SOCIOLOGY

Discovering Society

Second Edition

Jean Stockard

University of Oregon

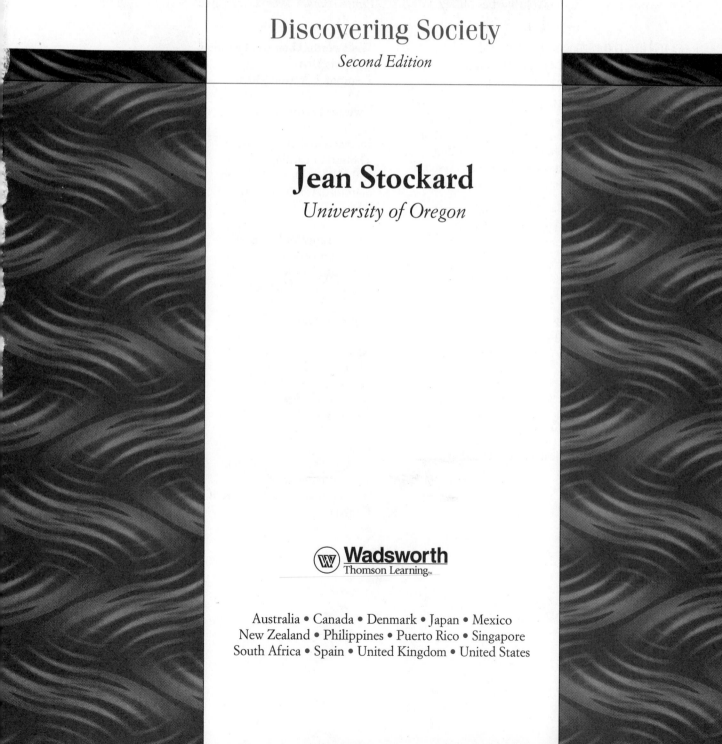

Wadsworth
Thomson Learning™

Australia • Canada • Denmark • Japan • Mexico
New Zealand • Philippines • Puerto Rico • Singapore
South Africa • Spain • United Kingdom • United States

Publisher: Eve Howard
Assistant Editor: Ari Levenfeld
Editorial Assistant: Bridget Schulte
Marketing Assistant: Kelli Goslin
Project Editor: Jerilyn Emori
Print Buyer: Karen Hunt
Permissions Editor: Susan Walters
Production Service: Robin Gold/Forbes Mill Press

Text Designer: Andrew Ogus ■ Book Design
Photo Researcher: Terri Wright/Terri Wright Design
Copy Editor: Robin Gold
Compositor: Forbes Mill Press
Cover Designer: Harold Burch
Cover Image: © 1999 Jenny Lynn
Cover Printer: World Color Book Services/Taunton
Printer/Binder: World Color Book Services/Taunton

For permission to use material from this
text, contact us by
Web: www.thomsonrights.com
Fax: 1-800-730-2215
Phone: 1-800-730-2214

Library of Congress
Cataloging-in-Publication Data
Stockard, Jean
  Sociology : discovering society / Jean Stockard.
—2nd ed.
  Includes bibliographical references and index.
ISBN 0-534-56521-2
  1. Sociology. I. Title.
HM585   99-36611
301—dc21

Wadsworth/Thomson Learning
10 Davis Drive
Belmont, CA 94002-3098
USA
www.wadsworth.com

International Headquarters
Thomson Learning
290 Harbor Drive, 2nd Floor
Stamford, CT 06902-7477
USA

UK/Europe/Middle East
Thomson Learning
Berkshire House
168-173 High Holborn
London WC1V 7AA
United Kingdom

Asia
Thomson Learning
60 Albert Street #15-01
Albert Complex
Singapore 189969

Canada
Nelson/Thomson Learning
1120 Birchmount Road
Scarborough, Ontario M1K 5G4
Canada

Contents

Preface

Even though sociologists often disagree on specific issues, almost all of us seem to believe, usually quite ardently, that sociology provides the best way to understand the world in which we live. Our passion for sociology reflects both our deep respect for the tools of our discipline—the theories and methods that we use to understand the world—and our earnest belief that through the dispassionate work of social science—the testing of theories with rigorous and sound methods—we can come to a better understanding of the world around us and help make it a better place in which to live.

Like most people who teach introductory sociology, I begin each term hoping that my students will come away from the course sharing my passion for the discipline. I hope that they will develop a sociological perspective and come to see the world in a radically new way. I hope that they will be able to use this new perspective to make sense of their daily lives and to be more useful and productive citizens. I also hope that they will develop respect for sociologists' research and that they will understand the excitement of participating in such a dynamic field.

Some students gravitate to sociology once they are exposed to its ideas. They flock around us after class, asking questions, sharing their own insights, and even begging for additional reading material. But other students, too many in my view, seem to have problems "getting it." The sociological perspective doesn't seem to make sense to them or to be particularly relevant to their own lives. They might have trouble making the connection between personal troubles and public issues that Mills defined as the essence of our sociological imagination. They might perceive the introductory course as a sequence of disconnected theories and concepts. They might be put off by what they see as jargon that has no relation to their own lives. They might have trouble understanding the significance of the technical details of the discipline or the reasons that theories and methods of research matter. Some students' problems with the field may also be more emotional in nature. They might be threatened by the ideas. The subject matter and the sociological approach might make them uncomfortable because it isn't necessarily easy to see one's world from a new perspective.

I have tried to write this book in a way that will help all students—both enthusiastic and reluctant—learn to use the sociological perspective. This text is designed to help convey an understanding and appreciation of sociology both as science and as a way of being engaged with the world. In response to the gap that often separates our own insiders' view of the field from our students' view, some of us may tend to focus on dramatic social problems and emphasize sociology's value in devising possible solutions while downplaying the more rigorous side of sociology as a science. Others may focus on theory and methods in order to present sociology as a valid social science. I believe that to serve our students well, we need to do both, to convey an understanding of both the passion and the dispassion of our discipline. I've tried to do so in a way any college student will find both accessible and inviting.

GOALS OF THIS TEXT

The three major goals of this book reflect the reasons we love the field and my attempt to help students understand both the passionate and dispassionate aspects of our discipline.

Understanding and Using Social Theory

First, I hope to show students how social theory helps us make sense of the world around us. I introduce theories and concepts within the context of real social issues and concerns to help students see how the need to understand a problem motivates our sociological thinking. In this way, I avoid presenting sociology as simply a sequence of definitions and terms and instead show how it can provide a powerful framework for comprehending our surroundings and actions. For instance, in Chapter 1, I introduce the views of the classical theorists as responses to the dramatic changes of the Industrial Revolution. In Chapter 3, I integrate the ideas of structural functionalism with a discussion of cultural universals and link the contributions of the Chicago School to a discussion of how both early and contemporary sociologists study racial-ethnic groups in the United States. In Chapter 5, I develop various theories of interaction in relation to real-life situations and problems.

At various points I also show how sociologists constantly use a variety of perspectives to understand an issue better, sometimes stepping back to a broader, macroanalysis, at other times taking a mesoperspective on groups and organizations. I show how some theories are very broad while others are more focused. I show how various theoretical perspectives sometimes complement one another and sometimes conflict. Throughout the book, I emphasize that sociologists can use a variety of theoretical approaches and that each of the many different theories that we use can be useful for understanding different problems and issues.

Theory is not always easy for students, especially in large doses, but it is central to our discipline and important for students to understand. To help students reach more sophisticated understandings, I gradually increase the sophistication of each chapter as the book progresses. I also often show students how theories used in one chapter relate to discussions in earlier chapters. For instance, the material on social interaction in Chapter 5 builds naturally on the discussion of socialization in Chapter 4, and the discussion of theories of deviance in Chapter 7 expands on the understandings of the classical theorists introduced in Chapter 1, as well as on material on socialization and interaction from Chapters 4 and 5. A feature at the end of each chapter called "Pulling It Together" provides a graphical way for students to see these theoretical linkages.

The Value of Research

Second, the book aims to help students understand why social research is important by continually giving examples of how research is used to develop, test, and modify theories that relate to real-world issues that affect us all. In the discussion of methodology in Chapter 2, I liken the process of social research to detective work, describing one research study in great detail. Each subsequent chapter describes at least one other study in detail and calls attention not only to the substantive findings but also to how they were achieved. Thus, the discussion of methodology is reinforced throughout the book, with consistent emphasis on developing students' critical analysis skills. Each of these featured research selections also illustrates how sociological research can be used to understand, and potentially address, social issues that face our country and countries around the world.

Throughout the book I emphasize that we have many different ways of obtaining knowledge of the social world and that the methods we choose may vary depending on the theoretical questions and substantive issues being addressed. Thus the research featured encompasses the entire span of methodologies used by sociologists, and I stress how a variety of perspectives helps us gain better understandings. Student skills in this area can be further developed by critical-thinking questions at the end of the Applying Sociology to Social Issues boxes and in the featured research studies. Each chapter also provides students with assignments and sources for exploring the Internet and the World Wide Web. In addition, the book is supplemented by a Web site created specifically to enhance the focus on research, critical analysis, and applications of sociology to the world around us. (See later.)

Insights into Daily Life and Social Problems

Third, throughout the book I hope to show students how sociology can provide a new perspective on both their daily lives and the social problems that face our society and world Each chapter begins by presenting a sociological issue at the level of individuals—in Mills's terms, at the level of private troubles. The chapters then use these stories as springboards to developing a sociological perspective on the topic, showing how these personal troubles are related to broader social issues. For instance, I use a discussion of ethnic identity as the basis for exploring culture and ethnicity in Chapter 3. I examine relationships among students from different social class backgrounds to explore issues related to social interaction in Chapter 5. Experiences regarding inequalities within schools and medical settings set the stage for the discussions of education and health and society in Chapters 15 and 16. Concepts related to each topic are presented within this context. I try to show students how personal troubles are public issues and how they can actively apply what they are learning to their own lives.

I also discuss how sociological findings can be used to develop policies to alleviate real-world social problems. Even though much of sociological work is not directly "applied," many of our analyses have direct implications for pressing social issues. To illustrate these implications, many of the featured research selections in the text deal with social issues. For instance, the featured research in Chapter 18 examines segregation in urban areas; Chapter 15 deals with student achievement in schools; and Chapter 6 looks at factors that influence delinquent and criminal careers. Beginning with Chapter 2, each chapter also includes a boxed feature entitled "Applying Sociology to Social Issues." These features directly establish the relevance of sociological analyses to social issues through discussing issues such as crime among street youth (Chapter 6), policies to combat poverty among children (Chapter 7), studies of methods that can be used to reduce racial-ethnic discrimination in organizations (Chapter 9), implications of research for helping children in disrupted families (Chapter 10), and ways that have been used to increase voter participation in societies around the world (Chapter 13.)

ORGANIZATION AND COVERAGE

This book includes all of the topics typically found in introductory sociology textbooks. Part One, "Doing Sociology," introduces students to sociological theory and methods. The writings of C. Wright Mills are used to introduce the notion of a sociological perspective, the vast range of sociological thought, and sociologists' interest in social issues. The writings of four major classical theorists—Marx, Durkheim, Weber, and Simmel—serve as an introduction to sociological theory and, just as important, to the intellectual and social motivations that have driven sociological thinking from its beginnings. This discussion allows the opportunity to present students with concepts basic to a wide variety of theoretical approaches, both microlevel and macrolevel. The notions of social structure and social action are also introduced, emphasizing that we not only are influenced by the world in which we live but also continually create this world. Each of these themes is expanded in subsequent chapters as the works of other sociologists are introduced.

Part One also examines the ways in which sociologists test theories. The goal is not to make students expert in methodology, but rather to convey a sense of the logic and importance of research methods, as well as the ways in which researchers try to ensure that their results are valid and the ways in which sociology differs from everyday thinking. In this manner students can become more critical consumers of the research that is reported both in the rest of the book and in the media.

Part Two, "Individuals and Society," includes chapters on culture and ethnicity, socialization and the life course, social interaction, and deviance and social control. Building on the theoretical ideas developed in Part One, these chapters show how the social world influences individuals' day-to-day actions and decisions, yet also how our actions influence the social world. I also illustrate how sociological theories can involve both macrolevel and microlevel perspectives. Because our society is becoming more heterogeneous and more conscious of racial-ethnic diversity and gender-related issues, and because so much of sociology (as well as my own research) relates to these areas, many examples concern social inequality and gender. Through such examples I hope to reinforce the notion that sociology can deal with students' own lives and social policy concerns.

In the chapter on culture and ethnicity, I use a macrolevel perspective to examine research on how cultures around the world are both similar and different. I also adopt a more microlevel perspective to examine the ways in which we come to see ourselves as part of our culture, focusing on issues related to ethnic identity. In the chapter on socialization, I discuss classical theory and the results of contemporary research on socialization and the development of self-identity, using many examples related to gender socialization. In the chapter on social interaction, I show how sociologists have examined social networks and how interactions involve social exchanges, social roles, and symbolic interactions, as well as nonverbal and verbal communication. I also discuss the influence of status characteristics on group interactions and the status attainment model to explore the ways in which social interactions are related to life chances and opportunities. Finally, in the chapter on deviance and social control, I show how various theories and ideas introduced in previous chapters can be used to understand deviant behavior. For instance, concepts introduced in the discussion of the life course are reinforced by examining the relationship of age to criminal behavior and influences on the course of "deviant careers" from youth to adulthood. This chapter also introduces a global perspective by examining variations in homicide rates around the world.

Part Three, "Social Inequality," deals directly with stratification based on social class, race-ethnicity, and gender. In these chapters I try to show how sociological analyses address issues that relate to students' own lives, both at home and around the world. For instance, in the chapter on social stratification, I examine variations in social inequality in countries around the globe. I also describe how sociologists have documented and explained stratification in the United States and what these analyses tell us about the possibility of attaining the "American

Dream" by examining issues of social mobility, poverty, and the power of the wealthy.

In the chapter on racial-ethnic stratification, I first describe the extent of such stratification in the United States and use a macrolevel analysis to examine sociologists' explanations of how racial-ethnic stratification has changed over the past two centuries. I then move to a more microlevel analysis, looking at research on how racial-ethnic stratification affects individuals and is continually reproduced through social interactions, again building on concepts and ideas introduced in earlier chapters. The featured research combines these levels of analysis to explore why racial-ethnic discrimination varies from one country to another in Europe and across time and place in the United States.

In the chapter on gender stratification, I document the nature of such stratification both cross-culturally and in the United States and then explore explanations of its existence. Here, the focus is on gender stratification in education, the family, the political world, and the economy, as well as the difficulties experienced by men and women who enter nontraditional fields. Each of the chapters in Part Three reinforces the notion that social structures, such as stratification, very much influence our lives, while at the same time we, as social actors, create and maintain these structures.

Part Four, "Social Organizations and Social Institutions," reviews research related to formal organizations and to six major social institutions: family, religion, the political world, the economy, education, and health and medicine. In order to help students see how sociologists' studies of these areas relate to their lives and concerns, I focus on specific issues. For instance, in the chapter on the family, I introduce basic concepts and theories by examining how families vary and change. In the chapter on the political world, I focus on issues of power and control. In the chapter on the economy, I focus on the role of the economy in fostering opportunities and constraints for individuals, groups, and societies. In the chapter on organizations, I look at how sociological concepts and theories can explain organizational change and success. In the chapter on education I look at issues regarding student achievement and effective schools; and in the chapter on health and medicine, I explore why societies and subcultures within societies vary in health and longevity.

As in other chapters, I show students how sociologists use different levels of analysis to look at these issues. For instance, at the macrolevel I examine cross-cultural variations in the family and changes in the family in the United States over the past two centuries. I also discuss economic relationships between nations, present varying perspectives on the political power structure, and explore why some nations have

healthier populations than others. At the microlevel I examine research on areas such as how changes in the family affect the lives of children, how individuals influence organizational change, and how social networks influence work careers. At the mesolevel I look at research on the role of political action committees in the political world, the ways in which schools can be organized to enhance the achievement of all children, and the reasons some religious groups grow in size while others get smaller.

Concepts and theories that specifically apply to each institutional area are introduced in the context of exploring the issues that define each chapter. These discussions often reinforce concepts and theories introduced in previous chapters. For instance, exchange theory and the concept of social networks, which were introduced in the chapter on social interaction, are used again in the analysis of why individuals choose to join religious groups and participate in political activities. Issues and concepts related to social stratification are reinforced in the discussion of the political world, the economy, and health and medicine.

The discussions also remind students how social issues and sociological analyses apply to global issues as well as their own lives. A discussion of global stratification—both its nature and explanations of its causes—is a prominent part of Chapter 14 on the Economy. Chapter 15 on Education includes an extended discussion of differences in literacy and schooling around the world as well as the development of common schools throughout the globe. Chapter 16 on Health and Society features research that examines world wide variations in health and how these variations are related to global stratification.

Finally, Part Five, "Social Change," includes chapters on population, communities and urbanization, social change, and social movements. Although global examples and research are featured in earlier chapters, the material in Part Five brings global issues, as well as environmental concerns, to the forefront. I believe that this is appropriate in that current social changes involve the world as a whole with intricate connections between the environment, global stratification, population growth, and urbanization. In addition, my experience suggests that students are often better able to apply sociological understandings to other cultures once they have developed the ability to apply them to their own world, and the chapters in this section often build on understandings developed in earlier chapters.

Chapter 17 on population focuses on the nature of demographic data and how these data can be used to understand how and why populations change. I explore how population changes produce opportunities and constraints for societies and their citizens, focusing on the different experiences of the least developed, the developing, and the highly developed countries. The linkage between population growth, economic conditions, and the environment is examined, both for less developed and more developed countries. Special attention is given to how social actions influence population changes around the world, looking at issues regarding immigration and at the influence of education on fertility.

Chapter 18 on communities and urbanization looks at the process of urbanization in both the United States and less developed countries and how this is related to environmental changes, such as desertification and deforestation. A great deal of attention is given to the issue of how communities can be sustained in urban settings, exploring both classical and contemporary work in this area. This question is then related to analyses of urban ecology and recent research on the very pressing issue of racial-ethnic segregation in urban areas in the United States.

Chapter 19 on technology and social change builds on understandings developed in previous chapters to show how social change involves the interaction of technological innovations, changes in social institutions, and alterations in the environment. Building on this theoretical base a long-range view of social change is presented from early hunting gathering societies to the current outpouring of new technology that allows us to build networks with others throughout the globe. Special consideration is given to issues regarding social change and development in the world's poorest societies.

Finally, in Chapter 20, I examine collective behavior and use research on the environmental movement to illustrate sociological analyses of social movements. From a mesolevel perspective I review research on why some social movement organizations are more successful than others. From a macrolevel perspective I examine theories regarding the development of social movements and describe contemporary cross-cultural research on how political cultures and history affect this development. Finally, from a microlevel perspective I review research on why individuals choose to participate in social movements and how social movements affect us as individuals. This analysis provides an appropriate way to end the text by showing students, yet again, how individuals continually create social structures through their social actions.

CHANGES IN THE SECOND EDITION

In developing the second edition of this book, I systematically reviewed issues from the last three years of general journals such as the *American Journal of*

Sociology, American Sociological Review, and *Contemporary Sociology,* as well as more specialized journals such as *Demography, Journal of Marriage and the Family,* and *Gender and Society,* looking for articles and books that were related to material covered in this text. This review took a lot of time, but helped me make sure that each chapter is as up-to-date and accurate as possible. All statistics, references, and recommended readings have been thoroughly updated. When the featured research studies have been replicated, information on the results of these studies has been added, helping to reinforce for students the importance of building knowledge and replicating results.

Although this careful updating and review might not be noticeable with a quick glance through the book, a change that should be striking for users of the first edition is the addition of a series of global maps that illustrate various issues discussed in the text. These maps depict the most recent worldwide data on variables such as homicide rates (Chapter 6), income inequality (Chapter 7), gross national product (Chapter 14), illiteracy (Chapter 15), life expectancy (Chapter 16), population growth rates (Chapter 17), extent of urbanization (Chapter 18), and telephone usage (Chapter 19).

An entirely new chapter on Health and Society (Chapter 16) has been added to the text, reflecting the widespread concern regarding this area both in the United States and in other nations. In this chapter, I use the issue of inequality in health care to examine research regarding social influences on individuals' health and well-being; access to health care, including issues regarding the funding of health care in this country and in other nations; and differences in health and longevity around the world. The featured research examines why the United States, the wealthiest nation on earth, is not the healthiest, and provides evidence that this result might be linked to social stratification and income inequality.

Chapter 19, "Technology and Social Change," and Chapter 20, "Collective Behavior and Social Movements," are also new to this edition, replacing a chapter in the first edition entitled "Social Change and Social Movements." The new chapter on technology and social change significantly expands upon material presented in the first edition by including an extensive analysis of the relationship between changes in social institutions, technology, and the environment as well as research on how technological innovations, both those in the past and those currently appearing, affect individuals' lives. Adding this new chapter also allowed me to expand upon issues associated with the notion of sustainable development as a way to deal with environmental problems. The new chapter on "Collective Behavior and Social Movements" expands upon the first edition by adding an extensive review of the collective behavior literature, a central part of many analyses of social movements, and features a new "Applying Sociology to Social Issues Box," on ending female genital mutilation.

Specific additions and changes to other chapters include the following:

Chapter 1: What Is Sociology?

- Added a comparison of sociology with other social sciences.

Chapter 3: Culture and Ethnicity

- Added very recent work on replications of the featured research involving the ethnic identity of Asian Americans.
- Altered the "Applying Sociology to Social Issues" feature to reflect forthcoming changes in the U.S. Census question regarding race/ethnicity.

Chapter 4: Socialization and the Life Course

- Added material regarding other studies of gender socialization to the featured research section.

Chapter 5: Social Interaction and Social Relationships

- Added material on replications and extensions of the featured research and how this work has been applied in schools.

Chapter 6: Deviance and Social Control

- Moved work on international comparisons that was a featured box to the text and transformed the data on homicide rates throughout the world into a map.
- Added a new "Applying Sociology to Social Issues" box on crime and youth on the streets.
- Expanded the discussion on cohort differences in criminal behavior to reflect very recent work.
- Altered the discussion of strain theory to include the most recent theoretical work in this area (general strain theory).
- Explicitly noted integrative work in contemporary theories of deviance.
- Added a discussion of replications of featured research work.
- Briefly discussed inequality and crime as a link to Part 3 on stratification.

Chapter 7: Social Stratification

- Added two global maps and a discussion of inequality around the world to demonstrate Lenski's analysis of variations in social stratification. This discussion introduces the Gini coefficient as a measure of inequality.
- Added comparison of income inequality in the United States with that in other countries.

- Moved the discussion of measuring poverty from a boxed feature to the text.
- Added a comparison of poverty among the elderly and the young.
- Focused the "Applying Sociology to Social Issues" box on childhood poverty and comparing policies in the United States with those in other countries.
- Separated the discussion of trends in poverty and trends in inequality into two distinct sections. The discussion of the Gini index is reinforced through the discussion of changes in inequality in the United States over time.

Chapter 8: Racial-Ethnic Stratification

- Added an entire section on altering racial-ethnic prejudice that links macro and microlevel analysis and introduces the concept of group threat as well as the notion of interaction effects. The new featured research uses both international and U.S. data.
- Moved the material on the contact hypothesis from the "Applying Sociology to Social Issues" box to the text.
- Added a new "Applying Sociology to Social Issues" box that shows how the contact hypothesis has been applied in actual organizations.

Chapter 9: Gender Stratification

- Added material on cohort variations in gender differences in income.
- Added international comparisons on income and occupational segregation, with a link to the new "Applying Sociology to Social Issues" box in Chapter 7.
- Added a new section on gender stratification in the political world, both in the United States and cross-culturally, with a new illustrative table.
- Updated material on gender stratification in the family and altered it to incorporate the most recent work on men's participation in household chores.

Chapter 10: The Family

- Altered the discussion of social institutions to include medicine.
- Changed some wording to more thoroughly represent current research findings.

Chapter 13: The Political World

- Added a paragraph on new voter registration law and the results of a Federal Elections Commission study of its effects.
- Changed the discussion of gender difference in voting to reflect most recent data.
- Expanded the featured research to include material from Clawson's most recent book. This includes an entire new section on soft money issues.

Chapter 14: The Economy

- Added discussion of trust, cooperation, and competition as key elements of all economic exchanges.
- Expanded discussion of the globalization of the economy and global stratification that is reflected throughout the chapter. Includes a map depicting GNP of countries throughout the world, a table listing various international economic organizations, and a discussion of the World Bank and IMF and their relation to global stratification.
- Added a brief discussion of institutionalization of workplace changes reflecting recent work on institutionalization.
- Added discussion of "megamergers" as the sixth wave of corporate mergers.

Chapter 15: Education

- Added a map of illiteracy rates of nations around the world.
- Added a new section on the influence of peers on academic achievement.

Chapter 17: Population

- Added a map that illustrates the "doubling time" of nations around the world and an explanation of this phenomenon.
- Added most recent material on how the AIDS epidemic is affecting populations in African countries.
- Revised and slightly expanded the discussion of the relationship of poverty, population growth, and the environment to include discussion of global warming and emissions of carbon dioxide.
- Greatly expanded the discussion of immigration to include a section on migration to the United States.

Chapter 18: Communities and Urbanization

- Updated discussion of urban/rural differences in growth in United States, highlighting recent growth in rural areas in some parts of the country.
- Added a map of urbanization around the world.

Chapter 19 Technology and Social Change

In the first edition, part of this chapter was included in Chapter 18, "Social Change and Social Movements." Additions to this edition include

- Expanded discussion of the relation between the development of technology, the environment, and changes in social institutions.
- Greatly expanded discussion of the effects of recent technological changes on individuals' lives and of changing political and economic systems throughout the world.

- Added new featured research section on how the invention of the telephone changed individuals' lives.
- Expanded discussion of issues related to sustainable development.
- Added new "Sociologists at Work" interview with Claude Fischer.

Chapter 20: Collective Behavior and Social Movements

In the first edition, part of this chapter was included in Chapter 18, "Social Change and Social Movements." Additions to this edition include

- Added extensive discussion of how sociologists define and conceive of collective behavior.
- Added explanations for why collective behavior occurs.
- Added new "Applying Sociology to Social Issues" box on ending female genital mutilation.

PEDAGOGICAL FEATURES

Several pedagogical features are incorporated throughout the book. These have been specifically designed to support the goals of the book and to help maximize students' understanding of the material. In developing these features, I have been guided by the recent work of cognitive psychologists, which suggests that we learn material most readily and easily when we can connect it with previously developed schema, that is, when we can connect it to mental scripts that we have already developed. Thus, I have tried to provide ways to help students develop their own "sociological schemas" and to link these schemas with knowledge and understandings they have already developed.

Building Concepts Tables

Almost every chapter in the book contains one or more "Building Concepts" tables. This feature is designed to summarize key concepts and ideas to help students focus on important notions and to understand key elements that differentiate various theories or concepts.

Common Sense versus Research Sense

This short boxed feature is designed to help students think about assumptions they might have about the social world and compare these to the results of research. Everyday notions regarding social life are presented and then compared with results of the featured research or of other research results discussed in the text.

Pulling it Together

To help emphasize and reinforce the continual building of knowledge that occurs throughout the text, a feature called "Pulling It Together" appears at the end of each chapter. It directs students to pages in other chapters where there is additional discussion of topics that have been covered, both those they might have already read and those that are in later chapters of the book.

"Featured Research" Sections

Each chapter, beginning with Chapter 3, has a "Featured Research" section that describes in detail the methodology and results of one research study. All of this research directly relates to issues discussed in the chapter and represents high-quality work that has been published in top journals and presses. The featured research ranges from classic studies, such as Elizabeth Cohen's work on how status characteristics affect group interactions, to contemporary work, such as Douglas Massey and Nancy Denton's analysis of racial-ethnic segregation in cities and Richard Wilkinson's analysis of the relationship of social inequality to the health status of nations. The studies represent all different types of research methodology and theoretical orientations, and they reinforce the explanations of research methodology given in Chapter 2. They also provide excellent examples of how sociologists merge their concern with social issues with dispassionate and careful scientific inquiry. Each "Featured Research" section ends with a few "Critical-Thinking Questions," which are designed to help students think about how sociological research is conducted, to develop their own critical-thinking skills, and to reinforce concepts and ideas introduced in the text.

"Sociologists at Work" Boxes

Each chapter has one or two "Sociologists at Work" boxes that present brief interviews with contemporary sociologists—sometimes the person whose work is described in the "Featured Research" section and sometimes a person whose research relates to other areas discussed in a chapter. The sociologists interviewed usually describe how they became interested in sociology and then discuss aspects of their work, such as what surprised them about their findings and how their colleagues reacted to their results. They also describe what implications and effects their work has had on public policy and how their findings may relate to students' lives. This feature helps to make sociological research come alive for students—not just the process by which it occurs but the way in which it can affect our social world.

"Applying Sociology to Social Issues" Boxes

Beginning with Chapter 2, each chapter includes an "Applying Sociology to Social Issues" box that describes how sociological research and theories can help address real-life social problems. For instance, in the chapter on socialization, the box describes how sociological insights are used to help develop effective programs to aid substance-abusing mothers and their children. In the chapter on social stratification, the box examines childhood poverty in the United States and compares policies in this country with those of other industrialized nations. In the chapter on communities and urbanization, the box reviews research on city residents' attitudes toward living in integrated neighborhoods. In Chapter 20, "Collective Behavior and Social Movements," the box explores how social movements helped end the practice of footbinding in China and the possibility that similar movements could end the practice of female genital mutilation in Africa.

Each of these boxes ends with a series of "Critical-Thinking Questions," which ask students to reflect upon the issues raised and the ways in which sociological understandings can be brought to bear on them. For instance, in the chapter on deviance and social control, I summarize research on "street kids" and ask students to think about how these findings would apply to their own community. In the chapter on racial-ethnic stratification I describe the efforts to improve racial-ethnic relationships within the United States Army and ask students to examine the extent to which the procedures and policies could be applied to groups in which they interact.

Internet Assignments and Sources

To help students use the latest technological advances in communication and data retrieval, a series of assignments and exercises in which students can use the Internet is included at the end of each chapter. These are designed to help students find sociological data on the Internet and to reinforce and apply theories and concepts presented in the text. To help the student and instructor grasp the benefits the Internet has to offer, a glossary of terms and a list of important sites to reference are included on the endpapers.

InfoTrac College Edition References

A special resource to students using Wadsworth books is the very large collection of articles within the InfoTrac College Edition collection. To help students access this material references are given at the end of each chapter to several articles that can supplement the material presented in the text.

Writing Style and Approach

In many ways, the most important pedagogical aid in this book is the style that I have tried to use—one that can both engage students and help them understand the sociological perspective. For instance, the chapter introductions, which describe the "personal troubles" of an individual, set the stage for subsequent sociological views of "public issues." These introductions also provide the basis for examples used throughout each chapter, showing students how sociological concepts and theories can provide insights into their own lives. Each chapter also includes an analogy or metaphor that summarizes the way in which sociologists look at an issue, such as the use of different camera lenses to portray different sociological theories and perspectives or Simmel's notion of a web to explain social networks and relationships. These metaphors can help students develop simple mental images that are part of their everyday life and pull together and understand the material that they read.

Because abstract theories and methodological techniques can often be difficult for students to understand, I try to give a number of real-life examples whenever difficult concepts appear, showing how sociology applies both to our own lives and to important issues in the world around us. I present ideas clearly and succinctly, building on previously developed concepts. Although I cover all of the material typically included in introductory texts, I do so in the context of explaining how sociologists look at particular issues, rather than simply presenting series of definitions and theories for students to memorize.

I have also included some material often not covered in other textbooks, material that I believe is central to a contemporary understanding of sociology and the social world. Thus, I have not shied away from presenting complex ideas, such as the distinction between age, period, and cohort effects; the difference between classical and contemporary Marxian interpretations of stratification; theories regarding racial-ethnic and gender segregation; the development of a new paradigm in the sociology of religion; and the meaning of segregation indexes. But I have tried to write about these and other complex ideas in a manner that peels away extraneous considerations and builds up a gradual understanding.

The order in which I have presented concepts and ideas has been carefully planned to help students gradually develop a fuller understanding of sociological theories and concepts. Thus, I gradually introduce theories and new terms. When a concept relates to an area that has been previously studied I make that connection in the text, thus reinforcing previous learning as well as helping students to develop and expand their understandings. For instance, after presenting the ideas of the four classical theorists in Chapter 1, I continue to refer back to their work in subsequent chapters, showing continuities in both methods and theoretical concepts. Similarly, the material in Chapter 5 on social interactions and relationships is used in later chapters in explanations of deviant behavior, racial-ethnic discrimination, decisions regarding religious affiliation, and participation in political activities and social movements. At the end of each chapter a feature called "Pulling it Together" describes many of these linkages as yet another aid to students (see description later).

Other Pedagogical Aids

Each chapter contains a series of tables, charts, graphs, photographs, and other graphics to illustrate or reinforce key concepts and data. Many of the photograph captions ask students to apply their sociological understanding and think about the portrayal in sociological terms. Summaries at the end of each chapter provide a succinct point-by-point review of major topics. All key concepts and terms are highlighted in the text, and a list of "Key Terms," with page references, is given at the end of each chapter to help students check their understanding. In addition, a complete glossary is provided at the end of the book. "Recommended Sources" at the end of each chapter give students additional resources with which to investigate key chapter topics.

ANCILLARY MATERIALS

To complete the goals and themes of the text, a standard set of supplements is available for both the student and instructor. They provide something for every learning style and something for every teaching style.

Wadsworth Classic Readings in Sociology

This set of classic readings in sociology includes a wide selection of articles that can supplement the material in the text. The reader includes excerpts from classic sociologists, such as Marx, Weber, Durkheim, Simmel, and Tönnies, to writers from just a generation or two ago, such as Goffman, Parsons, Davis and Moore, and Mills. More contemporary articles are written by Kozol, Ritzer, and Freeman. Using articles from the reader can provide students with an excellent taste of sociological writing and analysis by a wide range of authors.

InfoTrac College Edition

The extremely versatile InfoTrac College Edition system lets you create your own reader for your students. I have included a number of references to InfoTrac College Edition in the text, and you can also access this resource to develop a set of readings designed for your students from a very large set of journals from throughout the world.

Study Guide

For students, I have written a Study Guide to supplement the text, with chapter outlines, sample test questions, and exercises. I have also included advice on how students can do better in class and how they can adapt their study techniques to accommodate their own styles of learning.

Two special elements of the study guide have been especially designed to help students develop conceptual schema that incorporate their sociological learning. One of these elements is entitled "Sociology and You" and involves a series of questions designed to help students systematically apply the concepts introduced in the text to their own lives. Several sets of questions are provided for each chapter. The second special element is instructions for developing "concept maps," individually developed diagrams that help students think about and link together the theories, ideas, and concepts that they have studied. Students are instructed to continue the maps from one chapter to the next, gradually building up their understanding as they proceed through the text. When using these techniques in my own classes, I have found that they can significantly improve students' comprehension and grades.

Instructor's Resource Manual

Linda Heuser, of Willamette University, and I collaborated on an Instructor's Resource Manual that addresses the myriad ways we teach the introductory course. For those who teach writing-intensive classes or small, discussion-oriented classes, we provide suggestions for essay topics, discussion questions, and classroom exercises. For those who have large lecture sections, we have included several developed lectures for each chapter. Some of these lectures are designed to help students understand material that may be more difficult, others are designed to provide supplementary material. All of the lectures have been used successfully in my own classes.

Test Bank

Written by David Ford of The University of Central Oklahoma, the Test Bank provides a diverse set of questions, including multiple-choice, true/false, and completion questions with page references. The author has included a range of questions that test different levels of knowledge:recall/definition questions and content questions based on studies and examples provided in the chapter, as well as applied questions in which students apply the knowledge they learn in the chapter to hypothetical scenarios. The Test Bank is available in computerized Mac, Windows, and DOS formats, which allows professors to select questions, edit them, and add their own questions so that multiple versions of a test can be created and printed. Ask your Wadsworth/ITP representative about online testing options as well.

Transparencies

This set of 100 full-color transparency acetates includes graphs, charts, and other key images from the text.

Customized Videotapes

A customized collection of short videotapes giving a view of slices of society and societal interaction, as well as reflecting the applied research focus of the text, is available to adopters. Please contact your Wadsworth/ITP sales representative for details.

World Wide Web Home Page Resources

An extensive collection of resources for studying sociology can be found on the book's World Wide Web Home Page. This site can be accessed via the Wadsworth URL: http://www.wadsworth.com/wadsworth.html. This book-specific Web site offers numerous materials for both students and instructors. For students, there are links to Internet resources that are relevant to sociology, postings of views and events, and real-world examples analyzed sociologically to keep the book current and evolving. Multiple-choice questions for each chapter are available for self-quizzing along with the option of participating in discussion groups or collaborating with students in another part of the country on a research project. Instructors may also use the Home Page as a tool to encourage students to do extra-credit assignments, as well as a source for conducting research over the Internet to enhance and complement the focus of the book. The Web site will also be customized to the needs of different courses and instructors, providing a valuable service and a link among students of sociology everywhere. (Please contact your Wadsworth/ITP sales representative if you would like to post your syllabus or office hours or would like to sponsor a discussion group with other adopters.)

CONCLUSION

In writing this book, I focused on how to help students see how a sociological perspective can help them better understand and deal with the world around them. I have tried to convey the way in which we can carefully and dispassionately study the social world and how the results of our studies can be used to help us understand our own lives and to address pressing social issues, both at home and throughout the world. Thus I have tried to provide students with an understanding of the theories that we use and to show them how specialists in each area actually study real problems in their research. I hope that with these basic tools they will be able to do further study themselves. Even if they never take another sociology class (which is, in fact, true of many people who take our introductory classes), they will be able to enjoy reading monographs and trade books written by sociologists or simply to see the news of the world and our society with a fresh perspective. Those who do go on to further study will have the basic skills to explore the field in more depth.

I think it would be far easier to write a textbook for my fellow sociologists than for introductory students. That is, in fact, what we do in most of our professional writing. We don't have to help one another understand the basics of our discipline or convince one another of its importance and relevance. Though I would like to think that professional sociologists won't mind reading this book, my main hope is that students will enjoy it and, most important, that this book can convince at least a few students, and hopefully more than we have convinced in the past, that the sociological perspective can open up doors and vistas that they never dreamed existed, and that this perspective can help us deal with some of the most pressing issues that face our nation and our world.

ACKNOWLEDGMENTS

I am very grateful to the scholars who took the time to carefully read the first edition of this book and provide suggestions for this revision:

William Camp, Luzerne Community College
Juanita Firestone, University of Texas, San Antonio
George Klein, Oakton Community College
Sally Rogers, Rockland Community College
Edward Vaughn, University of Missouri, Columbia
Diana Wysocki, University of Nebraska, Kearney

In addition, these scholars reviewed the first edition of the book, and I am grateful for their assistance: Margaret Abraham, Hofstra University; David Ashley, University of Wyoming; Tim Biblarz, University of Southern California; Sampson Lee Blair, Arizona State University; Walter Carroll, Bridgewater State College; Karen A. Conner, Drake University; Michelle Curtain, Indiana University; Marlese Durr, Wright State University; Mohamed El-Attar, Mississippi State University; Jess G. Enns, University of Nebraska at Kearney; Michael P. Farrell, State University of New York at Buffalo; Marvin Finkelstein, Southern Illinois University; William Finlay, University of Georgia; David Ford, University of Central Oklahoma; Michael Goslin, Tallahassee Community College; David Hachen, University of Notre Dame; Ann Hastings, University of North Carolina at Chapel Hill; John Henderson, Scottsdale Community College; Frances Hoffman, University of Missouri; Darrell Irwin, Loyola University of Chicago; Michael B. Kleiman, University of South Florida; Cheryl Laz, University of Southern Maine; Sally Ward Maggard, West Virginia University; Donald B. Olsen, Kansan Wesleyan University; Harold W. Osborne, Baylor University; Mari Ruthi, Huntington College; Phil Rutledge, University of North Carolina at Charlotte; Anson Shupe, Indiana University–Purdue University; and Charles M. Tolbert, Louisiana State University.

I would also like to thank the staff of the Knight Library at the University of Oregon for their assistance in procuring books and especially Tom Stave for his help in locating the most recent statistical information. In addition, I appreciate the willingness of Claude Fischer, University of California, Berkeley, and William Cockerham, University of Alabama, Birmingham, for taking the time to help develop "Sociologists at Work" boxes that describe their work and careers.

The sociology team at Wadsworth has been continually helpful and supportive. Especially deserving of thanks are Robin Gold for her careful copyediting and guidance of the production process; Eve Howard for guiding the entire project through the editorial process; Jerilyn Emori for directing the production process; Ari Levenfeld for his work with myriad details; Andrew Ogus for his masterful design of the book; Bridget Schulte for tending to so many details with the supplements and day-to-day progress of the book; and, especially, Barbara Yien for her very helpful development of the photo program and new featured elements as well as her continual encouragement in the early stages of the revision process.

Finally, I owe an enormous debt to my family— my husband, Walt, and our children, Beth, John, and Tim. They provided useful advice and cheerful assistance with tedious details, but, even more importantly, they were always available for laughter and fun. I continue to dedicate this book to them with love and gratitude.

What Is Sociology?

The Sociological Imagination

Looking at Individual Experiences with a Sociological Perspective

Dispassionately Looking at the Passion of Life

A Passionate Science: How Sociology Developed

Responding to Social Ferment: The Classical Theorists

Sociology Today

Discovering Sociology: Three Key Themes

WHEN MARK FIRST arrived at Central University from a small town in another state, he knew no one. He had chosen Central because he thought he could win a spot on the track team running the hurdles, which he loved doing more than anything else.

During his first year Mark found some new friends and made the track team. However, all was not well in his life. His classes were much more difficult than he had thought they would be, and by the middle of the year he was in academic trouble. His girlfriend, who had gone to a different college, wrote that she had found someone else. Worst of all, Mark wrenched his knee so badly that the doctors said he would never compete in track again.

Mark continued to struggle with his schoolwork, and at the end of his freshman year, he was suspended from school for low grades. Embarrassed by his situation, he decided to move into a studio apartment and get a job. Unfortunately, with only a high school diploma, his job prospects were limited. Further, he now felt he had little in common with his school friends and teammates, who were caught up in campus life, and he rarely saw them. His parents still talked optimistically about his completing school and getting a good job, but he became increasingly discouraged and depressed.

By December the situation became too much for Mark to handle. With the last of his money, he bought a gun. He then locked himself in his apartment, loaded the gun, pointed it at his head, and pulled the trigger. The apartment manager heard the shot and called the police, but it was too late. Mark had committed suicide.

Mark's family and friends, as well as his former teachers and coaches, were horrified. In the weeks following Mark's death, they tried to understand why he decided to end his life. They looked at all that had happened to him in the previous year and probed what he must have been thinking and feeling. They tried to grasp why Mark, a healthy young man with many years of life ahead of him, had made this appalling choice.

It is hard to think of anything that is more personal and individual than deciding to take one's own life. Yet, unfortunately, an alarming number of people nationwide commit suicide each year. Why do they do this? Clearly it must result from deep personal despair and sadness.

Yet we can also look at suicide another way—as a sociologist would. Sociologists step back and examine not just one suicide, like Mark's, but a large number of suicides. Sociologists ask, Are people in some circumstances more likely to decide to end their lives than other people are? Is there more to explaining suicide than just the individual characteristics of particular people? Why do some people who face severe personal troubles decide to end their lives, whereas others do not?

ONE OF THE EARLIEST sociological studies examined why members of some groups are more likely than are members of other groups to commit suicide. In this chapter you will learn more about how these early sociologists began to look at events such as suicide in a different way—to look at events sociologically. When you step back from specific events and search for larger patterns in human behavior, you are taking a *sociological perspective.*

When you are used to seeing daily events from an individual perspective, looking at the world sociologically is similar to focusing the lens on a camera, so that things become much clearer and much more understandable. And once you have learned to apply this sociological perspective, you will never again look at either society or individual lives in the same way. You will have gained the ability to interpret social situations and events in ways you would not have thought of before.

Sociology is the science of society, or the scientific study of society and social interactions. When you think of science, you might envision someone in a white coat mixing chemicals in a laboratory or programming computers to send rockets into space. Or you might imagine a medical researcher studying retroviruses in search of a cure for AIDS. Sociologists don't study inert chemicals or program rockets or look for viruses; nor do they work in laboratories full of scientific instruments. Rather, sociologists use the scientific method to study the *social world,* the day-to-day interactions that we all experience with other people and the world around us. Sociologists' laboratory is society, from small groups such as families and groups of friends to organizations such as universities

A sociologist's "laboratory" is the social world. As you read this chapter, think about how a sociologist might study this setting.

and corporations and institutions such as religion and the government.

This is the world where people like Mark decide to end their lives in despair or resolve to keep on trying to deal with the problems they face. This is the world where you fall in and out of love, where you look for a job and are fired or promoted, where you join others in worship or wonder whether there is a God, where some people are poor and some are rich, where politicians win and lose elections, where wars are fought and peace is made, where children grow up and grandparents grow old. All these events involve emotions of sadness and joy, love and hate, faith and sorrow. And they are what sociologists study every day.

How, you might wonder, can anyone do this? Don't family problems, infighting at work, and brutal wars occur because particular individuals and groups can't get along? Don't individuals decide to commit suicide because of highly personal and unique circumstances and psychological problems? Doesn't religious faith develop out of each individual's own relationship with a supreme being? Aren't some people richer than others simply because they work harder or are luckier? In general, aren't important life events simply a result of individuals' personalities and idiosyncrasies?

Sociologists don't think so. They look at the world differently. They step back from the unique experiences of individuals and specific events such as suicides, marriages, or religious beliefs and concentrate on the broader processes that underlie all social life. Instead of asking why a specific individual might choose an action as drastic as ending his life, sociologists try to understand why suicides are more or less likely among specific groups of people or in certain social circumstances. Instead of asking why this or that particular marriage succeeds or fails, sociologists focus on the environment or social conditions in which marriages in general tend to endure or crumble. In short, sociologists look for evidence of social realities that can help us comprehend why many individuals behave as they do. Sociologists look for the ways in which the decisions of many different individuals reveal detectable *patterns*.

Through this research, sociologists often develop understandings that span many different situations, from our decisions as individuals to those of family units and even entire nations. These general explanations are social **theories,** broad systems of ideas that help explain patterns in the social world. Even though we can often easily explain why one person might have chosen to commit suicide, it is much harder to explain why suicides might be more common among some groups of people than among others. Yet, if a pattern exists-if, for instance, there is a trend toward more suicide attempts among young people-sociologists want to know why. Social theories provide a starting point for answering such questions. Social theories help sociologists explain why some aspects of the social world are relatively enduring. These theories also help illuminate how each of us participates in creating and changing our social world. In this chapter you will learn more about how sociologists look at the world and how their ideas developed.

Sociologists' work involves two basic elements: (1) developing possible answers to questions about how the social world works and (2) testing these answers, or theories, to see whether they are supported

Being stuck in a traffic jam can certainly seem like a personal trouble. But if you step back and take a sociological perspective, you can see how one particular traffic jam is simply part of public issues such as the development of cities, the ways people travel to and from work, and the pollution of the atmosphere.

by the facts or need to be modified. Before developing ideas about the social world, however, sociologists must discover what questions to ask. To do this, sociologists use a special perspective known as the *sociological imagination.*

THE SOCIOLOGICAL IMAGINATION

After Mark committed suicide, his friends and family examined every aspect of his last few days, looking for the crucial events that might have led him to lose all hope of having a happy or productive life. You may often look for such reasons for events in your own life. Imagine that you have just been fired from a job. You might legitimately attribute this unhappy event to a grumpy supervisor, to personality conflicts with other employees, or to your inability to get to work on time. But, in fact, you can view getting fired from a job-and even committing suicide-in a totally different way. You can look at such individual experiences sociologically.

Looking at Individual Experiences with a Sociological Perspective

More than thirty years ago, the sociologist C. Wright Mills tried to explain the way sociologists think by distinguishing between *personal troubles* and *public issues.*[1] As individuals we tend to think of our daily experiences, both positive and negative, as strictly personal. Yet, if we step back from these events and look at them from a sociological perspective, as public issues, we begin to see patterns and relationships that paint a different picture. The ability to shift our point of view from individual circumstances to social patterns is what Mills called the **sociological imagination.**

For instance, if you were fired from your job, you would certainly see it as a personal trouble. You might also try to understand why this unfortunate event happened to you rather than someone else. Suppose, however, that you were a sociologist who wanted to look at the issue of job loss. Although you might sympathize with the individual problems of someone who had just lost a job, you would also be more interested in the wider view. You would no longer look at one unemployed person and his or her personal problems. Instead, you would focus on entire groups of unemployed people and their experiences as a public or social issue. And in doing this, you might recognize an underlying pattern to the many stories of individual troubles. Personal problems and troubles can be linked to public issues and the wider social forces around us.

Looking for Patterns: The Nature of Social Structure When you consider the social issue of unemployment, rather than the personal problem of someone's job loss, you can see that unemployment affects vast numbers of people each year and that it is more likely to affect some people than others. For instance, African Americans are more than twice as likely as Euro-Americans to be unemployed.[2] Unemployment also is more common in some time periods and in some communities and industries than in others. People who live in inner cities are far more likely than those who live in suburban areas to be unemployed. Similarly, the unemployment rate for construction workers is typically two to three times the rate for those who work in finance or real estate.[3]

By looking at social issues, you begin to see unemployment as reflecting social patterns, rather than as simply individual problems of unlucky people. Sociologists call such patterns **social structure,**—the ways in which people and groups are related to each other and the characteristics of groups that influence our behavior.[4]

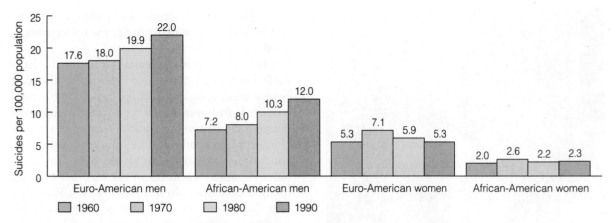

FIGURE 1-1 *Suicide Rates for Euro-American and African-American Men and Women, 1960-1990**

*The suicide rate tells us how many people out of each 100,000 in a particular group commit suicide in a given year.

Source: U.S. Bureau of the Census, 1994, p. 101; 1990, p. 86; 1975, p. 155.

The concept of social structure may seem strange at first, perhaps because you have never thought about how remarkably structured social life is. For example, each year, in what can only be imagined as terrible personal isolation, Mark and thousands of other people decide to end their lives through suicide. Yet, as you can see in Figure 1-1, from one year to the next, within any given society, the proportion of people in different groups who commit suicide actually varies only slightly. In addition, some people are more likely than others to commit suicide. For instance, Euro-American men have the highest rates of all, while African-American women have the lowest. Clearly the same people are not committing suicide each year, so something else must account for these regularities. Sociologists try to explain patterns such as these.

Creating Social Structure: The Importance of Social Action While sociologists look for the social structure that underlies individuals' lives, they also try to understand how we, as individuals, help to *create* and *change* this social structure through our day-to-day decisions. Sociologists often use the term **social action** to refer to these behaviors and decisions of individuals and groups. For instance, patterns in suicide reflect the decisions of individuals who decide to end their lives. Unemployment rates result from the decisions of employers to hire and fire individual workers. The sum of all these social actions create social structure.

Although social structure is a result of individual decisions, the relationship goes both ways. Aspects of social structure influence what individuals decide to do in obvious and subtle ways. For example, an employer's decision to fire workers is influenced not only by workers' performance but also by the demand for the company's products, competition from other firms, and the availability of other workers. In short, although the existing social structure can influence and limit our individual choices, the choices we make can also gradually alter the social structure.

Some sociological theories have been more concerned with understanding and explaining social structure; others have focused on social action and changes in social structure. Yet, to some extent, all sociological theories try to explain both social structure and social action, as well as the interactive process that links them. Throughout this book you will learn how different sociological theories have dealt with these issues.

Dispassionately Looking at the Passion of Life

In a sense doing sociology involves a paradox. On the one hand, because sociology is concerned with people's lives, sociologists often have strong views about what they study. Many sociologists have chosen their profession because they care deeply about social problems such as unemployment, poverty, and racism. They want to know more about why social problems exist and what they might do to eventually help lessen their impact. Similarly many sociologists are keenly interested in individuals' choices in such areas as family life, politics, and religion because they believe that these decisions are central to people's lives.[5] In short, sociologists often have strong feelings about their work, not only as scholars and scientists but as socially concerned individuals. As you will see in the chapters to come, this passionate engagement with social issues sometimes leads to fierce debates, but it also helps make sociology a thriving, changing, and exciting discipline.

On the other hand, sociology is a social science, based on a systematic study of the social world. More

Sociologists must step back from the emotions of everyday life to take a dispassionate and objective view of the social world.

than anything else, it is an **empirical** discipline; that is, it is based on experience and observation, not on our preexisting ideas. In other words, sociologists are committed to objectively understanding the world as it really is, not as they believe or wish it to be. This means that sociologists are not content to develop general ideas, or theories, about how social groups work. Sociologists are committed to *testing* these ideas in a disciplined way to see whether the ideas accurately describe reality. Just as the medical researcher in her white lab coat must be prepared to modify or abandon her pet ideas if the new drug she is testing doesn't work, sociologists want to ensure that their ideas about why unemployment occurs or which people are more likely to commit suicide are correct.

This is the paradox of sociology. Regardless of their passionate social concerns and cherished ideas, sociologists try to approach their studies with rigorous objectivity and careful procedures so that their explanations of the social world are as accurate and precise as possible. Of course the ideal of scientific objectivity is difficult to achieve, but sociologists believe that in the long run their commitment to this ideal will produce trustworthy results.

This chapter and the next deal with these two aspects of sociology: (1) sociologists' *passion* to understand more about how the world works and (2) their *dispassionate* or objective attempts to make sure that these understandings are correct-to abide by the rules of science. This chapter introduces some of the theories sociologists have developed to explain the social world by describing the work of some of the earliest sociologists. Chapter 2 discusses what social scientists call research **methodology,** the rules and procedures that guide research and help make it

valid. These two chapters may be the most important in this book, for they provide the underpinnings for all of the sociological enterprise.

As you will discover, social theories are far from static. Rather, they change and develop over the years as new ideas appear and older theories are tested, modified, and even discarded. Today sociology is theoretically diverse. To understand the wide variety of approaches used by contemporary sociologists, and to understand recent attempts to synthesize older theories, we need to see how sociological thought developed. In this process we will see that even the earliest sociologists tried to balance their passionate concern with the social world with dispassionate efforts to understand it.

A PASSIONATE SCIENCE: HOW SOCIOLOGY DEVELOPED

Throughout history people have developed ideas about how humans live together, or what sociologists call *social life.* Yet only in nineteenth-century Europe did the science of sociology as we know it today begin to appear. The discipline of sociology developed as social thinkers, responding to enormous changes in the world around them, followed the inspirations of such successful sciences as physics and chemistry.

In many ways this period can be seen as the culmination of two centuries of dramatic change in Western Europe. The explosion of technological innovation known as the Industrial Revolution had transformed the countryside, as hordes of peasants left their small rural communities to work in factories in large, dirty, poverty-ridden cities. Whereas for centuries the dominant economic class had been a feudal nobility in which membership was based on birth, now a new class emerged. These **capitalists** owned large industries, continually reinvested their profits to increase their wealth, and derived their power and social standing from their control of capital (money and other tangible resources). In addition, this newly powerful group had helped lead political revolutions to overthrow the aristocracy and produce massive changes in government. The Catholic church, which had been a dominant force in earlier centuries, no longer held religious sway over the vast majority of people. Instead, the Protestant Reformation, beginning in the sixteenth century, had spawned a large number of religious groups that competed for worshipers.

In addition, world exploration and trade had made people aware of many other cultures beyond their own borders. With increased travel and economic interaction, Western Europeans discovered

In the early years of the Industrial Revolution, women and children often were employed in factories. How would their lives be different if they still lived in rural villages? How might the personal troubles of these women and children be seen as public issues?

other peoples with lives strikingly different from their own. The world was no longer simple and straightforward. Where once people could live their entire lives within a small community with people very similar to themselves, they now might move several times and encounter many new and unfamiliar people, ideas, and beliefs. Meanwhile the traditional rules and sources of authority and social order-religious, political, and economic-continued to be replaced by new patterns.

The early sociologists were intrigued by these changes and the reasons that they had occurred. Many of these sociologists were also concerned with the way such forces as industrialization had affected their fellow citizens and wanted to know how society could continue in such times of social upheaval and change.[6]

At the same time, the physical and biological sciences were making rapid advances. The scientific method had led to such major technological discoveries as the mathematical basis of electricity and the nature of the circulatory system and cells. Clearly science was helping people learn more about the natural environment and their physical bodies. Could science also be used to study the processes at work in society? Over a span of two hundred years, political and social philosophers such as Thomas Hobbes, Jean-Jacques Rousseau, Georg Hegel, and John Locke had written influential treatises regarding the nature of human beings and their relationships with one another. Yet these philosophical writings were developed through logical reasoning and, unlike the discoveries of the scientists, had not been tested empirically.[7]

Inspired by the successes of physicists and biologists, the early sociologists set about creating a *social*

science. They began to develop theories, or ideas, about society that could be tested and modified as needed to more accurately reflect social reality. These sociologists recognized that they would need to use somewhat different research practices than the physical and biological scientists, but they wanted to apply aspects of the scientific method to test their theories about the nature of society. Controversy continues about how objective sociology can be. But all sociologists would agree that sociology is, at root, an empirical discipline that involves developing theories to explain the social world and examining the validity of these theories by looking at real life.

The social theorists we will discuss here, with one major exception, did not directly combine their sociological work with political activism. Yet many theorists hoped that a new science of society, called sociology, could suggest ways to create a more humane world.[8] Today, even as sociologists hope that their efforts will lead to better understandings of society, many also hope that their work will help alleviate social problems.

Responding to Social Ferment: The Classical Theorists

Four European social thinkers who began their work in the mid-nineteenth century are generally regarded as the founding theorists of sociology: Karl Marx, Émile Durkheim, Max Weber, and Georg Simmel. In their own ways, each of these theorists was able to step back from the social world and look at it in a completely new way. They introduced concepts that, even today, give us insight into the nature of social structure and social action.

Two definitions are necessary before we discuss the contributions of these thinkers. Each of these theorists used the term **society** to refer to the social world around us, and more specifically to groups of people who live within some type of bounded territory and who share a common way of life, or **culture.** Thus we might talk about the contemporary United States, Canada, or Japan-or ancient Greece or Rome-as a society. Each includes a group of people living within a specified territory who share a common culture.

Because each of these theorists viewed society from a distinct perspective, their ideas of social structure and social action often differed widely. To understand the diverse ways in which social theorists look at the world, imagine them as movie directors or as individuals wandering around a scene with video cameras. Each filmmaker can choose a different camera angle and perspective to look at the world. They might all focus on the same scene, yet by using diverse viewpoints and framing devices, they produce very different images. So, too, do social theorists use various perspectives when they look at the world around us, and thus they produce disparate descriptions of the nature of social structure and social action. Each angle has its own advantages, for each gives us a distinct viewpoint that highlights important aspects of our world. In addition, each theory can complement the others by suggesting a new way to see the same event. In this sense, as you will see throughout this book, no single theory can ever explain all aspects of the social world.

Karl Marx: Economic Relations as the Foundations of Society

Karl Marx is probably the most controversial of all the classical theorists, for he is best known as the intellectual father of communism. Virtually all his writings were directed toward radical political audiences and have inspired political figures throughout the world. Yet most of Marx's writings also involved sophisticated analyses of the nature of the social world, analyses that still intrigue and inspire sociologists.

Marx was born in 1818 in what is now Germany, the son of a lawyer and the grandson of rabbis. Shortly before Marx's birth, his father converted to a branch of the Lutheran church to maintain his legal practice, but religion was not practiced within the home. While growing up, Marx differed in many ways from other people in his community and even in his own family. This unique background might have helped him develop the ability to look at the world in fresh ways.[9]

Marx grew up surrounded by classic writings, encouraged by his father and, especially, by a neighbor-a widely read, wealthy government official who later

Marx suggested that in any society those who control the means of production are the dominant group. Who in this picture do you think controls the means of production? What influence might this control have on the lives of others who are pictured?

became his father-in-law. As a student at the University of Berlin, Marx became attached to the "Young Hegelians," a group of radical young philosophers interested in debating the fine points of the writings of the German philosopher Georg Hegel. After finishing his degree in 1841, Marx continued to pursue his interest in radical social and economic analysis and worked for a while as a newspaper writer and editor in Germany. He began to attract considerable attention for articles exposing the miserable conditions under which peasants and the poor lived, but because of governmental pressure, he lost his job and moved with his new wife to Paris. Later, after the French government objected to his presence there, he moved to London, where he lived most of the rest of his life. He never again held a regular, salaried job, instead living off meager fees from writing and the charity of his longtime collaborator, Friedrich Engels. He died in London in 1883 at the age of sixty-five shortly after the death of his wife and his oldest daughter.[10]

A prolific thinker and writer, Marx was greatly concerned with the conditions that workers of his day faced-the exploitation of children, the long work hours, the meager food supplies, and the low wages, especially when compared with the vast profits made by the owners of industry. In his writings he questioned why oppression and inequality existed and how this situation might change. Marx thought that through his writings he could help educate workers and others about the nature of the social world and that, once they understood this, they would join together to overthrow the existing order and create a more just society.

Marx believed he had found the key to social structure at any point in history in a single economic fact: the control of the **means of production,** the way in which people produce their living. Those who control the means of production, whether it be land, factories, tools, or capital, are the dominant group within any society. They can maintain their dominant position by controlling the services people receive and the ideas that are promulgated to the masses.

To examine this theory, Marx conducted extensive historical analyses of other societies. He used the results of this work to bolster his argument that the social structure inherently involves control and coercion of the less powerful by the more powerful. For instance, in the Middle Ages, feudal nobles not only owned the land on which the peasants worked but also regulated their medical care, education, and religious training, largely in ways that served their own interests. In the industrial society of Marx's own time, the **bourgeoisie,** the capitalists who owned the factories and mills, controlled the lives of the **proletariat,** the workers.

What set Marx apart from thinkers before and after him was his notion that the *economic relationships* within a society determine other aspects of the social world. He concluded that the economy, and especially the dynamics surrounding control of the means of production, make up the **base** of a society. Other areas, such as religion, family life, and political life, are all influenced by these economic facts and make up what he called the **superstructure.** For instance, because they owned the means of production, feudal lords controlled political decisions in their community, decided who would preside over the local church, and even influenced the ability of their serfs to marry and have families. Today wealthy citizens are much more likely than the rest of us are to contribute to political campaigns and to have access to high-level politicians. Thus, for Marx and his followers, the control of the means of production is the basis of social structure and the prime influence on social action.

Yet Marx did not believe that individuals are totally at the mercy of their relationship to the means of production. In fact, the potential for social action, through resistance and revolution, is central to his theory. Once again tapping historical evidence, he suggested that social change traditionally has occurred when the less powerful gained control of a means of production that was technologically superior to that currently used by the more powerful. Because the means of production and its control influences all other aspects of social life, this change affects everyone within the society, from the least powerful to the most powerful. When the less powerful people control superior technology, they can work together to alter the society. Thus the bourgeoisie overcame the agriculturally based feudal nobility in part because they owned the factories, which were the major mode of production after the Industrial Revolution. In addition, however, the transition from feudal agricultural to industrial economies involved the transformation of workers from serfs who were tied to their feudal masters to laborers who were more free to sell their skills to the owners of industry.

Marx predicted that the bourgeoisie would themselves be overthrown when the proletariat, or industrial workers, united to seize control of the means of production and ultimately create a truly egalitarian society in which all people would share the economic wealth equally.[11] In this respect Marx was not a particularly good prophet. Although his writings did inspire a number of revolutions throughout the world, these revolutions primarily occurred in poor agricultural societies, such as Russia, China, and Cuba, rather than in industrial societies. Moreover, these societies did not evolve into the idyllic egalitarian states he had envisioned. Instead, new privileged groups developed around a centralized and often totalitarian bureaucracy. Indeed, those societies that are most industrialized, including the various republics of the former Soviet Union and all of Eastern Europe, have recently rejected these governments, and many scholars now doubt whether the idyllic egalitarian societies Marx envisioned could ever emerge in modern industrialized societies.

Recent political changes, however, do not mean that Marxism will no longer influence social theory. Even within his own lifetime Marx's political predictions were consistently unfulfilled, yet Marxism as a theoretical perspective survived and continued to develop.[12] Even more important, Marx's view of social structure as a system in which the less powerful are controlled and coerced by the more powerful-and of social action as involving those with less power resisting and reacting to this situation-is an integral part of many contemporary sociological theories. Throughout this book you will see examples of the legacy of Marx's work, as sociologists examine issues such as inequality, conflict, and change.

Émile Durkheim: Social Differentiation and Integration Émile Durkheim was born in eastern France in 1858. His father and grandfather had been rabbis, and as a boy Durkheim planned to follow in their footsteps. However, in his teens, after a brief flirtation with Catholicism, he renounced religious involvements, became an agnostic, and devoted himself to a life of scholarship. Like Marx, Durkheim developed beliefs that set him apart from many of his colleagues and perhaps helped him gain the ability to look at his social world from a new perspective.

Durkheim finished his schooling in philosophy in 1882. Philosophy, however, left him dissatisfied

Even though we might not think about it, the collective consciousness affects our lives in many different situations. What do the behaviors of the people in this photo tell us about the collective consciousness?

because it seemed to avoid important questions about contemporary society and its methods did not allow its tenets to be empirically tested. Durkheim believed what was needed was a scientific study of society, a *sociology.* Because sociology was not yet taught in secondary schools or colleges, Durkheim initially continued to teach philosophy. After he published several articles about sociology, he was-at the relatively young age of twenty-nine-allowed to teach sociology courses within the philosophy department at the University of Bordeaux . He also regularly taught pedagogy or education courses, believing that through this applied field sociology could help change society for the better.[13]

Durkheim was particularly interested in what holds society together. Why is society so relatively unchaotic? How is this related to social change? Durkheim's answer involved examining the social relations between people, the ways in which these relations alter as society changes, and the ways in which they affect individuals' behaviors. Durkheim suggested that with industrialization and urbanization, societies were moving from what he called mechanical solidarity to organic solidarity. **Mechanical solidarity** refers to a society in which there is little *differentiation:* Individuals are all fairly similar to one another, with similar responsibilities, tasks, and behaviors. Today some small groups, such as strict religious sects, try to achieve this by requiring all members to abandon their previous lives, dress alike, and share all tasks, much like communities of monks and nuns in the past.

In such societies all individuals are part of, or *integrated* into, the same group, and so they share com-

mon views of what is right and wrong, important and unimportant. Durkheim called these common beliefs, values, and **norms**-rules defining behavior that is expected or required within a group or situation-the *conscious collective,* or simply the **collective consciousness.** In using this term Durkheim did not mean to imply that humans within a society have some type of "group mind," all believing as one. Instead, Durkheim suggested that people within a society develop agreements about what they should do and what is important. Beliefs, norms, and values cannot be seen or touched; they are ideas. Yet they are very real in our day-to-day lives, they continually influence our thoughts and actions, and they are central to social structure.[14]

In mechanical societies, because individuals are so similar to one another, all members share the collective consciousness, which results in relatively orderly and unchaotic social life. As societies become more complex, however, increased differentiation of tasks, responsibilities, and behaviors occurs, producing **organic solidarity.** For instance, the beginnings of differentiation can be seen in hunting and gathering societies, the simplest known tribal societies. In these groups men have certain tasks, such as hunting, and women have other tasks, primarily gathering and child care. Beyond this there may be relatively little differentiation. In contrast, within complex modern societies such as our own, there is much more differentiation, with people performing diverse types of work and exhibiting various behaviors. Individuals also may belong to any number of different groups, from families and work organizations to religious groups, political parties, and networks of friends. This more complex, organic society was the type that Durkheim believed was typical of nineteenth-century Europe.

You might think that, as a society becomes highly differentiated, the collective consciousness would weaken and that disorder and mayhem would result. Yet, in fact, this is generally not true. Within differentiated societies all the various tasks that people perform are necessary for the well-being of the society, just as in hunting and gathering societies both women's and men's tasks are necessary for the group's survival. Indeed, the simple fact of differentiation builds relationships between people. For example, you are unlikely to grow or gather all the food that you need to survive, build your own shelter, or even make all of your own clothes. To obtain these goods, you must depend on others who have specialized in these areas, such as farmers, grocers, builders, and tailors. From their varied relationships people within organic societies develop and maintain bonds with friends and families, co-workers, and neighbors. The collective consciousness develops from the interdependencies of these social groups, influences the behavior of the group members, and forms the basis of organic solidarity.

Durkheim insisted that the collective consciousness could be studied and understood apart from individuals-as something that applied to large groups and even whole societies. To demonstrate this, he conducted the first sociological study, on the paradox of suicide. As we saw earlier in this chapter, suicide is the most individual of choices, yet suicide rates appear to be relatively constant from one year to the next. Why?

For Durkheim the answer lay in the collective consciousness. Because the collective consciousness results from individuals' interactions within a social group, societies and segments within societies might differ in the types of interactions that their members have and in the strength of the norms that they experience. Sometimes the patterns of relationships are disrupted, so that previously stable and predictable relationships become more problematic. For instance, in times of economic crisis some wealthy people lose their money. They might still want to associate with their rich friends but might no longer be able to afford to participate in the same activities. The newly rich have a similar problem. All their old friends might be of modest circumstances, yet their newly acquired wealth might require them to behave like and associate with rich people. Similarly, students like Mark who have to drop out of school and work full-time because of problems with grades or money might be more attached to their school friends than to their new co-workers. And teenage mothers might still see themselves as young women who should be shopping and hanging out with their friends rather than dealing with the awesome responsibilities of motherhood.

Durkheim proposed that people who go through a period of temporary disequilibrium, like those just described, experience **anomie,** a situation of uncertainty over norms, or normlessness. They have no consistent, predictable set of norms, no constant collective consciousness to guide their actions. Durkheim reasoned that, because such individuals no longer had consistent, clear-cut norms to direct them, they would be more likely to commit suicide.

Sociologists are not content simply to develop theories, however plausible they may seem. What distinguishes Durkheim's work on suicide from mere speculation is his effort to *test* his ideas against observable facts. To test his deduction, Durkheim used a vast array of statistics gathered by various government agencies from throughout Europe. In the process he provided an excellent model for future sociologists by carefully testing his expectations in many different ways. He reasoned that the more tests he provided for his theory, the more faith he could have in his conclusions. The results of his study, as well as those of many other researchers in later years, have supported his theory.

For instance, Durkheim suggested that societies that had experienced economic crises or economic booms, resulting in more newly poor or newly rich people, would be more anomic and thus could be expected to have higher suicide rates than societies with relatively stable economies. To test this idea, he examined data spanning long time periods in a variety of countries. As predicted, he found that, because of what he called a "disturbance of the collective order," suicide rates are much higher during times of economic *change* than in other periods, even in disadvantaged societies where many people are quite poor. Perhaps surprisingly, suicide rates are actually quite low in countries that consistently have a great deal of poverty. Durkheim's theory also would suggest that people who experience discordant or unusual social positions would be more likely to experience anomie and thus would have higher suicide rates. For instance, it is relatively rare for a teenager to be a parent and rarer still for a teenager to be widowed. As Durkheim's theory would predict, however, teenagers in these positions indeed have higher suicide rates than other teens do.[15]

Durkheim, as well as other sociologists since his time, repeated this research process again and again and amassed considerable support for his notion that social relations influence both the collective consciousness and individuals' actions. Some sociologists have applied his insights to explain the large discrepancies in suicide rates between African-American and Euro-American men and women, as shown in Figure 1-1. As we will explore more in subsequent chapters, Euro-American men occupy a much wider range of jobs and positions within society than do Euro-American women or African Americans. African-American women face the most restrictions of people in these four groups. Why, then, do African-American women have much lower suicide rates? Drawing on Durkheim's insights, sociologists have suggested that the restrictions that African-American women face actually make their social relationships more predictable. Because African-American women have fewer choices and opportunities, they are much more likely to experience a smaller range of activities. Ironically, while these restrictions are much more likely to produce poverty and deprivation, they are also less likely to produce anomie, or normlessness. Based on Durkheim's reasoning, sociologists have suggested that the lower level of anomie contributes to the lessened probability of suicide.[16] In fact, these social conditions have persisted for many years, resulting in the patterns shown in Figure 1-1.

Durkheim remained a highly respected academic throughout his life, and his publications and teachings established sociology as a legitimate academic discipline in his country. In 1902 he moved to the Sorbonne in Paris, the most renowned French university. He died at the age of fifty-nine, broken-hearted over the death of his son in World War I.[17] His views of the primacy of social relations and collective norms and values to social structure, and the ways in which differentiation within a society affects social relations, remain important elements of sociological theory today.

Max Weber: The Importance of Social Action

Max Weber (pronounced VAY-ber) was born six years after Durkheim, in 1864, in Germany to Protestant parents. His father was a highly successful politician, and the family was economically comfortable. Weber received an excellent education in his early years and was extremely well read. Unlike Marx and Durkheim, who studied philosophy, Weber was trained as an economic historian, which he used to great advantage in developing his theories. He was fortunate to receive a university position at a fairly young age. This, coupled with a substantial inheritance he received in mid-life, allowed him the freedom to explore and develop his ideas without financial worries.

Despite these advantages, however, Weber's life held many contradictions. He was a somewhat frail boy and spent most of his time reading and studying. Nevertheless, in college, while he continued to pursue his intellectual interests, he also joined dueling fraternities, held his own in drinking bouts, and became a robust physical specimen. He was devoted to his parents and lived at home for many years, yet he also had violent arguments with them, especially with his father. Usually Weber worked extremely hard, far into the night. But he could also be paralyzed by bouts of depression, lasting for years, during which he could not work at all. And although he was very active politically, he was adamant about avoiding political pronouncements in his academic work and strongly chastised those who did not follow his example. Although it is impossible to know for sure how these contradictions in Weber's personal life affected his work, they might well have helped him develop the very broad approach to sociology that characterizes his writings and, like Marx and Durkheim, helped him step back from his own social world to look at it in a new way.

Although he was plagued by ill health for many years, Weber recovered sufficiently by the age of thirty-nine to produce numerous books and papers which still receive a great deal of attention. He continued his work until his death in 1920, at the age of fifty-six, from pneumonia.[18]

Weber wrote about many aspects of society, as you will see throughout this book. Two themes in his work, however, are central to the development of sociology. First, much of Weber's writings were in reaction to the writings of Marx. Weber began his university studies at about the time Marx died, when Marx's writings would have been widely available. Unlike Durkheim, Weber was always respectful of Marx's ideas, but he regarded Marx's analyses as too simplistic. Second, unlike both Durkheim and Marx, Weber gave central importance in his analyses to social action, the meanings that people attribute to their actions, and the way these actions and meanings affect the social order. He used the German term *verstehen* to refer to "interpretive understanding," a method of trying to grasp the meanings people attach to their actions. Weber stressed that part of a sociological perspective is comprehending the motives that underlie social actions, and this is best done by trying to understand other people's viewpoints. This does not mean that he thought individuals have a unique view of the world and their own life. Quite to the contrary, Weber believed that there are commonalities to people's experiences and interpretations, and that it was important for sociologists to understand these commonalities.

Weber's most famous work, *The Protestant Ethic and the Spirit of Capitalism,* illustrates these ideas. **Capitalism** refers to the free-enterprise economic system in which private individuals or corporations develop, own, and control business enterprises. Capitalism began to emerge in Western Europe in about the sixteenth century, eventually helped to stimulate the Industrial Revolution of the nineteenth century, and

BUILDING CONCEPTS

The Perspectives of Four Major Classical Theorists

The ideas of four classical theorists—Karl Marx, Émile Durkheim, Max Weber, and Georg Simmel—influence sociological thought and inquiry even today. Each of these theorists viewed society from a distinct perspective.

Theorist	View of Social Structure	
Karl Marx	Social structure involves coercion and control of the less powerful by the more powerful. Economic relations are crucial to understanding differential power.	
Émile Durkheim	Social structure is influenced by the degree to which individuals in a society share a collective consciousness. Social solidarity is influenced by relationships among people and the extent of differentiation within groups.	
Max Weber	An examination of what motivates social action is key to understanding social structure. Religious beliefs and practices, for instance, can influence economic relationships and social change.	
Georg Simmel	Social structure is a "web of group affiliations," created by microlevel and mesolevel interactions between people. Conflict is a necessary and often beneficial part of social life.	

today is the predominant global economic system. Why, Weber asked, did capitalism develop in Western Europe, but not in places like India and China, which were, in many ways, just as advanced at that time as Europe? China, for example, had a well-developed educational system, engaged in extensive international trade, built large cities, and made technological advances such as gun powder and book printing before other countries. Given all these advantages, why did China not develop the economic system of capitalism?

To answer this question, Weber tapped the concept of *verstehen*. Marx had suggested that economic relationships are the primary determinant both of historical changes, such as the development of capitalism, and of people's beliefs and values. In contrast, Weber suggested that changes in individuals' *motivations* and *beliefs* actually influenced the advent of capitalism. For capitalism to develop, individuals must be willing to reinvest their income to expand their businesses or create new products and markets. This requires that people not only work long, hard hours but also forgo everyday worldly pleasures and concentrate only on their long-term economic success. Weber called this orientation the "spirit of capitalism." In his study of economics and history, he found evidence that this spirit of capitalism was enhanced by the development of Protestantism, which began in opposition to Catholicism in sixteenth-century Europe.

Most of Weber's analysis focused on Calvinism, a branch of Protestantism that was a precursor to modern Presbyterianism. Calvinists believed in the doctrine of "predestination," the idea that people are fated from birth to be either saved from or condemned to hell. Unfortunately one could never know, while still on earth, if one was really saved, and no amount of good works or effort could alter this fate. Calvinists were, understandably enough, quite worried about whether they were saved or damned and would look for signs that they might be among those elected for salvation. One possible clue to membership in the chosen group was conducting one's life in a religious manner. In his readings of published sermons and ministers' guidelines for a godly life, Weber found that Calvinists were continually reminded that they should work hard; not waste time; avoid the pleasures of worldly life, such as the theater or even literature; and use their possessions for the glory of God. A good Calvinist would not spend money on lavish entertainments or possessions, but instead would work hard, reinvest the money in business, and see the expansion of that business as bringing glory to God. This "Protestant ethic," referred to in the title of Weber's book, was precisely the behavior that was needed for capitalism to develop and thrive.

Of course Weber recognized that not all capitalists were Calvinists. In fact, he suggested that strict

The Puritans who founded Plymouth Colony in New England were Calvinists. How might the success of Plymouth Colony be related to the "Protestant ethic"?

Calvinist beliefs were relatively short-lived as religious tenets, but that their legacy became part of the collective consciousness (as Durkheim termed it) and persisted for many years. Thus, Weber suggested, religious beliefs prompted actions that were necessary for the development of capitalism and altered the beliefs and values that make up the social structure. In contrast, although China and India had other characteristics that could have helped spur capitalism, their people did not have religious beliefs that corresponded to the capitalist spirit.

Whereas Marx argued that economic relations influence religious views as well as all other types of beliefs, Weber asserted that the world is much more multidimensional and complex and that sometimes beliefs can influence economic and social relationships. Weber also disagreed with Durkheim. He recognized the nature of the social structure of beliefs, values, and norms that Durkheim discussed. Yet, in contrast with Durkheim, he focused much of his own analysis not just on how this social structure influences individuals' actions but also on how individuals' actions influence social relations and social structure.[19] Weber's multifaceted view of the social world, his insistence that people's beliefs can influence their social actions, and his attempts to show the linkage between social action and social structure continue to influence sociology today.

Georg Simmel: Social Interactions as the Basis of Society Today it is not uncommon to hear sociologists call themselves Marxians, Durkheimians, or Weberians, signifying that they follow in the footsteps

Sociologists can take a macrolevel perspective that looks at entire societies and very large entities, such as this city, a mesolevel view that looks at a group or organization, such as this university, or a microlevel view that looks at individuals, such as these students.

of these early theorists. Rarely, however, do you hear sociologists refer to themselves as Simmelians, for, unlike the other three classical theorists discussed here, Georg Simmel did not try to develop overarching theories that explained the course of history or many aspects of society. Yet he introduced ideas that have continued to appear in sociology throughout this century, and, more than any of these other theorists, Simmel took a close-up view of the social world.

Born in Germany in 1858, Simmel was a contemporary of Durkheim and Weber. While he was still quite young, his father died. He was never close to his mother and is reported to have felt marginal and insecure from a young age. As with the other early theorists, this feeling of marginality may have helped Simmel step back from his world and look at it from a sociological perspective. With the support of a wealthy guardian (who also left him a substantial inheritance in later years) Simmel had the same type of

classical education as Weber did, studying history and philosophy at the University of Berlin. After receiving his doctorate, Simmel stayed at the university, unpaid except for fees from students who attended his lectures. He was a very popular lecturer and spoke and wrote on an extremely broad range of areas, encompassing both philosophy and the developing field of sociology.[20]

Sociological theories vary in their scope-that is, in how far they step back from the social world. These different orientations are similar to the different views you see when looking through the zoom lens of a camera. Some social theories, such as those developed by Marx, Weber, and Durkheim, tend to involve a distant, wide-angle view, focusing on broad areas of society. These perspectives are said to involve a **macrolevel** analysis. They can be used to explain such phenomena as unemployment, the persistence of poverty, the existence of various types of religious groups, and global conflicts and wars. Other theories zoom in to take a much closer look, focusing on narrower aspects of social life, such as our day-to-day activities and relations with other people. These perspectives are said to involve a **microlevel** analysis. They can be used to account for events as diverse as conflict within organizations or families, people's choice of marriage partners, and the rules children develop to guide their play activities. In between the macro- and microlevels is what is typically called a **mesolevel** analysis, one that focuses on social groups and organizations, such as individual families, classrooms, and work groups.

Simmel concentrated on micro- and mesolevel analyses. Basic to his work is the concept of **group,** which sociologists define as two or more people who regularly and consciously interact with each other. For instance, your family, the people you live with, the members of your stamp club, and your co-workers would all be groups. Simmel suggested that we can see society as a "web of group affiliations." He insisted that human interactions, or the relationships people have with one another, are the basis of large-scale social structure. Families, churches, political bodies, and nations are all ultimately based on these social interactions and relationships. From Simmel's viewpoint the web of social relations forms the social structure of society.

Recognizing, as Durkheim did, that modern societies are much more differentiated than earlier societies are, Simmel showed how this greater differentiation alters human relationships. In modern, industrial societies the webs of group affiliations are not as tightly connected as in premodern societies. For example, your friends at work might not be your friends at school, and neither set of friends might know any of your family members. Although this arrangement gives you much more freedom than you would have in a more tightly structured society, Simmel suggested that it can also lead to greater isolation and less regulation of your behavior.

Like Weber and other German scholars of his era, Simmel devoted much of his work to challenging the writings of Marx. Whereas Marx suggested that there could eventually be a truly egalitarian society in which conflict would no longer exist, Simmel claimed that conflict is a necessary and often useful part of social life. Even though he recognized that conflict can certainly be destructive to individuals and groups, he also suggested that it can be beneficial because conflict is almost always combined with cooperation, in both very small and very large groups. For instance, a husband and wife who are arguing might each try to enlist the support of friends. Likewise large-scale conflicts between nations involve the cooperation of armies and, often, the alliance of groups of nations.

Thus, although social structure is ultimately based on interaction, we should not see this structure as simply involving cooperative relationships; it also necessarily involves conflict. [21]

Despite the popularity of his work and the support of renowned scholars such as Weber, Simmel was not given a regular university position until shortly before his death. This is generally attributed to his broad interests not fitting into a single discipline, as well as to widespread anti-Semitism, or prejudice and discrimination against Jews, within the university system. Although this treatment no doubt helped reinforce Simmel's feelings of marginality and exclusion, he remained a well-respected lecturer and writer and, in conjunction with Weber, helped found the German Society for Sociology. Simmel died in 1918, at the age of sixty, just before the end of World War I, of cancer of the liver.[22]

Simmel bequeathed a rich legacy to future sociologists. Although much of his work focused on describing aspects of social interaction, he, like the other classical theorists, argued that individuals both are shaped by their social world and participate in shaping it. Society is developed through interactions of human beings, yet these interactions influence what we can and cannot do in our daily lives. As you will see in subsequent chapters, Simmel's ideas about the importance of interaction as the basis of social life and the role of conflict in the social world influenced both early and contemporary sociologists in the United States.

Sociology Today

Sociology began to develop in the United States, as in Europe, toward the end of the nineteenth and beginning of the twentieth centuries. A number of the early American sociologists studied in Europe, and many of them were heavily influenced by the works of Marx, Weber, Durkheim, and Simmel. In the chapters to come, you will read more about the theories of American sociologists.

Most contemporary sociologists work in universities and colleges and do research as part of their jobs. In this research they often study issues of concern to society-from poverty and racial inequality to family discord and educational achievement. A growing number of sociologists also are employed outside academe. Some work for government agencies, both in the United States and internationally, where they study issues as diverse as population growth, crime and delinquency, and child welfare. Some sociologists work in the emerging field of clinical sociology, where they help individuals, families, and organizations solve problems of daily living. Some work for corporations, helping them develop

BUILDING CONCEPTS

Macrolevel, Mesolevel, and Microlevel Analysis

Sociological theories vary in their scope, or the degree to which they step back from the social world.

Level of Analysis	Characteristics	Example
Macrolevel	A macrolevel perspective focuses on broad areas of society.	The theories of Marx, Weber and Durkheim tend to involve distant, wide-angle, macrolevel views.
Mesolevel	A mesolevel perspective focuses on social groups and organizations, such as individual families, classrooms, and work groups.	An example of a mesolevel perspective is Simmel's analysis of the role of conflict in promoting group cohesion.
Microlevel	A microlevel perspective focuses on narrower aspects of social life, such as our day-to-day activities and relations with other people.	Simmel took a microlevel perspective when he suggested that individuals in modern societies are more isolated and less regulated than they were in earlier eras.

more effective personnel practices and working with customer relations, marketing, and advertising departments. You will meet a number of sociologists in the "Sociologists at Work" feature of this book.

Contemporary sociology has much in common with the other social sciences, such as political science, economics, anthropology, and psychology. All the social sciences deal with events that relate to our daily lives. Like sociologists, other social scientists step back from looking just at their own lives or one person as an individual and try to get a broader view. But, in general, scholars in the other social sciences tend to focus on relatively discrete sectors, or parts, of the social world. For instance, economists look at monetary systems and economic relations, including areas as diverse as the cost of living and international rates of exchange. Political scientists look at governments and political systems in this country and around the world, ranging from studies of why people choose to vote for particular candidates to how governments rise and fall from power. Anthropologists have traditionally looked at relatively simple

and non-industrial societies, exploring how these worlds are both similar to and different from each other and our own. Psychologists study individuals and their behaviors, how people develop their unique personalities and life choices. Throughout this class you will see how sociologists also look at the economy and the political world, how they study societies other than our own, and how they examine individual behavior. Many sociologists have done research with social scientists in other fields; and interdisciplinary work involving social scientists from all these fields is becoming more common.

Yet sociology is different from these other social sciences precisely because it is so directly concerned with all of social life. Sociologists look at daily life from all types of angles. They don't just look at the economic or political world or at societies other than our own. They study individuals and their actions, but they also look at groups such as families, organizations, and even entire nations. In fact, part of what makes sociology so exciting and so much fun is that you can look at virtually anything people do from a sociological perspective. Sociologists study sport, medicine, governments, banks, business corporations, even pets and leisure activities. No other social science is so concerned with all aspects of life.

Building on the classical tradition, sociology in the United States has always been theoretically diverse. Like Simmel, today's sociologists look at the narrow microlevel of our interactions with others and the way these interactions affect our daily lives. Sociologists also look at the broad macrolevel of social reality and the mesolevel of organizations and groups. Like Marx, sociologists examine inequality and the reasons that some people have greater opportunities and benefits than others. Like Weber, sociologists try to understand the motivations for people's social actions. Today's sociologists also examine criminal behavior and areas such as the family, religion, education, and the political world, often drawing on the insights of Durkheim and the other classical theorists. They look at life in cities, in rural areas, and in countries around the world, and they try to understand why societies change and alter. Each subsequent chapter in this book features a detailed example of sociologists' work. You can get an idea of the wide range of issues sociologists examine by looking at Figure 1-2 and the list of questions addressed in these chapters.

In their research sociologists often use what the American sociologist Robert Merton has called middle-range theories. In contrast with the so-called grand theories of Marx, Durkheim, and Weber, **middle-range theories** deal with relatively limited areas of the social world. These theories often incorporate aspects of the various grand theories, but they are much more directed and applied toward specific research problems. They don't try to explain as many areas of the world as grand theories do, but, because they are more specific and limited, they can more easily be tested. In the next chapter you will see how sociologists tested the usefulness of two such middle-range theories in explaining the problem of domestic violence.

Some answers that today's sociologists have worked out to their questions about the social world are well developed. Others are much more tentative. This no doubt reflects the fact that sociology is a relatively young discipline, and there is much that sociologists do not yet understand about the world around us. The relative youth of sociology, however, also helps make it more exciting and fun for those who work in the field. There is so much to learn and discover that all sociologists can actively participate in developing greater knowledge of how our social world works.

Yet, just as the classical theorists often tried to show how their own ideas were superior to others, modern sociologists haven't always agreed with one another. At times sociologists have engaged in vocif-

erous arguments about what kinds of sociological questions should be asked and which theoretical frameworks provide the "best" answer. In more recent years, however, a number of sociologists have tried to demonstrate how various theories, often seen as competing in earlier decades, can yield more insights when used together. In essence they have said, "Hold on, people! No one theory is best. Sociology involves so much-the entire social world-that each of these theories can help us understand a different part of the puzzle. Each of these theories is simply using a different camera angle and frame-a different lens-to look at the same picture."[23] One sociologist who has helped formulate these integrative views is Janet Chafetz, whose work is described in Box 1-1.

You can see how sociologists use different lenses-or perspectives-by thinking about Mark, the young man described at the beginning of the chapter. Using a Durkheimian perspective, we could suggest that the many changes Mark experienced placed him in an anomic situation, one in which there were few norms to guide his behavior. Using a Weberian perspective, we might try to understand Mark's motivations for his actions. Using Simmel's perspective, we could examine Mark's relationships with other people and see that his web of social relations had been severely altered and that this could also have influenced his view of himself and provoked his tragic act. Each of these perspectives can, in its own way, help us understand not just Mark's behavior but also people's behavior in general-these perspectives allow us to step back and take a sociological perspective.

Even though each of the sociological theories presented in this book has its own unique features, they all tend to deal with three basic issues: (1) social structure, (2) social action, and (3) the linkage between the two. Contemporary social theorists stress that the association between social structure and social action is the key to understanding the relationship between various levels of social reality-between what we see on the micro-, meso-, and macrolevels. Social actions, or individual choices, lead to the structure of power relations, the control of the means of production, and the body of norms and values that make up Durkheim's collective consciousness. Regular patterns of social activity on the macrolevel of social reality, such as the number of people who commit suicide within a society, result from individual choices of people like Mark. These actions are influenced by the nature of the social structure, but they also, in turn, influence and create the social structure. In the chapters to come, we will continually return to different theoretical interpretations of the nature of social structure, social actions, and the relationships between them.

Chapter 2: *Methodology and Social Research*
What can police officers do to stop men from repeatedly abusing their wives and partners?

Chapter 3: *Culture and Ethnicity*
Why do people choose to see themselves as part of a particular ethnic group, and what does this group identity mean to them?

Chapter 4: *Socialization and the Life Course*
How do Little Leaguers' experiences on their teams help them develop a sense of masculinity?

Chapter 5: *Social Interaction and Social Relationships*
How can groups be structured to minimize the influence of race-ethnicity on group interactions?

Chapter 6: *Deviance and Social Control*
What can explain the probability that juvenile delinquents will continue in a life of crime as they get older?

Chapter 7: *Social Stratification*
Who are the wealthy in America, and how does their wealth translate into power and influence the rest of society?

Chapter 8: *Racial-Ethnic Stratification*
Why does racial-ethnic prejudice vary from one place to another and one time period to another?

Chapter 9: *Gender Stratification*
What happens when men and women enter occupations where they are in a minority, such as women becoming marines and men becoming nurses?

Chapter 10: *The Family*
How does growing up in a family with only one parent or with a stepparent influence children's lives as young adults?

Chapter 11: *Formal Organizations*
Why do some businesses thrive while others fail?

Chapter 12: *Religion*
Why have some churches lost members in recent years while others have gained members?

Chapter 13: *The Political World*
How do contributions to political candidates by business corporations affect the political process?

Chapter 14: *The Economy*
How do social relationships and interactions influence the work careers of film composers?

Chapter 15: *Education*
Why do some schools do a better job of educating children than others do?

Chapter 16: *Health and Society*
Why are people in some societies healthier than people in other societies?

Chapter 17: *Population*
Why is the education of women such an important factor in reducing population growth in developing countries?

Chapter 18: *Communities and Urbanization*
Why does racial-ethnic segregation persist in the United States, and what effects does this segregation have?

Chapter 19: *Technology and Social Change*
How did the introduction of the telephone change social life?

Chapter 20: *Collective Behavior and Social Movements*
How and why do environmental movements differ from one European country to another?

FIGURE 1-2 *Featured Research Questions by Chapter*

DISCOVERING SOCIOLOGY: THREE KEY THEMES

Throughout this book three central themes will be emphasized, all of which have been introduced in this chapter. First, *sociologists use a variety of theories, at several levels of analysis, to explain the social world.* These theories examine the nature of social structure and social action and involve macro-, meso-, and microlevel views of the social world.

Second, *to help formulate, test, and refine their theories, sociologists do scientific research.* Sociologists are committed to social science, to the objective testing of their ideas, and to using the results of these

BOX
1-1

SOCIOLOGISTS AT WORK

Janet Chafetz

In sociology the focus is on how you could use theoretical ideas to understand social life.

Janet Chafetz, University of Houston, has specialized in integrative theoretical work.

How did you become a sociologist, and why did you decide to study theory? My studies in history, both as an undergraduate and a master's student, focused on European intellectual and social history. While working on my M.A. at the University of Connecticut, I audited a graduate seminar in sociological theory that a friend was taking. I had never taken any sociology, but I found the seminar fascinating! At the same time, I was getting bored with a strictly historical approach and decided to change my major to sociology. In history the focus was on who influenced whom. In sociology the focus was on how you could use theoretical ideas to understand social life. I was far more interested in the latter than the former.

Students may wonder why sociologists think theory is so important. What would you say to a student who asks, "Why should I understand social theory?" Sociologists share subject matter and methods with many other fields of inquiry. What makes sociology unique is the perspectives and concepts we use to interpret and make sense out of the social world. Those concepts and perspectives are embodied in our theories-there is no sociology without sociological theory. Therefore students need to study the theory to learn the specific set of perspectives that we call sociological.

In your work you pull together the ideas of many social theorists. Why did you choose this integrative approach? I view theory as a kit bag of concepts and ideas that collectively help the sociologist to answer questions about the empirical world (now or historically, here or elsewhere). You employ whatever theoretical ideas you can to help you answer those questions. My view of research methods is the same: You employ whatever methods you can to help answer the question you're asking. The more theories and methods you can call upon and use, the richer and more complete your potential answers.

The students reading this chapter may have noticed that there are no women included among the classical social theorists. Could you comment on why this is the case and how the situation has changed in recent years? The founders of virtually all disciplines were men because women were effectively barred from pursuing scientific/academic careers. Moreover, the work of the few women who did manage to achieve such careers was largely ignored. There were some feminist social theorists in our discipline earlier in the century whose works are now being resurrected, but only a handful of feminist sociologists seem to know about them.

Since about 1970 feminist sociologists have been developing theories that reflect every theoretical tradition in sociology. I see signs that, gradually, mainstream sociology is becoming aware of feminist theories.

---

tests to refine their ideas to make them more accurate. In the next chapter you will learn more about the procedures sociologists use to systematically study the social world, and you will see the twists and turns that can occur in the research process. Each subsequent chapter also features an example of actual sociological research.

Third, *a sociological perspective can help us understand our own life experiences and begin to solve serious social problems.* When you apply your sociological imagination, you will begin to see your own

life and the lives of others in a new way. In addition, many sociologists hope that their efforts will help make the world a better place in which to live. Examples of sociological work that addresses social problems are given in the coming chapters in a feature called "Applying Sociology to Social Issues."

Critical-thinking questions that will help you apply and understand the issues presented in each chapter are found at the end of the "Featured Research" sections and at the end of the "Applying Sociology to Social Issues" boxes.

SUMMARY

This book has three major goals: (1) to present sociological theories, (2) to illustrate sociological research and methods, and (3) to show how sociology can be used to understand the world in which we live. The ideas of the four major classical theorists—Karl Marx, Émile Durkheim, Max Weber, and Georg Simmel—influence sociological thought and inquiry even today. Of central importance are the concepts of social action and social structure and the distinctions between micro-, meso-, and macrolevel perspectives. Key chapter topics include the following:

- Adopting a sociological perspective means stepping back to look for underlying patterns in social life. Sociological theories try to explain the nature of social structure and social action and the ways in which social structure and social action are related. Different theories look at society from different perspectives and at macro-, micro-, and mesolevels of analysis.

- As social scientists sociologists systematically test their theories through research. Even though sociologists often have passionate feelings about the areas they study, they try to study the world dispassionately and rigorously.

- The discipline of sociology developed in Europe and the United States at the end of the nineteenth and beginning of the twentieth centuries, as scholars like Marx, Durkheim, Weber, and Simmel tried to understand the many changes that had occurred within their social worlds and apply scientific methodology to test their theories.

- Marx asked why oppression and inequality exist and how they might be eliminated. He emphasized that social structure involves coercion and control of the less powerful by the more powerful

and that economic relations are crucial to understanding differential power.

- Durkheim asked what holds society together. He suggested that the collective consciousness—shared norms, values, and beliefs—is the basis of social structure and solidarity and that it is influenced by relationzships among people and the extent of differentiation within groups. He tested these theories with an empirical study of suicide.

- Weber wrote about many different areas of social life but was especially interested in the motivations that underlie social actions. In one study he asked why capitalism developed in Europe and not in other areas of the world and showed that religious beliefs and practices can sometimes influence economic relationships and social change.

- Like Weber, Simmel looked at many different areas of society, and like all the classical theorists, Simmel sought to identify the basis of social structure. Unlike the other theorists he took more of a micro approach, suggesting that interactions between people form the social structure and that conflict is a necessary and often beneficial part of social life.

- Modern sociologists often disagree about the adequacy of various theories. They also might develop middle-range theories to examine specific research questions. Sociology has much in common with other social sciences, but looks at more of the social world than any other discipline does. Recently, some theorists have tried to integrate various theoretical perspectives into more unified and global theories that recognize the multifaceted nature of social structure and its relationship to social action.

KEY TERMS

anomie 11

bourgeoisie 9

base 9

capitalism 12

capitalists 6

collective consciousness 10

culture 8

empirical 6

group 16

macrolevel 15

means of production 8

mechanical solidarity 10

mesolevel 16

methodology 6

microlevel 16

middle-range theories 18

norms 10

organic solidarity 10

proletariat 9

social action 5

social structure 4

society 8

sociological imagination 4

sociology 2

superstructure 9

theories 3

INTERNET ASSIGNMENTS

1. Using a search engine, enter a sociological theorist's name and see what types of Web pages can be accessed. For example, entering "Durkheim" might result in a link to the "Durkheim Pages," a Web site that contains information about Durkheim's life and scholarly work. By entering "Marx," you might find a link to the "Marx and Engels On-Line Library." If you enter "Max Weber," you might find a link to the American Sociological Association's Max Weber prize for a paper on work, occupations, or organizations. What type of information do you find on these theorists? Can you locate research or publications on these theorists or that are based on their work? Can you locate any on-line access to journals that focus on social theory? Does there seem to be any one theorist who is more widely referenced than others? What do you think this indicates about the influence of specific theorists on social research? (*Hint:* If you don't obtains results with one search engine, try another one.)

2. Enter several of the sociological concepts of key theorists into a search engine. What different types of information do you find? For example, enter the concept "anomie" and run a search. What links do you obtain? Do you find any links on academic research that uses the concept "anomie"? What types of sites do you locate that are not academically oriented? For example, you mighty find a link to the Web page of a band called "Anomie." Why do you think the band picked this name? Is this a good example of an application of Durkheim's theory? Can you find any other applications of sociological theory?

3. Enter "capitalism" into a search engine. What type of information do you retrieve? Can you readily identify which theorist's view of capitalism has influenced the creators of these pages? What other issues are included in these pages? How do some of these issues (for example, poverty, worker rights, the role of credit and debt in American society) relate to sociological theory? How are these pages examples of how we can use theory to help us understand social issues?

INFOTRAC COLLEGE EDITION: READINGS

Article A17589042 / Marxism, *The Columbia Encyclopedia,* Edition 5, 1993. This lengthy encyclopedia entry provides a useful and readable overview of the basic tenets of Marxism.

Article A53650181 / Riemer, Jeffrey W. 1998. Durkheim's "heroic suicide" in military combat. *Armed Forces & Society: An Interdisciplinary Journal, 25,* 103–113. Reimer tests some of Durkheim's theories regarding suicide with contemporary data regarding suicide in combat situations.

Article A20998291 / Garrett, William R. 1998. The Protestant Ethic and the spirit of the modern family. *The Journal for the Scientific Study of Religion, 37,* 222–234. Garrett builds on Weber's classic work by describing how the Calvinism and the

Protestant Ethic influenced the development of the modern family as well as capitalism.

Article A53857473 / Morgan, David. 1998. Sociological imaginings and imagining sociology: Bodies, auto/biographies and other mysteries. *Sociology, 32,* 647–657. Morgan, a British sociologist, discusses the "sociological imagination" and how it can be applied to specific areas sociologists study.

Article A14525743 / Hollinger, David A. 1991. Social sciences in U.S. history. *The Reader's Companion to American History,* Edition 1991, 1003–1007. This article gives an overview of the different social sciences, their development in the United States, and their relationship to each other.

FURTHER READING

Suggestions for additional reading can be found in *Classic Readings in Sociology,* bundled with this book. If you purchased a used copy of this book that does not include this custom-published reader, go to www.sociology.wadsworth.com for ordering information.

RECOMMENDED SOURCES

Chafetz, Janet Saltzman. 1990. *Gender Equity: An Integrated Theory of Stability and Change.* Newbury Park, Calif. : Sage. Chafetz (profiled in Box 1-1) provides a good example of the contemporary use of integrated theory.

Collins, Randall. 1994. *Four Sociological Traditions.* New York: Oxford University Press. Randall reviews sociological theory from the classics to the present time.

Coser, Lewis A. 1977. *Masters of Sociological Thought: Ideas in Historical and Social Context,* 2nd ed. New York: Harcourt Brace Jovanovich. Coser includes very readable biographies of all the classical sociological theorists.

Hughes, John A., Peter J. Martin, and W.W. Sharrock. 1995. *Understanding Classical Sociology: Marx, Weber, Durkheim.* Thousand Oaks, Calif.: Sage. This is a short, readable discussion of the life and writings of these three classical theorists.

Ritzer, George. 1995. *Sociological Theory,* 4th ed. New York: McGraw Hill. This is an extensive review of various sociological theories.

Turner, Jonathan H., Leonard Beeghley, and Charles H. Powers. 1997. *The Emergence of Sociological Theory,* 4th ed. Belmont, Calif.: Wadsworth. The authors trace the development of sociological theory from 1830 to 1930.

INTERNET SOURCE

The Dead Sociologists' Society site at Radford University. This very informative site provides biographical information and discussions of the writings of sociologists discussed in this chapter and other classical (and more recent) theorists.

Pulling It Together

We will return to the ideas that were presented in this chapter at many places throughout the rest of the book. You can also look up topics in the index.

Methodology and Social Research

*L*ORI WAS A PREGNANT twenty-five-year-old woman living with her boyfriend, James. James had lost his job in the lumber industry, and Lori's pregnancy made it difficult for her to hold a job. Every once in a while, from the time they first started dating, James would slap Lori, sometimes playfully and sometimes so hard that it left a mark. Although they had always argued and James had always had a temper, as Lori's pregnancy progressed and he couldn't find another job, James seemed to have an ever-shorter fuse.

To make things worse, James also was drinking more. He began to imagine that Lori was going to leave him and accused her of stealing his money and seeing other men. Lori denied his accusations, but it did no good. He started to beat her, not playfully and not just once, but over and over. The first few times he did this Lori locked herself in the bedroom. Later James would apologize and tell her how much he loved her. Lori always forgave him, but the beatings continued and even worsened. The neighbors noticed that Lori had strange bruises on her body, but they didn't ask any questions.

One day, however, James beat Lori so severely that the neighbors couldn't ignore her screams and called the police. The officers arrived at the scene in time to hear Lori and to see her bloodied and bruised body. They sent Lori to the hospital in an ambulance, afraid that she might lose her baby. As she left, James insisted that he loved her and that he never meant to hurt her.

What should the police do with James? Should they arrest him? Should they send him to a special counseling program for men who beat their partners? Or should they try to get him as far away from Lori as possible so that he can't abuse her again? How do we decide what policies the police should adopt?

*B*ASED ON A SAMPLING of cop/crime shows on television over the past thirty or forty years—from *Perry Mason* to *Murder She Wrote* to *Homicide: Life on the Streets*—one might suspect that police officers spend most of their time responding to intricately plotted murders and robberies. In reality police in the United States spend an extraordinary amount of time responding to domestic violence—situations in which one partner, most often a man, physically attacks and injures the woman he lives with and, sometimes, her children.

Domestic violence is dangerous and devastating for an alarming number of women and children. Recently three women in a small, rural county just to the south of where I live died at the hands of their husbands or boyfriends. One of these women was murdered in front of her two small sons.[1] Events like these are horrifying; unfortunately, however, they are not rare. The sad fact is, more than a quarter of all female murder victims are killed by a family member or other intimate.[2] Even if a woman is not killed, she might be severely injured, and she and her children emotionally scarred for the rest of their lives. Current estimates are that close to a million women are beaten by their husbands or boyfriends each year.[3]

Unfortunately a domestic assault is often not a one-time occurrence. One pattern identified by researchers is that such assaults tend to recur, even after police have been called to the scene. Worse, the chance of severe harm increases with each episode. Police officers around the country, as well as sociologists who study crime and violence, have long been concerned with this problem. How, they have asked, can we stop men from repeatedly abusing their wives and partners? Some sociologists have been specifically concerned with what police officers should do when a victim or witness of domestic violence calls for their help.

To answer this question, and others like it, sociologists do research. Through research they develop, test, and refine theories. They look at the empirical, or observable, world of everyday life and adjust their theoretical ideas so that they more accurately describe how the world really works.

In this chapter you will hear about how sociologists do research. Specifically you will see how one sociologist, Lawrence Sherman, and his colleagues, used research to test theories that can explain how police actions influence repeated bouts of domestic violence. In this chapter, I will introduce several technical terms that sociologists use in doing research. Instead of simply trying to memorize these terms and concepts, think about the logical processes involved in research and how they might be applied

Domestic violence affects many families each year. Police officers who are called to scenes such as this must decide what they should do to help make sure that it does not happen again.

Many people have opinions about the nature of society and about issues such as domestic violence. In contrast to everyday speculation, such as you can hear on television talk shows, sociologists carefully test their theories with research, following standard rules of methodology.

to other settings. Ask yourself how you might test ideas or theories you have about the social world. How would you set about finding answers to your questions? How could you ensure that your answers are valid and accurate?

Even if you believe you will never do social research yourself, considering such questions can make you a more critical consumer of sociological information. As a citizen, parent, or worker, you will frequently encounter claims about society that are allegedly based on research. How does the research you read about conform to the standards discussed in this chapter?

ASKING QUESTIONS: THE LOGIC OF SOCIAL RESEARCH

One of my favorite leisure activities is reading a fast-paced "whodunit." I love looking for the clues that will help me solve the mystery faster than the hero of the story does, and I am always impressed by the detective's logical powers and ability to put together all the disparate pieces of evidence needed to find the guilty party.

Social research is a lot like detective work. In the best detective stories the heroes and heroines have hunches about who committed a crime, and they systematically try to gather evidence to support or disprove those hunches. Instead of hunches, sociologists have theories. Sociologists use research to systematically gather evidence to test these theories. Detectives examine clues, whereas sociologists examine **data,** factual information about the social world that is used as the basis for making decisions and drawing con-

clusions. By looking at data, sociologists test the accuracy of their theories and often can get ideas that help them refine their theories or develop new ones. This dynamic interplay between theory and research is the core of the science of sociology.

Just as there are many sociological theories, there are also many different strategies or methods for investigating sociological questions. Each method has its own advantages and shortcomings. Just as detectives must use different strategies in different situations, so sociologists vary their techniques from one time to another, basing their choice on which technique seems the most appropriate for the problem at hand.

Nevertheless, the various research methods in sociology all follow certain basic rules of research **methodology** to help ensure that research results are valid, or true. To understand why these rules are important, think of how architects design buildings. When developing the plans for a new structure, architects can be extremely original, creating exciting spaces and innovative shapes. Yet, if the building is to be sturdy, it has to conform to the rules of sound building design. Similarly the logic of sociological research helps ensure that even very creative research is sound and adequately tests social theories. The commitment to follow methodological rules distinguishes sociologists' work from everyday speculations about the social world. It is what allows us to trust the results of sociological studies.

Steps in Research

The research process consists of three general steps: (1) asking questions, (2) looking at the real world by gathering data, and (3) using the data to answer the

FIGURE 2-1 *Steps in the Research Process*

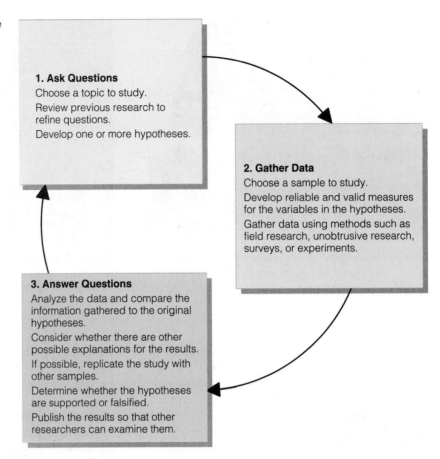

1. Ask Questions

Choose a topic to study.

Review previous research to refine questions.

Develop one or more hypotheses.

2. Gather Data

Choose a sample to study.

Develop reliable and valid measures for the variables in the hypotheses.

Gather data using methods such as field research, unobtrusive research, surveys, or experiments.

3. Answer Questions

Analyze the data and compare the information gathered to the original hypotheses.

Consider whether there are other possible explanations for the results.

If possible, replicate the study with other samples.

Determine whether the hypotheses are supported or falsified.

Publish the results so that other researchers can examine them.

questions. As you can see in Figure 2-1, each of these general steps contains a number of subtasks, some of which can be quite complex. Because sociologists do research in many different ways, how they carry out each of these steps, as well as the order of the steps, can vary from study to study. All research, however, contains these three basic elements.

The rest of this chapter describes each of these steps in detail. It is important to realize, however, that the research process is neither straightforward nor cumulative. For instance, when I begin a new research project, I always hope that it will develop smoothly into a series of clear questions and answers. But this almost never happens! Sometimes the problem becomes much more involved than I anticipated and leads to a much bigger and more complex research program. Sometimes I run into dead ends—my ideas simply don't pan out. I find results I didn't anticipate at all, so that my research questions and sometimes even my theoretical perspective change.

As we will see, Sherman's studies of how police behavior influences patterns of wife battering have also had their ups and downs and twists and turns. And, in fact, these curves in the road have helped to make his work much more important and interesting. In general the process of asking and answering

questions often leads sociologists to modify their original ideas and formulate new questions. Then the process starts over again.

Developing Questions and Hypotheses How do sociologists know what questions to ask? Often these questions come out of researchers' own interests. They might be curious about some particular topic, such as how people choose marital partners, or concerned about a social problem, such as domestic violence. Research questions also can be triggered by social theories, such as those reviewed in Chapter 1.

At first, a researcher's questions might be fairly general, such as "What causes repeated incidents of domestic violence?" Then the researcher reads intensely, looking at as much material as possible on domestic violence and other, relevant issues, such as other crimes and police procedures. In the early stages of research, the library is the sociologist's most important resource. Gradually, through reading and thinking about previous research and theories, the researcher narrows the research focus. For instance, Sherman moved from a rather general interest in domestic violence and police actions to an inquiry into what types of police procedures might help prevent men from repeatedly abusing their partners.

Specifics

Inductive

This autumn leaves fell.
Last autumn leaves fell.
The prior autumn leaves fell.

Deductive

Next autumn the
leaves will fall.

FIGURE 2-2   *Two Ways of Reasoning. Sociologists use both inductive and deductive reasoning in their research.*

Generalization

In the autumn leaves fall.

In the autumn leaves fall.

From their reading of previous research and theory, researchers also begin to develop certain expectations, or hypotheses, about how the world operates. **Hypotheses** are statements about the expected relationship between two or more variables. Hypotheses are like detectives' hunches, except that sociologists' hypotheses can develop not just from their own intuition but also from social theory and previous research. For example, Sherman was also interested in testing two very different theories, both "middle range" in nature, about criminal behavior. The first, known as deterrence theory, stems from the early work of Émile Durkheim. The second, known as labeling theory, derives from more of a microtradition, such as that started by Georg Simmel.

According to *deterrence theory,* greater social control and stronger enforcement of social norms deter violence. Quite simply, this theory suggests that people are less likely to misbehave when they know that they can't get away with it, when the social structure exerts strong control over behavior. On the basis of this theory, we might reason that repeated incidents of wife beating should diminish when police officers routinely arrest and detain suspected offenders. If men believe that they will be arrested whenever they abuse their partners, they will be *less* likely to repeat this behavior.

In contrast, *labeling theory* suggests that the process of being arrested and "labeled" as a violent criminal affects a person's self-image and subsequent behavior. According to this perspective, when people are repeatedly arrested and labeled by the police as violent, they come to accept this definition of themselves. This in turn can result in *more* aggression and violence, not less. Thus labeling theory would predict that men are less likely to continue to abuse their partners if the police do not categorize them as criminals, but instead provide some other type of treatment, such as counseling or separation. [4]

Sherman wanted to design a study that could test these theories. From deterrence theory he hypothe-

sized that offenders who are arrested will be less likely to abuse their partners again. From labeling theory he hypothesized that arrest could actually result in a greater tendency to repeat the abusive behavior. He and his associate, Richard Berk, then convinced the Minneapolis Police Department to help them conduct research to test these hypotheses. Later in the chapter you will find out more about how they gathered data and tried to answer their research questions.

Two Ways of Reasoning   When researchers develop hypotheses from theories, as Sherman did, we say that they use **deductive reasoning:** They deduce certain expectations from a theory. Sometimes, however, the process works in the other direction. That is, as researchers look at data, they develop research questions, gradually build up an understanding about what they are studying, and eventually arrive at hypotheses and theories to explain their observations. This process of developing theories from looking at data is called **inductive reasoning:** Researchers induce or derive first hypotheses and then general theories from specific observations of the world. We often see this type of reasoning in detective stories, as the hero or heroine sifts evidence and combs through witnesses' statements, rejecting one suspect after another and finally deciding that the one person we least suspected was actually the one "whodunit." Instead of starting out with a theory, the detective develops a theory of how and why the criminal committed the crime from looking at all the evidence. Figure 2-2 depicts these two ways of reasoning.

In reality, most sociological research—and most good detective work— combines inductive and deductive reasoning. Inductive reasoning is often used as researchers observe the world and create theoretical explanations of the phenomena they see. Deductive reasoning is then used to derive hypotheses or expectations from these theories that will allow researchers

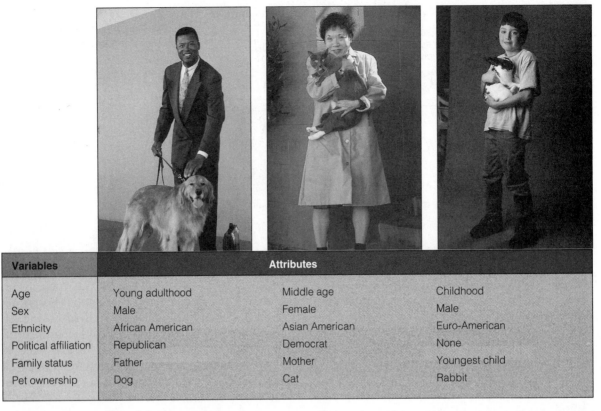

Variables	Attributes		
Age	Young adulthood	Middle age	Childhood
Sex	Male	Female	Male
Ethnicity	African American	Asian American	Euro-American
Political affiliation	Republican	Democrat	None
Family status	Father	Mother	Youngest child
Pet ownership	Dog	Cat	Rabbit

FIGURE 2-3 *Variables and Attributes*

to put their theories to the test. In looking at data and testing hypotheses, researchers again apply inductive reasoning to develop new questions and alter their earlier theories and expectations. In the chapters to come, you will see many examples of how researchers use both inductive and deductive reasoning in their work.

Variables, Correlation, and Causation

Even though most sociologists assume that individuals make choices about their daily lives, sociological theories are stated in terms of people in general rather than specific individuals. For instance, neither labeling theory nor deterrence theory tries to predict whether a given individual will return to beat his partner depending on how he is treated by the police. Instead, these theories predict the probability, or likelihood, that people who are arrested will be more or less likely than people who are not arrested to repeatedly abuse their partners. Nor do these theories try to explain all the details of why a particular man would choose to continually beat his partner. Instead, they try to explain why people who have had different experiences tend to choose different behaviors or actions. In general, sociological theories try to explain observed regularities or patterns in social events rather than individual behavior.

When we think about individuals, we often think about their individual characteristics, or what sociologists call **attributes.** For instance, you might be an African-American male in your twenties who is a registered Republican, and the person who sits next to you in class might be an Asian-American female in her fifties who is a registered Democrat. Each of these characteristics is an attribute. A sociologist who looked at your class would be interested not just in these individual attributes but also in the way they combine to make up variables that can be used to describe your class as a whole.

Variables are logical groupings of attributes (or, literally, things that vary or have different values). For example, in their work on domestic violence, Sherman and Berk categorized police behavior in cases of spousal abuse as involving the attributes (1) arrest, (2) referral to counseling, or (3) separation of the partners. Together these attributes make up a variable, "police action." Similarly they categorized the behavior of a man reported for wife beating as either (1) did not repeat the behavior or (2) did repeat the behavior. These two categories are attributes of the variable "repeated abuse by the suspect." Other examples of variables and attributes are shown in Figure 2-3.

No matter how they go about their work, detectives are always interested in both who the perpetrator is and why the individual committed the crime. In contrast, sociologists are interested in what happens and why at the level of social patterns. To explore these patterns, sociologists study *relationships* between variables. Two general types of relationships between variables are of interest to sociologists—those involving correlation and those involving causal relationships.

Correlation refers to the ways in which two variables might be related to each other in a predictable pattern. For example, if people who have blond hair also tend to have blue eyes, we could say that the variables eye color and hair color are correlated with each other. Similarly, as we will see in subsequent chapters, because African Americans tend to have lower family incomes than Euro-Americans do, sociologists say that race-ethnicity is correlated with family income.

Causal relationships are more complex than correlations, for they involve an association between variables in which one influences or causes the other. For instance, Sherman and his colleagues wanted to know how the actions of police officers influence the probability that men will continue to abuse their partners. Because they were interested in what could influence, or cause, a reduction in repeat offenses, what they were exploring was a causal relationship. In theories and hypotheses that posit causal relations, the variable that influences or causes another variable is called the **independent variable;** the variable that is influenced or caused by another variable is called the **dependent variable.** In Sherman and Berk's work, police behavior is the independent, or causal, variable, and suspects' behavior is the dependent variable.

How can social scientists demonstrate that an independent variable actually influences or causes a dependent variable? Philosophers have traditionally outlined three criteria whereby a causal relationship between two variables may be said to exist. First, it must be established that the independent, or causal, variable occurred before the dependent variable. For example, if a researcher is interested in how police actions affect offenders' abusive behaviors toward their partners, the police action must occur before the abuse.

Second, the two variables must be correlated with each other. When there are changes in one variable, the other variable should also change in predictable or patterned ways. For instance, if either labeling theory or deterrence theory is to have support, there must be an observable tendency for wife beaters who have been treated differently by the police to exhibit different kinds of subsequent behavior.

When sociologists talk about causal relationships that involve human behavior, they are not implying that individuals do not have free will. In fact, it is quite the opposite. As the early African American sociologist W.E.B. Du Bois (1868–1963) put it, sociology "is the science of free will." As you will see in the chapters ahead, sociology can help us understand why people choose their individual actions.

Third, the observed association cannot be explained away or be accounted for by the influence of other variables. For example, suppose researchers found that offenders who are arrested are more likely to abuse their partners again than are offenders who are not arrested. These results would support labeling rather than deterrence theory. Suppose, however, that in the city the researcher studied, police officers are much more likely to arrest men accused of domestic violence if they have extensive histories of violent behavior in other settings. If a man accused of domestic violence has no prior history of violent behavior, he generally is not arrested, but instead is separated from his partner or directed to some type of counseling program. In this case, even though men who have been arrested are more likely to repeat abusive behavior, they also are more likely to have personal histories of violent behavior. It could well be this history, rather than the police behavior, that accounts for their continuing wife abuse. It is impossible to tell which variable—the offender's prior history or the police behavior—actually is causing the continuing abuse. Consequently, before concluding that the association between arrest and later spousal abuse supports labeling theory, researchers would need to rule out the possibility that other variables, such as the offenders' history of violence, also influence this relationship. When a correlation between two variables only occurs because of the influence of a third variable, it is said to be a **spurious correlation.**

Human behavior is so complex and is influenced by so many factors that sociologists can never prove that their theories are true in the sense that a mathematician can produce a logical proof of a mathematical theorem. We can never know for certain that a causal relationship exists between two social variables, primarily because it is so difficult to ensure that we have ruled out all other possible explanations for an observed association between two variables.

But researchers can gather evidence and data to show that a causal relationship does *not* exist. When a proposed causal explanation survives empirical testing, gradually disproving alternative explanations can increase our confidence that the causal explanation is valid.

Philosophers call this line of thinking the logic of **falsification.** You cannot prove that a theory is true, but you can look for empirical evidence that it is false. For instance, if researchers repeatedly found that offenders who were arrested were less likely to continue their patterns of spouse abuse, they would come to have much less faith in this application of labeling theory, and they could say that it appeared to be false in explaining patterns of spouse abuse. In contrast, if they continued to find support for the theory's predictions, no matter what kind of situation they examined, they could say that the theory appeared to have explanatory value and had not yet been found to be false. And if there is no support for competing explanations, the theory gains in plausibility.

Research to test or develop a theory is a long-term process. A logical proof can establish the validity of a mathematical theorem, but no single study can establish the truth, or even the falseness, of a social theory, simply because it is impossible, in one study, to examine all social situations. To help counter this problem, researchers often try to conduct numerous studies of the same phenomenon to see if consistent patterns occur. Repetition of a study to see if the same results occur and if they hold in other settings is called **replication.** When similar results occur in a number of replications, researchers can have greater faith in their theories. Replication is a long and complex process; for example, Sherman and his colleagues have spent many years repeating their study in a number of cities and in a variety of ways. Yet, as we will see, this complex process is essential if sociologists are to develop theories that more fully and accurately explain the social world.

GATHERING DATA: OBSERVING SOCIAL STRUCTURE AND SOCIAL ACTION

The second major step in the research process is gathering data to answer research questions. To do this, sociologists must decide (1) what people or groups they will study, (2) what they will look at to answer their research questions and test their hypotheses, and (3) how they will make these observations. In sociological terms these aspects of data gathering are referred to as sampling, measurement, and observation, respectively.

Sampling: Choosing Cases to Study

Researchers seldom can study all the possible occurrences of a phenomenon. Instead, researchers must look at what are called **samples,** or subsets, of the larger group they are interested in. For instance, it would have been physically impossible for Sherman and his colleagues to study all the calls to police nationwide in cases involving domestic violence. Thus they chose to focus first on a police department in only one city, Minneapolis, primarily because Sherman worked there and knew the local police chief.

All studies have to begin somewhere, but because no one city is quite like any other, it would be very difficult for Sherman and Berk to suggest that any results they might find in Minneapolis would also occur in other cities. In sociological terms they could not say that Minneapolis was representative or typical of all other cities.

Ultimately sociologists want to generalize their findings to an overall **population,** the entire group or set of cases they could potentially study. For Sherman and Berk the relevant population might be all cities in the United States. The best way to ensure that a sample is representative is to use a **probability sample,** which generally involves some type of random selection process. In essence **random selection** ensures that each member of a population has an equal chance of being included in a sample. Mathematicians have developed ways in which findings from such random samples can be generalized to entire populations with a high degree of accuracy. These statistical procedures allow a researcher to observe relatively few people or situations and generalize from these observations to the larger population from which the sample was selected.

Measurement: Finding Real-Life Indicators of Theoretical Concepts

One of the most important steps in the research process involves selecting good measures of a variable. **Measures** are the way in which the concepts involved in a theory are translated into actual data. For example, without precise measures of the behavior of police and offenders, Sherman and Berk could not have tested their hypotheses.

Typically sociologists try to ensure that their measures have two key characteristics: **validity** and **reliability.** First, the measures must be *valid*—that is, they really measure the phenomenon they are supposed to be measuring. Second, the measures must be *reliable*—that is, the results will not vary if different researchers use the same measures on the same subjects at different times. Valid measures in the Sherman study should really describe the actual be-

Method	Description	Advantages	Disadvantages
Field Research	Researcher directly observes phenomena in their natural settings	Provides in-depth understandings of details of day-to-day lives and daily experiences of people	Can involve only a few cases, so it is difficult to generalize results to other settings
Unobtrusive Research	Researcher examines data that can be obtained without directly talking to or watching someone	Presence of researcher does not affect the data Allows study of sensitive situations and events that occurred in the past or in widely scattered places	Removed from day-to-day life, so it is difficult to assess the validity of data
Survey Research	Researcher examines data obtained by asking individuals questions through interviews or questionnaires	Data can be highly reliable and can be obtained from probability samples	Cannot provide in-depth information about subjects or about the context surrounding behavior
Experiments	Researcher uses experimental and control groups to test the effect of an independent variable on a dependent variable	High level of control allows the researcher to infer the presence of a causal relationship	Is often difficult logistically and ethically to use with sociological issues. Experimental setting can be highly artificial

FIGURE 2-4   *Advantages and Disadvantages of Different Research Methods Field Research*

haviors of both police and offenders. Reliable measures of these behaviors should yield the same results if someone besides Sherman and Berk used them or if the measures were used at different times of the day or in different settings.

Sherman and Berk worked very hard to ensure that their measures were both valid and reliable. For instance, to help increase the *reliability* of their measure of police behavior, they met regularly with the police officers to make sure that the officers were following similar procedures and behaviors when arresting a suspect or recommending separation or counseling. Further, to increase the reliability of the measure of suspect behavior, the people who interviewed the victims were trained to ask questions about repeated abuse in similar ways so that victims would give the same answers to whomever called.

To help increase the *validity* of the measure of abusive behavior, Sherman and Berk tapped several sources of information: data on re-arrests from police files and on repeated abuse from the interviews with the victims. Sociologists often try to use *multiple indicators,* or different measures of the same thing, to help increase the probability that they are really getting a valid measure of the concept they are trying to study.

From police records and interviews with the victims, Sherman and Berk determined, as best they could, whether the offender had again abused his partner. Also, using information from both sources, they calculated the number of days since the initial contact with the police that had elapsed without the offender again assaulting the victim, threatening to

assault her, or destroying her property. This variable could have a much larger number of attributes, ranging from 1 day, if the repeated abuse occurred the day after police contact, to 180 days, when the last six-month follow-up occurred. Measures such as these that involve real numbers are referred to as **quantitative data.**

Observation: Gathering Data on the Social World

Sociologists have devised several methods of gathering data about different areas of social life. Some of these methods work better for some types of problems, and some work better for others; but all of them allow researchers to examine the relationships between variables the researchers are interested in. In recent years, when possible, researchers have tried to use several approaches to the same problem to gather data. Throughout this book you will see detailed examples of how sociologists have used each of them. Figure 2-4 summarizes the principal methods we will discuss.

Field Research   One of the most straightforward ways to gather data is simply to observe the social world firsthand. The term **field research** refers to a variety of methods whereby a researcher directly observes behavior or other phenomena as they occur in natural settings. Researchers sometimes participate in the events or groups they want to study and carefully note details of the experience, a method called

participant observation. For instance, a researcher who wanted to study how police treat spouse abusers might train as a police officer and work with the officers on a regular basis.

Sometimes a researcher can't realistically participate in a setting. For instance, it might be difficult for a researcher to join a police force as an actual employee, and no reputable researcher would beat his wife simply to see how the police would treat him. In such cases a researcher could employ **nonparticipant observation,** observing as much of the ins and outs of the situation as possible without actually participating. For example, many researchers, including those working with Sherman, have ridden with police officers as they made their rounds and watched what happened when the officers confronted suspects. [5]

Whether or not field researchers participate in the behavior or events they study, they generally gather what is called **qualitative data,** measures of data that cannot be assigned real numbers. Instead, the data may consist of detailed reports of behaviors, quotes from research subjects, and observations of the setting. Unlike some research designs in which the research questions are answered only after all the data are gathered, field researchers often answer questions or analyze findings while gathering data. They typically use an inductive approach to research, developing their theoretical understandings and formulating new research questions as they gather more and more data.

Field research is probably the best way sociologists have of learning about the details of day-to-day life and people's daily experiences. The major advantage is the in-depth look field research provides at social life. The major drawback is that, because field research involves such a concentrated look at a setting, only one or a few cases can be examined at any one time. The findings can help a researcher develop hypotheses and theories, but often they are insufficient to permit generalizing to other settings or contexts.

Unobtrusive Research   Sociologists don't always look at people's behavior in real-life settings. Sometimes they conduct **unobtrusive research** whereby they obtain data without directly talking to or observing research subjects. For instance, researchers might examine official documents, such as court records and crime statistics, to find out how often people were arrested for domestic violence and what kinds of punishments they received.[6] In this way the experiences of large numbers of people can be examined without ever contacting them personally. Émile Durkheim used this technique in his study of suicide, described in Chapter 1.

Sociologists sometimes conduct unobtrusive research, *whereby they obtain data without directly talking to or observing research subjects. They might examine official document such as court records and crime statistics to gather data. Can you think of other sources of unobtrusive data?*

Researchers sometimes also look at documents such as books, newspapers, or magazines, or other types of material such as pictures, television shows, or movies, using a method called **content analysis.** For instance, a researcher interested in domestic violence and its resolution might systematically examine how writers have portrayed such situations in novels and short stories. A study like this might involve a quantitative analysis, such as counting the incidents of domestic violence and the ways in which they were handled. Or it might involve a more qualitative approach, such as looking at the emotions that are portrayed by the characters involved in the stories. Max Weber used a form of this method in his research into the development of capitalism when he analyzed the sermons of Calvinist ministers.

Two special types of unobtrusive research are **historical research** and **comparative research,** which are used to study events that occurred over a period of time or in many different settings. For instance, a researcher using historical research might examine court records and crime statistics, or novels and magazine articles, to see how reactions to domestic violence have varied over the years. Using comparative research, a researcher might compare statistics on domestic violence across a number of nations to see how experiences with domestic violence vary from one country to another. Max Weber's comparison of the development of capitalism in Europe, China, and India, described in Chapter 1, is a classic example of both historical and comparative research.

Unobtrusive research is especially useful when researchers cannot directly observe what they want to study. It is also well suited for gathering information about events that are widely scattered in time and space, and it can provide access to information about many different events and people. Because the researcher does not directly interact with the situations and people under study, unobtrusive research also helps ensure that the mere presence of the researcher doesn't affect the social situation. This might be especially important when you are studying sensitive issues such as domestic violence. A notable disadvantage of unobtrusive research is that, because it is relatively removed from the realities of day-to-day life, the information gathered might be less valid or trustworthy. For instance, crime statistics greatly underestimate the actual incidence of domestic violence because only the most serious incidents—and only a fraction of those—are actually reported to the police.[7] Similarly, books, movies, and magazine stories can show how writers and filmmakers depict domestic violence and other phenomena, but researchers can't always assume that these depictions accurately reflect the social world.

Survey Research Probably the most common way in which sociologists gather data today is through asking people questions. When sociologists systematically ask individuals a series of questions about a topic, through either interviews or written questionnaires, they are doing **survey research.**

There are many types of surveys. In *interviews,* the researcher asks people questions, either in person or, increasingly, over the telephone. Some interviews, especially in field research, are *unstructured*—they have no set format and amass data in an inductive manner. Most survey research, however, consists of *structured* interviews in which the questions have a set wording. A second type of survey research involves written questionnaires, sent through the mail or completed by respondents in the researcher's presence.

Structured interviews and questionnaires can yield highly reliable data, because the questions are always asked in the same way. Surveys can also yield information about a large number of variables, depending on the number of questions the researcher is able to ask. Perhaps the most important feature of

surveys, however, is that, when they are applied to random or probability samples, they can provide information about large populations. The results of public opinion polls, which are routinely reported in the media, are an example of survey research involving probability samples.

Surveys do have certain limitations. For instance, although they can yield reliable information about large numbers of people, they are not as effective as field research in providing in-depth information about people or the contexts in which people live.

Much of sociologists' knowledge about the actual incidence of domestic violence comes from surveys. As you might suspect, people often are reluctant to answer truthfully about such matters. Survey researchers try to figure out ways to ensure that they are getting valid, or accurate, responses to their questions. Figure 2-5 shows a portion of the "Conflict Tactics Scale," a series of questions that the sociologist Murray Straus and his colleagues have developed to ask both women and men about their experiences with domestic violence. They have selected several probability samples of people from across the country and questioned them about the extent to which they have experienced domestic violence and even if they have inflicted violence on their partners.

You might notice that the questions ask about tactics that range from the mundane (discuss an issue calmly) to the deadly (use a knife or gun). Straus and his colleagues have developed scoring procedures for the scale that result in three different measures: (1) the extent to which couples tend to "reason" about their problems with each other, (2) the extent to which they use "verbal aggression" and threats, and (3) the extent to which they resort to "physical aggression" or violence. Based on this research we now know how widespread domestic violence actually is, affecting millions of women and men each year, and what a small proportion of these incidents are ever reported to the police. [8]

Experiments When you think of experiments, you probably envision scientists in white lab coats trying out new medicines or working with exotic chemicals. In elementary school science classes you probably learned about the techniques and logic of experiments. Perhaps you planted radish seeds in three different containers. You placed one container in a sunny window, another in a dark closet, and the third in a corner of the room away from direct sunlight. You gave all the plants the same amount of water and you measured them each week to see which one grew the most. In this experiment the dependent variable is how tall the plants grew in a few weeks. The **experimental** or independent **variable** is the amount of

FIGURE 2-5
The Conflict Tactics Scale

Source: Straus and Gelles 1990, pp. 29–47.

The following is an example of how sociologists use survey research to examine domestic violence. After reading the introductory material, the interviewer reads the first subpart of the question and waits for a response before asking about the next tactic on the list.

No matter how well a couple get along, there are times when they disagree, get annoyed with the other person, or just have spats or fights because they're in a bad mood or tired or for some other reason. They also use many different ways of trying to settle their differences. I'm going to read some things that your partner might do when you have an argument. I would like you to tell me how many times (once, twice, 3–5 times, 6–10 times, 11–20 times, or more than 20 times) in the past 12 months he

a. Discussed an issue calmly

b. Got information to back up his side of things

c. Brought in, or tried to bring in, someone to help settle things

d. Insulted or swore at you

e. Sulked or refused to talk about an issue

f. Stomped out of the room or house or yard

g. Cried

h. Did or said something to spite you

i. Threatened to hit or throw something at you

j. Threw or smashed or hit or kicked something

k. Threw something at you

l. Pushed, grabbed, or shoved you

m. Slapped you

n. Kicked, bit, or hit you with a fist

o. Hit or tried to hit you with something

p. Beat you up

q. Choked you

r. Threatened you with a knife or gun

s. Used a knife or fired a gun

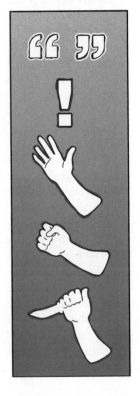

light each plant received. If the plant on the windowsill grew more than the others, you would know that the sunlight was responsible because that was the only difference in the plants' treatment. This design is called an experiment because the only difference in the three plants' treatment is related to the independent or experimental variable, and any difference that occurs in their growth must be attributed to this and not to other conditions. In general an **experiment** is a research technique that uses control groups and experimental groups to assess whether a causal relationship exists between an independent and a dependent variable.

The primary advantage of experiments is that they can fulfill all the criteria needed to establish causality. By controlling the way in which the groups under study, whether plants or people, are exposed to the independent variable, experiments try to guarantee that the only difference between the groups involves the independent variable. This element of control also ensures that the independent variable occurs before the dependent variable.

Despite this ability to help establish the presence of a causal relationship, experiments have several drawbacks. Sociologists actually use them only occasionally, mainly because it is so difficult (and often unethical) to experiment on their fellow human beings. In addition, sociologists are interested in many aspects of social life that are simply not amenable to experimental manipulation. Sociologists are concerned with how variables such as geographic location, the circumstances of our births, and our day-to-day interactions influence individual choices and behaviors. However, such factors are not readily amenable to experimental control. After all, children cannot be randomly assigned to parents, and citizens cannot be randomly assigned to neighborhoods.

Occasionally, however, it is possible to conduct an experiment in a real-life setting; this is called a **field experiment.** Lawrence Sherman and his colleagues chose to conduct a field experiment in their study of the relationship between police actions and repeated incidents of spouse abuse in Minneapolis. They focused only on cases of what the police call

Sociologists are often interested in studying situations that aren't amenable to experimental manipulation, such as the way these residents of the Dutch village of Ochen have reacted to a flood that threatens their community. What research techniques might be used to study a situation such as this?

"simple" or "misdemeanor" assault, cases in which the woman has not suffered serious injury or had her life seriously threatened, and incidents that the officers believed had occurred in the previous four hours. Previous studies using observations from field research and analysis of police documents suggested that when police officers are called to a home where domestic violence has occurred, they quickly look things over. Officers try to see if anyone has been injured or if people have been drinking. They talk to the man and the woman involved. If the woman wants to sign an official complaint against her partner, or if the man is disrespectful to the officers, they tend to arrest him.[9] But unless someone has been seriously injured or killed, the decision of whether to arrest the man or to take some other action is left to the officers' discretion. Officers must use their own best judgment to decide what type of police action would be most likely to prevent the man from violently attacking his spouse again.

Sherman and Berk's experimental design called for a slightly different procedure. Each time police officers were called to a situation of "simple" domestic assault, they were to do one of three things: (1) arrest the suspect, (2) recommend that the partners receive "mediation," a form of counseling to deal with the causes of the dispute, or (3) recommend that the partners be separated for a period of time. The officers' actions were determined not by their gut instinct in the field, but by a pad of color-coded report forms, with each color corresponding to a different action. The forms were randomly arranged, so that each would be equally likely to be used by an officer in each situation. This random arrangement was designed to ensure that the treatment, or independent variable, was not related to any other possible characteristic of the suspects or victims.

Within a few days of the police action, members of Sherman and Berk's research staff interviewed the victims in person, and the researchers continued to have telephone contact with the victims at two-week intervals for the next six months. The purpose of these interviews was to determine whether the suspect had assaulted the victim again, or, in sociological terms, to obtain measures of the dependent variable "repeated abuse by the suspect." The researchers also monitored police records for the six-month period to see if there were any additional reports of domestic violence involving the suspect.

As mentioned earlier, it is difficult to study human beings experimentally, and even well-designed field experiments can run into problems. In this case, for the results of the study to be considered valid, it was extremely important that the officers' response be determined randomly, as the experimental design called for. However, because this experiment involved actual police officers responding to real-life situations, this behavior could not be controlled perfectly. In 18 percent of the cases, the officers did not do what was prescribed by their color-coded pads. This almost always involved a situation in which they chose to arrest the suspect rather than recommend counseling or separation, generally because the suspect was rude to the police or tried to assault one or both of the officers, because weapons were involved, or because the victim insisted that the officers arrest the suspect.

Sherman and Berk also ran into difficulties with the follow-up interviews with the victims. Despite their best efforts the research team was able to locate and

FIGURE 2-6 *Results of Sherman and Berk's Study of the Relationship Between Police Behavior and Repeated Wife Beating in Minneapolis*

Source: Adapted from Sherman and Berk, 1984, pp. 265–267.

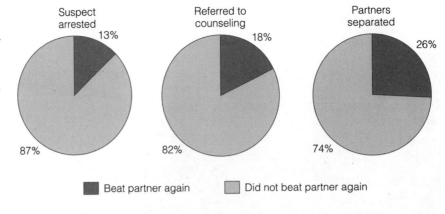

interview only 62 percent of the women immediately after the contact with the police. Less than half of the women completed all twelve of the interviews scheduled for the follow-up period. Sherman and Berk tried to determine if the people they did interview were any different from those they failed to contact. Fortunately they could find no systematic differences.

Clearly, from a research standpoint, it would have been preferable if the police had strictly followed the experimental guidelines and if the researchers had managed to contact all the victims for the six-month follow-up. On the whole, however, the researchers concluded that they had sufficient accurate information to proceed with answering their research questions by analyzing the data they had obtained.

ANSWERING QUESTIONS: DATA ANALYSIS AND THEORY REVISIONS

The third major step in the research process is answering questions. This step involves examining the data and comparing the results to hypotheses generated from previous research and theory.

Analyzing the Data

Sherman and Berk's research team spent more than a year gathering data in Minneapolis. During this time police officers participating in the study were called to deal with 314 men accused of spousal abuse. Even though 60 percent of the men had had prior contact with police regarding domestic violence, only 5 percent had ever been arrested for domestic assault. The experimental design called for the police to arrest men suspected of domestic assault one-third of the time. If the prior arrest records of the men in the experiment are a good indication of typical police practices in Minneapolis, the experiment called for quite a change in police action. Thus, the researchers rea-

soned, the experiment provided a good opportunity to see if, as deterrence theory predicts, arrest can lead to a reduction in abusive behavior.

Sherman and Berk's analysis involved sophisticated statistical techniques designed to help rule out the possibility that other variables might actually be causing a relationship between police behavior and repeated domestic abuse. The results depicted in Figure 2-6 were compiled from the records of the Minneapolis police. The pie charts show the relationship between the independent variable, police behavior, and the dependent variable, suspect behavior. For instance, in the first chart we see that 13 percent of the suspects who were arrested (the first attribute of the independent variable) beat their partners again (the first attribute of the dependent variable). In each pie chart, you can see the outcomes associated with a particular police action.

If you study the data, you can see that arrested suspects were less likely to abuse their partners again, a finding that supports deterrence theory rather than labeling theory. However, this relationship is probabilistic and far from "perfect." For instance, a large proportion of the men who were referred to counseling (82 percent) or who were simply separated from their partners (74 percent) did not beat their wives again, while some of the men who were arrested did do so (13 percent). But on average these results support deterrence theory by suggesting that men who were arrested were less likely to repeat their violent behavior and those who were not arrested were more likely to do so.

Sherman and Berk found similar results when they examined the data obtained from interviews with the suspects and when they looked at how long the suspects managed to avoid repeating or threatening assault. Because not all domestic assaults are reported to the police, the victims tended to report more repeated incidents in the interviews (29 percent compared to 18 percent in the police records). But the deterrence effect of arrest remained, with only 19 percent of those whose partners had been arrested reporting repeated incidents.

Based on their findings, Sherman and Berk suggested that there was little support from these data for labeling theory and much more support for deterrence theory, at least as it could be applied to the relationship between police actions and repeated domestic violence. The research results were published in the leading sociological journal, the *American Sociological Review*. Research findings reported in journals such as the ASR are evaluated by other sociologists to see if they meet the standard rules of research methodology discussed in this chapter, a procedure called **peer review.** Sherman and Berk's research clearly met the usual methodological standards.

Asking More Questions

Sherman and Berk's conclusions seem amazingly clear-cut. Labeling theory was discounted, and deterrence theory was the "winner." If police departments want to reduce the probability that spouse abusers will repeat their violence, the results of the Minneapolis study suggest that police should arrest the abusers.

The vast majority of sociological studies are read only by other sociologists (but see Box 2-1). Sherman and Berk's work was an exception. The media quickly published the findings and the recommendations for police action, and the publicity appears to have resulted in a sharp change in police policy. In 1984, the year the research findings were published, only about 10 percent of city police forces nationwide instructed their officers to routinely arrest men accused of domestic violence. By 1986 almost half of the cities had such policies, and many of them reported that Sherman and Berk's research had influenced this change.[10]

Such a quick reaction to research results rarely occurs in sociology, but it is not uncommon in fields such as medicine. A few years ago, for example, a research report suggested that eating oat bran could help reduce blood cholesterol levels. Within weeks bakeries were pushing oat muffins and oat bread, and the makers of Cheerios proclaimed that their cereal was a natural source of oat bran. As it turned out, however, the early faith in the curative properties of oat bran was misplaced. Later research, which replicated the first studies, showed that oatmeal is still a healthful breakfast food, but it is not a magical preventive or cure for high cholesterol levels. How would replications affect the lessons derived from Sherman and Berk's work?

Replicating the Minneapolis Study In their 1984 article Sherman and Berk stressed that their study should be repeated elsewhere. No two cities in the country are the same. Not even the twin city of St. Paul could be said to be identical to Minneapolis. Cities vary in size, in the types of jobs people have, in the characteristics of their citizens, and so on. For instance, the largest minority group in Minneapolis is Native Americans, whereas in most large cities the largest minority group is African Americans. Similarly police departments can differ greatly in how they respond to domestic violence and how they treat suspects. For example, in Minneapolis most arrestees were released within a short period of time, but this does not always happen in other areas. For all these reasons it was important to see whether the results obtained in Minneapolis also appeared in other cities.

At least partly because of all of the publicity surrounding Sherman and Berk's article and the hope that the results could help with the problem of domestic violence, nineteen police departments participated in plans for replications of Sherman and Berk's work that were presented to the National Institute of Justice in 1986. This federal agency agreed to sponsor studies in Omaha, Nebraska; Charlotte, North Carolina; Milwaukee, Wisconsin; Dade County, Florida; and Colorado Springs, Colorado.

The replications all involved experimental designs similar to that used in Minneapolis, where people accused of domestic violence were either arrested, separated from their victims, or referred to counseling. Each of the studies generally involved more people than the Minneapolis study did and included improvements that helped increase the probability that the police officers followed the experimental design and that the victims cooperated in follow-up interviews. For instance, to better control police decisions, officers in Milwaukee were instructed to call the research office at police headquarters after they reached a domestic assault scene. Members of the research staff then told them, on a predetermined random basis, what type of action they should take. This resulted in less than 2 percent of the officers acting in a way that did not conform with the experimental design. Similarly, in Milwaukee, improved techniques of contacting victims resulted in more than three-fourths of the victims completing follow-up interviews, compared with less than half in Minneapolis.

Most important, each of the replications involved a much longer follow-up time period, allowing the researchers to track suspects' behaviors for as long as two years after their initial involvement in the experiment. Thus, the researchers could see whether the deterrent effect of arrest lasted beyond the six-month period in the first study.

The only contact most people have with social research is through their local newspaper or television newscasts. Most of these reports are quite short, designed to grab viewers' attention and to present some supposedly new and exciting data. If research isn't exciting, it can't be news!

Unfortunately, when social research is reduced to a thirty-second TV sound bite or a three-paragraph newspaper article, a great deal can be lost. Further, much of the social science research reported in the media can be misleading because important details are omitted or because the research has not been replicated or does not meet sound methodological standards. Being aware of some of the pitfalls in interpreting research reported in the media can make you a more intelligent consumer of sociological information.

The following are the kinds of things I look for when reading media reports:

■ What kinds of questions were the researchers trying to study? Journalists usually aren't very interested in sociological theories, but they often can give you hints as to what the general research question was.

■ Where was the research reported? If it is published in a journal (a professional magazine) or a book, it is more likely to have been reviewed by other social scientists and to have met rigid methodological standards.

■ What size sample was used? Were only a few people studied? Or was a large sample used? Generally, though not always, research that involves larger samples is likely to be more accurate.

■ How representative or typical of the rest of the population was the sample? If the research was done on only a certain group of people or in only one locale, the results might not apply elsewhere.

■ What kind of research design was used—experiments, field research, unobtrusive research, or surveys? Does the report indicate why the chosen techniques were appropriate?

■ Are there other possible explanations for the research results that the researcher did not consider or that could only be discovered through replications? What kind of evidence does the researcher give to show that these results can't be explained by something else?

Sherman and Berk's work in Minneapolis was widely reported in the media, and one of these accounts is reprinted here.

Obtaining New Results and Revising Theories
As sometimes happens in a good detective story, the replications "thickened the plot." Even though the results from the Minneapolis study seemed straightforward, the picture after the replications was more complex. In fact, the results varied from one city to another. Results from Colorado Springs and Dade County showed a long-term beneficial effect of arrest and thus supported deterrence theory. In contrast, the long-term data from Omaha, Charlotte, and Milwaukee showed that, after about a year, suspects who were arrested were more likely than those who received other treatments to again abuse their partners. This finding directly contradicts deterrence theory and the Minneapolis data.[11]

In the years since the replication data began to become available, Sherman and his associates have tried to understand why these differences appeared. One suggestion is that arrest might work as a deterrent far better for some kinds of suspects than for others. In particular, individuals who have more ties with others, such as marriage and a job, might be much more likely to be deterred by the stigma of arrest than will those with fewer ties with others. Note how this hypothesis relates to the work of Durkheim and Simmel. It suggests that people who have stronger social bonds have a higher stake in conforming to social norms and thus should find arrest to be more of a deterrent to further violence. If people don't have these ties, arrest isn't as embarrassing and thus doesn't tend to serve as a de-

BOX
2-1

**APPLYING SOCIOLOGY
TO SOCIAL ISSUES**

Continued

*Domestic Violence: Study
Favors Arrest*
by Philip M. Boffey

Despite the conventional psychological wisdom that domestic disputes are best resolved through mediation, a pioneering study of police tactics in domestic assault cases has concluded that the best way for the police to prevent repeated acts of violence in the home may be simply to arrest men suspected of assaulting their wives or lovers....

The study was conducted by the Police Foundation, a private research organization based in Washington, with the help of the police department in Minneapolis. It was supported by a grant from the National Institute of Justice, a unit of the Federal Justice Department.

Over a 10- to 12-month study period, some 30 to 35 cooperating police officers in Minneapolis used three different tactics, selected at random, to handle cases of "moderate" domestic violence, defined as

simple assaults that did not cause severe or life-threatening injuries. The three tactics were arrest, advice or mediation, and ordering a violent spouse to leave for eight hours. The suspects were then followed through police reports for six months to see if the violence was repeated.

As it turned out, only 10 percent of those arrested generated a new official report of domestic violence within six months, compared with 16 percent of those given advice or mediation and 22 percent of those ordered out of the house for eight hours. The arrested suspects manifested significantly less violence than those who were ordered to leave and less violence than those who were advised but not separated, says a preliminary report by Lawrence W. Sherman,

director of research for the Police Foundation.

The findings "suggest that police should reverse their current practice of rarely making arrests and frequently separating the parties," the report said. "The findings suggest that other things being equal, arrest may be the most effective approach and separation may be the least effective approach." . . .

Source: Article excerpted from *The New York Times,* Tuesday, April 5, 1983, pp. C1, C4. Copyright (c) 1983 by The New York Times Company. Reprinted by permission.

CRITICAL-THINKING QUESTIONS

1. Based on what you have learned about their study, how accurate was this media report?

2. How well does it address each of the issues outlined earlier?

3. Is this kind of account sufficient to enable you to form an opinion about the policies police should follow in dealing with domestic abuse? Why or why not?

terrent. Thus the results could differ because the sample of suspects varied in their social ties from one city to another. For instance, arrest might have appeared to produce a long-term deterrent effect in Colorado Springs and Dade County simply because more of the suspects in those cities had social bonds that would make arrest an embarrassment.

To test this new hypothesis, Sherman and his associates decided to look at the way in which arrest affected the probability of repeated spouse abuse among people with different types of social bonds. Figure 2-7 shows the results using the data from Dade County. The figure is similar to Figure 2-6, but the pie charts measure three variables rather than two. The dependent variable is still the suspects' subsequent behavior.

To simplify the charts, the independent variable of police behavior has only two attributes: "arrested and "not arrested." The new variable, employment status, was chosen to measure the suspects' stake in conformity. It is called a **control variable** because it allows us to see the association between police actions and suspects' subsequent behavior once the influence of employment status is controlled or removed.

As you can see, the results depicted in Figure 2-7 support the new hypothesis. Among suspects who were employed, those who were arrested were only about half as likely as those who were not arrested to repeat the offense. This supports deterrence theory. Among suspects who were unemployed, and who thus presumably had fewer social bonds with others

Suspect Unemployed

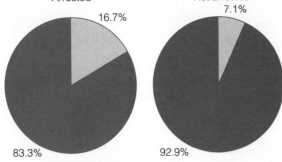

Suspect Employed

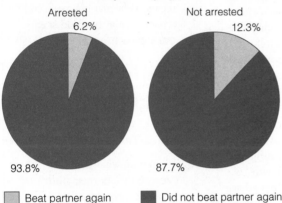

☐ Beat partner again   ■ Did not beat partner again

FIGURE 2-7 *Relationship Between Police Behavior and Repeated Wife Beating, Controlling for Employment Status, in Dade County, Florida*

Source: Adapted from Pate and Hamilton, 1992, p. 695.

and would be less likely to be embarrassed by arrest, the relationship was just the opposite—those who were arrested were more likely to repeat the offense. Sherman, Berk, and their associates have looked at data from other cities, often using complex statistical techniques, and have found results that replicate those shown in Figure 2-7. Deterrence theory is supported, but only for some people. [12]

You can see now how the work of Sherman and his associates fits the cyclic model of theory and research depicted in Figure 2-1. In his initial work with Berk, Sherman designed research questions to test the adequacy of deterrence theory and labeling theory. The answers they developed appeared to strongly support deterrence theory. Sherman and other colleagues then asked additional questions as they replicated the work in other cities. This time the answers suggested that deterrence theory applied to some suspects, but not to others.

But the story doesn't end here. In fact, the findings from these replications have complicated the

search for ways in which police actions can effectively deter domestic violence. The field experiments of Sherman and his associates can't explain why labeling theory seems to work with some suspects, but not others; nor can they explain the process by which arrest deters some suspects from repeat offenses. In Chapter 6 we will return to the issues raised by their work. Meanwhile sociologists will continue to test these ideas, gradually discovering more about how police actions might help reduce domestic violence and other crimes, as well as more about deterrence and labeling theory in general. This is part of what makes the field so exciting for sociologists: Our understanding of the world is constantly developing and changing based on our research.

ETHICAL ISSUES IN RESEARCH

All researchers need to be aware of ethical issues. For instance, chemists must be concerned with the potential toxicity of substances they work with; zoologists must be concerned with the health and safety of the animals they study. Because sociologists study real people and real-life settings, they often face special ethical concerns. Researchers need to be sure that they will not hurt the people and groups that they study, and they need to make sure that their own beliefs and emotions about their research topics don't undermine their objectivity.

Protecting Research Subjects

To do their research, sociologists often need to observe real people and interact with them in ways that might interfere with their normal way of life. No matter what type of methodology is used, a researcher runs the risk of making people uncomfortable, invading their privacy, or upsetting their daily routine.

To help guard against such intrusions, sociologists and other social scientists have developed codes of ethics to guide researchers. The American Sociological Association, the professional association to which most sociologists in the United States and Canada belong, developed the code for sociologists. Universities and research institutes where sociologists work also have committees that review research plans before they are carried out to make sure that they won't harm subjects.

Practices that try to keep participants as safe and unthreatened as possible and that maintain their privacy are common to social research. For instance, Sherman and his associates excluded from their study incidents in which the woman's life had been seriously threatened or in which she had suffered se-

BOX
2-3

SOCIOLOGISTS AT WORK

Lawrence Sherman

Men who have little left to lose can be desperate; men with a stake in society can be turned around.

The featured research in this chapter was conducted by Lawrence Sherman, University of Maryland, and his colleagues.

How did you get interested in studying the area described in the featured research? I spent thousands of hours riding with police in patrol cars, answering 911 calls. The most frequent type of call was the domestic disturbance. As I watched police deal with the situations and read the substantial literature on the problem, I realized no one was asking the key question: What effect does the police action have on violence after the police leave?

What surprised you most about the results you obtained in your research? The fact that police action could make any difference at all. Most criminologists supported the doctrine that government does not matter and that crime is determined by social forces. The Minneapolis experiment was the strongest evidence against that view.

What was the reaction to your research among your colleagues? Did it surprise you in any way? I was surprised at the extent to which my colleagues who supported the "police don't matter" doctrine attacked the publicity that the research results got. Their view was almost in favor of censorship—that the public was unprepared for these results and they should be kept within the profession until they had been replicated.

What research have you done since completing the study described in the chapter? Does it build on your previous research? I am now focusing on cases of more serious injury. Unfortunately we have found that domestic homicides cannot be predicted by the prior number of police calls, either in Milwaukee or in Melbourne, Australia. But we can say that women are at the greatest risk in the first several days after a police call, which supports the use of shelters and even portable alarms for women during the high-risk window.

How does your research apply to students' lives? For the too many students who have current and past experiences of domestic violence in their lives, the research is very practical: If the batterer is employed, try to get him arrested; if he is not, try to get as much distance as possible and do not rely on the police. Men who have little left to lose can be desperate; men with a stake in society can be turned around.

Much of your work has been directly oriented toward helping police and criminal justice professionals. Can you describe specific situations in which police officers or departments used your work? On a different subject, my finding that 50 percent of all crime occurs at only 3 percent of all addresses has helped police departments to look for their crime "hot spots" and to focus more resources on those locations. The evidence from New York and other cities is that this has helped them to reduce crime.

rious injury because arrest would generally be the only proper course of action in these cases. In addition, the researchers tried to be sensitive to the victims' concerns and fears in the weeks following their experiences and to choose interviewers who were similar to the victims in race ethnicity and the like. The researchers also repeatedly assured the victims that their responses would be kept confidential and that their names would not be revealed to others.

Separating Personal Views and Research Findings

Researchers need to be concerned not only with participants' well-being but also with their own feelings and emotions. Sometimes in a whodunit the detective falls in love with the villain. If the detective's vision is clouded by emotion, it becomes much harder for him or her to correctly interpret the evidence and recognize who really committed the crime. Similarly social researchers and laypeople alike must guard against "falling in love" with their favorite ideas of how the social world works. In Chapter 1 some theories undoubtedly made more sense to you than others. In the chapters to come, as you explore the various issues sociologists study, you will likely be more interested in some issues than in others; you will also find yourself more attracted to some theories and explanations of the world than to others.

The very passion that animates sociologists can potentially interfere with their commitment to the dispassionate methods of science. Just as a good detective must be able to set aside emotion and biases, so must good social researchers see beyond their own passions and theoretical preferences to evaluate what the empirical data actually are telling them. Following the rules of research methodology described in this chapter is the best way to ensure that this happens. Sociologists must choose research designs and measures that can provide the best tests of the theories and hypotheses they are studying. Sociologists must try to think of alternative explanations for their findings and do their utmost to falsify all possible explanations, especially their own favorite hypotheses.

The practice of peer review is an important part of this process. Often the sociologists who review a research report before its publication have different theoretical preferences than the authors. Thus they can spot places where the researchers might have been biased toward their own favorite views. These reviewers can also help ensure that researchers use the best methods and procedures and that they consider other possible explanations for their results.

Lawrence Sherman's work illustrates how the careful social scientist can maintain an open mind during the research process and how important such an open mind is to advancing theory and research. When the results from the Minneapolis study first came in, Sherman and his colleagues easily could have concluded that deterrence theory was supported and moved on to new areas of research. Or, if they were less ethical in their approach and simply wanted to promote deterrence theory, they could have designed replications that would have supported their initial findings.

Thankfully, they didn't do this. Although they clearly reported, in a peer-reviewed article, that their findings supported deterrence theory and even offered policy recommendations based on this work, they made no secrets of the study's limitations. In addition, several other researchers critiqued the original work and suggested ways to improve it.

After analysis of these shortcomings, Sherman and his associates improved their design and replicated the study in several different cities. When these analyses failed to support their earlier results, they could have downplayed the new findings and tried to explain them away. Instead they objectively evaluated the new results and used them to develop both a new understanding of the limits of deterrence theory and new theories about the relationship between police action and domestic violence. To learn more about Sherman and his work, see Box 2-2.

Doing research with honesty and integrity, following standard rules of methodology, and maintaining an open academic community are the primary ingredients in advancing sociology. In the chapters to come, you will see many examples of how the ongoing research process has led to new understandings of the social world.

SUMMARY

Research is the way in which social scientists develop, test, and refine their explanations of the social world. The dynamic interplay between theory and research is at the core of sociology. The work of Lawrence Sherman and his colleagues illustrates the research process and shows how well-executed investigations can add to our knowledge of important social issues. Key chapter topics include the following:

- To be accurate and trustworthy, research should follow established procedures, or methodological rules.

- The research process consists of three basic steps: asking questions, gathering data, and answering these questions. Sociologists can use either inductive or deductive reasoning in their work.

- Sociologists look at relationships or correlations between variables. To show that an independent variable causes a dependent variable, researchers must show that the independent variable occurs before the dependent variable, that the two variables are correlated with each other, and that no other variable can account for or explain this correlation. Hypotheses can never be proved, but they can be falsified.

- To ensure accurate results, sociologists use representative samples of the group under study and measure variables in ways that are both valid and reliable. Sociologists use many different kinds of research methods, including field research, unobtrusive research, survey research, and experiments. Only experiments can establish causal relationships.

- Through data analysis researchers try to answer their research questions. Replication of results is essential to ensure that similar results appear with different samples. In the process of doing studies and replications, researchers often revise theories and develop new research questions.

- All researchers must be concerned with ethical issues. Because sociological research often involves work with humans, researchers must be careful to protect these people from harm and balance research interests with concern for subjects. Sociologists also need to be careful that their personal feelings don't affect the accuracy of their research.

KEY TERMS

attributes 30
causal relationships 31
comparative research 34
content analysis 34
control variable 41
correlation 31
data 27
deductive reasoning 29
dependent variable 31
experiment 36
experimental variable 35
falsification 32

field experiment 36
field research 33
historical research 34
hypotheses 29
independent variable 31
inductive reasoning 29
measure 32
methodology 27
nonparticipant observation 34
participant observation 34
peer review 39
population 32

probability sample 32
qualitative data 34
quantitative data 33
random selection 32
reliability 32
replication 32
sample 32
spurious correlation 31
survey research 35
unobtrusive research 34
validity 32
variables 30

INTERNET ASSIGNMENTS

1. Using a search engine, locate the American Sociological Association's (ASA) home page. From this Web page locate information about the journals that are sponsored by ASA. What theory, method, or substantive issues do each of these publications emphasize? Which journals does your college or university library subscribe to? Look through some recent issues of these publications and see what research methodologies and techniques are used in the articles. Do you see any trends? That is, is one method or technique used more often than others? What does this tell you about the methods sociologists use?

2. Several research centers have pages on the Web. Using a search engine, locate some of these organizations. For example, some of the best known centers are: the Survey Research Center at Princeton University, the Population Study Center at the University of Michigan, the Data and Program Library Service at the University of Wisconsin–Madison, the Public Opinion Laboratory at Indiana University–Purdue University, Inter-University Consortium for Political and Social Research (ICPSR), the Gallup Organization, and the Roper Center. What types of data do these organizations have? How many different data sources can you locate that contain information on work, employment, and earnings? How many can you find that contain information on social attitudes (for example, on abortion, welfare, politics, and so forth)?

One of the best known and widely used public opinion surveys is the General Social Survey. Try to locate information on this data set and publications that contain analyses of these data.

3. As you know, sociologists use many different ways to gather data. Try searching for material on these data gathering techniques by entering terms such as "qualitative research," "content analysis," survey research," or "social statistics" into a search engine. What kinds of results do you get? You might find discussion groups or listserves. What kinds of issues are discussed in these groups? Do the topics vary from one type of data gathering strategy to another? You might find research firms. What types of studies do these firms do? You might also find descriptions of ongoing research or entire research reports. Compare these descriptions to the steps of the research process described in this chapter and assess how well each part of the process has been described.

TIPS FOR SEARCHING

Search engines that are categorized by topic, like Yahoo, are useful when using terms that have many meanings. For example, entering "survey" or "data" in an unrestricted search might lead you to discussions of geological surveys or electronic data transfers. Searching with these terms under the category of social science or sociology will make your search more efficient.

INFOTRAC COLLEGE EDITION: READING

Article A21036269 / MacCoun, Robert J. Biases in the interpretation and use of research results. *Annual Review of Psychology, 49,* 259–288. MacCoun examines debates regarding bias in social science research and possible ways to deal with it.

FURTHER READING

Suggestions for additional reading can be found in *Classic Readings in Sociology,* bundled with this book. If you purchased a used copy of this book that does not include this custom-published reader, go to www.sociology.wadsworth.com for ordering information.

RECOMMENDED SOURCES

Babbie, Earl. 1997. *The Practice of Social Research,* 8th ed. Belmont, Calif.: Wadsworth. Babbie provides an entertaining and easily understood discussion of research methods.

Becker, Howard S. 1998. *Tricks of the Trade: How to Think About Your Research While You're Doing It.* Chicago: University of Chicago Press. This is an entertaining discussion of the ins and outs of research from a noted practitioner of field research.

Cohen, Bernard P. 1989. *Developing Sociological Knowledge: Theory and Method,* 2nd ed. Chicago:

Nelson-Hall. Cohen provides a more philosophical look at the relationship between theory and method.

Sherman, Lawrence W. 1992. *Policing Domestic Violence: Experiments and Dilemmas.* New York: Free Press. This is a detailed account of the ongoing research into domestic violence and police policy.

Zeisel, Hans. 1985. *Say It with Figures,* 6th ed. New York: Harper & Row. This is a classic and informative guide to both presenting and reading data.

INTERNET SOURCES

American Communication Association Social Science Communication Research Page. This site offers links to research-related sites in many disciplines, including sociology, history, linguistics, psychology, public health, and statistics.

The Institute for Social Research at the University of Michigan is a long-established and very large university-based research center. Its Web site describes many different social science research projects.

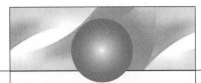

Pulling It Together

We will return to the ideas that were presented in this chapter at many places
throughout the rest of the book. You can also look up topics in the index.

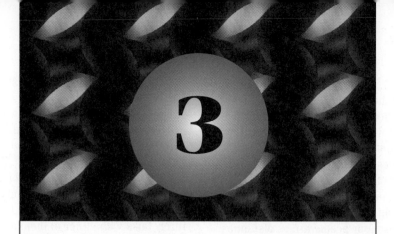

3

Culture and Ethnicity

MARIA, CRAIG, SHIRLEY, and Chan are all avid fans of the Los Angeles Dodgers baseball team. They share other characteristics as well: They are all well educated and smart; they all read the *Los Angeles Times* each morning; they all love to go to movies; and they all live near Sunset Boulevard, the street immortalized in the 1950 movie of the same name. Yet Sunset Boulevard is a twenty-four-mile-long thoroughfare, and in many ways the lives of these four people are as distant as the two ends of the street.[1]

Maria's ancestors settled near what is now the eastern end of Sunset Boulevard in the late 1700s. Maria teaches in a local elementary school in a community called Echo Park. She cares deeply about her students but finds her job increasingly difficult. Almost 40 percent of the children in Los Angeles public schools speak only a little English, and more than eighty different languages are spoken by the district's students. These students come from families that differ not only in the languages they speak but in the clothes they wear and the foods they eat; not only in their religious beliefs but in their views about learning and their attitudes toward women, schools, and government. Some of the students' parents have fled oppressive governments, and some of the parents have very little schooling themselves. At times, all the parents seem to share is a desire to make a new life and raise their children. The atmosphere in the community is often tense, and these tensions sometimes erupt into violence. Maria worries about her students' safety as well as her own.

Craig lives a few miles west of Echo Park in Hollywood. Like many young people in Hollywood, Craig hopes to enter the entertainment business. He majored in English and has written several unsold screenplays. Currently he is working as a waiter in a restaurant that caters to people who work in films. Most of his friends are connected to the business: aspiring actors, writers, producers, and technicians. The large gay community in West Hollywood also appeals to Craig. He often felt uncomfortable in college because of his sexuality, but here he feels as if he belongs. Craig loves the vibrant atmosphere of Hollywood. Sometimes he goes to the corner of Sunset and Vine to look at all the stars on the sidewalk memorializing movie stars.

If Craig looked long enough, he would find the star that celebrates Shirley's career. Shirley was a Hollywood success—so successful that she now lives in retirement in Beverly Hills, just west of Hollywood. She often watches tapes of the movies in which she starred, plays bridge with her wealthy friends, and occasionally travels abroad. Although Shirley was born into a Jewish family in New York City, she carefully hid this fact from her fans. Recently, however, she has begun to think again about her Jewish heritage and customs. Shirley's maid and gardener both live in Echo Park, where Maria teaches, but Shirley rarely talks with either of them and thus knows little about their lives.

Chan is a student at UCLA, which borders Sunset Boulevard to the west of Beverly Hills. Chan's parents immigrated to the United States a few years before he was born. They have always encouraged him to do well in school and make no secret of their hope that he will complete a doctorate in physics. He has done well in his classes, but he worries about his social life. His freshman class is as ethnically varied as Maria's school: more than half are Hispanic, Asian, or African American. Chan would like to get to know some of the Hispanic and African-American students, but, so far, his only contacts with them have been in class. Somehow the students from different backgrounds don't spend much time socializing with one another.

The similarities and differences in the lives of these four people all reflect what sociologists call **culture**—the complex pattern of living that we as humans have developed and that we pass on from generation to generation. All four share a culture that is common to well-educated residents of the United States, yet in many ways they live in different worlds.

If we broaden our view beyond the Sunset Boulevard area, we can readily see that the world is filled with cultural groups. Like the residents of Los Angeles, all these groups have much in common. They all have families and children; they all construct shelters and gather food; they all have strong beliefs about their way of life. Yet their customs, beliefs, and rituals vary greatly. Often their cultural differences seem to overwhelm all their commonalities, and sometimes the result is violent confrontations between cultural groups.

How do these similarities and differences arise? How do they affect the people who live within a given cultural group? To what extent do people in the United States belong to different cultures, and what implications do these differences have for our society?

*A*S YOU MIGHT REMEMBER from Chapter 1, one historical condition that prompted the development of sociology in Western Europe was the widespread discovery and exploration of new lands and peoples. Explorers and world travelers brought back descriptions of religious practices, family arrangements, beliefs, and customs that differed markedly from the European way of life. Early European sociologists were fascinated by the vast variety of ways in which people around the world live. Sociologists continue to be curious about why societies develop different cultures and why similarities exist across cultures. Sociologists also are interested in how our cultures influence the way we see ourselves and how we make decisions such as what type of religious beliefs we will adhere to or how important higher education is to us.[2]

Questions about culture have special significance for our own society. With its relative freedom and abundant resources, the United States has been a magnet for people from all over the world. We constantly hear about issues that arise from our cultural diversity, from movements to declare English the official language on ballots and in the schools to hate crimes against members of ethnic groups, gays and lesbians, or new immigrants. In addition, the globalization of the world's economy brings us into contact with unfamiliar cultures, creating opportunities for either increased understanding or tragic misunderstanding and even conflict. The more we comprehend how culture affects our experiences and worldviews, the better we can deal constructively with the challenges of living in a culturally diverse society and a multicultural world.

This chapter explores culture from a sociological perspective. We begin at the macrolevel, looking at what cultural similarities and differences exist around the world, why these similarities and differences occur, and how cultures change. We then move to a more microlevel view and examine how we come to see ourselves as part of a particular culture.

CULTURAL UNIVERSALS AND VARIABILITY

To better understand the diverse ways in which human beings live, I sometimes imagine a roomful of artists, all of whom are attempting to depict the idea of "home." All artists would almost certainly sketch some type of shelter and its environs. Yet the similarity might end there. The shelters could have very different shapes, from straw huts to elaborate multistoried mansions. The surrounding environments might range from acres of sunlit farmland to blocks

These women are members of just one of many different cultures around the world. How might their lives be similar to your own? How might they be different?

of gloomy tenements. The colors might range from somber browns and grays to vivid yellows and reds. In a similar manner all societies seem to share certain cultural traits or characteristics, but specific forms can vary dramatically.

Simply because we are all members of the same biological species, human beings share certain capacities. Probably the most important thing that distinguishes us from other animals is that we have extensive language systems. Scientists now know that many animals communicate with each other; for example, bees can tell one another where to find honey, and monkeys can warn their young about an approaching eagle or snake. Human language, however, allows us to do much more because of our ability to use symbols for things we can't hear, see, or touch. For instance, we can't touch love or hunger or fear; we can't literally hear or see pride or power, compassion or loyalty. Yet these terms have meaning for us, and we can communicate their meanings to others who share our language. In short, language allows us to communicate with **symbols**—in this case, words—that represent concepts, or abstract ideas. More generally a symbol can be anything—an object, a gesture, a word, a drawing—that people use to represent something else. This unique and marvelous capacity to develop and share symbols has enabled humans the world over to develop their complicated cultures and ways of life.[3]

All humans also share certain needs. Two stand out as most important in a discussion of culture. First, unlike some animals, young human children are incapable of caring for themselves, but instead must be nurtured for many years before they can fend for themselves. Second, because of our highly developed cognitive capacities, humans need to attach some meaning to their activities. We want to have reasons for the things we do; we want to make sense of the world we live in.

All known societies have developed ways to deal with these two needs. All societies have rules about who should do child care and other important tasks, and all have developed some type of family structure for the care of children.[4] All societies also have beliefs about what is important, and all have various symbols and rituals that reinforce these beliefs.[5] In a sense the existence of family ties and belief systems can be seen as the basis of the common humanity of all people.

Throughout the world we can see **cultural universals,** such as norms about the family and values regarding individual behavior, that are characteristic of virtually all societies. Yet the form or nature of these structures can vary a great deal from one group to another. Some societies define the family as only a mother and father and children whereas others define it much more broadly. Similarly some cultures value independence and ingenuity whereas others value cooperation and patience. Sociologists and other social scientists have looked at these commonalities and differences and developed theories to account for both the universal patterns and the variability.

In their studies of cultural commonalities and differences, social scientists have identified several **cultural elements,** basic aspects or characteristics of a culture. The cultural elements of a group can be seen as providing a roadmap or blueprint of its way of life—a way of understanding what members do and why. Some of these elements involve what is called **material culture,** the various physical objects that people make and that they attach meaning to, such as their food, clothes, houses, and hairstyles. Other elements are termed **nonmaterial culture,** the way of thinking of a group, including norms, values, ideology, folklore, and language.

Caring for Children and Each Other

Much of our knowledge of different cultures comes from the work of anthropologists who have spent many years living with groups of people around the world. Many of these groups were preliterate and pre-industrial; that is, they had no written means of communication and generally had relatively simple means of subsistence, such as hunting and gathering food or farming without the use of a plow or irrigation. Because today virtually all societies have had at least some contact with other groups, anthropologists and sociologists believe that the most reliable information about different cultures comes from studies conducted before such contact was common. One ambitious anthropologist, George Murdock, devised a way to systematically accumulate this type of information so that it could be summarized and compared. This work, available in most college and university libraries, is called the Human Relations Area Files, or HRAF.[6] Sociologists have used materials from these files to study large numbers of societies, using comparative research techniques not unlike those first employed by Max Weber in his studies of religion and economic life in Europe, China, and India.

Gender-Based Division of Labor Reviews of data from the Human Relations Area Files show that in all known societies women have primary responsibility for child care. This cultural universal no doubt reflects the fact that women, not men, have babies and can breast-feed. Throughout most of human history the only way for infants to survive was through nursing, and only women are able to do this. Similarly, in all known societies hunting and waging war are primarily men's responsibilities. In the few societies where women do hunt, the activity doesn't interfere with their ability to care for their children.[7] For instance, among the Agta, a small society in the Philippines, women regularly hunt for meat even while caring for young children. They can do this because they live in a tropical area with abundant game and can easily return to their camps with their catch in less than an hour.[8] Women also have been known to participate in warfare on occasion, but this appears to be only on an emergency basis, as in guerrilla wars or desperate last-ditch battles.[9]

These rules about what men and women can do in a society are examples of **norms,** cultural rules that define expected or required behavior. The early American sociologist William Graham Sumner defined three types of norms. **Folkways** are norms governing the customary ways of doing everyday things, such as wearing clean clothes, waving to neighbors, or walking on the right-hand side of the walkway. **Mores** are norms that are vital to the society and seen as morally offensive if violated, such as norms against child abuse or murder. **Laws** are norms, either folkways or mores, that are codified, or written down, so that they may be enforced by some group.

What elements of material culture, the physical objects people make and use, are shown here?
What elements of nonmaterial culture, such as beliefs, norms, and values, can you observe?

Norms are found in all cultures and in all situations. As we noted in Chapter 1, collectively norms help to make life predictable and form part of the basis of the social structure. Whereas Agta women find nothing remarkable about their hunting activities, Aleut women in Alaska and northern Canada, as well as many other women throughout the world, wouldn't dream of participating in a hunt. Norms apply to entire societies, such as the Agta and Aleuts, but also to **subcultures,** groups or cultures that exist within a broader culture. Thus norms guide the lives of Maria in Echo Park, Craig in Hollywood, Shirley in Beverly Hills, and Chan at UCLA in their daily interactions with the people around them. Shirley assumes that she will have a maid and a gardener and that her manager will supervise them; Chan assumes that he should study hard to be a physicist. Whatever the norms of your culture or subculture, you have been immersed in them from birth. You simply accept them as unremarkable, as the way life should be.

Norms regarding child care, hunting, and warfare appear in all societies, but other norms about men's and women's activities vary a great deal. In some societies women care for fowls and small animals, and in others men do. In some societies men care for dairy animals, and in others women do. Together the norms about the activities of men and women form a **gender-based division of labor,** rules about what tasks members of each sex should perform. All societies appear to have a gender-based division of labor, as well as an aged-based division of labor. Men and women do different types of work, and children and adults are also assigned different tasks.[10]

Families and Kin Groups Besides norms regarding the gender-based division of labor, all societies also have rules about whom a person may marry and who is responsible for the welfare of each individual child. The **nuclear family**—a family group consisting of a mother, a father, and their children—exists in virtually all societies and is seen as the basic unit for child rearing. All societies also recognize a broader group of relatives, or what social scientists call **kin.** However, just as artists can depict "home" in many ways, societies have developed many different ideas of who should be considered kin and what obligations they have toward children.[11]

For instance, even though you and many people you know might not have grown up with both a mother and a father in the home, you probably still think of such an arrangement when you hear the word *family.* That is because the normative family in our society is the nuclear family. In addition, when you think of grandparents, you probably think of the parents of both your mother and father. Your aunts and uncles are the sisters and brothers of both of your parents. Your kin, whether on your mother's or father's side of the family, are generally equally likely to give you presents or leave valuables to you after they die. As an adult you would not be allowed to marry close relatives on either side of your family.

If, however, you had been born in another society, your notion of family and kin would be very different. For example, if you had been a child in a traditional Navajo family (before the Navajo way of life was altered by contact with Spanish and Anglo immigrants to the American Southwest), you would have lived in a *hogan,* or house, with your mother and father, much like our own version of the nuclear family. This hogan, however, would not be set off by itself, but would sit beside other hogans belonging to your maternal grandmother and your mother's sisters. In addition, although you would know your father's relatives and consider them kin, your most important relatives would be those on your mother's side of the family. You would say that you belonged to your mother's, and not your father's, lineage or clan. Your mother's sisters and your maternal grandmother would be very much involved in rearing you and would care for you if your own mother died. You would inherit property only from your mother's relatives. If your father were to die before you, you would not receive his property because you would not belong to his clan. Instead, the property would go to his brothers or his mother's brothers, who would belong to his clan. In adulthood you would not be allowed to marry someone who belonged to your mother's clan, even if in our society he or she would be seen as only very distantly related to you, perhaps through several great grandmothers.[12]

If you had been born into a traditional Chinese family before the nineteenth-century Sino-Japanese war, your living arrangements also would differ. Navajo society is said to be *matrilineal,* because it "counts kin" or views family relationships by concentrating on the mother's side of the family. In contrast, the traditional Chinese family was *patrilineal.* Your mother would have left her own family, perhaps at a young age, and gone to live with your fa-

ther and his family. She would be expected to exhibit total devotion and service to her new in-laws. Although you would recognize your mother's relatives as kin, you would rarely see them. Your most important relatives would be those on your father's side of the family, not just because you would probably live with them, but also because you would regularly honor them and even their ancestors.[13]

The traditional Navajo and Chinese families are just two examples of the many types of kin structures found throughout the world. Along with these kinds of structures, all cultures also have developed strong beliefs about why these traditions are important.

Making Sense of the World

If you did not grow up in a traditional Navajo or Chinese home, you might have difficulty imagining what those societies are like or understanding why their members would regard their own culture as simply the way life should be. In fact, we all have beliefs about what life should be like. As you will remember from Chapter 1, these general standards about what is important to a group are called **values.** Values are another basic element of all cultures. Whatever culture you belong to, values provide standards that help you choose your actions and make decisions in your daily life.[14]

For instance, in the United States people typically value individual ingenuity and independence.[15] Heroes are famous inventors such as Thomas Edison, the Wright brothers, and George Washington Carver or daring explorers such as Lewis and Clark, Amelia Earhart, and John Glenn. Individuals who leave home to seek adventure are admired and honored and the family system, which de-emphasizes extensive ties with kin, facilitates such independence and daring. In contrast, traditional Chinese society valued cooperation, courtesy, patience, and self-control. Chinese households often included a large number of people living in close quarters. Thus these values helped keep conflicts and disruptions to a minimum and also helped ensure that family members would be available to do the work required to maintain the family's land and farms.[16]

Cultural values and norms are supported by belief systems. All cultures appear to have developed marvelously imaginative descriptions of why people live the way they do. These descriptions can be embedded within **folklore,** myths and stories that are passed from one generation to the next, or within religious tenets and **ideologies,** complex and involved belief systems. These ideologies and folklore are also generally supported by **rituals,** cultural ceremonies that often mark important life events.[17]

For instance, Navajo legends describe the behaviors of the gods, telling of their marriage customs and the rituals they performed at birth, puberty, and death. These behaviors of the gods are seen as models that good Navajos should follow.[18] Navajo rituals help reinforce clan ties and norms such as those regarding marriage and kin obligations. Similarly Chinese myths and rituals support the importance of the family and its continuation through sons. Although not a religion in the sense of Judaism or Christianity, Confucianism embodies many of the ideas of "filial piety," the notion that children should serve their parents with great devotion and respect.[19] In traditional Chinese society families regularly observed feasts and ceremonies throughout the year to commemorate dead ancestors and help family members remember and maintain family ties.

Like people throughout the world, the individuals described at the beginning of the chapter regularly participate in rituals and subscribe to ideologies and folklore that help guide their behaviors, even though they might not be conscious of doing so. For example, Maria is a devout Catholic and regularly attends mass with her family. She views her work with the children in Echo Park as a means of serving God. Although Shirley once tried to hide her Jewish ancestry, she now regularly attends synagogue and celebrates Jewish holidays. When she does this, she feels connected not just to her family but to other Jews throughout the world. Craig has heard all the Hollywood stories about overnight stars. He firmly believes that if he continues to work hard, he, too, will be discovered and will someday have his own sidewalk star. And from his first day at UCLA, Chan has been exposed to campus folklore. Fellow students have told him stories about one-third of all students flunking a certain elementary physics class, no matter how hard they work, and about the ghost that haunts an old lab. Listening to such stories, and retelling them to other students, makes him feel that he belongs, that he is linked to his fellow students and to campus life.

We all have beliefs about our world and participate in ceremonies and rituals that help us feel connected to others. These actions and beliefs are so much a part of our daily lives that we rarely think about them or acknowledge their presence.

Explaining Cultural Variations and Regularities

Why do all societies have norms, values, and ideologies? Why have different cultures "painted pictures" that, while always including families and rituals, can look so different from one part of the world to another? Two theoretical orientations that have addressed these questions are structural functionalism and cultural materialism.

Structural Functionalism   Like the classical sociologists discussed in Chapter 1, the American sociologist Talcott Parsons (1902–1979) was fascinated by questions such as, What holds society together? Why do societies remain relatively stable and not simply fall apart? To answer these questions, Parsons developed the theory of **structural functionalism,** which looks at the **function,** or part, that social structures play in maintaining the society. Like the classical sociologists, Parsons conceived of **social structure** as the patterns that underlie social life, as social phenomena that are relatively stable. A particular social structure can be a type of behavior, such as the gender-based division of labor or the norms that govern students' behavior. A social structure might also be the way in which society is organized, such as the nature of kin groups and families or the relationships between businesses and consumers. To Parsons the important point was simply that a structure is a part of society that has been present for a relatively long period of time. Parsons suggested that the key to understanding why these structures exist is to determine what part or function they play in either maintaining or altering the society.[20] Although few sociologists today call themselves structural functionalists, the ideas introduced by Parsons and other theorists are commonly used by sociologists in their study of many different aspects of society, including culture.

Structural functionalists would argue that cultural norms and values, such as those regarding the division of labor and kin relations, are part of the structure of a society. The most obvious function of these norms and values is to guarantee child care and mutual support within groups. The presence of families and kin ensures that children will be cared for, even if something happens to their biological parents. In a traditional Navajo family the child's clan relatives—that is, those related to the child's mother—would assume the responsibility; in a traditional Chinese family the father's relatives would do the same. The traditional gender-based division of labor can also be explained in this manner. Simply put, if infants, and ultimately a society, are to survive, women are less expendable than men. Thus women hunt extensively only when the activity won't interfere with their childbearing and child-rearing tasks, and they risk themselves in battle only in emergencies.

Structural functionalists would suggest that folklore, ideologies, and rituals help support these structures by giving people reasons for their lives, by helping them to make sense of their world. We have no way of disproving ideologies or folklore, and because they are learned at such an early age and generally accepted by all who surround us, we simply accept them as truth. Rituals often invoke the supernatural, beings beyond the level of human life. Thus the rituals not only reinforce cultural practices, as the Navajo myths and Chinese rituals do, but also suggest that powerful, nonhuman forces have decreed that life should be the way it is.

These ideas of the function of kin relations, the division of labor, rituals, and ideologies might seem rather obvious. In fact, another structural functionalist, Robert Merton (b. 1910), called those functions that are easily seen and obvious **manifest functions.** Merton argued that structures can also have what he called **latent functions,** functions that are less obvious, often unintended, and generally unnoted by the people involved.[21] To understand why structures persist, we need to understand both their manifest and latent functions. For example, each cultural element we discussed has the important latent function of binding members of the society to one another and promoting social solidarity.

You can perhaps grasp this more easily by recalling the work of the classical sociologists Émile Durkheim and Georg Simmel. Durkheim noted how differentiation within a society, which would include the gender-based division of labor, actually makes people more dependent on one another, thus binding them closer together. This effect can be viewed as a latent function of differentiation. Similarly Simmel noted how a "web of group affiliations" underlies society. Through kin relations, rules about marriage, rituals, and ideologies, humans reassert their bonds with one another. Even though we might not think about it consciously, each time we engage in some type of cultural ritual—perhaps attending a wedding or a funeral—we reinforce our relationships with and common commitment to one another and, ultimately, to society as a whole. Structural functionalists assert that these latent functions of cultural elements are just as important as the more manifest functions are in maintaining the structures of a society.

Structural functionalism is sometimes criticized for giving an overly rosy view of a society, for any analysis of the functions that structures play can imply that all parts of a society continually work together to make everything function smoothly. In fact, conflicts abound in our everyday lives, much like in the neighborhood of Echo Park where Maria teaches. In addition, some aspects of society can harm not just individuals but the nature of group life itself. Merton recognized this possibility and introduced the idea of **dysfunctions,** or negative conse-

quences of a structure of society for the whole of society. He suggested that structures can be both functional and dysfunctional and that it is important to determine the relative balance of these negative and positive consequences.[22]

You can see this by looking at our own culture as well as others, for many norms and values that you hold are not always as functional to or supportive of group life as they seem. For instance, along with their independence, people in the United States have generally valued freedom and ease of movement.[23] As a result, a large proportion of the population owns cars, and we have developed extensive road and highway systems. These are functional in large urban areas in that they make it easy for us to travel from our suburban homes to jobs many miles away. Highway systems also allow us to easily travel long distances to visit relatives or take vacations. Yet these highway systems are also dysfunctional because, while they encourage the use of automobiles, they discourage the use of mass transit, thus contributing to the depletion of the earth's oil and gas reserves. Similarly, the traditional Chinese family system can be seen as functional in ensuring the maintenance of a cooperative and integrated kin system across generations. Yet it also helped perpetuate vast inequalities between the sexes and the subordinate position of women.

Cultural Materialism   To a structural functionalist, characteristics of a culture continue to exist simply because they are functional, or beneficial, for a group. But why do these traits appear in the first place? And why do we see certain characteristics in some societies and other characteristics in others? Cultural materialists take a step backward and look at how the physical environment in which people live influences their way of life.

Anthropologist Marvin Harris (b. 1927) has been the most eloquent proponent of this view.[24] According to **cultural materialism,** societies develop differently primarily because they exist in different environments and thus tap different resources, or materials. To understand this perspective, simply think of the different diets or houses that people around the world have. For instance, Polynesians probably traditionally built houses out of grass and plant material because these were readily available and the climate did not require that they have protection against cold. Similarly, because they often lived in areas with abundant plant and animal life, Polynesians consumed plants as well as fish and animals. In contrast, the Inuit lived in a much harsher climate. Igloos built of blocks of ice provided extensive protection from the cold, and the Inuit relied virtually exclusively on hunting for food.[25]

BUILDING CONCEPTS

Structural Functionalism and Cultural Materialism

Structural functionalism and cultural materialism are two approaches to explaining cultural variations and regularities.

Perspective	Characteristics	Example
Structural functionalism	Examines the function, or part, that social structures play in maintaining social order.	The presence of families and kin is a social structure that functions to guarantee child care and mutual support within groups. The presence of families and kin across societies is an example of *cultural regularity.*
Cultural materialism 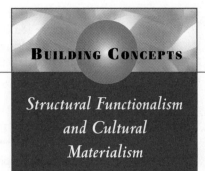	Examines how the physical environment in which people live influences their way of life.	Polynesians traditionally built houses out of grass and plant material in part because they were readily available and the climate did not require protection against harsh weather. Inuit, on the other hand, built igloos to provide extensive protection from the cold. The difference in habitat between Polynesians and Inuit is an example of *cultural variation.*

Other cultural variations, such as the different types of family structures, can also be understood in relation to the material environment. For example, groups that farm crops or herd animals require the cooperative labor of many people to ensure a steady food supply. In this type of society, **extended families,** whose members live together and cooperate in food production, help the society survive. The families of both the traditional Navajo, with their herds and small gardens, and the Chinese, with their small farms, are typical of such groups. In contrast, in hunting and gathering societies a mother and father can generally obtain enough food without extensive help from others to sustain themselves and their children—in these societies smaller kin groups are more functional. Interestingly, our own and other industrialized societies more closely resemble these latter groups. We typically rely only on members of our nu-

clear family for support and often live many miles from extended kin.

Understanding Cultural Dynamics

It is important to remember that cultures are always dynamic and changing. You can easily see how much U.S. culture has changed over the past century by visiting your local museum or watching reruns of old television shows or broadcasts of old movies on television. The world depicted there often differs greatly from that of today, not just in elements of material culture such as clothes, hairstyles, cars, appliances, and music, but also in elements of nonmaterial culture, such as the words we use and the values we espouse. Some of these changes—such as electric lights, automobiles, and Levis—result from **cultural innovation,** the process whereby new cultural

Some people have suggested that the Internet provides a way to bypass cultural gatekeepers. Think about your experiences with the Internet. How have cultural gatekeepers affected what you have seen on and contributed to the net?

elements are created. Others—such as pizza parlors, reggae music, and punk hairdos—come from **cultural diffusion,** the process whereby elements of one culture or subculture spread from one society or culture to another.

When one aspect of a culture changes faster than another, the result is what is called **cultural lag.**[26] Often material elements of culture such as technology will change more quickly than nonmaterial elements such as values and norms. For instance, genetic researchers are rapidly developing the capacity to identify carriers of many inherited diseases and conditions, such as forms of breast cancer, diabetes, and mental retardation. Yet medical ethicists and the general public have not yet formed clear norms about how this knowledge should be applied—who should be tested, who should know the results of the tests, or what should be done once the results are in. This lack of congruence can be extremely problematic for both medical personnel and affected laypeople.

In addition, not everyone has the same influence on how cultures change. We can all create new cultural elements, from poetry and songs to machines and technology. Yet only a few of these innovations ever become widely known and integrated into the general culture. To actually reach the public, new products and ideas must successfully pass **cultural gatekeepers,** powerful organizations and individuals that control the entrance of new cultural elements into a society. These gatekeepers—newspaper and magazine editors, art gallery owners, film producers, book publishers, manufacturers—have the power to determine what kinds of products we can buy, what art and movies we can see, and what kinds of ideas we can read about. Each of us can decide whether to buy a particular book or see a particular movie, and

some cultural innovations that make it past gatekeepers never really catch on. Ultimately, however, the range of cultural elements that we are exposed to is determined by these gatekeepers.[27]

CULTURE AT THE MICROLEVEL: CULTURAL IDENTITY

When we compare one culture to another or trace the way cultural elements change over time, we generally take a macro approach—a wide-angle, long-distance view that encompasses broad features of social life. Yet we can also take a much more micro or close-up view of the relationship between culture and individuals. One way to do this is by examining people's perceptions of themselves as belonging to or being part of a subculture, much as Maria calls herself Mexican American, Shirley calls herself Jewish, Chan calls himself Chinese American, and Craig calls himself gay.

A Nation of Subcultures

The United States is populated by the descendants of people from a wide variety of cultures. A trip to your local greeting card store during the winter holidays can illustrate this diversity, for there you would find missives appropriate for a wide range of cultural groups. You could send a religious Christmas card to your devout Christian friends, a Happy Hanukkah card to your Jewish friends, a Chinese New Year card to your Chinese friends, and a Kwanza greeting card to your African-American friends who celebrate the newly created winter holiday of Kwanza. Similarly birthday parties for Chinese immigrant children in Echo Park will no doubt differ from the birthday celebrations of Chan and his friends at UCLA or those of Shirley and her friends in Beverly Hills.

Sociologists use the term *subculture* to refer to many different types of groups, from the gay community in West Hollywood and the campus community at UCLA to the various ethnic groups that live in Echo Park. Later in this book we will discuss the culture of schools, the culture of organizations and work groups, and subcultures of crime and violence. In this chapter, however, we will focus on subcultures that are related to the national origin of individuals or their ancestors.

Several terms are commonly used to describe and classify people according to their national, cultural, or biological origin. Occasionally the meanings of these terms are slightly different or more precise in sociology than in common usage. For example, sociologists use the term **ethnic group** to describe people who have a common geographical origin and biological heritage and who share a number of cultural elements, such as language; traditions, values, and symbols; religious beliefs; and aspects of everyday life, such as food preferences. In addition, sociologists say that people belong to an ethnic group if they define themselves as special and if other people recognize this distinctiveness.[28] The classical theorist Max Weber emphasized this subjective nature of ethnicity. According to Weber an ethnic group exists because people *believe* that it exists; members believe that they have a common culture and a common heritage.[29] We will see the importance of this subjective aspect of ethnicity a little later in this section.

Sociologists commonly use the term **white ethnics** to refer to people who identify themselves as belonging to a group with European origins (other than English), such as the German, Irish, French, or Italian. Although the terms used vary somewhat from one author to another, many sociologists use the term **people of color** to refer to people whose ancestors or who themselves came from non-European areas of the world and who can be identified through the color of their skin. This term is often applied to people of Hispanic origin, such as Mexicans, Puerto Ricans, Cubans, and Central and South Americans; to people of Asian origin, such as Chinese, Japanese, Filipinos, Koreans, Laotians, and Vietnamese; to African Americans; and to Native Americans or American Indians.

Social scientists emphasize that the social fact of group membership is much more important than physical characteristics. What is significant is not that individuals resemble one another in some way, but that they share a common culture—that they share values, beliefs, and ideologies and that they believe they are part of the same group. Thus, in addition to the term *people of color,* sociologists often use the term **racial-ethnic group** to signify subcultures that can be distinguished on the basis of skin color and ethnic heritage. For instance, because of the color of her skin, Maria would probably be called "white" by others who saw her. Yet Maria sees herself as part of the Mexican-American subculture, not only because of her ethnic heritage but also because of her religious beliefs and her links with others in the Mexican-American community. We turn now to a discussion of how people come to see themselves as part of such communities.-

Cultural Identity and Ethnocentrism

The Navajo refer to themselves as Dine, which translates as "the people." Most of their myths and folklore describe the origins of the Dine and the elements of their culture.[30] If you were born into a traditional Navajo family, you would come to see yourself as part of the Dine and as separate from those who were different, such as the Utes and Hopis who lived nearby or the Anglos and Mexicans who were entering your people's territory. If you were born in the United States or have lived here most of your life, you probably call yourself an American. You can recite the Pledge of Allegiance, you know the national anthem, and you call the war of independence that began in 1776 the "Revolutionary War" rather than the "Colonial War," as it is known in Canada and Great Britain. When you travel to other countries, the people there may view you as an American because of the way you talk and dress, and, as you witness other ways of living, you become even more aware of your own "Americanness."

No matter where you were born in the world, you would come to see yourself as belonging to or being part of a culture or cultural group; as sociologists put it, you would come to have a **cultural identity.** All people seem to identify with their own culture as they come to see the norms and values of their group as natural, to accept its ideologies and believe its myths and folklore.[31] This occurs with subcultures as well as larger cultural groups. Thus, although you might see yourself as an American, you may also see yourself as part of a student subculture or gay subculture or as an African American, Chinese American, or Irish American. The feeling of belonging specifically to an ethnic group is often referred to as **ethnic identity.**[32]

Feeling part of one's culture or ethnic group can help a person feel secure and connected to others, but it can also have a more negative impact. Social scientists use the term **ethnocentrism** to refer to the belief that one's own culture or way of life is superior to that of others. At least some degree of ethnocentrism is probably inherent to all ethnic groups and to a sense of ethnic identity. If you see your own culture as natural, then you might also tend to see other cultures as not just different but unnatural, inferior, and even immoral.[33] In many ways the battles between ethnic groups around the world, as well as within the United States, can be interpreted as a consequence of ethnocentrism, as the ultimate representation of the belief that one group's way of life is so superior to another's that it must be forcefully imposed. Some ethnic conflicts become so extreme that the goal is to eliminate the other group rather than merely impose one's way of life; examples include the genocidal Nazi campaign against the Jews in World War II or the recent campaigns of the Serbs against the Bosnians and Albanians in the former Yugoslavia.

Ethnocentrism also exists in the United States in many forms. In its extreme form ethnocentrism leads to ethnic violence, the so-called race-motivated crimes in which members of one racial-ethnic group batter members of another group for no reason other than the color of their skin. But less extreme ethnocentrism is a part of everyday life. For example, Maria feels the tension and danger as she teaches in Echo Park. Chan notices that members of different racial-ethnic groups have few contacts with one another on campus. Occasionally they might even exchange hostile words with each other and suggest that members of one group aren't as worthy as they are.

The problems associated with ethnocentrism have plagued the United States since the first immigrants landed. Thus, not surprisingly, these problems, as well as the experiences of ethnic groups and individuals' cultural or ethnic identities, have long fascinated American sociologists. By reviewing their work, you can learn more about how culture affects your view of yourself, how members of various subcultures in the United States have different experiences, and also how you, as an individual, help create the culture in which you live. Using your sociological imagination can even be an antidote to ethnocentrism itself.

Understanding Ethnic Groups and Subcultures: The Chicago School

Just as in Europe, the early sociologists in the United States were concerned with the problems of industrialization and urbanization. Because of its unique history, however, the United States faced different problems than many of those found in Europe, and American sociology reflected these concerns. In particular, American sociologists traditionally have grappled with issues related to the massive immigration to the United States and the legacy of African-American slavery.

Sociology departments began to appear in universities and colleges in the United States in the late 1880s and early 1890s, but the one at the University of Chicago was destined to be the center of sociological theory and graduate training until the 1930s. The faculty and graduate students at what would come to be called the "Chicago School" epitomized the tension between a passionate desire to understand and alleviate social problems and a dispassionate need to develop a science of society—a tension that continues within sociology. Much of the Chicago School's work concerned the large immigrant and growing African-American populations in Chicago. At the same time, these sociologists wanted to develop techniques of looking at the world that would validate their observations and to create and test theories that could be used in many other settings.

The ancestors of today's white ethnics often came to the United States in "steerage," the cheapest form of sea transport. Richer passengers traveled on the higher decks. How might these on-ship distinctions be related to the group boundaries that these immigrants experienced after their arrival in the United States?

Much of the Chicago School's work involved the techniques of field research whereby sociologists personally observe social situations and use inductive logic to form theories.[34] One of the Chicago School's most influential members, Robert Park (1864–1944), used the research of his students and colleagues to develop a theory about how relations between racial-ethnic groups evolve over long periods of time. Park called this theory the "race relations cycle."

Park suggested that when ethnic groups move to a new area and try to earn a living, **competition** with groups that are already there is almost inevitable, and conflicts are common. The lengthy battles between the colonists in the New World and the Native Americans, and the discrimination and even violence experienced by European immigrants to the United States, both illustrate this process.

Eventually, however, Park suggested that these ethnic groups reach a mutual **accommodation:** They develop norms regarding the ways in which they will interact with one another. For instance, when African Americans were brought to this country as slaves, very strict rules of etiquette regarding the slave–master relationship developed. Rules such as prohibitions against sitting or eating in the master's presence helped maintain distance between the white owners and the black slaves.[35] Other forms of accommodation might involve norms regarding the types of jobs members of various ethnic groups hold and the places where they live. For example, Italian immigrants tended to do manual labor, whereas people of Slavic descent often worked in mines.[36] Irish immigrant women often worked outside the home as domestic servants, whereas Italian women were much more likely to take boarders into their homes or, sometimes, to work in factories with other immigrant women.[37] In New York City, Italian immigrants lived in the "Little Italy" section, which was even further subdivided into enclaves that housed immigrants from different areas of Italy, such as Apulia, Naples, and Sicily.[38] Similar patterns were found throughout the country.[39]

Sociologists say that these patterns of accommodation help to mark **group boundaries** by defining who does and who does not belong to a given group. Of course these patterns also heighten ethnocentrism, as members of each group have only limited contacts with people from other subcultures.

Although this pattern of accommodation might continue for many years, eventually, Park suggested, members of the groups will begin to interact more with one another, group boundaries will begin to weaken, and, ultimately, **assimilation,** or a virtual merger of the groups, will occur. Building on Simmel's view of society as a "web of group affiliations," Park regarded it as inevitable that members of various ethnic groups would eventually develop friendships with one another, perhaps first through schools and work, but then through neighborhood associations, and finally through family and kin relations. As the affiliations between members of different ethnic groups increased, the older patterns of accommodation that maintained group boundaries would decline and people would become less likely to define themselves by their ethnic group membership. In other words they would be assimilated to the broader society, and ethnic identity would be less important to them.[40]

A distinction introduced by another early American sociologist, Charles Horton Cooley (1864–1929), can help to clarify this process. Cooley distinguished between **primary groups,** those that involve intimate, face-to-face interaction, and **secondary groups,** those that involve larger numbers of people and more distant social relations.[41] Your family and friends might both be considered primary groups because they include only a few people and are characterized by long-lasting intimate relationships. Your classroom and college or university would be considered secondary groups because they include more than just a

few people who come together not for intimate long-term relationships, but because they share a common goal of getting an education. In addition to primary and secondary groups, sociologists sometimes speak of **aggregates,** large groups of people who actually have no relationship to one another except that they might happen to be in the same place at the same time.

Collectively all the various racial-ethnic groups within the United States can be seen as an aggregate, a group of people who live within the country's boundaries. When members of these groups belong to the same work organizations or attend the same schools or churches, they belong to the same secondary groups. When they become friends or members of the same family, they belong to the same primary groups. Park's theory of the race relations cycle suggests that as members of different ethnic groups become members first of the same secondary groups and then of the same primary groups, their differences will diminish and they will become assimilated to the larger society.[42]

This view has been called the "assimilationist perspective" or the "melting pot theory" because it suggests that eventually the various ethnic groups that make up the country will melt down into one undifferentiated community. This perspective was propounded for many years in the United States by essayists such as the nineteenth-century philosopher Ralph Waldo Emerson.[43] However, Park and others in the Chicago School were the first to explain the sociological reasons underlying these changes—to develop a theory and testable hypotheses. According to Park's theory, once people from different groups develop secondary and then primary relationships with one another, the group boundaries that separate their cultures will break down.

We should note that Park's theory applies primarily to members of white ethnic groups. Park realized that African Americans and other people of color were treated differently because of their color. Consequently, he believed, the group boundaries separating them from other groups might be much more difficult to break down than are the boundaries between white ethnic groups.[44]

Although Park's ideas have been highly influential, they have been challenged. Other sociologists argued that Park and his followers underestimated the strength of ethnic identity. This approach is often termed "cultural pluralism" because it suggests that plural, or several, cultures continue to exist within the larger society. According to this perspective the assimilation Park predicted is not inevitable. Today members of various ethnic groups have been in the United States for several generations. They are no

a.

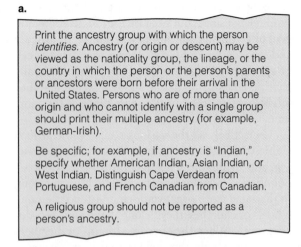

Print the ancestry group with which the person *identifies.* Ancestry (or origin or descent) may be viewed as the nationality group, the lineage, or the country in which the person or the person's parents or ancestors were born before their arrival in the United States. Persons who are of more than one origin and who cannot identify with a single group should print their multiple ancestry (for example, German-Irish).

Be specific; for example, if ancestry is "Indian," specify whether American Indian, Asian Indian, or West Indian. Distinguish Cape Verdean from Portuguese, and French Canadian from Canadian.

A religious group should not be reported as a person's ancestry.

b.

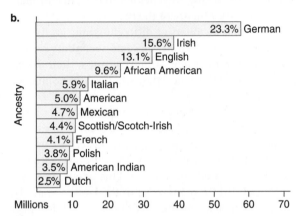

Ancestry	
23.3%	German
15.6%	Irish
13.1%	English
9.6%	African American
5.9%	Italian
5.0%	American
4.7%	Mexican
4.4%	Scottish/Scotch-Irish
4.1%	French
3.8%	Polish
3.5%	American Indian
2.5%	Dutch

Millions 10 20 30 40 50 60 70

c.

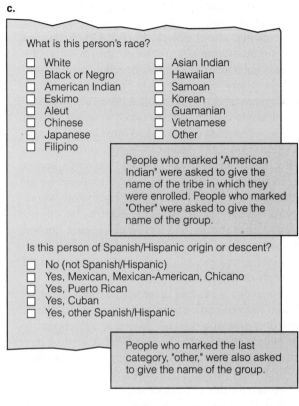

What is this person's race?

☐ White
☐ Black or Negro
☐ American Indian
☐ Eskimo
☐ Aleut
☐ Chinese
☐ Japanese
☐ Filipino
☐ Asian Indian
☐ Hawaiian
☐ Samoan
☐ Korean
☐ Guamanian
☐ Vietnamese
☐ Other

People who marked "American Indian" were asked to give the name of the tribe in which they were enrolled. People who marked "Other" were asked to give the name of the group.

Is this person of Spanish/Hispanic origin or descent?

☐ No (not Spanish/Hispanic)
☐ Yes, Mexican, Mexican-American, Chicano
☐ Yes, Puerto Rican
☐ Yes, Cuban
☐ Yes, other Spanish/Hispanic

People who marked the last category, "other," were also asked to give the name of the group.

FIGURE 3-1 *Ethnic Identity in the United States*

longer confined to their traditional jobs and neighborhoods, and they have married descendants of other ethnic groups. Cultural pluralists assert, however, that ethnic identity will continue to be important, perhaps because people need to feel that they are part of a group. Even if the various mechanisms of accommodation described by Park have declined, true assimilation, at least in the way people see themselves in relation to an ethnic group, will not occur.[45] Even people who become friends and belong to the same primary group might continue to see themselves as Irish or as Italian.

It is now more than a half-century since members of the Chicago School studied the neighborhoods of their city and Park first developed his theories of assimilation. The group boundaries that separated various white ethnic groups have virtually disappeared, although many boundaries between the white majority and racial-ethnic minorities remain. By this time, if Park's view of a race relations cycle were correct, relatively few descendants of the white ethnic immigrant groups should have a strong ethnic or cultural identity. Has that happened? Do people in the United

States still have strong ethnic identities? What do these ethnic identities mean to them? And what about people of color? How does their situation differ from that of descendants of the white ethnics?

Ethnic Identity in the United States Today

Every ten years the U.S. government conducts a **census**—a complete count of a group of people—of all the people living within its borders. The data collected by the Census Bureau are closely studied by policymakers, businesspeople, social scientists, and interested citizens. One census question that has particularly intrigued sociologists involves ethnic identity, and data from that question can be used to examine the basic hypothesis derived from Park's theory of the race relations cycle.

In both 1980 and 1990, in a random sample of about 20 percent of households nationwide, the person filling out the census form was asked to describe the ancestry of each person living in the household. The full text of the question used in 1980 is shown in Figure 3-1a. Note that the question is **open-ended,** meaning that respondents can give whatever responses they desire rather than having to check specific categories. The respondents were told not to use a religious group, such as Jewish, as their ancestry because the law prohibits the Census Bureau from gathering data on religion.

As you can imagine, given the millions of people living in the United States, all kinds of ancestries and their combinations were listed. To make sense of these responses, the Census Bureau combined a number of answers into one category. For instance, people who listed their ancestry as a particular ethnic group in what is now Italy (such as Sicilian, Tuscan, or Calbrian) were all categorized as Italian. In addition, the Bureau usually listed only the first two ancestries mentioned for each person and listed triple ancestries only if they were among the seventeen most common ones used in previous surveys. (The most common triple ancestry is English-German-Irish.)[46] Some citizens simply stated that they were "American," perhaps because their families have been in the United States for many generations and they don't know or care about their ancestry. Still, the vast majority did report one or more ethnic ancestries, with German, Irish, and English the most common responses. Figure 3-1b shows the twelve most common responses to the question about ancestry in the 1990 census.

In responding to the census question about ancestry, about 10 percent of the population identified themselves as African American, 4.7 percent as Mexican, and 3.5 percent as American Indian. However, the census also asks separate questions that can indicate whether citizens would be considered a person of color. These questions are shown in Figure 3-1c. Note that responses to these questions are **closed-ended,** meaning that respondents have only certain possible options from which to choose their answers.

Interestingly the answers that people gave to these closed-ended questions didn't always match those they gave to the open-ended question regarding their ethnic origin. For instance, in 1990 only about 2 million people defined themselves as American Indian in the closed-ended question, but almost 9 million reported that they were of American Indian ancestry. Most of these 9 million people are probably defined by other people, and see themselves, as "white"; yet they have some ancestors who were Native Americans and thus choose to think about themselves in this way.[47]

In this sense ethnic identity can be seen as a *choice.* At least to some extent, people can choose to see themselves as Native American, as African American, as Italian or Irish or Polish. Yet people might also ascribe such an identity to others, especially to people of color. Based on your observations of or interactions with someone, you might decide that the person is African American or Scottish or Middle Eastern. How do people make these decisions? To what extent can people in the United States choose to identify with a particular ethnic group? How do the results of studies of ethnic identity conform with Park's theory of the race relations cycle?

Common Sense versus Research Sense

COMMON SENSE: The values held by any given ethnic group are what distinguish them from other ethnic groups.

RESEARCH SENSE: The Mary Waters study found that many values thought to be distinctive to one ethnic group were actually shared across many different ethnic groups.

FEATURED RESEARCH STUDY

CREATING ETHNIC IDENTITIES

Sociologist Mary Waters wanted to study these questions, but she was dissatisfied with the quality of the information she could get from the census. The responses given there could tell her the number of people who claimed certain identities, but the data couldn't tell her *why* they chose to see themselves this way and what this identity meant to them.

Waters decided to try to answer these questions through field research in two suburban communities, one outside of San Jose, California, and one outside of Philadelphia, Pennsylvania. These suburbs were similar to many others throughout the country. Most

of the residents had attended college and held professional jobs, such as lawyer, engineer, nurse, or teacher. Most of the residents were also white. To control for, or eliminate, the possibility of religious differences influencing her results, Waters chose her sample from members of a Catholic parish in each community. Because many people identified as white ethnics, such as Poles and Italians, are also Catholic, she hoped that this would yield a number of people who would identify themselves as a member of an ethnic group. She also limited her sample to whites. As noted previously, people of color have many different experiences than

Mary Waters found that many of the special feelings that people have about their ethnicity revolve around families, food, holidays, and special events. How might the activities of these families be related to the development of their ethnic identity?

white ethnics, and limiting the sample in this way helped Waters to focus on the questions she wanted to answer without introducing complexities that would make the research results difficult to interpret.

Ethnic Identity: The Choices of White Ethnics

Waters conducted long interviews (from one to three hours) with sixty people, both men and women, and sometimes parents and their adult children. The interviews began with Waters showing the people a copy of the census question regarding ethnic identity and asking them how they would answer the question and then why they gave this answer. She followed up with a whole series of questions about their family history and ethnicity. In the tradition of the Chicago School, Waters developed her conclusions inductively. She mulled over respondents' answers, she asked new questions, and she gradually developed conclusions about what ethnic identity meant to these descendants of immigrants. As you will see, she found that the choices available to white ethnics are quite different from those available to people of color.

Virtually all of Waters' respondents, whether they mentioned just one ancestry or several, chose to emphasize certain aspects of their heritage. Some people simply knew more about the background of certain of their ancestors. Others identified with ethnic groups on the basis of their last names. Some people justified their allegiance to an ethnic group, as well as the way in which they classified other people into ethnic groups, by the way they looked. They had stereotyped views of what Italians or Poles or Irish should look like, and they chose their own ethnic identity—and identified others'—based on these stereotypes. Amazingly, however, respondents often had totally different ways of describing the same

groups of people. For example, one person said that he could always tell the Irish by their pale skin and dark hair; another said that the Irish always had ruddy complexions and red hair. Both people claimed to always be right in their judgments![48]

At the same time, the respondents' ethnic identity was often quite fluid. Some people reported that they emphasized one part of their ethnic heritage with one side of their family, another part with other relatives, and a third part with friends. Thus they chose not just what ethnic identity to emphasize, but also when to emphasize it.

Some ancestries were much more popular than others. For instance, when faced with the possibility of emphasizing their Scottish or their Italian heritage, respondents were more than *1200* times as likely to identify with their Italian background.[49] What a change from the early twentieth century, when Italians were often stereotyped as dirty criminals in contrast with earlier settlers, such as those of Scottish descent.

In general, for Waters' subjects, ethnic identity did not have a negative connotation, but rather brought them pleasure. It gave them a feeling of belonging, of being part of a group; it helped them feel special and interesting.

After talking with and watching her respondents for many days, Waters decided that these special feelings are related to the cultural activities and beliefs that the subjects associate with their ethnicity. Most of these activities involved their families and larger kin groups as well as various holidays. They ranged from using phrases of their ethnic language, such as swear words or terms of endearment that they remembered their parents or grandparents using, to cooking ethnic foods on special occasions and following customs associated with rites of passage such as weddings or funerals. These cultural practices helped them to remember their families, feel connected to kin, and justify their behaviors and actions.

Yet, as Waters continued to talk with her respondents, she found that their descriptions of these cultural practices were remarkably similar, no matter what they saw as their particular heritage. As she put it, "Many people reported that meals after funerals were a specifically ethnic tradition in their families: They were variously described to me as an Irish, Italian, Polish, Portuguese, and Slovenian custom."[50] They also believed that their own ethnic groups each had certain unique values and behavioral traits, but these values were also remarkably consistent from one group to another.

Waters' report of this finding illustrates how her work was often inductive in nature: her conclusions gradually developed as her study progressed. As she put it, "I noticed the similarities after I conducted interview after interview in which the same qualities were mentioned, but each time with a different ethnic label attached."[51] Whether they reported that they were Irish, Italian, Slovenian, Polish, or any other group, subjects always believed that their ethnic group differed from others in three ways: (1) they highly valued the family, (2) they highly valued education, and (3) they were extremely faithful to God and loyal to their country. In fact, however, these are values held by most well-educated people in the United States. The respondents simply assumed that their values and experiences reflected their ethnic heritage rather than this much broader community.

Gradually Waters came to the conclusion that these descendants of immigrants were not members of separate ethnic groups, in the sociological sense, as their forebears had been. Instead, their ethnic identity was "symbolic." They chose to identify with one or another part of their European ancestry, but that choice had little real effect on their everyday lives, and they could assume this identity when they wanted to and forget it at other times.

Given the cultural similarities across the ethnic identities and the fact that virtually all of her subjects had descended from a variety of white ethnic groups, Waters and others have suggested that a broad new ethnic group of Euro-Americans has replaced the older, separate white ethnic groups.[52] A participant in a recent St. Patrick's Day parade in Cincinnati epitomized this view when he told a television reporter, "It's a great day for the Irish and the Germans!" Divisions between the Irish and Germans that once existed in the United States are now meaningless, and he can celebrate both strands of his ethnic heritage by marching with fellow celebrants in the parade.

This development of a new Euro-American ethnic group supports Park's theory of a race relations cycle. Since the early twentieth century the boundaries between the various white ethnic groups have broken down. Descendants of the earlier immigrants share not just secondary group ties through school and work but also primary group ties through their families and friends. They have assimilated with one another.

Yet, given this assimilation, why does ethnic identity persist, even if only in a symbolic form? Why is ethnic identity still important to today's descendants of immigrants? Waters' work suggests an answer. Like people throughout the world, these white ethnics can reaffirm their ties to their family and extended kin through the cultural practices they associate with their ethnic identity. These practices help them feel part of a larger group of people, to feel connected to others.

Although these effects certainly can be functional for the individuals involved, Waters cautions that they can also be dysfunctional, for ethnic identity always carries the possibility of ethnocentrism and feelings of distrust and disdain toward those outside the group. If Euro-Americans now represent a white ethnic group within the United States, people of color are those outside that group. Moreover, people of color are not nearly as free as white ethnics to choose their ethnic identity. To learn more about Mary Waters' views, see her comments in Box 3-1.

CRITICAL-THINKING QUESTIONS

1. How might you replicate Mary Waters' study in your own community? What kind of sample would you use?

2. Would you ask questions other than those she asked? Would you use additional methods of gathering data? How might these changes affect your results?

BOX
3-1

SOCIOLOGISTS AT WORK

Mary Waters

Sociology is a way to both study and improve the condition of poor and oppressed people in our society.

Mary Waters, Harvard University, conducted the research featured in this chapter.

Where did you grow up and go to school? I grew up in the Flatbush section of Brooklyn, New York, and I think that is where I developed my interest in race and ethnic relations. My neighborhood was a mix of Irish and Italian Catholics and Reform, Conservative, and Orthodox Jews. I knew some black people who were Protestants but I honestly never met a white Protestant until I went away to college to Johns Hopkins University in Baltimore. I did not realize until I started studying ethnicity at college that the demographics of the United States were so very different than what I had experienced in Brooklyn.

Why did you become a sociologist? My work-study job as an undergraduate was working for the Center for Social Organization of Schools, a research institute studying the sociology of education. I learned a great deal about methodology and about the excitement of doing research in real settings—in this case the Baltimore city schools. I decided to go to graduate school in sociology both because I liked research and because I thought (and still think) that sociology was a way to both study and improve the condition of poor and oppressed people in our society.

Has the study been replicated by you or others? What were the results? At the same time I was doing my study, my colleague Richard Alba at SUNY Albany was doing a random survey sample study of white later-generation people in the Albany area. His study found very similar results but with a random sample. This was very important and reassuring because my subjects were chosen via snowball sample, so the sample could have been biased in some way that affected the results. The fact that Alba found very similar answers with a much larger group of people also shows the neat ways in which

different methods in sociology can often work together to give a better picture of a phenomenon.

How does your research apply to students' lives? White college students tend to unconsciously assume that their black classmates have ethnic identities that are similar to their own. They tend to come from backgrounds where they have never lived closely with black people and may have few black friends, while black college students, because of the demography of middle-class black life in the U.S., tend to have had many white classmates and know a lot about white people. The white students are therefore more curious about the black students than vice versa. The white students often assume that black students will think about identity issues much the same way they do—that ethnic and racial identity is an intermittent symbolic thing that is mostly a source of tradition, pride, and pleasure, with few social costs. Black students do have identities that bring them tradition, pride, and pleasure, but they also suffer from discrimination and unequal treatment because of their skin color. Many misunderstandings develop because whites don't appreciate this difference.

For example, white students in the dorms often have never been around black people when they are fixing their hair and will ask black students all sorts of questions. Some black students will find it insulting that whites who barely know them want them to educate them about their hair. Yet white students think to themselves, "If someone asked me about my hair, I would not be upset. Why are they making such a big deal about it?" I think my work on symbolic ethnicity should help white and black students to understand how different their notions of identity can be and what it means, and this might help them to better understand each other in social settings.

Ethnic Identity: Limited Choices
for People of Color

Symbolic ethnicity is an option only for some groups of people. Consider two people, Jack and Jerry, both of whom had a great-grandfather from Ireland. Suppose that Jack's other ancestors also came from Europe whereas Jerry's other ancestors came from Africa. Jack and Jerry would technically have the same amount of Irish heritage, and if Jack called himself Irish American, chances are no one would bat an eye. If, however, Jerry tried to call himself Irish American, he would probably face disbelief and scorn. That is because, as a person of color, Jerry has limited options. People will call him black or African American, no matter how he might see himself and no matter where his great-grandfather came from. In other words, Jerry and other people of color don't have the option of symbolic ethnicity.

The difference between the ethnic identity of whites and African Americans is manifested in many ways. For example, descendants of white ethnic groups hold similar kinds of jobs, earn comparable incomes, freely marry each other, and live together in the same neighborhoods. Some of Waters' respondents noted that their parents might have preferred that they marry someone of similar ethnic and religious backgrounds, but many of them did not do so. The biggest prohibition regarding potential mates involved color: their parents, and they themselves, believed that they should not marry a person of color. Traditionally, few African Americans have married outside their own racial-ethnic group. In addition, neighborhoods are often highly segregated, with people of color, especially African Americans, relatively unlikely to live in neighborhoods with those of Euro-American descent. For instance, the community of Echo Park, where Maria teaches, has many residents with different ethnic heritages but very few African-American residents. As we discuss in subsequent chapters, there are also sharp differences between white ethnics and many people of color in income and type of employment.

Even though group boundaries that separate African Americans from other racial-ethnic groups are perhaps the strongest within our society, the ascribed nature of ethnic identity affects other people of color as well, although the nature of this effect might vary from one group to another. For instance, the sociologist Mia Tuan replicated Mary Waters' work by studying Asian Americans, a racial-ethnic group that is sometimes termed a "model minority" because of its high levels of education and occupational success. Tuan interviewed Asian Americans in California whose families had lived in the United States for several generations, asking about their experiences and identification with their ethnic heritage. She found that, like the white ethnics in Waters' study, these Asian ethnics displayed a great deal of flexibility in cultural traditions they continued to keep in their personal lives with most traditions involving food and, sometimes, holidays. At the same time, however, interactions with others made it clear that they did not have the freedom to choose whether or not to identify with their ethnic heritage. Even though all the respondents were born and raised in the United States, others would often assume that they were from other countries, ask where they were from, and even try to speak to them in Chinese or Japanese. Thus, even though the people in Tuan's study had "made it" in terms of their educational and work experiences and a large proportion had married Euro-Americans, they still felt that others did not consider them to be "real Americans."[53]

In short, the assimilation that Park's theory predicts appears to have occurred for white ethnics, but the group boundaries that separate white ethnics and people of color are still evident. The tensions that surround relationships between racial-ethnic groups within the United States are no doubt related to this fact. Euro-Americans and people of color share some secondary group relationships but few primary group ties. Park's theory and the work of Mary Waters suggest that building such relationships is a key step in improving intercultural relationships.

Cultural elements that make up the social structure are not fixed in stone. These elements, including our definitions of ethnic groups and group boundaries, are always developing through social actions, as individuals contribute the brushstrokes and tints that help to create the total reality of our culture. (See Box 3-2.) Through their interactions Maria and other residents in Echo Park create their subculture, Craig and his friends continually re-create the subculture that defines Hollywood, and Shirley and Chan help create their subcultures. In addition, each of them contributes to and constantly re-creates the broader culture in which we all live. In turn, these cultures and subcultures influence their daily behaviors, the ways in which they see themselves, and their views of other people. In the next chapter we look more closely at how individuals learn the behaviors expected of them within their cultures and develop views of themselves as unique individuals.

BOX
3-2

APPLYING SOCIOLOGY TO SOCIAL ISSUES

Measuring Ethnic Identity and Race-Ethnicity

The questions asked in the U.S. census change over time, but only after careful review by social scientists and interested citizens. Changes that do occur often reflect changes in our culture. One change that will occur for the census in the year 2000 involves the way people are asked about their race-ethnicity and the way in which people of *multiracial* ancestry will be able to report their heritage.

The impetus for the proposed changes is that a growing number of people have trouble giving just one answer to the current question. For instance, General Colin Powell, whose parents immigrated to the United States from Jamaica, is of African, English, Scottish, Arawak Indian, and Jewish descent.[54] In Jamaica, as in many cultures worldwide, he would be seen as multiracial; in the United States he is seen as African American. This American tradition is a remnant of the post–Civil War era when so-called one-drop laws decreed that a person with even one African-American ancestor would be considered black.

The change that will occur in the year 2000 census involves the wording of the item shown in Figure 3-1c. Because the question required respondents to choose only one answer, a number of citizens argued that it forced people to deny part of their heritage. In response to these concerns the Census Bureau held hearings throughout the country through the mid 1990s to gather information about citizens' views about how the Census should measure race and ethnicity. National opinion polls found that almost half of all African Americans and more than a third of all Euro-Americans supported the addition of a "multiracial" category to the list of options. In addition, almost half of both groups said that the Census Bureau should simply stop collecting information on race and ethnicity.[55]

After a great deal of debate and thought the Census Bureau finally decided to alter the instructions that go with the question in 3-1c to allow Americans to mark as many racial categories as they believe apply. For instance, the golfer Tiger Woods is of African, European, Thai, and American Indian descent and calls himself "Cablinasian" to denote his ancestry. To answer question 3-1c, he could now mark several boxes rather than choosing only one. In general, this change in the wording of the census question will allow the Census Bureau to count people who identify themselves as falling in each individual category and in all of the various combinations.

Those who supported changing the census question suggest that the revision will more accurately reflect individuals' ethnic identity. To use sociological terms, advocates of this change suggest that the answers will be more valid. Critics of the change are not sure that validity would be increased and also suggest that it could have other far-reaching consequences—from affecting the way that congressional districts are drawn, to determining federal benefits for cities and states, to drawing attention away from the injustices that face people of color.[56]

CRITICAL-THINKING QUESTIONS

1. Should the census have altered the way people respond to the list in Figure 3-1c? Will such a change help you or some of your friends answer the question?

2. What effects do you think this change will have? Will the responses be more valid? Will it help change the way people in the United States think about race-ethnicity and ethnic identity? Will it increase the probability that individuals will identify as multiracial? Will it affect ethnocentrism? Will it affect relationships between racial-ethnic groups?

3. Does the fact that such a change has occurred suggest that greater assimilation of racial-ethnic groups is occurring in the United States? Or will it simply draw attention away from difficulties experienced by people of color?

SUMMARY

The theory and research discussed in this chapter have provided ideas that help us understand both the common and distinctive elements of cultural experiences. Looking at culture from both macro- and microlevel perspectives helps us appreciate how our culture influences how we see ourselves (our cultural identity) and how we, as individuals, help to create culture. Key chapter topics include the following:

- Human beings share certain basic capacities and needs: extensive language systems that allow the use of symbols for communication, a protracted period of care as children, and the need to make sense of their experiences.

- All societies have developed certain kinds of structures to address these needs. These cultural universals include (1) a gender-based division of labor, (2) some form of a family and kin structure, and (3) values, ideologies, beliefs, rituals, and folklore and myths. However, the exact nature or form of these structures varies greatly from one society to another.

- According to Parsons' theory of structural functionalism, these social structures help maintain the social order and thus the entire society. Merton suggested that these functions can be manifest or latent. Parts of a society can also be dysfunctional and have negative consequences for the group.

- According to Harris's theory of cultural materialism, societies develop different types of family forms and other cultural elements primarily because they exist in different environments and thus have access to different resources, or materials.

- Cultures change through both cultural innovation and cultural diffusion. Some parts of a culture might change more quickly than others, and some organizations and people serve as gatekeepers with the power to influence which elements become widely accepted.

- Cultural identity refers to individuals' identification with a culture and their acceptance of its elements as simply natural.

- Cultural elements, such as rituals and norms, help unite people within the United States. At the same time, the United States contains many subcultures, such as those based on race and ethnicity.

- According to Park's theory of the race relations cycle, as members of various groups develop more relationships within primary groups, group boundaries disappear and assimilation occurs.

- Data gathered by the U.S. Census Bureau indicate that, for many people, ethnic identification involves a choice regarding which ancestry to emphasize. As Mary Waters showed, this choice can be influenced by several factors, ethnic identity carries no real cost today for white ethnics, and white ethnics' identity with an ethnic group is largely voluntary and symbolic. In contrast, people of color do not have this choice, for their ethnic identity is generally ascribed by others.

- Divisions between white ethnic groups have largely disappeared and have been replaced with a common Euro-American ethnic identity. In contrast, boundaries between people of color and Euro-Americans remain largely intact.

KEY TERMS

accommodation 60
aggregates 61
assimilation 61
census 62
closed-ended questions 63
competition 60
cultural diffusion 58
cultural elements 52
cultural gatekeepers 58
cultural identity 59
cultural innovation 57
cultural lag 58
cultural materialism 56
cultural universals 52
culture 50
dysfunctions 56

ethnic group 58
ethnic identity 59
ethnocentrism 59
extended families 57
folklore 54
folkways 52
function 55
gender-based division of labor 53
group boundaries 61
ideologies 54
kin 53
latent functions 56
laws 52
manifest functions 56
material culture 52
mores 52

nonmaterial culture 52
norms 52
nuclear family 53
open-ended questions 62–63
people of color 59
primary groups 61
racial-ethnic group 59
rituals 54
secondary groups 61
social structure 55
structural functionalism 55
subculture 53
symbols 51
values 54
white ethnics 59

INTERNET ASSIGNMENTS

1. One of the best ways of learning about cultures other than your own is to interact with people from another culture. Using search engines, locate discussion groups and other Web sites that focus on intercultural exchange. In your discussions, exchange information about what each of you finds to be the important aspects of your culture. How do your responses differ? How are they similar?

2. Subcultural variation occurs within cultures. Locate newsgroups and listservs that focus on particular group interests (for example, groups based on race-ethnicity, sexual orientation, gender, regional location, age/birth cohort, and so forth). What can you learn about the cultures of particular groups from their discussion groups? Can you identify unique aspects of these cultures? Choose a group that you do not belong to (for example, if you are African American, look at an Asian American group). How does the culture of this group differ from your own?

3. Using a search engine, locate information on journals and research groups that focus on studying culture. For example, one journal that contains research on cultural issues is the *Journal of Material Culture.* What issues do researchers who are interested in culture focus on?

4. A lot of media attention has been focused on generation gaps, most recently between Baby Boomers and Generation X/Thirteenth Generation. Using a search engine, locate information on, about, and for members of these groups. Are these groups culturally different? What indicates this?

TIPS FOR SEARCHING
The areas of "American studies" and "cultural studies" might be good to search. Although these are interdisciplinary between the social sciences and the humanities, Web sites under these areas yield a lot of information about culture and cultural diversity.

INFOTRAC COLLEGE EDITION: READINGS

Article A20565299 / Logan, John R., Fuqin Bian, Yanjie Bian. 1998. Tradition and change in the urban Chinese family: The case of living arrangements. *Social Forces, 76,* 851–883. Logan and his colleagues look at how the contemporary Chinese family retains traditional elements and has also changed some of these elements.

Article A19362368 / Smolicz, J. J. 1997. Australia: From migrant country to multicultural nation. International Migration Review, 31, 71–187. Smolicz describes the policies Australia has used to change from a monocultural nation to a multicultural nation that incorporates a very large number of different ethnic groups.

Article A19145911 / Rowan, Mike. 1997. Pacific revival. (cultural revival in the islands of the Pacific; includes related articles on Melanesian cargo cults and land inheritance) *Geographical Magazine, 69,* 24–26. This short article describes how modern development disrupted the traditional cultural life of Pacific Islanders and, more recently, promoted a cultural revival.

Article A19178124 / Bunis, William K., Angela Yancik, and David A. Snow. 1996. The cultural patterning of sympathy toward the homeless and other victims of misfortune. *Social Problems, 43,* 387–403. Burris, Yancik, and Snow examine the way in which sympathy toward the homeless and victims of famine varies from one cultural setting to another.

Article A20565331 / Charney, Paul. 1998. A sense of belonging: colonial Indian cofradias and ethnicity in the valley of Lima, Peru. *Americas, 54,* 379–408. Charney explores the development of ethnic identity among an indigenous group in the area of Lima, Peru.

Article A20142576 / Perlmann, Joel. 1997. Multiracials, intermarriage, ethnicity. *Society, 34,* 20–24. Perlmann explores the controversies and debates over the wording of census questions regarding "race" and "ethnicity."

Article A20142578 / Light, Ivan, and Cathie Lee. 1997. And just who do you think you aren't? (Americans' ethnic and racial identity). *Society, 34,* 28–31. Light and Lee examine why Americans sometimes give conflicting answers to the census questions regarding their ethnic and racial identity.

FURTHER READING

Suggestions for additional reading can be found in *Classic Readings in Sociology,* bundled with this book. If you purchased a used copy of this book that does not include this custom-published reader, go to www.sociology.wadsworth.com for ordering information.

RECOMMENDED SOURCES

Davis, Murray S. 1993. *What's So Funny? The Comic Conception of Culture and Society.* Chicago: University of Chicago Press. Davis analyzes American culture by looking at jokes.

Driedger, Leo. 1996. *Multi-Ethnic Canada: Identities and Inequalities.* Toronto: Oxford University Press. Driedger provides extensive discussion of issues related to ethnic identity throughout all regions of contemporary Canada.

Epstein, Jonathon S. (ed.). *Youth Culture: Identity in a Postmodern World.* Malden, Mass,: Blackwell. This is a series of essays describing many different aspects of contemporary youth culture.

Glazer, Nathan, and Daniel P. Moynihan. 1970. *Beyond the Melting Pot: The Negroes, Puerto Ricans, Jews, Italians, and Irish of New York City,* 2nd ed. Cambridge, Mass.: M.I.T. Press. This is a classic statement of the cultural pluralism perspective.

Harris, Marvin. 1979. *Cultural Materialism: The Struggle for a Science of Culture.* New York: Random House. Harris gives a solid overview of the basic tenets of cultural materialism.

Johnson, Benton. 1975. *Functionalism in Modern Sociology: Understanding Talcott Parsons.* Morristown, N.J.: General Learning Press. Johnson provices a short and very readable summary of the basic tenets of structural functionalism.

Jordan, Glenn, and Chris Weedon. 1995. *Cultural Politics: Class, Gender, Race, and the Postmodern World.* Oxford, U.K.: Blackwell. This book examines from a postmodern perspective how issues surrounding power affect the way in which culture is defined.

Mathews, Gordon. 1996. *What Makes Life Worth Living? How Japanese and Americans Make Sense of Their Worlds.* Berkeley: University of California Press. Mathews examines similarities and differences in the ways Japanese and Americans find meaning in their lives.

Nagel, Joane. 1996. *American Indian Ethnic Renewal: Red Power and the Resurgence of Identity and Culture.* New York: Oxford University Press. Nagel describes reasons underlying the recent increase in American Indian ethnic identity.

Tuan, Mia. 1998. *Forever Foreigners or Honorary Whites? The Asian Ethnic Experience Today.* New Brunswick, N.J.: Rutgers University Press. Tuan gives a very readable report on a replication of Mary Waters' work that examines the ethnic experiences of Asian Americans.

Waters, Mary C. 1990. *Ethnic Options: Choosing Identities in America.* Berkeley and Los Angeles: University of California Press. Waters provides a very readable detailed report about the research featured in this chapter.

INTERNET SOURCE

American Sociological Association. Members of the American Sociological Association (ASA) interested in the study of culture have a research network on the Web with information about current research activities, forums, meetings, and discussion groups.

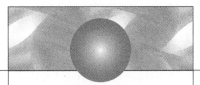

Pulling It Together

You can read more about many of the topics covered in this chapter in
other places in this book. You can also look up topics in the index.

Socialization and the Life Course

Becoming Part of the Social World

The Interaction of Biology and Environment

Social Networks and Social Roles

Targets and Agents of Socialization

*Early Socialization: Interactions
with Parents and Peers*

*Theoretical Perspectives on Learning
and Development*

Developing a Notion of the Self

Aspects of the Self: A Developmental View

Social Roles and Self-Identity

FEATURED RESEARCH STUDY

**Developing Gender Schemas:
The Influence of Peer Groups**

Socialization Through the Life Course

Developmental Change or Age Effects

History or Period Effects

Peer Group or Cohort Effects

ANGELA AND ANDY were born to different families on the same day. By about six months of age, they were playing with anything that was handed them, from kitchen spoons to push toys, dolls, and trucks. Andy's parents, however, were a little uncomfortable when he played with his sister's dolls and tried to distract him with blocks and toy cars. Meanwhile Angela's parents would take her older brother's Ninja Turtle figures away from her and give her toy dishes to play with instead.

By their first birthdays both Angela and Andy spent several hours with other children each day, but they weren't yet capable of playing involved, imaginary games with other children. Sometime after their second birthday, however, this changed. At their day-care centers Andy would generally play with the other boys while Angela would play with the other girls. They now shunned toys that they thought were inappropriate to their sex and held firm beliefs about what boys and girls should do at day care.

From infancy both Angela and Andy showed their own personalities and interests. For instance, even as a baby, Andy loved music, and his parents could always comfort him by singing softly. By age four he would spend hours watching old musicals with his mother and mimicking the dancers by doing his own version of a "soft shoe." Shortly after his sixth birthday he announced that he wanted to learn to tap dance like Fred Astaire. His mother, who had briefly been a professional dancer herself, promptly enrolled Andy in a children's dancing school.

On their first visit to the school, Andy was so excited that he literally soft-shoed his way across the parking lot. When he and his mother entered the studio, however, Andy realized that he was the only boy in the room. All of the other students were girls! "Mom," he whispered, "I've changed my mind. I don't want to take dancing lessons."

Unlike Andy, Angela was indifferent to music, but she loved physical activity and sports. By age six she was playing touch football with her older brothers. When her brothers joined a flag football team, Angela begged her parents to let her play, too. When she was seven, her father registered her for a neighborhood team.

On Saturday morning as they walked to the first practice, Angela was elated. She was unprepared, however, for what greeted her. "Who's this," one of the boys said, "a girl? Girls can't play football."

"I can too play!" Angela insisted and braved the boys' catcalls to join in the throwing and running routines.

The coaches all encouraged Angela, and some of the boys were nice to her, too. But she was the only girl on the team, and several of the boys purposely avoided throwing her the ball, taunting her with comments like, "Why don't you go home and play with your Barbies?"

By the end of the practice, Angela was fighting back tears, and on the way home she decided that playing on the football team wasn't such a good idea after all.

Andy and Angela now have finished high school. Andy hardly ever watches old musicals anymore, and he only does his Fred Astaire routine for his family or close friends. He has maintained his interest in music, though. He played the saxophone in his high school band and is thinking about becoming a jazz musician. Angela was an "all-league" girls' soccer player in high school, where she also became active in politics and women's issues. Now she wonders whether she'll get the chance to pursue athletics in college. Unfortunately the only women's sports at the local college are gymnastics, basketball, and volleyball—none of which interests her very much. She is toying with the idea of trying out as placekicker for the football team—"just to see what would happen," she says with a sly smile.

Angela and Andy have much in common. Both were lucky enough to be born into loving families. Both showed individual characteristics from early infancy—Andy's love for music and dance, Angela's attraction to sports—that they continue to display as young adults. Yet both have changed over time as they encountered new experiences that influenced their developing views of themselves, including their identity as male and female. How much would their stories differ if they had had different experiences when they were growing up? Might Andy have become a dancer if there had been a few boys in that dance class years ago? Might Angela have earned an athletic scholarship if she had had more opportunity to pursue sports? What if Andy had been born a girl, or Angela a boy?

Like Angela and Andy, we all have a sense of who we are and how we fit into the world. We have beliefs about what is suitable for us to do in life and values that guide our actions (or that sometimes bring us feelings of guilt). All of us have enduring qualities that we think of as being part of ourselves, yet all of us change over time. And we all likely ask "what if" questions from time to time. What if we had been born into a different family or even in a different culture or a different time in history? What if we had had different experiences growing up? To what extent would we be the same people we are today?

*I*NFANTS LIKE ANGELA AND ANDY do look and act very much alike. Yet not only do young children quickly begin to look very different from each other, but their actions become more distinctive and they come to see themselves as separate, unique individuals. In the process they learn to act in ways that they believe are appropriate for them—as a girl or a boy, a daughter or a son, a sister or a brother, as students, and as friends.

We continue to learn new roles throughout our entire lives. Sociologists call this process **socialization,** a very broad concept that refers to the way in which we develop, through our interactions with others, the ability to relate to other people and to play a part in society.[1] At the same time children develop a view of themselves—how they differ from others and how they are the same person over time. Sociologists refer to this process of coming to see oneself as a distinct person with a clear identity as the development of the **self.**[2]

To better understand the process of socialization and the development of the self, I sometimes think about pilots of boats and the different types of waters they can explore. Some boats navigate streams that flow gently through rolling meadows. Other boats venture down swift, dangerous rivers that roar past steep, rocky banks. Some boats have strong, capable crews whereas others do not. The pilots all steer their boats through the waters and must choose how they will react to changing conditions. But the way in which their trip unfolds depends not just on their own capabilities and skills as a pilot but also on the environment of the river and the ability of their crew.

When sociologists look at people's lives, they try to understand how both our unique capabilities and our social environment influence our self-perceptions and our actions. Just as the nature of a river and the quality of the crew influence a boat pilot's journey, the specific social environment in which we live plays a large role in shaping the choices available to us, the way we behave, and the way we come to see ourselves. At the same time, just as boat pilots choose how they will respond to each new situation, so do we, within the limits of our capabilities and resources, choose how we will respond to the particular opportunities and constraints presented to us. Many sociologists hope that a greater understanding of socialization and the development of the self can suggest ways to help all people, and especially those in more challenging environments, make the most of their capabilities.

In this chapter we focus first on how we develop in childhood and then on how socialization continues throughout our lives and how it may vary for people born at different periods in history. Throughout this discussion you will see how our views of ourselves and our behaviors reflect both our social environment—the social structure in which we live—and our social actions.

BECOMING PART OF THE SOCIAL WORLD

Socialization includes a variety of aspects, including how children come to learn what it means to be "good," whether there is a God, what "democracy" means, and what friends are. One experience we all

Socialization occurs in all cultures. What might the young women pictured here be learning?

Even though all of these children are participating in the same choir, it is clear that they are also unique individuals. How might the individual characteristics of each of these children affect the way that they react to their common experiences in the choir?

have in common is what sociologists call **gender socialization,** the process of learning to see ourselves as male or female and learning the roles and expectations associated with our sex group. Gender socialization is a particularly relevant topic today in politics, family life, and many other areas. Politicians discuss the "gender gap" in voting, and sex scandals involving people in power lead to fierce debates over the ways in which men and women should relate to one another. At the personal level many people struggle with differing expectations about the "proper" behaviors of men and women in dating, courtship, marriage, and child rearing. Throughout this chapter we will focus on this aspect of socialization to illustrate the process of socialization and the development of the concept of self. We turn first to the interaction of biology and environment.

The Interaction of Biology and Environment

As noted in Chapter 3, humans are totally dependent on others at birth and remain so for many years as they learn about their social world. At the same time, all infants have certain biological capabilities that enable them to develop socially. At birth they appear to be somehow "ready" to respond to other people, to make eye contact, and to imitate facial and hand movements. The familiar game of peek-a-boo illustrates this phenomenon, as infants from very early ages begin to smile and coo at the sight of a human face. Even babies who are born blind will smile at the sound of a human voice.[3] Like Andy and Angela we begin communicating with others and developing socially the moment we come into the world.

Of course newborns can't get their own food, change their diapers, or discuss politics. No matter how much parents might try to speed up their babies' development, an infant usually isn't capable of sitting up before about the age of six months or understanding or using words until around the first birthday. This fact underlines an important point—the experiences that children have and the nature of their interactions with others depend a great deal on their age and their maturation or developmental level.[4] As we will see later in the chapter, this notion of *developmental preparedness* has been used to understand how children learn to think and how they learn to play a role in the social world.

Biology also helps to explain why no two infants are identical. Like all babies you were born with certain temperamental tendencies. You might have been more irritable, active, and sensitive to touch than your brothers and sisters. Or perhaps you cried more or slept less than others in your family.[5] Scientists think that these differences result from basic neurological patterns in the brain. Such temperamental differences might well appear throughout your life, as you continue to be more easily bored or excitable, or to need more or less sleep, than most other people.

At the same time, however, the social environment in which you grew up influences these characteristics.[6] For instance, you might have learned to control your tendencies toward excitability or irritability so that you don't continually place yourself in embarrassing situations. Or perhaps you were quite shy when you were young, but your friends and family helped you develop skills for dealing with new situations so that you no longer seem to be shy, even though you might still feel shy underneath. We all have certain personality traits that probably are related to our neurological makeup, but the ways in which we deal with and present these characteristics to others are very much influenced by the experiences we have as we grow up—by our socialization process.

Gender socialization provides a good illustration of the interaction of biological and environmental influences. Males and females obviously differ in their biological characteristics, with different sex organs at birth and different body shapes, hormonal secretions, hair patterns, and voice characteristics in adulthood. They also differ in their growth patterns and susceptibility to a wide variety of diseases and physical conditions. For example, girls reach puberty earlier than boys; females more often suffer from diabetes, phlebitis, and migraine headaches; males more often suffer from hemophilia, heart attacks at early ages, and most forms of cancer.[7]

Yet, surprisingly, on average males and females have relatively few temperamental differences. For many years social scientists have studied phenomena such as children's activity levels and their tendency to aggress against others or to display nurturant behaviors. These studies of gender differences in personality traits have been replicated many times, making it possible to compare results over a wide variety of reports. Replications of studies help scientists gain confidence in their conclusions. When a lot of studies have been conducted of a phenomenon, scientists analyze and combine the results of these studies to see what patterns emerge, in what are called **meta-analyses.** This is what researchers have done with the hundreds of studies of gender differences. In general, these meta-analyses show that boys and girls on average are very similar in their tendencies to be timid and to try to control or influence others, in their intellectual abilities, in their dependency on others, and in their desire to be sociable with other people.[8]

Only two broad sets of gender differences seem to consistently appear. Studies in both the United States and other countries suggest that boys are more likely than girls to express hostile feelings or be aggressive toward others and to be more physically active, especially in more restricted, stressful, or unfamiliar settings. These results don't suggest that girls are more passive, on average, than boys. Instead, their style of activity differs, as any observer of childhood birthday parties can see. Boys are much more likely to run around, while girls are much more likely to play quietly. Girls are also more likely to exhibit nurturant behavior toward younger children—to want to interact with them and to respond to their needs. There is a great deal of overlap between the sexes in these tendencies: Many girls are more aggressive and physically active than a large number of boys; many boys are more nurturant and helpful toward young children than a large number of girls. Yet on average boys tend to be more physically active and aggressive than girls, and girls tend to be more interested in young children than boys are.[9]

No one knows for sure why these differences appear. Because they exist at such early ages, however, a number of scholars believe that that these differ-

ences might result at least partly from different neurological patterns that develop prenatally along with sex-related physical characteristics. The evolution of the species can help to explain these differences. Contemporary humans are only a few generations removed from the time in which all people lived in hunting and gathering bands. It takes many generations for living organisms to alter their biological characteristics to conform to new environmental demands, and thus humans today are probably virtually identical genetically to their hunting and gathering ancestors.[10] Females' propensity toward nurturance and males' propensity toward aggression and physical activity might well reflect a biological heritage from our hunting and gathering ancestors.

Even though, as we discussed in Chapter 3, these differences were quite functional at one time, they probably aren't that important in modern industrialized societies. In fact, in many ways they might be dysfunctional. Parents and teachers seem to recognize this when they try to help boys who are overly aggressive and active learn to control their actions, and studies show much smaller sex differences in aggression among adults than among children.[11] Many parents also try to encourage both their sons and daughters to be supportive of all younger children. Even though young boys might not be as interested in babies as their female counterparts, at later ages both males and females are fully capable of caring for young children and responding in a nurturant manner to infants.[12] This development involves their social roles and social networks.

Social Networks and Social Roles

When you were born you entered a **social network,** a term sociologists use to describe patterns of social relationships, or the linkages between individuals formed by social interactions. In Chapter 5 we will discuss how this concept applies to many areas of our lives; here we will focus on how our social networks and relationships with others influence our socialization.

Quite likely the most important first bond or relationship within your social network was with your mother, for, as we noted in Chapter 3, in virtually all cultures and settings the mother is primarily responsible for child care. As with Angela and Andy, as you got older the number of people in your social network increased to include brothers and sisters, adults who helped care for you, playmates, and, eventually, classmates and teachers. These people with whom you interact and who are emotionally important to you are called **significant others.** They are people whom you care about and whose opinions of you matter to you.

Each time you enter a social network, you hold a **social status,** a term sociologists use to describe positions that we hold within the social structure. Attached to every social status are **social roles,** or expectations, obligations, and norms that are associated with a particular position in a social network. A social status is a position that you occupy; a social role is the behaviors associated with a status. Thus throughout our lives we occupy social statuses, and we learn the social roles associated with these statuses.

The first social role you learned was that of baby. You might also have learned other family roles, such as little sister or brother, grandchild, niece or nephew, and cousin. In play groups you learned the role of friend, and in school you learned the role of student. As you've gotten older, you have learned a wide variety of roles, perhaps including those of worker, spouse, and parent. In each of these situations you learned the behaviors—the expectations, obligations, and norms—that were associated with your status. For many sociologists, especially those who take a more macrolevel view, socialization involves learning the role behaviors that are associated with our positions in social networks.

What agents of socialization do you see in this photo? What messages might young women be getting from these agents?

Targets and Agents of Socialization

When you learn new roles, a sociologist would say that you are the **target of socialization.** The source of your learning is called the **agent of socialization,** the person or group that provides information about your new social role.[13] Figure 4-1 lists several agents of socialization, including people, the media, and organizations. All of these agents of socialization transmit messages about how we should play certain roles.

Because all of us have different social networks, we encounter different agents of socialization. We have different parents, family members, and sets of friends, and we attend different schools and churches. We also often read different magazines and books and watch different television shows and movies. Because we belong to various networks, we learn to play various roles. For instance, virtually all children attend school and hold the status of student, but parents differ a great deal in the expectations or roles they attach to this status. Some assume that their children will attend college and take college preparatory courses in high school. Others might prefer that their children take vocational courses to prepare for employment after high school. These expectations can have lasting influences on our lives.

The terms *target* and *agent* of socialization make me think of someone shooting an arrow full of information toward a child who then magically absorbs it. It is important to realize that this *isn't* what these

terms mean at all. Instead, by using these words, sociologists are simplifying their analysis of the socialization process and looking at just certain aspects of it. For example, consider the interaction of mother and baby. You could say that the baby was the target and the mother the agent of socialization as the baby learns more and more about the social world—what to call people and things, how to hold a spoon, how to interact with others. At the same time, however, the mother is the target and the baby the agent of socialization as the mother learns how the baby likes to be held, what can calm the baby, and what can make the baby laugh. The baby is teaching her the role of mother. As you encounter agents of socialization throughout your life, you will often be both the target and the agent of socialization. Yet, to simplify matters, sociologists generally look at one part of socialization or role learning at a time.

Early Socialization: Interactions with Parents and Peers

In your early interactions within the family, as the target of socialization, you learned both what it was like to be nurtured and how to nurture others. Many social scientists believe that the early relationships with

People	Media	Organizations
Mother and father	Television	School
Sisters and brothers	Movies	Church
Grandmothers and grandfathers	Books	Community groups,such as Boy Scouts or Campfire Girls
Aunts and uncles	Radio	Sports teams
Teachers	Video games	
Friends		

FIGURE 4-1 *Some Common Agents of Socialization*

your parents are very important in later life because in these relationships you learn what it is like to be loved by others and how to love others in return. The style or content of these interactions can vary a great deal from one child and family to another. Some children like a lot of cuddling and holding; others do not. Some mothers and fathers talk to their children a great deal; others use more nonverbal means of communication. Some parents adhere to rigid schedules of care-giving activities such as feeding and sleeping; others follow no time schedule. In some families both mothers and fathers will be heavily involved in child care, and the baby will become heavily attached to both; in other families only the mother will be significantly involved in the baby's care.

Whatever the style of interaction, however, infants become very attached to their mothers (and sometimes their fathers) by the age of three to four months. Similarly infants who have been placed in foster homes at birth and are moved to adoptive homes after this time often show signs of emotional disturbance after the move because they have become so attached to their foster mothers.[14] This special mother-child attachment develops even if a child's mother works and the child is left with a babysitter for a good part of each day. The critical factor doesn't appear to be the number of hours of contact with the mother, but rather is the strength of the emotional tie.[15]

Besides the family, the most important agent of socialization in childhood is probably the peer group. Throughout childhood the social networks of boys and girls tend to be relatively separate and distinct.[16] Think back on your own childhood and your best friends. Like most children you probably played with others from the same sex group much more often than with those from the other sex group, unless your parents or teachers somehow arranged for you to get together. This simple fact has a large impact on gender socialization, for it means that much of children's learning about what it means to be male or female actually comes from their interactions with their same-sex peers. Later in this chapter we will

look more closely at the nature of peer group socialization, but first we turn to a more microlevel analysis of socialization and social roles.

Theoretical Perspectives on Learning and Development

In studying the process of socialization, some social scientists have tended to emphasize how agents of socialization affect children's development—a social learning perspective. Others have tended to focus more on how children actively participate in the process—a cognitive learning perspective. As in most areas we will look at, viewing this issue from both perspectives helps us better understand what actually happens in the socialization process, and most contemporary scholars use a combination of both perspectives.[17]

Social Learning   Those who take what is broadly termed a **social learning perspective** suggest that we learn behaviors by being *reinforced* or rewarded for those that conform to norms and by *modeling* or imitating those that we think conform to the roles we want to fill. Although this perspective recognizes that children think about the world around them as they respond to the things that parents and other agents of socialization do, these thought processes are not central to the theory. Instead, the theory tends to emphasize the active role of agents of socialization in shaping children's behavior through reinforcement and modeling.[18] This theory suggests simply that children will behave in ways for which they have been rewarded and that they will imitate the behaviors of significant others.[19]

To what extent is social learning theory supported by the evidence? Research suggests that parents' behaviors influence children's behaviors in very general ways. For instance, if mothers are more sensitive to and accepting of their infants and rarely express anger toward them, the babies tend to be more strongly attached to their mothers. Boys whose fathers have rejected them or been negligent or aggres-

Socialization is a two-way street. What might the baby and the mother pictured here be learning?

What would social learning theory suggest that the boy shown here might be learning?

sive are more likely than other boys to be delinquent.[20] Children also imitate adults' behavior in many areas, from waving bye-bye, to dressing up in mom's and dad's clothes, to speaking and gesturing like their parents.[21]

With more specific behaviors related to those that are seen as typical of males and females, it is harder to find strong support for the ideas of modeling and reinforcement. Although the differences aren't large, the research evidence suggests that adults, like Angela's and Andy's parents, tend to promote at least some gender differences in toy choice. But stereotypically "feminine" mothers are no more likely than other mothers to have daughters who act in stereotypically feminine ways, such as being very interested in dolls or their hair or clothes. Likewise stereotypically "masculine" fathers are no more

likely than other fathers to have sons who are interested in stereotypically masculine activities, such as cars or sports. By the time children are old enough to attend nursery school, almost all boys and girls, like Angela and Andy, choose to play with different toys, in different activities, and with members of their own sex. Experimental attempts to alter boys' and girls' gender-typed behavior through changing the kinds of rewards that teachers and parents give or through changing the models that children see have been unsuccessful. Changes might occur in the short term, but once the experiment ends, the children revert to their gender-typed activities.[22] Because the research has given only partial support to social learning theory, many social scientists have looked for other explanations of how children learn social roles.

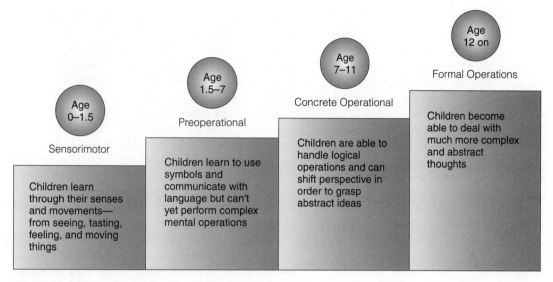

FIGURE 4-2 *Piaget's States of Cognitive Development*

Cognitive Development Whereas social learning theory emphasizes almost automatic processes rather than children's own thinking, other approaches to understanding socialization assign far more significance to children's thoughts and actions. These theorists focus on the child's **cognitive development,** or emerging ability to understand and interpret the world, and the ways in which these changing developmental abilities affect the socialization process.[23]

Much of this work is derived from the ideas of the Swiss psychologist Jean Piaget (1896–1980).[24] Based on his extensive observations of young children, Piaget concluded that all children go through a series of discrete stages in their ability to think about and understand the world around them. (Figure 4-2 depicts these four stages.) In other words, depending on their stage of cognitive development, children see the world in qualitatively different ways; they have different sets of rules for interpreting the world. By the same token children who have not yet advanced to a given stage cannot perform certain cognitive tasks, just as babies cannot learn to sit up until their bodies have reached a given stage of muscular-skeletal development.

Piaget used the term **scheme** to refer to the mental framework or set of rules we use to understand how the world works. A scheme is a cognitive structure, a set of ideas in our minds. As children get older, their schemes become more sophisticated and complex.

Piaget called the first stage, from birth to about eighteen months of age, the **sensorimotor stage,** the time when children learn through their senses and their movements—through seeing, tasting, feeling, and moving things. Children at this stage, like Angela and Andy, will play with anything they are handed because they are simply exploring the object's physical characteristics and do not see it as a specific object, such as a toy or a tool. The major cognitive milestone within this stage is the discovery of what Piaget called *object permanence,* the realization that objects still exist when they are out of sight. For instance, very young children don't understand that when you cover a toy with a blanket, the toy is really still there. That is why they take such delight in playing peek-a-boo. Babies are surprised to find that a person who hides her face actually reappears when she takes her hands away. By the time a child is almost a year old, however, peek-a-boo is a bore. If you were to cover your face with your hands, the child might reach over and remove your hands to show that your face hadn't disappeared. Children know now that objects remain even when they are out of sight.

Gradually children's cognitive capabilities become more developed and more complex. In the **preoperational stage,** from about eighteen months to seven years of age, children learn to use symbols and to communicate with other people through language, but they do not yet have the mental flexibility to perform mental operations that involve complex relationships. At least at the beginning of this stage, they closely identify words and symbols with the objects they represent. A child at this stage might truly believe that a storybook monster lives under her bed or that a cartoon character is a real person. Importantly for social development, children's thinking at this stage is also very egocentric in the sense that they find it difficult to step out of their own shoes and adopt someone else's perspective. For this reason they have trouble appreciating how others might view them and their actions. Nevertheless, the acquisition of language enables them to begin incorporating the various elements of their culture as their own. Language also allows them to think about their daily experiences, that is, to talk to themselves as well as to others.

As children become better able to handle logical operations, between the ages of seven and eleven, they enter what Piaget called the **concrete operational stage.** Children at this point can learn to read and do arithmetic, measure items, and understand the inner workings of plants and animals. Unlike preoperational children they also can shift perspectives and see themselves as others might see them. In addition, they can begin to grasp more abstract ideas. A younger child will insist that the monster in her closet is real, but by the age of seven or so she understands that the monster was only in her mind and has no material existence. Similarly, a young child might think that a tall, narrow glass holds more water than a short, wide glass containing the same amount of water, but children who have reached this third stage of cognitive development realize that the amount of water stays the same no matter what container it is in. Despite the ability to perform mental operations, however, children's thinking at this stage is still closely tied to concrete objects—to things they can see or touch.

In Piaget's final developmental stage, beginning about age twelve, children become capable of more complex and abstract thought; this is the **formal operations stage.** Instead of solving simple arithmetic problems by manipulating objects and numbers, older children can do complex mathematics involving formulas and proofs. They become capable of understanding hypothetical situations and abstract lines of reasoning. For instance, it probably would have been impossible for you to comprehend the various sociological theories discussed in this book when you were a child or maybe even when you were in high school. But now that you are older and have more developed cognitive capacities, you are capable of understanding theoretical ways of looking at the world.[25] Similarly, whereas younger children find it very difficult to imagine that the real world can be any different than they think it should be, older children are more flexible. They can begin to comprehend how people in various cultures can have norms and values that differ from their own.

What's most important here is to recognize that cognitive and social development go hand in hand. The developmental psychologist Lawrence Kohlberg (1927–1987) has expanded Piaget's work by looking at how children's developing cognitive capacities affect their views of the social world.[26] One area he

studied was children's views of **gender roles,** the norms and expectations associated with being male or female.[27] Based on extensive interviews with boys and girls of varying ages, Kohlberg suggested that the most important element in children's developing gender roles is their belief that they are a boy or a girl. Once they come to realize that they are male or female and that their sex cannot change, they then try to figure out what kinds of behaviors and actions are appropriate to their own sex. Note how, in contrast to the social learning perspective, Kohlberg's cognitive orientation emphasizes the active work of the target of socialization in figuring out the social world and deciding what he or she should do.

When young children are in Piaget's preoperational stage, their cognitive schemes allow them to see the world only in simplistic ways. This applies to their views of gender as well, so young children tend to see gender-typed behaviors as moral obligations. You can easily see this when you suggest to a four- or five-year-old boy that he play with Barbies or to a girl of that same age that she play with an action figure such as Ninja Turtles or GI Joes. They are often offended and embarrassed at the thought. Similarly Andy wanted nothing to do with a dance class made up of girls, and Angela retreated from the boys' football team when it became clear that they thought girls shouldn't play. Kohlberg, and others who emphasize the cognitive orientation, suggest that preschoolers place such a strong emphasis on gender typing because they aren't able to conceive of the world as a place without rigid divisions between categories, such as male and female. They try to adopt the various behaviors that they think are appropriate for their own gender group. Because of the relative simplicity of their thought processes, however, some of these behaviors are exaggerated far beyond what really exists in the adult world.

Within a few years these same children develop more flexible cognitive schemes. For instance, by age ten or so, in Piaget's concrete operational stage, a boy might say, "Well, a boy could play with Barbies, but people would think he was weird." By adolescence, now capable of formal operational thinking, the same boy might be able to distinguish between people's biological sex and their preferences for certain kinds of play or other activities. As adolescents become capable of more complex and abstract thinking, they may become more accepting of a range of behavior in themselves and others.[28] For example, Angela was willing to consider defying convention to try out for football in college. Similarly, even though Andy performs his Fred Astaire routines only for friends and family, he isn't embarrassed to tap dance simply because he's a male.

BUILDING CONCEPTS

Theoretical Perspectives on Learning and Development

Some social scientists emphasize how agents of socialization affect children's development—a social learning perspective. Others focus more on how children actively participate in the process—a cognitive learning perspective. Examining the issue from both perspectives helps us better understand the socialization process.

Perspective	Characteristics	Example
Social learning perspective	Suggests that children will behave in ways for which they have been rewarded and that they will imitate the behaviors of significant others (for example, parents and other adults).	Children learn to wave bye-bye as a result of watching their parents wave and the reward they receive when they start to wave.
Cognitive learning perspective	Explains children's behavior by focusing on a child's emerging ability to understand and interpret the world.	A child's realization that he or she is a boy or girl results in the child's then trying to figure out what kinds of behaviors and actions are appropriate to his or her own sex.

Of course our cognitive schemes develop within our social environments. The cultures in which we live, the agents of socialization with whom we interact, and the models and reinforcements they provide influence our cognitive schemes. But as we develop more flexible cognitive schemes, we can become better able to look beyond rigid definitions of what we should do and consider a wider range of behavior. This developing ability to choose our own behaviors and actions is related to our growing conception of ourselves as unique individuals. Several sociologists have studied how this self-view develops and how it interacts with social influences as we define our roles in society.

DEVELOPING A NOTION OF THE SELF

When sociologists talk about the *self,* they are referring to a stable sense that we have of our own identity and of how we should and do interact with others. Of course, psychology has much to say about

how a sense of self develops as we grow older and interact with others. Sociology contributes to this understanding by showing how our sense of self is closely related to our relationships with others in the world around us. Although you might think of your "self" as your most private and unique being, by applying your sociological imagination you can see how much of the self is really a *social* creation.

Sociologists have tended to look at the self in two general ways. Some have focused on children's development, especially in the early years, whereas others have taken a slightly more macrolevel perspective and looked at how the multiple social roles we fill contribute to our sense of self. We will examine both of these approaches and then apply them to the specific aspect of the self we call *gender.*

Aspects of the Self: A Developmental View

Many of sociologists' ideas about the development of the self derive from the work of George Herbert

Mead (1863–1931), a philosopher by training, who taught social psychology classes at the University of Chicago.[29] Many of his students were sociologists who used his insights in their own work and later developed their own variations and elaborations of his ideas.[30] Mead suggested that we can conceive of the self as involving two different aspects, the "I" and the "me."[31] The "I" represents the active part of the self, as when we initiate actions and decide to do things. For instance, you might say, "*I* decided to skip class today," or "*I* am going to buy a new car." The "me" represents the reflective part of the self, as when we make ourselves the object of our thoughts. For example, you might say, "My math teacher will be upset with *me* if I don't finish my paper, so I will skip sociology class to finish it," or "When my friends see my new car, they'll be impressed with *me*."[32]

Mead regarded the "I" and the "me" as always engaged in a kind of rapid-fire internal dialogue in our minds.[33] We initiate behaviors and we think about these behaviors and their possible actions; these thoughts then lead us to perhaps change these actions or to continue with them. As we think and "talk to ourselves" about our relationships and actions, we mentally assume the role of other people, as if we were engaged in conversations with them. This process of being able to think about, or reflect on, our own actions is called **reflexive behavior.**[34] We engage in such reflexive behavior many times each day as we continually think about what we have done or plan what we might do.

Mead realized that children only gradually develop the ability to engage in reflexive behavior, to look at themselves as objects. At first, Mead suggested that children are in the **play stage,** where they "play at" or assume various roles one at a time. For instance, young children might copy or model the activities of significant others, such as their parents: pretending to shave or vacuum or wash dishes or tend a baby. In so doing, they are practicing or playing at the activities of one role occupant. They don't, however, recognize that these roles are intertwined with others. For example, very young children will pretend to rock and feed dolls, but they don't understand how to "play house" with other children because they aren't yet able to understand how the roles of mommy, daddy, and baby go together. Thus very young children engage in "parallel play," doing the same activities side by side, but without really interacting with each other in coordinated activities.

As children grow older, they enter what Mead called the **game stage** and begin to understand how different roles go together. Mead suggested that children now become able to take the role of the other. They can think about what other people must be thinking and apply this understanding to develop their own behaviors and actions. Mead used the example of a baseball game to illustrate this concept. Although even young children love to play with balls, baseball as a game requires that teammates work together and understand one another's roles and those of their opponents.

You can see how this ability to take the role of others gradually develops by watching children of various ages try to play baseball. In the community in which I live, children in the early elementary grades play tee-ball, hitting the ball from a tee. In the later elementary grades they play baseball and softball, hitting a ball thrown by a pitcher. In both games, however, they are expected to run around the bases and to get opposing players out by fielding the ball and throwing to the appropriate base. The tee-ball games are a lot of fun to watch, but clearly the players often have no idea of what is happening on the field around them. When a ball is hit to them, they dash in to get it but then don't know what to do with it. The parents and coaches may yell loudly, "Throw it home!" or "Throw it to third!" yet the child will look around blankly, not quite understanding what everyone else is doing in the game. In contrast, older children are generally quite aware of what is happening all over the field. When a ball is hit to them, they can anticipate the moves of their opponents. They know where the ball should be thrown and can even execute double plays.

More generally you "play the game" all the time as you consider what others with whom you interact might be thinking as you plan your own behaviors. But you only developed the ability to do this as you became older and more able to understand the complexities of both your own and others' roles. For instance, suppose you worked the counter in a fast-food restaurant. In your interactions with customers, other counterpeople, cooks, and managers, you would know how each of these people were related and what their varying expectations of you were. In contrast, when you were younger, you might have played "McDonalds" by eating pretend french fries or flipping plastic hamburgers on a toy stove, but you were not yet able to grasp the full complexity of the various roles.

As children interact more and more with other people, Mead suggested that they develop a notion of the **generalized other,** a conception of the expectations and norms that all people have, not just the individuals with whom they interact. This more generalized view is the basis of the "me," the standards you use in your internal conversations with yourself.

When you accept the norms that other people also hold, sociologists say that the process of **inter-**

With the aid of masks these children can play at taking the role of the other. What other ways do children use to develop the ability to take the role of others?

nalization has taken place. Internalization has occurred when you engage in behavior without thinking about possible rewards and punishments for it. Through internalization you develop a **conscience,** an internal set of ethical and moral principles that guide your actions. Children who develop normally become increasingly capable of self-control. Because they have a conscience, they can conform to behavioral standards without either fearing punishment or expecting rewards for doing so.[35] Moreover, these standards become an aspect of what we think of as "the self."

Social Roles and Self-Identity

From our interactions with others and our internal conversations with ourselves, we gradually develop a **self-identity,** a set of categories we use to define ourselves.[36] Although we might often think of our identity as something private and uniquely our own, sociologists point out that our views of ourselves reflect our relationships to other people.[37] That is, our self-identity reflects our role relationships with other people.

To study people's views of themselves, sociologists and psychologists often look at the **self-concept,** the thoughts and feelings that individuals have about themselves. One way they try to measure self-concept is the "Who Am I?" test.[38] With this measure respondents write down fifteen different answers to the question "Who Am I?" You might want to try writing down your own responses to this question before reading further.

Figure 4-3 shows the responses to the "Who Am I?" test of nine-year-old Josh and college sophomore Arlene. Sociologists have found that people typically write three different kinds of self-descriptions, which reveal different aspects of the self-concept.[39] First, and most common, are *role identities.* For instance, Josh sees himself as a boy, as Louis's little brother, and as being in the third grade. Arlene describes herself as a person, a member of the human race, a daughter and sister, a student, a creator, and a music enthusiast. All of these descriptions involve social roles. The second major category involves *personal qualities* such as temperament, body image, and preferences and tastes. Josh sees himself as having big ears, liking to eat, and being capable of beating up Andy. Arlene sees herself as loving people, enjoying nature, and always changing. Third, but least common, are *self-evaluations,* judgments about our personal qualities. For example, Josh says he is "sometimes a good sport," and Arlene says that she is lonely.

Clearly your view of yourself is complex. You fill many different roles, you see yourself in a variety of ways, and not all of these aspects of your identity are equally important to you. Some sociologists suggest that we actually have a **hierarchy of identities** in which some self-identities are more important to us than others.[40] For instance, former President Lyndon Johnson is reputed to have said, "I'm an American, a Texan, and a Democrat—in that order."[41] In your responses to the "Who Am I?" questionnaire, suppose you said that you were a student, a parent, a woman, and an African American. Maybe the most important of these self-identities is that of parent. When you discuss your role of student with others, or when you think about your schoolwork, you

FIGURE 4-3

*Sample Responses to the
"Who Am I?" Test*

Source: Michener,
DeLamater, and Schwartz,
1990, p. 86.

I am...
a boy
do what my mother says, mostly
Louis's little brother
Josh
have big ears
can beat up Andy
play soccer
sometimes a good sport
a skater
make a lot of noise
like to eat
talk good
go to third grade
bad at drawing

Josh (a nine-year-old boy)

I am...
a person
member of the human race
daughter and sister
a student
people lover
people watcher
creator of written, drawn, and
    spoken (things) creations
music enthusiast
enjoyer of nature
partly the sum of my experiences
always changing
lonely
all the characters in the books I read
a small part of the universe,
    but I can change it
I'm not sure?!

Arlene (a college sophomore)

might talk about the ways in which your schooling can help you get a better job and so improve the lives of your children. Or perhaps you identify most closely with your African-American identity and are more inclined to get involved in issues of racial discrimination than in women's issues.

The salience, or importance, of our self-identities is often closely related to the nature of the relationships we have with other people and our social roles.[42] Consider Mark, the young man described in the beginning of Chapter 1, who committed suicide after losing his spot on the track team to an injury and flunking out of school. In Chapter 1 we used Durkheim's analysis to show how suicides like those of Mark reflected *anomie,* or a situation of normlessness. We can also interpret Mark's situation from a more microlevel perspective by using the idea of hierarchy of identities. Mark's identity as an athlete was highly salient to him, and his role as a student was crucial to maintaining ties with his school friends. Even though there was much more to Mark than being an athlete and student—among other things he was a son, a brother, a young man embarking on adulthood, and so on—when he was forced to leave school and find work, he found it difficult to build up new relationships and a new identity. Mark's anomie resulted from the loss of his old social networks and the lack of new roles and a new identity to replace them.

The salience of a particular self-identity also depends on the particular situation in which you find yourself. For instance, you might not think of yourself as a son or a daughter much when you are in class, but if you are at a family reunion, this role is probably very salient to you. Sometimes sociologists use the term **situated self** to refer to the subset of self-concepts or self-identities that apply to one's self-views and behaviors in particular situations.[43] For instance, you might see yourself as a competent, independent adult student and worker during the week, but at a family reunion you might instead see yourself as the baby of the family. At a party your view of yourself as shy and withdrawn will probably be more important than your self-image as a competent worker. In other words the aspects of your self-identity that are most important to you at any one time depend very much on the situation in which you are interacting. In a sense we all have many "selves."

As we have seen, one of the first elements of children's developing sense of self is their view of themselves as either male or female. This aspect of the self is referred to as **gender identity,** a gut-level belief that you are a male or a female. This belief apparently develops by the time that children begin to talk. As children acquire a gender identity, they also develop a **gender schema,** a cognitive framework that they use to organize information around them as relevant to one sex or the other—and that helps them understand the world in gender-typed ways.

An important way in which children develop their ideas of gender-appropriate behavior, as well as other notions about the social world, is through interactions with friends and playmates. Here we will use the concept of gender schemas to illustrate the role of peer groups in children's socialization.

DEVELOPING GENDER SCHEMAS:
THE INFLUENCE OF PEER GROUPS

Gary Fine, a sociologist at the University of Georgia, wanted to investigate the subculture of preadolescent boys. To do this, he used the method of participant observation to study boys' interactions and participation on Little League baseball teams in five different communities.[44] Of course, he couldn't actually be a Little Leaguer himself. Instead he obtained permission from the organizations, coaches, and teams to hang around their games and practices. He purposely decided not to take the role of coach or umpire, preferring, as much as possible, to be seen as the boys' friend.

This process invariably took time. Whenever Fine first began to hang out with a team, the boys weren't quite sure what to make of him. Gradually, however, they came to accept him—especially when he explained that he was "writing a report for school," something they could all relate to. They gave him nicknames, teased him, and gradually told him more and more about their activities. They wouldn't let him observe their pranks or attend boy-girl parties, but they would readily answer his questions about these events, and they talked freely with one another while he was present.

Like all field researchers Fine had multiple sources of data. When he observed the games and practices, he often served as scorekeeper and, in the process, took copious notes of the boys' interactions and behavior and sometimes tape-recorded their conversations. In addition, he distributed questionnaires to and conducted in-depth interviews with the players, their parents, and the coaches.

Remember that the term subculture means a group or culture within a broader culture. Fine believed that these preadolescent boys form their own subculture, a group with distinctive norms and values. Although coaches and parents influenced the norms and values of the players, Fine saw many ways in which the Little Leaguers themselves defined what it meant to be a good team player and also what it meant to be masculine.

Fine approached his observations by trying to understand the "moral code" of the boys' culture. In one sense this is probably similar to what Durkheim might have done, for in effect Fine was trying to understand the *collective consciousness* of the Little League teams—the values and norms that they held in common. In contrast to what Durkheim probably would have done, however, Fine took a more micro perspective and focused on the boys' actual behaviors and interactions. From these observations he

identified four basic parts of the boys' definitions of what good or "proper" behavior includes—four elements of their "moral code," all of which involved the display of gender-appropriate emotions.

First, it was important for the boys to show that they were tough. Even when they were hurt or intimidated, they expected one another to hang in there and continue to play. Second, the boys felt that it was important for them to control their emotions, something some of them had more difficulty doing than others. For instance, some tended to show anger more frequently and to cry and mope when things didn't go their way, and some were more sensitive to pain than others. When they were unable to keep their anger or tears in check, their teammates might be sympathetic but clearly communicated that this shouldn't be done. Similarly, even though they were sympathetic when teammates were hurt, the clear-cut norm was that they should try to play if at all possible. Third, the boys expected one another to be competitive and to want to win. Finally the boys valued group unity and loyalty. For example, it was improper for a boy to "rat" or "tattle" on others.

These notions of toughness, unemotionality, competitiveness, and loyalty parallel the stereotypical conceptions of masculinity found among adults and older adolescents in the United States. Research has shown that in our society people consistently characterize men as more aggressive, independent, and competitive, but less emotional and excitable, than women.[45] Thus the moral code of the Little Leaguers' subculture in many ways reflected stereotypes found throughout the culture.

Because preadolescent boys do not have the cognitive complexity of older boys and grown men, many of the Little Leaguers' definitions and portrayals of what they perceive to be masculine behavior are no doubt exaggerated. Fine noted two general themes in particular. The first was a preoccupation with sex. As he put it,

> In "hanging around" with preadolescent boys (particularly those eleven and twelve years old) I was impressed by the time devoted to talk about sex and members of the opposite sex.... Males maturing in our sexualized society quickly recognize that sexual prowess is a mark of maturity. One must convince others that one is sexually mature, active, and knowledgeable. Boys must try to be men. Although the techniques change, males throughout the life course are subject to the same demanding requirements.[46]

Fine studied the subculture of preadolescent boys. How does this picture illustrate both material and nonmaterial elements of this subculture?

The boys' activities in this area took a wide variety of forms. For instance, they sometimes teased one another about girls they thought their friends had crushes on. They talked extensively about sexual behavior and tried to impress one another with their knowledge. Some of the conversations promoted an identification of masculinity with heterosexuality by degrading homosexuality. Most of the boys had only vague ideas of what homosexuality was, but they often ridiculed others by calling them derogatory homosexual names. Their actual contacts with girls were also exaggerated. In one of Fine's questionnaires, more than half of the boys claimed to have had a girlfriend, but in reality most of the contact occurred in groups or through go-betweens.

Fine found that a second theme, closely related to the sexual talk and behavior, was "the perceived need of preadolescent males to demonstrate they are men by being daring and bold." They did this with insults, often sexually oriented, directed at one another, or with pranks directed at others whom they generally did not know. In one community the boys delighted in "mooning" passing traffic, in another they liked to "egg" houses, and in others they enjoyed ringing doorbells and then running away. Typically the boys would perform these pranks with a partner and later describe in detail their heroics to their friends in order to demonstrate their daring and toughness.

Fine's work on the preadolescent subculture illustrates both the nature of gender socialization and the development of the self. From their interactions with their parents and coaches, but especially with their friends, the boys developed a moral code of behavior—ideas of what it meant to be a Little Leaguer and a member of their peer group. This code involved norms and values that emphasize toughness, unemotionality, competitiveness, and loyalty. The boys internalized these expectations or incorporated them into what Mead would call the generalized other, their views of what should be done. Conformity to these expectations was reinforced through social learning—through the feedback the boys gave to one another when they failed to display these characteristics or when they were especially tough or competitive. Conformity was also reinforced by the boys' sense that following these norms was related to proving their masculinity. Through their interactions and behaviors they tried to define and bolster their gender identity. If they could display sexual knowledge and show that they were truly daring and brave, the boys could show that they were really boys. (To learn more about Fine and his work, see Box 4-1.)

Other scholars who have studied gender socialization have also found that children's interactions with each other contribute to the development of their self identity and views of appropriate role behavior. They have also documented the presence of several different subcultures among children.[47] For instance, the sociologist Barrie Thorne examined boys' and girls' interactions on school playgrounds. In an important extension of Fine's work she documented variation in the ways that children define what it means to be masculine and feminine and how these variations are related to the peer groups in which they participate. For instance, some of the young boys she observed sorted boys into "nice boys," "machos," and "show-offs." A college age woman that she interviewed described girls in her youth as "fast," "pretty," "smart, " or "tough."[48] All children appear to develop views of what it means to be a boy or a girl though their interactions with other people, but the precise nature of this meaning can vary from one subgroup to another..

All of us continue to develop our sense of masculinity or femininity and other aspects of our self-identities throughout our lives. But the nature of our self-views and the roles that we learn are very much influenced by the particular historical context in which we grew up. Sociologists have tried to understand more about these influences by looking at socialization throughout the life course.

 BOX
4-1

SOCIOLOGISTS AT WORK

Gary Fine

There is an active, somewhat disturbing subculture among preadolescents.

Gary Fine, University of Georgia, conducted the featured research in this chapter.

When you were growing up, was there anything that made you want to become a sociologist? I grew up in New York City (Manhattan) and I remember walking down Park Avenue in seventh or eighth grade. It's lined by big apartment houses on either side, and I was suddenly struck by the huge number of people in those buildings leading lives that rarely intersected. In the same geographical area one person could be receiving an award while another was being diagnosed with cancer. Did these people interact at all, on the street corner or at the local market? The idea of all of those activities in a small physical space intrigued me.

What interested you about studying groups, and Little Leaguers in particular? I wanted to study relationships among people, the development of group life, and how groups create culture and tradition. Little League teams were an ideal small group to study because they have a task orientation and a socio-emotional orientation. Plus, I like baseball and enjoy being with preadolescents.

What surprised you the most about your research findings? I was surprised at the underside of preadolescent culture, the "dirty play" that was so evident, much more so than I remembered. I realized that kids are less angelic than parents believe. There is an active, somewhat disturbing subculture among preadolescents.

If you could repeat the study, what would you do differently? It would be difficult to do the study now, with parents' fears about strangers hanging out with their kids. If I could, I would expand the ethnic and class background of the kids and delve more into the history of preadolescent sports.

What kinds of social issues does your research address? An ongoing social issue involves kids' participation in organized sports and the question of whether it's too competitive. In my work there's not a lot of evidence that this is the case, and, in fact, Little League looks like a reasonably good way for kids to spend the spring and summer. More broadly, the work demonstrates the strength of the preadolescent need to differentiate along ethnic and gender lines through the sexually explicit, racist, and sexist talk I reported. While adults should criticize the talk when they hear it, they shouldn't expect the talk to disappear right away. By condemning it, adults plant the seeds for kids to reject these attitudes at a later stage. So adults/parents should know that addressing this kind of talk may be effective in the long run.

How do you think sociology affects your students' lives? I want my intro students to develop the sociological imagination. I want them to go home for Thanksgiving dinner and see it in a new way, understanding its group dynamics, status issues, family and economic structures, and interaction patterns. My research on groups can help students reflect on any group—fraternity, sorority, work group, family—because every group develops its own idioculture.

CRITICAL-THINKING QUESTIONS

1. How do the results of Fine and others illustrate the theories that were discussed earlier in this chapter?

2. Compare Fine's observations and conclusions with your own observations of the subculture of preadolescent boys. How are they similar; how are they different? How many different subcultures can you identify?

3. How might Fine's conclusions have differed if he studied girls rather than boys? What might explain these differences?

4. What other methods, besides field research, could a researcher use to study the cultures of young boys and girls? What might be the advantages and disadvantages of using such methods?

Socialization Through the Life Course

Throughout our lives we enter and leave social networks, we learn new roles, and we leave roles behind. Sociologists use the term **career** to refer to a sequence of roles that we enact during our lifetimes. Although you might think about a career in relation to work, you can also talk about your career as a friend, a son or daughter, a parent, or a student.[49] For instance, you might change jobs many times and move from one locale to another. Or you might be single and childless for a number of years, then marry and have a family, and then divorce a spouse or have your children leave home. The moves from one role to another are often called **role transitions.**[50]

The roles we enact are very much influenced by our age. In addition, the historical period in which we live and even the era in which we were born influence our lives. For example, your grandparents might not have been able to play Little League ball when they were young, simply because it didn't exist. Similarly you might be majoring in an area of study, such as computer science or women's studies, that hadn't been developed when your grandparents, or even your parents, were in school. Sociologists have tried to evaluate the changing impact of age and historical context on our lives by looking at three basic concepts: the effects of age, period (or history), and cohort.

Developmental Change or Age Effects

No matter how hard we might wish it, we are powerless to change our age. All societies distinguish between activities that people of different ages should do. They have **age-based norms,** expectations and rules about behaviors that should occur at various ages. When these age-based norms influence our behaviors and actions, sociologists say that there is an **age effect.**

Part of the reason that age is important simply reflects physiological change—our bodies and our physical abilities change as we get older. There is, of course, a great deal of variation in the ages at which these physiological events occur. For example, some children enter puberty at age ten or eleven, and others much later in the teen years. Women from approximately age fifteen to forty-five are biologically capable of bearing children. Men are able to father children until the end of their lives, whereas women can enter menopause (the end of their childbearing years) from their late thirties to well into their fifties.

As we get older, we usually have less energy and more health problems than we did at younger ages.

Some people do not feel the detrimental effects of aging and start to slow down their activities until they are well into their eighties; others feel the impact much earlier. In addition, age can affect our participation in various roles in different ways. For instance, we can generally expect to continue to paint, write, and play music far longer than we will be able to play a game of pick-up basketball or go downhill skiing.

Biology only sets the outside parameters for our activities, however. Age-based norms specify ranges of ages in which it is generally expected and acceptable to attend school, work, get married, and retire. You were required to begin school by the age of six or seven, even though you might still have been quite physically and emotionally immature. Similarly in most states you can't legally drop out of school until the age of sixteen whether you want to or not. And even though young people might be fully capable physiologically of sexual relationships and parenthood, most states prohibit marriage until the age of sixteen and even then require parental permission until the age of eighteen.

These age-based norms are socially defined, and they can vary a great deal from one society to another. In some societies women marry and bear their first child while in their early teens. In the United States, however, teenage parenthood violates age norms regarding when we should have children. As a result teenage parents often have difficulty in other areas of their lives, such as finding jobs and finishing school.[51]

Drawing on the concepts of life careers and age-based norms, sociologists often talk about **normative life stages,** periods during the life course when people are expected to perform certain activities.[52] People in the United States typically talk about four life stages after childhood: (1) youth or adolescence, (2) young adulthood, (3) middle age, and (4) old age. These stages, along with ages at which they generally occur, are summarized in Figure 4-4. Note that the ages associated with each stage overlap because the borders between stages are somewhat fuzzy. Each of us may enter and leave various life stages at different chronological ages.

It is important to realize that these descriptions involve broad generalities, or modal activities. The word **modal** (derived from mode) simply means the most common or frequently occurring. That is, many analyses of the life course focus on what the largest number of people within a group tend to do, rather than on the variety of choices that individual group members can make. This focus helps to simplify the analysis. Yet not everyone follows the modal course, and as we will discuss shortly, in some historical eras the amount of variation is much greater than in oth-

Age
16–23

Youth and
Late Adolescence

The stage in which young people are traditionally expected to finish school, leave the parental home, and begin to establish work careers and intimate relationships. Because this period of life involves so many changes and is often a time of searching and trying out new roles, it is sometimes said to involve an "identity crisis," a period of self-reflection in which young people try to understand more about who they are.

Age
18–40

Young
Adulthood

The time when formal schooling is completed and roles associated with work and family careers become more important as many people marry and begin to raise children. The major challenge facing people in this stage is balancing work and family commitments

Age
35–70

Adulthood, Maturity,
Middle Age

The stage in which children get older and begin to leave home and the struggles involved in balancing work and family demands tend to lessen. Both men and women begin to feel encroaching signs of age, report experiencing more health problems, and often have parents and other elderly friends and relatives who are in failing health.

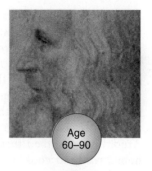

Age
60–90

Late Maturity,
Old Age

The stage in which people begin to retire from work and suffer from some type of disability that requires drastic changes in activities. Changes generally involve unavoidable losses, such as jobs, significant others through their deaths, and health and energy. The major challenge of old age is learning to cope with the new identities and roles associated with these losses.

FIGURE 4-4 *Normative Life Stages*

ers. For instance, college students today are much more likely than were previous generations of students to have already worked for a number of years and to have had families. Thus, if some of the conditions described in Figure 4-4 don't seem to apply to you or someone you know, remember that they reflect merely the modal patterns within our society.

The existence of age-based norms helps to make role transitions somewhat easier than they might otherwise be, for we can prepare for new roles. Through a process that sociologists call **anticipatory socialization,** we think about and rehearse the actions, emotions, and skills that might be involved in these new roles.[53] This anticipatory socialization involves conversations with ourselves (Mead's reflexive behavior) as well as the development of new schemas and new self-identities. For instance, you probably thought a lot about what college would be like. You might have talked to people you knew who had attended your school to try to learn more about what would be expected of you and what your classes and social life would be like. You might have thought about how you should act to impress your teachers and what you might do to make new friends.

Role transitions are much harder on us when they are unexpected because we haven't been able to en-

gage in anticipatory socialization. For example, if a close relative or friend has been terminally ill for a long time, you are able to think about what life will be like when he or she dies. Even though you still grieve deeply when death comes, the transition is generally easier to cope with than it would have been if death had been sudden and unexpected. Similarly we can experience great difficulty with role transitions associated with other unexpected circumstances such as suddenly having to get a job, becoming disabled, losing a spouse through divorce or death, or being forced to retire from a job.

History or Period Effects

Even though you would encounter age-based norms no matter when you were born, the exact nature of these norms depends on the historical context. This can be seen in the way three different television shows portrayed the experiences of teenage boys and their families set in a particular historical era: *Happy Days* in the 1950s, *The Wonder Years* in the late 1960s and early 1970s, and *Family Ties* in the 1980s. Many of the variations in the plots and situations depicted in these shows reflect differences in the historical eras in which the fictional families lived.

For example, consider how the mothers are portrayed in the shows. Reflecting different norms about gender roles, neither Mrs. Cunningham in *Happy Days* nor Mrs. Arnold in *The Wonder Years* worked outside the home. Even though a large number of women worked outside the home in the 1950s, 1960s, and early 1970s, most mothers of school-age children were housewives. Richie Cunningham and Kevin Arnold just assumed that their mothers would be home, ready to cook and clean for them. Subsequently, however, norms surrounding women's roles in the home changed a great deal. Thus, although she loved Alex and took care of him, Mrs. Keaton on *Family Ties* also had a full and demanding career outside the home. In fact, historical changes in the last half-century or so have affected people of both sexes and all ages, not just mothers. For instance, all young people, and especially women, are much more likely to go to college now than in earlier years. Women, and especially those with children, are much more likely to be employed outside the home. The divorce rate has risen considerably, and many young people wait until they are older to marry. Both men and women have a higher chance of living into old age.[54]

Sociologists refer to the influence of the norms and events of a particular historical time on individuals lives a **period effects.**[55] "The Wonder Years" took place during the Vietnam War era, and many episodes dealt with tensions surrounding the changes that occurred during this time. For instance, Winnie, Kevin's girlfriend, lost her brother in the war, and, to his parents' dismay, Kevin's older brother, Wayne, once tried to enlist in the army. Conflicts in the Arnold household often revolved around lifestyle issues, as when Kevin's sister lived in a commune and became a vegetarian. In contrast, the Cunningham family rarely had political arguments, but political discussions in the Keaton family usually involved the conservative Alex, who idolized President Ronald Reagan, opposing his more liberal parents. Even though Richie, Kevin, and Alex all had experiences that teenage boys commonly face, such as problems with friends, teachers, and family members, the content and nature of these problems were heavily influenced by the times in which they lived.

These period effects influence all people, no matter what their age. For instance, everyone who was old enough in the 1960s to be aware of world events remembers the assassinations of John F. and Robert Kennedy, Martin Luther King, and Malcolm X, as well as the turmoil surrounding the Vietnam War. These events undoubtedly affected the way people living in these periods saw themselves and the world around them. Yet the impact of these events often varied for people of different ages and at different life stages. For instance, middle-aged people had lived through the end of World War II, a time of unusual national unity and unprecedented dominance for the United States in world affairs. These people also had established family and work careers. In contrast, adolescents and young adults had no direct memory of the war. They were just finishing their schooling and entering their work careers, and they had not yet started families. In many ways they were still trying to develop their identities. During that era parents and children commonly had passionately conflicting views on the war in Vietnam, as well as on values concerning work, sex, and marriage; civil rights; and many other areas of life. Sociologists try to understand these different impacts of historical events by looking at groups of people with similar experiences.

Peer Group or Cohort Effects

Sociologists use the word **cohort** to refer to a group of individuals who have some characteristic in common. For instance, people who are born in the same year are referred to as a **birth cohort.** They all experienced the same historical events when they were the same age, and these experiences during the same developmental period might have influenced their attitudes and behaviors, through what is called a **cohort effect.**[56]

Let's imagine, for example, the experiences of the birth cohorts of the fictional TV characters Richie Cunningham, Kevin Arnold, and Alex Keaton. Richie was probably born around 1940, Kevin about 1957, and Alex about 1968. Figure 4-5 illustrates the life course of these birth cohorts. People in Richie's cohort were preschoolers when World War II ended, were teenagers during the Eisenhower administration, and were launched into adulthood and beyond draft age before the Vietnam War escalated in the 1960s. They can expect to retire from work around the year 2005.

Kevin and others in his cohort were only six years old when President John F. Kennedy was assassinated. By the time they were thirteen, they had witnessed Martin Luther King's and Robert Kennedy's murders, they had seen race riots in cities nationwide on television, and they had experienced the Vietnam War on the nightly news. The civil rights, environmental, feminist, and anti-war movements were all in full swing during their teen years. News of the Watergate scandal during the Nixon administration permeated their high school years, and Nixon left the presidency a few months after they finished high school. In contrast to members of Richie's cohort, they experienced a great deal of political turmoil and unrest during their growing-up years.

1956: Elvis Presley's first hit

1963: John F. Kennedy assassinated

1968: Martin Luther King assassinated

1974: Richard Nixon resigned

1980: The Iran hostage crisis

1986: The Challenger explosion

1990: The Berlin Wall fell

FIGURE 4-5 Life Course of Richie's, Kevin's, and Alex's Birth Cohorts. *The diagonal lines in this collection of photos represent the life course of the fictional characters Richie, Kevin, and Alex, showing the age of each in different years. From examining this set of pictures, you can see the way historical events might affect members of different birth cohorts. For instance, when Elvis had his first hit in 1956, Richie was a teenager but Kevin wasn't even born. When Richard Nixon resigned as president, Richie was in his 30s, Kevin was 17, and Alex was only 6. Try plotting your life on this chart by finding the year of your birth and drawing a diagonal line that parallels the other diagonals. How have the historical events pictured affected you? Now draw a line for someone you know from a different birth cohort. How have the influences of these historical events been different for them?*

Alex and his cohort were only babies when Robert Kennedy and Martin Luther King were killed, and only young children during the Vietnam War. During their teen years President Ronald Reagan became the first president since Dwight Eisenhower in the 1950s to serve two full terms in the White House. Although Alex's cohort certainly grew up in a more violent world than did Richie's, as teenagers they witnessed less political turmoil and conflict than did Kevin's cohort, and the political atmosphere was less charged and more conservative.

Because age-related norms have changed greatly over the years, individuals in different birth cohorts have also been exposed to different role expectations. Richie and his cohort grew up in an atmosphere in which it was uncommon for women with young children to work outside the home or even for people to choose not to marry or have children and in which most people who went to college did so right out of high school. Alex and his cohort lived in a strikingly different world. To the extent that these growing-up years influenced people in Richie's, Kevin's, and Alex's birth cohorts to have different attitudes and experiences throughout their lives, a sociologist would say that there was a cohort effect on their life course. You can use the idea of age, period,

and cohort effects to understand how your life experiences differ from people who are older and younger than you.

Whatever birth cohort we might belong to, our lives are affected by the historical periods in which we live. For instance, Angela and Andy were children during a period of greater technological sophistication than any previous time in history. Unlike their parents they might not be able to even remember what it was like to live without a VCR, a microwave, and a computer. They also faced fears at a young age that other birth cohorts did not. They saw technology fail when they were quite young and the Challenger space shuttle exploded. They worried about AIDS and cried when Ryan White, who was a little older than they, finally succumbed to the disease.

We all also go through role transitions as we age. As Angela and Andy approach the end of adolescence, they have reached a bend in the river, a new stage in their life careers. Although much of their identity and part of their life course are set, many role transitions and new challenges await them in young adulthood. How they negotiate the future bends in the river will depend partly on their biological heritage and partly on the way circumstances and

Angela, Andy, and the boys in Gary Fine's study were all quite lucky. Even though their leisure activities and other behaviors were constrained by expectations associated with their roles as males or females, they were healthy and they were raised by parents who encouraged their interests and activities. To return to our analogy of the riverboats, these children were pilots who had capable crew members, a smooth stream, and nonperilous riverbanks. Not all children are so fortunate.

Many people who study socialization are interested in how we can promote children's healthy development. Two basic preconditions seem important: (1) the health of the child at birth and (2) the health of the environment. Children do best when they are born without physical or mental limitations and when they are raised in supportive and stimulating environments.

In recent years the problem of children born to alcoholic and drug-addicted mothers has become much more apparent to the general public. Babies who have been exposed to drugs and alcohol in the womb often suffer from a number of impairments, including retardation, malformation of body parts, and difficulties concentrating and interacting with others. After birth they also often suffer painful withdrawal symptoms as their bodies adjust to the absence of the drugs or alcohol.[57] Even though special educators have developed many ways of helping these and other children born with disabilities, such children clearly start off their lives at a disadvantage because of their physical problems.

At the same time, even if children of drug-addicted and alcoholic mothers could be born without physical and mental impairments, their substance-abusing mothers would not be able to provide the quality of care that other mothers can. In fact, a great deal of evidence now suggests that children born into such situations are far more likely than others to eventually have drug- and alcohol-related problems of their own unless effective interventions can alter their environment.[58]

People—such as educators, social workers, and medical personnel—who work with mothers who have chemical dependency problems often use intervention programs that try to get the mothers to stop using alcohol or drugs as early in the pregnancies as possible so that the physical effects of these substances on their infants are minimized. This often involves helping the mothers to break ties

their own social actions help to shape their environment. But much of their future experience, and that of their children, also will be shaped by their early socialization. They both were lucky enough to grow up in loving, nurturing families that encouraged them to develop their special talents. (For a discussion of children who aren't so lucky, see Box 4-2.) In the years to come, the opportunities and constraints they face will be very much influenced by the interactions they have with other people. In the next chapter we examine how sociologists have studied these interactions.

SUMMARY

Through socialization and the development of a sense of self, we learn to be part of society. One aspect is gender socialization, the process of learning to see ourselves as male or female and to adopt the roles associated with this status. Socialization continues throughout the life course as we move from childhood to adulthood to old age and undergo associated role transitions. Key chapter topics include the following:

- Human children depend on the care of others for a long time after birth and are responsive to other people from an early age. Temperamental differ-

**APPLYING SOCIOLOGY
TO SOCIAL ISSUES**

Continued

with those who promoted their alcohol and drug use and to develop new relationships with people who are not substance abusers. To provide more supportive environments for the children, they work with the mothers to keep them drug- and alcohol-free after giving birth. They help the mothers develop new social networks that can provide the support the mothers need to break old habits and develop new ones. They help the mothers learn new roles—such as responsible worker and mother. They also help mothers learn how to care for their babies and to enact the role of competent mother. On a day-to-day basis they encourage the women to develop different self-views and self-images. To do this they might use models of women who have broken the cycle of drug and alcohol dependence, as well as reinforcements for successfully enacting positive behaviors. Much of their work with the

children involves helping them overcome their physical limitations and problems. People who work in special education have become very skilled at the use of reinforcements and modeling techniques that work with children at different cognitive levels.[59]

Sometimes sociologists evaluate the effectiveness of intervention programs such as those used with drug-affected mothers through what is called **evaluation research.** Evaluation researchers can use any of the various techniques described throughout this book, but their work is focused not so much on developing theory as on helping practitioners develop more effective ways of dealing with social problems. Thus far evaluations of programs dealing with

drug-affected mothers and babies indicate that it is very difficult to change the mothers' behaviors. However, few if any programs use all of the various intervention techniques mentioned here, and programs that use more of the methods tend to be more effective.[60] That undoubtedly occurs because only a combination of approaches allows changes in both social structures and social actions. However, the fact that some mothers can overcome their addictions and change the way in which they relate to their children shows how socialization occurs throughout life.

**CRITICAL-THINKING
QUESTION**

1. If you were in charge of a program to help substance-abusing mothers and their children, which interventions would you implement first? Why? Which would you give lowest priority? Why?

---

ences between infants may be biologically based, but they are also greatly influenced by the socialization process. Few basic temperamental differences exist between males and females, and those that may exist are modified through socialization.

■ Socialization can be seen as involving the learning of social roles within social networks. The types of roles we learn and the support we have from others are largely determined by the nature of the social networks we enter. The family and the peer group are the most important agents of socialization in childhood.

■ A social learning perspective on socialization emphasizes how children's actions are influenced by receiving rewards from and by imitating the behaviors of significant others.

■ Cognitive perspectives on socialization, largely influenced by the work of Piaget, emphasize children's developing ability to interpret the world around them and the ways in which children's own interpretations of this world influence the socialization process. Kohlberg suggested that children's views of gender roles change as their cognitive development changes.

■ Mead's analysis of the development of the self involves two aspects, the "I" and the "me." The "I" represents the active, decision-making part of the self, and the "me" represents the reflective part. Through social interactions children develop a sense of the generalized other as a guide to behavior.

- Our self-identity or self-view develops from our interactions with others and our adoption of social roles. Some identities are more salient than others, often depending on the situations in which we find ourselves.

- Peer groups are important influences in childhood socialization and the development of the self. Gary Fine's participant observation of Little League teams showed how boys, through their interactions with one another, actually create the male preadolescent subculture, their own distinctive norms and values, and their definitions of masculinity.

- Throughout our lives we acquire new roles and leave roles behind through series of role transitions.

- Age-based norms are influenced by both physiological changes associated with aging and social expectations of when we should engage in various activities. Events that affect our lives at particular historical periods are called history effects and influence people of all ages. In contrast, cohort effects refer to common influences on groups of people who have experienced an event at a particular point in the life course.

KEY TERMS

age-based norms 90	gender identity 86	role transition 90
age effect 90	gender roles 82	scheme 81
agent of socialization 78	gender schema 86	self 75
anticipatory socialization 91	gender socialization 76	self-concept 85
birth cohort 92	generalized other 84	self-identity 85
career 90	hierarchy of identities 85	sensorimotor stage 81
cognitive development 81	internalization 84–85	significant others 77
cohort 92	meta-analyses 77	situated self 86
cohort effect 92	modal 90	social learning perspective 79
concrete operational stage 82	normative life stages 90	social network 77
conscience 85	period effect 92	social role 78
evaluation research 94	play stage 84	social status 78
formal operations stage 82	preoperational stage 81	socialization 75
game stage 84	reflexive behavior 84	target of socialization 78

INTERNET ASSIGNMENTS

1. There are several agents of socialization, in addition to parents and the family as primary socializing agents. Using a search engine, locate information on parenting and on children's education. What trends/themes can you identify in the information you retrieve? How are children being socialized today? For example, what does the information you find say about disciplining children, entertaining children, and teaching particular skills to children (for example, reading, math, and so forth)? Do you think that what is considered important for parenting and child rearing has changed from when you were a child? Why or why not? What impact do you think early socialization has on children as they become adults?

2. Much attention has been given to violence and the media. Some argue that as an agent of socialization, the violence contained in the media has a detrimental effect. Using a search engine, locate information on this issue from several sources (for example, social policy, government, academic research, discussion) and compare the information. What reasons can you find for limiting or not limiting violence on television, in movies, in music, and in other media? Is the issue of censorship central in any of this information? Why do people feel strongly about their positions? Is this an indicator of the importance of socialization in society?

INFOTRAC COLLEGE EDITION: READINGS

Article A19155710 / Godwin, Glen J., and William T. Markham. 1996. First encounters of the bureaucratic kind: early freshman experiences with a campus bureaucracy. *Journal of Higher Education, 67,* 660–692. Godwin and Markham describe how the process of socialization enables college freshmen to learn the ins and outs of campus bureaucracies.

Article A20997422 / Settersten, Richard A., Jr. 1998. A time to leave home and a time never to return? Age constraints on the living arrangements of young adults. *Social Forces, 76,* 1373–1401. Settersten uses survey data to explore age-based norms regarding where young adults should live.

FURTHER READING

Suggestions for additional reading can be found in *Classic Readings in Sociology,* bundled with this book. If you purchased a used copy of this book that does not include this custom-published reader, go to www.sociology.wadsworth.com for ordering information.

RECOMMENDED SOURCES

Chudacoff, Howard P. 1989. *How Old Are You? Age Consciousness in American Culture.* Princeton, N.J.: Princeton University Press. Chudacoff provides a fascinating look at the development of age-based norms in the United States from the mid-1800s on.

Elkin, Frederick, and Gerald Handel. 1989. *The Child and Society: The Process of Socialization,* 5th ed. New York: Random House. This is a very readable textbook covering all aspects of socialization.

Fine, Gary Alan. 1987. *With the Boys: Little League Baseball and Preadolescent Culture.* Chicago: University of Chicago Press. Fine provides a highly entertaining account of his research.

Howard, Judith A., and Jocelyn Hollander. 1997. *Gendered Situations, Gendered Selves.* Thousand Oaks, California: Sage. Howard and Hollander include a detailed examination of many areas related to gender socialization.

Light, Paul C. 1988. *Baby Boomers.* New York: Norton. Light provides a detailed look at the lives of the birth cohort born after the end of World War II, often called the "baby boom."

Lopata, Helena Znaniecka. 1996. *Current Widowhood: Myths and Realities.* Thousand Oaks, California:

Sage. This is an in-depth examination of the final stage of the life course for many women.

Stockard, Jean. 1999. "Gender Socialization" in Janet S. Chafetz (Ed.) *Handbook of Gender Sociology.* New York: Plenum. This is a complete review of contemporary literature on gender socialization.

Stockard, Jean, and Miriam M. Johnson. 1992. *Sex and Gender in Society,* 2nd ed. Englewood Cliffs, N.J.: Prentice-Hall, chapters 6–11. This is an in-depth examination of material related to gender socialization and the life course.

Thorne, Barrie. 1993. *Gender Play: Girls and Boys in School.* New Brunswick, N.J.: Rutgers University Press. Thorne reports the results of her very interesting field study of children on school playgrounds.

Whiting, Beatrice Blyth, and Carolyn Pope Edwards. 1988. *Children of Different Worlds: The Formation of Social Behavior.* Cambridge, Mass.: Harvard University Press. This is a fascinating report about several comparable research projects on childhood socialization conducted in a wide variety of countries.

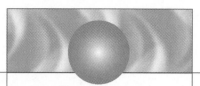

Pulling It Together

Material related to socialization and the life course is found in several other chapters of your text. You can also look up topics mentioned in this chapter in the index.

TOPIC	CHAPTER	
Age, period, and cohort effects	6:	Deviance and Social Control, pp. 131–132
	8:	Racial-Ethnic Stratification, p. 202–203
	17:	Population, pp. 452–453, 461–464
Family interactions	10:	The Family, pp. 257–266
	15:	Education, pp. 402–404
	17:	Population, pp. 470–472
Gender socialization	9:	Gender Stratification, pp. 216–239
Peer groups	6:	Deviance and Social Control, p. 140
	12:	Religion, pp. 320–321
Self-identity	3:	Culture and Ethnicity, pp. 59–60, 62–65
	6:	Deviance and Social Control, pp. 137–138
Role transitions and life course changes	6:	Deviance and Social Control, pp. 145–147
	10:	The Family, pp. 259–261
Social networks	5:	Social Interaction and Social Relationships, pp. 101–104
	6:	Deviance and Social Control, pp. 140–144
	10:	The Family, pp. 257–258, 265–266
	12:	Religion, pp. 320–321
	13:	The Political World, pp. 339–352
	14:	The Economy, pp. 361–371, 376–385
	16:	Health and Society, pp. 425–428
	18:	Communities and Urbanization, pp. 487–492, 495–503
	20:	Collective Behavior and Social Change, pp. 535–536, 544–545
Field research	2:	Methodology and Social Research, pp. 33–34
	3:	Culture and Ethnicity, pp. 63–65
	9:	Gender Stratification, pp. 235–237
	14:	The Economy, pp. 380–383

INTERNET SOURCE

The Center for the Future of Children. This Packard Foundation site offers multidisciplinary research on issues affecting children, such as divorce and health.

Social Interaction and Social Relationships

Linking Social Actors: Social Networks

Opportunities and Constraints in Networks

Social Capital

Creating Social Structure Through Interactions: Three Perspectives

Balancing Costs and Benefits: Exchange Theory

Norms Associated with Our Place in Society: Role Theory

Taking Others' Roles: Symbolic Interaction Theory

Producing Social Structure Through Interactions: Nonverbal and Verbal Communication

Actions and Social Ties: Nonverbal Interaction

Words and Social Ties: Verbal Interaction

Interactions in Groups: The Influence of Status Characteristics

Expectation States Theory

FEATURED RESEARCH STUDY

Overcoming Status Generalizations

Status Attainment: Social Networks, Social Interaction, and Life Chances

RITA GREW UP in a small town in northern Michigan. Her father drove trucks for the local mining company, and her mother worked as a waitress in a small cafe. Because her family didn't have the money or time to take vacations, Rita and her two brothers rarely left their hometown while growing up, and none of them ever traveled outside the state.

From early childhood Rita was a voracious reader. By the end of high school, she had not only read almost every book in the town's library but decided to go to college and learn to be a librarian. With the urging of one of her favorite teachers, she applied to private schools in the East as well as to Michigan State. Somewhat to her surprise, her grades and exceptional test scores brought several scholarship offers. Despite some anxiety about leaving her family and friends, she decided to accept a scholarship from Greenville College, a small, private, and very expensive school in New England.

Rita's first days at Greenville were difficult ones. She was one of just a handful of scholarship students on campus. Most Greenville students came from wealthy New England families and had attended private schools. When Rita met her assigned roommate, Michelle, her eyes nearly popped out. Her parents had sometimes commented ruefully on "how the other half lives," and now she could see what they meant. Rita had never seen such a collection of cashmere sweaters and silk blouses, not to mention Michelle's laptop computer, CD player, golf clubs, and skis. "How," she wondered to herself, "will I ever fit in here? I've never known people like this before in my life! What must Michelle think of my clothes and my old portable typewriter?"

For a week or so Rita wondered whether she'd made a mistake in choosing Greenville. Although Michelle seemed pleasant enough, she was always running off with a small group of friends she knew from prep school, leaving Rita alone to read or take long walks on campus.

One day, as Michelle was about to go out, Rita impulsively asked whether she could go along.

"Why, sure," Michelle replied, raising an eyebrow. "I would have asked you before, but you seemed all caught up in your books."

"And I just assumed I'd be intruding," Rita answered.

"Don't be silly! It'll be fun for my friends to meet you. Come on, you can tell us all about life in Michigan. We've certainly used up all of our own stories by now!"

Once the ice was broken, Rita and Michelle began to become friends. As Michelle and her pals introduced Rita to still other students, some of her homesickness began to ebb away. Still she occasionally felt a bit like an interloper, especially when the talk turned to ski trips, European vacations, and other experiences she had only read about.

Despite her discomfort Rita soon found that she loved the academic part of life at Greenville. Unlike the social scene the classroom environment was very familiar. She knew how teachers expected her to behave, and she had little trouble earning her professors' respect.

Although it took time, by her junior year Rita felt much more at home at Greenville. Part-time jobs gave her enough money to share in a few ski vacations and even a holiday trip to Mexico. Eventually both she and her widening circle of friends came to see one another as much more alike than different, in part because so much of their lives revolved around the school.

Meanwhile Rita's career plans changed dramatically. Noticing her articulateness in class and her skill in logical argument, several of her professors encouraged her to consider taking up law. After some soul-searching Rita decided to abandon her ambition to become a librarian. Thanks to her excellent record and test scores—as well as the strong letters of recommendation from professors—she was accepted to an Ivy League law school.

Once again Rita was faced with adjusting to a new environment, a task that was complicated by resentment from some male students when her work was singled out for praise. After her experiences at Greenville, however, the transition to law school didn't seem so difficult. Meanwhile she kept up with friends from college, some of whose parents held executive positions in major corporations. When she obtained her law degree, these connections helped her land a job as a staff lawyer with an insurance firm in Boston, where her work is once again attracting notice. Even though she is still surrounded by books, her life today seems a world away from the tiny rural town of her youth.

Although you might not have had as many dramatic role transitions in your life as Rita experienced, you undoubtedly have often changed directions in large or small ways, entering new positions at school or work or in your community. In what ways did those changes involve developing relationships with new acquaintances—students, teachers, co-workers, bosses? How did you develop those relationships, and in what ways have they influenced your life course? How are social worlds like the one at Greenville constantly created and re-created by people's day-to-day interactions? How do our daily interactions affect our later life chances?

You can look at this group of skydivers in many different ways, such as the overall pattern they form as they fall, the way that pairs of divers hold on to each other, how they come together to form a pattern, how they separate at the end so that their parachutes don't get entangled, and so on. In the same way, sociologists look at social interactions from a number of different perspectives.

WE CREATE OUR EVERYDAY LIVES through social interaction—the ways in which we communicate with and relate to other people. These interactions have a strong influence on our life course and the opportunities and constraints that we face throughout our lives. Much like Rita and Michelle and the other people at Greenville College, we continually create the social structures in which we live through our interactions with one another. As Simmel put it, linkages or relationships between people—the web of social relations—are the basis of society. Without these connections there would be no social life; there would be no society. Moreover, through these interactions society itself changes. For instance, people fall in love and form lasting relationships, workers might decide to protest their working conditions and strike against their employers, politicians can negotiate peace with other countries or fail at these negotiations and declare war.

When my children were younger we would sometimes take a magnifying glass with us when we went on hikes. Simmel's notion of a "web of social relations" and the different sociological perspectives on interaction often make me think of what we would see when we looked at a spiderweb. When you examine the web from a few feet away, you can see many strands that link one part to another and help keep the web stable and strong. As you move closer, you can focus in on different parts of the web, such as the nature of the threads and the pattern of the connections. If you get still closer and use the magnifying glass, you can look at the intricate connections that link one part to the next. And if you are lucky, you can actually see the spider spinning and creating the web.

A sociological perspective of the world around you can also involve these different views. You can take a long-distance view and focus on the links or connections that bind groups and individuals together. You can move in closer and look at how these connections stay strong and why they exist. Or you can zoom in and examine how people actually build these relationships from one minute to the next. Thus, as with all issues in sociology, you can look at social interactions from a macroperspective, from a microperspective, or from some vantage point in between. You can focus on social structure—how stable and predictable patterns of interaction are and how social structure influences our interactions with others. Or you can focus on social action—how we actually create and change this social structure through our day-to-day relationships. In this chapter we discuss each of these perspectives. We also look at how characteristics, such as gender and race-ethnicity, can influence our interactions and how our interactions with others influence our life chances in adulthood. Just as some spiderwebs are much more extensive and developed than others, so too do some people have much more extensive and supportive interactions than others.

LINKING SOCIAL ACTORS: SOCIAL NETWORKS

In describing webs of social relations, sociologists often use the term **social networks,** patterns of social

FIGURE 5-1 *Network of Colleges in Greenville's Athletic Conference*

Note: Because all the schools compete in athletics, those ties are omitted from the diagram.

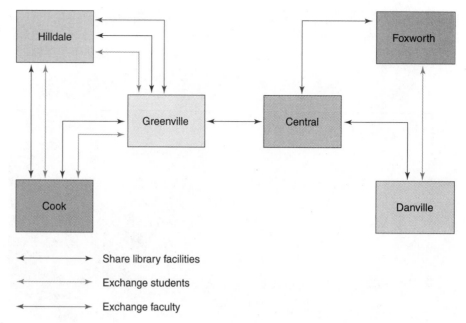

→————————→ Share library facilities

←————————→ Exchange students

←————————→ Exchange faculty

relationships connecting organizations, groups, or individuals. Thus you can think about social networks as occurring on various levels of analysis: between large bodies such as nations or corporations, between smaller groups such as church congregations and kin groups, and between individual people.[1] Whether a network describes individual people or larger groups, sociologists call its elements **social units** or **nodes,** and the relationships between these units **social ties.**[2]

Opportunities and Constraints in Networks

Your college or university is a social unit that is tied to other colleges and universities through athletic competitions, the sharing of library facilities, faculty and student exchanges, and a variety of other means. Your school might routinely play basketball against a nearby college several times a year but compete against one several hundred miles away only once a season or maybe even once every few years. Similarly your school might have many relationships with the nearby school—the sharing of library books and journals, the exchange of faculty and students, the joint use of facilities, and academic and athletic competitions. With other schools your college occasionally might exchange faculty and students but otherwise have very few associations. In turn, these other schools have ties not just with your college but with others as well. By tracing these ties, you could map out the entire web of associations, or the social network, that links these colleges and universities.

Figure 5-1 illustrates patterns such as these with the colleges in the athletic conference that includes

Rita and Michelle's school, Greenville College. Because the schools are in the same conference, they all compete with one another athletically. Yet they vary in their other contacts, with each school tied more closely to some than to others. For instance, the ties between the schools on the left of the diagram (Greenville, Hilldale, and Cook) are relatively strong. A sociologist would say that this part of the network is relatively *dense,* because it has so many ties.[3] In contrast, the other three schools, shown on the right side of the diagram, have many fewer relationships with one another, so this part of the network is far less dense.

Although the network ties shown in Figure 5-1 are between large organizations, they have a direct impact on individual lives. They provide opportunities for faculty and students at the various schools while imposing constraints or limits. For instance, Greenville College has ties with libraries at three other schools, while Danville City College has library exchanges with only Central State. As a result a student at Greenville probably has access to a much wider range of books and periodicals than does a student at Danville. Similarly all the schools except Central State have a student exchange program with at least one other school. Thus Central students who want to take courses at other schools face constraints that they might not if they were elsewhere.

The notion of social networks and their associated opportunities and constraints can also be applied to individuals. You are linked to many other people in your life: friends, family, co-workers, schoolmates. These people in turn are linked not just to you but to many other people. By stepping

We are linked with other people through social networks, much as telephone lines can link one household to another. Do you think that social networks in this rural area might differ from those in more heavily populated, urban areas? In what ways? We will look at this question more closely in Chapter 19.

back from your everyday life and taking a long-range view, you could see the linkages you and your acquaintances and friends have with other people, much like the linkages between colleges illustrated in Figure 5-1. The ties that you have with other people, and in turn the ties that your friends and acquaintances have with others, help provide you with both opportunities and constraints within your daily life.[4]

Some of your linkages with other people are quite strong, involving primary group relationships. These ties are very important to you in your everyday life. When something bad happens, when you are sad or lonely, or when you have good news to share, you probably call a friend or a close family member. The people with whom you have strong ties provide emotional support in your daily life, especially in times of stress, and they also help you celebrate when good things happen.[5]

Other ties in your network are weaker: casual friends and acquaintances on campus and in your neighborhood, and co-workers you rarely see away from the job. These weak ties in your social network might not provide emotional support, but they can help you find resources and new opportunities.[6] For instance, if you are new to a community and want to find an auto mechanic or a doctor, you will probably ask around for recommendations. You are much more likely to trust a doctor or mechanic who has been recommended by someone you know, even only casually, than one whose name you picked out of the phone book. Similarly, if you need to find a job, you will probably ask many acquaintances whether they are aware of any possibilities. If you

know more people—that is, if you have many weaker ties—your chances of finding someone who knows where you might find a job are higher. The popular idea of "networking" to find a job refers to this process of using your social ties to broaden your opportunities.

Social Capital

Sociologists use the term **social capital** to refer to the resources or benefits that people gain from their social networks. The term *capital* is often used to refer to a supply of money or wealth. By adding the modifier *social,* sociologists suggest that we also have noneconomic resources—we accumulate advantages and opportunities from our social ties, our relationships with other people.[7]

Some of us have more social capital than others. If you have a dense network of strong ties, you will be well supported in times of emotional crisis. If, however, you have few strong ties, these times of crisis can be much more difficult. Similarly, if you have a large network of weaker ties, you will have many opportunities and advantages in tapping resources within your environment and community. If, however, you have only a few relationships with others, your supply of social capital is limited and your opportunities are more restricted. For instance, Rita's career opportunities as a lawyer were greatly enhanced by the friendships she made at Greenville. In contrast, her brothers, who attended the local community college, could not develop social networks that included people with the material advantages of most of the Greenville students and thus did not enjoy the same opportunities. In general, sociologists have found that the nature of our social networks and social capital has a large effect on our life chances in school, at work, and in our communities. We will return to these ideas at several points in later chapters.

When you apply the concept of social networks to an examination of the linkages you have with others, you can begin to understand an important aspect of the social structure in which you live. Although we all have a part in creating it, this structure of linkages between people and groups in turn affects our choices. It provides the framework that helps to define the opportunities we enjoy and the constraints that hinder us. How is this framework constructed? What influences the relationships we form, and what keeps them going? To use the analogy of the spiderweb, how do we spin the individual threads in our webs of social relationships? We turn now to three explanations of how social networks are established and maintained.

CREATING SOCIAL STRUCTURE THROUGH INTERACTIONS: THREE PERSPECTIVES

Of the many theories that have tried to explain how we establish social ties, three general explanations have been most important. Each of these was developed in the United States in the first half of the twentieth century. The first, exchange theory, began as a reaction against what was seen as an overly macrolevel view of the world in the structural functionalist perspective. The other two, role theory and symbolic interactionism, developed out of the often inductive and fieldwork-oriented tradition of the Chicago School.

Each of these perspectives represents a slightly different take—a different camera angle—on the same scene. Exchange theory focuses on the dynamics of interpersonal relationships and the reasons underlying individuals' choices; role theory looks at the social norms that influence people's actions; and symbolic interactionism looks at the thought processes of social actors. All of these perspectives can help us understand how social structure is created by individuals through their everyday interactions with others.

Balancing Costs and Benefits: Exchange Theory

George Homans (1910–1989) taught at Harvard beginning in the 1930s, the same period in which Talcott Parsons developed the tenets of structural functionalism. However, Homans was not a fan of structural functionalism or macrolevel explanations of the social world. Neither did he think that we can learn much about society by looking only at the ideas and beliefs of individual people, as he thought psychologists did. Instead, Homans suggested, we can best understand society by examining interactions between people, specifically the patterns of rewards and costs that occur within these relationships.

Inspired in part by the earlier work of Simmel, Homans developed these ideas into what became known as **exchange theory.** According to exchange theory, social action is an ongoing interchange, or exchange, of activity between rational individuals who decide whether they will perform a given action based on its relative rewards or costs. By "rewards" and "costs" Homans was referring not only to money and other goods but also to nonmaterial things such as love, esteem, approval, and affection. Given this understanding of rewards and costs, he believed that all people are rational actors who look after their own interests.[8]

This perspective can help us understand the interactions between Rita and Michelle described at the beginning of the chapter. Rita was reluctant to approach her new roommate because she felt different from her and feared rejection, a cost she didn't want to pay. Michelle might have felt that getting to know Rita would cost her time to interact with her own friends. Ultimately, however, Rita decided that the potential rewards of approaching Michelle, such as making new friends, outweighed the possible costs of rejection. Michelle then rewarded her with a friendly response, and so Rita continued the interaction.

Exchange theorists suggest that before we decide what to do and say, we weigh the possible costs and benefits on a mental scale. Guiding these decisions is the notion of **distributive justice,** or the **norm of equity,** the belief that things should be "fair." That is, we tend to believe that we and others should receive rewards that equal what we contribute to a relationship or interaction.[9] Much the way diplomats negotiate terms of a peace treaty, trying to produce agreements that each side believes is fair, so too do each of us, in our daily lives, decide what we will do and how we will do it based on what we think is fair or equitable.

Social exchanges almost always involve an element of power, for some people might derive more benefit from a particular relationship than others do. When sociologists use the word **power,** they mean that one social element (either a group or a person) can get another element to do what it wants. You have greater power in a relationship when the other person is more likely to accede to your wishes than the other way around. Exchange theorists say that those who depend more on a relationship than the other party does, those for whom the relationship is more important, have less power within that relationship. Thus the idea of power doesn't make sense unless we think about it as a relationship between two or more elements of a social network. You cannot be powerful all by yourself.

Power differentials are often obvious. For instance, as individuals employees almost always have less power than employers; students have less power than teachers; prisoners have less power than guards. Those with less power will comply with the requests of those with more power because the former depend on the resources the more powerful parties provide, such as paychecks, grades, or, in the case of prisoners, even food and water. However, when the demands of the more powerful parties become unreasonable, those with less power will object in whatever ways they can. For instance, underpaid workers might band together to demand higher wages, having more power than their employer only when they act as a group. Sometimes the less powerful are forced to use less obvious means of resistance. For instance, under slavery African Americans would feign illness in the fields or on the auction block to reduce the labor they could perform or their selling price. Even though these actions did not end slavery, they helped to restore some sense of equity within the relationship by diminishing, at least to some degree, the strength of the more powerful slave owners.[10]

But power also can be less obvious. Relationships that appear to be equitable usually involve a balance of power, a situation in which both parties are equally strong or in which some type of compromise has been reached. Power is also a factor in relationships that we ordinarily think of as based on affection and loyalty, such as families and friends. The fact that power exists in these relationships often becomes apparent only when the compromises that keep these situations equitable are disrupted.[11]

A few years ago there were stories in the media about mothers who went on strike against their families. For many years they might have cooked and cleaned for their families and seen the situation as equitable. But one day they simply got tired of doing all the work and announced to their families that, until other family members decided to pitch in, they would not do any more chores. In the exchange of household duties, these mothers were more powerful than either their children or husbands, for they were much more experienced and skilled at cooking and cleaning. As a result the family soon capitulated. The children and husband began to help more around the house, and a sense of equity was restored.

What do you think the woman pictured here has considered in her decision to go on strike? What power differentials might be involved in this situation?

You can use exchange theory to analyze all types of social relationships, including those in the workplace, in politics, in leisure activities, between nations and corporations, and even between the police and suspected criminals. Consider what happens when the police arrest several suspects in a crime. Typically police meet with the suspects separately in hopes of getting one to confess and "rat on" his or her confederates. The police have more power in this situation, because they can threaten the suspects with imprisonment if they don't cooperate and promise leniency if they turn "state's evidence." At the same time, however, the suspects can use their knowledge of the crime (assuming they committed it) to bargain for more lenient treatment. Both the police and the suspects try to balance out the relative costs and benefits of their actions. The police ask themselves how much the potential sanctions should be reduced for the information; the suspects ask themselves how much the reduced sanctions compensate for ratting on their confederates.[12]

Our relationships with others, even those between criminal suspects and police, are dynamic.

Even though we might not realize it, we constantly weigh the costs and benefits of our actions before deciding what we will do. Through relationships based on social exchange—at home, at school, at work, in our communities—we establish linkages with other people, the individual threads in the web of our social network. We build relationships and the patterns that underlie social life. Using the perspective of exchange theory, we can see everyday exchanges between people as a basis of social structure.

Norms Associated with our Place in Society: Role Theory

Exchange theory looks at the linkages between units in a social network, the relationships that connect each position in the web of social relationships. Some sociologists focus not only on the relationships between units but also on the nature of the units themselves and the norms associated with them. This perspective involves the ideas of status and roles, terms that we encountered in our discussion of socialization in Chapter 4. These concepts were first developed and applied by early sociologists such as Robert Park and other members of the Chicago School.[13]

Roles and Statuses These early American sociologists recognized that all of us occupy various kinds of **statuses,** or positions within the social structure. For instance, in your school some people are professors, some are students, some are administrators, some are clerical and support workers. Sociologists often talk about two types of statuses. One is **ascribed status,** positions that we attain through circumstances of our birth and that we can do nothing to change, such as race-ethnicity, gender, or family economic status. The other is **achieved status,** positions that we attain through our own efforts, such as our education or occupation.

People in each status, whether achieved or ascribed, are expected to do different kinds of things. Consider the achieved statuses of professor and student. Professors teach, do research, and evaluate students' work; students go to classes, study, and learn. Similarly, the ascribed status of gender also has associated expectations, with males often expected to behave differently than females. Together these norms or expectations associated with statuses are called **roles.** Roles define the obligations and expectations associated with a given status—that is, what others expect us to do as a student, an employee, a mother or a father, a daughter or a son.[14]

Just as exchanges involve two people and social networks involve linkages between units, social roles are always *reciprocal.*[15] That is, for every defined so-

cial role there is a reciprocal role, one that has a corresponding set of expectations. For instance, you can't play the role of student without also having someone who plays the role of teacher; you can't be a parent without having a child; you can't be a husband without having a wife; you can't be a co-worker without a work partner. Each of these sets of reciprocal roles involves corresponding sets of expectations and obligations. For example, parents are obliged to take care of their children, and children are expected to obey their parents.

Collectively the many people within our society, all interacting in various complex networks of reciprocal roles, may be seen as creating the patterns that underlie social life, or social structure. From the perspective of **role theory,** we can say that life is relatively predictable from one day to the next, one situation to the next, because we generally act in ways that conform to our social roles.[16] In each of our interactions, as we act in ways that conform to social roles, we *legitimate* them; we help promote the view that they are valid. In other words, by fulfilling social roles we help to make them real. In this sense we are constantly constructing and maintaining the social structure.

Roles can also be seen as a way to simplify our understanding of the world and to smooth our interactions with others. Recall, for instance, how Rita felt comfortable in the college classroom environment. She understood and easily fulfilled the obligations associated with her role as student. Similarly, suppose you are standing outside a class at the beginning of the term and you start to talk to the person beside you. Knowing whether that person is a professor or a student is very important in guiding your actions. To a fellow student you might say, "Have you heard anything about how hard this class is? Have you heard whether it is interesting?" But you probably wouldn't ask these questions of your professor. Likewise you would be unlikely to ask a student a technical question about the prerequisites for the course or the contents of the syllabus.

Just because we fulfill roles does not mean that we all behave in exactly the same manner. Quite to the contrary! Each student plays the role of student in a more or less unique way. Some students are anxious; others are relaxed. Some study hard from the first day of class; others wait until the last minute to cram for tests. In all your roles you have choices about how you will behave, how and even whether you will fulfill the obligations and expectations associated with being a friend, a worker, a child, or a parent.[17] Nevertheless, the statuses we hold and the roles we play greatly influence our social behavior. Although we might think of ourselves as doing whatever we please, most of the time we act in accordance with the demands of our social roles.

Although roles help to simplify and define social structure, they also create conflict. You can't always fulfill all of your assigned roles or even meet all the expectations associated with a single role. For instance, sometimes the norms or obligations associated with the statuses you fill are incompatible; sociologists call this situation **role conflict.** Suppose you are the supervisor of a work crew and one of your friends is assigned to the crew. You now are placed in a situation in which you must play the role of both supervisor and friend. Each of these roles contains a variety of behavioral expectations. As a friend you are expected to support your pal when difficulties arise; as a supervisor you are expected to treat all workers equitably and to make sure that the work gets done. If your friend starts to mess up on the job, you are faced with a conflict: Do you support your friend, or do you treat your friend as you would other co-workers? The legitimate and usual expectations associated with the roles you are fulfilling simply don't match up. Because the conflict is inherent to, or part of, your roles, it is often said to be *structural* in nature.[18]

Sometimes, too, things don't work out as planned. Your performance of your role is flawed, and something goes awry. The sociologist Erving Goffman (1922–1982) was especially interested in these situations. By looking at such situations, he developed an analysis of how we play our social roles and continually create social structure.

Goffman's Dramaturgical Theory In embarrassing moments, when people aren't meeting the usual expectations of their roles, others can show remarkable tact. Perhaps your professor appears for class with his fly unzipped or with her slip showing. As a student you might quietly suggest that the professor zip his pants, or you might ignore the lace showing below the skirt. The professor would try to maintain an air of dignity, and class would go on. Similarly President George Bush once suffered from the stomach flu at a diplomatic dinner and, in full view of television cameras, threw up on the prime minister of Japan. Yet the diplomats at the state dinner quickly helped the president; they tried to cover up his embarrassment, and the dinner proceeded as if nothing had happened.

In a variation of role theory called **dramaturgical theory,** Goffman uses the metaphor of a drama to explain how individuals play social roles and thus produce social structure. We all play our parts, and if something starts to go awry, we try to cover it up or help to make it right so that the "play" can go on.

Much like an actor will stage-whisper forgotten lines to a fellow performer, so we cover up the failure of others to perform their expected roles. In this way each of us, through our everyday actions, conspires to keep the social structure intact.[19]

Dramaturgical theorists sometimes talk about the **frame** of interaction, the setting in which interaction occurs and the norms that apply to a given setting. Just as a movie is made up of a series of pictures, or frames, we continually interact in different situations. Once we understand the frame of interaction, we can adapt to the appropriate roles for the setting. For instance, in the status of student you might interact in many different frames, ranging from the classroom to a professor's office to informal discussions with friends in a dorm. Each of these frames includes a set of norms that indicate what behaviors are appropriate and what roles you should play.[20] In the classroom or a professor's office, the student role involves expectations of attentiveness to the professor and a scholarly demeanor; in the dorm the student role might involve raucous bantering with fellow students and even joking about teachers and assignments.

Goffman described how part of the drama of role playing involves the ways in which each of us uses **impression management,** manipulating the impression or view that others have of us and giving out cues that we hope will guide an interaction in a particular direction. Impression management can involve the way we try to look, the way we act, and even the things we choose to use or place around us, much as actors use props. For instance, a few years ago several books instructed job aspirants how they could "dress for success." These books described in great detail the types of clothes that one should wear to give the impression that one was a suitable candidate for career advancement. Trial lawyers often engage in impression management as they feign anger or outrage simply to win the sympathy of a jury. They plan when they will lose their temper, how long they will display their anger, and when they will recover— all calculated to make a tactical point. Professionals in many fields, such as law, medicine, counseling, and even music teaching, display copies of diplomas on their office walls, and professors line their offices with books to bolster the impression that they possess superior knowledge.[21]

We might be much more concerned with impression management at some times than at others. Goffman distinguishes between two types of frames or settings that affect our behaviors. Some behavior occurs **frontstage,** in front of an audience or group that we especially want to impress or influence. Here impression management is important. Other behavior

What do you think is happening in this classic photograph by Henri Cartier-Bresson?
What roles are being played? How are the people pictured using impression management?

occurs **backstage,** where impression management is not needed. Here we aren't as concerned about the impressions that we want to convey and may even display behaviors that are quite contrary to those displayed frontstage.[22] For instance, waiters in a restaurant might be calm, courteous, and gracious while serving customers but yell at one another and at the kitchen staff, slop food around, and even make fun of the customers when out of their sight. Impression management can be stressful, and only backstage can people let their hair down and release some of the tensions that build up while they are frontstage. For example, students can make fun of their teachers, doctors can joke about their patients, and teachers can laugh about the stupidity of their students.[23]

Applying role theory, we can see how the patterns that underlie social life are based on the statuses we hold and the roles associated with these statuses. Each day, as we play our social roles, we continually reconstruct the social structure, legitimate the roles we play, and make the social structure real. Role theory illuminates the linkage between social action and social structure. Our social actions are highly influenced by the structures in which we live—the statuses

we hold and the roles attached to these statuses. At the same time, our social actions—the ways in which we play our social roles—continually re-create and sometimes modify the social structure.

Taking Others' Roles: Symbolic Interaction Theory

The notion of social roles can help us understand the expectations and obligations that guide our day-to-day decisions. Other sociologists from the Chicago School have tried to explain what happens when we act out our social roles. These sociologists often drew on the work of George Herbert Mead, some of whose ideas were introduced in the previous chapter, in developing the theory of symbolic interactionism.[24]

According to **symbolic interactionism,** social interaction involves a constant process of presenting and interpreting symbols through thinking about what another person is trying to communicate through the use of symbols. This perspective has two basic tenets. First, symbols are basic to social life. As you will remember from our discussion of culture in Chapter 3, a primary reason that humans have been

BUILDING CONCEPTS

Exchange Theory, Role Theory, and Symbolic Interactionism

Exchange theory, role theory, and symbolic interactionism all help us understand how social structure is created by individuals through their everyday interactions with others.

Theory	Characteristics	Example
Exchange theory	Focuses on the dynamics of interpersonal relationships and how individuals decide how to act by evaluating the costs and benefits involved.	A suspect arrested in a crime might decide to "rat on" a co-conspirator if she believes the police will grant her leniency in return for the information.
Role theory	Looks at the social obligations and expectations associated with the statuses people hold.	When children obey their parents, they are fulfilling their social roles and legitimating the roles both of their parents and themselves.
Symbolic interactionism	Looks at how individuals continually interpret communications (symbols) presented by others and base their actions on these interpretations.	When someone calls to you in a loud voice, you interpret this action and base your response on this interpretation.

able to develop such a marvelous array of cultures is that we have very extensive language systems. The words in these languages are symbols, representing not just concrete objects in the world around us but also thoughts and emotions. In addition to words, symbols include expressive gestures such as shaking a fist, nodding one's head, or smiling. Symbols also can involve pictures; items that have religious or patriotic meanings, such as a flag or a cross; and even colors, such as the traditional black to symbolize mourning or white to symbolize purity.

Second, in our interactions with others we continually present and interpret symbols. When someone says something to you or displays some other type of symbol, such as smiling or nodding, waving a flag, or dressing in black, you attribute meaning to the person's act. A symbolic interactionist would say that you *take the role of the other person*. You think about what that person must be thinking or trying to

tell you, and you interpret the symbol. Your response, the action that you take, results from your interpretation of this symbol.

Suppose you are reading the newspaper in class. Your instructor notices this, catches your eye, and frowns at you. This action is a symbol, and you will probably interpret it to mean that the instructor is annoyed. You then must decide what type of action you will take in response to this symbol. If you want your instructor to see you as a hard-working, responsible student, you will probably put the newspaper away and try to look very attentive during the rest of the class. Your instructor in turn could interpret this action as meaning that you are paying attention to the lecture and might reward you with a smile. This simple interaction follows a chain: One person presents a symbol; the other interprets the symbol's meaning and then responds based on that interpretation.

You can see how important role taking is when you think about what happens when you misinterpret the meaning of symbols. Suppose you weren't really reading the newspaper in class because you were bored, but instead were searching for something you had recently read so you could ask a question related to the day's lecture. You would probably feel hurt that your instructor misunderstood and might try to explain after class what you were really doing. Your instructor might feel bad for mistaking your actions. Each of you would try to help the other understand your views and would probably see this mutual understanding as important in allowing you both to continue your roles as helpful professor and diligent student.

Misinterpretation of symbols can have potentially disastrous effects in many situations. Today many businesses develop relationships with other corporations throughout the world. Similarly governments often negotiate over policy issues ranging from the oceans to air travel, immigration, regulation of imports, and deportation of suspected criminals. In each case people from different cultures must communicate with one another regarding difficult and sensitive areas. If negotiations are to go smoothly, participants must correctly interpret what each party is saying and maintain mutual respect. Yet, because people from different cultures use symbols that might have no common meaning, the chances for miscommunication are increased. For instance, in many Asian cultures it is considered polite to present and receive gifts with both hands. In contrast, Muslims believe that the left hand is unclean and should not be used to pass objects. Thus they could be highly offended by the Asian way of gift giving or receiving. In some cultures, such as the United States, people cross their legs when they are relaxed and comfortable; in Korea such an action would be seen as unpardonably rude. In the United States and Europe people stand up when they want to honor someone; in Tonga people sit down when they are with a person considered superior to them.[25]

Applying the ideas of symbolic interactionists, we can see that our ability to assume others' roles—to interpret the meaning that underlies the communication or symbols others give us—forms the basis of society. Our ability to understand what others mean when they talk to us and to think about how they must be seeing us is central to cooperative relationships, to linkages with others, to social structure or the patterns that underlie social life.[26] Through this process of continually taking the role of the other, anticipating what others expect, and then responding to those expectations, we build up and maintain, or break down, our ties and linkages with other people.

How can the body language shown here tell you what might be happening in this basketball game?

PRODUCING SOCIAL STRUCTURE THROUGH INTERACTIONS: NONVERBAL AND VERBAL COMMUNICATION

In looking at the social world with a social network perspective, we can see how we are linked with other people and how they in turn are linked with still others. Exchange theory, role theory, and symbolic interaction theory all can help us understand how we produce these linkages through our everyday actions. Yet we can also use our social magnifying glass to get even closer to the web of social relationships. We can look at the actual content of interactions—what people do (their nonverbal interactions) and what people say (their verbal interactions)—to gain further insight into how we spin our webs of social relations.

The theories of interaction discussed in the previous section were generally developed in the early twentieth century. In contrast, work on nonverbal and verbal interactions has developed most rapidly in recent years, aided in large part by the technology of photography, audiotape and, most recently, video-

:-)	smile	:-o	uh oh!
:-(	chagrin, sad	:-<	very sad
;-)	wink	[:-)	wearing a Walkman
:c)	pig-headed	:-9	yum
:-,	hmmm	:-#	lips are sealed
8-)	wearing glasses	:-I	grim
:-&	tongue-tied	:-[	pouting
I-O	yawning	:-II	mad
:-D	laughing	>-II	very mad
:-/	skeptical	8-o	surprised

FIGURE 5-2
"Emoticons" as Paralanguage

tape. Through these media social scientists can record communications and repeatedly study them to locate common patterns and characteristics.

Actions and Social Ties: Nonverbal Interaction

Nonverbal communication refers to all the ways in which we send messages to others without words, including posture and movements, facial expressions, clothes and hairstyles, and manner of speaking. Contrast the way a computer-generated phone message "talks" to you—in a flat, expressionless voice—with the way you communicate in face-to-face interaction. You change volume and tone of voice, emphasize certain words, look at others or avoid eye contact, gesture, slouch, and so on. In speaking, your tone can be deadly serious or playful. You might sound hostile or comforting. These aspects of your speech are referred to as **paralanguage,** nonverbal aspects of speech, such as tone of voice and emphasis on words. With different paralanguage—for example, saying "Thanks a lot" in a sarcastic rather than sincere tone of voice—you alter or even reverse the literal meanings of your words. Paralanguage is so important in our communications that people who communicate via computers have developed symbols—often called "emoticons"—to add to their messages indicating how the message should be interpreted. Figure 5-2 shows some emoticons. Different symbols attached to the same computer messages could result in entirely different meanings.

You also enhance or change verbal meanings with your **body language,** your physical movements, postures, and gestures. Do you shake your fist or do you hold your hands open in a welcoming gesture? Do you stand far away from someone or do you move closer and establish eye contact? These actions are symbols that provide information to others, helping them to interpret your meanings and intentions.[27]

When we talk with other people, we can often control our paralanguage and body language, deliberately adopting a friendly tone or edging away from someone to create emotional as well as physical distance. We find it more difficult, however, to control our facial expressions, whether we are feeling anger or surprise or joy. It is far harder to lie with our face than with our words.

A number of studies by Paul Ekman and his associates suggest that facial expressions and their association with certain emotions are universal throughout all cultures. These researchers sorted through a large collection of still photos and found several pictures that portrayed people displaying different emotions, including happiness, sadness, anger, surprise, disgust, and fear. The pictures showed both men and women; all of the people shown were white; and, in most of the studies, all were adults. Ekman and his associates also tried to make sure that the pictures were "pure," that is, that they displayed only one emotion, rather than a mixture of feelings. Being careful to randomize the order in which they showed the pictures and using samples of students from the United States, they found a great deal of agreement when they asked students what emotion each picture depicted.

Ekman and his associates then replicated this study by showing the same pictures to people in many different cultures and asking them to describe the affect, or emotion, being portrayed. They showed the pictures to people from industrialized countries or regions, such as Brazil, Germany, Greece, and Hong Kong, and to people from isolated villages in New Guinea and Borneo, who would be very unlikely to have been exposed to media from the United States or even to have seen the kinds of people shown in the pictures. No matter where the respondents were from, almost all of them described the same emotional responses as the students in the United States had when they saw the pictures.[28] Other research shows that children who have been blind from birth smile, frown, and show anger on their faces in the same way as sighted children do.[29] All these findings suggest that humans have a biological predisposition to certain facial expressions. Our central nervous systems are wired in such a way that when we have an emotional response, we display our emotions in our faces. Because this is a physiological response, it is very difficult to control or alter.

Even though people in all cultures display emotions on their faces in similar ways, the type of interactions that produce a certain emotion, and the extent to which people display their emotions, can vary a great deal from one culture or subculture to another. For instance, among young children in the United States, especially boys, bodily sounds such as emissions of gas and loud belches are the source of great merriment; in other groups, such as secretarial pools or church socials, such noises usually produce shock and embarrassment.

Similarly people are much more likely to smile at strangers in some cultures than in others. If you grew up in the United States, you probably often smile at people you meet casually, such as sales clerks and fellow students, without thinking about it. If others fail to respond in like manner, you might well think that they are rude and unfriendly. Yet this type of display is unthinkable in other cultures and can, in fact, be quite disconcerting to international students newly arrived in the United States. Consider the comments of two students, one Arab and one Japanese:

> When I walked around the campus my first day many people smiled at me. I was very embarrassed and rushed to the men's room to see if I had made a mistake with my clothes. But I could find nothing for them to smile at . . .

> On my way to and from school I have received a smile by non-acquaintance American girls several times. I have finally learned they have no interest for me; it means only a kind of greeting to a foreigner. If someone smiles at a stranger in Japan, especially a girl, she can assume he is either a sexual maniac or an impolite person.[30]

Cultural myths, folklore, and life experiences also help to define what elicits such emotions in the first place. If you grew up in traditional Ireland, for example, you might be terribly frightened at the thought of a "banshee," a female spirit whose wailing is said to warn family members of an approaching death. If, however, you grew up in contemporary California, you would probably laugh at this notion. In war-torn areas of the world, children might shrink in terror when they hear an approaching plane, fearing a bombing attack. This reaction can continue for many years after war has ceased. In contrast, children who have always lived in peaceful areas might look toward an oncoming plane with excited and happy expressions.

Because nonverbal communication is so automatic, we almost never think about our paralanguage, our body language, or our facial expressions. Yet these actions are critical to the building of our social world and our linkages with other people. They are part of the symbols we display to others, part of the way we play our social roles, and part of the costs and rewards in interactions. To use the analogy of the spiderweb, our nonverbal interactions are part of the way in which we spin our web of social relationships.

Words and Social Ties: Verbal Interaction

Although nonverbal interactions contribute significantly to human communication, most of the content of our exchanges with others is verbal. This content involves words and the meanings that speakers and listeners attach to these words. Some social scientists specialize in studying these communications, using what they call **conversation analysis** to examine patterns in verbal interactions. They have recorded hundreds of everyday conversations in a wide variety of settings on video- and audiotape. Then, much as you might look at a spiderweb through a magnifying glass, they scrutinize these conversations, reviewing the tapes many times to look at what people say, how they say it, and especially at what seems to be similar and what seems to differ from one conversation to the next. In this process researchers have identified a number of rules and skills that we all use in conversing with others, often without realizing it. These rules and skills help us maintain our ties with other people and so contribute to the building and maintaining of social structure.[39]

For example, we have rules or norms that govern the way we start conversations. According to one of these norms, we begin conversations with strangers with a question, something we are less likely to do with people we know well. If you look closely at how you open your own conversations, you will probably notice very similar sequences. On seeing a casual acquaintance, you might say, "Hi, how are you?" When seeking help at a grocery store, you might say, "Excuse me, but could you tell me where to find the soy sauce?" While standing in line outside a movie theater, you might say to the stranger next to you, "Boy, isn't this weather awful?"[32] With your best friend, on the other hand, you might begin a conversation with,

"I have great news! I got tickets to the ballgame!" Thus the norms governing how we start conversations allow for different kinds of relationships. If you were to address a stranger or a casual acquaintance in the way you addressed your friend, you would probably make that person very uncomfortable.[33]

Another regularity conversation analysts have identified involves what they call **turn taking,** the process whereby first one person and then the other talks. In fact, we learn to accomplish this so skillfully that we might not be aware of consciously taking turns. We seem to have certain cues or signs, both verbal and nonverbal, that help us know when it is our turn to speak and when it is the other person's. For instance, when you are through saying something, you generally will look directly at your listener, pause in your speech, stop moving your hands, or tack something fairly meaningless onto your last sentence, such as "you know." If you don't want to let the other person talk, you may avoid looking at him or her, tense up your hands, and raise your voice. If you want a turn at talking, you probably indicate this by taking a deep breath, tensing or moving your hands, moving your head away from the speaker, and making fairly loud sounds to indicate your interest such as "un-huh" or "Yes."[34]

Throughout our conversations we continually let the other person know how things are going. Symbolic interactionists describe how we base our actions on our interpretations of another person's responses. Conversation analysts have tried to explain this process by examining what they call **back-channel feedback,** the verbal and nonverbal techniques we use to let others know whether we understand what they are saying. If you grasp what another person is saying, you might nod your head, utter little sounds like "um-hm," and even move your body in a rhythmic motion. If you are talking and your listener isn't responding in these ways, you might add phrases such as "you know" at the end of sentences or ask a question to try to pull your listener back into the conversation.[35] For instance, when I am lecturing to a large introductory class and students suddenly aren't maintaining eye contact or nodding in the way they usually do, I ask them, "Do you want me to give another example?" or "Should I go over these points again?"

Sometimes, of course, conversations don't continue in the way we expect them to. Just as the dramaturgical theorists describe how we try to cover up embarrassing moments so that we can all continue to play our social roles, so, too, conversation analysts point to ways in which we try to repair things that go awry in conversations. Suppose your instructor uses a word you don't understand or refers to a concept

whose meaning you can't quite remember. At first, you might pretend that you understand what she is saying, but you soon realize that you can't keep up your end of the conversation because you can't figure out what she is talking about. Your instructor might also realize that you are having trouble, and both of you will try to repair the situation. You might gather up your courage and say, "Can you explain more what these words mean?" Or she might say, "Do you want me to tell you a little more about the background details on this?" If the two of you don't repair the conversation, you really cannot continue to communicate meaningfully. You will not be ascribing the same meanings to the words each of you say.[36]

Conversation analysts have helped people in a variety of different areas develop better communication skills. For instance, some have studied the interactions that occur when people make emergency 911 phone calls.[37] Researchers have also worked with people who handle customer questions and complaints, such as those who staff toll-free hot lines. Using their understandings of conversation regularities, or patterns, these sociologists have helped workers learn to communicate better. The result might be improved customer relations and, for 911 operators, faster response time and more lives saved in emergencies.[38]

Conversations are complex and subtle performances involving many skills. From a sociological viewpoint these performances are important because we build up obligations and relationships, our social networks, through our conversations. In our verbal and nonverbal interactions we have exchanges with others, we act out our social roles, and we interpret others' responses. In this way we help create the patterns that make up social life, and this social structure in turn influences our interactions.

The cultural elements we discussed in Chapter 3, such as norms and values, ultimately develop from our interactions with one another. So, too, do other aspects of social life, such as ethnic identity and the roles that are attached to other statuses that we hold, for example, our gender and our occupations.[39] In turn, these cultural elements and status characteristics influence our interactions, especially in groups. We turn to this topic next.

INTERACTIONS IN GROUPS: THE INFLUENCE OF STATUS CHARACTERISTICS

Many of our daily interactions occur in **groups,** associations of people who are engaged in some common activity beyond a casual conversation and who have some relatively stable social relationship. Groups can

How can you tell that this is an informal group? How might the status characteristics of people in this group affect their interactions?

range in size from just two people, such as a married couple or two roommates, to several people who carpool together or work in the same office or attend the same class, to employees of a large hospital or a department store. The key element is that the group members have social ties to one another and that they interact in some way that is more than transitory. If you think about your daily life, you will realize that you spend a lot of time in various groups. For this reason sociologists have long been interested in group interactions.

Sociologists usually distinguish between **informal groups,** those that have no specified goals or formalized norms and no set membership, and **formal task groups,** those that have specific goals, established norms, and recognized membership.[40] For instance, a group of friends who go out to dinner or the movies together would be an informal group. A class group assigned to do a project together or a group of politicians and high-level government employees charged with providing options to deal with a foreign policy crisis would be a formal task group.

Sociologists have looked at many different characteristics of groups. One of these is *power,* the ways in which one or more members of a group have greater influence or control over group activities than the other members do. Numerous studies have revealed something you've probably experienced yourself: In all groups, both formal task groups and informal groups, some people talk more than others and have more influence on group decisions than others. These power differences within the group represent a **power structure,** for they are part of the underlying pattern of relationships or social struc-

ture of the group.[41] How do these power structures develop? Why do some people in a group acquire more power than others?

Expectation States Theory

Joseph Berger, a sociologist at Stanford University, and his associates, tried to answer this question.[42] According to their **expectation states theory,** group members' beliefs about what other members can do influence group interactions and the power structure within the group. Central to this theory is the notion of **status characteristics,** the statuses that people hold and the evaluations and beliefs (or characteristics) that are attached to these statuses. Status characteristics include such aspects as our race-ethnicity, occupation, gender, age, education, and even physical attractiveness.

In attempting to explain group dynamics, Berger suggests that some characteristics are **specific status characteristics**—that is, they are closely related to the group's given task. For instance, a presidential foreign policy advisory group charged with developing recommendations after an allied nation had been invaded might pay considerable attention to the ideas of members who have the specific status of military official.

Group power structures also are influenced by what Berger calls **diffuse status characteristics,** those that go beyond specific task-related skills and reflect more general statuses, such as occupation, education, gender, and race-ethnicity. Within society as a whole, these diffuse status characteristics are generally associated with various cultural stereotypes. For

Based on expectation states theory, what would you predict about the interactions in this formal task group? If the participants knew about how diffuse status characteristics can affect interactions, what might they do to try to improve the group's effectiveness?

instance, within the United States many people assign higher status to others on the basis of their occupations and levels of education. In addition, males tend to be assigned higher status than females, and Euro-Americans tend to be assigned higher status than people of color. To a large extent, as we will discuss in subsequent chapters, people who hold these more "favorable" status characteristics also tend to have higher levels of income and more powerful positions within organizations.

Through what Berger and his associates call **status generalization,** these diffuse status characteristics influence group interactions even when they have no apparent connection with the task at hand. Because certain statuses are stereotyped as more favorable, we tend, rightly or wrongly, to generalize all types of favorable characteristics to people who hold these diffuse status characteristics. In other words, people's status characteristics influence the expectations others have of their ability to perform a given task. Rita might have experienced this phenomenon in law school when some of her male classmates reacted as if a woman shouldn't excel in legal work.

These **expectation states,** or beliefs about what others can do, then influence the interactions within task groups, and specifically the group's power structure. Given expectation states theory, we would hypothesize that people with less favorable status characteristics, such as children, people of color, and women, would have less influence on group decisions than would adults, Euro-Americans, and men.

These diffuse status characteristics are most likely to influence group interaction when the participants don't know one another well and have little information about one another's specific status characteristics.[43] For instance, even the faculty at her law school might have assumed initially that Rita would

have less to contribute than her classmates to class discussions, not only because she was a woman, but because her social background was less impressive to them than that of the other students. In time, however, her consistently excellent work overcame these expectations.

Several studies support expectation states theory. For example, a common stereotype is that, compared with men, women talk all the time and dominate conversations. In reality, however, this generally doesn't happen. Social scientists have examined many instances of male-female interaction, often using the conversation analysis techniques described in the previous section. As expectation states theory would suggest, most of these studies show that in mixed-gender groups, the men tend to talk far more than the women do. The topics that men introduce result in extended conversations much more often than do topics introduced by women,[44] and men also are more likely than women to control turn taking through interrupting other speakers.[45] Even though both women and men interrupt each other, one study showed that women allowed men to have twice as much uninterrupted speaking time as men allowed women.[46]

These findings appear in studies of both informal groups and formal task groups. Moreover, similar patterns appear in nonverbal behavior. For instance, in groups composed only of women or only of men, people tend to see the person who sits at the head of the table as the group leader and to look to that person for leadership and ideas. In mixed-gender groups, however, this doesn't happen as much. Even if a woman is sitting at the head of the table, both men and women participants tend to regard one of the men in the group as the leader.[47]

Many sociologists find this support for expectation states theory disturbing, for diffuse status characteristics aren't necessarily an indicator of which group members have the specific skills that are important to the work of the group. Whenever diffuse status characteristics influence who has more impact on a decision, the group's decisions might not be as sound as they could be. For instance, suppose that the presidential advisory group dealing with a foreign policy crisis includes a person from the Department of State who understands better than anyone else the international political ramifications of the alternatives being considered. Suppose also that this person is a woman. If her opinions and ideas are given less attention because of this diffuse status characteristic, the advisory group runs a great risk of reaching a less-than-optimal decision, one potentially affecting the lives of many people. The same point applies to other status characteristics besides gender. For example, the advisory group might discount the views of a person of color because of his racial-ethnic background or the views of a nonveteran because of his lack of military experience. Whatever particular diffuse status characteristic influences group interactions, the potential is there for the group to be less effective and productive than it could be, for it has not used all the available skill and talent.

Common Sense versus Research Sense

COMMON SENSE: Women are more talkative than men and dominate conversation.

RESEARCH SENSE: Studies have shown that in many instances of male-female interaction, men tend to talk far more than women do. How would "expectation states theory" explain this?

FEATURED RESEARCH STUDY

OVERCOMING STATUS GENERALIZATIONS

Because diffuse status characteristics can make groups less effective, some sociologists have asked how status generalizations can be minimized. That is, how can groups be structured so that diffuse status characteristics are less likely to influence interactions? Over the years social scientists have suggested a number of solutions to this problem. Currently the most promising method seems to involve giving group members information that contradicts the stereotypes or expectations associated with diffuse status characteristics. In other words, status generalization can be overcome if the group members know more about specific status characteristics, the actual skills that group members bring to a task.

One of the most influential studies in this area was conducted by Elizabeth Cohen and Susan Roper.[48] Previous work had shown that diffuse status characteristics influence group interactions even among students in junior high schools. In particular, Euro-American students tend to dominate when African-American and Euro-American students work together on a task, even when the students have similar family backgrounds and attitudes toward school. Cohen and Roper devised an experiment to test ways to counter this status generalization.

The researchers assigned junior high–age boys to four-person groups, with two African-American and two Euro-American students in each group. These groups were then divided into three experi-mental conditions, as shown in Figure 5-3. All of the African-American students were taught how to build a radio. For students assigned to two of the conditions (B and C in Figure 5-3), the African-American students' competence in radio building was demonstrated through a videotape to both the Euro-American and the African-American students and through high praise by the experimenters in front of all the students.

The African-American students in these two groups then taught the Euro-American students in their groups how to build a radio, and their skills in teaching were again reinforced through reviewing videotapes. Thus, in these two experimental conditions (B and C), the African-American students' skills at radio building were clear to all the study participants and were reinforced not just by the experimenters' statements but also by their skill as teachers. In condition A the African-American students taught staff members how to build a radio but initially had no contact with the Euro-American students assigned to their group.

The experimenters then attempted to manipulate students' expectations in the following way. The students in condition C, but not in conditions A and B, were told that the skills involved in building radios and teaching other students how to do it would also influence their ability to play a decision-making game called "Kill the Bull." In other words the students in condition C were told that the African-

	African-American students taught radio building to Euro-American students	Students told radio-building skills would generalize to game
Condition A	no	—
Condition B	yes	no
Condition C	yes	yes

FIGURE 5-3 *Cohen and Roper's Research Design for the Study of Altering Expectation States*

American students' demonstrated skill in radio building also applied to another specific area. If Cohen and Roper had developed a way to counter status generalization and the influence of the diffuse status characteristic of race-ethnicity, the students in condition C would be expected to have more equitable interactions and group power structures than those in the other two conditions.

As the students played the game, the experimenters measured the amount of influence each student showed in the group by counting the number of ideas they suggested and the actions they initiated. After the game was finished, the students were also asked, in individualized questionnaires, to choose which of their group members appeared to be a leader.

The results were striking. Students in condition A, in which the African-American students had not demonstrated their radio-building skills to the other students, behaved the way expectation states theory would predict. Euro-American students dominated the interactions and were more often chosen as leaders. In contrast, in both conditions B and C, the African-American students were much more active participants and were chosen much more often as leaders. The effects were most striking in condition C, where the students had specifically been told that radio-building skills were relevant to the game. In this condition almost two-thirds of the students picked an African-American student as leader, and an African-American student was generally the most active participant in each group.

These results suggest that it might indeed be possible to counter status generalization, but that doing so requires that all group members be aware of the specific status characteristics that relate to the task at hand. Diffuse status characteristics don't always have to influence group interactions, but to counter their influence, group members must have access to other information.

Cohen has spent many years extending the work described here and applying it in the educational field. Based on replications and extensions of this work, she and her colleagues have developed curriculum and instructional techniques that teachers can use to counter diffuse status expectations in the classroom. Today's schools often include children from different racial-ethnic backgrounds, with different language capabilities, and with different levels of ability and, unfortunately, many children are not doing well in school. The techniques developed by Cohen and her associates, which are based on expectation states theory, are helping to change this situation. They provide concrete ways in which teachers can help students realize their own capabilities and the capabilities of others without being distracted by diffuse status characteristics. As a result, the achievement of all students, and especially those who are most at risk, appears to be enhanced.[49] To read more about Elizabeth Cohen's work, see Box 5-1.

If social groups are to maximize their production, the skills and talents of all people must be recognized and used. Expectation states theory can help explain why some people's contributions aren't adequately encouraged in group interactions. Cohen and Roper's study, and Cohen's subsequent work, help show how this situation can be altered, for we continually create our social world through our daily interactions. Many sociologists hope that we can develop patterns of interaction that encourage the contributions of people based on their actual skills and qualifications, rather than on irrelevant status characteristics, and thus build a better society.

Sociologists have also examined other ways in which social interactions influence our lives. Instead of focusing on how status characteristics influence interactions, they have looked at how interactions and social networks influence the statuses that we achieve in adulthood.

CRITICAL-THINKING QUESTIONS

1. How might Cohen and Roper's study be altered to examine the effect of diffuse status characteristics other than race-ethnicity?

2. Can you think of how sociologists might use methods other than experiments to test hypotheses derived from expectation states theory?

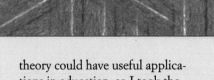

BOX
5-1

SOCIOLOGISTS AT WORK

Elizabeth Cohen

I took the theory from sociology and applied it to the problem.

Elizabeth Cohen, Stanford University, conducted the research with Susan Roper featured in this chapter. Professor Cohen has spent many years studying the effect of status characteristics on social interactions and using the results from this research to help classroom teachers more effectively teach students from all kinds of backgrounds.

How did you become a sociologist, and why are you now in the School of Education? Partly by chance: My undergraduate degree from Clark University was in psychology. When I was ready to do my graduate work, the man who was to become my husband was in Harvard's Department of Social Relations. Psychology at Harvard was all about running experiments on dogs and rats, and that was not my interest, so I switched fields to sociology and received my degree from the Department of Social Relations.

My husband started working in the Stanford sociology department and at that time it was unthinkable that a man and wife be in the same department, so I shifted to education. At the beginning, education was only a variable I used in my dissertation!

How did you get interested in studying the area described in the featured research? There were real issues of theoretical weakness of applied research in education. Status characteristics theory was a powerful tool being developed in the sociology department at the time, and I wanted to show that this

theory could have useful applications in education, so I took the theory from sociology and applied it to the problem of interracial interaction among schoolchildren.

Sociology can help us understand events in our own lives. How does your research apply to students' lives? An example I have seen is a woman entering a male-dominated workplace such as an engineering firm. An important strategy for her is to write a smashing technical memo as early on as possible. This addresses the expectations of those who have high status in the firm and prepares them to recognize her competence.

What kinds of applications has your research had in the world of education? Through the Program for Complex Instruction, thousands of teachers are using applications of status characteristics theory to treat status problems today in California, Arizona, Massachusetts, Sweden, the Netherlands, and Israel. The program is designed to help teachers of academically and ethnically heterogeneous classrooms to teach at a very high intellectual level. Students work in small groups on a multiple ability curriculum, which gives them multiple ways to succeed through group tasks. For example, one project might be for a group to build a Gutenberg printing press. A person with spatial abilities will make a major contribution to this project, while a good reader will make another kind of contribution. In schools, it is necessary to use both a multiple ability curriculum and a special status treatment in order to equalize interaction in classroom groups.

STATUS ATTAINMENT: SOCIAL NETWORKS, SOCIAL INTERACTION, AND LIFE CHANCES

The influence of our environments and social networks continues throughout our lives, and many sociologists have been interested in how these influences work. Given the importance of our occupations and incomes in adulthood, much of their work has focused on how

our early life experiences and interactions influence the kinds of achieved statuses we obtain in adulthood, such as how much education we have, what kinds of jobs we get, and how much money we earn. The result of their work is known as the status attainment model. This model was first developed in the 1960s by Peter Blau and Otis Dudley Duncan and was later modified and extended by William Sewell and Robert Hauser.[50] One version of this model is shown in Figure 5-4.

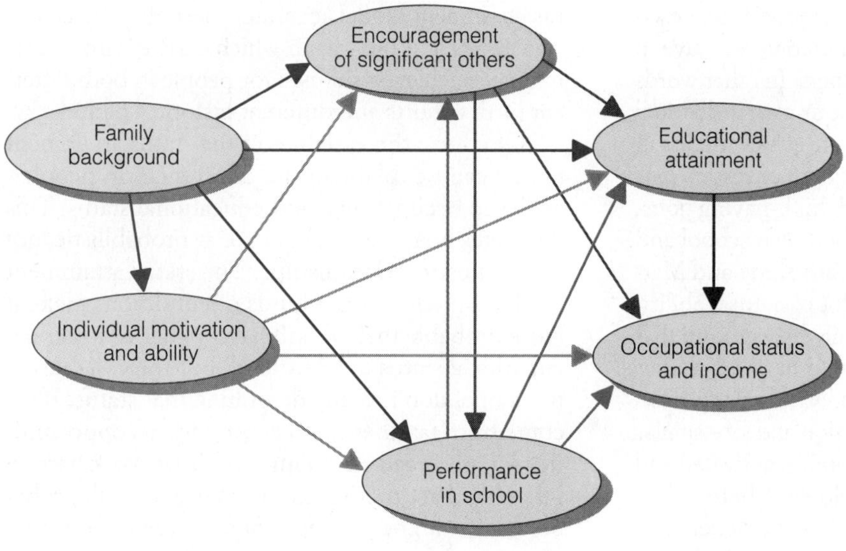

FIGURE 5-4 *Status Attainment Model*

Diagrams like the one in Figure 5-4 are called **causal models,** graphic devices that illustrate sociological theories involving causal relationships between variables. To read causal models, you simply start at the left-hand side of the graphic and follow the arrows. By following these arrows you can see that they describe role transitions in educational and work careers throughout the life course and how our interactions at earlier life stages affect our later experiences.

As you can see, the **status attainment model** suggests that variables such as family background, individual motivation and ability, and interactions with significant others influence the educational levels and occupations and incomes we have in adulthood. For example, if our parents have more education and more money, we are much more likely to have contacts with people who will encourage and support us in school. If we work hard and if we have more natural ability than others, we are also likely to receive more encouragement and to do better in school. In turn, successful experiences in school and the encouragement of significant others influence how long we choose to stay in school, what sociologists call **educational attainment.** For instance, it is unlikely that you would be in college if you had not done well enough in high school to meet the entrance requirements and if your parents, or other people who were important to you, had not expected or encouraged you to attend. Finally your educational attainment influences the kind of job you will have and the amount of money you will earn in adulthood—that is, your **occupational status** and income, respectively. People who have higher educational attainment tend to have higher incomes than those with less education.

Notice how the status attainment model incorporates many of the different ideas and theories about interaction discussed earlier in this chapter. It suggests that our social networks and social capital—the opportunities and constraints that we obtain from our relationships with parents, teachers, and others who are important in our lives—influence our later opportunities. It also suggests that our interactions with these significant others influence our life decisions. For example, when our teachers or parents let us know what they think and expect of us—what kind of students they think we are, how far they expect us to go in school—we interpret these communications and then make our own decisions and act on them based on our interpretation of their views. All these influences occur through both verbal and nonverbal communications in our daily lives.

One of the most important theoretical aspects of the status attainment model is the lines that are missing from the graphic. If you look closely, you can see that no line directly connects family background with occupational status and income. Instead the influence is indirect, going through the other variables in the model. Our family backgrounds influence the kinds of jobs and levels of income we have in adulthood, but this influence comes from the impact that our parents' education and income have on how much schooling we obtain and how well we do in school. Because educational attainment comes between family background and ability and eventual adult status in the model, it is called an **intervening variable,** a variable that comes between a dependent variable and an independent variable in a causal relationship. The status attainment model suggests that our social capital and family interactions influence our work careers only to the extent that they influence our educational careers.

Another important aspect of the model is that each of the influences on the jobs and income we have in adulthood is *independent* of the others. In other words *both* our family background and our own individual motivation and ability influence our achievements in school. For instance, suppose that both of Andy's parents attended college and hold high-paying jobs, whereas Matt's parents only finished high school and have lower-paying jobs. Even if both Andy and Matt are highly motivated and have a lot of natural ability, Andy enjoys benefits from his family background that Matt doesn't and, as a result, would be predicted to have a higher-paying job in adulthood. Similarly, however, if Russell's parents had education and jobs similar to Andy's parents but Russell was more motivated and able, he would be predicted to do even better than Andy. Thus the model suggests that the benefits or handicaps of our family background, individual characteristics, and interactions with others *add together* to influence the nature of our life course.

Sociologists have tested the status attainment model with large samples of people, following the steps of the research process outlined in Chapter 2. Researchers have gathered data to measure variables, such as family background, occupational status, and income, and used complex statistical techniques to see if these data support the theory. By the early 1980s, more than five hundred different studies had examined the model, and many more have been completed since then. In general these replications support the model. The studies suggest that the sta-tus attainment model accurately describes life course transitions and the way in which early environments influence achieved statuses for people in both different birth cohorts and different historical periods.[51]

Of course the variables in the status attainment model can't explain all the differences in people's achieved occupational and educational status. This illustrates how sociologists look at probabilistic, not deterministic, relationships. The status attainment model can tell us what kinds of conditions make it more probable that we will have higher or lower occupational status in adulthood, but the variables in the model don't totally determine that status. If we come from families that can provide us opportunities for advanced schooling, and if we work hard in school and are motivated to achieve, we will probably have a higher occupational status in later life. However, other factors can also affect our work careers. In many ways the status attainment model shows how both social structures and social actions affect our interactions and our life course. Yet, at the same time, each of us acts out our social roles in ways that help influence how these opportunities and constraints affect our lives.[52] (See Box 5-2.)

Like all of the theories discussed in this chapter, the status attainment model focuses on how individuals conform to norms—how people exhibit behaviors that others expect of them. In fact, however, sometimes just the opposite occurs: people don't conform to norms. The next chapter looks at this other side of social life.

SUMMARY

By looking at our interactions with others both from a macrolevel, social network perspective and from more microlevel viewpoints, we can see how our nonverbal and verbal communications help construct our relationships with others. We can also see how our interactions within groups are affected by the status characteristics of group members. Key chapter topics include the following:

- Social networks link elements of society, including large organizations, smaller groups of people, and individuals. Some network ties are stronger than others. Social networks influence our opportunities and constraints and provide us with social capital.

- Three theoretical perspectives have contributed to our understanding of how we form social ties. Exchange theory focuses on the relative costs and benefits of interactions and suggests that people act based on their judgment of an action's relative costs and benefits. Not all people have equal investments in relationships, and those with more to lose tend to have less power in a relationship.

- Role theory suggests that social roles help us simplify our understanding of the world and our interactions with others. Social roles are always reciprocal. Role conflict occurs when roles we occupy contain conflicting or incompatible expectations. Goffman's dramaturgical theory, a variation of role theory, describes how individuals work together to ignore lapses in role performance, thus maintaining the social structure.

- Symbolic interactionism focuses on the meanings we attach to our communications with one another. When we decide how to respond to the symbols, or communications, presented by others, we mentally put ourselves in the other person's position and then base our actions on our interpretation of these symbols. Misinterpretation of symbols can have potentially disastrous effects in social interactions.

BOX
5-2

**APPLYING SOCIOLOGY
TO SOCIAL ISSUES**

*Status Characteristics
and Effective Group
Interactions*

The research discussed in this chapter suggests that our ascribed status characteristics can affect our interactions in groups as well as affect our life chances and opportunities. Expectation states theory and the work of Cohen and Roper suggest that groups might not use all the skills and talents of group members because diffuse status characteristics affect the group's interactions. As a result, groups can lose valuable input of some of their members. The status attainment model shows how our family background influences our life course, independent of our own ability and motivation. People from less advantaged backgrounds, even with higher levels of ability and motivation might not attain positions that use all of their capabilities.

In studying these issues, sociologists have often wondered how the impact of ascribed characteristics could be lessened. How might groups change their inter-

action patterns to try to counter the effect of diffuse status characteristics? What could be done to minimize the effect of family background on status attainment so that all individuals have the same opportunities?

Status attainment is a process that occurs over the life course, and as you will remember from the status attainment model, the effect of family background on occupational status and adult income is indirect, channeled through educational attainment. How can the effect of family background on educational attainment be altered? How can children from all types of families gain equal access to education and receive equal encouragement to succeed? We will return to

these issues later in this book. For now, consider ways in which schools, families, and communities can provide interactions and social capital that enhance the opportunities of all children. In your community do all children have the same type of interaction and encouragement to do well in school? What long-term effects might these practices have?

**CRITICAL-THINKING
QUESTIONS**

1. Think about the groups in which you interact—at school, at work, in organizations to which you belong. Do diffuse status characteristics affect interactions in these groups? Are there ways in which the interactions could be altered to ensure that the contributions of all members are considered?

2. How might Cohen and Roper's findings help the groups you participate in be more equitable?

■ Sociologists have also looked at social relationships at the microlevel by studying both verbal and nonverbal communication. Nonverbal communication includes the paralanguage and body language that affect interactions. Research suggests that facial displays of emotion are universal from one culture to another, although the stimuli that can produce a given emotion vary among societies.

■ Conversation analysts have identified norms that underlie conversations, including the way we start interactions, take turns in speaking, let others know how the conversation is proceeding through feedback, and repair the conversation when it starts to go awry. Our daily conversations are complex and subtle performances that help to create social structure.

■ Social structure also includes various kinds of groups, both informal and formal. Group interactions and power structures can be influenced by both the specific and diffuse status characteristics of participants. According to Berger's expectation states theory the process of status generalization can lead to less effective decision making when groups are influenced by diffuse rather than specific characteristics. Elizabeth Cohen and Susan Roper's research indicates that status generalization can be altered if specific information about the characteristics of group participants and their relevant skills is given to all group members.

■ The status attainment model shows how our social networks and interactions with significant others influence our educational achievement and our occupations in adulthood.

KEY TERMS

achieved status 106	formal task groups 114	role 106
ascribed status 106	frame 107	role conflict 107
back-channel feedback 113	frontstage 107	role theory 106
backstage 108	groups 113	social capital 103
body language 111	impression management 107	social networks 101
causal model 119	informal groups 114	social ties 102
conversation analysis 112	intervening variable 119	social units 102
diffuse status characteristics 114	nodes 102	specific status characteristics 114
distributive justice 104	nonverbal communication 111	status 106
dramaturgical theory 107	norm of equity 104	status attainment model 119
educational attainment 119	occupational status 119	status characteristics 114
exchange theory 104	paralanguage 111	status generalization 115
expectation states 115	power 104	symbolic interactionism 108
expectation states theory 114	power structure 114	turn taking 113

INTERNET ASSIGNMENTS

1. How do sociologists study social networks? What kinds of research and information on networks and network analysis are available on the World Wide Web? Can you locate any on-line journals that are directed toward the study of networks and communication? What issues are discussed in these electronic publications?

2. In what ways do sociologists look at on-line communication and the World Wide Web as social networks? What discussion groups exist that address the influence of computer technology on the way we interact and form social ties? What are considered to be the positive aspects of this technological revolution in communications? What are the negative aspects? Do you think the Internet and the World Wide Web have influenced your ability to communicate and form ties with others? How?

TIPS FOR SEARCHING

You might want to look at on-line networks that have been established for people with specific interests. For example, many networks (mailing lists, on-line discussion groups, lists of resources) are specific to certain groups (racial-ethnic groups, political groups, gay/lesbian/bisexual groups, and so on). Access some of these networks and determine the purpose and goals of the groups.

INFOTRAC COLLEGE EDITION: READINGS

Article A20151735 / Dunn, Thomas P., and Sandra Long Dunn. 1997. The graduate assistant coach: role conflicts in the making. *Journal of Sport Behavior, 20*, 260–272. Dunn and Dunn use role theory to examine the experiences of graduate students who serve as assistant coaches to college teams.

Article A21261649 / Shanahan, Michael J., Richard A. Miech, and Glen H. Elder, Jr. 1998. Changing pathways to attainment in men's lives: Historical patterns of school, work, and social class. *Social Forces, 77,* 231–257. The authors explore how the status attainment process has changed across historical periods.

Article A16659998 / Lal, Barbara Ballis, 1995. Symbolic interaction theories. (Theories of Ethnicity). *American Behavioral Scientist, 38,* 421–442. Lal summarizes symbolic interaction theory and discusses its usefulness in analyzing relationships between racial-ethnic groups.

Article A18888993 / Ekman, Paul. 1996. Why don't we catch liars? (Truth-Telling, Lying and Self-Deception). *Social Research, 63,* 801–818. In this article Ekman, one of the pioneers in the study of nonverbal communication, applies his knowledge to the detection of liars.

Article A19284904 / Wagner, David G., and Joseph Berger. Gender and interpersonal task behaviors: status expectation accounts. *Sociological Perspectives, 40,* 1–33. In this article Wagner and Berger (one of the developers of status expectation theory) illustrate the application of this approach.

FURTHER READING

Suggestions for additional reading can be found in *Classic Readings in Sociology,* bundled with this book. If you purchased a used copy of this book that does not include this custom-published reader, go to www.sociology.wadsworth.com for ordering information.

RECOMMENDED SOURCES

Cohen, Elizabeth G. 1994. *Designing Group Work: Strategies for the Heterogeneous Classroom,* 2nd ed. New York: Teachers College Press; and an accompanying videotape entitled "Status Treatments for the Classroom," and Cohen, Elizabeth G., and Rachel A. Lotan (eds.). 1997. *Working for Equity in Heterogeneous Classrooms: Sociological Theory in Practice.* New York: Teachers College Press. These works, developed by the sociologist whose research was featured in this chapter, provide an outstanding example of how sociologists' insights can be applied to help people in the real world. The result of her years of research, Cohen's explicit practical guidelines for classroom teachers advise how to counter expectation states in the classroom. The edited volume with Lotan provides the research evidence that the techniques actually work.

Howard, Judith A., and Jocelyn Hollander. 1997. *Gendered Situations, Gendered Selves.* Thousand Oaks, Calif.: Sage. Howard and Hollander provide an extensive and readable review of theories of social interaction and how they apply to issues related to gender.

Jencks, Christopher. 1979. *Who Gets Ahead? The Determinants of Economic Success in America.* New York: Basic Books. This book includes several analyses of variables that influence occupational status and income in adulthood, all based on national samples of men from varying birth cohorts and the status attainment model.

Kerckhoff, Alan C. 1990. *Getting Started: Transition to Adulthood in Great Britain.* Boulder, Colo.: Westview. Kerckhoff analyzes status attainment in Great Britain with comparisons to the United States.

Lemert, Charles, and Ann Branaman (eds.) 1997. *The Goffman Reader.* Malden, Mass.: Blackwell. This is an extensive collection of the writings of Erving Goffman plus a detailed and informative introduction.

Michener, H. Andrew, and John D. DeLamater. 1993. *Social Psychology,* 3rd ed. San Diego: Harcourt Brace Jovanovich. This well-written, easy-to-understand social psychology text provides many more details on the ideas presented in this chapter.

Samovar, Larry, and Richard Porter. 1996. *Intercultural Communication: A Reader,* 8th ed. Belmont, Calif.: Wadsworth. Numerous short articles relate many issues covered in this chapter to communication and interaction between people from different cultures.

Stephan, Walter G., and Cookie White Stephan. 1996. *Intergroup Relations.* Boulder, Colo.: Westview. The authors use sociological insights to analyze intergroup relations.

Wiggins, James A., Beverly B. Wiggins, and James W. Vander Zanden. 1994. *Social Psychology,* 4th ed. New York: McGraw Hill. This is another well-written, easy-to-understand social psychology text.

INTERNET SOURCES

Current Research in Social Psychology (CRISP). This electronic journal publishes articles and research in the subfield of social psychology.

Social Science Information Gateway. A searchable database of social science resources useful for many disciplines.

Pulling It Together

Additional discussion of material covered in this chapter can be found in
many other chapters in this book. You can also look up topics in the index.

Deviance and Social Control

WHEN JASON WAS IN first grade, he and his pal Justin loved to play cops and robbers. They would spend hours reenacting stories they had seen on television or making up their own, taking turns playing the roles of police detective and fugitive criminal.

In time, however, the game stopped being fun, partly because in their neighborhood a number of gangs were active, and "cops and robbers" was anything but make-believe. One day after school, the two friends were horrified to see a teenage boy gunned down in a drive-by shooting. The screams of the boy's companions and the puddle of blood on the sidewalk were all too real. As if by an unspoken agreement, Jason and Justin never played cops and robbers again.

As the years passed, the young friends gradually went their separate ways. Justin, always the more athletic of the two, played sports in junior high school and dreamed of going to college on an athletic scholarship. By the time he was in high school, however, he had to give up sports and work part-time to earn money for clothes and other things his parents couldn't afford.

Although he was bitterly disappointed about having to give up organized sports, Justin did play in a late-night basketball league at the local community center. There he met a number of police officers who participated in the program, which was designed to keep teenagers and young men off the streets at night. One officer in particular became almost a second father to Justin. Partly because of the officer's encouragement, Justin went on to study police science at the local community college, and eventually he became a police officer himself.

Meanwhile Jason's life was filled with troubles. In middle school he was bored and frustrated, while at home his mother and father were constantly yelling at each other or at him. He began skipping classes, hanging out with other kids on the street, and swiping small items from stores. Before long he got caught shoplifting a candy bar. That night Jason's father slapped him hard across the face, shouting that he was a good-for-nothing delinquent in the making.

Jason was twelve when his father left for good after a fight that left his mother, Cindi, with a bloody and swollen face. At that time Cindi was pregnant with her fourth child, and by the time the baby was born, Cindi was living on welfare. Preoccupied with the new baby and her own troubles, Cindi paid even less attention to Jason than before. She hardly seemed to notice whether he attended school, and she didn't question him when he said that the brand-new stereo he brought home was a gift from a friend.

The truth was that Jason and his friends had graduated from shoplifting candy to breaking into cars and stores. When Jason was fifteen, he and a couple of buddies were sent to a juvenile detention center for several weeks after being caught trying to hot-wire a car. When he was released, Cindi encouraged him to go back to school and straighten out his life. For a few weeks he tried to do as she suggested, but he was so far behind academically that he only felt stupid. Before long Cindi was ignoring him again, while his street pals were taunting him for wasting time in school. Soon he dropped out for good.

Over the next few years Jason learned quite a lot about how to support himself primarily through petty crimes. Unlike most of his friends he avoided violent and high-risk crimes, and the few times he was arrested, the charges were dropped. He prided himself on being too smart to get involved in anything that would mean serious prison time. His acquaintances who had "done time" often told him that, sooner or later, he would, too, but he laughed them off.

In his early twenties Jason's luck ran out. He had taken to carrying a gun, partly for his own protection and partly as a last resort if he ever did get into a spot he couldn't escape any other way. One night he and some friends were hanging out when one of them dared Jason to "roll" an elderly and apparently well-to-do man who was walking down the street. Jason tried to shrug them off, but they wouldn't let up. "Come on," they urged him. "We need the cash. What's the matter? Are you afraid of a little old man—even with that big gun in your pocket?"

Finally Jason gave in and accosted the old man. To his amazement the man began to struggle, and Jason pulled out his gun to threaten him. The man struck at Jason's arm, the gun went off, and the man crumpled to the street.

For a moment Jason was too stunned to move, but then he and his friends scattered. A few hours later a couple of police officers found Jason huddled on a bench in the city park. Armed with a witness's description of the shooter, they handcuffed him and read him his rights. As they escorted Jason to the patrol car, a backup unit drove up, and an officer got out. Glancing up at the new arrival, Jason felt his mouth drop open. "Justin?" he said wonderingly. "Is that you? Man, I haven't seen you in years."

Justin gazed sorrowfully at his childhood pal. "Man, what happened to you?" he asked. "How did it ever come to this?"

Jason shrugged and shook his head. "Cops and robbers, man," he replied. "That's what it's all about."

Given their backgrounds and upbringing, was it inevitable that Jason and Justin would turn out so differently? Are children from broken homes or other troubled circumstances fated to embark on a life of deviance and crime? Can anything be done to deter youths from this course? How concerned should society be with this problem?

*I*N PREVIOUS CHAPTERS we have looked at how we come to be part of the society in which we live—how we identify with our cultures, how we are socialized to become part of our social world, and how we interact with others around us. Much of our discussion has focused on why social structures exist and how our social actions both are influenced by and help create these patterns of social life. Yet the experiences of Justin and Jason remind us that there is another side to the story. Not everyone adopts the prevalent societal values. In particular, despite the power of socialization, in every society many people do not conform to social norms, including the special type of norms encoded in a society's laws. In our own society, lawbreaking has long been an issue of grave concern. People worry about crime, especially violent crime; politicians debate what to do about it; social critics argue that the high levels of violent crime in the United States are a sign of something deeply disturbing about our society. And both the public at large and professionals who study crime passionately debate what causes criminal behavior.

Sociologists use the term **deviance** to describe behavior that violates social norms, including laws. Just as they are interested in why people generally conform to social norms, sociologists are equally interested in why many people do *not* conform to these norms. What causes deviance? Why does deviant behavior appear in all societies? Is deviance more common in some parts of society than others? If so, what does that tell us about why deviance occurs? And, from a more microlevel perspective, why are some individuals more likely than others to violate social norms?

Like many other topics explored in this book, many people have strong views about deviance. People commonly frown on "nonconformists," whether the latter are simply taking too many items through the supermarket express lane or are willfully breaking society's laws. People wonder what is "wrong" with these "deviant" individuals and tend to attribute their behavior to their personalities or bad character.

Sociologists, too, may be very concerned about deviant behavior, especially when, as in the case of serious crime, it causes social harm. But in asking why deviance occurs, sociologists do not focus exclusively on the unique qualities of individuals. Instead, researchers step back and use their sociological imagination to investigate patterns that can help us understand deviance from a fuller sociological perspective. For example, recall from Chapter 2 how Lawrence Sherman and his associates examined the effects of different police actions on the behavior of men who engaged in domestic violence. Although we didn't use the term *deviance* in that chapter, their study is an example of investigating one type of deviant behavior from a sociological perspective.

In this chapter we explore sociological research into and theories about the nature and causes of deviant behavior. After exploring the meaning of deviance in more detail, we focus on crime in the United States to illustrate the ways in which sociologists look for patterns in deviant behavior and the kinds of questions raised by these patterns. We then turn to a range of sociological theories that address these questions. Finally, we take a close look at a study that uses a life course analysis, such as that introduced in Chapter 4, to explain why some people continue patterns of deviant behavior throughout life while others take alternate paths.

It is important to realize how much deviance is simply a part of all the different aspects of the social world that we have examined in previous chapters. In a sense the study of deviance is similar to looking at the negatives of the pictures that we take with our sociological camera. Much of the work we explored in previous chapters has tried to understand the orderly patterns that structure social life and the ways in which individuals become part of the social world. In contrast, in this chapter we will look at why there always appears to be at least some resistance to this social structure and why some people are more likely than others to exhibit this resistance.

DEFINING DEVIANCE

Because norms apply to all areas of our lives, the concept of deviance is one of the broadest sociological ideas. Recall that norms include folkways, mores, and laws. Accordingly deviance can involve rude or unusual behavior, such as wearing formal evening clothes to class or talking loudly during a movie. Deviance can involve choices that are inconsistent with cultural practices, such as wearing grass skirts and skimpy thongs to work. Or it can involve behavior that is proscribed by law, such as drug use, assault, burglary, rape, and murder.

Most important, deviance is *socially defined*. Whether a given act is deviant can vary from one social situation to another and from one culture to another. For instance, cannibalism is unknown in most societies but is practiced in certain circumstances in a few societies. Similarly, the ritual sacrifice of animals in mainstream religious services is unheard of in our society but has been common in many cultures. Even within societies definitions of deviance can vary from one subculture to another. For example, peyote is an integral part of the religious ceremonies of some Native American groups in the southwestern United States, but other Native American groups see its use as deviant.

Deviant behavior appears in all settings and all societies. What might be considered deviant in this setting? Where and when might these behaviors not be considered deviant?

Within a society, definitions of what is deviant can also change over time. For instance, not so long ago it was considered deviant for women to wear pants to work or even while shopping, and smoking was considered fashionable rather than a serious health hazard. Similarly, sexual behaviors that were once widely condemned, such as premarital intercourse and homosexual relationships, are now much more widely accepted.

Deviance is also situation-specific. For instance, if you were to go to work without any clothes on, your behavior would be deviant, but such attire (or lack of it) would not be at all deviant on a nude beach. Likewise aggressive behavior is expected in some situations, such as wars or football games, but is totally out of place in most other settings, such as at work or in the family. In general, what is seen as deviant depends very much on the social situation. Deviance is defined by the nature of a group's norms.

Deviant behavior is ever present, or at least potentially present. Wherever norms exist, there is also the possibility, and usually the reality, that they will be broken. Thus, although the definition of what is deviant can vary from one society to another, from one historical era to another, and from one group to another, it is always potentially part of the social world and daily interactions.

In addition, deviant behavior never exists without **social control,** efforts to help ensure conformity to norms. Usually we don't even notice these attempts because they are integral to our everyday lives. For instance, even though laws prohibit public nudity, you probably never think about these laws when you get dressed in the morning, even if

it is very hot outside. Sociologists say that this reflects **internal social control,** or control over one's behavior that is based on internalized standards—in this case that you should wear clothes. This internalization is part of the development of the conscience, as discussed in Chapter 4. If, however, one day you decided to leave your house totally naked, you would undoubtedly encounter **external social control,** or attempts by others to control your behavior. Your neighbors might ask about your health and sanity; police officers would wrap you in a blanket and perhaps take you to jail or to a hospital for observation.

No matter what the source of social control—whether yourself or others—it involves **sanctions,** social reactions to your behavior, generally reflecting attempts to control the behavior. Sanctions can be negative, with someone showing disapproval of your actions by frowning at you, making negative remarks, gossiping, or even, in the case of the police, arresting and detaining you. Or sanctions can be positive, such as offering compliments or praise, nodding in approval, or giving you awards and money.[1]

Sociologists have looked at behavior that has been defined as deviant in many different settings, such as sexuality, alcohol and drug use, mental illness, unusual religious beliefs and actions, and atypical lifestyles. Researchers have tried to understand why these behaviors, which might be seen as perfectly normal in some societies and some historical periods, have been considered deviant. How prevalent are the deviant behaviors? Has their incidence changed over time? Why do they appear? Why are some people more likely than others to choose these behaviors? In this chapter we will attempt to answer these questions as they relate to many different types of deviant behavior, but especially to crime.

PATTERNS OF DEVIANCE: CRIME IN THE UNITED STATES

Real-life stories of crime saturate our newspapers and television news shows, and graphic depictions of crime are a staple of television and movies. It's no surprise that so many people fear crime or have such strong feelings about crime and criminals. But, sociology is an empirical discipline. That is, sociologists want to know what the reality of crime is. They draw on valid data to tell them how much crime of different kinds occurs, who commits crimes, and who the victims of crime are. Researchers try to identify patterns in the data and the underlying social realities that help explain these patterns.

In this section we examine crime in the United States from a macrolevel perspective. First, we look at how crime is defined and measured and at what empirical data tell us about crime. How much crime is there in the United States? How has the amount of crime changed over time? Second, we review changes in the patterns of crime and explanations for these changes. Can age, period, and cohort effects, introduced in Chapter 4, help us understand changes in crime rates? Finally, we look at patterns of victimization. Who are the likeliest victims of crime? What are your chances of being a crime victim at some point in your life?

Defining and Measuring Crime

Unlike other types of deviant behavior, **crime** involves actions that society explicitly prohibits and sanctions by official means.[2] Laws, the most formal and codified type of social norms, specify illegal actions and the punishments associated with these acts. Officials, such as police officers and judges, enforce these laws and dispense sanctions.

Like other forms of deviance, crime is socially defined, and the definitions vary from place to place as well as over time. For example, one community might have an ordinance specifying that ice cream cannot be eaten outdoors in the downtown area (because spilled ice cream makes a mess on the sidewalk), while the same behavior is perfectly legal in the neighboring town. In the United States it was once legal to own slaves, whereas during Prohibition in the 1920s it was a crime to manufacture or distribute alcohol.

Moreover, at any given time, people in the same society often disagree about what acts should be defined as crimes. Many people argue that certain currently illegal actions, such as prostitution or possession of marijuana, are "victimless crimes" or crimes in name only, because they believe that these behaviors do no social harm. Others believe that actions that are now legal, such as abortion, should be criminalized. Certain behaviors, such as murder, however, are so fundamentally threatening to a society that they are defined as crimes in virtually all times and places.

When sociologists study crime, then, they are looking at the special class of "deviant" behavior that is defined as illegal in a given society or community at a given time. The most common sources of information about crime are official reports. Not all crimes are reported to the police, however, and sociologists have used other means to ensure that their information about crime is as valid as possible.

Using FBI Data: The Uniform Crime Reports

The most common data source on crime for the news media and the general public is reports of crime made to the police and compiled by the Federal Bureau of Investigation (FBI). Each month local law enforcement agencies send the FBI information on the crimes reported in their jurisdiction, whether or not an arrest has been made. These data are compiled into a larger document called the *Uniform Crime Reports* (UCR).[3]

The UCR divides information on crime into two major categories: Index crimes and other crimes. Index crimes are a set of eight serious crimes used in the Uniform Crime Reports as the basis to calculate crime rates. **Index crimes** include two categories of serious offenses: **violent crimes** (murder, forcible rape, robbery, and aggravated assault), all of which involve direct contact with a victim, and **property crimes** (nonviolent offenses such as burglary, larceny-theft, motor vehicle theft, and arson), which do not involve direct victim contact. The "other" category includes a broad range of offenses, such as fraud, embezzlement, vandalism, prostitution, illegal gambling, and illegal use of weapons. Because reports of Index crimes are thought to be the most reliable measure of crime, and because they are considered much more serious, these are the crimes that policymakers and social scientists most often examine.[4]

Getting Comparable Data: Calculating Crime Rates
If you were interested in knowing what your community's crime rate is, you would probably use data from the UCR. For instance, the UCR for a recent year showed that about 2,200 violent crimes were reported to the police in Ann Arbor, Michigan, but slightly less than 2,000 violent crimes

In the United States millions of dollars are spent each year constructing prisons, largely because Americans believe that imprisonment will help deter crime. As you read this chapter, think about what a sociological perspective might suggest about the effectiveness of prisons in combating crime.

were reported in Anchorage, Alaska. Does this mean that Ann Arbor is a more violent community than Anchorage?

Actually, the answer is no, for these raw numbers are misleading. Ann Arbor has a population of about 494,000, whereas Anchorage has a population of only about 242,000. To accurately compare the amount of crime across time or locations, we need to take into account the size of the relevant population. To do this, sociologists use a **crime rate,** a calculation of how frequently crimes occur for every 100,000 people within a population:

$$\text{Crime rate} = \frac{\text{Number of reported crimes}}{\text{Total population}} \times 100,000$$

Using this formula we can see that Ann Arbor's crime rate is 445.3 and Anchorage's crime rate is 826.4. Thus Anchorage has almost twice as much violent crime per capita as Ann Arbor.

Obtaining More Accurate Measures of Crime

For a variety of reasons, many crimes are never reported to the police. Crime victims are sometimes too embarrassed or ashamed to report a crime, or a crime might be relatively minor and not worth the trouble to report. Some victims may have their own reasons for avoiding contact with the police. Social scientists know that serious underreporting of crime will undermine the validity of the UCR data. Remember that *validity* refers to the extent to which a measure actually represents what it purports to measure. If a significant number of crimes are never reported to the police, then the UCR is not a completely valid measure of the incidence of crime.

As you will recall, one way to increase the validity of a measure is to obtain multiple indicators—to look at it from a variety of angles. To try to get another measure of the incidence of crime, the U.S. government has, since 1972, conducted an annual survey of a representative sample of people and households around the United States to determine the extent to which they have been victimized by crime. This survey, called the *National Crime Victimization Survey* (NCVS), asks about the respondents' experiences with six different Index crimes: rape, robbery, assault, larceny-theft, burglary, and motor vehicle theft. (Murder and arson are omitted because better measures are available for these crimes from medical personnel and firefighters.)

Figure 6-1 shows UCR and NCVS data for the numbers of violent and property crimes reported in both sources. As you can see, the numbers from the victimization survey are much higher than those from the UCR in all categories except motor vehicle theft. (Most car thefts are reported so that owners can get their losses reimbursed by insurance companies.) A scholar who was interested in estimating how much crime occurred in a certain year or city therefore would get different answers depending on whether UCR or NCVS data were used.

Despite this underreporting problem, however, UCR data are still valuable for certain purposes if they are used with care. One area in which they have been especially useful is in studying trends in crime rates over time. Sociologists who have studied crime statistics in both the UCR and NCVS suggest that the two sets of data tell similar stories about the rise and fall of crime. When reports of crime in the UCR go up, reports from victims in the NCVS also tend

Violent crimes	UCR	NCS
Rape	102,220	167,550
Robbery	618,950	1,298,750
Aggravated assault	1,113,180	2,478,150
Total	1,834,350	3,944,450
Property crimes		
Burglary	2,712,800	5,482,720
Larceny-theft	7,879,800	23,765,790
Motor vehicle theft	1,539,300	1,763,690
Total	12,131,900	31,012,200

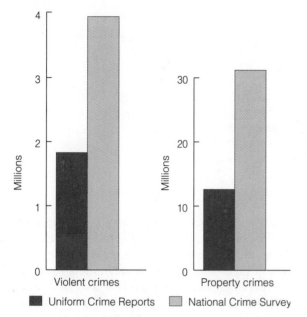

■ Uniform Crime Reports　　■ National Crime Survey

FIGURE 6-1 *UCR and NCVS Data on the Number of Six Index Crimes for the United States, 1994*
Source: McGuire and Pastore, 1997, pp. 208, 306

to go up, and when one goes down, the other goes down as well. In other words, the two data sources seem to reflect similar trends in the crime rate over time. This has led a number of sociologists who study crime to conclude that UCR data can be a valid indicator of changes in the incidence of crime in the United States over time, as long as we remember that the UCR figures often underestimate the true incidence of crime.[5]

A second reason for using UCR data involves sociologists' concern with public policy. Because the UCR data come from police agencies, they reflect the burden placed on law enforcement officials. It might be quite appropriate for sociologists who want to help these agencies become more effective to focus on these official data.

Changes in Crime over Time

What about changes in crime over time? Are crime rates rising? Are they falling? Questions such as these are often the subject of fierce social and politi-

cal debate. We hear people citing increases in crime rates as signs that the moral and social fabric of society is unraveling. By the same token, when crime rates decrease, politicians and others are quick to take credit for instituting get-tough measures or improved policing. In fact, there might be entirely different reasons for changes in the crime rate.

To illustrate this, refer to Figure 6-2, which shows the U.S. murder rate from 1943 through 1997 (Murder rates are one of the most valid elements of the UCR because murders are quite likely to be reported to authorities.) A quick glance at Figure 6-2 shows you how much the rate has fluctuated over these years. Rates were usually quite low through the 1950s and early 1960s, but then grew rapidly, fluctuated around this higher level from the early to mid 1970s, and then fell in the late 1990s to levels about equal to those of the late 1960s. Similar patterns also occur with other Index offenses.

Why have these changes occurred? Was the United States simply more crime-prone in 1980 than in earlier and later years? Or are there other explanations for these changes?

The answer, interestingly enough, involves the notions of age, period, and cohort effects, which were introduced in Chapter 4. The data in Figure 6-2 suggest that there are large **period effects,** or influences of the prevailing norms and events on individuals' lives. There is much more murder today than there was from the late 1940s through the early 1960s. Scholars and political commentators have suggested several historical factors that might have contributed to these differences.

However, a different picture emerges when we consider that people are most likely to commit crimes when they are in their late teens and early twenties. In fact, about half of all arrestees in the United States are under the age of twenty-five.[6] From 1947 through the early 1960s, there simply weren't that many people in this crime-prone age group compared with the rest of the population. Beginning in 1946 (at the end of World War II), the United States experienced a "baby boom," an unprecedented increase in the number of births, that continued until about 1960. As a result the number of people in the most crime-prone age group expanded very rapidly beginning in the early 1960s, as those born in the late 1940s and early 1950s became teenagers. After 1960 the United States entered a "baby bust" in which many fewer young people were born. This resulted in a much smaller group in the crime-prone age bracket beginning in the early 1980s.

Figure 6-2 also shows the percentage of people between the ages of fifteen and twenty-four from 1943 to 1997. As you can see, the shape of this graph is remarkably similar to the one for the murder rate.

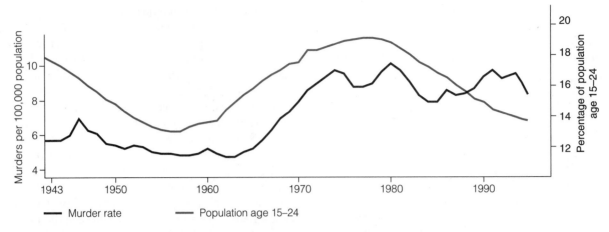

FIGURE 6-2 *Murder Rates and Percentage of Population Age Fifteen to Twenty-Four, 1943–1997*

Source: Federal Bureau of Investigation, 1992, p. 58; 1976, p. 49; Archer and Gartner, 1984; U.S. Bureau of the Census, 1985, pp. 9*-*16; 1994, p. 14; 1995, pp. 15, 199; 1998, pp. 15; Bureau of Justice Statistics, 1999, Web site, Homicide trends in the United States, Homicide victimization rates, 1950*-*97: www.ojp.usdoj.gov/bjs/homicide/homtrnd.htm

In other words, the changes in the murder rate essentially mirror the changes in the proportion of the population in this crime-prone age group for most of the years shown. This indicates that a large part of the change in crime over time in the United States can be explained as an **age effect,** whereby age-based norms influence our actions. This age effect results from a shift in what sociologists call the age structure, alterations in the proportion of the population in the most crime-prone age group.[7]

But this is only part of the story. It now appears that there is also a **cohort effect** such that individuals who are exposed to similar influences because of the era in which they were born exhibit different patterns of criminal behavior. That is, the data suggest that some birth cohorts—people born in certain periods of time—may be more or less prone to crime and deviant behavior throughout their life span than are people in other birth cohorts, even when historical conditions are considered.[8] For instance, my colleague, Robert O'Brien, and I have found that birth cohorts that are large relative to others and cohorts that have a larger proportion of people born in single parent families are more prone to violent behavior—both toward others and toward themselves. Recent data indicate that cohorts that are just now reaching young adulthood might be much more prone to violence than earlier birth cohorts are. Among youths age fourteen to seventeen, the murder rate more than doubled from the mid 1980s to the mid 1990s, while the rate for the population as a whole remained fairly constant. At the same time, the rate of suicide among

young people also increased dramatically while the rate for older people stayed the same or declined.[9] These changes can be seen in the right hand portion of Figure 6-2 where the two lines in the graph are less likely to be parallel. In a later section we will look at a possible explanation for these cohort differences.

In this discussion we have focused on one type of pattern revealed by crime data—changes in crime rates over time. But there are many other patterns to investigate and questions to ask if we are to understand criminal behavior. Later we will discuss the principal theories sociologists have advanced to explain crime and other forms of deviance. Before doing that, though, let's look at one other type of data—patterns in crime victimization.

Patterns in Victimization

Not all residents of the United States are equally likely to be victimized by crime. For instance, perhaps not surprisingly, crime rates are lower in rural areas than in urban areas. Perhaps even more surprising, older people, who tend to have an understandable fear of crime, are far less likely than younger people to be victims of crime.[10] Some of the largest differences, however, involve the status characteristics of race-ethnicity and gender.

To illustrate this phenomenon, consider the data in Figure 6-3 on homicide rates. These data come not from the UCR, but from official reports of cause of death filed by doctors and other medical personnel. Whenever someone dies, the government records the cause of that person's death in a Bureau of Vital Statistics. From that information sociologists can determine how many people die as a result of homicide, as well as from other causes. Émile Durkheim gathered the data for his research on suicide from such government records almost a century ago.

Like the crime rates reported previously, the data depicted in Figure 6-3 show how many homicides occurred for every 100,000 people in each group. This figure is referred to as a *homicide rate*. It is comparable to the murder rate reported by the UCR, except that it reports the rate of occurrence for victims rather than for perpetrators. In a recent typical year the homicide rate was 7.8 for Euro-American males in the United States, 2.7 for Euro-American females, 11.1 for African-American females, and 56.3 for African-American males. This last figure is not a typographical error. African-American males are more than five times as likely as African-American females, seven times as likely as Euro-American males, and twenty times as likely as Euro-American females to be murdered!

These figures become even more shocking when we consider the age of the victims. Among people between the ages of twenty and twenty-four, the homicide rate was 18.2 for Euro-American males, 4.2 for Euro-American females, 17.2 for African-American females, and 155.5 for African-American males. A young African-American man is more than *thirty-seven* times as likely as a young Euro-American woman to be murdered![11]

These striking patterns illustrate how a number of variables must be considered if we are to paint an accurate picture of a social phenomenon—in this case crime victimization. Here we see the role played by environment, age, gender, and race-ethnicity. And still other variables might come into play as well, such as socioeconomic status (which, among other things, affects the kind of neighborhood individuals live in). Notice, too, how the interaction of different variables must be considered. For example, the variable gender interacts with the variable race-ethnicity, so that we get quite different results for Euro-American versus African-American males, as well as for females in both groups compared with males and to each other.

We can pursue similar lines of inquiry about people who commit crimes and the societies in which crime occurs. That is, we can look for the social factors that seem to be related to the occurrence of crime or other forms of deviance and the ways in which these factors interact to produce the picture we observe. Let's turn now to some principal theories sociologists have advanced to explain patterns in crime and deviance.

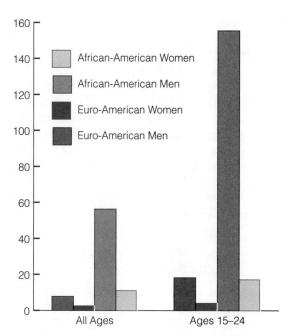

FIGURE 6-3 *Homicide Rates by Gender and Race-Ethnicity, 1995*
Source: U.S. Bureau of the Census, 1998, p. 109

THE INEVITABILITY OF DEVIANCE: THE CLASSICAL THEORISTS

Deviant behavior has attracted the attention of sociologists ever since the days of Durkheim, Marx, and Weber. These classical theorists tended to view deviance from a macrolevel perspective. More recently, several contemporary middle-range theories have focused on more specific patterns. Because studying deviance is similar to looking at the "flip side" of orderly social life, these explanations build on many of the ideas presented in previous chapters. As with the other areas we have explored, no one of these theories, by itself, is sufficient to explain the vast range of deviant behavior in a society, but each contributes to our understanding of this phenomenon.[12]

The classical theorists were fascinated by something that most people take for granted—namely, that social life is relatively predictable and orderly, so that societies maintain themselves even while constantly changing. These theorists were also intrigued by enduring patterns in deviance. Why is it that deviance occurs in all societies, even though it might be defined very differently from one society to another? Why don't deviance and crime ever disappear? In simple terms, do all societies have crime because all societies have criminals, or do all societies have criminals because something in the very nature of the social structure produces them?

Simply asking such questions is an example of applying the sociological imagination. Instead of bemoaning the fact that "deviants" seem to crop up in every society, these thinkers examined the nature of society itself for clues about why deviance occurs. As you might imagine, given their quite different theoretical perspectives, their explanations of why deviance occurs in all societies concentrated on different aspects of the social world. But they all contributed valuable insights into why deviance is an inescapable part of social life.

Durkheim: Deviance and the Collective Consciousness

In Chapter 1 we examined Émile Durkheim's view that the basis of social structure is the **collective consciousness**—the beliefs, values, and norms that members of a society hold in common. Durkheim saw deviance as an integral part of this social structure, as a necessary and ever-present part of any society. In this sense Durkheim analyzed the *functions* of deviance, much as we discussed the notion of social functions of other aspects of society in Chapter 3. According to Durkheim deviance helps societies maintain themselves by reconfirming the collective consciousness and the common basis of the society. Condemning and shaming those who do not conform to accepted norms reinforces the cohesiveness and the norms of the group.[13]

To illustrate his contention, Durkheim suggested that we imagine a "society of saints," a group of people who, to most of us, would seem completely without sin. Deviant and criminal acts such as we know them, from killing someone to speeding on the highway to telling lies, would be unknown. Within such a society, Durkheim suggested, other actions and emotions would be defined as deviant, including those that you and I would see as everyday occurrences, such as feeling jealousy or greed or praying eight hours a day rather than ten. Members of this society would be scandalized by those who had not prayed as many hours as they should or who confessed to harboring unworthy thoughts. By the simple fact of recognizing and treating these behaviors as deviant, the other society members could identify the group's norms more clearly, thus reinforcing the collective consciousness. In short, Durkheim suggested that we can never have a society that is completely free of deviance, because deviance always is redefined according to the prevailing social norms.

Besides reinforcing a society's norms, Durkheim suggested that deviance has a second function: De-viance can help facilitate change, both by showing that it is possible to behave or think in unconventional ways and by setting the foundation for later changes in the collective consciousness. Durkheim used the experience of the Greek philosopher Socrates as an example. Socrates taught his students by asking them questions and, on hearing their answers, asking them more questions. This technique, which some of your teachers may use in class discussions, is called the Socratic method. Unfortunately Athenian leaders disliked and feared Socrates' methods, which they believed encouraged disrespect for traditional customs and beliefs. Because his behavior was seen as deviant and even criminal, Socrates was brought to trial and the ultimate form of social control, execution, was applied. Yet, despite his fate, Durkheim suggested, through his "crime" Socrates both introduced new ways of thinking into the society and laid the foundation for later changes in the norms governing thought and expression. Later, Socrates' ways of teaching and thinking were both accepted and widely practiced.[14]

Similarly only a few decades ago mothers of young children who worked outside the home were considered deviant (though not, fortunately, criminal), especially when their husbands could support their families. These women often faced strong disapproval from relatives and neighbors and were condemned as bad mothers. As more and more young mothers entered the labor force, however, these views changed. Today mothers are just as likely as other women to be employed outside the home, and they are much less likely to face criticism when they do so. Thus "deviant" behavior can itself gradually change social norms and the way deviance is defined.

Marx: Social Control and Maintenance of Power

Like Durkheim, Karl Marx was fascinated by the question of why societies remain so orderly and what social structures produce these patterns. In contrast with Durkheim, however, Marx suggested that social structure was based on the oppression of some people by others. Marx believed that a critical element of society is social control, the way in which the powerful maintain their position of authority within societies. One way in which powerful groups enforce control is by defining deviance and influencing the sanctions for engaging in deviant behavior. So, for example, in Marx's time child labor and other forms of worker exploitation were not regarded as criminal, but stealing from the rich was dealt with very harshly. Thus, whereas Durkheim saw social control as reinforcing the values, norms, and beliefs of all of society, Marx saw it as solidifying the advantages of the powerful.[15]

Examples supporting the Marxian perspective abound. For instance, Socrates was labeled a criminal and sentenced to death because the power structure in Athens was threatened by his activities. If his students and admirers had been politically and economically powerful, Socrates might have been honored rather than executed.

Similarly, before the Civil War, millions of African Americans were kept as slaves by wealthy and politically powerful plantation owners in the South. Not only was "deviance" defined quite differently for slaves than for free persons—slaves could be whipped mercilessly simply for talking back to their masters—but other norms also supported the slave system. Even in many northern states, for instance, abolitionists who openly advocated ending slavery or who helped fugitive slaves escape to freedom were castigated as dangerous and ungodly revolutionaries (deviants, in other words) who had no right to speak in public gatherings. Often they were assaulted and even murdered by outraged supporters of slavery and "Southern rights." That social norms defined the abolitionists' activities as deviant and even evil helped to maintain the slave economy, and for years state and federal laws supported these norms.

You can see similar examples today. For instance, as many social critics note, well-to-do criminals in the business world who defraud people of millions of dollars are treated less harshly by the criminal justice system than are many street criminals from impoverished areas who commit small-scale thefts. Similarly, although well-dressed people can walk freely through office buildings and upscale department stores, shabbily dressed people will likely be detained by police or security guards. In fact, many cities have passed laws and ordinances that prohibit homeless people from staying in parks or "loitering" in business areas, even though citizens with jobs and money are not prohibited from doing the same things.

The idea of social change is central to the Marxian perspective. Marx suggested that the potential for resistance and revolution is present in all social situations because the less powerful can always act against the status quo. Whereas Durkheim focused on how deviant behaviors can lay the foundation for the development of new norms, however, Marx focused on how oppressed groups can join together to act against more powerful groups, directly challenging their control and bringing about social change. For instance, abolitionists not only introduced ideas that challenged the norms surrounding the system of slavery but joined together to directly challenge the status quo and eventually to end slavery altogether.

Keeping the ideas of Weber in mind, in what ways is this judge's authority legitimated?

Weber: Legitimation of Social Control

Although Durkheim and Marx, each in his own way, looked at how deviance and social control help maintain the social structure, Max Weber was more interested in the reasons that people allow this social control to exist. Weber insisted that the way we see the social world—our beliefs and interpretations—as well as our actions help shape and build the social order. Using the method of *verstehen,* or "interpretive understanding," Weber tried to explain how we develop our common views of what is and is not appropriate behavior and why we accept the authority of others in enforcing societal norms. In other words Weber implicitly accepted both Durkheim's view that members of a society share common values and norms and Marx's notion that some people in a society dominate others. But, he asked, *why* do we let this happen? For instance, why did Socrates and his followers accept the authority of government officials and not question their right to sentence Socrates to death? Why do you adhere to the dress code when you go to a fancy restaurant? Why do you stop speeding when you see a police car in the distance?

Weber suggested that social control is effective because we accept domination and authority as *legitimate,* that is, as justified or proper. According to Weber, domination works only if both the rulers and the ruled believe in the rulers' right to their authority. These beliefs help maintain the stability of the group. For instance, because motorists accept the authority of the city government to control traffic and of the police to enforce traffic laws, drivers and pedestrians alike can normally be expected to obey traffic lights. Because diners accept the authority of restaurant owners to impose dress codes, the formal ambiance of fancy restaurants can be maintained.

BUILDING CONCEPTS

Perspectives on Deviance and Social Control

Marx, Durkheim, and Weber all contributed valuable insights into why deviance and social control are an inescapable part of social life.

Theorist	Perspective on Deviance and Social Control	Example
Marx	A critical element of society is social control. One way in which powerful groups enforce control is by defining deviance and influencing the sanctions for engaging in deviant behavior.	Socrates was labeled a criminal and sentenced to death because the power structure in Athens was threatened by his activities.
Durkheim	Deviance helps societies maintain themselves by reconfirming the collective consciousness. Condemning and shaming those who do not conform to accepted norms reinforces the cohesivenesss and norms of the group.	A few decades ago, mothers of young children who worked outside the home were considered deviant, especially when their husbands could support their families. The norm was for women to stay at home, and women who worked outside the home faced strong disapproval.
Weber	For social control to be effective, authority must be accepted as legitimate.	Motorists must accept the authority of police and traffic laws before they can be expected to obey traffic lights.

Weber distinguished three ways in which authority can be legitimated. First, **traditional authority** may be accepted simply because it is an integral part of the social structure and people cannot conceive of any other system. For instance, a Jewish community might accept the authority of its rabbi, or Catholics the authority of the pope, or pupils the authority of their teacher. Second, individuals can gain domination over others through **charismatic authority** based on their personal qualities. This charismatic authority is the opposite of traditional authority because it is rooted in the unique characteristics of a particular leader, rather than in the status the person happens to occupy. Religious leaders such as Moses, Jesus, and Mohammed, who gained followers through their personal magnetism and inspirational qualities, are often cited as charismatic leaders, as are political figures with less lofty aims but with similar power to enthrall others, such as Adolf Hitler. Third, **legal authority** is based on written rules that provide a systematic way of maintaining the authority of people who occupy a certain position or office. Governments throughout the world have extensive legal codes that define, often in great detail, the actions that we may or may not take and that empower officials to enforce the rules and impose prescribed sanctions when they are violated.[16]

Weber's three categories of authority involve what he called **ideal types,** concepts or descriptions of phenomena that might not exist in a pure form in the real world but that define basic aspects of a given situation. The three types of legitimate authority are conceptually distinct, but in real life they may overlap. For instance, traditional authority, such as that of the pope, can be reinforced by church law and enhanced by a particular pope's charismatic saintliness. And even if we rarely encounter ideal types, Weber believed that such conceptual distinctions can illuminate social actions and structures. Other sociologists have also used ideal types to describe the world, and throughout this book you will see examples of such typologies.[17]

Like Marx and Durkheim, Weber suggested that deviance and social control help to explain not only the stability of a society but social change as well. According to Weber, social control is effective only when authority is accepted by those who are ruled. When the legitimacy of this authority is challenged by the development of another system of authority, social change can result.[18] For instance, the charismatic leader Mahatma Gandhi challenged England's legal domination of India in the mid-twentieth century. The efforts of Gandhi and his followers eventually led to the collapse of English rule in India and the establishment of the Indian state. Similarly changes in the law have often resulted in changes in traditional patterns of domination. For example, husbands' traditional dominance of wives in the United States has eroded substantially in the twentieth century as new laws have given women rights to vote, to own property in their own names, to sue for divorce, and to prosecute abusive spouses.

Collectively the works of Durkheim, Marx, and Weber can help us see that deviance appears in all societies as the necessary counterpart to orderly social patterns, much like the negative of a photograph. To these classical theorists, deviance doesn't appear because some people are inherently "bad" or "nonconformist." Instead, deviance is simply part of the social structure. Yet these theories can't tell us why some people are more likely than others to violate norms. They can't explain why Jason pursued a life of crime and Justin became a police officer. More contemporary sociologists have explored these issues, and we turn now to their work.

EXPLAINING DEVIANT ACTIONS: CONTEMPORARY MIDDLE-RANGE THEORIES

Whereas Marx, Durkheim, and Weber all took a very broad, macrolevel view of society, contemporary sociologists usually zoom in closer. In contrast with the grand theories developed by the classical theorists, contemporary sociologists often propose middle-range theories—that is, theories that deal with relatively limited areas of the social world and that are directed toward specific research problems. Build-

ing on many of the ideas introduced in previous chapters, contemporary theorists have explored deviant behavior by looking at (1) our social interactions, (2) our positions within society, (3) the norms and values of subcultures, and (4) our linkages to and bonds with other people. Each of these perspectives provides another perspective or camera angle by which to understand deviant behavior. Although in earlier years the theories often developed in opposition to other viewpoints, contemporary theorists have shown how, when used together, these theories can provide a more complete understanding of why individuals engage in deviant or criminal behavior.[19]

Deviance as the Result of Interactions: Labeling Theory

We first encountered labeling theory in Chapter 2, when we reviewed Lawrence Sherman and associates' research on preventing repeated incidents of domestic violence. According to **labeling theory,** definitions of deviant behavior develop from social interaction, and the key element in becoming deviant is how others *respond* to people's behavior, rather than how people actually behave. Sherman hypothesized that the process of repeatedly being arrested and "labeled" a spouse abuser affects individuals' self-image and later behavior such that they come to accept this definition of themselves. Thus labeling theory would predict that men would be less likely to continue to abuse their partners if the police did not categorize them as criminals but, rather, provided some other type of treatment, such as counseling or separation. Sherman and Berk's original research did not support this hypothesis, but subsequent replications provided support for labeling theory for some suspects but not others.

Sociologists have used labeling theory to explain the development of deviant behavior in many situations besides domestic violence—from juvenile delinquency to drug addiction and mental illness. The theory is based on the ideas of identity formation and the notion of the self and on the theory of symbolic interactionism developed by George Herbert Mead, which we discussed in Chapters 4 and 5, respectively. Labeling theory suggests, perhaps surprisingly, that the key element in becoming deviant is not how people *actually* behave, but how others in their social networks *respond* to their behavior and, especially, whether others label them as deviant. Once individuals are defined as deviant, their interactions with others might change in ways that help confirm and support this definition. To put it another way, we become what other people say we are. The deviance that results from a change in identity caused by labeling is called **secondary deviance,** precisely because it is secondary to the original or primary first offenses.[20]

For instance, recall how Jason's father called him a "good-for-nothing delinquent" after he was caught shoplifting. Labeling theory would suggest that the fact that the police and Jason's father labeled him in this way actually led him to alter his self-image. This change in self-identity then promoted more deviant behavior, or secondary deviance. Similarly, suppose that in high school you occasionally cut class and hung out with friends. If you had been caught, you might have been taken to the principal's office and officially declared a truant. The administrators and teachers at your school would then have watched your behavior much more closely, and your parents and fellow students would have become aware of your status as a truant. Through these interactions they would have communicated to you and to one another that you were a truant. As you interpreted these interactions, through what symbolic interactionists call *reflexive behavior,* you would have begun to see yourself as a truant and thus been more likely to act in that way. If, however, your teachers had not noticed you were missing or simply ignored your behavior, you wouldn't have been labeled a truant, and your interactions with significant others at school and home wouldn't have changed. Thus you would have had little reason to redefine your self-image.

Labeling theory stresses that definitions of deviance are the product of social interaction. Whether a person is seen as deviant can vary from one setting to another, just as the definition of deviance varies from one situation to another. When we look at deviance from the perspective of labeling theory, we see how relative the definition of deviant behavior can be. Consider, for example, homosexuality. In various times and places, including the United States, homosexuality has been considered a mental illness or has been illegal; in other societies, such as ancient Greece, it was widely accepted. Similarly, in some times and places, youths seen drinking alcohol or roaming the streets after a certain hour might be designated as juvenile delinquents—as deviant—but just a few years later or in another community, these actions might be accepted as normal.

In addition, as Sherman's work showed, not everyone who is labeled as deviant comes to see him- or herself in that way and increase his or her deviant behavior. This probably reflects the fact that each of us has many components to our self-identities. Recall from Chapter 4 how our self-identities reflect our role relationships with other people and our views of ourselves. We actually have a hierarchy of identities, in which some identities are more important to us than

How might labeling theory explain the appearance and behavior of rock star Marilyn Manson?

others. If you were labeled a truant but other aspects of your identity were much more important or salient to you—such as son or daughter, good athlete, strong student, and friend—then seeing yourself as a truant might be an insignificant part of your self-image. Because Jason did not have these other salient identities, his label as a troublemaker might have become an important aspect of his self-image, and this self-image might then have influenced his decision to continue to act in these troubling ways.

Similarly Sherman and his associates found that labeling was much more likely to result in continued domestic violence among men who didn't have strong bonds with others. One way to interpret this finding is that the label of deviant was more salient for these men because they had fewer other strong and salient identities. For those whose most salient or important identities were those of worker, spouse, or parent, a label of spouse abuser and deviant would be more likely to be rejected.

Positions within Society: Strain Theory

Labeling theory focuses on how our interactions with others influence our self-identities and our behavior. In contrast, Robert Merton, the sociologist who coined the term *middle-range theory,* wondered how our positions within society affect our ability to follow social norms.

Merton believed that all societies have two types of related norms: (1) culturally defined *goals* and (2) institutionalized *means. Culturally defined goals* are the goals that are valued within society. For example, Merton noted that in the United States a major goal is financial success. Novels, movies, television, and stories about lottery winners reflect this preoccupation with wealth. *Institutionalized means* refers to culturally approved ways of achieving these cultural goals. In the United States the appropriate means of achieving wealth include working hard in school and getting a good, well-paying job.

Merton suggested that most people both value financial success and follow the socially accepted means of trying to attain this goal. Yet, Merton noted, many people who accept this goal don't have the means to achieve it. Thus, according to **strain theory,** deviant behavior results when individuals accept culturally defined goals but lack the institutionalized means of attaining them. Merton called this frustrating situation "structural strain" and saw it as resulting from individuals' places in the social structure. A logical response would then be to use *illegitimate* means, such as fraud or theft, to achieve the cultural goals.[21]

The significance of Merton's theory is its suggestion that people choose deviant actions because of their place within social structures, not because of some personality flaw. Those with access to institutionalized means of attaining cultural goals are least likely to choose deviant behavior, and those who lack access to such means are much more likely to do so. Thus, for example, Merton's theory implies that there should be a strong correlation between poverty and crime. Because of the strain between cultural goals and institutionalized means, disadvantaged individuals will be much more likely than more privileged individuals to choose illegitimate means of obtaining wealth. According to this perspective Jason may have been more likely than others to commit petty crimes because his mother was very poor and he did not have access to more legitimate means of gaining money.

Do the data support this view? Sociologists have tested this hypothesis many times over the last fifty years, and the evidence has been mixed at best. For instance, the data seem to suggest that poor people are more likely than others to be arrested and convicted of crimes that involve robbing and stealing. Yet there are other ways to obtain money illicitly, and those with access to institutionalized means are much more likely to use these illicit techniques— overcharging for goods and services, stealing from their employers, developing fraudulent banking schemes, or selling junk bonds. For instance, even though he was extremely well educated and financially comfortable, the former Wall Street executive Ivan Boesky developed an elaborate scheme to earn money from stock deals. First, he bribed investment bankers to give him advance information about sales of corporations. Armed with this information, he then bought shares of these companies at a low price, held onto them until after the sale occurred and the price rose, and then sold them at a huge profit. By using these illegal methods, Boesky made more than $200 million dollars in only ten years.[22]

Sociologists refer to nonviolent crimes such as these, which generally involve fraud and deception and are committed in the workplace, as **white-collar crime.** (The term *white collar* is used to designate people in executive and professional jobs.) Sometimes entire corporations engage in white-collar crime. For example, stores might use misleading marketing practices, advertising a very cheap product to draw customers into the store and then telling them that the product is sold out and substituting a more expensive one. Companies might engage in fraudulent accounting procedures to cheat the government out of taxes. Corporations might charge much more for a product than it actually costs to make, either because they are the only group that makes the product or because they have conspired with other manufacturers to keep the price at an unreasonably high level. In this sense, the presence of white-collar crime undermines support for Merton's theory; after all, these corporate officials have access to institutionalized means of gaining wealth but still resort to deception and fraud.[23]

As you can imagine, this lack of support for Merton's formulation led many sociologists to attack strain theory and suggest that it be abandoned.[24] Relatively recently, however, a few sociologists have begun to revise and expand the original perspective in what they term **general strain theory.** These general strain theorists suggest that the sources of structural strain can be much broader than Merton suggested and that the response that individuals choose is influenced by their perceptions of the fairness of the situation and the nature of their social networks. Individuals pursue many goals besides economic success—such as having happy family relationships, good friends, and enjoyable leisure pursuits and work lives. Yet, sometimes people do not attain the goals that they expect, or they believe that their experiences have not been just or fair. In other words, they believe that the *norm of equity* has been violated. One of the possible reactions to this distress can be deviance. For instance, workers who believe that they are underpaid might steal from their workplace; students who believe that their efforts have not been properly rewarded by their teachers might vandalize a school.[25]

Of course, such deviant or criminal behavior is only one possible response to strain and stress, a response that is actually only adopted by a small minority. General strain theorists suggest that how people interpret and respond to stressful events reflects their relationships with other people.[26] Other theories have looked much more closely at how our relationships with others influence our decisions to engage (or not engage) in deviant and criminal behavior.

The Importance of Peer Groups: Differential Association and Subcultures of Deviance

In Chapter 4 we noted the importance of peer groups as agents of socialization. Through our association with our peers—people with whom we share common interests and statuses—we learn social roles and develop our self-identities. Through these associations we can also learn about deviant behaviors. Theorists who have explored these ideas have suggested a variety of approaches that involve the ideas of differential association and subcultures of deviance. **Differential association theory** focuses more on how people learn or acquire deviant roles from their association or relationships with others engaged in deviant activities; **subcultural theory** looks more at the groups or contexts in which these roles are learned.[27]

Both perspectives suggest that it takes a while to learn a deviant role and that the role is most easily learned from "established deviants." In other words we learn how to be deviant in the same way we learn other social roles—through training by skilled, experienced individuals, as well as through the more subtle processes of anticipatory socialization. For example, you might be very curious about what it would be like to try some illicit drug. But if you don't know anyone who uses the drug and is willing to teach you how to obtain and ingest it, you will find it difficult to do so. Just as in other areas of social life, your contacts with other people, your social networks, are an extremely important influence on both the opportunities and constraints that you face.[28] If you have more contacts with people who engage in various types of deviant behavior than with people who do not, differential association theory suggests that you will be more likely to engage in these deviant behaviors yourself.[29] Because Jason had a large number of friends who were engaged in deviant behavior, he could easily learn how to hot-wire and steal cars. In contrast, Justin had a different peer group and didn't become involved in such activities.

In testing this theory, researchers frequently have looked at gangs, groups of people, often young,

who identify with one another and share a goal of engaging in deviant acts. The first studies of gangs were conducted by sociologists from the Chicago School, who were fascinated by the groups of urban youths within some ethnic neighborhoods who joined forces and developed a common identity as troublemakers, thieves, or some other type of criminal.[30] In more recent years sociologists have studied contemporary gangs and found similar arrangements in cities throughout the country. Their research reveals that these gangs tend to develop ways to mark their group identity—often with certain colors and styles of clothes, hand signals, terminology, and mannerisms that are said to be unique to their group. In other words, like all cultures, they develop their own norms, values, ideologies, and rituals that members can use to signify their membership within the group.[31]

Just as we described in Chapter 3 how people develop ethnic identities, so, too, do gang members come to identify with the gangs with which they are affiliated. Through their interactions with others in the group, they come to see themselves as gang members: they view the gang's norms and values as natural, accept its folkways, and believe its myths and ideologies. Subcultural theories of deviance suggest that there are subcultures or groups whose cultural elements support lifestyles that are deviant in the context of the larger culture. For people who belong to those subcultures, what appears to be deviant behavior to those *outside* the group is actually conforming behavior *within* the group.[32] Jason and his friends may be seen as forming such a subculture. When they stole cars and robbed people, they were conforming to their subcultural norms, demonstrating their allegiance to the group, and reaffirming their identities as group members.

Relationships and Social Bonds: Control Theory

Even if, like Justin and Jason, you grew up in a neighborhood containing a large number of gangs and many people in your neighborhood participated in them, you may well have not chosen to do so. Not everyone who is exposed to gangs joins them or adopts their lifestyle. Why then, some sociologists have asked, do certain individuals choose deviant behaviors? Why are some young people more likely than others, even with the same type of economic circumstances and the same neighborhood environments, to participate in deviant activities? Sociologists who have addressed this issue have turned to some of the ideas we explored in earlier chapters regarding individuals' bonds with

Teacher Mary Kay LeTourneau's affair with her 13-year-old former student drew widespread attention for its degree of deviance from societal norms. How might control theory explain her actions?

other people and the ways in which we internalize the norms of the groups to which we belong. Their ideas show the legacies of the classical theorists, especially Durkheim, and involve the notion of social control.[33]

Remember Durkheim's work discussed in Chapter 1 regarding suicide. (Suicide can be considered a form of deviant behavior because it violates social norms regarding the sanctity of life.) Durkheim was especially interested in anomic situations, those in which patterns of relationships have been disrupted or are unstable and in which no consistent, predictable set of norms exists. Durkheim found that individuals in these anomic situations were more likely than others to commit suicide. As a result of his research, Durkheim insisted that our adherence to society's norms depends on our attachment to others within the society. As he put it, "We are moral beings to the extent that we are social beings."[34] Similarly **control theory** suggests that our connections to others within society are the major influence on our desire to conform to society's norms. Attachments to others are the basis of social control.

External Social Controls The distinction between two types of social control—external and internal—that was made early in this chapter can help illuminate how our relationships with others deter

deviance. External control derives from our attachments to and involvement with others through social groups—from the extent and nature of our social networks. The more you are tied to conventional, and not deviant, others through your social networks, the stronger your ties to conventional norms will be. In other words, the more social capital you have—the more you are associated with groups and people that promote nondeviant behavior—the less likely you are to behave in a way that they would not approve of and thereby risk the dissolution of your social networks. In a sense this perspective on deviance is the opposite of the differential association and subcultural theories. Whereas these latter theories focus on how attachments to deviant peers promote deviant behavior, control theory emphasizes how our attachments to conventional, or nondeviant, groups and people help prevent deviant behavior.

Thus control theory would predict that if Jason had had more conventional ties to others—perhaps through success at school, strong attachments to family members, and involvement in church and community groups—he would have been less likely to engage in criminal behaviors. In contrast, Justin's involvement in the midnight basketball league and his friendship with the police officer helped build up social capital that would be expected to promote more conventional behavior. Similarly, recall the results of Sherman's research on domestic violence described in Chapter 2. As the replications of his original study with Berk showed, being arrested was most likely to serve as a deterrent to further incidents of spouse abuse when the men had ongoing ties with others, such as being married and having a job. These social ties functioned as a means of social control, and the stigma of arrest was so great for the men with these ties that they were more likely to be deterred from further violent behavior.

Control theory also has been used to explain the cohort differences in crime that were described earlier in this chapter, especially why members of "boom" cohorts and today's younger cohorts have been more crime-prone. Essentially large birth cohorts and cohorts with fewer two-parent families place pressure on a society that weakens controls over deviant behavior. When a baby boom occurs, a society suddenly has to deal with many more young people than previously. Schools might have to be built; there might not be enough teachers, counselors, and scout leaders; families are larger and might be hard-pressed to give children as much attention as they once did. In adulthood these baby boomers might have fewer employment opportunities because there is more competition for jobs. In short, people in such "boom" cohorts might have more difficulty maintaining the conventional ties to family, work, and community that people in "bust" cohorts take for granted. Similarly, when there are more single-parent families, there tend to be fewer adults available to supervise children and to provide needed attention than in earlier years. As a result, like members of the "boom" cohorts, members of these recent young cohorts appear to have higher crime rates and higher rates of self destruction.[35]

Finally, the notion of external social control has been used to explain differences in crime rates found around the world. As you can see in Map 6-1, nations differ a great deal in the extent to which citizens engage in various criminal activities. The United States stands out as a particularly dangerous nation because only a few nations (some in South and Central America and some that were part of the former Soviet Union) have higher homicide rates than the United States. Countries that are culturally and economically most similar to the United States, such as Canada, Australia, New Zealand, Japan, and all of the European countries, have homicide rates that are a fraction of the United States' rate.

Part of these differences occur because of the greater availability of guns in the United States, as well as because of a tradition of interpersonal violence that dates back to the American frontier period. About two-thirds of all murders in the United States are committed with firearms. Most other countries have strict gun control laws, and most experts believe that such laws could substantially lower our murder rate.[36]

Yet variations in external social controls also can help explain these different rates. The sociologist Freda Adler, who is profiled in Box 6-1, decided to investigate this possibility by comparing countries such as Japan, Switzerland, Saudi Arabia, Costa Rica, and Nepal—all of which have very little crime—with the United States. After an extensive examination of these societies, she suggested that these low-crime societies differ from others primarily in the nature of social control. Most of the low-crime countries have very strong extended family systems, which exert a great deal of pressure for conformity. Cultural norms, religious practices, and legal regulations support these strong family ties. Misbehavior by one family member can bring shame to the others. Adler suggested that these societies were characterized by **synnomie,** a term she coined to represent the opposite of anomie. That is, each of these societies has a very strong collective consciousness and a high degree of cohesion, promoted by kinship structures that provide strong social networks and ample amounts of social capital for their members.[37]

Internal Social Controls   Although the development of external social control involves relationships with social networks, the development of internal control involves the concepts introduced in Chapter 4 regarding internalization and the development of the conscience. If you have strongly internalized certain norms, you will be unlikely to violate them, and if you do violate them, you will feel guilty and will be unlikely to repeat the behavior. Most social control is actually internal control. The reason that most of us are law-abiding most of the time is that we simply don't think of disobeying the law or we know that if we did do so we would feel very bad afterwards. Our actions are constrained, or controlled, by our views of what we should or shouldn't do. These internalized views of right and wrong, and the sanctions we apply to ourselves, are probably the most important influence on our decisions from one moment to another, and thus our probability of committing deviant acts.[38]

Some people have more strongly internalized norms and thus are more committed to them than others. Over the past half-century sociologists and developmental psychologists have conducted numerous studies of adults and juveniles who have been heavily involved in delinquency, crime, and aggressive behavior. These accumulated results suggest that children's relationships with their parents are the most important influence on the development of internal social control. When parents and children have strong emotional attachments to one another, and when the parents both supervise and discipline their children, the chances of subsequent aggressive, delinquent, and criminal behavior are reduced. In other words, the most important aspects of child rearing in the prevention of delinquency and aggressive behavior are affection, supervision, and discipline.[39] Thus Jason would have been far less likely to behave as he did if his parents had been able to exhibit a better balance of affection, supervision, and discipline in their relationships with him.

Because internal social control has its roots in early childhood, it is not surprising that there is a strong correlation between behavioral problems in childhood and deviant behavior in adulthood. Several studies have found strong associations between aggressive and antisocial behavior in childhood and deviant behavior in adulthood, including spouse abuse, fighting, traffic violations, and criminal activities.[40]

Yet this association is far from perfect. As you will remember from the discussion earlier in this

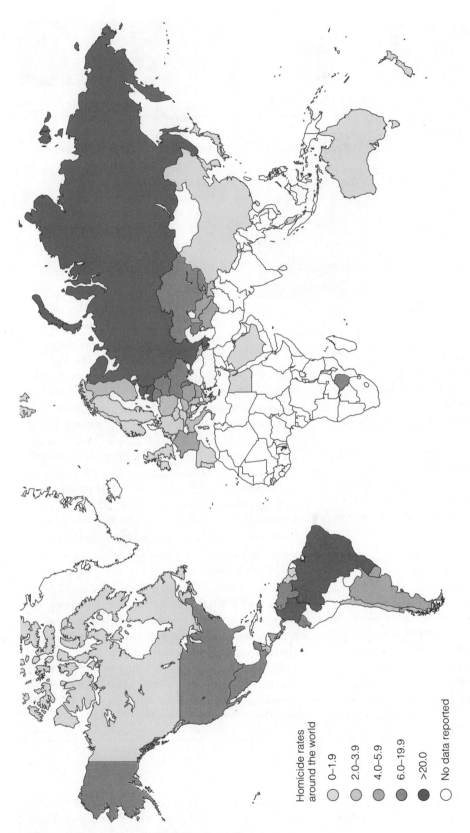

MAP 6-1 *Homicide Rates Throughout the World*

Note: All rates represent the number of homicides (people who were murdered) per 100,000 people in the population.

Source: United Nations, 1997, pp. 486–507.

BOX
6-1

SOCIOLOGISTS AT WORK

Freda Adler

As a youngster it occurred to me that everyone around me was an amateur sociologist.

Freda Adler, Rutgers University, has looked at crime in many different countries.

Why did you become a sociologist? As a youngster it occurred to me that everyone around me was an amateur sociologist. People were guessing and surmising about their groups and their society and their relations with each other. I simply wanted to take the guesswork out of those surmises and learn whether science could provide me with the tools to find answers that made sense.

How did you get interested in studying the range of crime rates in different countries? As a consultant to the United Nations, I was asked to explain why some countries have extraordinarily low crime rates. What is it that these countries have in common? What distinguished them from countries with high crime rates? I went to the countries under study and lived with people in their homes, farms, desert tents, and skyscrapers; worked with them in their police agencies, courts, and prisons; and met with them in their schools, places of assembly, and factories in order to learn firsthand how their social controls work.

What surprised you most about the results you obtained in your research? What surprised me most was how the people in the highly diverse societies under study—rich and poor, urban and rural, dictatorships and democracies, monarchies and republics, industrialized and agricultural—nevertheless could find agreement on how to live harmoniously amongst themselves.

What research have you done since completing the study described in the chapter? Does it build on your previous research? My research interests have been quite varied. For example, I have been involved in the study of maritime crime. There, too, the question arises whether the concept of synnomie or its opposite, anomie, can explain high or low rates on or affecting the world's waterways, from drug trafficking to smuggling to modern piracy. Similarly, the concept of synnomie has been useful to me in the study of female criminality and crimes against the environment.

What do you think the implications of your work are for social policy? Have any of these implications been realized? My study demonstrates that social solidarity and informal social control mechanisms must not be underestimated in crime control. In my work with government officials and agencies at the national level, I find great willingness to accept the need for strengthening informal social controls. Obviously politicians vary widely on the extent to which such social policies should be implemented by legislation; however, funding has been provided for programs such as Head Start, drug rehabilitation facilities, and various communal organizations.

How does your research apply to students' lives? Ask yourself what it is that keeps you from punching your friend anytime he or she angers you. Is it the fear of arrest, and maybe a jail term, or is it the conventional values you have learned from your family, social groups, and community? It appears that the informal social controls of family, religion, schools, and workplaces are much more effective than the criminal justice system.

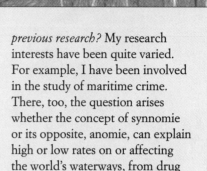

chapter, people are most likely to commit crimes when they are in their late teens. Beginning in their early twenties, the probability of their engaging in deviant and criminal behavior declines dramatically.[41] In addition, many young children with behavioral problems do not exhibit problematic behaviors when they get older, and most adolescents who get in trouble don't continue these behavior patterns in later years.[42] Two sociologists, Robert Sampson and John H. Laub, wondered whether control theory might be able to explain these continuities and changes in behavior over the life course. In the next section we look at their research in detail.

Lillie was a twelve-year-old living on the streets of Seattle in 1983. Based on what you have read so far in this book, what do you think might have happened to her in later years? What might have been done to change her life chances?

DEVIANCE AND CRIME OVER THE LIFE COURSE

In Chapter 4 we discussed how sociologists who study the life course use the term *career* to refer to a sequence of roles that we enact throughout our lives. Our discussion there focused on family, work, and educational careers, but we can also think of criminal or deviant careers.

Building on control theory, Sampson and Laub were especially interested in learning more about how the quality and strength of individuals' social ties might modify their criminal careers. Sampson and Laub offered two basic hypotheses. First, based on the work described in the previous section, they suggested that antisocial behavior in childhood, such as extensive misbehavior, temper tantrums, and juvenile delinquency, is linked to a variety of troublesome behaviors in adulthood, such as criminality, deviance, and problems in school, at work, and in personal relationships. In other words, they suggested that observable patterns of deviant and problematic behavior in childhood will likely also be apparent in adulthood. Second, the researchers suggested that individuals' bonds with others in their family, school, and work environments will influence criminal behavior over the life course. In other words, the strong association between childhood and adult deviance can be modified through strong ties to work and family. Together these two propositions suggest that there is *continuity* between childhood and adult behavior but that criminal or deviant ca-

reers can be *modified* through alterations in other aspects of the life course. Specifically Sampson and Laub predicted that abundant social capital in family and work relationships can greatly lessen the chance that a person will continue to engage in deviant behavior.

Gathering an Archived Data Set

To test these hypotheses, Sampson and Laub used data that had been gathered by Sheldon and Eleanor Glueck beginning in 1939 and continuing into the 1950s. The Gluecks were especially interested in how families influenced criminal careers and how criminality in adulthood could be predicted from earlier life experiences. Although they published the results of their studies in numerous books and articles, their work was largely ignored by sociologists during their lifetimes.[43] Fortunately their data, as well as copies of all their writings and correspondence, were preserved in the archives of the Harvard Law School library, so that Sampson and Laub were able to consult them.

When Sampson and Laub read some of the Gluecks' books and articles, they realized how much of the Gluecks' data could be used to test hypotheses based on control theory. An especially relevant data set was a **longitudinal study**—a study that involves data that have been collected at different times—

designed to understand more about the long-term outcomes of boys who had been juvenile delinquents. Because sociologists had not been interested in the questions raised by the Gluecks until relatively recently, no one else had gathered a similar set of data.

To develop this data set, the Gluecks and their research team obtained samples of white boys born in Boston between 1924 and 1935. One part of the Gluecks' group, called the delinquent sample, included five hundred boys, ages ten to seventeen at the time of the first data collection, who were committed to correctional schools in Massachusetts. The other part of the group, called the nondelinquent sample, included five hundred boys of the same age who attended Boston public schools. All the boys were from neighborhoods where poverty and delinquency were common. The Gluecks made the two groups as similar as possible on other variables that could be related to delinquency, including age, race-ethnicity, general ability, and neighborhood type.

The Gluecks and their research team followed the subjects for an average of eighteen years, collecting data at three different times, in what are often called **waves of data collection.** The first wave occurred when the subjects were an average of fourteen years old, the second at an average age of twenty-five, and the third at an average age of thirty-two. Ninety-two percent of the subjects were followed throughout the entire period, a figure that is considered very high even today. The resulting data set is often called a **panel study,** a special kind of longitudinal data set that includes information on the same people over a long period of time.

The Gluecks obtained data in many different ways, thus increasing the reliability of the information. They interviewed the subjects, as well as their families, employers, teachers, neighbors, and criminal justice and social welfare officials who knew them. They also combed the official files of both public and private agencies that might have had contact with the subjects or their families.

During their lifetimes the Gluecks did very little analysis of these longitudinal data. Sampson and Laub decided to try to reanalyze the data gathered years before using more sophisticated statistical techniques.

Testing Control Theory

Both of Sampson and Laub's hypotheses were strongly supported. First, no matter which measures of childhood and adult behavior were used, they found strong evidence that antisocial behavior in childhood was associated with troublesome behavior in adulthood. For instance, three-fourths of the boys who were official juvenile delinquents, but only one-fifth of the boys who were not, had been arrested between the ages of seventeen and twenty-five. Similar results appeared when other measures of childhood and adult behavior were used. Remember that the boys came from the same neighborhoods and were matched on race-ethnicity, ability, and age. This matching helps researchers have even more confidence that these differences in adult behavior are related to their different characteristics in childhood.

Second, like virtually all the relationships sociologists study, the association between childhood and adult antisocial behavior was far from perfect. Some men who were delinquent as children did not participate in criminal and other deviant activities as adults; others who were not identified as delinquent or antisocial in childhood did turn to these behaviors in adulthood. As Sampson and Laub hypothesized, the nature of the social bonds that the men had in adulthood was a major influence on their behavior. Those who had stable jobs and strong marital ties were far less likely to engage in adult deviant behavior, whether or not they had a childhood history of problems. In other words, supportive social networks and conventional work and family careers in adulthood helped alter the path of the men's criminal or deviant careers. These findings applied for both the delinquent and nondelinquent samples, and the correlation was so strong that it could not be attributed to chance.

Like all good sociologists, Sampson and Laub were not content with these findings, even though the findings strongly supported their hypotheses. Instead, Sampson and Laub continued to think about other possible explanations for their findings, about ways that their results might be falsified. For instance, they knew that the subjects differed a great deal in the severity of their antisocial and deviant behaviors in childhood and adolescence. Some of the boys had been involved in quite serious incidents of violent behavior, whereas others had only engaged in relatively minor crimes such as truancy and petty theft. Perhaps the most delinquent subjects were more likely to have job instability, higher rates of crime, and less happy marriages, while those who were guilty of only minor crimes, but who had still been classified as delinquent, had much more stable family and work careers.

To answer this question, Sampson and Laub conducted several complicated statistical analyses in which they examined the effect of adult social bonds on adult criminal and deviant behavior once the seriousness of earlier delinquency and crime was controlled. The results continued to support their basic

Kip Kinkel was fourteen years old when he shot his parents and several fellow students at Thurstone High School in Springfield, Oregon, in May 1998. Based on what you have read in this chapter, how would you sentence Kip Kinkel?

hypotheses that both childhood antisocial behavior *and* adult roles influence adult deviance and crime. Even subjects who were involved in very serious criminal activity as juveniles could alter these careers when they had supportive social networks in adulthood. Strong marital ties seemed to be especially important. No matter how much juvenile crime the subjects had engaged in or how successful they were in their jobs, they were less likely to engage in criminal activities and other deviant behaviors in adulthood if they were strongly attached to their spouses. Just as studies have found that strong attachments to parents in childhood help deter juvenile crime and delinquency, the results of this study indicate that strong and loving attachments to a partner in adulthood help deter adult crime.[44]

Sampson and Laub looked at the Gluecks' data in a lot of different ways. No matter what type of analysis they tried or which measures they used, they always obtained the same kind of results: Early childhood differences in delinquency and deviance were related to adult behaviors. But other life course events, especially those that involved strong support from marital partners and job stability, could alter

these connections, even when the seriousness of the childhood deviance was taken into account.

Results that support these findings have been obtained in replications of Sampson and Laub's work. Because it would be very expensive today to gather a data set like that the Gluecks developed, other researchers have looked at the influence of social support and attachments over shorter time periods. For instance, a study of adult prisoners found that, in the two to three years before being incarcerated, virtually all the inmates had committed a number of crimes. This supports Sampson and Laub's finding of continuity in criminal behavior over time. Yet, the inmates' rate of involvement in crime was not always constant over this time period. Instead, involvement tended to vary depending on the nature of their social relationships. When the inmates were in school, employed, or living with their wives, they were less likely to engage in criminal behavior.[45] Another study, described in Box 6-2, examined the role of social support in diverting young street people from criminal careers.

These findings echo the conclusions we reached in our study of socialization and the life course. The social networks and interactions, or *social capital,* that we have in childhood certainly influence the opportunities and constraints that we face in our lives. Yet, even if our early experiences are not especially supportive or the types that could, by themselves, predict later success, we can, with the right kind of social support, take alternative life paths. Thus Sampson and Laub's results suggest that Jason could have altered his criminal career if he had been able to develop the kinds of supportive social ties that Justin did.

Of course, some people are much more likely than others to be able to have supportive social ties, and a key influence is economic well-being. Although not all people who are economically well off have abundant social capital, it is often far easier for families with economic resources to find ways to provide this support. All the subjects in Sampson and Laub's study grew up in poor neighborhoods. In general, people who live in poverty are far more likely to be involved in crime, both as victims and as perpetrators.[46] In the next section of this book we look more directly at variations in economic well-being and access to power, the phenomenon of social stratification.

CRITICAL-THINKING QUESTION

1. Based on Sampson and Laub's results, what types of policies might you recommend to officials who are concerned with combating crime? Why?

BOX
6-2

**APPLYING SOCIOLOGY
TO SOCIAL ISSUES**

*Street Kids,
Social Capital,
and Deviant Behavior*

The phenomenon of "street kids," young people who have left their homes and live on the streets, is found throughout the world, both in rich countries such as the United States and Canada, and in poorer, developing countries such as Brazil. These young people must often fend for themselves and daily look for food and safe shelter. Studies of homeless youth find that they often resort to shoplifting, robbery, prostitution, or the sale of drugs to help support themselves. Although street kids are only a small proportion of all young people, they are involved in a substantial proportion of crimes committed by the young.[47]

The sociologists John Hagan and Bill McCarthy studied the lives of street youth in two Canadian cities: Vancouver, British Columbia, and Toronto, Ontario. Both are large metropolitan areas with substantial numbers of young people who live on the streets with no permanent address. Throughout one summer Hagan and Mc-

Carthy's team of youthful research assistants interviewed several hundred of these street youth three different times in a *panel study.* Their questions were designed to find out as much as they could about the young people's lives before they came to the streets as well as their experiences while they were homeless. For comparative purposes, the researchers also gathered information for many of the same questions from a sample of young people who were in school and not living on the streets.

Hagan and McCarthy's analysis of these data supports each of the contemporary theories regarding deviant behavior that have been described in this chapter. As control theory would predict, the young people on the street were far more

likely than students living at home to come from families that did not provide optimal levels of affection, supervision, or discipline. As generalized strain theory would predict, these aversive home environments prompted many of these young people to leave home. Among those living on the streets, some youth were more likely than others to engage in deviant and criminal behavior. As generalized strain theory would predict, the street youth were much more likely to turn to criminal activity when they needed food or adequate shelter and had no means of legitimate employment. As differential association theory would predict, the activities of friends and associates were also a strong influence on an individual's own criminal behavior. Young people were much more likely to turn to crime if they had friends who were also involved in criminal activity. Finally, Hagan and McCarthy found, as labeling theory would predict, that being officially labeled and known as a criminal to

SUMMARY

Sociologists have long been interested in deviance—in why people do and do not participate in various deviant behaviors and why there are variations in the extent of crime and deviance in different times and places. Deviance is ever present, or at least potentially present, and never exists without social control, or efforts to ensure conformity to norms. Key chapter topics include

- Deviance can appear anywhere and can take many different forms, although the definition of what constitutes deviant behavior will vary from one society or situation to another.

- The principal sources of crime data are official reports. The *Uniform Crime Report* is developed by the FBI from data compiled by local police agencies, and the *National Crime Victimization Survey* asks a representative sample of households about crime they have experienced each year. Because many

crimes are not reported to the police, the NCVS reports many more crimes than the UCR.

- The crime rate has varied a fair amount over the past fifty or so years. A major reason for this variation is that most crimes are committed by young people, and the proportion of young people in the population has also varied over this time. Variations in the crime rate might also reflect a cohort effect.

- Not all people in the United States are equally likely to be victimized by crime. African-American males are much more likely than Euro-Americans or African-American females are to be victims of violent crime.

- Among the classical theorists, Durkheim suggested that deviance helps a society reaffirm its goals and bind members of the society more closely. Marx emphasized how social control reflects the divi-

the police intensified criminal in-volvement. This was especially true for youth who had suffered abuse from their parents while they were at home. For these young people contacts with the police appeared to feed into a spiral of ever increasing secondary deviance.[48]

The processes that influenced whether or not individuals would engage in crime and the level of violent crime that the street kids committed were the same in both Vancouver and Toronto. (Remember, however, that Canada has a much lower rate of violent crime than the United States, as shown in Map 6-1.) Yet the two cities differed in the level of nonviolent crimes. Street kids in Vancouver were far more likely than those in Toronto to be involved in nonviolent offenses involving drugs, theft, and prostitution. Hagan and Mc-Carthy suggest that these differences reflect differences in the access to social capital that the two cities provide for young people on the streets. Toronto tends to have a

"social welfare orientation," with a number of overnight shelters and a relatively high level of support services. In contrast, Vancouver tends to use a "crime control model," providing few resources or support. As a result, Vancouver youth are more likely to turn to various criminal means to obtain support for their daily needs.[49]

Furthermore, just as Sampson and Laub found that paid employment helped to keep former delinquents away from a life of crime, Hagan and McCarthy found that having a job was an important way to help young people move off the streets, even though their jobs tended to be low-paying. As they put it,

We found over the course of the summer during which we tracked street youth in Vancou-

ver and Toronto that those youth who were able to find employment were more likely to begin moving away from the street: They spent less time hanging out, panhandling, searching for food and shelter, using drugs, stealing with other youth, or pursuing other kinds of criminal activities. Youth who began to move along this track and away from the street were not enamored with the employment they found, but they nonetheless saw these entry-level jobs as a place to begin transitions to more conventional lives apart from the street.50

CRITICAL-THINKING QUESTIONS
1. What do Hagan and Mc-Carthy's results suggest can be done to reduce crime among street youth? Could these policies work in your community?

2. Are there implications for reducing crime among older populations as well?

sions within societies between the powerful and the less powerful. Weber showed how social control must be legitimated by members of a society.

■ Labeling theory suggests that when people are labeled and treated by others as deviant, they will come to see themselves as deviant and then increasingly act in ways that conform to this label.

■ The original version of strain theory suggests that people who value cultural goals but lack the institutional means to obtain them will experience frustration and therefore will use illegitimate means to obtain these goals. Strain theory has been criticized for its failure to account for white-collar crime. The more recently developed generalized strain theory suggests that sources of strain can involve many areas of social life and that people can respond to strain in different ways.

■ Differential association and subcultural theories of deviance emphasize that we learn deviant behaviors

from our associations with others and that subcultures exist in which the group norms support behaviors that are seen as deviant in the larger culture.

■ Control theory describes how our associations and linkages with conventional peers and groups serve as a form of external social control. Children who have strong emotional attachments to parents who also supervise and discipline them are most likely to develop strong internal social control. Control theory can also account for cross-cultural variations in crime.

■ Robert Sampson and John Laub used data gathered many years ago by Sheldon and Eleanor Glueck to test hypotheses developed from control theory. Sampson and Laub found that a strong association exists between childhood behavior problems and adult deviancy and crime, but that this strong linkage can be modified by stable marriages and occupational careers.

KEY TERMS

age effect 132

charismatic authority 136

cohort effect 132

collective consciousness 134

control theory 141

crime 129

crime rate 130

deviance 127

differential association theory 140

external social control 128

general strain theory 139

ideal types 136

Index crimes 129

internal social control 128

labeling theory 137

legal authority 136

longitudinal study 145

panel study 146

period effect 131

property crimes 129

sanctions 128

secondary deviance 138

social control 128

strain theory 139

subcultural theory 140

synnomie 142

traditional authority 136

violent crimes 129

waves of data collection 146

white-collar crime 139

INTERNET ASSIGNMENTS

1. Look for deviance and social control on the Internet. Browse through several chat rooms dealing with different types of topics. How is deviance defined within these rooms. How does it vary from one room to another? What types of social control mechanisms are used. To what extent is social control internal? To what extent is it external? How can you tell?

2. Using a search engine, locate a Web site (for example, Bureau of Justice Statistics) that contains information about crime rates in the United States. Obtain selected crime rates (for example, violent crime rates, property crime) for several states and for different cities. You might want to choose cities and states from different regions of the country. What patterns do you see? Can you use the theories on deviant behavior discussed in this chapter to help you explain these patterns?

3. Try to locate on-line publications or discussion groups that are devoted to discussions of deviant behavior. What issues are being discussed? Try to find opposing viewpoints on these issues. Can you identify any "new" forms of deviant behavior (for example, new musical forms, body art, computer hacking, and so forth)? Who considers these to be deviant and who accepts them as viable behavior?

What trends do you see? Try to use the ideas of the classical theorists (Durkheim, Marx, and Weber) to help you interpret this material. For instance, what functions might the deviance fulfill? For whom? Who defines the behavior as deviant? What types of power differentials are involved? What kinds of social control mechanisms are present and how is this social control legitimated?

4. Using a search engine, locate Web sites that discuss gang behavior. What types of information can you retrieve? From these sites can you determine any similarities among gangs? What types of differences can you identify? What are the motivating factors behind gang formation? How does this information relate to the theories discussed in the text? What types of social control mechanisms are used to counter gang behavior? Based on your readings, which of these mechanisms do you think would be most effective? Why?

TIPS FOR SEARCHING

The following key words are useful when you search for the topics of deviant behavior: crime, crime statistics, crime victims, criminal justice, criminology, delinquency, gangs, juvenile crime, law, violence.

INFOTRAC COLLEGE EDITION: READINGS

Article A20575292 / Haley, John O. 1998. Apology and pardon: learning from Japan. *American Behavioral Scientist, 41,* 842-868. Haley examines the role of apologies and pardons in the reduction of Japan's crime rate since World War II and

explores the possibility that similar practices might be effective in the United States.

Article A53286529 / Gorman-Smith, Deborah, Patrick H. Tolan, Rolf Loeber, and David B. Henry. 1998. Relation of family problems to patterns of delinquent involvement among urban youth. *Journal of Abnormal Child Psychology, 26,* 319–329.Gorman-Smith, Tolan, Loeber, and Henry explore the relationship of family interactions to juvenile delinquency using many of the ideas found in control theory.

Article A19487818 / Heimer, Karen. 1997. Socioeconomic status, subcultural definitions, and violent delinquency. *Social Forces, 75,* 799-834. Heimer uses a large data set to examine the ways in which families, peer subcultures and prior behavior all influence the probability that young people will engage in delinquent behavior.

Article A20870808 / Sullivan, Richard, Marcia Fenner Wilson. 1995. New directions for research in prevention and treatment of delinquency: a review and proposal. *Adolescence, 30,* 1–18. Sullivan and Wilson review juvenile delinquency prevention programs and propose research that could test their relative effectiveness.

FURTHER READING

Suggestions for additional reading can be found in *Classic Readings in Sociology,* bundled with this book. If you purchased a used copy of this book that does not include this custom-published reader, go to www.sociology.wadsworth.com for ordering information.

RECOMMENDED SOURCES

Adler, Freda, Gerhard O. W. Mueller, and William S. Laufer. 1997. *Criminology,* 3rd ed. New York: McGraw-Hill. This is an extensive criminology text written from a sociological perspective.

Akers, Ronald L. 1998. *Social Learning and Social Structure: A General Theory of Crime and Deviance.* Boston: Northeastern University Press. Akers provides a contemporary view of differential association theory.

Decker, Scott H., and Barrik Van Winkle. 1996. *Life in the Gang: Family, Friends, and Violence.* New York: Cambridge University Press. Decker and Van Winkle report on an extensive field research study of gang members and their social relationships in St. Louis, Missouri.

Erikson, Kai T. 1966. *Wayward Puritans: A Study in the Sociology of Deviance.* New York: Macmillan. Erikson provides a classic application of Durkheim's theory of deviance.

Gottfredson, Michael R., and Travis Hirschi. 1990. *A General Theory of Crime.* Stanford, Calif.: Stanford University Press. Gottfredson and Hirschi give an extended development of the basic ideas of control theory.

Hagan, John, and Bill McCarthy. 1997. *Mean Streets: Youth Crime and Homelessness.* New York: Cambridge University Press. This book provides a full report of the research described in Box 6-2.

Henry, Stuart, and Werner Einstadter (eds.). 1998. *The Criminology Theory Reader.* New York: New York University Press. This reader is a collection of the most important articles appearing in the journal *Criminology* from the mid 1980s to the mid 1990s.

Sampson, Robert J., and John H. Laub. 1993. *Crime in the Making: Pathways and Turning Points Through Life.* Cambridge, Mass.: Harvard University Press. Sampson and Laub give book-length treatment of the research featured in this chapter.

Short, James F., Jr. 1997. *Poverty, Ethnicity, and Violent Crime.* Boulder, Colo.: Westview. Short provides a very readable and carefully developed analysis of the relationship between poverty, race-ethnicity and crime that includes many statistics.

INTERNET SOURCES

Bureau of Justice Statistics. The source of many of the statistics presented in this chapter, including both UCR and NCVS data.

Gray Areas. This on-line publication contains information on deviant and controversial behavior and issues.

The National Criminal Justice Reference Service (NCJRS). Provides links to many sites concerned with criminal justice policy and data.

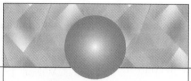

Pulling It Together

Material related to topics discussed in this chapter are found in a number of other places in this book. You can also use the index to find related material.

TOPIC	CHAPTER
Social networks and social capital	4: Socialization and the Life Course, pp. 77–79
	5: Social Interaction and Social Relationships, pp. 101–104
	7: Social Stratification, pp. 000. 179–180
	8: Racial-Ethnic Stratification, pp. 200, 203–204
	10: The Family, pp. 257–258, 265–266
	11: Formal Organizations, pp. 283–283, 287–288
	12: Religion, pp. 320–321
	13: The Political World, pp. 339–352
	14: The Economy, pp. 361–371, 376–385
	15: Education, pp. 402–404, 408–412
	16: Health and Society, pp. 425–428
	18: Communities and Urbanization, pp. 487–492, 495–503
	20: Collective Behavior and Social Movements, pp. 535–536, 544–545
Age, period and cohort	4: Socialization and the Life Course, pp. 90–95
	8: Racial-Ethnic Stratification, pp. 202–203
	17: Population, pp. 452–453, 461–464
	18: Communities and Urbanization, pp. 502–503
Symbolic interactionism and self identity	4: Socialization and the Life Course, pp. 83–85
	5: Social Interaction and Social Relationships, pp. 108–110
	8: Racial-Ethnic Stratification, pp. 204–207
	9: Gender Stratification, pp. 233–237
	12: Religion, pp. 320–321
	14: The Economy, pp. 382
	16: Health and Society, pp. 425
Subcultures	3: Culture and Ethnicity, pp. 53, 58–67
	11: Formal Organizations, pp. 284–285
	12: Religion, pp. 320–321
	18: Communities and Urbanization, pp. 490–492
Life course analysis	4: Socialization and the Life Course, pp. 90–95
	10: The Family, pp. 259–261
	14: The Economy, pp. 380–385
	15: Education, pp. 404–406

Social
Stratification

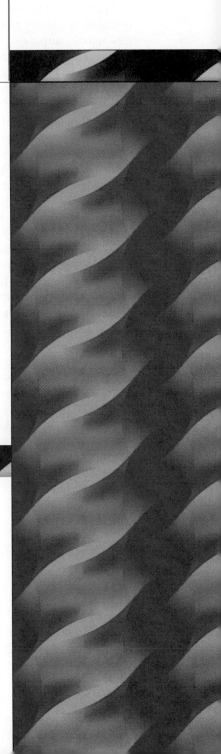

WE FIRST MET RITA in Chapter 5. She had left her small hometown in northern Michigan to go to Greenville College, an elite private school in New England. Her father was a truck driver for the local mining company, and her mother worked as a waitress in a local cafe. Her new friends at Greenville seemed to come from another world. Some of them spent more money on clothes and vacations than many families in her hometown had to spend on food and rent. Her roommate Michelle's father was the president and chief executive officer of a large computer and software manufacturing firm. Her friend Lori's great-grandfather had struck gold in the late nineteenth century and established a highly profitable gold mining and jewelry manufacturing corporation. The monthly checks Lori received from trust funds that her grandfather and father had established for her were large enough that she would never have to work to earn a living.

Rita also felt out of place because most of the other students had attended prep schools and seemed to know one another from schools, summer camps, or resort communities where they spent the holidays. Rita wasn't quite sure how to behave around her new friends or even which fork to use when she went out to dinner at an elegant and high-priced restaurant with Michelle's family.

Rita rarely talked about her feelings with Lori or Michelle. Even when she told them about life back home, they didn't seem to grasp how different their experiences were from those of people like Rita and her family. Of course Rita knew that she wouldn't mind being more comfortable financially, and after college her job as a staff lawyer provided a lifestyle that was luxurious by the standards she'd grown up with. But she would never have the financial security that Michelle and Lori had—and she was still unsure about which fork was for what. Although she sometimes dealt with wealthy clients, they clearly considered her a kind of legal plumber or high-priced servant. Money alone wasn't enough to make her part of their world. In a way she was in between worlds now. "Home" had lost a lot of its meaning.

One day, as Rita walked from the subway to her office at the law firm, she was stopped by a young woman in a ragged coat who was holding out her palm, hoping for a handout. "Homeless," the woman muttered. "Got three kids. Can you help?"

Usually Rita brushed right by panhandlers, never quite knowing whether to believe their stories. This time, though, she fumbled in her purse for some change. Maybe it was because the woman's eyes reminded Rita of her mother's, making her think of her worries that her parents would have too little money to be comfortable in their old age. On an impulse she opened her wallet and took out a five-dollar bill. "Here," she said, thrusting it at the woman. "I hope this helps." With that she hurried off, not waiting for a thank you.

All that day the encounter with the woman lingered in the back of Rita's mind. Were Rita's parents and the homeless woman to blame if they had to struggle or even beg money from strangers to get by? Rita knew people who were quick to say that anyone who worked hard enough could get ahead, just as she had done. But was that really true? Could her own parents have done what she did, in becoming a well-paid professional? And if money was all that mattered, why did she still feel so out of place in Boston society?

Of course, Rita had wondered such things before, but today the questions seemed more real and urgent than usual. That night she wrote her parents a newsy letter as an excuse to send them a check "to buy a little something with." At the last minute she added a postscript: "Dad and Mom, I've been thinking about how hard you've always worked. It just doesn't seem fair that you didn't have the same opportunities as me. So I hope you don't mind if I share a little of my good fortune sometimes."

Rita smiled ruefully. "Good fortune," she thought. "Was that what it all came down to, in the end?"

How can anyone be without a place to live in a country of such wealth? Why are there such huge differences in people's life chances? Are some people just lucky to be born into the "right" families? And who decides which families get to be the "right" ones in the first place?

$\mathcal{R}$ITA'S EXPERIENCES REFLECT **social stratification,** the organization of society in a way that results in some people, like Lori and Michelle and their families, having more and some people, like Rita's family, having less.[1] Social stratification also involves **social class,** groups of people who occupy a similar level in the stratification system. Sociologists would say that Lori and Michelle come from different social classes than Rita.

Americans like to think that they live in a "classless society"—that they all have pretty much the same opportunities and that they all can have an equal influence on their world. You might have heard about the "American Dream," the idea that each of us has a right to "success" and the opportunity to "better our station in life" and reach positions of power. Even if we are born into a poor family, the American Dream allows us to do whatever we want in life—to be just as successful as Lori's great-grandfather or Michelle's father. No one needs to be trapped in poverty; anyone can move up—all that is needed is hard work.[2]

Unfortunately, when we take a sociological view of our society, and other modern industrialized societies, the American Dream often doesn't match reality. We saw this in our discussion of the status attainment model in Chapter 5. The model shows how people who come from families with more material advantages are more likely to have higher incomes in adulthood, even if individuals from poorer families are equally smart and work equally hard. When we use the sociological imagination to look at society, we see that many people have virtually no chance of "making it big." A line worker in a factory can work just as hard and be just as dedicated to his job as the company president. But no matter how hard the factory worker toils, he is highly unlikely to ever have a job that pays much more than he is currently earning, let alone reach a position of power.

This chapter explores issues related to social stratification in the United States. We first look at how sociologists have studied stratification and what they have found. We then consider theoretical explanations for stratification. Why are societies stratified, and why are some societies more stratified than others? Finally we look more closely at the American Dream. How do people move about in the American stratification system? How unequal is our society? Can people escape poverty? How unequal is the influence wielded by rich and poor people in our society?

As you read about stratification, you might picture systems of stratification as mountain ranges. Mountains come in many different shapes. Some, like those in the Rockies or the Cascades, are tall and peaked, with large bases and very little room at the top. Others, like the Appalachians, are flatter and

How do the people shown here represent different social classes? What elements of the photo led you to this conclusion?

more rounded. Similarly stratification systems vary in their dimensions. Some societies have a great deal of inequality, with a small elite group at the top and a large group of people with many fewer resources and few chances to move up. Other societies have much less inequality and, sometimes, more chances for people to move from one level to another.

Social stratification is clearly much more than a sociological abstraction. Stratification and inequality affect people's life chances, and they raise profound questions of economic and political justice. By trying to understand the reality of stratification from a scientific perspective, many sociologists hope that they can contribute to creating a world where inequality is lessened and more people have the opportunity to use their talents and skills to their fullest capacity.

SOCIAL STRATIFICATION IN THE UNITED STATES

Sociologists who have studied stratification in the United States have found it to be a complex phenomenon that involves several variables. The first studies tended to focus on single communities, looking at the way in which social hierarchies developed within individual cities and towns. More recently sociologists have examined stratification on a societywide basis.

Stratification in Local Communities: Yankee City

Some of the earliest studies were conducted by W. Lloyd Warner and his students and colleagues. They

FIGURE 7-1 *The Six Social Classes of Yankee City*
Source: Warner and Lunt, 1941, p. 88; Gilbert, 1998, p. 27.

1. Upper-upper (1.4%) The old-family elite, who had inherited wealth that had been in the family for more than one generation and who lived in large houses in the best neighborhoods.

2. Lower-upper (1.6%) The newly wealthy, who usually were slightly richer than the old-family elite but who had not yet attained the same polish and self-assurance.

3. Upper-middle (10.2%) Business and professional people who were moderately successful but not as well-off as those in the lower-upper group. Some education and social skills were necessary to belong to this group, but ancestry wasn't important.

4. Lower-middle (28.1%) Smaller business owners, schoolteachers, and foremen in industry, who belonged to the local churches and lodges.

5. Upper-lower (32.6%) The "solid, respectable" working people, who kept their houses clean, worked hard, and stayed out of trouble.

6. Lower-lower (25.2%) The "lulus" or disreputable people, who often dug clams and waited for public relief.

began their work in the early 1930s in a New England town of 17,000 people, which they called "Yankee City." It was a fairly stable coastal community with a long history. Warner and his research team lived in Yankee City for several years, familiarizing themselves with the community.[3] One of their principal goals was to understand how the town's stratification system was structured—who were the "notables" and "nobodies," who was at the top, middle, and bottom of the stratification "mountain"—and what influenced where people fell within this system.[4]

Using techniques of field research popularized by the Chicago School, Warner and his associates observed the interactions and behaviors of the town's residents, combed birth and marriage records, and examined tax reports and other official records revealing ownership of property and transfers of income. They carefully read the local newspaper and any other documents about the town and its history that they could find.[5] Gradually they developed a card file that summarized information on each of the 17,000 residents, including their social networks, their lifestyles and occupations, their income levels, their ethnic heritage, and even the doctors they called when they were sick. From all this information Warner and his associates developed a picture of the class structure of Yankee City.

At first, Warner expected that the residents of Yankee City would use economic criteria to differentiate their fellow citizens. Most of the early interviews supported this hypothesis, as residents talked about

"the big people with money" and "the little people who are poor."[6] But as time went on, Warner and his associates realized that membership in a social class reflected not just people's wealth or their occupation and education, but a whole combination of variables, including their lifestyle, the people they associated with, and their parents and even their ancestors.[7]

Eventually Warner and his associates decided that Yankee City's class structure consisted of six separate groups, as shown in Figure 7-1. These groups are what sociologists call **prestige classes,** aggregates or clusters of people who possess similar characteristics, who are perceived by their fellow citizens as being similar, and who are accorded similar levels of respect and esteem. Warner suggested that these groupings reflected both the objective reality of the residents' associations and activities and their subjective views of themselves and one another. The groupings should be seen as *ideal types,* in the sense that Weber used the term, for they are a sociologist's description or representation of the class structure and probably provide more clear-cut and sharper definitions than most Yankee City residents would have given.

Notice that with this notion of prestige class, class status is associated with families, not just individuals. As a child you had the status of your parents. As a married couple, you and your spouse share your class status. Of course people can move from one class group to another, a process that sociologists term **social mobility.** For instance, a young woman from a family Warner would classify as upper-lower, perhaps like Rita, might attend college, enter an

These young girls are learning the proper way to eat artichokes. Where might they fit into the class structure of Yankee City? What opportunities and constraints might they face in their lives because of this class placement?

upper-middle-class profession, and eventually marry into a family of higher status. However, mobility can be downward as well as upward. For example, someone from a family in the lower-middle-class group in Yankee City might have dropped out of school and taken a job in a factory, thus moving into the upper-lower-class group.

Warner and his associates replicated the Yankee City study in a number of other communities and consistently found that they could differentiate the residents into prestige classes.[8] The exact number of classes sometimes varied, as did the proportion of the population found in each class group. In addition, communities with diverse racial-ethnic populations typically had sharp divisions between Euro-Americans and people of color, with the Euro-Americans enjoying higher status and greater economic rewards. Social class groups, however, typically appeared within each racial-ethnic classification, leading to parallel, although unequal, stratification systems.[9] (We look at racial-ethnic stratification in more detail in the next chapter.)

Measuring Prestige and Social Class Today

The techniques that Warner and his associates used in the 1930s and 1940s within single communities would be very hard to replicate in larger cities. To describe the stratification systems in larger cities and the nation as a whole, contemporary sociologists use survey rather than field research. In placing people in social classes, researchers typically look at their education, occupation, income, and, sometimes, friendship patterns.

If they have to pick only one variable to use as a measure, sociologists generally use occupation.[10] Over a number of years and in many different countries, sociologists have asked people to describe jobs that they think are "better" or "worse" than others to find out the type of prestige people ascribe to various occupations. The results are strikingly consistent. In different times and in different societies, some occupations, such as physician, nuclear physicist, and scientist, are continually ranked at the top of the prestige scale; and other occupations, such as garbage collector, street sweeper, and shoe shiner, are continually ranked at the bottom.[11] An example of these rankings, which are commonly called occupational prestige scores, is given in Table 7-1.

Occupational prestige scores are highly associated with income and education.[12] People in jobs with more prestige generally earn more than people in occupations farther down the scale. Sociologists often use occupational prestige scores as a quick and easy way to figure out where people fall within the stratification system.

Besides looking at occupational prestige scores, sociologists today, like Warner, also refer to the social class to which an individual belongs. To use the metaphor of a mountain, social class categories divide the social mountain into discrete layers, much as a geologist might identify different strata of minerals and rocks beneath the surface of a hill. A typical way in which sociologists have used the information of occupational prestige scales, along with other data, to group people into discrete social classes is shown in Table 7-2. In making the distinctions between these social class groups, sociologists have considered whether people

TABLE 7-1   *Occupational Prestige Scores*

SOURCE	OCCUPATION	SOURCE	OCCUPATION
94	U.S. Supreme Court justice	72	Policeman
93	Physician	71	AVERAGE
92	Nuclear physicist	71	Reporter on a daily newspaper
92	Scientist	70	Bookkeeper
91	Government scientist	70	Radio announcer
91	State governor	69	Insurance agent
90	Cabinet member	69	Tenant farmer who owns livestock and machinery and manages the farm
90	College professor		
90	Member, U.S. Congress	67	Local labor union official
89	Chemist	67	Manager of a small store in a city
89	U.S. Foreign Service diplomat	66	Mail carrier
89	Lawyer	66	Railroad conductor
88	Architect	66	Traveling salesman for a wholesale concern
88	Country judge	65	Plumber
88	Dentist	63	Barber
87	Mayor of a large city	63	Machine operator in a factory
87	Board member of a large corporation	63	Owner-operator of a lunch stand
87	Minister	63	Playground director
87	Psychologist	62	U.S. Army corporal
86	Airline pilot	62	Garage mechanic
86	Civil engineer	59	Truck driver
86	State government department head	58	Fisherman who owns his own boat
86	Priest	56	Clerk in a store
85	Banker	56	Milk route man
85	Biologist	56	Streetcar motorman
83	Sociologist	55	Lumberjack
82	U.S. Army captain	55	Restaurant cook
81	Accountant for a large business	54	Nightclub singer
81	Public school teacher	51	Filling station attendant
80	Building contractor	50	Coal miner
80	Owner of a factory that employs about 100 people	50	Dock worker
78	Artist whose paintings are exhibited in galleries	50	Night watchman
78	Novelist	50	Railroad section head
78	Economist	49	Restaurant waiter
78	Symphony orchestra musician	49	Taxi driver
77	International labor union official	48	Bartender
76	County agricultural agent	48	Farmhand
76	Electrician	48	Janitor
76	Railroad engineer	45	Clothes presser in a laundry
75	Owner-operator of a printing shop	44	Soda fountain clerk
75	Trained machinist	42	Sharecropper who owns no livestock or equipment and does not manage farm
74	Farm owner and operator		
74	Undertaker	39	Garbage collector
74	City welfare worker	36	Street sweeper
73	Newspaper columnist	34	Shoe shiner

Source: Hodge, Siegal, and Rossi, 1964, pp. 290–292.

TABLE 7-2   *A Contemporary View of the American Class Structure*

CLASS, PERCENT OF HOUSEHOLDS	SOURCE OF INCOME, OCCUPATION OF MAIN EARNER	TYPICAL EDUCATION	TYPICAL HOUSEHOLD INCOME, 1995
Privileged Classes			
Capitalist 1%	Investors, heirs, executives	Selective college or university	$1.5 million
Upper-middle 14%	Upper managers and professionals, medium-sized business owners	College, often post-graduate study	$80,000
Majority Classes			
Middle 30%	Lower managers, semi-professional, nonretail sales workers	At least high school, often some college	$45,000
Working 30%	Operatives, low-paid craftsmen, clerical workers, retail sales workers	High school	$30,000
Lower Classes			
Working Poor, 13%	Most service workers, laborers, low-paid operatives, and clerical workers	Some high school	$20,000
Underclass, 12%	Unemployed or part-time workers, many dependent on public assistance and other government transfers	Some high school	$10,000

Source: Gilbert, 1998, p. 286.

earn their money by working for a wage or from investments, how extensive their educational credentials are, how much independence and freedom they enjoy at work, and how stable and consistent their participation in the labor force is. For example, sociologists estimate that less than 1 percent of the households in the United States belong to the elite capitalist class, whose members can live off their investments. In contrast, the working poor and the underclass together outnumber capitalist families by about 25 to 1. In between are the privileged upper middle class (who would be regarded as very wealthy by most of the world) and the middle and working classes, who together make up most of the population.[13]

As noted previously, all societies have some type of stratification system, although some are more unequal than others. But why does stratification exist? Why do some people live in poverty while others are very rich? Why does stratification persist even though particular individuals and families move up or down in the social class ladder? These questions have intrigued sociologists since the beginning of the discipline.

WHY DOES STRATIFICATION EXIST? FOUR THEORETICAL PERSPECTIVES

As with most areas of social life, sociologists have a variety of perspectives on stratification. Each high-

lights different explanations of why stratification exists, why it varies from one type of society to another, and why it persists.

Economic Exploitation: Marxian Views

Karl Marx believed that the key to understanding social order was recognizing who controlled the means of production. Because Marx saw the material basis of society as the key to social structure, he believed that social stratification *is* social structure—they are one and the same thing.

Notions of conflict permeate Marx's view of social structure and stratification. According to Marx stratification occurs because the more powerful exploit the less powerful—that is, some people are wealthy *because* other people are poor. But this very exploitation has the potential to arouse the lower classes and provoke revolutionary change. In this sense the stratification "mountain" is not permanent and unchangeable; rather, it is a temporary equilibrium between dynamic forces. If the equilibrium is upset, a volcanic eruption can blow off the mountaintop and rearrange the layers. Let's examine both Marx's original ideas and a contemporary sociologist's updated version that more accurately reflects modern society.

A Dual Class Structure: Classical Marxism A key component in Marx's original theory is that those

How would the people pictured here be classified in Marx's view of the stratification system? To what extent do you expect these people had class consciousness in the sense Marx used the term?

who control the means of production maintain their position both by exploiting people's labor and by controlling other aspects of their lives, such as religion, family life, and politics. This division between those who control the means of production and those who do not is central to Marx's view of social stratification. He suggested that in the European societies and the United States of his day there were two major social classes: the *bourgeoisie,* who owned the factories and mills (the means of production), and the *proletariat,* or workers.

As we saw earlier, the concept of social class can involve nonmaterial qualities such as occupational prestige. But for Marx the definition of social class reflected his belief that society is built on divisions, conflict, and change related to the means of production. According to Marx groups of people with similar positions in the production process—such as workers, owners, peasants, or feudal lords—have the potential to become social classes because people in each of these groups have similar experiences in the work process, especially in either exploiting or being exploited by other people.

However, similar experiences aren't enough. To truly become a class, people within the group need to develop what Marx called **class consciousness,** awareness of their common interests and concerns and of the fact that they are in conflict with another class group. In addition, and most important in Marx's view, "separate individuals form a class only in so far as they have to carry on a common battle against another class."[14] Thus the proletariat would become a social class when members realized that

they had common interests that opposed those of the bourgeoisie and when they joined together in conflict against them.

Marx recognized that industrial societies contain groups besides the bourgeoisie and proletariat. However, he felt that these other groups lacked the necessary sense of common purpose to qualify as a true class. For instance, the society he lived in included a fair number of people we would today call "upper middle class" and "middle class," including self-employed professionals, such as doctors, lawyers, and businessmen. Yet Marx thought that these people lacked sufficient consciousness to form a class. He predicted that the gap between rich and poor would widen as capitalism became more established and that the middle class would shrink and eventually become part of the proletariat. The worsening conditions would lead to stronger class consciousness and, eventually, struggles to help produce a more equitable society.[15]

Marx died more than one hundred years ago, and economies have changed greatly since that time. For instance, contrary to his prediction, the middle class in industrialized countries has become much larger over the past century. Yet Marx's idea that social stratification is the basis of social structure and that it involves a continual process of economic exploitation remains persuasive to many sociologists. When we look around a busy city or even a small town, we can immediately see that some people are able to exploit others, whereas others are much more likely to be exploited. How, some sociologists have asked, could the basic insights of Marx be applied to modern societies?

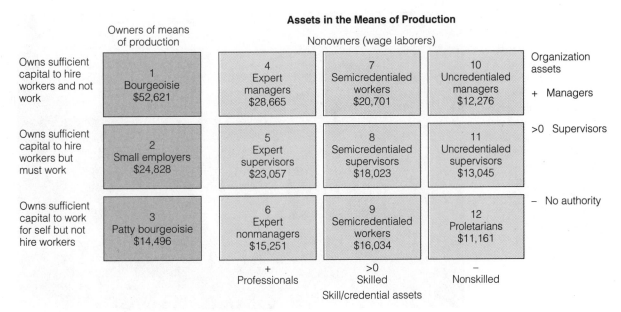

Assets in the Means of Production

FIGURE 7-2 *Wright's Neo-Marxian Formulation of Social Classes*

Note: Dollar figures are the mean gross annual individual income from all sources, before taxes in 1980.

Source: Grimes, 1991, p. 185; Wright, 1989a, p. 35; Wright, 1997, p. 155.

Multiple Methods of Economic Exploitation: A Contemporary Marxian View Erik Wright, perhaps more than any other contemporary Marxian sociologist, has tried to describe the class structure of the United States and other industrialized countries. In his analyses he has tried to retain the basic elements of Marxian thought while giving an accurate picture of modern societies. In particular, he has focused on the class placement of those whom Marx would see as belonging to neither the bourgeoisie nor the proletariat.

Wright suggests that in modern societies stratification (or economic exploitation) involves three dimensions: (1) one related to property, or whether we own the means of production; (2) one related to authority, or the level of control we have in organizations; and (3) one related to skill or expertise, and the extent to which our skills are in scarce supply. In short, stratification in today's world is based not just on ownership of the means of production but also on our ability to control others through positions in large organizations and our degree of access to technical skills and knowledge. Because Wright's formulation is an expanded and modified version of traditional Marxian notions of economic exploitation, it is termed a **neo-Marxist** approach (neo means "new" or "recent").[16]

Figure 7-2 summarizes Wright's conceptualization of social classes. It includes information he developed, based on national surveys, of the average annual income (from all sources including interest and nontaxable income from bonds and other investments) of people in each class group. The two class groups that were central in Marx's theory are in opposite corners of the diagram, representing the large differences in their ability to exploit property, authority, and expertise. The bourgeoisie, located in the upper-left-hand corner, own the means of production and have enough money that they can hire workers to labor for them and not have to work themselves. (Lori and her family would be in this group.) The proletariat, located in the lower-right-hand corner, do not own the means of production and have no organizational or skill assets. (Rita's parents would be in this group.) The bourgeoisie have annual incomes that are more than four times greater than the proletariats' incomes.

Those in the other cells generally hold what Wright has sometimes called "contradictory" positions and represent the middle class group that Marx felt would inevitably join with the proletariat. On some dimensions these groups of people can be seen as exploiters, but on other dimensions they are exploited. Their incomes fall between those of the bourgeoisie and the proletariat.

Wright shares the Marxian concern for realizing social change and, especially, for transforming what he and other Marxists see as the repressive nature of capitalism. He suggests that the multiple class positions in the middle of Figure 7-2 illuminate sources of class conflict and class struggle. Whereas Marx predicted struggles primarily between the bourgeoisie and the proletariat, Wright suggests that class struggles can occur in a variety of ways, depending

If it weren't for the French revolution, Henri de France, pictured here with his wife, could have been King of France and owner of the marvelous grounds of Versailles. Even though Henri has lost the power associated with his hereditary title and much of his family's great wealth, he is still accorded status or prestige by others and maintains a lifestyle that differentiates him from those of lower status. How does his life illustrate Weber's theory of social stratification?

on how different groups join with one another. For instance, the expert managers, whose incomes are second only to the bourgeoisie, might try to use their organizational and skill assets to join the bourgeoisie. Or they might be induced to align with the class interests of the bourgeoisie through salary bonuses, stock plans, and other incentives. Wright and other scholars suggest, however, that such practices might also prove to be so expensive in the long run that they would be discontinued. The expert managers might then see greater advantages in aligning themselves with other class groups.

Just as Marx's original predictions regarding class struggles could only be tested through the course of history, so, too, these current speculations can only be assessed as time passes. Whatever the future of class struggle might be, many sociologists find the Marxian and neo-Marxian framework useful in analyzing social stratification, especially its focus on economic exploitation, the centrality of the notion of class conflict, and the importance of class consciousness.

Bargaining Power in the Economy: Weber's Views

Long before Erik Wright modified Marx's analysis of social class, Max Weber expanded on Marx's work, and many contemporary sociologists have continued to develop Weber's ideas in this area. Weber agreed with Marx that stratification involves economic exploitation, but he saw stratification as multidimensional and multisided.

Weber described three basic ways in which stratification appears. The first involves Marx's notion of

class. Weber, however, focused on how class struggles reflect attempts to obtain a monopoly or control over aspects of the economy, such as land, credit and loans, needed skills, or specific industries. Groups that have a strong monopoly over things that others most want or need become the dominant classes. Those who have only a partial monopoly over these most highly valued things or who have monopolies over less valued things have less dominance. Those who have no monopoly at all are at the bottom levels of the class hierarchy. Like Marx, Weber used the term *class* to describe these groups and suggested that our social class determines our life chances.

According to Weber the second dimension of stratification involves **status,** or *prestige,* by which he meant communities or social networks of people with similar lifestyles and viewpoints. This perspective is very similar to that used by Warner in his description of prestige classes in Yankee City. Status or prestige groups arise from economic distinctions. People who share an economic monopoly typically have social relationships with one another. They also develop similar customs and lifestyles that distinguish them from other people and usually come to see themselves as better than others within the society. In turn, these differences in lifestyles and viewpoints can reinforce economic distinctions. For instance, owners of exclusive clothing stores will look only for the "right kind" of person to serve their customers, and members of private clubs have "blackball" rules to bar membership for those who "wouldn't fit in."

Weber called the third dimension of stratification *party,* by which he meant what we call **power.** Earlier we defined power as the ability to compel others to

BUILDING CONCEPTS

Why Does Stratification Exist?

Sociologists have a variety of perspectives on why stratification exists. Here are three views.

Perspective	View of stratification	Example
Marx	Stratification occurs because those who control the means of production exploit others (for example, workers) in order to maintain their own positions.	Marx suggested that in his day, there were two major social classes: the "bourgeoisie," who owned the factories and mills, and the "proletariat"—workers who were exploited by the bourgeoisie.
Weber	Stratification involves differences in property, prestige, and power, and these dimensions can be independent of each other.	The President of the United States might not command a million-dollar salary, but his position of power affords him power and prestige.
Functionalist	Stratification exists because societies give greater rewards to those who fill difficult (and more highly valued) positions.	Surgeons are paid more than orderlies because the role of a surgeon is seen as more important and requires far more time spent in training.

do what we want them to do, even if they might not want to. As with status, Weber suggested that power can operate independently of property. For instance, politicians in the United States, such as presidents Jimmy Carter and Bill Clinton, have often gained positions of political power without having the great economic wealth that would place them within the bourgeoisie of the Marxian typology. Bureaucrats in large organizations, such as corporate executives, can have a great deal of power over the lives of their workers and can influence legislation and community affairs without owning a large part of the corporation. The heads of large government agencies such as the Internal Revenue Service or Federal Bureau of Investigation wield a great deal of power over all citizens while receiving only modest salaries when compared with executives in private corporations.

People who hold such powerful positions may eventually become rich, but, in contrast to the traditional Marxian view, they can become powerful with-

out first being wealthy. Nevertheless, as we will see later in this chapter, power often arises out of the advantages that both class and status provide. Owners of large industries have more power than the rest of us and can use their economic clout, or class in Weber's terminology, to wield power over people such as politicians and bureaucrats.[17]

Supply and Demand of Special Skills: A Functionalist Perspective

A third perspective of stratification stems from the ideas of structural functionalism. As you will remember from our discussion in Chapter 3, structural functionalists suggest that relatively stable aspects of a society continue to exist because they perform a function—they help maintain the social order. Given that stratification is present in all societies, structural functionalists naturally try to explain its persistence by asking what function it performs.

Many different workers are needed to complete the construction of a house. Some of these workers earn more than others do. How would the functionalist theory of stratification account for these differences?

The most famous attempt to apply this perspective was developed by Kingsley Davis and Wilbert Moore in the 1940s.[18] They asked why some occupations are more highly rewarded than others. For instance, why are doctors paid more and given more respect than sanitation workers? Why are professional football players paid more and held in greater esteem than nursery school teachers? More generally, why do all societies seem to have a stratification system in which some roles, whether they are that of shaman or scientist, are more highly rewarded and valued than others are? The answer, according to Davis and Moore, involves four simple reasons.

First, some positions are more important to a given society than others. For instance, the position of shaman or priest might be very important to a society because only the person in that role can perform religious rituals that are deemed necessary for the society's survival or spiritual well-being. Similarly in our own society, the role of surgeon is seen as much more important than that of hospital orderly because only the surgeon can perform delicate and difficult medical procedures.

Second, some positions are inherently more difficult than others because they require skills that few people possess or a great deal of special training. For instance, it takes many years to learn to be a skilled surgeon but little time to learn to be an orderly.

Third, because some positions are more important and more difficult than others are, societies need to motivate people to aspire to these positions and to stay in them once they have attained them. Given the difficulty involved in becoming a surgeon or a shaman, as well as the importance of these positions

to a society, there must be some way to induce people to want to fill these occupations.

Fourth, as a result of this need to motivate people to enter these hard-to-fill slots, societies have developed systems of stratification, or unequal rewards. Positions that are more important or that are harder to attain have higher rewards. Those that aren't as important or that don't require special skills or training are less highly rewarded.

As with all sociological theories, Davis and Moore's functionalist explanation has been subjected to a great deal of comment and criticism. One basic question involves the issue of functional importance. How can we determine that one position is more important to a society than another? For instance, public health research suggests that effective sanitation practices—such as providing clean drinking water, good sewage systems, and regular garbage collections—do much more to lower the number of deaths in a society than does the provision of highly trained doctors. How can we then say that the role of physician is functionally more important than the role of sanitation worker?

Part of the answer appears to involve the notion of *replaceability,* how hard it would be to replace either the position or the people who hold it.[19] For example, a sanitation worker is easier to replace than a physician. In simple terms, the functionalist theory is what economists call a "supply and demand" argument. People and positions that are in short supply and high demand—that are less replaceable and very important to the survival of a society—will be more highly rewarded. Those that are in large supply and low demand—that are highly replaceable and not as important—will be less highly rewarded.[20]

However, this argument doesn't address a more difficult logical issue. According to the notion of replaceability, physicians and professional football players earn more money than sanitation workers and orderlies because there are fewer people who can do their jobs. In other words the supply is low. But this doesn't address the question of why the *demand* is high. Clearly some things are important to all societies and thus in greater demand, such as health, food, shelter, and water. But other things, such as the type of entertainment that societies value, clearly vary from one time to another and from one society to another.

For instance, skilled soccer players might be just as scarce as skilled football players are, yet they are not in great demand in the United States (though they are highly esteemed in other countries). Davis and Moore's functionalist argument suggests that the role of football player is important because only the person in that role can perform it, but it can't tell us why it is more important than the role of soccer player. The argument seems to be that it must be important because it is highly paid, and it is highly paid because it is important.[21]

Finally, structural functionalism is sometimes criticized for painting an overly rosy picture of a society. The functionalist theory of stratification is no exception. Davis and Moore point out that a causal explanation of a phenomenon, such as stratification, does not necessarily mean it is good or just.[22] Nevertheless, their theory could be interpreted as implying that systems of stratification are fair and that societies operate better when certain positions carry greater rewards than others.

Other sociologists working within the functionalist tradition have described three general *dysfunctions* of stratification systems.[23] First, to the extent that systems of stratification prevent talented people from being discovered and encouraged in their efforts, they limit the potential for a society to maximize its productivity. Recall how the status attainment model, discussed in Chapter 5, shows how both our family background and our innate intelligence or talent influence our adult status. No matter how smart you are, if you are born into a family with few resources, you will have more difficulty attaining a high status in adulthood than if you had been born into a more well-to-do family.

Second, these theorists suggest that stratification systems actually work against greater integration of a society by encouraging hostility, suspicion, and distrust among citizens. The history of immigrant groups within the United States provides ample evidence of this process. Members of newly arrived groups generally held jobs that offered the lowest levels of re-

wards. The tensions that existed between the various ethnic groups of European ancestry lasted for many years, and divisions between members of different racial-ethnic groups remain even today.

Third, stratification systems can work against the possibility of social change because they help maintain the power of the elite. The elite, in turn, will quite likely want to maintain the privileges they gain from the status quo.

Societal Complexity and Technology: Lenski's Views

Although functionalist theory can explain why stratification systems reward those who have skills desired by the society, it can't explain why some societies have very elaborate systems of stratification whereas others have relatively little. The sociologist Gerhard Lenski tried to understand more about these differences by examining data from the Human Relations Area Files regarding many different types of societies. (The HRAF were first described in Chapter 3.) Lenski suggested that differences in the nature of stratification were related to variations in the technology available to societies and their divisions of labor.[24] Table 7-3 summarizes some of these differences among four types of societies: (1) hunting and gathering, (2) horticultural, (3) agrarian, and (4) industrialized.

Hunting and Gathering Societies Societies that obtain their food through hunting animals and gathering plants and that use the simplest, most primitive tools and work techniques are known as **hunting and gathering societies.** Most of the members' activities are oriented toward simply getting enough to eat because they generally have no way to preserve food for a long period of time. The groups typically are quite small, averaging about fifty people, and they often must move from place to place to find sufficient food. Examples of such groups include the Australian aborigines, the Punan of Borneo, the Pygmies and Bushmen of Africa, a number of groups in the Amazon basin of South America, certain Native American groups from the Great Plains area to the Pacific, and some of the Inuit.[25]

Because they are nomadic and have to carry their belongings from place to place, these groups produce very little surplus, or extra, food or goods. As a result everyone has just about the same amount of personal property. In addition, apart from the fact that men are assigned to hunting and women to gathering, the tasks assigned to group members are not very different. For all these reasons, there is relatively little social stratification in these societies.[26]

TABLE 7-3   *Lenski's Analysis of Technology, Division of Labor, and Stratification of Society*

TYPE OF SOCIETY	LEVEL OF TECHNOLOGY	DIVISION OF LABOR	AMOUNT OF ECONOMIC SURPLUS	POLITICAL DOMINATION BY A FEW OVER THE MASSES	EXTENT OF STRATIFICATION
Hunting and gathering	Simple tools for gathering food and hunting animals	Minimal, except for that based on age and sex	Minimal	Rare	Minimal
Horticultural	More advanced than hunting and gathering Includes digging tools, hoes, some irrigation and fertilization of crops	Greater than in hunting and gathering societies	Greater than in hunting and gathering societies	Greater than in hunting and gathering societies	Greater than in hunting and gathering societies
Agrarian	More advanced than horticultural societies Includes metallurgy, plows, and harnessing of animals	Greater than in horticultural societies	Greater than in horticultural societies	Extensive	Much greater than in other types of societies
Industrialized	Highly advanced	Extensive	Extensive	Less than in agrarian societies	Less than in agrarian societies

Horticultural Societies   In contrast to hunting and gathering societies, **horticultural societies** are based on a gardening economy. Examples include the Iroquois and Zuni in what is now the United States, several groups in New Guinea such as the Kapauku and Garia, South American groups such as the Boro and Jivaro, as well as the Incas and Aztecs, and a number of groups in sub-Saharan Africa.[27] The ability to cultivate land means that members of a horticultural society can produce a surplus of food, live in one place, and begin to accumulate goods and wealth beyond the basics needed for survival. It also means that they have a greater division of labor and more stratification. Thus they are much more likely to develop strong political systems, which often include hereditary chiefdoms or kingdoms, classes of people who have accumulated much more wealth than others, larger differences between shamans and ordinary people, military prowess and special ceremonies associated with war, and, as an outcome of war, slavery.

These characteristics, especially the development of kingdoms and slavery, are more common in the advanced horticultural societies that use hoes and can produce a greater surplus of goods. For instance, the Incas were a highly advanced horticultural society, with extensive irrigation systems and very productive croplands. Within just a century they managed to conquer many of their neighbors and build a society of some four million people.[28]

This point illustrates the key element of Lenski's theory. He suggested that when societies develop technology to the point that an economic surplus can occur, differences in power determine both how these surplus goods and services will be distributed and who will have more privileges. Although he agreed with Davis and Moore and other functionalists that the division of labor is an important element of stratification, Lenski suggested that "the distribution of rewards in a society is a function of the distribution of power, not system needs."[29] In this sense Lenski's theory sees differentials in economic property, in power, and in prestige as just as central to the development of stratification as Weberian and Marxian theories do.

Agrarian Societies   The patterns of stratification seen in horticultural societies are even more pronounced in **agrarian societies,** societies with intense agricultural production made possible by the plow. These societies first appeared in the fertile river valleys of the Middle East about five to six thousand years ago and subsequently spread throughout much of Asia, Europe, and North Africa. In addition to developing the plow, a tool that allowed them to break the ground and plant much larger crops, these groups also learned to harness and domesticate animals and developed the skill of metallurgy, which enabled them to forge iron plowshares, rather than having to use wooden ones. Thus, in contrast with horticultural societies, they could produce much more food than they needed and had even more leisure time to spend on ceremonial activities and political pursuits.[30]

The superior technology of agrarian societies allowed them to produce a wide variety of new tools and instruments of warfare—wheeled vehicles, food storage equipment, chariots, and armor. As a result, the division of labor was much more complex than that found in either the horticultural or hunting and gathering societies. People held many different occupations, not unlike our own society today. In addition, all the large agrarian societies were formed by forcibly subjugating and often enslaving other groups.[31]

All these factors contributed to the growth of the **state,** the organized monopoly or control of the use of force in a society, or what we often think of as the government. In agrarian societies the state became a powerful mechanism for administering vast territories, collecting taxes, and protecting the interests of the rulers. The rulers and their close associates had much more power and wealth than did others in the society, often taking more than half of the society's annual income for themselves. Together with the increased division of labor, this development made agrarian societies much more highly stratified than horticultural groups.

Industrialized Societies The onset of the Industrial Revolution in the eighteenth and nineteenth centuries brought dramatic changes to agrarian societies. Whereas previously most of the energy to produce food and goods had been provided by animals and men, with the Industrial Revolution machines powered by steam, oil, gas, wood, water, and coal, and eventually nuclear and solar energy, did most of the work. The result was an enormous change in the productivity of human beings. With the aid of machines, a single farmer could now plant and harvest hundreds of acres. As a consequence only a small proportion of the workers in modern industrialized societies need to be involved in producing food, freeing many people to do other things.[32]

You might think that these changes would produce even greater stratification and inequality in complex industrialized societies than in agrarian societies. In fact, however, exactly the reverse has happened.[33] Differences in wealth are much smaller in industrialized societies than in agrarian societies, and political power is spread among more people. We can see this pattern by looking at contemporary societies.

Countries today differ in the extent to which they have industrialized, as illustrated in Map 7-1. In some countries, especially those in Africa, South America, and Asia, more than three-fourths of the labor force works in agriculture. In others, such as most European countries, the United States, Canada, Japan, New Zealand, and Australia, less than 10 percent of the labor force works on farms, largely because farming has become so mechanized that only a relatively few people are needed to produce large quantities of food. As Lenski's theory would predict, these more industrialized countries have more democratic and stable governments. They also tend to have somewhat greater economic equality.

If a society had total economic equality, all people would receive the same proportion of the society's income. In reality, of course, this never happens —stratification exists and some people have more than others. Sociologists and economists often use a formula called the **Gini index** to measure income inequality within societies. The index can range from zero, which would indicate that everyone receives an equal share of the country's income, to 100, which would indicate that all the income is received by just one person. Data gathered by economists who work for the World Bank indicate that this coefficient varies from a low of about 25 in several European countries to values close to 60 in a number of Latin American and African nations (see Map 7-2). By comparing Maps 7-1 and 7-2 you can see that, as Lenski's theory would predict, countries that have more people employed in agriculture generally tend to have greater income inequality. For instance, in Belgium, Denmark, and Sweden, which are all highly industrialized, the Gini index is 25. In contrast, in societies such as Kenya, Paraguay, and Guatemala, where more than half of the populace works in agriculture, the Gini index is near 60.

Why was the trend toward greater stratification reversed with industrialization? The answer, interestingly enough, still revolves around the nature of technological change and the division of labor. As industrialization progresses and technology becomes more complex, jobs become more specialized and require more skills. Workers are better educated and better trained, and they become less replaceable. As the functionalist approach would suggest, when workers are less replaceable (that is, when they are in shorter supply), they tend to be rewarded more highly. Elites might still want to amass the huge profits that they did in agrarian societies, but they are forced to share their wealth with the skilled workers they need to do the work.

In addition, as the growing complexity of work has required that workers be more highly educated, they have also acquired greater political rights. The practice of slavery, the belief in the "divine right of kings," and the exorbitant taxation of agrarian societies have also disappeared. Thus the very complex division of labor and very large surpluses found in

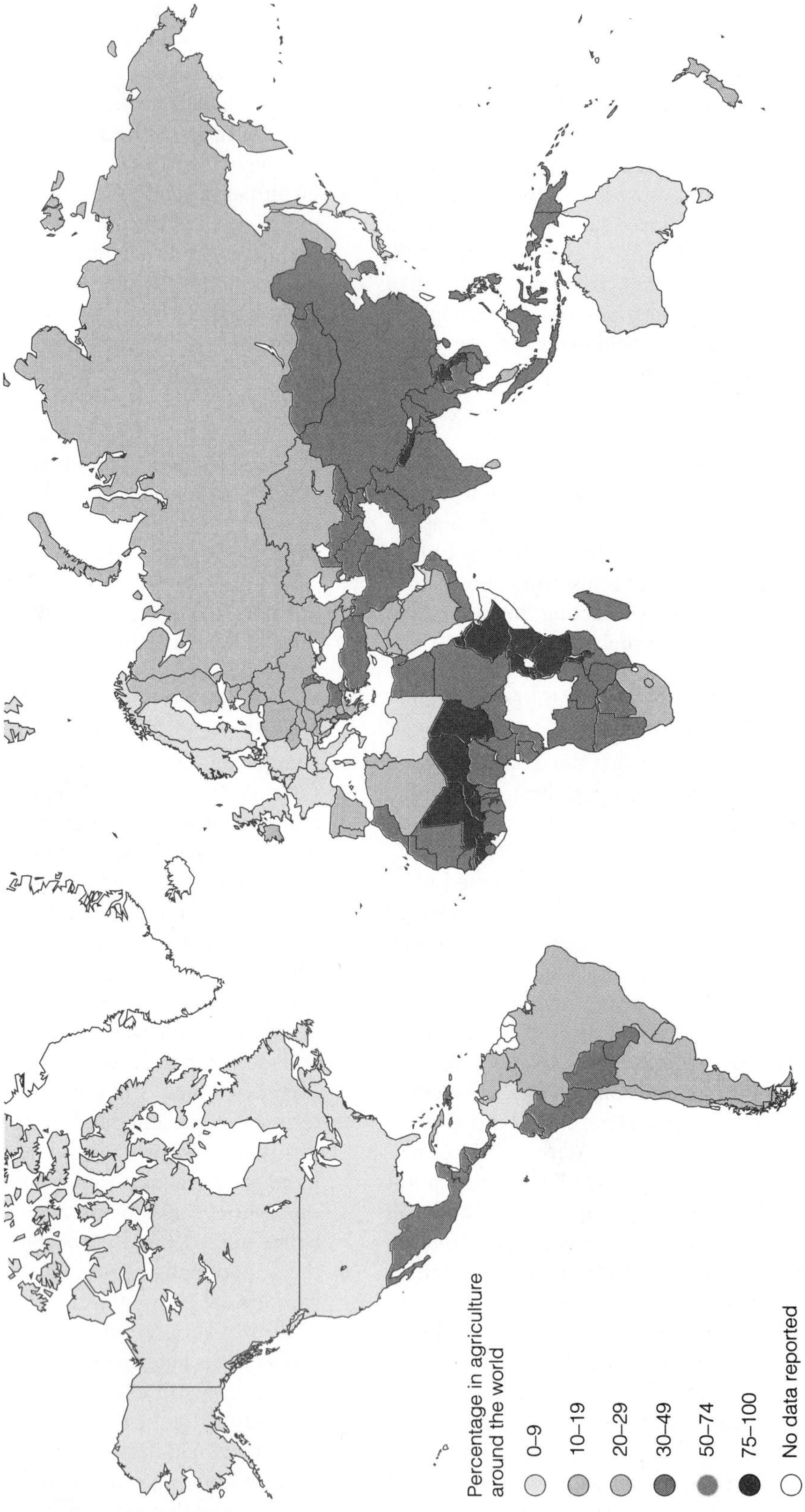

MAP 7-1 *Percentage of the Population Employed in Agriculture in Countries Around the World*

Source: World Bank, World Development Indicators, CD-ROM, 1998, Tables 2-5 and 2-8

Percentage in agriculture around the world

0–9

10–19

20–29

30–49

50–74

75–100

No data reported

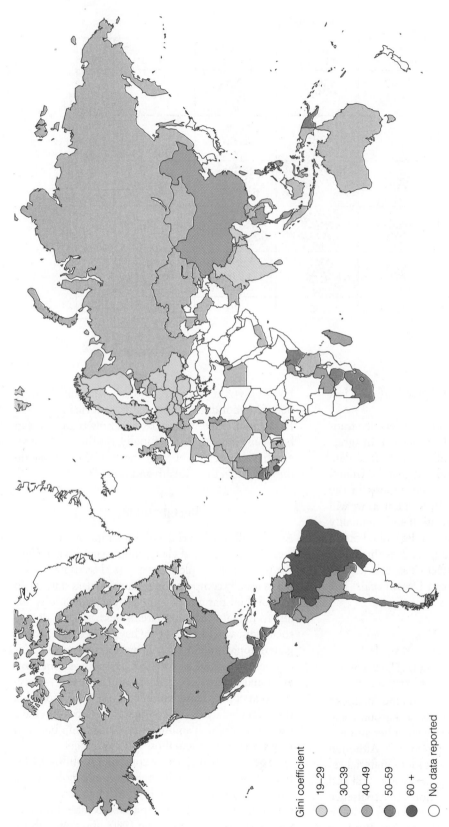

MAP 7-2   *Income Inequality Around the World as Measured by the Gini Index*

Source: World Bank, World Development Indicators, CD-ROM, Tables 2-5 and 2-8

Son's Occupation

Father's Occupation	Upper white-collar	Lower white-collar	Upper manual	Lower manual	Farmer	Total
Upper white-collar	125	30	24	30	1	210
Lower white-collar	45	22	23	29	1	120
Upper manual	74	24	80	59	3	240
Lower manual	78	29	69	102	2	280
Farmer	28	15	34	50	23	150
Total	350	120	230	270	30	1000

FIGURE 7-3   *Example of a Mobility Table*

industrialized countries actually helped produce less stratification.[34]

Of course, industrialized countries remain stratified. The mountain might not be as steep as in agrarian societies, but it certainly still exists. In fact, data gathered by the World Bank indicate that the United States is one of the more unequal nations in the world, with a Gini index of 40. In addition, as we will see in the next section, stratification and inequality might even have been increasing in the United States in the last few decades. What impact does this stratification have? How does it affect us as individuals? To what extent is the American Dream real? The next section deals with these issues.

THE AMERICAN DREAM: MOBILITY, POVERTY, WEALTH, AND POWER

In many ways Rita's life seems to fulfill the American Dream. Despite her working-class background, she went to a good school, did very well in her studies, and landed a well-paying, prestigious job. Although she grew up in a family without a lot of material resources, she managed, through hard work, skill and intelligence, and college connections, to experience **upward mobility**—to move to a social class position that was higher than her parents' position. But how common is Rita's experience?

In this section we explore specific sociological issues related to the American Dream. First, to what extent can we move up the stratification hierarchy?

How easy is it to move, and why do these moves occur? Second, how unequal is American society? How much poverty is there, and how big is the gap between rich and poor? Finally, do the wealthy really have more power than the rest of us? How do they maintain their wealth and power?

Social Mobility

As you will recall, social mobility simply means movement from one social class group to another. To find out how much people move across class strata, sociologists usually compare adults' current class status with that of their parents. If our class status as adults differs from that which our parents held, as Rita's does, a sociologist would say that we have experienced **intergenerational mobility.** If we have the same position as our parents, we are said to have experienced **intergenerational succession.** Because, until the last twenty-five years or so, many mothers did not work outside the home, analyses of mobility and succession have usually involved examining the relationship between men's occupations and those of their fathers.

Sociologists frequently construct **mobility tables** to illustrate patterns of intergenerational mobility and succession. For example, Figure 7-3 shows the relationship between the occupations of one thousand sons who were working in the 1980s and their fathers. Five different categories of occupations are shown: (1) "upper white-collar," or administrative and professional positions that usually require a college education; (2) "lower white-collar," or office, clerical, and

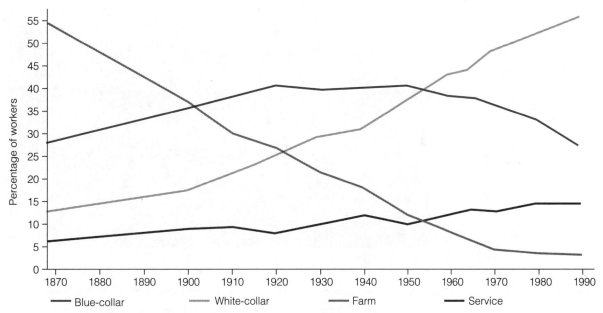

FIGURE 7-4   *Change in the U.S. Occupational Structure, 1870–1990*
Source: Gilbert, 1998, p. 65

sales jobs that may require less education; (3) "upper manual," or jobs in the high-paying skilled crafts; (4) "lower manual," or jobs that require less training and generally pay less; and (5) "farmer."

There is a high level of occupational succession in the United States, as shown by the relatively high numbers on the diagonal of Figure 7-3. Most farmers are the sons of farmers, and most manual workers (also often called blue-collar workers) are the sons of manual workers.[35] Yet there is also a great deal of mobility, especially into white-collar fields. About half of the workers in white-collar jobs in the late 1980s had fathers who held blue-collar jobs, and less than a fifth of the sons of farmers worked on farms.[36] Many people have been able to move into higher-status jobs than their parents held because they have received more education than their parents did.

Even if they had wanted to, many sons of blue-collar and farmer fathers would not have been able to take the same types of jobs as their fathers because many of these jobs have disappeared while the number of white-collar jobs has increased. For instance, only a third of the fathers in Figure 7-3 were employed in white-collar jobs, while almost half of the sons were. Put another way, children of blue-collar workers and farmers who have received appropriate levels of education could obtain white-collar jobs simply because there were more of these jobs available.

These changing opportunities reflect a change in the **occupational structure,** the set of occupations available within a society at a given time. We can understand more about these changes by taking a very

broad macrolevel perspective on the **labor force** (or **work force**), the people employed within a society, and sometimes those actively seeking work. For example, after about 1870 the United States experienced rapid industrialization. As Figure 7-4 shows, this process of industrialization was accompanied by dramatic declines in the proportion of the work force involved in farming and a corresponding dramatic increase in the proportion employed in white-collar jobs. The representation of blue-collar workers was more variable, rising to a peak of more than 40 percent from the 1920s to the 1950s and then falling to only 27 percent in 1990. During this period the overall occupational structure of the United States (and other industrializing countries as well) changed from one in which farmers predominated, to one in which blue-collar workers were most common, to one in which white-collar workers are now most common.[37] The intergenerational mobility that resulted from these changes in the occupational structure is called **structural mobility.**[38]

For many years structural mobility allowed people to move up the class ladder, to hold higher-status, higher-paying jobs than their parents had. In recent years, however, the nature of structural mobility has changed. Even though the number of white-collar workers has continued to increase in the past quarter-century, the new jobs have not been in the most highly paid professions, such as the law and medicine. Instead, the growth has occurred among the less well-rewarded professions—such as teachers, librarians, social workers, accountants, nurses,

A poverty-level income affects virtually every area of a person's life, including where one can afford to live.

and middle-ranking rather than top-level managers. At the other end of the spectrum, the proportion of both unskilled laborers and highly skilled blue-collar and crafts workers has declined, while the proportion of low-paid, unskilled service workers, such as janitors and hospital attendants, has grown.[39] Consequently, even though upward intergenerational mobility is still more common than downward mobility, the probability that a young person will be downwardly mobile is growing, especially for men in their twenties and thirties with low levels of education.[40]

These structural changes can help us understand the lives of individual workers and families. The American Dream, the belief that upward mobility comes to those who work hard, is a genuine aspect of American culture. But when the number of high-status, well-paying jobs stops growing, it won't matter how much training we have or how hard we work or what kinds of jobs our parents had. The American Dream will be harder for most people to attain, and downward mobility will become more common, as it has in the United States in the past two decades.

Poverty and Inequality

These structural changes are also related to the probability that people in the United States live in poverty. At first glance defining what it means to be poor would seem to be pretty simple. It means that you don't have enough money! But how much is "not enough"? Social scientists and policy makers have not always agreed on an answer, and we look first at the controversy about how to measure poverty and then at trends in the prevalence of poverty and inequality over time.

Measuring Poverty In the early 1960s the U.S. government developed an official definition of poverty. This definition derived from a practice that

had long been used by charitable groups to determine how much money would be needed to fill a "market basket" of cheap but nutritious foods that would keep a family alive and also provide other essentials, including rent money. The **poverty line,** or income level under which families are officially defined as being poor, was then set at three times the cost of this "market basket" of food, assuming that families would need a third of their income to purchase food and the remainder to pay for other essentials, such as housing and clothes. Because prices of food and other goods change over time, the government adjusts the poverty line regularly by increasing it in proportion to the change in the Consumer Price Index, a measure that the government uses to chart inflation.[41]

Yet this definition of poverty was also influenced by political considerations. Government nutritionists who developed the market basket of foods used in setting the original poverty level in the early 1960s actually presented two different types of market baskets to government officials. One was an emergency diet designed only for short-term use by a family; the other, which cost 25 percent more, would provide the long-term nutrition a family needed. The government chose to use the lower level because the number of people found to be in poverty using the higher income level was embarrassing to the administration.

Because a substantial number of people have incomes below the poverty level for more than just a few weeks, many people believe that the higher figure would be more appropriate. This higher number (which works out to 125 percent of the official poverty figure) is sometimes reported in government statistics.[42] Although slightly more than 13 percent of the population lived below the official poverty line in 1997, almost 18 percent had incomes below this higher 125 percent figure.[43]

The official poverty line is subject to another type of problem, one that involves changing standards of

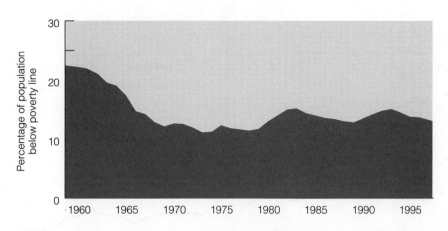

FIGURE 7-5 *Official Poverty Rates, 1959–1997, for Total Population*

Source: Dalaker and Naifeh, 1998, page C-2.

living and definitions of what constitutes a "good life." Sociologists use the term **reference group** to indicate a group of people that we look to, or use for reference, in evaluating ourselves and our position in life. Thus, if you lived in a society in which no one had a telephone or had graduated from high school, you wouldn't think that it was important for you to do so. But if most people around you had a telephone and an education, you would see these as part of a minimum standard of living. In other words, our definition of what constitutes poverty depends on our reference group and can change over time. For instance, in 1908 charity workers assumed that a family of four needed twenty-two pounds of high-protein food per year, but by 1960 charity workers assumed that a family of four needed fifty-five pounds. Similarly in 1908 the same family was thought to need four rooms to live in but no bath; by 1960 this family was assumed to need five rooms and a full bath.[44]

From 1908 to the 1960s, the subsistence budget used by charitable agencies was always close to 40 percent of the average income of families of four across the country.[45] This figure is close to the amount that people seem to think is needed to live on. Since 1946 the Gallup Poll (a survey that is regularly given to a representative nationwide sample of respondents) has asked people, "What is the smallest amount of money a family of four needs to get along in this community?" The average response has generally been from 50 to 55 percent of the average family income. Thus, when people try to determine what is needed to keep someone from being poor, they tend to give a similar answer—about half of the average family income.

The official government poverty figures don't work this way. Because the official poverty line is updated each year by simply adjusting the 1960s figure for inflation, it does not take into account changes in the standard of living and lifestyles that have occurred during that time. Although the poverty line was about equal to one-half of the average income

when it was first set in the 1960s, it is now only about 38 percent of this figure. Family incomes have grown since the 1960s, but our definition of poverty has not kept pace.[46]

Thus many sociologists suggest that a better way to measure poverty—and one that would reflect the way in which people actually define poverty—would be to use a relative measure, one that takes into account the society's general standard of living. A measure that is often used is one-half of the median income. (The **median** is the midway point in a distribution, the point at which 50 percent of the cases are larger and 50 percent are smaller. For instance, in 1960 the median income was about $6000. Half of all families had incomes higher than this amount and half had incomes that were lower.)[47] Using this standard in the 1990s would reveal that close to a quarter of the families in the United States were poor, the proportion sociologists generally see as belonging to the working poor and underclass (see Table 7-2).[48]

Trends in Poverty Even though there are some concerns about how well the official poverty line really measures poverty, this measure can help us see how many individuals experienced poverty over the past three decades and how these numbers have changed. Figure 7-5 shows the change in the **poverty rate**—the percentage of the total population that lives in families with incomes under the official poverty level—in the United States from 1960 to the late 1990s. As you can see, the poverty rate declined markedly in the 1960s and remained low during the 1970s, primarily because the government instituted a number of programs to aid the poor. These include food stamps, which people can redeem for food at the grocery store; Medicare, which covers medical costs for the elderly; and Medicaid, which covers medical costs for some of the poor.[49]

Figure 7-6 shows the poverty rate from 1960 to the late 1990s for two subgroups of the total population: children and the elderly. In the early 1960s the elderly were more likely to be poor than were

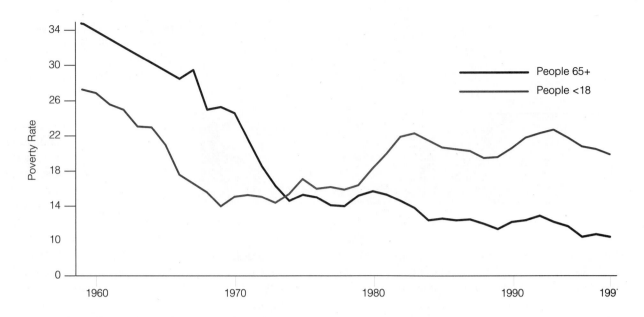

FIGURE 7-6 *Poverty Rates of People Under 18 and People 65 and Older, 1959 to 1997*
Source: Dalaker and Naifeh, 1998, p. C-6.

children or people in the general population. However, the poverty rate of the elderly fell rapidly during the 1960s and throughout the 1970s and 1980s so that today older people are less likely than are those in any other age group to live below the poverty line. This was not the case for children, whose poverty rate rose rapidly during the 1980s and has remained around this higher level ever since. A child in the United States is now almost twice as likely as a person older than 65 is to live in poverty. The rate of childhood poverty in the United States is substantially higher than that in most other industrialized societies. As discussed in Box 7-1, this situation has concerned both policy makers and social scientists.

It is not only jobless people who are poor. In the late 1990s more than 40 percent of all poor people over the age of 16 worked at least some time during the year, and 10 percent of these poor adults were full-time, year-round workers.[50] In recent years an increasing number of workers simply haven't been able to earn enough to rise above the poverty line.[51] This increase in poverty corresponds to the change in employment opportunities described in our discussion of structural mobility. As the proportion of jobs paying higher wages has declined, more people have sunk into poverty, even if they are working.[52] In fact, if the dollar amounts are adjusted for inflation, the real wages of male workers actually declined between 1970 and the mid 1990s.[53]

Trends in Inequality Looking at the amount of poverty within a society represents one possible cam-

era angle, but another viewpoint gives a slightly broader perspective. This involves the notion of *inequality,* the gap between the rich and the poor. As you will remember from the discussion of Lenski's work, societies can vary a great deal in how much inequality they have. In addition, the amount of inequality within a society can vary over time.

Figure 7-7 shows the *Gini index,* the measure of income inequality introduced in Map 7-2, for families in the United States over the past half-century. As you can see, inequality within the United States gradually declined through the 1970s, and families at all income levels enjoyed growing prosperity. In the early 1970s, however, inequality began to increase. For most people incomes ceased growing as they had before, but for the richest segments of the population, incomes began to surge. In fact, even though more and more people fell below the poverty line after the mid 1970s, the proportion of families with incomes greater than $100,000 (adjusted for inflation) doubled! And even as the average earnings of individual workers fell, the income of the richest 1 percent of all families, most of whom belong to the capitalist class, rose by more than 50 percent. During this same period corporate executives' salaries increased sharply, so that the wage gap between executives and ordinary workers became a chasm. In the mid 1970s the average chief executive officer of a corporation earned 40 times as much as the average factory worker; by the mid 1980s he (and only rarely she) earned 190 times as much as the worker.

The increase in poverty among children during the last few decades appears to be related to two trends: the decline in well-paying jobs for people with few skills and an increase in the number of children who live with only one parent. Many adults with low levels of education and training can no longer earn enough to support their families. As a result, two incomes are now much more often needed to support a family, and children who live with only one parent are much more likely than other children to live in poverty.[76] These trends have been especially troublesome for children of color. The poverty rates for African American and Hispanic children are more than twice as high as those for Euro-American children.[77]

Sociologists have documented several long-term effects of living in poverty. Although some children who experience poverty do very well in adulthood, on average poor children tend to have more problems throughout their lives than do other children. Based on extensive longitudinal studies, sociologists have concluded that, when compared with other children, children who grow up in poverty tend to have poorer health, do less well in school, have more problems in their interactions with others, and have lower incomes and occupational status in adulthood.[78]

The decline in well-paying jobs for less well educated workers and the increase in single-parent families has appeared in several industrialized countries. If we only consider how much money parents earn, about as

many children would be poor in many other industrialized countries as in the United States. Yet, in fact, most of these other countries have instituted programs that result in actual child poverty levels that are only about a third or less of those in the United States.[79] Only Australia, among all the industrialized countries, has childhood poverty rates that are similar to those found in the United States.[80]

The United States does not ignore the poor. Both the federal and state governments have developed extensive programs of cash welfare assistance, food stamps, housing subsidies, unemployment insurance, and various tax credits and deductions to help poor families and individuals stay afloat. Why then have other countries been so much more successful in keeping children out of poverty?

The answer appears to be two-fold. First, other countries appear to be somewhat more generous in programs that are directed toward the poor, such as those used in the United States. Second, they have developed programs designed to promote the health and well-being of *all* children, and it is these more general programs that appear to be most central to reducing the childhood poverty rate. For instance, most industrialized countries have a "family allowance" or

"child allowance" system whereby all parents, no matter how wealthy they may be, are given a set amount of money for each child to help with the expenses of child raising. All these countries also have extensive systems of government provided medical care, ensuring the health of mothers and children and shielding families from unexpected medical costs. Many countries also provide extensive programs of child care and early childhood education. This reduces the costs of child care for working parents, and it also helps ensure that all children have access to high quality child care.[81]

Of course, such programs cost money. For instance, the French, who have an extensive program of medical care, income supplementation, and early childhood education, spend about 59 percent more per person on family related costs through their taxes than do people in the United States. Because people throughout French society, and not just the poor, benefit from these programs, they receive broad support, and even when there have been economic downturns the family and child-oriented programs have not been targeted for elimination.[82]

CRITICAL-THINKING QUESTIONS
1. Could such extensive programs be developed in the United States? Would there be support for paying for these programs?
2. Should such programs be available to all children? Why or why not?

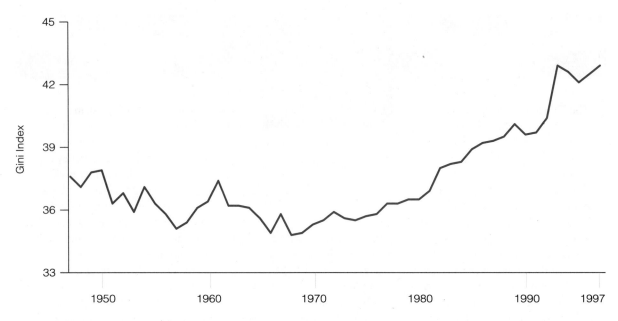

FIGURE 7-7   *Gini Index for Family Incomes, 1947–1997*
Source: U.S. Bureau of the Census, 1998a, p. C-32.

One reason for these changes is declining wages for workers with few skills and low levels of education. Between 1980 and the mid 1990s, the median wage of workers with a high school education fell by 6 percent. In contrast, the earnings for college graduates rose by 12 percent. This increasing inequality also reflects changes in families as the number of single-parent households has increased. In addition, many more women have entered the labor force, and women with higher earnings tend to marry men with higher earnings, thus widening the gap between high-income and low-income households.[54]

In short, the material wealth of the United States is the envy of much of the world, but this wealth is distributed in strikingly unequal ways. The stratification mountain in our society comes to a tall, narrow peak representing a relatively small class of people who command great wealth. Who are these very rich people, and how are they able to maintain their wealth? We turn now to an examination of how one sociologist looked at this very powerful group.

Common Sense versus Research Sense

COMMON SENSE: The American Dream tells us that every person can be as powerful as any other if they just work hard enough.

RESEARCH SENSE: Michael Allen's study of the corporate rich finds many factors, such as social and cultural capital, that enable the wealthy to wield and maintain power in ways others cannot.

FEATURED RESEARCH STUDY

THE WEALTHY IN AMERICA

In addition to the idea that everyone who works hard can be upwardly mobile, the American Dream tells us that every person—rich or poor—can be as powerful as any other. Yet the theories we reviewed earlier in this chapter, as well as our own everyday observations, suggest that members of the elite classes in a society have far more power than other people. What about the rich in America? How do they influence the rest of society? Does their wealth translate into power, as

Marxian and Weberian theories would suggest? We explore these questions using the research of the sociologist Michael Allen. (To learn more about the work of Michael Allen, see Box 7-2.)

Defining the Capitalist Class

To understand the experiences of the wealthy, it is important to distinguish between income and

BOX
7-2

SOCIOLOGISTS AT WORK

Michael Allen

The use of deception in a study raises serious ethical and legal issues.

Michael Allen, Washington State University, conducted the research on the corporate rich featured in this chapter.

How did you get interested in studying the corporate rich? I had distant relatives who were wealthy (a great uncle was one of the founders of Dole Pineapple). They visited our family once a year and made a great impression on me. My father was a truck driver and my mother was a receptionist, and I was always fascinated by the class distinctions within my own family. My father's father was a judge and had been a state senator in California. My mother's father was a successful businessman.

What surprised you most about the results you obtained in your research? I was surprised and impressed by the methodical and deliberate manner in which the members of these families worked to maintain their economic and social advantages and to transmit these advantages to their children and even their grandchildren. The creation of networks of trusts, to avoid taxes, was the most obvious example of the care and attention given to the perpetuation of wealth by these families.

What was the reaction to your research among your colleagues? Did it surprise you in any way? Sociologists are uncomfortable with data that they cannot quantify easily and analyze using statistical methods. I think many sociologists remain unconvinced of some of the findings of my research because it is based largely on qualitative historical comparisons. Moreover, sociologists like to believe, as do all Americans, that the rich are basically the same as everyone else. They are not. They are different inasmuch as they are treated differently by everyone from a very early age. Robert Coles once suggested that wealthy children possess a sense of "entitlement" that others do not possess.

Did you run into any problems in conducting your research? Without exception, my requests for background interviews were summarily rejected by even the most liberal and outspoken members of these families. Perhaps I might have obtained more cooperation if I had been willing to deceive those individuals about my intentions, but the use of deception in a study raises serious ethical and legal issues. In the end, I decided to limit my analysis to information that was available, although not always readily accessible, from various public records. I have never regretted this decision. In fact, I am now convinced that public documents are the best source of systematic and reliable information on the historical evolution of corporate rich families and their fortunes.

What research have you done since completing the study described in the chapter? Does it build on your previous research? After completing the book, I did go on to publish a couple of articles based, at least in part, on data collected for the book, including three articles dealing with the campaign contributions of wealthy families and the impact of campaign finance reforms on those contributions. I have since moved off to study less problematic subjects like the organization of cultural production in the television and film industries.

What do you think the implications of your work are for social policy? Who benefits from the "minimal state" promoted by many Republican members of Congress? Many of the social programs funded by the federal government were intended to redress the inequalities of wealth inherent in a capitalist society. I find it curious that most Americans fail to realize that they live in a class society, or at least in a society with extremes of wealth and poverty. It is the poor and even large portions of the middle class that will lose from the reduction of federal programs (such as student aid), not the rich or the big corporations.

How might these young members of the capitalist class be accumulating social and cultural capital through attending this party?

wealth. **Income** refers to how much money a person receives in a given time, such as $20,000 per year. **Wealth** refers to a person's assets that result from the accumulation of income, such as houses, cars, real estate, and stocks and bonds. Assets are commonly measured by what is called *net worth,* which is simply the total value of a person's assets minus any outstanding debt. In some forms, such as rental properties or stocks and bonds, assets become *capital*—that is, they produce income. The ownership of large amounts of these income-producing assets is the defining feature of the capitalist class.[55]

In the United States (and in other countries as well), wealth is much more concentrated than income. For instance, the top 1 percent of income earners in the United States receive about 16 percent of the total annual income. In contrast, the top 1 percent of wealth holders control almost one-third of all of the wealth. Furthermore, the net worth of this top 1 percent is roughly equal to the net worth of the bottom 90 percent![56] If we exclude ownership of cars and homes, which makes up the net worth of most people, and focus on other assets, such as corporate stock, rental properties, and other real estate holdings, we see that a mere 1 percent of all households owns almost one-half of the stock and almost two-thirds of the small business assets. The *average* net worth of these households is more than $5.5 million![57]

These are the people who make up the capitalist class. Members of the capitalist class, like Lori and her family, don't need to hold jobs to have income (even though many of them do) because they earn plenty of money from the businesses, stocks, bonds, and other assets they own. In addition, unlike monthly income, which most people have to spend each month simply to keep afloat, wealth can beget more wealth. Wise investors not only live off their assets but also reinvest the money to increase their net worth even more. Because this wealth generally remains stable or increases, large portions of it are passed on to future generations. Thus the capitalist class comprises not just wealthy businesspeople but their spouses, children, grandchildren, and even generations of descendents. Just as Warner found that membership in the upper class in Yankee City was dependent on having the proper ancestors, so, too, membership in the capitalist class often depends on having wealthy ancestors who were canny enough to pass on their fortune to future generations.

Note that when sociologists use the terms *capitalist class* and *upper class,* they are not necessarily referring to the same people. In the tradition established by Warner and his colleagues, the upper class comprises a social network of people who have very high status or prestige. Although they are all wealthy, their family fortunes can be relatively modest, and they might even have to work for a living. In contrast, some members of the capitalist class, despite their incredible wealth, may not be welcome in upper-class circles because they are perceived as lacking the proper manners or connections. As one observer of the upper class in Philadelphia put it, "Inherited money is better than made money,"[58] and those with "made money" might well be looked down on by those who grew up in wealthy surroundings.[59]

In our discussion we will use the term corporate rich synonymously with capitalist class. Allen and other researchers who study the very wealthy often use this term because most of the wealth in the United States comes from the ownership of stock in corporations.[60]

Studying the Capitalist Class

Allen wanted to understand more about how members of the capitalist class enter the upper class and how these wealthy families survive with their wealth intact or even increase it over the years. To guide his work, he used the theoretical ideas of Pierre Bourdieu, a contemporary French sociologist who has long been interested in stratification.[61] Like Weber, Bourdieu envisions stratification as involving several dimensions and suggests that a family's status rests on three different forms of capital (or resources).

First, and most obvious, is **economic capital,** income-producing wealth. Bourdieu sees economic capital as most important because it allows an individual to accumulate the other types of capital. Second is **social capital,** the networks and contacts that reflect the family's place in the social structure. Third is **cultural capital,** by which Bourdieu means the ways in which people behave, dress, and talk, and the ways in which these manners and styles differentiate those in one class group from those in another. Bourdieu suggests that members of the capitalist class are able to reproduce their position in society by accumulating not just economic capital but also social and cultural capital. Because many capitalist families had relatively modest origins, they might not have the social and cultural capital of the upper class, but their economic capital allows them to gain these advantages for their children. As they do so, their children can be accepted in the upper class. In Allen's words, "Nothing succeeds like money, manners, and connections!"[62]

Members of the capitalist class are often the targets of individuals and organizations seeking money from them for business or charitable reasons or even for nefarious schemes such as extortion. Thus they try very hard to preserve their privacy and not let others know about the extent of their wealth.[63] Given their penchant for privacy and secrecy, how could a sociologist, especially one who was not part of the capitalist class himself, study this group?

At the beginning of his research, Allen tried to obtain background interviews with various members of the capitalist class, but without exception they declined. Thus Allen resorted to the techniques of **historical research,** a method that involves the analysis of existing documents and materials without contacting the people or groups who are the object of the study. Much of the information he obtained on the economic capital of the corporate rich came from reports filed with the Securities and Exchange Commission, the federal agency that regulates stock transactions. Allen also examined many other public records, including the regular reports of corporations, probate and court records filed at the time family members died, and newspaper and magazine articles about the families. He found obituaries, published biographies, family histories, probate records, and other sources that genealogists often use to trace family trees.

Altogether Allen spent many years searching through corporate records, newspapers, and courthouses across the country.[64] Eventually, however, by combining and collating the information about several hundred corporate rich families, Allen learned a great deal about how they developed and maintained their economic, cultural, and social capital and how this was related to their powerful position within the United States.

Maintaining Economic Power and Wealth

The defining element of corporate rich families is what Bourdieu calls economic capital, Weber called class or property, and Marx called the control of the means of production. The United States, like other countries, has fairly hefty estate and income taxes. Estate taxes in particular are designed to try to "level the playing field" a little, to control the transfer of large amounts of money from one generation to another. Yet many loopholes exist, especially through *trusts,* accounts in which money is designated for the benefit of a particular individual or group. The wealthy often use trusts to protect their fortunes for at least two generations. Consider the family of William du Pont, one of the original partners in the Du Pont company, one of the world's largest producers of chemicals and related products. Prior to his death in 1928, du Pont converted his $35-million estate into a trust, with the principal to go to his grandchildren on the death of his children. When his son, William du Pont, Jr., died in 1965, his 60 percent share of the trust was worth more than $400 million. Because the assets were in a trust, his children (the first William's grandchildren) paid no estate or gift taxes on the fortune and didn't have to do so until they created new trusts for these assets for their own children and grandchildren.[65] Through these and other means, the corporate rich maintain their economic capital and pass it down to succeeding generations, thus helping to maintain their position in the capitalist class.

Accumulating Social and Cultural Capital

As we have already noted, money alone does not buy a family's acceptance in the upper class. Through his research Allen discovered how members of corporate rich families manage to move into the upper class by accumulating what Bourdieu calls social and cultural capital. Both of these are related to what is known in the Weberian tradition as prestige. For example, Henry Ford, who created Ford Motor Company, was certainly part of the corporate rich, but he was not accepted into the upper class. In fact, he despised their ostentatious lifestyle. In contrast, his son, Edsel, very much wanted to be accepted into the upper class and worked hard to build up his family's social and cultural capital so he could attain this goal. He married the daughter of a prominent upper-class family in Detroit; built a massive summer lodge in Seal Harbor, Maine, near where many upper-class families summered, so his children could play with their offspring; and sent his sons to elite private schools and universities.[66] Attending these schools and playing with upper-class children produced social capital, or social networks, that included upper-class children. These associations also helped produce cultural capital, or knowledge of how the upper class behaves. In this way, even though he might still have been a little suspect given his own upbringing, Edsel Ford paved the way for his children and grandchildren to be fully accepted as members of the upper class.

Wielding Wealth and Power

What does the existence of the corporate rich class mean for the rest of us? Why should we be interested in the corporate rich? The answer to this question involves the third dimension of status outlined by Weber—*power*. As Marx suggested, the corporate rich are able to translate their economic capital into a great deal of power, much of which is generally invisible to the rest of society.

Allen identified a variety of ways in which the corporate rich use their wealth to influence our lives. One of the most obvious is through large contributions to political candidates, most often Republicans. It is not unreasonable to expect that candidates who receive these donations subsequently will vote in ways that favor the interests of the donors.

Besides making donations, the corporate rich directly lobby government officials to pass legislation for their own benefit. Allen reports several instances, including one that involves the Gallo family:

> Ernest and Julio Gallo and their four children are the owners of a large wine company worth roughly $600 million. When the House Ways and Means Committee began considering a tax-reform package in 1985, one of the proposed changes in gift and estate taxes was the imposition of a new "generation-skipping tax." . . . Under this law, for example, the Gallo brothers and their wives would have been able to transfer a total of $4 million in company stock to their grandchildren before they had to pay generation-skipping taxes. Ernest and Julio Gallo, who planned to distribute some of their stock in the family company to their twenty grandchildren, wanted a more generous exclusion. [A] temporary version of the "Gallo Amendment,"

as it came to be known, was included in the final tax law. The passage of this amendment may have been aided by the fact that Ernest and Julio Gallo and their wives had contributed over $276,000 to various federal campaigns since 1977. In any event, the Gallos [had] until 1990 to transfer a total of $84 million worth of stock in the family corporation to their twenty grandchildren without paying any generation-skipping taxes.[67]

In addition, several members of wealthy capitalist families have used their family wealth to finance their own political campaigns. Most notable of these, as Allen documents, are members of the Rockefeller family. This family's wealth originated with John D. Rockefeller, who founded the Standard Oil Company in 1870. With his partners in Standard Oil, Rockefeller once controlled 80 percent of the oil-refining capacity of the United States and was widely condemned as an "utterly ruthless 'robber baron.'"[68] Yet, with the passing of time, Rockefeller's descendents were able to transcend this characterization, and three of them have had prominent political careers. John D.'s grandson, Nelson, served four terms as governor of New York, tried (and failed) four times to gain the Republican nomination for the presidency, and finally was appointed to the vice presidency by President Gerald Ford in 1974 after Richard Nixon's resignation. Winthrop, another grandson, became governor of Arkansas. Most recently great-grandson Jay has served as West Virginia state representative, governor, and U.S. senator. In just two of his campaigns, he spent more than $20 million of his own money.[69] Similarly between 1958 and 1972, Nelson and members of his family contributed $17 million to his various electoral campaigns.[70] Clearly only members of corporate rich families can afford to finance high-spending campaigns that greatly increase their influence over the political process.

Allen also documented how the corporate rich exert power in more subtle ways. For instance, more than half of the newspapers, most of the radio and television stations in large metropolitan areas, and most of the book- and magazine-publishing firms in the United States are owned by large media corporations. Many of these corporations are largely controlled by descendents of the families that first founded them. For example, a handful of corporate rich families, such as the Hearsts, Scrippses, Pulitzers, Coxes, and Newhouses, own almost all of the stock in the major media corporations.[71] The media can influence public opinion by what they choose to report and how they do so. Newspapers also regularly endorse candidates for public office and offer editorial opinions on public issues. Even though we generally assume that the people hired as

This gate is just one way that the wealthy separate themselves from the rest of the population. What are other means that the wealthy use to reinforce and maintain their class position?

editors represent their own views in the editorial page, corporate rich families who control media corporations have at least occasionally imposed their political preference on editors. For example, in 1972 all forty-three of the newspapers owned by Cox Enterprises and all forty of the newspapers owned by the E. W. Scripps Company provided editorial endorsements of President Nixon.[72] The fact that Nixon won a landslide reelection notwithstanding, such uniform endorsement by so many editors is unlikely to have occurred simply by chance.

A final way in which the corporate rich exert their power is even more subtle, involving their philanthropic and charitable activities. The corporate rich can avoid paying taxes by placing part of their wealth in foundations and other tax-exempt organizations. Almost all of the large charitable foundations in the United States were originally founded by members of corporate rich families.[73] They and their descendants generally serve on the boards of directors of these organizations and influence, if not control, the ways in which foundation funds are allocated. Traditionally they have provided money for cultural projects; for various types of civic affairs at the national, state, and local level; and for research into social problems and policy solutions. In contrast, governments generally have granted only relatively small amounts of money to such projects and, in recent years, have reduced their contributions even more. Thus the foundations have a clear and strong influence on what types of projects will be considered and developed.[74]

Although boundaries between the social classes in the United States are often somewhat vague, Allen concluded that the corporate rich families "belong to a distinct and coherent social class [and that they] form the core of the capitalist class in America."[75] They share economic interests related to their concern over the profitability of their corporations, and these interests differ from those of large portions of the population. They use this economic capital to help promote national policies that can help maintain their interests, as well as to support lifestyles and social networks that maintain their distinctive position as a social class.

Descriptions of social class groups sometimes make it sound as though stratification simply exists as an inevitable part of society. Yet it is important to realize that individuals continually act in ways that help to maintain both their class position and the stratification system, as the actions of the corporate rich demonstrate. Rita may have gotten to know members of the capitalist class while at college, but she is unlikely ever to belong to that class herself. She has experienced a lot of upward mobility in her life, but the chances are slim that she can ever enjoy the power and wealth that the corporate rich take for granted.

CRITICAL-THINKING QUESTIONS

1. Do you think that Allen would have obtained different results if his study had dealt with a different historical era? With a different society? Why or why not?

2. Can you think of ways, besides historical research, that you could study the capitalist class? What might be the advantages and disadvantages of using such methods? What research questions might you ask?

SUMMARY

Systems of social stratification, or inequalities based on social class, exist in all societies. However, across different times and societies, the extent and nature of stratification varies widely, as do theoretical explanations about why it occurs and how it is maintained. Key chapter topics include the following:

■ Even though Americans often think that they live in a "classless society," the sociological view reveals a great deal of inequality and stratification.

■ The early sociologists studied social stratification by examining the social hierarchies in single communities. Contemporary sociologists often measure social stratification through occupational prestige.

■ Several theoretical perspectives offer insights into why stratification exists. The Marxian view focuses on economic exploitation. Although traditional Marxians concentrated on only two class groups, Wright's contemporary neo-Marxian approach involves more categories. A Weberian view stresses that stratification involves differences in property, prestige, and power. A functionalist explanation suggests that stratification exists because societies give greater rewards to those who fill difficult social roles. Finally,

Lenski's theory focuses on technology, the division of labor, a society's ability to produce surpluses, and the relative power of the members of a society.

■ Although the American Dream encourages us to believe that anyone can be upwardly mobile, the reality is far more complex. Much of the upward mobility in the twentieth century has been structural mobility resulting from changes in the occupational structure, but in recent years the amount of this upward mobility has declined.

■ The number of people who have incomes under the poverty line has varied over the years. Since the 1970s, both the proportion of people living in poverty and the amount of income inequality in the United States have increased. Poverty has decreased among the elderly, but it has increased among children.

■ Michael Allen's research describes how the capitalist class—less than 1 percent of the total population—maintains its powerful position within the stratification system by accumulating economic, cultural, and social capital. In addition to controlling a disproportionate share of economic resources, this select class wields great political power.

KEY TERMS

agrarian societies 166

class consciousness 160

cultural capital 179

economic capital 179

Gini index 167

historical research 179

horticultural societies 166

hunting and gathering societies 165

income 178

intergenerational mobility 170

intergenerational succession 170

labor force 171

median 173

mobility table 170

neo-Marxist 161

occupational prestige scores 157

occupational structure 171

poverty line 172

poverty rate 173

power 162

prestige classes 156

reference group 173

social capital 179

social class 155

social mobility 156

social stratification 155

state 167

status 162

structural mobility 171

upward mobility 170

wealth 178

work force 171

FURTHER READING

Suggestions for additional reading can be found in *Classic Readings in Sociology,* bundled with this book. If you purchased a used copy of this book that does not include this custom-published reader, go to www.sociology.wadsworth.com for ordering information.

INTERNET ASSIGNMENTS

1. Using a search engine, locate a Web site that contains census information on income levels in the United States (for example, Section 14 of *Statistical Abstracts*). How has disposable income in the United States changed since 1980? How have individual wages and family income changed during this same period of time? (Look for data that adjust for inflation by reporting the amounts in "current dollars.") What can you conclude about inequality in the United States?

2. Using the same or a similar Web site, locate information on poverty and public assistance by state. What percent of the population of selected states are below the poverty threshold? What percent receive public assistance income (for example, AFDC, SSI, and so forth)? Are the proportions consistent among states or are there variations? How might you explain these findings?

INFOTRAC COLLEGE EDITION: READINGS

Article A53643942 / Garis, Dalton. 1998. Poverty, single-parent households, and youth at-risk behavior: an empirical study.*Journal of Economic Issues, 32,* 1079-1081. Garis reports research that links poverty to delinquency and other at-risk behavior.

Article A20779525 / Pressman, Steven. 1998. The gender poverty gap in developed countries: causes and cures. *Social Science Journal, 35,* 275–287. Pressman looks at poverty among female headed households in 15 industrialized nations and how different policies address this issue.

Article A19956527 / Lichter, Daniel T. 1997. Poverty and inequality among children. *Annual Review of Sociology, 23,* 121–142. Lichter documents the extent of poverty among children and the lifelong effects that such poverty can have.

Article A53552958 / Burtless, Gary. 1999. Growing American Inequality. *Brookings Review, 17,* 31–32. Burtless documents changes in income inequality in the United States and discusses the implications of this trend.

RECOMMENDED SOURCES

Allen, Michael Patrick. 1987. *The Founding Fortunes: A New Anatomy of the Super-Rich Families in America.* New York: Dutton. Allen provides a fascinating, in-depth, easy-to-read report on his research into the corporate rich.

Bendix, Reinhard, and Seymour Martin Lipset (eds.). 1966. *Class, Status, and Power: Social Stratification in Comparative Perspective,* 2nd ed. New York: Free Press. This is an older but still classic collection of theoretical works related to stratification, such as those by Marx and Weber.

Davis, Allison, Burleigh B. Gardner, and Mary R. Gardner. 1941. *Deep South: A Social Anthropological Study of Caste and Class.* Chicago: University of Chicago Press. This gripping and extensive description by students of Warner of the stratification system of a pre–World War II community in the Deep South provides important insights into how social stratification appeared within both the Euro-American and African-American communities and how these two separate social stratification systems were related.

Frank, Robert H., and Philip J. Cook. 1995. *The Winner-Take-All Society: How More and More Americans Compete for Ever Fewer and Bigger Prizes, Encouraging Economic Waste, Income Inequality and an Impoverished Cultural Life.* New York: Free Press. Two economists discuss how cultural norms that support inequality can be found throughout many areas of social life.

Gilbert, Dennis. 1998. *The American Class Structure: In an Age of Growing Inequality,* 5th ed. Belmont, Calif.: Wadsworth. In a contemporary examination of stratification in the United States, Gilbert discusses a wide range of both theoretical and empirical work.

Hacker, Andrew. 1997. *Money: Who Has How Much and Why.* New York: Scribner. Hacker provides an accessible description of the extent of inequality in the United States.

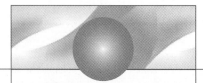

Pulling It Together

Listed below are other chapters where you can find additional discussion or examples
of some topics covered in this chapter. You can also look up topics in the index.

INTERNET SOURCES

The RAND Corporation. This Web site focuses on
public policy research, including many issues related
to stratification.

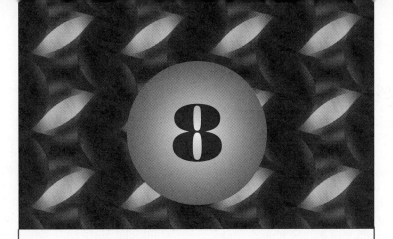

Racial-Ethnic Stratification

**Racial-Ethnic Groups and Stratification
in the United States**

Racial-Ethnic Groups in the U.S. Population

Racial-Ethnic Stratification and Social Stratification

Social Structure and Racial-Ethnic Stratification

*The Pre–Civil War South: The Plantation Economy
and Racial-Caste Oppression*

*The Civil War to World War II: Industrialization,
Segregation, and a Split Labor Market*

*World War II to the Present:
Advanced Industrialization and a Segmented
Occupational Structure*

**Individual Discrimination: Social Actions
and Racial-Ethnic Stratification**

Prejudiced Attitudes

Discriminatory Actions

Prejudice and Discrimination in Everyday Life

**Prejudice and "Group Threat":
Linking Macro- and Microlevels of Analysis
to Diminish Prejudice**

FEATURED RESEARCH STUDY

**Group Threat and Racial-Ethnic Prejudice:
The Work of Lincoln Quillian**

Prejudice in Europe

Trends in the United States

*Decreasing Group Threat in Everyday Interactions:
The Contact Hypothesis*

THE SUMMER BEFORE Joe Johnson started college, he spent a weekend at the Johnson family reunion at Grandpa John's modest home outside of Chicago, where the Johnson clan assembled every summer without fail. Unlike most families Joe knew, the Johnsons could trace their heritage back many generations, and they were fiercely proud of it.

One of the reunion rituals involved listening to the family elders tell stories of days gone by. These days it was Grandpa John who did the honors. But when John began talking, Joe could close his eyes and picture his great-grandfather, Peter, holding forth when Joe was small. Ancient and frail, old Peter had had a surprisingly strong, clear voice that seemed to carry across the long years.

Old Peter's mother and father, Joe's great-great-grandparents, had been born into slavery in the Civil War–era South. Albert Johnson, Peter's father, grew up to become a sharecropper, working a plot of land on the plantation where his family had been slaves in return for part of the crop and a ramshackle house to live in. In many ways sharecropping wasn't much different from slavery. It was backbreaking work on land Albert and his wife, Pearl, didn't own, and it barely kept their family fed.

As a child Albert had learned to read when he spent several months in a "freedom school" taught by Northerners just after the war. Intensely proud of his literacy, he read passages from the Bible, the only book he owned, to his family each night. When Albert died, Peter inherited the Bible along with his father's pride in being able to read it.

Like his father Peter worked as a sharecropper for several years until World War I broke out. As word reached the South that there were jobs to be had up North, he and his wife, Ethel, decided to move to Chicago. Peter found work in a factory while Ethel worked as a maid and cook for well-to-do white families. In some ways life was better in the North. People weren't being lynched, as they sometimes were down South, and the family had enough income to buy a few luxuries, like the big console radio that Grandpa John still kept in his living room. Yet, in practice, the North was nearly as segregated as the South, and white workers were often hostile to blacks who migrated to the cities and competed with them for jobs.

In 1921, around the time that Grandpa John was born, the Johnsons started having annual get-togethers to reunite the northern and southern wings of the family. Peter and Ethel wanted their children to know about their roots as much as they wanted them to have opportunities their ancestors had never known.

Encouraged by his parents, Grandpa John became the first Johnson to complete high school. When World War II began, John enlisted in the army and served in an all-black unit in Europe, winning a medal for valor.

After the war John married Joe's grandmother, Mary, and took college courses at night while working at the post office. In time he worked his way up to local postmaster, one of the few white-collar jobs open to African-American men at that time. It was a decent enough job, but John always felt he could have been a business executive if only he'd had the chance.

John and Mary were devout Baptists. Through their church they became active in the civil rights movement in the 1950s. They spent some of John's vacations in the South, staying with relatives and marching with Martin Luther King, Jr., in places like Selma and Montgomery.

After Joe's father, Andrew, was born, his parents took him along on these trips. At this point in Grandpa John's story, Andrew chimed in, telling how he could still remember visiting this strange and frightening part of the country where he couldn't use water fountains and bathrooms that were designated "white only" and where police turned fire hoses on peaceful marchers.

In the late 1960s Andrew became the first Johnson to finish college. He then was accepted into law school, and when he received his degree and passed the bar, he declined a few offers from private law firms to work as a public defender. Things had improved in some ways thanks to the civil rights movement, but African Americans in Chicago were routinely abused by police and the criminal justice system. Andrew wanted to do what he could to protect the rights of his neighbors, whether they were guilty of crimes or not. He felt proud of the fact that he was carrying on, in his own way, his parents' struggle for civil rights.

While he was still in law school, Andrew had married Joe's mother, June, a schoolteacher. By the time Joe was born, in 1980, the family was financially secure. Although Andrew worried about the boys being too sheltered, he agreed when June proposed sending Joe and his brothers to private schools where academics and discipline were paramount. Every generation of Johnsons had enjoyed a better life than the one before, and Andrew hoped the same would be true for his sons.

As the familiar family saga came to an end, Joe sensed the approving looks of his relatives. Here he was, about to carry on the family tradition by going to an even better college than his father had attended. He knew he was the favored child in the eyes of many of his older relatives because he had done well in school and stayed out of trouble. In contrast, several of his cousins had gone to dilapidated schools where the teachers seemed to be just putting in time and no one learned much. A few of them had dropped out of school altogether and taken jobs in fast-food restaurants and gas stations. One cousin was already a mother twice over at the age of seventeen.

Joe wasn't sure how he felt about his past and his future. In many ways he shared the Johnson family's pride in and respect for their ancestors. But sometimes the old stories didn't seem all that relevant. Things may have been better than they used to be, but he was still hassled by cops for no reason, and too many African Americans' opportunities seemed severely limited because of their "heritage." They lived and went to school where they did because of housing patterns that had been established years

ago, back when Grandpa John got his "big break" working for the post office. Even if segregation laws and lynch mobs were a thing of the past, there still seemed to be invisible lines drawn in society, and African Americans usually were on the wrong side.

What will Joe's experience be at a "select," predominantly white college? If he does well, will he be respected for his achievements, or will some people think he's getting special treatment because of his color? Should he wear his family heritage like a badge of honor, or should he treat the past as something irrelevant to who he is? Will there ever come a time when the color of one's skin simply doesn't matter in America?

*T*HE EXPERIENCES OF JOE and his family illustrate what sociologists call **racial-ethnic stratification,** the organization of society such that people in some racial-ethnic groups have more property, power, or prestige than do other groups. You are probably most familiar with racial-ethnic stratification as it exists in the United States. In fact, tensions, conflict, and exploitation of one group by another seem to appear in all societies that include more than one group. The dreadful slaughter of the Tutsis by the Hutus in Rwanda; the bloody civil war between the Serbs, Croats, Bosnians, and ethnic Albanians in the former Yugoslavia; the periodic violence in Sri Lanka, India, Burundi, South Africa, Sudan, Lebanon, Israel, Spain, Russia, and many other countries—all are examples of this phenomenon.[1]

Sociologists use the term **racial-ethnic group** to signify distinctions between groups of people based on both physical characteristics and ethnic heritage. Although you often encounter the term *race* in everyday conversations and popular writings, social scientists usually try to avoid the term in their scholarly writings because it implies that we can reliably distinguish between groups of people based on their physical characteristics. In actuality, however, attempts at such racial classifications have never produced consistent results. Instead, many different kinds of classification schemes have appeared depending on the objectives of the people doing the classifying. Scholars today agree that there are no pure physical "races." What is much more important, especially from a sociological standpoint, is the

social meaning that people attach to the differences between groups.[2]

The study of stratification along racial-ethnic lines is inseparable from the study of discrimination and prejudice. **Discrimination** refers to differential treatment accorded to a group of people based solely on ascribed characteristics—on who they are, not what they know and can do. Discriminatory acts can range from making derogatory ethnic references, to ignoring someone trying to speak to you, to denying someone a job or housing or education, to committing acts of violence and aggression.[3] Joe and his family have experienced all of these types of discrimination, as have countless other people around the globe and throughout history.

Discrimination is often related to **prejudice,** preconceived hostile attitudes, arbitrary beliefs or feelings, toward a group of people simply on the basis of their group membership.[4] Prejudice refers to people's attitudes and beliefs, whereas discrimination refers to their actual behaviors. As we will see later in the chapter, not all discrimination is caused by prejudice, and not all prejudiced people act in discriminatory ways. Although we'll examine both discrimination and prejudice in this chapter, in discussing stratification we are primarily concerned with discrimination, whatever its causes may be.

Although both prejudice and discrimination are nearly universal in human experience, they represent a special issue for Americans. Prejudice, discrimination, and severe racial-ethnic stratification directly contradict our culture's deeply held beliefs about freedom and equality. This contradiction between our ideals and the reality of racial-ethnic stratification was called the "American dilemma" by the Swedish social scientist Gunnar Myrdal.

In 1937 the Carnegie Corporation (a charitable foundation funded by a bequest from the extremely wealthy industrialist, Andrew Carnegie) asked Myrdal to direct "a comprehensive study of the Negro in the United States, to be undertaken in a wholly objective and dispassionate way as a social phenomenon."[5] In other words they asked Myrdal to do a sociological study of racial-ethnic stratification in the United States.

In 1938 Myrdal came to America and embarked on a car tour of the southern states. Writing about this trip more than twenty years later, Myrdal noted, "I was shocked and scared to the bones by all the evils I saw, and by the serious political implications of the problem." He suggested to the Carnegie Corporation that they "give up the purely scientific approach and instead deal with the problem as one of political compromise and expediency."[6] But his sponsors rejected this notion and asked that he continue to approach his study as a social scientist—with rigor and dispassion while not ignoring the moral implications of his work.

Myrdal did so, enlisting the assistance of many American social scientists, both Euro-Americans and African Americans, and both southerners and northerners. The team of researchers gathered information on racial-ethnic stratification in all areas of American life—work, schools, churches, politics, housing, and many others. As Myrdal and his staff continued to ponder this "American dilemma," Myrdal sensed that this was a moral issue dating back to the arrival of the first African slaves in the British colony of Virginia in 1619. How could a society that valued freedom and equality so deeply allow such intense racial-ethnic inequality to exist?

Myrdal's study was conducted at a time when segregation was still legal in the southern states. African Americans were compelled to attend their own schools, to use different public facilities than whites, to ride in the back of public buses, and to sit in designated areas in theaters. Yet, despite all the changes of the past half-century, the American dilemma continues to affect all of us.[7] Hardly a day goes by that racial-ethnic stratification is not in the news, whether the topic is unemployment among young African-American males in the cities or among Native Americans on reservations, the quality of education in inner-city public schools, or the "white backlash" against Affirmative Action programs or immigration policies. Issues like these stir people's emotions, from rage over injustice to resentment

How do these African–American children, living in poverty in sight of the United States Capitol, illustrate the "American dilemma"?

over the "special treatment" that some believe is received by members of minority groups. For many people—social scientists like Myrdal and others as well—racial-ethnic stratification raises questions that are not simply factual or scientific, but moral as well.

Even though we might all have these emotional reactions and feel these tensions, we still can step back and look at racial-ethnic stratification in a sociological manner. We can analyze how racial-ethnic differences appear within the stratification mountain, producing large fault lines that separate groups—limiting opportunities for some and enhancing opportunities for others, while producing tensions, stresses, and strains in everyday lives. We can also begin to understand more about the American dilemma—how our cultural values of equality and freedom painfully clash with the day-to-day realities of inequality and stratification. Many sociologists hope that the insights they achieve through studying racial-ethnic stratification will ultimately help mitigate this situation and contribute to a resolution of the American dilemma.

In this chapter we explore some of this work. We begin by looking at racial-ethnic groups in the United States and the extent to which members of these groups vary in their social class placement. We then discuss sociologists' explanations of racial-ethnic stratification. We first apply a macrolevel perspective to take a long-range, historical view of changes in racial-ethnic stratification in the United States. We then adopt a more microlevel perspective to see how racial-ethnic stratification also involves the behaviors of individual people and small groups. Finally, using both macro- and microlevel views, we look at the possibility of changing prejudice and discrimination, exploring why racial-ethnic prejudice varies from one country to another, from one time period to another, and from one situation to another.

RACIAL-ETHNIC GROUPS AND STRATIFICATION IN THE UNITED STATES

In Chapter 3 we discussed the wide variety of racial-ethnic groups in the United States. From that discussion you know how difficult it is to determine an individual's ethnic background because so many of us are descended from ancestors from a number of cultures and parts of the world. You also will recall that race-ethnicity is not just an attribute of people of color. We can all be defined by our race-ethnicity, and patterns of racial-ethnic stratification affect the lives of every American. However, some people, particularly Euro-Americans, have a great deal of choice about how salient their ethnic identity is to them and even which ethnic identities they might choose. People of color do not typically have these options.

In talking about racial-ethnic stratification, sociologists often use the term *minority group*. Even though the word *minority* implies that a group constitutes less than one-half of a given population, the sociological definition is actually much more specific and involves the nature of a group's relationships with other groups. According to sociologists any subculture can be considered a **minority group** if it is subordinate to another group or groups within the society—that is, if it has less power, privilege, wealth, or prestige. Thus even though black South Africans represent a numerical majority in that country, in sociological terms they would be considered a minority group.[8]

A primary source of information about minority groups and racial-ethnic stratification is the data collected by the U.S. Census Bureau and other government organizations. These classifications will be modified somewhat beginning with the census in the year 2000 as people are allowed to describe themselves as having more than one racial-heritage. As discussed in Box 3-2, "Measuring Ethnic Identity and Race-Ethnicity," this illustrates how people's

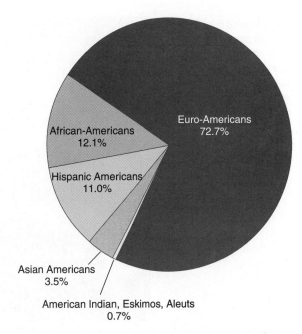

FIGURE 8-1 *Racial-Ethnic Groups in the United States*
Source: U.S. Bureau of the Census, 1998, p. 19.

preferred terms for describing their ancestry can change over time. Nevertheless, the government data give us a way to characterize the makeup of the U.S. population as a first step toward exploring racial-ethnic stratification.

Racial-Ethnic Groups in the U.S. Population

Figure 8-1 shows the distribution of the major racial-ethnic groups counted by the Census Bureau. Euro-Americans make up a little less than three-fourths of the population of the United States. African Americans compose 12 percent of the population, and Hispanic Americans make up 10.7 percent. Hispanic American is a global term the Census Bureau uses to designate people who migrated from or whose ancestors migrated from Mexico, Puerto Rico, Cuba, or Central and South America—countries and regions that represent a broad range of distinct ethnicities, languages, and cultures. About 60 percent of Hispanic Americans are of Mexican ancestry, about 12 percent Puerto Rican, and about 5 percent Cuban.[9] The number of Hispanic Americans is growing rapidly, and the Census Bureau estimates that they will outnumber African Americans by 2005.[10]

Like Hispanic Americans, Asian Americans are another highly diverse group whose representation is growing rapidly as a result of recent immigration.[11] In 1970 people of Chinese and Japanese heritage made up the most populous ethnic groups among Asian Americans, but by the late 1990s Chinese were

The children in this school represent some of the racial-ethnic diversity found in the United States. As you read this chapter, think about how racial-ethnic stratification might affect the life course of these children.

less than 25 percent of all Asian Americans, with Filipinos at 19 percent, and people of Japanese, Asian Indian, and Korean descent each a little more than 10 percent.[12]

American Indians, or Native Americans, are the smallest racial-ethnic minority group today, although the number of people who identify themselves as American Indian to the Census Bureau has increased phenomenally over the past two decades. American Indians trace their heritage to a large number of distinct groups. The largest tribe today is the Cherokees, with more than 300,000 enrolled members. The largest reservation is the Navajo, which covers parts of Arizona, New Mexico, and Utah and is home to more than 140,000 people. A large proportion of American Indians do not live on reservations.[13]

Because of the historical patterns of immigration and later movement throughout the country, members of various racial-ethnic groups aren't evenly dispersed across the United States. At the beginning of the twentieth century, 90 percent of all African Americans lived in the South, much like Joe's ancestors, but today only slightly over half of all African Americans live there. More than a third of all Asian Americans live in California, the first port of entry for most immigrants from Asia. More than 9 percent of the population of Los Angeles and almost 15 percent of the population of San Francisco are of Asian descent.[14] Similarly almost 80 percent of all Mexican Americans live in the two border states of California and Texas. Large Puerto Rican communities are found in the northeastern United States, and more than two-thirds of all Cubans reside in the Miami, Florida, area.[15] American Indians most often live in the western part of the

country, with Oklahoma, California, Arizona, and New Mexico having the largest populations.[16]

These residential patterns obviously affect the possibility of contacts between members of different groups. For instance, Euro-Americans who live in the South are much more likely to be acquainted with African Americans than with Asian Americans. Similarly many Mexican Americans probably have never met any Cuban Americans or Puerto Ricans, even though the Census Bureau classifies members of all three groups as Hispanic. In addition, these residential patterns influence the opportunities and life chances of members of different racial-ethnic groups. For instance, for many years the quality of schooling available in the South lagged far behind that available in other areas of the country. African Americans suffered most from low-quality education, because the practice of segregation meant that their schools were even worse than those whites could attend. In contrast, Asian Americans benefited from the higher quality of schooling found in the West.[17] Finally, these patterns affect the nature of racial-ethnic tensions and concerns that characterize different regions of the country. Issues in the eastern and southern parts of the nation more often focus on relations between African Americans and Euro-Americans. In the western United States, concerns and issues more often involve a larger number of different racial-ethnic groups.[18]

Racial-Ethnic Stratification and Social Stratification

As we pointed out in the discussion of social stratification in Chapter 7, sociologists often use variables

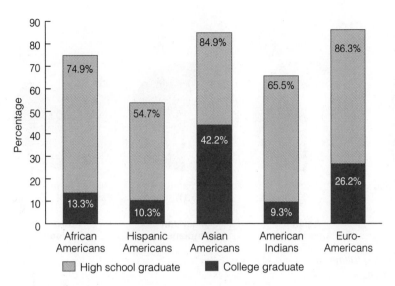

FIGURE 8-2 *Educational Attainment by Race-Ethnicity for Persons Age Twenty-Five and Older*

Note: Data for Hispanics, African Americans, Asian Americans, and Euro-Americans are for 1997; Data for American Indians are from 1990.

Source: Day and Curry, 1998, p. 2; U.S. Bureau of the Census, 1997, p. 50

such as education, occupation, and income to determine the social class to which we belong. To understand more about racial-ethnic stratification in the United States, we can look at differences among various racial-ethnic groups in educational attainment, occupation, and income.

As you will recall from our discussion of the status attainment model in Chapter 5, education is an important influence on occupational status and income. Figure 8-2 shows the differences among racial-ethnic groups in the average amount of education they have received. Not only are Asian Americans more likely to have a college education than all the other groups, including Euro-Americans, they also score higher than any other racial-ethnic group on mathematics aptitude tests and are second only to Euro-Americans on verbal tests. However, some Asian groups have much more success in this area than others. For instance, even though Chinese Americans born in the United States are more than twice as likely as Euro-Americans to have graduated from college, both Filipinos and Vietnamese are less likely than Euro-Americans to have done so, even when they were born in this country.[19] The situation is strikingly different for the other three groups. Although almost three-fourths of African Americans have completed high school, only two-thirds of American Indians and slightly more than half of all Hispanics have this level of education. Far fewer have completed college.

It is important to realize, however, that the data shown in Figure 8-2 include all adults and that the situation has improved somewhat in recent years, affecting the situation of some people in younger cohorts. For instance, by the mid-1990s 86 percent of African Americans and 61 percent of Hispanics age twenty-five to twenty-nine had completed high

school. Unfortunately these trends have not extended to higher education. Euro-Americans in this age group are still more than twice as likely as African Americans and three times as likely as Hispanics to have completed college.[20] Hispanics appear to be severely disadvantaged in educational attainment when compared with other racial-ethnic groups. Almost thirty percent of Hispanics age sixteen to twenty-four have dropped out of high school compared with 13 percent of African Americans and 7 percent of Euro-Americans in this age range.[21]

A central element of the status attainment model is the idea that educational attainment translates into occupational status. Figure 8-3 shows the occupational distribution of workers in four major census categories; Figure 8-4 shows the annual income, unemployment, and poverty status of members of each group. As we would expect from their high educational level, Asian Americans are more likely than are those in any other group to hold white-collar jobs. This is especially true for those of Chinese, Korean, Japanese, and Asian Indian descent. As shown in Figure 8-4, Asian Americans also have average higher incomes, even though their poverty rate is substantially higher than that of Euro-Americans.[22] The situation is very different for members of the other groups. Hispanics, American Indians, and African Americans are all less likely than Euro-Americans to hold white-collar jobs. They also have substantially higher unemployment rates and lower family incomes, and they are much more likely to live in families with incomes below the official poverty line.

The differences become even greater when we consider net worth—the standard way in which wealth is measured. Euro-Americans are much wealthier, on average, than minority families are. The mean, or average, net worth of Euro-American families is more than

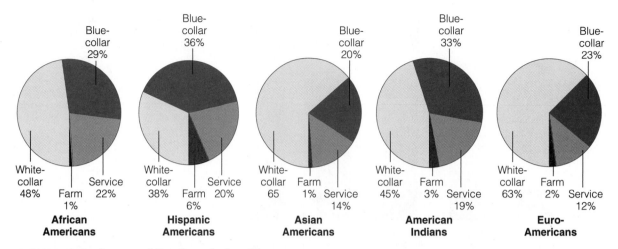

FIGURE 8-3 *Occupational Distribution by Race–Ethnicity*

Note: Data are from 1990 for American Indians and 1997 for the other groups.

Source: U.S. Bureau of the Census, 1993a, p. 45; DeBarros and Bennett, 1998; Hooper and Bennett, 1998; and Reed and Ramirez, 1998.

four times as large as that of all the other groups *together*. By the early 1990s half of all Euro-American families had a net worth of almost $46,000; the corresponding figure for other families was less than a tenth of this amount![23] These numbers are so strikingly different because virtually all people in the capitalist class—the extremely wealthy members of the corporate rich—are Euro-American, while a very large proportion of the very poor underclass and working poor are members of minority groups. The former own most of the wealth in the country, and the latter own very little or none.

All these data suggest that racial-ethnic stratification is an ongoing reality in the United States. People of color, except for some Asian Americans on some dimensions, tend to be situated lower on the stratification mountain than Euro-Americans are. In a sense racial-ethnic stratification acts like a huge fault line that runs through the stratification mountain, keeping part of the population, by virtue of their race-ethnicity, from the highest realms. Why does this stratification exist, especially within a society, such as the United States, that values equality and freedom? In the next section we look at how one sociologist has studied changes in racial-ethnic stratification over time to try to answer this question.

SOCIAL STRUCTURE AND RACIAL-ETHNIC STRATIFICATION

One of the most influential sociologists who studies racial-ethnic stratification from a macrolevel perspective today is William Julius Wilson. Using both historical accounts and contemporary data, he shows how racial-ethnic stratification is influenced by people's re-

lationships to the means of production, or *property* in the Weberian sense, and by their relative *power,* especially in the political arena.[24] Although his analysis has focused on African Americans, some of Wilson's ideas can be applied to other groups as well.

Wilson distinguishes three historical eras with different economic and power relationships that have influenced racial-ethnic stratification as it relates to African Americans: (1) the period preceding and immediately following the Civil War (the time when Joe's great-great-grandparents were in slavery and then freed and when his great-grandparents Peter and Ethel were born), (2) the time of large industrial expansion and growth that began in the late 1800s and continued until the 1930s and 1940s (an era that spans the adulthood of Joe's great-grandfather Peter and grandfather John), and (3) the decades following the end of World War II, marked by advanced industrial growth and the civil rights movement (the period in which Joe's father grew up and when Joe was born). We use Wilson's macrolevel analysis to look at racial-ethnic stratification in each of these periods, giving special attention to the implications of his analysis for the situation today.

The Pre–Civil War South: The Plantation Economy and Racial-Caste Oppression

Before the Civil War almost all African Americans lived in the South, and the vast majority were slaves, much like Joe's great-great-grandparents. Most of them lived on huge sugar, tobacco, and cotton plantations, providing a source of cheap labor. Only a very small proportion of free whites actually owned large plantations, however, and only about a fourth owned any slaves at all. Most of these owned fewer than twenty.

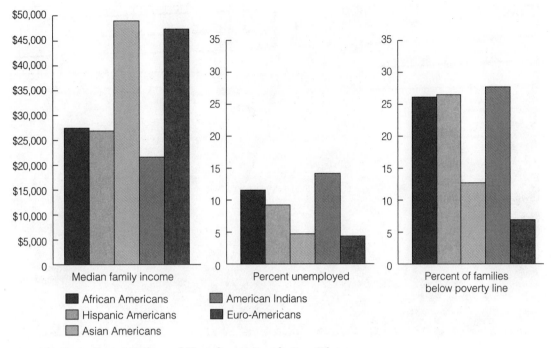

FIGURE 8-4 *Income, Poverty, and Unemployment Rates by Race–Ethnicity*

Note: Income and poverty data are from 1996 for African Americans, Asian Americans, and Euro-Americans and from 1990 for American Indians. Unemployment data are for 1997 for African Americans, Asian Americans, and Euro-Americans and from 1990 for American Indians. If unemployment is calculated only for American Indians living on reservations, the figure is actually much higher.

Source: U.S. Bureau of the Census, 1997, p. 51; U.S. Bureau of the Census, 1993a, p. 44; DeBarros and Bennett, 1998; Hooper and Bennett, 1998; and Reed and Ramirez, 1998.

Despite their small numbers the owners of large plantations managed to influence politics and law in ways that helped maintain their wealth. The requirement that men must own property to vote (of course women weren't allowed to vote), combined with manipulations of the ways in which slaves were counted in determining the allocation of legislative seats, ensured that most whites who did not own slaves had little voice in the political process. Thus, plantation owners wielded enormous political power, which they used to maintain the system of racial-ethnic stratification that was central to the plantation economy.

Wilson suggests that racial-ethnic stratification in the rural South during this period had a castelike orientation. (A **caste system** is a system of stratification like that traditionally found in India, where the family into which you are born determines your social status and the types of occupations you can hold.) Because of the nature of the slave-based economy, most contact between African Americans and Euro-Americans involved interactions between the wealthy masters and their slaves. An elaborate system of norms, duties, obligations, and rights developed that served to "keep people in their place." At the same time, African Americans were portrayed as ignorant, uncivilized, and childlike, and thus in need of this help and protection. In short, racial-ethnic stratification in

the pre–Civil War South resembled what a Marxian analysis would suggest exists: Wealthy plantation owners used their wealth, or ownership of the means of production, to control other areas of the society—politics, the law, and even morality—in ways that perpetuated their privileged position.

The Civil War to World War II: Industrialization, Segregation, and a Split Labor Market

The northern United States began to industrialize before the Civil War, and this process intensified after the war's end. The South was slower to industrialize, but eventually many agricultural workers were displaced by machines. From his macrolevel perspective Wilson suggests that the relationship of various racial-ethnic groups to the means of production and their access to political power during this time can help explain their economic status and the nature of the norms that governed their lives.

Most African Americans, like Joe's ancestors, continued to work in the fields after the Civil War, but the increasing use of machines in agriculture eventually forced many of them out of work. At the same time, reports of jobs in more highly paid industries lured many people, like Joe's great-grandfather Peter

*The slave-based economy of the plantation era involved a system in which virtually every
aspect of a slave's life could be controlled by his or her owner. How did these experiences
of African Americans differ from the experiences of other racial-ethnic groups?*

and his family, to the North. When European immigration was strongly curtailed in the early twentieth century, industrialists even actively recruited blacks.

As more African Americans moved to the North, they began to compete with whites, especially the newly arrived immigrants, for jobs. Wilson and others suggest that the effect of the economy on racial-ethnic stratification in this period can be understood using the notion of a split labor market, a term introduced by the sociologist Edna Bonacich.[25] The **split labor market** includes three class groups: (1) business owners and employers, who, like the plantation owners, control and own the means of production; (2) higher-paid labor, workers employed by the business owners and who might be demanding better wages; and (3) cheaper labor, people willing to work for less than the higher-paid labor, often because they have such a low standard of living that even very low wages will be an improvement. Thus African Americans, like Joe's great-grandfather Peter, could be enticed to factory jobs in the North at wages lower than those of immigrant workers because even these jobs were an improvement over their lives in the rural South.

Clearly these cheaper workers represented a threat to the economic well-being of the higher-paid workers. In response, the more highly paid workers often banded together to exclude the new workers or prevent their taking new jobs. At times these efforts turned violent, resulting in deadly riots—in East St. Louis in 1917 and in twenty-one other cities in 1919, including Chicago, where Joe's family lived. In these clashes, scores of people were killed, hundreds were injured, and many buildings were destroyed.

Despite the fact that they were willing to work for lower wages, African Americans remained locked out of the higher-paying jobs. This reflects differences in *power*. Waves of European immigrants had settled in northern cities many years before large numbers of African Americans came north. Over the years these immigrants developed a great deal of power through "political machines" that controlled neighborhoods and even cities and states. This political power often translated into occupational mobility and economic betterment for members of immigrant groups, who reaped the benefits of contracts and employment in areas ranging from construction to transport to industry. In contrast, even though they were growing in numbers, African Americans were cut off from these political resources and thus didn't have the means to undercut the power of more highly paid labor.[26] As a result the work force in both the North and the South was highly segregated. Euro-American workers monopolized skilled positions, often through restrictive apprenticeship and training programs, and highly segregated housing patterns appeared throughout the North.

In the South, the plantation owners' voting power dissipated with the end of the Civil War, but

Asian Americans have a high rate of business ownership. In fact, Asian Americans of Korean, Asian-Indian, Japanese, and Chinese descent are more likely than Euro-Americans to own businesses. How might a decision to open a small business be related to the realities of a split labor market, racial-ethnic hostilities, and exclusion?

after the brief period known as Reconstruction, new systems of political control developed. Because the end of the Civil War and the emancipation of the slaves also pitted African Americans against working-class whites in economic competition, the same type of hostility that appeared in the North was also apparent in the South. Euro-Americans reasserted political power in the South by preventing African Americans from voting through a variety of legal maneuvers and developing the so-called Jim Crow system. This involved a very complex system of laws that required segregation in almost all areas of social life—housing, work, schools, health care, churches, leisure activities, and transportation.[27]

Thus, in both the North and the South, economic competition between Euro-American and African-American workers, combined with the greater political power of Euro-Americans, resulted in ethnic hostility and strong patterns of segregation. Although the paternalistic caste system such as that found on plantations tended to disappear after the Civil War, a castelike stratification system was perpetuated through segregation, this time promoted by Euro-American workers instead of by wealthy plantation owners.

Bonacich has shown how economic competition characterizes relationships involving other racial-ethnic groups besides African Americans and how this competition can often result in **exclusion,** attempts to remove the lower-paid group from the labor market entirely.[28] This process is evident in the country's long history of restrictive immigration laws. For instance, workers in the western United States successfully fought for the exclusion of Chinese and

Japanese labor in the late nineteenth century. The restrictive immigration laws of the early twentieth century were aimed at European immigrants. Many of the current pressures to curtail immigration are directed toward Mexicans and other Hispanics, as well as southeast Asians. Like the patterns of segregation forced on African Americans, these exclusionary attempts have been strongly supported by workers who were threatened by competition and often have been accompanied by hostilities and virulent stereotyping and prejudice.

World War II to the Present: Advanced Industrialization and a Segmented Occupational Structure

Until around the end of World War II, racial-ethnic stratification experienced by African Americans was perpetuated by restrictive laws and policies, most notably the laws that supported the practice of segregation. Some affected the entire country, such as the practice of maintaining segregated units in the armed forces and excluding African Americans from labor unions. Others were limited to the southern United States, where schools, drinking fountains, restrooms, restaurants, and so on were segregated and where African Americans' opportunities to vote were severely restricted. These practices clearly affected individuals' life chances, but perhaps even more important, they provided a constant reminder of the power and dominance of the white majority.

Beginning with Franklin Roosevelt's presidency in the 1930s, however, things began to change, at least for African Americans. (Ironically Japanese

Americans faced new discrimination during this period; many lost their jobs and property and were placed in internment camps during World War II.) This is also the period in which the Carnegie Corporation commissioned Gunnar Myrdal to study racial-ethnic stratification, bringing the issue to the attention of people throughout the country.

Changes in the Status of African Americans

Roosevelt appointed several African Americans to positions within his administration and, in 1941, issued an executive order that prohibited discrimination in federal employment. (This executive order is what helped Joe's grandfather John get his job in the post office.) Roosevelt also helped bring about changes in laws governing collective bargaining practices, and labor unions began to accept and even recruit African Americans. Following World War II President Harry Truman ordered an end to segregation in the armed forces, and in 1954 the Supreme Court, in the landmark *Brown v. Board of Education of Topeka* case, declared the practice of segregated schooling unconstitutional.

Gradually over the next two decades, through the courageous efforts of many people like Joe's grandfather John, the barriers to voting and the practices of segregation disappeared.[29] Meanwhile, changes in the economy (described in Chapter 7) created many more white-collar jobs, and increasing structural mobility opened new opportunities for many workers and their families. Wilson suggests that these economic and political changes have resulted in a pattern of racial-ethnic stratification that is strikingly different from prior patterns—one that reflects both the legacy of previous decades of economic and political inequality and more recent political and economic changes.

Census Bureau data on the economic well-being of African Americans reveal a complex picture of gains in educational attainment, occupational status, and income of individual workers, coupled with a discouraging lack of change in rates of unemployment and poverty and in family incomes.[30]

As Figure 8-5a shows, the proportion of African Americans who finish high school now is almost the same as the proportion of Euro-Americans who do so, a striking change from earlier years. The differences in college graduation rates have also declined slightly, although Euro-Americans are still much more likely to have higher levels of education.

In the past half-century African Americans have also been more likely to be employed in white-collar jobs. During that time all members of society have benefited from structural mobility, but as Figure 8-5b shows, the increase in the representation of African Americans in white-collar jobs is even larger than that for whites. This increased representation is no doubt related to a change in individual incomes. During the past half-century, as their educational and occupational status has increased, the wages of individual African-American workers have moved closer to those of Euro-American workers, especially for those who are employed full-time. For instance, in 1940 a young African-American man who had graduated from college earned, on average, about 61 percent what his Euro-American counterpart did; by 1980 the figure had increased to 87 percent.[31] Figure 8-6 depicts this trend for all men and women since 1959.

Changes also are evident in the political world. A half-century ago the vast majority of African Americans, especially in the South, found it extremely difficult, if not impossible, to vote, and very few people of African-American heritage held elective office. Today there are more than eleven thousand local elected African-American officials.[32] In addition, African Americans serve in both houses of the U.S. Congress, on the Supreme Court, in important military positions, and on presidents' cabinets. By the early 1990s more than three hundred U.S. cities had African-American mayors. Even though this representation is still far less than the proportion of African Americans in the population, it represents a tremendous change from a half century ago.[33] White ethnic groups no longer control political machines in most major cities.

Despite this progress, the reality in other areas is less encouraging, which indicates how complex the phenomenon of racial-ethnic stratification really is. For example, since 1950, when data first become available, the unemployment rate of African Americans relative to Euro-Americans actually has worsened. As Figure 8-7a shows, in 1950 slightly less than twice as many African Americans as Euro-Americans were unemployed, but since the 1980s African Americans have been more than twice as likely to be out of work.

In addition, even though individual workers' wages have become more equitable over time, this gain has not translated into more equal family incomes. As Figure 8-7b shows, in 1950 the average African-American family had a family income that was a little more than half that of the average Euro-American family. This figure increased to 61 percent in 1970, but by 1996 it had dropped to only 56 percent. In addition, as Figure 8-8 shows, there have been very few changes in the relative poverty rates of the two groups since a large number of federal programs designed to combat poverty were established in the late 1960s. Through the mid 1990s, African-American families were almost three times as likely as Euro-American families to live in poverty.

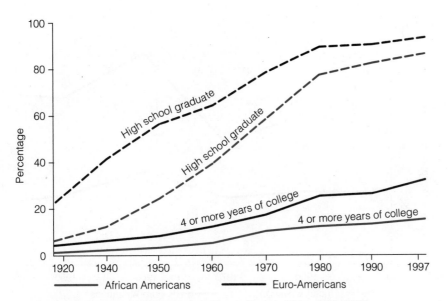

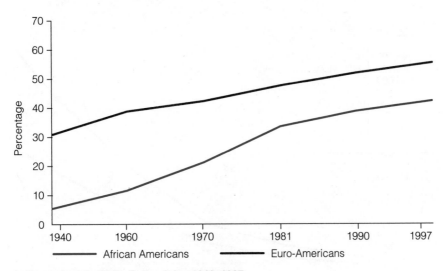

FIGURE 8-5 *Educational and Occupational Achievement of African Americans and Euro-Americans*

Source: Education, 1997, p. 17; Marger, 1994, p. 256 and DeBarros and Bennett, 1998.

a **Educational Attainment, Age Twenty-Five to Twenty-Nine, 1920–1997**

b **Percentage in White-Collar Jobs, 1940–1997**

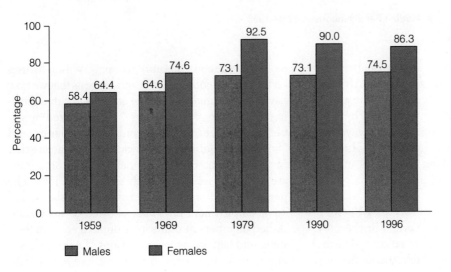

FIGURE 8-6 *African-American Median Income as a Percentage of Euro-American Median Income (full-time workers), 1959–1996*

Source: O'Neil, 1992, p. 144; U.S. Bureau of the Census, 1995, p. 477; DeBarros and Bennett, 1998.

FIGURE 8-7 *Unemployment Rates and Median Family Incomes of African Americans and Euro-Americans*

Note: Part a: Through 1980 the category "African Americans" includes other people of color; however, more than 90 percent of this category is African American. *Part b:* Through 1960 the category "African Americans" includes other people of color.

Source: Marger, 1997, p. 249, 246; DeBarros and Bennett, 1998.

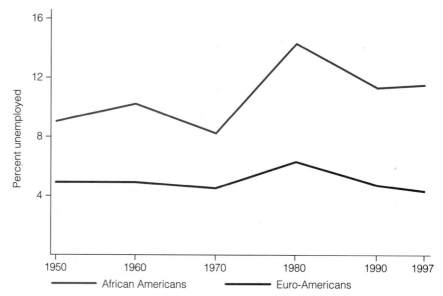

a **Unemployment Rates, 1950–1997**

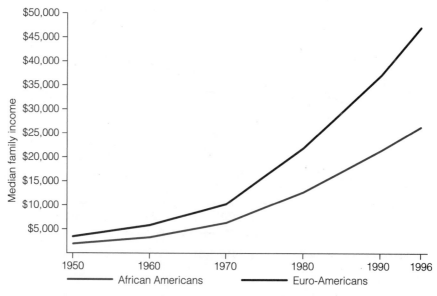

b **Median Family Incomes, 1950–1996**

Together these trends show a complex pattern of changes in racial-ethnic stratification, with African Americans showing significant gains in some areas while stagnating or declining in others. Why has racial-ethnic stratification changed in this way? To answer this question, Wilson again applies a macrolevel perspective, looking at changes in the economy and the distribution of power.

A Segmented Occupation Structure and Racial-Ethnic Stratification The pre–World War II economy, which included primarily blue-collar and farm jobs, had a split labor market in which African Americans were largely relegated to low-level jobs. When looking at the contemporary economy, Wilson focuses on the types of jobs available and suggests that we now have a **segmented occupation structure,** with a growing proportion of "good jobs," characterized by stable employment, benefits, and higher salaries, and "not-so-good jobs," which lack these characteristics. Most important, in Wilson's view, African Americans are no longer relegated solely to low-level jobs. The passage and enforcement of civil rights legislation and equal employment laws have allowed educated and skilled members of minority groups to enter higher-status and better-paying occupations.[34] As you can see in Figure 8-5b, African Americans, like Joe's father, who have the education and skills to fill these newly

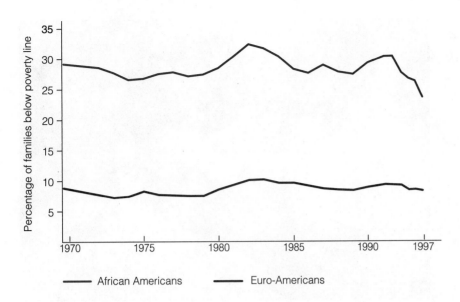

FIGURE 8-8 *Families Below the Poverty Line, 1970 to 1997*

Source: U.S. Bureau of the Census, 1997, p. 479; Dalaker and Naifeh, 1998, p. vii.

created white-collar positions benefited from the structural mobility that the changes in the occupation structure allowed. Thus, over time, the proportion of African Americans in the middle class has grown.

Yet, even as white-collar occupations have become more common, many low-skilled and unskilled positions have disappeared. Whereas once people with few skills and little training could still find some kind of work, now those who lack the requisite education often can't find any jobs at all, especially jobs that pay enough to support a family. Because of the legacy of discrimination in schooling and other areas, African Americans, as well as members of other racial-ethnic minority groups such as Hispanics and American Indians, are much more likely to be among this group. Thus, just as the changes in the occupational structure can account for African Americans' increasing presence in white-collar jobs, so, too, can it account for their increasing unemployment relative to Euro-Americans. Jobs for people with fewer skills simply have become far less common.

These alterations in the occupational structure also help account for the lack of changes in relative family incomes and poverty rates of Euro-Americans and African Americans. The reasoning is a little complex and involves two different ways in which the economic changes have affected families. The first involves shifts in the labor force participation of women. African-American women have always been much more likely than Euro-American women to work outside the home because their husbands earned such low wages. Thus median family income figures were as close as they were in the 1950s because relatively few Euro-American families had two wage earners. Beginning in the late 1960s and accelerating in the 1970s, however, many Euro-American women also entered the labor force, often because their husbands' wages were falling relative to inflation. Even though the individual incomes of African Americans were rising during this period, the fact that Euro-American families now more often had two incomes tended to keep the income gap between the African-American and Euro-American families about the same.

The second reason for the very small improvement in relative family incomes and poverty rates involves shifts in the composition of African-American families. Since the late 1960s the divorce rate has risen for all families, but especially for African Americans. In addition, many more women are choosing to have children outside of marriage. (As we will discuss in Chapter 10, both of these changes appear to be related to changes in job opportunities that resulted in the higher unemployment rate of African-American men.) As a result many more families now than in 1950 have only one parent, usually the mother, in the home. In the late 1990s, among African-American families with two parents in the home, the average family income was more than 80 percent of that of Euro-American families with two parents, and less than 10 percent of these families had incomes that placed them below the poverty line. In contrast, almost 40 percent of all African-American families with a mother but no father (and more than 30 percent of all families in this situation) lived below the poverty line.[35] Clearly having only one breadwinner greatly increases the chance that a family will live in poverty. Because African Americans are more likely than Euro-Americans or Asian Americans are to have single-parent families, a greater proportion of African-American families are poor.

What about political changes? Just as white ethnic groups controlled political machines in past

It is important to remember that racial–ethnic stratification appears throughout the world. It is not uniquely American or a problem of the past. Blacks in South Africa, such as the people shown here, first got to vote only in 1994.

years, could not the increased political power of African Americans, particularly in large cities, help create jobs and spur economic well-being? Again, using a macrolevel perspective, Wilson locates the answer in the changing economy and the changing locus of power. He suggests that in the post–World War II economy the locus of power has moved from politicians to corporations and other large economic enterprises. It is no longer city officials and party bosses who call the shots; it is corporate executives. African Americans have gained political positions just as these positions have become less powerful.

Most important, Wilson suggests that the status differences between African Americans and Euro-Americans today reflect stratification based on *social class* more than stratification based on race-ethnicity. Today there are no Jim Crow laws; in fact, a number of laws guarantee equal opportunity and nondiscriminatory treatment. Yet, there is still what sociologists call **structural discrimination.** This type of discrimination can't be traced to prejudiced attitudes or to laws or norms that promote segregation or exclusion, but rather results from the normal and usual functioning of society (the social structure). In other words the source of stratification today might be differences in education and family resources rather than a deliberate policy of discrimination. Of course, these differences in education and family resources can be traced to the fact that African Americans and other minority groups historically have been more likely to experience inadequate schooling and low incomes. But, Wilson claims, these problems are not unlike those faced by all poor people. Poor children from all racial-ethnic groups tend to live in areas with less adequate schools and to have less social capital than other children do. For this reason, Wilson sug-

gests, we should focus directly on issues related to poverty rather than those related to color.[36]

Wilson also stresses that the inner cities, which are overwhelmingly populated by very poor people of color, have their own special problems. Because of extensive interethnic hostility, for many years African Americans of all class backgrounds lived in **ghettos,** highly segregated neighborhoods populated primarily by people of one racial-ethnic heritage. As nondiscriminatory housing policies were enacted into law and economic and housing opportunities expanded, however, many middle- and working-class African Americans moved out of the ghetto into neighborhoods that were often still segregated but included many fewer low-income people. At the same time, the economy changed, especially in northern cities, in ways that drastically decreased the numbers of jobs available to people with low levels of education, the same people who were less likely to leave the ghetto.[37]

These two changes resulted in many inner-city neighborhoods experiencing extreme social problems. Poverty has become much more concentrated, so that almost all the inhabitants of some neighborhoods are poor and unemployed. This means that there are few employed people who can provide linkages into social networks that might lead to jobs or teach job skills. The high unemployment rate means that few men can afford to marry and support a family. Such areas also tend to attract criminals and thus to have high crime rates. Wilson suggests that this concentration of poverty and the isolation of residents of these neighborhoods is part of a "vicious cycle" that results in extremely high rates of violent crime, unemployment, female-headed families, out-of-wedlock births, and welfare dependency. To learn more about Wilson and his work, see Box 8-1.

BOX
8-1

SOCIOLOGISTS AT WORK

William Julius Wilson

I'm not scandalized by the fact that there is unflattering behavior in the inner cities.

William Julius Wilson is one of the most famous sociologists currently studying racial-ethnic stratification. He is at the Kennedy School of Government at Harvard University and previously taught at the University of Chicago. In an interview with Eric Bryant Rhodes published in the journal *Transition,* **Wilson spoke about his work and its implications.**

Since the 1920s, the University of Chicago has been at the forefront of investigation of the black ghetto. How do you situate yourself and your work in relation to this tradition? Ever since we began the Urban Policy and Family Life Study, people have been saying that our research has revived the Chicago School's approach to the neighborhood, combining surveys with ethnographic research. . . . In those earlier Chicago studies there was also an integration, not only of methods, but also of ideas and approaches. They straddled different disciplines, and I think that my work is in that tradition. My work is not just read by sociologists. It's read by economists, political scientists, developmental psychologists, anthropologists, historians. It lacks the kind of narrow focus that would reflect a particular methodology or a particular theoretical orientation.

What principles do you think should serve as guidelines when researchers are faced with situations where making an honest and accurate representation of inner-city residents may be embarrassing to the black community? We have sometimes been reluctant to describe certain patterns of behavior that I think are pathological—by which I mean ultimately destructive for the individual, and for the families, in those neighborhoods. You do a disservice to the black population, and to the inner-city population, by not acknowledging and explaining those patterns of behavior. If we don't make the effort to analyze the causes and contexts of unflattering types of behavior, we leave a vacuum for people like Newt Gingrich who are going to describe inner-city pathologies without making reference to their broader structural determinants.

I think we also have an obligation to ensure that other black people do not get a distorted idea of the families, institutions, and individuals in these communities. I'm not scandalized by the fact that there is unflattering behavior in the inner cities. You should expect it, given the obstacles that these people have to overcome. In these kinds of environments, some children are going to drop out of school and become involved with drugs. Of course, some children, despite everything, are going to be highly successful. The question is why are there still so many who don't live up to the expectations of mainstream society. In describing these neighborhoods, we need to talk about those who fail as well as those who succeed, the better to see the strains on those who are trying to live up to mainstream norms and expectations, how damn difficult it is—and why so many of them slip down into the "underclass."

There is a large amount of good will and concern for the problems of the poor within the black middle class. What kinds of programs do you think are most effective at harnessing this good will? Community-based programs are probably the best solution. One of the greatest problems facing low-income blacks is alienation. You can see this in the increasing alienation of blacks from the political process. As we develop a progressive agenda, we must attempt to pull together groups that represent different ethnic groups and different class groups—working class, middle class, lower class. . . . There are progressive multiracial organizations that are designed to address the problems of these low-income neighborhoods and to get people to participate more in the ongoing life of the community: programs to increase black voter registration, increase black voter turnout, and involve poor blacks in the educational system. There are people who are working with schools and attempting to improve the learning environment in a lot of homes in these neighborhoods, so that the home environment reinforces what the children learn in school. Middle-class blacks can get more involved in these basic institutions of society.

Wilson's analysis of the nature of today's racial-ethnic stratification has not gone unchallenged. In fact, undoubtedly because this issue is so emotionally tinged, the challenges to his work (and to that of others who have examined these questions) have often been quite harsh.[38] For instance, some critics contend that Wilson's picture of a growing economic equality among educated African Americans and Euro-Americans is too optimistic. Even among college graduates, African Americans have higher unemployment rates and smaller accumulations of wealth, or net worth, than Euro-Americans do.[39] In addition, although conditions did improve relatively rapidly for African Americans from 1940 to 1970, since then the rate of improvement slowed considerably as the chances of structural mobility diminished.[40]

Others worry that by taking a macrolevel perspective Wilson has downplayed the day-to-day reality of individual discrimination and the effects it has on each of us, as well as on the perpetuation of racial-ethnic stratification. Many sociologists have taken a more microlevel view of racial-ethnic stratification.

INDIVIDUAL DISCRIMINATION: SOCIAL ACTIONS AND RACIAL-ETHNIC STRATIFICATION

Individual discrimination involves actions of individuals or small groups rooted in both prejudice and discrimination. Joe and many other members of racial-ethnic minorities often experience both. Several sociologists have examined these phenomena, often with the explicit hope that their insights can help improve interactions between members of different racial-ethnic groups and diminish the incidence of such actions.

Prejudiced Attitudes

Virtually all racial-ethnic groups in the United States have faced prejudice at some time. But the prejudice directed against people of color has been much more extensive and long-lasting than that directed against Euro-Americans. For instance, American Indians often have been stereotyped as hostile savages on television shows and in movies, and the nicknames of sports teams such as the Florida State Seminoles and the Atlanta Braves, as well as the chants and war whoops of their fans, only perpetuate these stereotypes.[41] Similarly Hispanics, especially those of Mexican origin, traditionally have been stereotyped in the media and elsewhere, as lazy, shiftless, and villainous. Even advertising has traded on these stereotypes. For example, for a number of years, until

protests forced it to cease the practice, the Frito-Lay Company used the Frito Bandito to advertise its corn chips.[42] Asian Americans faced severe legal restrictions on where they could live and work and were also legally barred from entering the country through the Oriental Exclusion Act passed in 1924.[43]

The most severe prejudice, however, has been experienced by people of African-American heritage. As Wilson noted in his historical research, to help perpetuate slavery and its economic advantages, plantation owners promoted views of African Americans not only as heathens and savages but also as inferior to whites in very basic ways, such as their intelligence and initiative. These negative stereotypes became even more pronounced after the Civil War and helped to justify the intense segregation and the restrictive regulations of the Jim Crow period.[44]

Identifying Historical Trends in Prejudiced Attitudes   How much prejudice is there today? Have the changes in African Americans' economic, political, and educational status since the 1940s and the passage of civil rights laws been accompanied by changes in the attitudes of the Euro-American majority? Are Euro-Americans more likely now to support equal treatment and to want to engage in relationships with people of other racial-ethnic backgrounds? Fortunately there are data that can answer these questions, for social scientists began studying prejudice in the late 1930s and early 1940s. Beginning then, and continuing until today, polling firms, such as Gallup and the National Opinion Research Center, have conducted large-scale surveys of representative samples of Euro-Americans that asked questions about their attitudes toward African Americans and their relationships with them.

Some of these questions and the responses obtained over time are shown in Figure 8-9. Analyses of data from surveys that were conducted over a long period of time, such as those in Figure 8-9, are called **trend analyses.** A trend analysis examines data collected at various time points but uses a different sample each time. One way to think of a trend analysis is to see it as a series of snapshots of something taken at different time periods.

Trend analyses of questions regarding prejudice and discrimination, such as those in Figure 8-9, suggest that Euro-Americans' support for the general principles underlying nondiscrimination has increased dramatically since Myrdal wrote *The American Dilemma* in the early 1940s.[45] By the 1970s and 1980s responses to some of the questions "topped out," with very few people giving prejudiced answers. For instance, in 1982, when asked, "Do you think white students and black students should go to the same

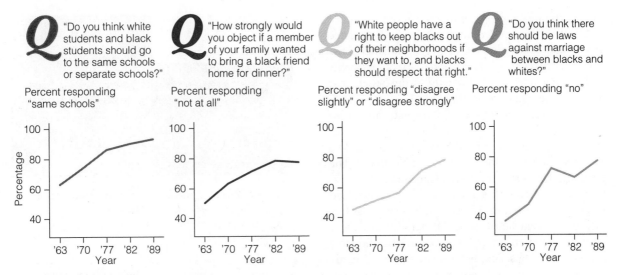

FIGURE 8-9   *Trends in Euro-Americans' Attitudes on Racial Equality and Integration*
Source: Marger, 1994, p. 276.

schools or to separate schools?" 90 percent of the respondents said "same schools," and by 1989, 93 percent did so.[46] Similar results occur with questions regarding **social distance,** the types of social ties or interactions people are willing to establish with others.

Why have attitudes changed so much over time? Did prejudice decrease simply because people got older and more tolerant (an age effect)? Because of historical changes and a generally less prejudiced society in more recent years (a period effect)? Or because people who were born more recently are less likely to be prejudiced, whereas older, more prejudiced people are dying off (a cohort effect)? Several sociologists have examined the survey data to answer these questions. They have found that we don't seem to have any systematic changes in prejudice as we get older, apart from changes that appear in all of us as a result of the historical period in which we live. In contrast to this lack of age effects, both period and cohort effects are evident. Americans of all ages and cohorts seem to express less prejudiced attitudes now than they did previously. In addition, however, younger cohorts are less prejudiced than older cohorts, a fact that contributes more to the overall declining trends in prejudice than do period or historical effects.[47] In other words the fact that there is less prejudice now than twenty years ago is less a result of changes in everyone's attitudes than it is a result of younger cohorts being less prejudiced and older, more prejudiced cohorts dying off.

Even though prejudice has declined markedly over the past half-century, a substantial minority of the population still holds prejudiced views. Why are some people more likely than others are to exhibit prejudice?

Explaining Variations in Prejudice   Education and region of residence are two of the most important microlevel predictors of how prejudiced someone is. People with more education have consistently been more likely than are those with less education to give nonprejudiced answers to survey questions. However, this difference is most apparent when the questions involve general principles, such as whether Euro-Americans and African Americans should attend the same schools. It tends to disappear in response to specific questions about social distance, such as when the integration involves contact with more than a few African Americans or involves the respondent's own family, or when the questions involve social policies such as special government spending or policies.[48] For instance, college-educated individuals are less likely than other people to say that they would object if their children went "to a school where a few of the children are blacks" or "where half of the children are blacks." But when asked how they would feel about sending their children to a school "where more than half of the children are blacks," they actually object somewhat *more* than people with less education.[49]

In addition, for as long as these surveys have been conducted, northerners have been more likely than southerners to give nonprejudiced answers. However, this difference has declined rapidly in recent years, primarily because younger cohorts of white southerners have been changing their views much more rapidly than their northern counterparts. Among younger cohorts today there is little difference in the prejudiced attitudes of northerners and southerners.[50]

Another reason that some people are more likely than others to be prejudiced involves the idea of social networks and the notion of reference groups, which was introduced in Chapter 7. **Reference groups** are the people we look to for standards of behavior. They tend to serve as models for our thoughts and actions, and we tend to exhibit the beliefs that others in the reference group hold.

A study conducted several years ago shows how racial-ethnic prejudice is related to the views of our reference groups. Leonard Pearlin taught at The Women's College of the University of North Carolina (now the University of North Carolina at Greensboro) in the early 1950s. Even though the college did not yet admit African Americans, he knew that faculty and students at the school generally had a more favorable opinion toward integration than did most southerners. Thus, simply by virtue of attending the college, these students were exposed to more accepting ideas. But, Pearlin hypothesized, merely being exposed to these new ideas probably wouldn't result in less prejudiced attitudes. Instead, changes in the students' social relationships—their reference groups—would be more important.[51]

To test this hypothesis, Pearlin and some of his students administered a survey to a random sample of women in the school. The survey included a number of questions designed to measure the nature of their prejudiced attitudes, as well as questions about their relationships with their parents, friends from their home community, and new friends and teachers in college. In support of his hypothesis, Pearlin found that students who continued to see their pre-college associations as their most important reference group were the most prejudiced, and those who more strongly identified with their new college associations were more accepting.[52]

Pearlin's study and the trend studies using data from national polls show how prejudice isn't constant and unchanging, but instead can vary over time and situations. In addition, several studies have shown that how prejudiced we are is not a perfect predictor of how much discrimination we will exhibit. Instead, our behavior seems to reflect both specific situations and general attitudes.

Discriminatory Actions

Robert Merton developed a typology, shown in Figure 8-10, that can help us see how there is no necessary association between prejudice and discrimination. The descriptors in the figure represent ideal types, so they don't perfectly reflect the behavior or views of any one person. But, as ideal types do, they provide conceptual distinctions that can help us understand more about the relationship between prejudice and discrimination and the ways in which situations influence our actions.

		Prejudiced?	
		Yes	No
Individual discriminator?	Yes	Active bigots	Fair-weather liberals
	No	Timid bigots	All-weather liberals

FIGURE 8-10 *Merton's Typology of the Association Between Prejudice and Individual Discrimination*

Source: Merton, 1949.

Two of the categories in the typology reflect *consistency* between attitudes and behaviors. One consistent group is "active bigots," people who openly and consistently behave in discriminatory ways and who hold prejudiced attitudes. Members of the Ku Klux Klan and neo-Nazi groups would be examples of this type. The other consistent group is "all-weather liberals," people who accept principles of social equality and also consistently act in nondiscriminatory ways, even when they might be in situations where others would expect them to do otherwise.

The other two types display an *inconsistency* between attitudes and behaviors. "Timid bigots" are people who might express prejudiced attitudes but who, when placed in a situation in which nondiscriminatory actions are expected, conform and don't act in a discriminatory way. For instance, before civil rights laws were enacted in the 1960s, Euro-American southerners who traveled to the North were often placed in situations in which they were expected to use the same restaurants, restrooms, and transportation facilities that all racial-ethnic groups used. Even though they might have preferred segregated facilities, they acquiesced and used the integrated services. The second inconsistent group is "fair-weather liberals," who express liberal views but who, when placed in situations in which they believe that others expect them to discriminate, will do so. For instance, many Euro-Americans in the pre–civil rights South did not share the intense prejudice of their neighbors, but to preserve harmony, they did not challenge the practices of segregation and Jim Crow.[53]

Note how each of these types of actions involves individual choices and decisions. We can use the theories regarding social interactions that were first introduced in Chapter 5 to illuminate what happens in such situations. In that chapter we discussed *role theory,* which stresses the social norms that influence our actions; *symbolic interactionism,* which focuses on our thought processes as we choose our actions; and *ex-*

change theory, which looks at why individuals choose to act the way they do. Both the timid bigots and the fair-weather liberals in Merton's typology are no doubt responding to role expectations. The timid bigot is in a role in which nondiscriminatory actions are expected, whereas the fair-weather liberal is in a situation in which discriminatory actions are the norm. In each case individuals interpret the expectations that others have for them, and their discriminatory actions (or lack of them) reflect their views of both others' expectations and the situation. At the same time, according to exchange theory, they balance the costs and benefits associated with any action. Active bigots and all-weather liberals believe that any costs associated with acting on their prejudices (or lack thereof) do not outweigh the benefits. In contrast, the fair-weather liberals and timid bigots see the costs of acting in ways that would conform to their attitudes as greater than any benefit they would gain.

Prejudice and Discrimination in Everyday Life

If you have experienced prejudice and discrimination, you know how hurtful it can be. If you haven't had such experiences, you might find it hard to fully empathize with people who have. The sociologist Joe Feagin has tried to document such experiences and explore the extent to which individual discrimination occurs today. He was especially interested in whether African Americans who have "made it," who are firmly in the middle class, experience discrimination or whether "money talks" and discrimination ceases to be a problem for those who have achieved middle-class status.[54]

Feagin's respondents included corporate managers and executives, health care and other professionals, government officials, college students, journalists and broadcasters, clerical and sales people, and entrepreneurs. Like Joe Johnson and his parents, all of the respondents had incomes that would place them well within the middle class. More than 90 percent had some college education, and more than a third had graduate training.

Feagin's data came from a larger study that featured in-depth interviews of African-American respondents from twelve major U.S. cities. Even though the interview questions dealt with education, employment, and housing, several subjects volunteered accounts of discrimination they had experienced while shopping or dining in restaurants or while driving or walking down the street. Feagin was fascinated by these digressions and by the importance the respondents placed on the events. He decided to focus on where the events occurred, how the respondents reacted, and, especially, how their middle-class status affected their response.

Feagin's respondents reported a variety of discriminatory encounters. The most common discrimination experienced in stores or restaurants was having clerks refuse to serve them or receiving poorer service than other customers. For instance, a news director at a major television station reported how she and her boyfriend waited to be seated in a restaurant but no one would help them, even though the restaurant was almost empty and waiters and waitresses were everywhere.[55] Similarly a president of a financial institution recounted how he twice took his staff to a fancy new restaurant and waited over a half-hour to be waited on, even though others around him were being served much more quickly. A utility company executive described how he and his son were asked to go outside a small neighborhood store to buy a snow cone, even though Euro-Americans were being served snow cones inside.[56]

The most common encounters reported on the street were threats and harassment from police. A man now working as a television commentator recalled two incidents that occurred when, as a college student, he was conducting interviews for a survey firm. He was dressed in a suit, driving a company car, and carrying a batch of survey instruments. Yet twice he was stopped by several white officers, spread-eagled on his car, and accused of "running through" the neighborhood.[57] A female professor at a major white university in the Southwest gave the following account:

> When the cops pull me over because my car is old and ugly, they assume I've just robbed a convenience store. Or that's the excuse they give: "This car looks like a car used to rob a 7–11 [store]." And I've been pulled over six or seven times since I've been in this city—and I've been here two years now. Then I do what most black folks do. I try not to make any sudden moves so I'm not accidentally shot. Then I give them my university I.D. so they won't think that I'm someone that constitutes a threat, however they define it, so that I don't get arrested.[58]

Feagin was especially interested in how his middle-class respondents reacted in these encounters. He found that the most common response was some type of verbal reply. For instance, although her boyfriend initially wanted simply to leave the restaurant, the television news director asked to speak to the restaurant's manager. As she recounted the event,

> I said, "Why do you think we weren't seated?" And the manager said, "Well, I don't really know." And I said, "Guess." He said, "Well I don't know, because you're black?" I said, "Bingo. Now isn't it funny that you didn't guess that I didn't have any money" (and I opened up my purse) and I said, "because I certainly have money. And isn't it odd that you didn't guess that it's because I couldn't pay for it because I've got two American Express

cards and a Master Card right here. I think it's just funny that you would have assumed that it's because I'm black." . . . And then I took out my card and gave it to him and said, "If this happens again, or if I hear of this happening again, I will bring the full wrath of an entire news department down on this restaurant." And he just kind of looked at me. "Not [just] because I am personally offended. I am. But because you have no right to do what you did. . . ."[59]

Other times, especially with the police, the respondents chose to simply accept the encounter without challenging the officers. For instance, the television commentator who had been stopped while working as a survey researcher in college was clearly upset and angry over his treatment but felt he had no choice but endure it and walk away. The professor who has been stopped by the police several times routinely pulls out identification that will neutralize the encounter and allow her to leave.

Just as we can use various sociological theories to explain our actions when we choose whether to discriminate, so, too, can these theories be used to explain our actions when we experience discrimination. Remember that symbolic interactionism stresses how our behavior results from our interpretations of others' communications. Interpreting actions that might be discriminatory can be problematic in that we need to decide if it really is discrimination or if everybody is treated the same way. We all interpret our daily interactions with others, but we often do so unconsciously. The situation is different for those of us, like Joe, who regularly, or even occasionally, encounter individual discrimination. As one of Feagin's respondents put it:

You have to decide whether things that are done or slights that are made are made because you are black or they are made because the person is just rude, or unconcerned and uncaring. So it's kind of a situation where you're always kind of looking to see with a second eye or a second antenna just what's going on.[60]

Once you interpret an event as discriminatory, you then must decide how you will respond. From an exchange theory perspective this involves evaluating the costs and benefits of various actions. For instance, Feagin's respondents often decided that the benefits of some type of verbal response or other challenge to discrimination encountered in public places like restaurants or stores outweighed the possible costs, such as hostility or embarrassment. In contrast, they generally decided that the costs associated with challenging police authority were too great and that the benefits associated with simply withdrawing or acquiescing might be greater.

Finally the types of roles we hold can strongly influence our potential responses. The advantages that middle-class status gave to Feagin's respondents are obvious. A news director for a television station could threaten to expose a restaurant on the news. The owner of the financial firm was a personal acquaintance of the attorney retained by the owners of the restaurant where he had been discriminated against. The educational, financial, and career achievements of these middle-class African Americans gave them the freedom, expertise, and resources to effectively challenge discrimination—advantages that are not enjoyed by many members of racial-ethnic minorities. Yet even these people, many of whom lead fairly sheltered lives because of their occupations and income, experience an individual cost of coping with the possibility of discrimination and the accumulation of incidents. In fact, both the cumulative nature of discriminatory incidents and the fact that such incidents might occur at any time often produce a good deal of stress for people who experience them. One member of Feagin's sample eloquently described these costs:

. . . if you can think of the mind as having one hundred ergs of energy, and the average man uses fifty percent of his energy dealing with the everyday problems of the world—just general kinds of things—then he has fifty percent more to do creative kinds of things that he wants to do. Now that's a white person. Now a black person also has one hundred ergs; he uses fifty percent the same way a white man does, dealing with what the white man has [to deal with], so he has fifty percent left. But he uses twenty-five percent fighting being black, [with] all the problems of being black and what it means. Which means he really only has twenty-five percent to do what the white man has fifty percent to do, and he's expected to do just as much as the white man with that twenty-five percent. . . . So, that's kind of what happens. You just don't have as much energy left to do as much as you know you really could if you were free, [if] your mind were free.[61]

The experiences of Feagin's respondents show how the racial-ethnic fault lines in our society affect everyday interactions, even for those who have climbed far up the stratification mountain. Even though many individual members of minority groups have experienced social mobility, they cannot escape the harsh realities of racial-ethnic stratification. Feagin's findings also show how pervasive the American dilemma is and how—more than a half-century since Myrdal wrote his report—it is still with us. Cultural values and traditions of equality and freedom have helped produce the changes described by Wilson and have helped Feagin's respondents attain their middle-class status. At the same time, however, the reality of racial-ethnic stratification and individual discrimination continues to affect the lives of people of color, like Joe Johnson and his family, in very painful and disturbing ways—ways that contradict the very notions of equality and freedom.

PREJUDICE AND "GROUP THREAT": LINKING MACRO- AND MICROLEVELS OF ANALYSIS TO DIMINISH PREJUDICE

How can these patterns of racial-ethnic stratification and individual discrimination and prejudice be changed? One way to answer this question is to look at situations and societies with different amounts of prejudice and discrimination. For instance, why are immigrants and migrant laborers treated with much

more disdain and hostility in some European countries than in others? Why have African Americans historically faced more prejudice and discrimination in the southern states than in other parts of the country? Why do some communities or work groups have little strain between racial-ethnic groups, and others have much more? Microlevel variables, such as education and age, can explain why some individuals are more prejudiced than others are. Yet, can they also explain large differences in prejudice and discrimination between countries and, within countries, between different regions? Can they fully explain why prejudice and discrimination are more apparent in some groups and situations than in others? Two strands of research—both of which link macrolevel and microlevel analyses—have addressed these issues and provide the basis for hypotheses regarding the improvement of racial-ethnic relations both in this country and around the world.

FEATURED RESEARCH STUDY

GROUP THREAT AND RACIAL-ETHNIC PREJUDICE: THE WORK OF LINCOLN QUILLIAN

Lincoln Quillian, a sociologist at the University of Wisconsin in Madison, hypothesized that differences in prejudice from one area or time period to another could be explained by the notion of **group threat**—the extent to which members of the dominant group perceive that minority groups threaten their well-being.[62] When members of a dominant group believe that others are threatening their way of life they will be more likely to respond with prejudice and discrimination. When they feel relatively secure and unthreatened, such prejudice will be far less likely to appear. Specifically, Quillian hypothesized that prejudice would be more extreme in areas and time periods when a minority group was large relative to the dominant group and when economic conditions were more precarious. To test this hypothesis he used **secondary analysis,** analyzing data that has already been gathered by other researchers. Secondary analysis allows researchers to use very high-quality data sets and provides an excellent way to replicate studies by seeing if the same results occur with different samples and in different settings. In his tests of this hypothesis Quillian first used data that had been gathered in Europe and then looked at data from the United States.[63]

Prejudice in Europe

In the late 1980s a very large survey was conducted with samples of about 1,000 citizens in each of the

12 countries that constituted the European Economic Community at that time (Belgium, Denmark, France, Germany, Greece, Ireland, Italy, Luxembourg, Netherlands, Portugal, Spain, and the United Kingdom). Several questions asked respondents about their views toward "people of another race" and "people of another nationality," trying to capture different elements of prejudiced attitudes, much like the questions shown in Figure 8-9.

Because the questions did not specify the nationality or "race" of those to whom the questions applied, the respondents could potentially be thinking of any group when giving their answer. Other questions on the survey revealed that in Britain most respondents were thinking of South Asian immigrants, in France they generally thought of Arabs, and in Germany they most often thought of black Africans and Turks. In the rest of the countries, respondents most often thought of black Africans. This variability is important in testing Quillian's hypothesis. If the notion of group threat is valid, it should be supported no matter what particular mix of nationalities or racial-ethnic groups occurs within a country.

Quillian used two different measures of his central concept of group threat: One was the proportion of residents within each nation who were not citizens of a country in the European Economic Community. The other measure was a five-year average of the gross domestic product (GDP) in the years just before the survey was given. His hypothesis led him to expect that countries with larger proportions of noncitizens and with lower economic wealth would experience more group threat and thus have higher levels of prejudice.

Quillian also had information on several microlevel variables such as the respondents' education, their age, income, satisfaction with life, recent changes in their employment and economic status, and how much daily contact they had with people of other races and nationalities. To make sure that the levels of prejudice within each country really reflected the macrolevel concept of group threat he needed to also see to what extent these microlevel variables influenced the responses of each individual. Many sociologists today use very complex statistical models to sort out the effects of a variety of variables, and Quillian used such methods to examine the separate effects on prejudice of the macrolevel measures of group threat and the individual, microlevel variables.

As other researchers have found, Quillian's results indicated that individual level variables generally are related to prejudice—both toward people from other countries and toward people of other races. The largest effects were from age and education. More highly educated people and younger people reported less prejudiced attitudes. Yet, in general, the differences in prejudice expressed by older and younger people or by those who had more or less education were not very large. Even more important, age and education couldn't explain why people in some countries expressed more prejudice than people in other countries did. Societies that had a lot of overall prejudice weren't more likely to have older people or people with less education than societies where there was less prejudice.

As a second step, Quillian looked at the effect of the macrolevel variables. How, he asked, might differences across societies in the size of the immigrant population and economic conditions—the amount of group threat—be related to differences in prejudice? In contrast to the results with the individual level variables and in direct support of his hypotheses, Quillian found that the macrolevel variables had strong effects. Countries where there were more immigrants and where economic conditions were less favorable tended to have more prejudice.

In addition, these two variables worked together to increase prejudice. As he explained it, "The effects of percent non-EEC immigrants and economic conditions intensify each other, so that prejudice is more likely when there is both a large foreign presence and poor economic conditions than would be expected by the sum of their additive effects."[64] Sociologists call this an **interaction effect**—the influence of one variable changes depending upon the status of another variable.

Quillian also found an interaction effect between the microlevel and macrolevel variables. Individual level variables, such as education, had a greater impact in societies where there was more of a group threat. To illustrate this pattern, he compared Ireland, a country with low threat because of a small immigrant population and fewer economic problems, with Belgium, a country with a larger immigrant population and more economic problems at the time of the study. He found that individual variables such as age, education, and social class, had more of an effect in Belgium than in Ireland. In other words, in Belgium, a context where prejudice was more likely, these individual level variables had more influence on whether individual people expressed prejudiced attitudes.

Still, because these individual variables had only a small overall effect on prejudice, their impact remained much smaller than that of the macrolevel factors. In other words, all things being equal, Quillian's analysis of the European data indicated that the chance that individuals would express prejudice was influenced more by the social structure in which they lived than by their own personal, individual characteristics.

Trends in the United States

But could such patterns explain variations in prejudice in the United States? Can the amount of prejudice that we feel and express be explained more by the social structures in which we live than by our own personal characteristics? Quillian decided to answer this question by examining the extensive survey data on prejudice discussed earlier in this chapter and portrayed in Figure 8-9.[65] Could his hypotheses regarding group threat help explain why prejudiced views have changed over time and why they have varied from one region of the United States to another?

As with the European study, Quillian used a sophisticated statistical analysis that allowed him to separate the influence of macrolevel and microlevel variables and found results that replicated his analysis of the European data. Although microlevel variables, such as increasing levels of education, can help explain the decrease in prejudice found over the years in both the North and South, they can account for only a portion of this change. Instead, macrolevel variables related to perceptions of group threat can account for changes in prejudice over time and variations from one part of the country.

As in Europe, economic vulnerability is an important contributor to perceptions of group threat. In regions where per capita income is higher, views regarding social distance are less prejudiced and opposition to social policies regarding racial-ethnic equality is lower. Similarly, in time periods when per capita income has grown, prejudice had decreased. These results have occurred regardless of how well educated individual citizens might be or how old they are. When economic times have been good, prejudice has been less likely to appear among people of all ages and levels of schooling. For instance, after World War II and from 1972 to 1985 the country experienced strong economic growth. Quillian's analysis suggests that this economic well being can account for the relatively sharper decline in prejudice during this time period, over and above any change in individuals' characteristics.

At the same time, when minority groups are a larger presence in the population, prejudice and opposition to policies that are proposed to address inequities are more common. This aspect of group threat can explain the traditional differences between the North and South in prejudiced attitudes. Even though residents of these regions have similar levels of education and people within the same age range, the African American population has historically been a much larger proportion of the total population within the South. As a result, residents of that area might perceive a greater threat than do those in other regions of the country.

In general, Quillian's analyses support his hypothesis that perceptions of group threat can increase the level of prejudice within a society or a time period. Particularly, his results imply that increased economic well-being of all members of a society is important in diminishing the possibility that prejudice will occur. Economic well-being might be especially important when minority groups represent a larger proportion of the population.

CRITICAL-THINKING QUESTIONS

1. Based on Quillian's work how would you expect prejudice toward Hispanics, Asian Americans, and American Indians to vary across different regions of the United States? Why would these variations occur?

2. Think about William Julius Wilson's analysis of the segmented labor force and current patterns of racial-ethnic segregation discussed earlier in this chapter as well as the discussion of income inequality in Chapter 7. Based on those analyses, what might you predict about future trends in prejudice within the United States and among sub-groups within the United States?

Quillian's analysis focused on entire societies and large regions within the United States. What about everyday situations such as the organizations and work groups in which people interact on a daily basis. How can prejudice be reduced in these situations? The answer, developed over the last half century, involves the notion of reducing group threat by promoting contact in situations with norms that promote equality and nondiscrimination.

Decreasing Group Threat in Everyday Interactions: The Contact Hypothesis

Most social scientists now agree that the most effective way to minimize racial-ethnic stratification is to focus on changing social structures, rather than try to change individuals' prejudiced attitudes. Much of this work has been influenced by what has come to be known as the *contact hypothesis,* which developed from the separate work of a number of sociologists and psychologists.[66] Basically this hypothesis suggests that individual discrimination and prejudice toward members of a minority group will diminish once members of the two groups have direct interpersonal contact. However, not just any type of contact will do. In fact, if the groups are competitive with each other, enhancing perceptions of group threat, conflict can actually increase. This happened in the post–Civil War era when minority groups were seen as competing with white ethnic groups for jobs and can also be seen today in the animosity displayed toward many immigrant groups in the United States and Europe.

The **contact hypothesis** suggests that greater interpersonal contact between two racial-ethnic groups will lead to less discrimination and prejudice when the following five conditions hold: (1) the two groups have equal status, (2) the attributes of the minority group disconfirm common negative stereotypes about it, (3) members of the two groups have sufficient contact to really get to know one another, (4) members work together to achieve some common goal, and (5) the norms of the situation in which contact occurs support equality and nondiscrimination.[67]

The contact hypothesis has been tested in many different settings and has received a great deal of support. Consider a sports team, an endeavor where individual skills are obvious. Some of us can shoot baskets and catch balls very well and some of us have much more trouble doing so. Many sports teams are designed so that the team members all have skills that are roughly equal. In addition, members of teams have a lot of contact with each other through their daily practices, and they are clearly working toward a common goal—winning. When the coaches and players support norms that support equality and nondiscrimination, researchers have found that team members can form strong interpersonal bonds that cross lines of race-ethnicity.

The contact hypothesis can also help explain the results that Elizabeth Cohen and Susan Roper obtained in their research on expectation states theory, as described in Chapter 5. They were able to counter diffuse status characteristics by equalizing the status of group members and having them engage in a cooperative task. Similarly, the contact hypothesis can be related to Merton's typology of the relationship between discrimination and prejudice, as shown in Figure 8-10. Many people fall into the "timid bigot" category. When placed in situations in which nondiscriminatory actions are expected, they will act that way. Based on these results many sociologists now suggest that the

most efficient way to diminish ethnic stratification is to diminish perceptions of group threat. This can best be done by promoting contact between members of different groups in settings that provide equality and norms of fair treatment and justice.

Clearly the American dilemma is still with us. Even though many members of racial-ethnic minorities have attained higher levels of education and income, deep cleavages between racial-ethnic groups span the stratification mountain of our society. Individual prejudice and discrimination still exist and affect the lives of many individuals. Yet, many sociologists continue to hope that as we learn more about why racial-ethnic stratification occurs we can also learn how it can be diminished. One of the most impressive successes in these efforts in the United States is described in Box 8-2, Applying Sociology to Social Issues.

SUMMARY

Racial-ethnic stratification, or inequalities based on race-ethnicity, is an ongoing problem in the United States. Despite occupational, political, economic, and educational advancements, racial-ethnic minorities continue to lag behind Euro-Americans and to experience both structural and individual discrimination. Key chapter topics include the following:

- Racial-ethnic stratification appears throughout the world and involves both prejudice and discrimination. The long-standing and deep-seated problem of racial-ethnic stratification in the United States was termed the "American dilemma" by the Swedish social scientist Gunnar Myrdal.

- The United States contains a wide variety of racial-ethnic groups. Euro-Americans constitute about 73 percent of the population, African Americans about 12 percent, Hispanics close to 11 percent, Asian Americans a little over 3 percent, and American Indians about 1 percent. Members of various racial-ethnic groups tend to live in different areas of the United States because of historical patterns of settlement and movement.

- Except for some Asian Americans on some dimensions, people of color tend to have lower levels of education, lower status occupations, lower incomes, and less wealth than Euro-Americans.

- William Julius Wilson outlines three historical eras with different economic and power relationships that have influenced racial-ethnic stratifica-

tion in the United States with respect to African Americans. The first stage, which occurred before the Civil War, involved a caste system in which powerful plantation owners controlled both the political and economic systems in the South. The second stage, which lasted until World War II, involved the growth of a split labor market. Euro-American workers used their political power to promote patterns of segregation that relegated African Americans to lower-level jobs and to segregated housing and education. The third stage, which began after World War II, involved the end of formal, legal supports for segregation, a shift to a segmented occupational structure, and the emergence of African Americans in political roles. The result has been complex changes in patterns of racial-ethnic stratification, with a growing African-American middle class but a continuing pattern of African-American poverty.

- Wilson suggests that these contemporary patterns reflect social stratification more than racial-ethnic stratification. Critics of Wilson have countered that, by focusing on structural discrimination, he has downplayed the importance of individual discrimination.

- Although a substantial minority of individuals still express prejudiced attitudes, trend analyses show that prejudice has declined considerably in the United States in the past half-century, primarily because younger cohorts are less prejudiced than older cohorts.

Quite possibly your college or university campus bears the marks of racial-ethnic division and even, perhaps, animosity. You might have few friends from racial-ethnic backgrounds that are different from your own. Only if you attend an historically Black college will there be a substantial number of higher administrators or teachers who are African American.

The situation would be very different if you belonged to the United States Army. The Army is, in fact, more integrated than any other large organization in the United States. It has more African Americans at high levels of achievement and authority than any other large organization and is the only place in American life where Euro-Americans are routinely given orders by African Americans. In addition, daily life in the army is highly integrated. Euro-Americans and African Americans routinely live together, work together, and socialize together—and do so because they want to. Although racial-ethnic tensions and hostilities can and do appear, they are reported to be far less frequent or serious than in other areas of American life.[68]

Relationships between racial-ethnic groups were not always this good in the Army. Until 1950 the armed forces were segregated. Even after African Americans served in the same units as other soldiers, they were largely absent from the leadership ranks. The 1970s were an especially difficult time with numerous incidents of racial hostility and widespread perceptions of discrimination. In 1973, at the end of the Vietnam War only 3 percent of all Army officers were African American.

Yet, about this time the situation began to change. The number of African American officers almost quadrupled, and African Americans obtained the highest leadership positions in both the military and civilian sides of Army command. By the 1990s African Americans constituted almost a third of all senior non-commissioned officers, 11 percent of all commissioned officers, and 7 percent of all generals.[69]

How did these changes occur? Charles Moskos and John Sibley Butler, sociologists who have extensively studied the army, suggest that these changes support the five conditions under which the contact hypothesis holds. First, members of different racial-ethnic groups have equal status—rank, not color, determines how one is treated. Second, the attributes of the minority group disconfirm common negative stereotypes: African Americans hold positions of power and authority and do so because they have demonstrated their skills and competence and met well-established and accepted standards. These standards are the same for all, no matter what their color. Third, members of the two groups have sufficient contact to really get to know one another: African Americans constitute more than a quarter of all personnel in the Army and units throughout the forces are purposely integrated so that squads of

soldiers or officer staffs never consist of only one racial-ethnic group.[70] Fourth, members work together to achieve a common goal—to defend the United States and its interests—and this goal is continually stressed. Finally, norms support equality and nondiscrimination. Army regulations clearly prohibit discrimination and have established a variety of procedures to ensure that discrimination does not occur. Army authorities believe that racial-ethnic prejudice and discrimination interfere with the Army's mission of combat readiness and thus must not be part of daily army life.

Could other organizations emulate the Army's success? Moskos and Butler suggest that they could. One suggestion they have made involves a national service program, in which young civilians from all racial-ethnic backgrounds could participate. It could be a means of providing services to needy groups and regions of the country while promoting extensive contact between racial-ethnic groups as well as the training and educational benefits that the Army gives its enrollees.

CRITICAL-THINKING QUESTIONS
1. Do you think such a national service program could help counter prejudice and discrimination? How would it have to be organized to attain this aim?
2. Can the contact hypothesis suggest ways in which relationships between racial-ethnic groups could be improved at your college or university? In your community?

- People tend to express less prejudiced attitudes when their reference groups are less prejudiced. Merton's typology of the relationship between prejudice and discrimination shows how social situations influence whether people act in discriminatory ways. These relationships can be explained using the ideas of exchange theory, role theory, and symbolic interactionism.

- Joe Feagin's interviews with middle-class African Americans demonstrate the nature of individual discrimination and show how theories of social interaction can explain people's responses to discriminatory actions.

- Lincoln Quillian's research shows how differences in individual prejudice from one society to another, one region to another, or one time period to another are influenced by perceptions of "group threat," the belief that the presence of a minority group will harm one's well-being.

- The contact hypothesis suggests that individual discrimination and prejudice will decrease when members of a minority and dominant group have direct interpersonal contact under situations that promote equality and nondiscrimination.

KEY TERMS

caste system 193

contact hypothesis 209

discrimination 187

exclusion 195

ghetto 200

group threat 207

individual discrimination 202

interaction effect 208

minority group 189

prejudice 187

racial-ethnic group 187

racial-ethnic stratification 187

reference group 204

secondary analysis 207

segmented occupational structure 198

social distance 203

split labor market 194

structural discrimination 200

trend analysis 202

INTERNET ASSIGNMENTS

1. Locate the Statistical Abstracts Web site and look in Section 1 on "Population." How have the populations of specific racial-ethnic groups (white, African American/Black, Asian and Pacific Islander, American Indian–Eskimo–Aleut, Hispanic) changed between 1980 and 1990 (census years)? Can you find information about changes since the 1990 census? What implications might population shifts have for greater social equality among racial-ethnic groups?

2. Peruse several discussion groups that focus on racial-ethnic issues. Can you find evidence of concern about inequality and discrimination among the topics? How can the sociological theories used in the text help explain the acts of individual discrimination described in these discussions?

3. Locate the Web sites of several diverse academic institutions (public universities, private universities,

liberal arts colleges, community colleges, private high schools, public high schools, and so forth). Do any sites contain information on multiculturalism or racial-ethnic diversity? What patterns do you see? What might account for these patterns? Do any describe policies used to promote better relations between racial ethnic groups.

4. Locate the Bureau of Labor Statistics Web site. See if you can use data from this site to update the information on wages, unemployment, and occupational categories of different racial-ethnic groups given in Figures 8-3 and 8-4.

TIPS FOR SEARCHING

Many colleges and universities have programs or departments that specialize in particular racial-ethnic studies. Many have Web sites that discuss their curriculum and often contain links to pertinent information on racial-ethnic issues.

InfoTrac College Edition:
Readings

Article A53953545 / Hogan, Richard, and Carolyn C. Perrucci. 1998. Producing and reproducing class and status differences: Racial and gender gaps in U.S. employment and retirement income. *Social Problems, 45,* 528–548 Hogan and Perrucci document a growing racial gap and a declining gender gap in incomes in the last two decades.

Article A20505586 / Thernstrom, Abigail, and Stephan Thernstrom. 1998. Black progress: How far we've come—and how far we have to go. *Brookings Review, 16,* 12–17 Thernstrom and Thernstrom document both the areas in which the status of African Americans has improved and the areas in which it has not improved.

Article A20505587 / Patterson, Orlando. 1998. Affirmative action: opening up workplace networks to Afro-Americans. *Brookings Review, 16,* 17–24. Patterson describes the role of social networks in promoting employment opportunities and success and the role of affirmative action in developing network ties.

Article A20505591 / Loury, Glen C. 1998. An American tragedy: the legacy of slavery lingers in our cities' ghettos. *Brookings Review, 16,* 38–43. Loury traces the problems of poverty and segregation in cities to the historical legacy of African American slavery.

Further Reading

Suggestions for additional reading can be found in *Classic Readings in Sociology,* bundled with this book. If you purchased a used copy of this book that does not include this custom-published reader, go to www.sociology.wadsworth.com for ordering information.

Recommended Sources

Marger, Martin N. 1997. *Race and Ethnic Relations: American and Global Perspectives,* 4th ed. Belmont, Calif.: Wadsworth. Marger discusses the experiences of a wide variety of different racial-ethnic groups, as well as theories that can account for these experiences.

Miller, Norman, and Marilynn B. Brewer (eds.). 1984. *Groups in Contact: The Psychology of Desegregation.* Orlando, Fla.: Academic Press. The authors provide a summary of the ideas of and evidence for the contact hypothesis.

Moskos, Charles C., and John Sibley Butler. 1996. *All That We Can Be: Black Leadership and Racial Integration the Army Way.* New York: Basic. This is an extensive discussion of the research described in Box 8-2.

Myrdal, Gunnar. 1962. *An American Dilemma: The Negro Problem and Modern Democracy,* 20th anniversary ed. New York: Harper and Row. This republication of Myrdal's original work includes his reflections twenty years after the original publication.

Oliver, Melvin L., and Thomas M. Shapiro. 1995. *Black Wealth/White Wealth: A New Perspective on Racial Inequality.* New York: Routledge. This is an extensive analysis of racial-ethnic differences in the accumulation of wealth.

Patterson, Orlando. 1997. *The Ordeal of Integration: Progress and Resentment in America's "Racial" Crisis.* Washington, D.C.: Civitas. This book is a series of essays on contemporary racial-ethnic relations and policy by a leading African American sociologist.

Thernstrom, Stephan, and Abigail Thernstrom. 1997. *America in Black and White: One Nation, Indivisible.* New York: Simon and Schuster. The authors provide an in-depth analysis of the historical and current nature of racial-ethnic stratification in the United States.

Wilson, William Julius. 1980. *The Declining Significance of Race: Blacks and Changing American Institutions,* 2nd ed.; 1987. *The Truly Disadvantaged: The Inner City, the Underclass, and Public Policy.* Chicago: University of Chicago Press; and 1997. *When Work Disappears: The World of the New Urban Poor.* New York: Knopf. These books provide three explications of Wilson's general theoretical perspective.

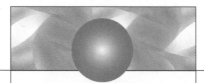

Pulling It Together

Listed below are other chapters where you can find additional discussion or examples
of some topics covered in this chapter. You can also look up topics in the index.

INTERNET SOURCES

Asian American Resources, African Americana, Native Links, and Latino Web. These Web sites are specifically aimed toward members of or those interested in particular racial-ethnic groups. Such sites are excellent sources of information not only on what types of inequality exist but on what people are doing about this.

9

Gender Stratification

Comparing Gender, Racial-Ethnic, and Social Stratification

Gender Segregation and Stratification
Gender Stratification in Preindustrial Societies
Gender Stratification in Contemporary Societies

Explaining Gender Segregation and Stratification
The Structure of the Occupational World
Social Actions of Employers and Employees
A Psychoanalytic View of Gender Stratification

FEATURED RESEARCH STUDY

Testing Psychoanalytic Theory

CARLA WAS A "MILITARY BRAT." She had lived on military bases in Europe, Japan, and the Middle East as her father, an officer in the Air Force, was transferred from one place to another. During her last two years of high school, Carla lived with her father in Virginia, near his latest posting at the Pentagon. Her mother had died in Japan the year before, and Carla had taken over many of the household chores, as well as the role of "hostess" when her father entertained Air Force brass. Although she tried to emulate her mother's restrained cordiality, she hated the way the men complimented her looks or the food and then ignored her while they "talked shop." Even her father didn't seem to notice her interest in his work or the way she studied his pilot manuals in her spare time.

When Carla was accepted into the Air Force Academy, she was so thrilled she couldn't sleep. The next morning she told her father the secret she had been keeping for the last few years—she was going to be a test pilot, something even he had never done. At first he tried to talk her out of it, citing the terrible risks involved in flying experimental aircraft and the difficulties she would face in a "man's world." But in the end he surrendered to her determination and gave his blessing.

Throughout the summer Carla prepared for life at the Academy by exercising strenuously and brushing up on her math. She knew from her admissions materials that she would be taking physics and calculus during her first year. She also knew that basic cadet training was a grueling experience in the August heat that would test her will and stamina as well as her physical preparedness.

Upon arriving at the Academy, Carla was issued a set of uniforms—both skirts and dress slacks—and given a short military haircut, just above her shirt collar. Thanks to her summer regimen, she survived basic training with only a few blisters to show for it. Afterward she proudly marched in the acceptance parade signifying that she was officially a part of the Cadet Wing.

By this time Carla was already becoming fast friends with her roommate, a young woman from New York who hoped to fly in space someday. Still, her first year at the Academy challenged her determination to become a test pilot. Even after growing up on military bases, she found it hard to adjust to the strictly regimented lifestyle. She had to follow a complex code of behavior that forced her to think about how she sat, walked, and addressed older cadets and instructors.

Classes were no pushover either, but Carla worked diligently, and her grades in calculus and general physics put her at the top of her squadron.

In the spring semester she was allowed to enroll in "soaring training," an optional course in which she learned to fly a sailplane. The first time she glided solo, she was exhilarated. After that she told herself she would make it through her four years at the Academy no matter what.

Other aspects of Academy life were more troublesome, however. From her first days at the Academy, Carla overheard some of the men making comments about her and other women cadets, sometimes whispering "ratings" of the female cadets' bodies when they walked past. When she earned a silver wreath for making the commandant's list in military performance, the comments got nastier. "Trying to show us how tough you are?" one male cadet sneered. "Don't worry," she replied with exaggerated sweetness. "I'm sure you can make the list yourself next time." Fortunately not all the men were biased against the women cadets, but even the friendly ones seemed vaguely threatened by women who could perform as well as they.

As her fourth and final year at the Academy approached, Carla was thrilled to learn that she had been accepted into the undergraduate pilot training program, which meant that she would be learning to fly a propeller-driven plane solo. When she excelled in the course, she felt she was closing in on her dream of becoming a test pilot. Then one day another cadet, Eric, threw her a curve. Eric had been a pillar of support during her Academy days, so she was all the more shocked when he took her aside and tried to persuade her to change her career plans.

"Look," Eric said earnestly. "You know I think you're a wonderful pilot. But this test-pilot routine isn't for everybody. What I mean is, well, a lot of the guys just don't feel right about having a woman around or maybe flying with her as pilot in some new tin can they're working the bugs out of. And besides"—Eric paused uncomfortably—"I kind of care about you, if you know what I mean. It's been hard to show you that, because you're not like the kind of woman I ever thought I'd—well, never mind that. The point is, women ought to have babies and not go risking their lives that way for no reason. What kind of mother would you be? I mean, if you have to fly couldn't you pilot transport planes or something? You wouldn't be hassled nearly as much, and you could leave the really risky stuff to those crazy flyboys. After all," he added with a grin, "they're more expendable, you know?"

Carla was devastated. Being a test pilot had been her dream since high school. She had put up with the catcalls and insults, she had survived every test the Academy threw at her, and now her fellow pilots didn't want her? Was this the reason that there were so few women doing hazardous flight duty—that one way or another, they just got discouraged or frozen out even if they could do the job? And what was she to think if even Eric, who was finally confessing the interest in her that she'd always suspected he had, believed the things he did? With a shock, she realized that it was *because* he cared for her that Eric thought of her as a woman first and a fellow cadet and aspiring pilot second. He was trying to *protect*

her! If her best male friend couldn't get past his ideas of what a woman should be like, was there any point in fighting it anymore?

As she pursues her military career, will Carla always confront such attitudes? Will women continue to struggle to enter fields where men work? Why is it such a big deal when women try to enter jobs that men traditionally have held? Is it just as hard for men to enter jobs that women usually hold?

*C*ARLA'S EXPERIENCES ILLUSTRATE **gender stratification,** the organization of society such that members of one sex group have more access to wealth, prestige, and power than members of the other sex group do. Carla's experiences also illustrate **gender segregation,** the restriction of members of each sex group to different statuses and roles, such as the belief that men but not women should be fighter pilots or that women but not men should be nurses. At the same time, this example shows the complexity of gender stratification—how it is intertwined with occupational roles, family lives, and personal relationships.

Like the other forms of social inequality we have studied (racial-ethnic and social stratification), gender stratification affects all of us. Women like Carla, trying to break into a male-dominated field, are not the only ones who feel the effects of gender-based inequalities. For example, today most families in the United States depend on two incomes, yet women typically earn about 74 percent of what men earn, even when they work full-time—a discrepancy that affects the entire family, not just the woman worker.[1] Although most mothers work outside the home and many fathers are taking on increased responsibility for child rearing, child care and domestic chores remain primarily women's responsibility, creating stresses and strains for the entire family. Only a small proportion of employers have developed "family-friendly policies" that allow for parental leave, flexible schedules, child care, and other types of support that help both mothers and fathers to participate in both paid work and family chores.[2]

Many other issues related to gender stratification and gender segregation affect the lives of both men and women. It's important to remember that not only women but also men "have" gender. That is,

men as well as women might have more limited choices available to them because of the sex-typing of many occupations. For instance, men who want to be nurses or secretaries or elementary school teachers often face disbelief and discouragement. And, as we shall see, traditionally "female" occupations pay less well than traditionally "male" occupations, which can further limit the choices of both sex groups. Parents who have strong aspirations for their daughters, men and women who are affected by issues like sexual harassment in the workplace, individuals who struggle to redefine relationships between women and men in a more equitable way— all are affected by gender stratification.

Just as we all have emotions regarding social stratification and ethnic stratification, so, too, do most of us probably have feelings related to gender stratification. Perhaps you feel uncomfortable when the issue comes up or can't understand what all the ruckus is about. Or perhaps you have experienced gender discrimination and felt frustrated because others haven't taken you seriously. These experiences are understandable because we confront aspects of gender stratification every day. Yet they are so embedded in our culture that we sometimes find it difficult to step back and look at them scientifically.

Actually, sociologists themselves "discovered" gender stratification only relatively recently. When I took my first sociology class almost thirty years ago, no one talked about gender stratification. Sociologists recognized that males and females had different roles and that all societies had a gender-based division of labor. They also realized that all societies, to varying degrees, exhibited **male dominance,** cultural beliefs that give greater value and prestige to men and to their roles and activities.[3] With a very few exceptions, however, sociologists simply accepted these facts. They didn't

How does this photo of the United States Senate reflect gender stratification? What might you hypothesize about other political bodies? Would you expect gender segregation on school boards or city councils? Why or why not?

see them as *problematic,* as worthy of attention. Like most members of society, most sociologists simply assumed that gender stratification and male dominance were a natural part of the social order.

In the late 1960s this situation began to change, largely because of the influence of **feminism,** an ideology that directly challenges gender stratification and male dominance and promotes the development of a society in which men and women have equality in all areas of life. In addition, beginning in the late 1960s, an unprecedented number of women began entering graduate training in sociology. Many of them were feminists and were active in the **feminist movement,** a social movement to promote the interests of women, much as the civil rights movement promoted racial-ethnic equality.

Some of these new entrants, along with women who had entered the profession in previous years, had experienced discrimination in their own work lives. For example, female sociologists were typically paid less than male sociologists, for it was assumed that the men had to support a family. Women whose husbands were employed by a given university could generally not be hired by that institution because of what were called "nepotism" rules. Thus, for many years, a large number of women with graduate training as sociologists did not teach in colleges and universities, but instead held part-time or insecure positions as research assistants. Even today, female sociologists, especially those with families, are unlikely to be employed in the most prestigious universities, often earn less than their male counterparts, and tend to specialize in different areas than men.[4]

Inspired by the feminist movement, many of these sociologists began to look at their lives and these experiences in a new way. To use the terminology of C. Wright Mills, what they had seen as private troubles, they now began to view as a public issue, a system of gender stratification. More important for sociology, they moved beyond their political concerns and passion for attaining gender equality and began to examine gender stratification scientifically. They tried to step back and understand what gender stratification is and why it exists.

In this chapter we explore this work. We begin by examining the ways in which gender stratification differs from racial-ethnic and social stratification. We then look at the reality of gender segregation and stratification in both preindustrial societies and the contemporary United States. Finally we review explanations for gender-based inequalities.

COMPARING GENDER, RACIAL-ETHNIC, AND SOCIAL STRATIFICATION

Gender stratification differs from racial-ethnic and social stratification in several ways. Members of different social classes and different racial-ethnic groups often have social networks that have few interconnections, especially in highly stratified societies. This lack of social ties between various groups reflects the fact that both racial-ethnic and social stratification are highly related to economic competition and to differences in control of property and levels of prestige and power.

The relationships between the two sex groups are exactly the opposite. Our most intimate and loving attachments, at least in childhood, involve relationships with members of both sex groups—with mothers and fathers, grandmothers and grandfathers, brothers and sisters. Moreover, men and women work together to promote the economic benefit of their families, although men often have greater control of the family's assets than women do. They share their class status, and its accompanying benefits or disadvantages, with other members of their families. In general men and women rarely compete directly with each other for economic benefits.

In addition, your **gender identity,** the deepseated belief that you are male or female, is usually a very salient aspect of your **self-identity,** the way you conceive or think about yourself. From very early childhood you know that there are two sex groups, and the realization that you are a male or a female affects your self-concept and your interactions with other people from a very early age. Your racial-ethnic identity and your class identity might also be salient to you, but you almost certainly developed an understanding of class and racial-ethnic differences much later than you recognized the presence of sex differences in the world.

Finally, although some societies have very little variation in class status or in race-ethnicity, no societies have only one sex group, and all societies have different roles for males and females. Thus sociologists sometimes say that all societies are *gendered,* a characteristic that can be seen in virtually all areas of social life.

It's important to realize that the differences in the way males and females are treated are *social* facts, not products of biology. The terms *gender* and *gendered* are used by many sociologists to underline this socially defined reality. Whereas the word *sex* refers to the biological differences between males and females, the word *gender* refers to the different roles and expectations that people have for members of the two sex groups—the social expectations that are attached to sex differences.[5]

Because gender stratification is unique in important ways, the image of a stratification mountain from Chapter 7 needs to be modified to accurately depict gender stratification. Within families, women and men share their social class. Yet, within their families and within society at large, males tend to enjoy greater prestige and authority in many areas of daily life. Thus gender stratification might be seen as involving groups of men and women arranged at various levels on the stratification mountain. Within groups, or social classes, they are linked together, just as mountain climbers are bound together by ropes, but the men usually have the lead position and enjoy a superior status. These gender differences permeate the relationships of the climbers in each group, who continually interact with one another even if they have little or no contact with the groups above and below them.

Perhaps you think that gender stratification isn't as real or dramatic as other types of stratification. After all, we know that today, to a greater extent than ever before, women are pursuing careers, becoming entrepreneurs, moving into areas previously dominated by men, and being elected to political office. A few years ago it would have been unheard of for a woman like Carla to even attend the Air Force Academy, let alone train as a test pilot. In addition, laws have been passed to prevent sexual harassment and to punish it when it occurs. In fact, one of the complaints of "angry white males" is that women—along with members of racial-ethnic minorities—are getting special treatment in employment and other areas of life. So just how unequal *are* the sex groups today? To answer this question, let's see what sociologists have discovered about gender segregation and stratification.

GENDER SEGREGATION AND STRATIFICATION

Gender segregation and stratification have appeared throughout history. For instance, even though women appear throughout the Bible, their role is usually minor, such as Mary Magdalene or, in the Old Testament, Ruth. Even the Declaration of Independence asserts that "all *men* are created equal"; and it wasn't until 1920, almost one hundred and fifty years after the Declaration of Independence was written, that women in the United States secured the right to vote.

In the nineteenth century many passionate abolitionists who worked to secure freedom (and even the vote) for African-American slaves recoiled from the idea that women were entitled to economic independence or the right to participate in political affairs.

But is gender stratification largely a matter of history? To find out, let's look first at preindustrial societies and then examine patterns of gender stratification and segregation in contemporary times.

Gender Stratification in Preindustrial Societies

All cultures have a gender-based division of labor, rules about what tasks members of each sex group should perform. By looking at how these tasks differ from one society to another and how they are related to gender stratification, sociologists hope to get a better idea of how gender stratification works, not just in the United States but in other settings as well. Most of the work in this area has involved data from the Human Relations Area Files (HRAF), first described in Chapter 3, which summarize information about many different preindustrial societies.

Sociologists who have examined data from the HRAF have found that, even though all societies have a gender-based division of labor, activities aren't always gender-typed in the same way. For instance, in some societies women take care of small animals and fowls; in other societies men do. In some societies men take care of dairy animals; in others women do. The major exceptions involve child care and hunting. Sometimes women assist with the hunt and sometimes men participate in child care, but in no known society are only women assigned hunting and men assigned child care. Likewise, although women participate in war, planning and actually carrying out acts of war are the province of men.[6]

Some of these role assignments might well be explained by physical differences between the sexes. After all, only women can breastfeed infants, and men might be more expendable and thus be sent to war, much as Carla's friend Eric suggested that the male cadets would be better suited for test-pilot work. Yet other differences in roles cannot be explained by these characteristics. For instance, men are almost always assigned the task of making musical instruments, but it is hard to imagine how they might be better equipped physically than women to do this.[7]

Of course, simply because a society has gender *segregation* doesn't necessarily mean that it will also have gender *stratification*. Men and women could do different tasks and be assigned to different roles, but all the work could be equally valued and rewarded. In other words, a society could be *gendered* without having gender stratification or male dominance.

Interestingly enough, this doesn't seem to happen. No matter how various societies assign specific tasks to men and women, the tasks assigned to men tend to be defined as more worthwhile and necessary than those assigned to women. For instance, in some parts of New Guinea, men grow yams, which are a very prestigious food that is eaten at feasts, while women grow sweet potatoes, which aren't anything special at all! Similarly, among the Iatmul, another New Guinea group, both men and women fish for food, but the women's fishing is simply viewed as their work, while the men's fishing is seen as an exciting expedition.[8]

Anthropologists have never found a society in which women as a group control the lives of men in the political or economic world. In other words researchers have never found a "female-dominant" society, or a society in which the system of gender stratification favors women rather than men.[9] Yet there is a lot of variation from one society to another in the extent to which gender stratification and male dominance exist. In addition, within a given society some aspects can be relatively egalitarian whereas others are not. That is, women might be given a fair amount of freedom or authority in one area of the society, but not in another.[10]

The Twana, an American Indian group that lived in the Hood Canal region, just west of what is now Seattle, Washington, illustrate this phenomenon. In some areas the Twana showed a great deal of gender inequality. Men provided most of the food because women were forbidden to have contact with canoes and weapons or to engage in river net fishing. Men built and owned the houses, acted as the informal leaders of the community, and held almost all of the specialized roles, such as canoe maker or hunter of sea mammals. There also were strong sexual double standards. Young boys were encouraged to seduce women, but young women were discouraged from responding. In cases of adultery among adults, the woman was always considered at fault and might be divorced or even killed. In contrast, a husband's infidelity generally was seen as harmless. In other areas of their lives, however, the Twana showed a great deal of gender equality. For instance, a woman could inherit property from her parents and then pass it on to her children. Relationships between husbands and wives were fairly harmonious, and both shared in child care; males and females had an equal voice in the choice of marriage partners; and wife beating apparently was neither common nor approved. Women were responsible for most of the domestic work, but men would help out by doing some of the cooking and by dressing skins and fur. In addition, women and men participated together in secret societies, and older women would some-

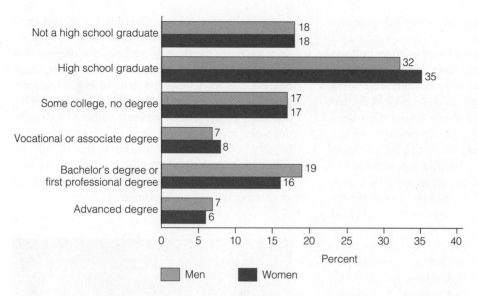

FIGURE 9-1 *Educational Attainment of Women and Men Ages Twenty-Five and Over, 1997*

Source: Day and Curry, 1998, pp. 1–6.

times even direct younger men on their vision quests (a search for a guardian spirit).[11]

Thus, although at least some degree of gender stratification seems to exist in all societies, its extent and characteristics can vary a great deal from one group to another. But what about the United States? How egalitarian is our own society? And to what extent do gender segregation and stratification vary from one area of the society to another?

Gender Stratification in Contemporary Societies

Sociologists who study gender stratification in the United States have often compared women's and men's education, occupations, income, and family lives. As in other cultures, researchers have found that there is more segregation and stratification in some areas than in others, and the extent of this gender stratification is greater for some groups of people than for others. Much as teams of mountain climbers might walk side by side part of the time but lag behind one another part of the time, so does gender stratification in social statuses and roles vary across situations and social groups, even within the contemporary United States.

Education As you will recall from previous chapters, in modern industrial societies such as the United States, the education we obtain is an important predictor of our adult status. As Figure 9-1 shows, for all adults in the United States, men are slightly more likely than women to have received higher degrees. Sociologists suggest that this pattern reflects gender stratification.

Yet you might have noticed that the differences in men's and women's educational attainment are small, especially when compared with the differences between members of various social classes or racial-ethnic groups. This reflects one basic way in which gender stratification differs from racial-ethnic and social stratification. Men and women share many of the life chances that reflect their race-ethnicity and social status. Thus upper-middle-class women almost always have more education than working-class men; and Euro-American women, on average, have more education than African-American, Hispanic, and American Indian men. When families can afford to do so, they tend to give both their sons and their daughters equal levels of education.

Higher education can be seen as providing the status or prestige dimensions of social stratification that Weber described. Even though families might not expect their daughters to hold the same types of jobs that their sons will, or even to work at all, they know that attending college will provide the young women with the social and cultural capital to find "appropriate" marriage partners—that is, someone from a similar prestige class. Even if they don't marry someone they meet in college, the women's college experiences will enable them to enter jobs and social networks later in life where they will meet others of similar class backgrounds.

Note that the data in Figure 9-1 include all people over the age of twenty-five and thus group together several different birth cohorts. Since the mid 1980s more than half of all college students have been women. Part of this increased representation of women comes from older women being more likely than older men to return to school. But in recent years it also reflects a change in men's behavior. Since 1988 women who have just finished high school have been more likely than men of that age to go on to college.[12] In fact, as shown in Figure 9-2, among Americans in their early twenties, women are more likely than men to have gone to college.

This pattern also appears with younger cohorts, especially when we look at students' plans for the future. By the early 1990s, among students who graduated from high school, women were more likely than men to expect to go to graduate school. This is exactly the opposite of the situation twenty years earlier, when men were more likely to plan advanced study.[13] In short, gender stratification, as measured by the number of years of education men and women have, is becoming much smaller and could, if present trends continue, disappear or even reverse itself in the future.

Despite their similar levels of education, males and females start to specialize in different subject areas in the later years of high school and especially when they get to college. For example, as Table 9-1 shows, at the undergraduate level women (who receive 55 percent of all degrees granted) earn the vast majority of the degrees given in home economics, health sciences, education, and psychology but only 18 percent of the degrees in engineering. These differences appear at all degree levels but become even more striking at the doctoral level, which men are still more likely than women to complete. For instance, almost half of all bachelor's degrees in mathematics go to women, but one-fifth of the doctoral degrees do. Similarly slightly more than one-third of the bachelor's degrees in physical sciences are attained by women, but less than a quarter of the doctoral degrees are earned by women. The only areas in which more than half of all doctoral degrees go to women are education, English language and literature, foreign languages and literature, health sciences, home economics, and psychology. If you think about your women and men friends in college, you will probably realize that they often have chosen different majors.

Nevertheless, just as gender differences in college attendance patterns have changed in the past few years, so have the areas women choose to study.

Many of the changes are phenomenal! As shown in Table 9-2, undergraduate women today earn more than ten times the share of architecture degrees that they received in 1960, six times the share of business degrees, and forty times the share of engineering degrees! Similarly, as Figure 9-3 shows, until the 1970s less than 10 percent of all recipients of degrees in dentistry, medicine, and law were women, but this began to change dramatically during that decade. By 1996, 36 percent of all graduating dentists, 41 percent of all new medical doctors, and 44 percent of all graduating lawyers were women.

What do you want to major in? Are the people in your chosen major predominantly male or female, or equally represented by both sexes?

FIGURE 9-2

Educational Attainment of Men and Women Ages Twenty to Twenty-Four, 1997

Source: Day and Curry, 1998, pp. 1–6.

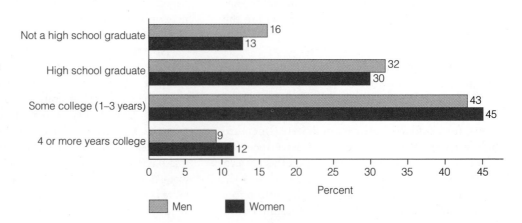

TABLE 9-1 *Percentage of Degrees Granted to Women by Level and Area of Study, 1995–1996*

DISCIPLINE	BACHELOR'S	MASTER'S	DOCTORAL
All fields	55	56	40
Agricultural sciences	43	44	27
Architecture and related programs	36	41	32
Biological sciences, life sciences	53	53	42
Business management and admin. services	48	37	29
Computer and information sciences	28	27	15
Education	75	76	62
Engineering	18	17	13
English language and literature	66	64	62
Foreign languages and literatures	70	67	56
Health professions and related sciences	82	79	57
Home economics	89	83	72
Mathematics	46	39	20
Philosophy and religion	34	35	30
Physical sciences	36	32	23
Psychology	73	72	66
Social sciences and history	48	46	38

Source: U.S. Department of Education, National Center for Education Statistics, 1998, pp. 15–23.

TABLE 9-2 *Percentage of Bachelor's Degrees Granted to Women in Selected Areas of Study, for Selected Years*

DISCIPLINE	YEAR		
	1959–60	1975–76	1995–96
All fields	35%	45%	55%
Architecture and environmental design	3	19	36
Biology and life sciences	25	34	53
Business and management	7	20	48
Computer and information sciences	—	20	28
Education	71	73	75
Engineering	0.4	3	18
Foreign languages	66	76	70
Mathematics	27	41	46
Physical sciences	12	19	36
Psychology	41	54	73

Note: No data available for computer science for 1959–1960.

Source: Snyder and Hoffman, 1995, pp. 250, 294–304; U.S. Department of Education, National Center for Education Statistics, 1998, pp. 15–17.

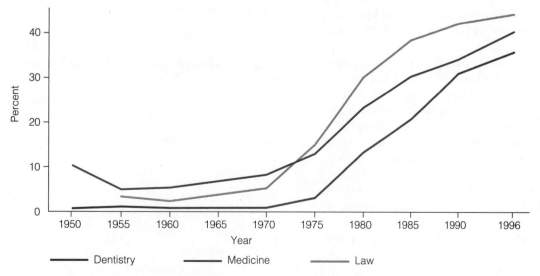

FIGURE 9-3 *Percentage of Medical, Dentistry, and Law Degrees Granted to Women, 1949–1996*

Note: Data before 1955–56 for law are not given because they are not comparable with those for later years.

Source: Snyder and Hoffman, 1995, p. 278; and U.S. Department of Education, National Center for Education Statistics, 1998, p. 8.

Yet, although many more women have entered these traditionally male-typed fields, few men have gone into traditionally female-typed fields, such as education, foreign languages, health sciences, and home economics. In addition, women's entrance into traditionally masculine areas of study has been largely limited to students at four-year colleges and graduate professional schools. Vocational education is still extremely sex-typed, and this has changed very little during the past twenty-five years. For instance, almost all the students studying to be dental hygienists, child care workers, and clerical workers are women. Almost all the students studying to be carpenters, welders, and plumbers are men.[14]

Note that these different patterns of change in gender segregation in education result from alterations in the educational specialization of women, not men. In addition, they are largely restricted to the middle class and have not appeared in training for working-class jobs.

Occupations Because men and women specialize in different areas in school, we could expect to find a similar specialization in their occupations. This does, in fact, occur. Sociologists use the terms **gender segregation of the labor force** and **occupational gender segregation** to refer to the gender-based division of labor in the occupational world, or the phenomenon of men and women holding very different jobs. The closer they look at the labor force, the stronger this segregation appears to be.

Table 9-3 shows the proportion of men and women who hold occupations in each of the major categories used by the Census Bureau. As you can see, women are four times as likely as men to hold administrative support and clerical positions; men are nine times as likely as women to hold precision craft and repair jobs and four times as likely to work in farming, forestry, or fishing.

But even these figures tend to obscure some gender-based occupational differences. For instance, you may have noticed that women and men are about equally represented within the various white-collar jobs, as listed in the first four categories in Table 9-3. Within these categories, however, they fill very different positions. Table 9-4 shows the percentage of women found in a wide variety of professional specialty occupations. As you can see, women make up the majority of dieticians, librarians, nurses, social workers, and teachers. In fact, in recent years almost half of all women professionals were in these jobs. Men make up the vast majority of the clergy, dentists, engineers, and physicians. The same pattern appears when we look more closely at other broad census groupings. For instance, even though women

TABLE 9-3 *Distribution of Men and Women in Major Occupational Categories, 1997*

OCCUPATIONAL CATEGORY	MEN	WOMEN
Executive, administrative, and managerial	15%	14%
Professional specialty	13	17
Technicians and related support	3	4
Sales	11	13
Administrative support, including clerical	6	24
Service	10	17
Precision production, craft, and repair	18	2
Operators, fabricators, and laborers	20	8
Farming, forestry, and fishing	4	1

Source: Bureau of Labor Statistics, 1998, Table 10.

TABLE 9-4 *Distribution of Women in Selected Professional Specialty Occupations, 1997*

OCCUPATION	PERCENTAGE
Architects	18
Chemists, except biochemists	26
Clergy	14
Computer systems analysts and scientists	29
Dentists	17
Dieticians	89
Economists	52
Editors and reporters	51
Engineers	8
Lawyers	27
Librarians	80
Operations and systems researchers and analysts	40
Pharmacists	46
Physicians	26
Psychologists	59
Public relations specialists	66
Registered nurses	94
Social workers	69
Teachers	
Prekindergarten and kindergarten	98
Elementary school	84
Secondary school	58
College and University	43

Source: Bureau of Labor Statistics, 1998, Table 11.

and men are about equally likely to be employed in sales, women tend to work as clerks in retail stores whereas men tend to work as manufacturing sales representatives.

The extent of this occupational segregation becomes even greater when we look past the job titles the Census Bureau uses and examine how employers classify jobs and the work they assign to their workers. Two sociologists, William Bielby and James Baron, studied the experiences of more than 60,000 workers in four hundred different firms over a twenty-year period. They found that only 10 percent of these workers had job titles that were assigned to both men and women. When they looked at what these workers actually did on the job, Beilby and Baron found that men and women rarely worked together, side by side at the same job. For instance, the women might work day shift while the men with the same job title would work the night shift; the women might work in one part of the plant while the men worked elsewhere.[15]

You have probably noticed patterns of segregation in the profession of teaching. The Census Bureau reports that about 43 percent of all college teachers and 58 percent of all high school teachers are women. But if you look more closely, you will find that they are in very different specialties. Generally women are found in languages and literature and in some of the social sciences and natural sciences, more rarely in mathematics or the physical sciences.

Patterns of gender-based occupational segregation have changed over time, almost always because women have entered fields that were once predominantly male-typed, not because men have taken female-typed jobs. For instance, women now make up more than one-fourth of all chemists, computer systems analysts, lawyers, and pharmacists, even though they were rarely found in these fields twenty-five years ago. Yet many jobs remain overwhelmingly female, such as bank tellers, child care workers, dental hygienists, secretaries, and nurses.[16]

In addition, the entrance of women into previously male-dominated fields doesn't automatically mean that gender segregation has disappeared or even declined dramatically, or that women enjoy the same benefits that men do. For instance, women have been increasingly employed in some fields just as the nature of the work has changed and the amount of individual discretion and opportunities for advancement and high pay have declined. For example, the field of insurance adjusting went from only about 30 percent female in 1970 to 75 percent female in 1994. Adjusters decide whether insurance claims should be granted and how much money should be paid out. Since 1970 much of the work of adjusting has become highly computerized. As women were hired into this field, they were primarily employed as "inside adjusters," where they used this computerized decision-making process. Men, however, continued to be employed as "outside adjusters," making individual decisions about claims using their own knowledge and judgment. Similarly women went from 28 percent of all bus drivers in 1970 to 55 percent in 1989. But men still make up the majority of drivers in metropolitan transit districts, while women are primarily employed as part-time school bus drivers.[17]

Common Sense versus Research Sense

COMMON SENSE: Workers who have more education will have higher paying jobs.

RESEARCH SENSE: Research has shown that women consistently earn less than men with comparable (and sometimes lower) levels of education.

Income The patterns of gender-based occupational segregation translate into differences in income, because women tend to work in areas that pay less. For instance, outside insurance adjusters earn more than inside adjusters; metropolitan transit drivers earn more than school bus drivers. At most colleges and universities the people who teach foreign languages and literature earn much less than those who teach physics and chemistry. (Sociology instructors usually fall somewhere in between.) Not surprisingly, the areas in which teachers earn less are those in which there are more women.

These patterns can also be seen when we look at census data on the annual incomes of men and women workers in the United States. Although we would expect that workers who have more education would have higher-paying jobs, women consistently earn less than men with comparable levels of education, as Figure 9-4 shows. In fact, on average, women who have an associate degree earn less than what men who have graduated from high school earn. And women who have graduated from high school earn $2000 less each year than men who have dropped out of high school.

Why does this happen? Is it simply because men and women hold different types of jobs? To control for the type of work that men and women do, the data in Table 9-5 show median wages in each of the broad census categories of occupations. Once again, women earn less than men, no matter what general type of job they hold.

FIGURE 9-4
Median Income of Men and Women Ages Twenty-Five and Older Working Year-Round, Full-Time, by Years of Schooling, 1997

*Women's income as a percentage of men's.

Source: U.S. Bureau of the Census, 1998b, pp. 31, 33

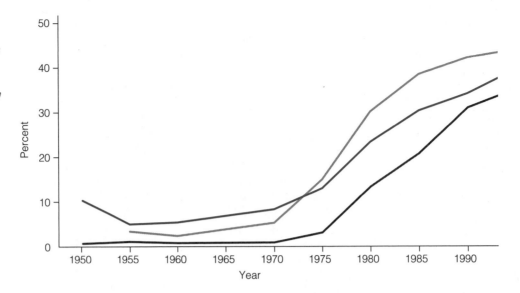

In the last few decades, several laws have been passed mandating "equal pay for equal work." Because gender segregation of jobs is so extensive, however, most men and women aren't doing the same job, and thus the laws can be interpreted as inapplicable. For instance, if women work day shifts and men work night shifts, the men's higher wages for essentially the same tasks can be justified because they are working at night. In the few instances in which women and men are doing substantially the same work, as in the category of "farming, forestry, and fishing," the income gap is quite small. The problem is, of course, that gender segregation of the labor force is so pervasive that relatively few women and men hold exactly the same jobs. There just aren't that many women who work in farming, forestry, or fishing. In addition, both men and women in this category earn far less than other employees.

On the other hand, the gender gap in pay is much smaller for younger workers than for older workers and appears to be diminishing for younger cohorts. For instance, in 1962 women between the ages of 25 and 34 who worked full-time throughout the year earned only 62 percent of the average man's salary. By the late 1990s this figure had increased to 83 percent.[18] Part of the decrease in the wage gap has occurred because younger women have been more likely to enter jobs traditionally held by men, which tend to pay more highly. Part of the decrease, however, has also resulted from the fact that the wages of young men with relatively little education have declined.[19]

Gender segregation of the labor force and gender differences in pay also appear in other societies. For instance, throughout the industrialized world women tend to be overrepresented in clerical, service and low-prestige sales and professional areas while men

TABLE 9-5 *Median Income of Men and Women Working Year-Round, Full-Time, by Occupation, 1997*

OCCUPATION	MEN	WOMEN
All jobs	$35,248	$26,029
Executive, administrative, and managerial	50,149	33,037
Professional specialty	50,402	35,417
Technicians and related support	37,705	27,576
Sales	35,655	21,392
Administrative support, including clerical	29,442	22,474
Service, except private household	22,359	16,120
Precision production, craft, and repair	31,496	21,649
Machine operators, assemblers, and inspectors	26,969	17,683
Transportation and material moving	28,227	21,024
Handlers, equipment cleaner, helpers and laborers	21,475	15,774
Farming, forestry, and fishing	17,394	17,301
Private household workers	—	12,648

Note: Median income not given for men as private household workers because the base number is less than 75,000 people.
Source: U.S. Bureau of the Census, 1998b, pp. 28–29.

are more likely to be employed as administrators, managers, and in production work. As in the United States, these occupations have different levels of pay, and women earn less than men, even when they have similar levels of education and work full-time throughout the year.[20] Yet, as described in Chapter

7, other industrialized societies tend to have both higher levels of government benefits for families and less overall inequality of incomes. These conditions then produce gender differences in income that are somewhat smaller than in the United States.[21]

It is important to realize that the gender segregation of occupations and women's lower wages result from social actions.[22] They reflect the choices that workers and employers make. Women and men choose to enter the occupations that they do; employers choose to pay the wages that women and men earn. At the same time, these choices reflect the nature of the social structure. The jobs we choose to pursue reflect our beliefs about what type of work we would enjoy (which involves our self-identity), our social networks, our education and social class, the level of pay associated with different jobs, and our perceived chances of getting these jobs. Simply knowing that few men work in a field such as nursing often is enough to discourage a man from training for that occupation. Similarly knowing that few women are welders is sufficient to keep many women from considering that line of work. The actions of people like Carla, who aspires to be a test pilot despite the fact that few test pilots are women, can and do gradually change the social structure. But for every Carla, there might be thousands of women and men whose realistic choices are limited by gender stratification and segregation.

The Political World   Gender segregation and stratification also appear in the political world. But, as in other areas, these patterns have gradually changed, during the last two decades. For instance, for many years after women in the United States gained the right to vote men were more likely than women to cast ballots. Since the 1980s, however, women have been more likely than men to participate in elections.[23] Even so, women have been far less likely than men to hold elective or appointive political offices. Although their representation has risen sharply in recent years, they are still very underrepresented, especially at the highest levels of government. As shown in Table 9-6, by the late 1990s, women held less than 10 percent of the seats in the United States Senate or in governor's offices and only about one-quarter of the elected state legislative and lower executive positions. Women tend to fare best in elected posts at the local level, where they hold about a third of all positions.[24]

Similar patterns appear throughout the world. Even though women can now vote in almost all countries, they are vastly underrepresented in political office, both elected and appointed. Data gathered by the United Nations indicate that by the mid 1990s women held only about 10 percent of the seats in national legislatures or parliaments around the world,

TABLE 9-6 *Women's Representation in Elective Offices in the United States, 1999*

United States Senate	9%
United States House of Representatives	13%
State Governors	6%
All state-wide elective offices (for example, Governor, Lt. Governor, State Treasurer, State Superintendent of Schools, etc.)	28%
State Legislatures	22%

Source: Center for the American Woman and Politics, 1998a, b.

were head of state or head of the government (a position such as President or Prime Minister) in only a dozen countries, and held less than 6 percent of all ministerial or cabinet posts. Women's representation in these cabinet posts corresponds to patterns of gender segregation in the world of work, with women officials much more likely to assume responsibility for education, social welfare and health than for political affairs or the military.[25]

Yet, the participation of women in political roles varies from one society to another. Although there is no society in which women constitute a majority of office holders, their representation in national legislative bodies varies from more than 30 percent in countries such as Denmark and Sweden to around 10 percent in countries such as Austria, France, Israel, and the United States.[26]

Why are women in the United States less likely than women in several other modern democracies to hold political office? The answer does not appear to involve individual prejudice against women candidates. Studies have found that, when all else is equal, voters are just as likely to cast their ballots for and contribute campaign money to women as to men candidates. Instead, most scholars now believe that women's underrepresentation reflects two aspects of the structure of the American political system that keep women and men from having equal representation. The first involves incumbency. In the United States incumbent office holders, people who already hold a political office, are very likely to be re-elected. Very few women are incumbents, and when women do run for office they often face incumbents; this, rather than their gender, appears to contribute to their defeat. This can explain why women's representation has gradually increased over the years as they, too, join the ranks of the incumbents. The second aspect involves the structure of the representation system. Legislative seats in the United States are filled through a representational system with one candidate elected from each given geographic area, and this one-winner-take-all system does not appear to increase the probability of women winning office. Women are more likely to gain office in countries that have a proportional representation system in which seats are allotted to a party based on the number of votes that party receives within the election.[27]

Family Life Just as gender segregation appears in education, occupations, and the political world, it can also be seen in the family. Fathers and mothers, and brothers and sisters, typically have different tasks and responsibilities. For example, sociologists have found that women are much more likely than men to be responsible for cooking, cleaning the house, and taking care of children. Men are more likely to be responsible for household repairs, yard work, and pet care. Sons are more likely to take out the garbage and do yard work. Daughters are more often asked to help with the cleaning and cooking, much as Carla did for her father after her mother died.[28] The tasks associated with women's family roles are much more time-consuming than those of men, and thus women are generally saddled with more of the family work. At the same time, however, fathers often miss out on many of the interactions with their children that mothers enjoy.

Sociologists also have studied how families make decisions, focusing especially on the relative power that husbands and wives wield in household decisions. Researchers have found that both husbands and wives influence family choices but that, especially when big purchases are involved, the husband generally has more influence.[29] Much as one member of a team of mountain climbers can be in the lead, husbands tend to take precedence over wives in major family decisions.

Until recently this situation was largely accepted as the way life should be. This might reflect the fact that only since about 1980 have a majority of women worked outside the home for pay, as Figure 9-5 shows. But today three-fourths of all married women with school-age children work outside the home, and more than half of all married women with children under age three are employed.[30] These changing patterns of employment have prompted some equalization of housework. Compared with the 1960s, men devote more hours to household chores and women devote less.[31] The changes have been especially strong in the area of child care. As more women are working outside the home, many men are spending more time taking care of children, especially in families in which the parents work different shifts and can't afford child care.[32] Mothers, however, generally retain the ultimate responsibility for organizing and managing both child care and housework and spend many more hours in these tasks than do fathers.[33]

Are the household chores in your family gender-segregated? Why do you think this gender segregation does or does not appear?

Even though women who work outside the home still do the majority of the housework, the fact that they work for pay and contribute to the family's economic resources has helped mitigate other ways in which gender stratification occurs in the family. For instance, women who work outside the home have more power in the family and more influence on major family decisions.[34]

Despite these benefits, there can also be dysfunctions associated with these recent changes in gender segregation and women's and men's roles. Many of these involve issues related to child care when both parents are in the labor force. Unlike most European countries, the United States does not provide paid maternity leave for working mothers, nor does the government provide preschools and day care for young children. Finding appropriate and affordable child care is often very difficult for families, especially for those in the working class, where both parents must work to make ends meet and where the family income is usually quite low.[35]

Clearly the gender segregation of the labor force underlies much of the stratification observed in families. Women earn less than men because they work in different areas, and this can translate into their having less power within the family. In addition, business or the government might be more likely to provide child care if more women held high-paying,

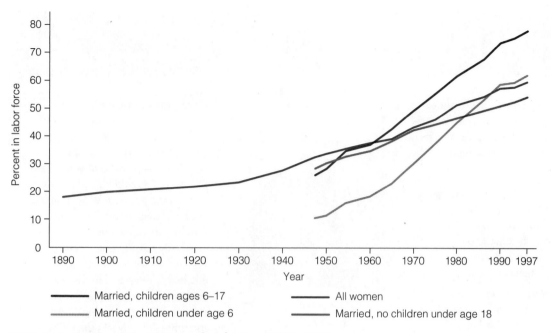

FIGURE 9-5 *Women's Participation in the Labor Force, 1890–1997*

Note: Before 1947 the proportion is based on all women ages fourteen and older. After 1947 it is based on all women ages sixteen and older.

Source: U.S. Bureau of the Census, 1975, pp. 131–134; 1994, pp. 401–402; 1995, p. 406; 1997, pp. 398, 404; 1998, pp. 404, 409.

influential jobs. But what factors explain why gender segregation and stratification occur in the first place?

EXPLAINING GENDER SEGREGATION AND STRATIFICATION

Most explanations for the gender segregation of the labor force and women's lower incomes use the notion of **gender discrimination,** differential treatment of women because of their sex. You might think that such discrimination rarely occurs anymore, but the experiences of Christopher and Julie, two college graduates in their late twenties, tell us otherwise. They were sent by the ABC television show *Prime Time Live* to apply for jobs advertised in the help-wanted columns. The show's producers wired Christopher and Julie for sound and rigged hidden cameras to document what happened to them. Christopher and Julie were armed with resumes showing similar managerial experience and similar educational backgrounds. However, when they applied for a job as territory manager for a lawn-care firm, the company's owner gave Julie a typing test and offered her a receptionist's position at about six dollars an hour. Christopher was given an aptitude test and offered the territory manager job at substantially more than Julie's starting wage.[36] If Christopher and Julie had taken these positions, they would have been embarking on different career paths—Julie toward additional low-paying cler-

ical jobs, Christopher toward lucrative, upper-level management positions. Of course Christopher and Julie are only two people, but cumulatively such experiences produce the social patterns of gender stratification discussed in the previous section.

But why does such discrimination occur? Why do employers make these decisions? Why do male cadets object to females, such as Carla, learning to become a test pilot? Why did Eric suggest that she might consider a specialty other than testing planes? Since women and men live together in families, wouldn't men want to have women—or at least their sisters and wives, or the sisters and wives of their friends—earn as much as men and enter fields that they really wanted to pursue? Why are some men more likely than others to be sensitive about these issues?

Some theories of gender discrimination take a macrolevel perspective, looking at differences in the kinds of jobs that men and women typically hold and the structure of the occupational world. Other explanations take a more microlevel perspective, focusing on the actions of employees and employers. Still other explanations move in even closer to look at the motivations that underlie gender segregation and stratification—the reasons that people choose to discriminate. In this section we focus on these explanations and on research designed to test them. To learn more about the work of a sociologist who studies gender discrimination in occupations, Paula England, see Box 9-1.

BOX
9-1

SOCIOLOGISTS AT WORK

Paula England

I decided to examine whether it was true that women's jobs were less skilled and this explained the pay gap.

Paula England is a professor of sociology at the University of Arizona who has done extensive studies of gender discrimination in the world of work.

Where did you grow up and go to school? I grew up in the suburbs of Minneapolis, the first of four children born during the baby boom to a professor of industrial relations and a homemaker.

Why did you become a sociologist? During high school, I wanted to become a social worker. I majored in sociology and psychology at Whitman College during the late 1960s. The social movements of the times, together with what I was learning, made me see being a social worker as putting Band-Aids on a system that needed fundamental changes in inequalities. I decided to do graduate work to study the systemic causes of inequalities I had come to believe were at the root of a lot of suffering. Besides, I enjoyed ideas and the college environment.

How did you get interested in studying occupations and gender stratification in the economy? I was becoming interested in feminism. But none of the sociology courses I took in college discussed gender. I went to graduate school at the University of Chicago. Although none of the faculty members there taught courses on or did research on gender, I did a dissertation on gender. It began my path of research. I focused on the sex gap in pay.

What has surprised you most about the results you obtained in your research? I realized that the main problem couldn't be lack of equal pay for men and women in the same job (even though that occurs sometimes) because jobs were so segregated by sex that men and women often didn't work in the same job. I learned that in the United States employed men and women averaged about the same years of education. At first I assumed that the problem must be that women were in menial, less skilled jobs either because their socialization channeled them to these jobs or because employers discriminated in

hiring. Instead of studying how women and men got into the jobs they hold (a very important question!), I decided to examine whether it was true that women's jobs were less skilled, and this explained the pay gap. What I discovered was that women's jobs often require lots of the kinds of skill—especially cognitive skill obtained through education—that is generally highly rewarded in the labor market. Yet, women's jobs requiring a certain amount of education and cognitive skill pay less than different male jobs requiring the same or less of such skill. So, for example, secretaries make less than welders, psychiatric social workers much less than chemists, nurses less than mechanics, day care workers less than lawn mowers. Also, some skills required in women's jobs are hardly rewarded at all. For example, the social skills required to do a good job of caring for people—skills of empathy, patience, communication—are often not even recognized as skills. The issue of the underpayment of female occupations relative to their demands has come to be called "comparable worth" or "pay equity." The general point that I began to realize is that whatever activities women do tend to get devalued because of their association with women. Thus I realized there were really two separate parts to sexism: (1) keeping women out of respected, rewarded "male" roles and (2) devaluing whatever activities women do. It seemed to me the second form of sexism was more subtle, often not recognized, and that changing it would be harder. This is what I have emphasized in my research.

What are your plans and hopes for the future? In the future I am interested in figuring out what it would take to have a society that encouraged people to care for each other, inside the family and outside the family, and that rewarded caring more. Perhaps some of you reading this book will do the research in the future necessary to answer or reformulate this question.

The Structure of the Occupational World

When looking at the **labor market,** or set of jobs within a society, sociologists and economists often distinguish between the primary and secondary labor markets. The **primary labor market** consists of "good" jobs with good wages and benefits, comfortable working conditions, job security, and chances for advancement; the **secondary labor market** consists of "bad" jobs with low pay, few if any benefits, little job security, and few chances for advancement. As you know from previous discussions, members of the working poor and racial-ethnic minorities are more likely than others to hold jobs in this secondary market. But, within each class and racial-ethnic group, women are also more likely to hold jobs in the secondary market, and some theorists point to this *segmented* (or divided) *labor market* as a major reason that women have lower incomes and less prestigious occupations than men. As a result, this theory is often referred to as **labor market segmentation theory.**[37]

Besides looking at divisions within the labor market, some theorists have focused on the **internal labor market,** the series of jobs that people might hold within an organization throughout their work careers. For instance, to become a test pilot, Carla must work her way through a series of positions within the U.S. Air Force. Several studies of the jobs that women and men hold suggest that men are much more likely than women to take jobs that have more possibilities of advancement. Women tend to enter fields, such as clerical work, school teaching, and nursing, that have very short career ladders and few opportunities for advancement. For instance, when Christopher and Julie applied for jobs with the lawn-care firm, Christopher, but not Julie, was offered a job that would allow him to advance to upper-level management.[38]

Explanations for gender stratification that focus on labor markets suggest that gender inequality can result simply from the everyday functioning of the occupational world. Once women and men are sorted into different entry-level positions and different labor markets, "business as usual" will result in differences in pay, job security, and opportunities for promotion. But, of course, individual employers and employees make decisions that produce these labor markets. Employers hire workers, and employees choose to pursue certain occupations. Other theorists have taken a more microlevel view of gender stratification to look at these decisions.

Social Actions of Employers and Employees

One popular explanation of gender segregation and stratification is human capital theory. Much as the term *social capital* refers to the resources we derive from our social relationships, the term **human capital** refers to the resources that we have as individuals and workers, such as our education, skills, and work experience. **Human capital theory** suggests that different occupations and incomes reflect different amounts of human capital, and thus that women have lower incomes than men because they have less human capital. According to this theory, if women would choose to develop as much human capital as men, then gender disparities in income would sharply decline. Certainly people who have more human capital tend to have greater success in the labor market. People with more education and job training earn substantially more than people with less training do. But human capital differences hardly account for gender stratification. Even when men and women have similar levels of work experience, training, and occupational status, men still tend to earn more.

Another popular explanation of gender stratification is the theory of **compensating differentials,** which suggests that employers compensate their employees for working in undesirable working conditions by paying them more. Women earn less than men because they choose jobs that don't have these undesirable working conditions. In fact, however, there is little support for this theory either. For instance, one extensive study of many different jobs found that the jobs women hold are better than men's only in how much vacation time they receive and how dirty the job is. Men's jobs are better, on average, in areas such as how many hours they have to work each week, how much on-the-job training they receive, how repetitive the work is, how much choice they have in work hours, how closely they are supervised, and how secure the position is. Interestingly, the gender differences in these other conditions actually were greater than the differences in pay.[39]

A third theory, called the model of **statistical discrimination,** looks at the social actions of employers. This theory suggests that employers have beliefs, or stereotypes, about the relative stability and productivity of men and women workers, which they rely on when hiring employees. For instance, they might believe that, on average, women will be more likely to leave their jobs than men or that men, on average, tend to be better pilots. Perhaps because of his belief that young women might be less devoted to the work force than young men, the owner of the lawn-care firm offered Julie a job as a receptionist rather than as a territory manager. Similarly some of Carla's classmates might have discouraged her attempts to pursue test piloting because of their views of the relative abilities of men and women pilots. When these

BUILDING CONCEPTS

Macrolevel and Microlevel Theories on Gender Segregation and Stratification

Here are some examples of micro- and macro-level theories on gender segregation and stratification. Can you think of arguments both for and against each of these theories?

Level of Analysis	Focus	Theory
Macrolevel	Segmentation of the labor force	Women are more likely than men to hold jobs in the "secondary labor market," where incomes are lower and occupations are less prestigious.
Macrolevel	Internal labor markets	Men are more likely than women to take jobs with more possibility of advancement.
Microlevel	Human capital theory	Different occupations and incomes require different levels of "human capital"—that is, education, skills, and work experience. Gender stratification exists because women have not acquired as much human capital as men.
Microlevel	Theory of compensating differentials	Employers compensate their employees for working in undesirable environments. Men earn more than women because they have less favorable work environments.
Microlevel	Statistical discrimination	Employers hold stereotypes about the relative stability and productivity of men and women workers, which influences their hiring decisions. These stereotypes affect all workers, including those to whom the stereotypes don't apply.

average characteristics don't apply to individual women, statistical discrimination is said to have occurred. Julie might have been just as good a territorial manager as Christopher, and Carla could be as good a test pilot as any of her classmates. But because of beliefs about women as a group, they were denied these opportunities.

All these theories, like others we have looked at in this book, assume that the motivations for human behaviors and social patterns are rational. For instance, employers might be acting rationally when they employ statistical discrimination simply because their decisions might, on average, tend to be correct. The theory of compensating differentials suggests that women rationally choose to enter jobs that pay less because the jobs provide other rewards that compensate for the lower pay. The segmentation of the labor market can also be seen as rational. After all, dividing labor up among members of a group is an efficient way to get tasks completed and, as we know from the discussion of Durkheim's work in Chapter 1, can even help to bind members of a group more closely together.

But it is hard to understand why one type of work should be seen as more valuable than another, especially when the tasks are so similar. Why should growing yams be more important than growing sweet potatoes for the tribe in New Guinea? Why should it be acceptable for women to fly transport planes but not to test fighter jets? Why do women, but not men, typically work as receptionists? In other words, why do so many of the rules about what women and men should do seem irrational?

Yet, when you think about it, you will realize that all of us behave irrationally at times. Even the mighty Indiana Jones is afraid of snakes! Because irrational behavior occurs every day, it would seem important to understand its origin. Fortunately one theoretical tradition, psychoanalysis, takes irrational behavior seriously. Psychoanalysts acknowledge that we often behave in ways that don't make sense and search for reasons that we do so. A number of sociologists have used ideas from this tradition to try to account for the motives underlying gender stratification.[40]

A Psychoanalytic View of Gender Stratification

Psychoanalysis was founded by the famous psychiatrist Sigmund Freud (1856–1939). Many of Freud's original ideas, especially about the differences between men and women and the way young boys and girls develop, are no longer given much credence.[41] But Freud did introduce three notions that still are widely accepted. The first is the concept of the **unconscious,** the idea that we have thoughts and impulses that we aren't consciously aware of. The second is the concept of **repression,** the notion that we repress, or push out of conscious awareness, ideas that are uncomfortable or painful to think about. The third is the idea that our *early experiences* in the family, especially our emotional relationships with our parents, have a deep and lasting influence on how we see ourselves and other people and how we behave as we grow older. Beginning in the 1920s, some of Freud's students and followers used these concepts to develop ideas about how boys and girls come to define themselves as masculine and feminine and why gender segregation exists.[42]

According to psychoanalytic theorists a child's early relationship with his or her mother is very important. Using the ideas of reciprocal roles introduced in Chapter 4, this relationship can be seen as one in which we learn both what it is like to be loved and how to love other people. These theorists also suggest that because this very strong early tie is almost always with a woman, our first identification, or self-view, is feminine rather than masculine.

As we grow older and become more independent, we need to lessen the very strong ties that we had with the mother during infancy. We also learn what it means to be a male or a female. For a girl, this is relatively easy because her mother was the first person with whom she identified. But, these theorists note, achieving gender identity is harder for a boy because in the process he must reject his first identity as feminine. In addition, because fathers and other male figures often aren't such a central part of young boys' lives as their mothers and other women are, it might be hard to develop a strong idea of just what masculinity involves.

Because boys know most intimately what is feminine, they come to define masculinity as being "not-feminine." In their behaviors and relationships with others, they devalue what is feminine and deny their attachment to the feminine world. In psychoanalytic terms, they repress the feminine identification developed in their early relationship with the mother. As a result, boys' gender identity tends to be somewhat more tenuous, or less firm, than girls' gender identity.

Gary Fine's study of Little League ballplayers that was reported in Chapter 4 provides several examples of the outcome of this process. Remember how these boys developed subcultural norms that emphasized toughness, unemotionality, competitiveness, and loyalty. Remember also how they were often preoccupied with demonstrating their masculinity to one another, often through talking about sex (about which they actually knew very little) and performing daring feats. Through these interactions the boys tried to show themselves and one another that they were really boys and to define what masculinity meant. (To learn about the work of another sociologist who has studied masculinity, Michael Kimmel, see Box 9-2.)

Theorists who favor the psychoanalytic perspective go on to suggest that these early childhood experiences can help explain why gender segregation is so pervasive. Essentially these theorists assert that the tenuous nature of men's identity provides an unconscious motivation for maintaining gender segregation and for devaluing women's contributions.[43] The late anthropologist Margaret Mead described this process, using her knowledge of gender segregation in many different societies:

SOCIOLOGISTS AT WORK

Michael Kimmel

There's a certain fear that asking questions about masculinity shows you're not a true man.

It's important to remember that gender stratification involves both males and females and that gender identity involves notions of both masculinity and femininity. Michael Kimmel, who teaches at SUNY-Stony Brook, was one of the first contemporary sociologists to study masculinity and how it relates to gender stratification.

How did you get interested in studying masculinity and gender issues? My intellectual concerns have always been guided by my heart and my feelings; I've always done research on things that moved me, what I considered to be pressing moral and emotional issues.

For my dissertation, I was thinking a lot about the Vietnam war. I had two questions: What was it about some countries that made them want to control other nations, and how did other countries besides the United States do this? I spent two years in Paris archives trying to figure out the similarities and differences among France, England, and Germany; my dissertation focused on seventeenth-century France and England. This is the time of the origin of the nation-state, and I wanted to study this period to continue my thinking about the historical origins of American involvement in Vietnam.

My political involvement continued with my first teaching job at UC Santa Cruz, where I did volunteer work with Santa Cruz Men Against Rape. We were working to prevent date/acquaintance rape and to think about the feminist issues raised by these actions. As well, the women's movement was really taking off in my and my friends' lives.

When I went to Rutgers University, I wanted to teach about masculinity, and I taught the first course on men in New Jersey in 1983: Sociology of Male Experience. The course was immediately filled to overflowing; it really struck a chord. In the course we took feminist issues seriously and used them to look at masculinity. Out of my political commitment, my academic interest grew and I started doing research in this area in 1984.

Areas I've studied include sexuality and the history of men who have supported feminism in the United States. Two of my recent books have focused on the historical development of manhood in the United States and the debate between feminist men and the so-called mythopoetic men's movement.

How unusual is it for a man to study gender stratification? How do you think this has affected your work? It is still somewhat unusual for a man to study gender; look at the composition of courses on gender. There's still this idea that "real men" don't study gender. Men are confused about masculinity and what it means, yet at the same time there's a certain fear that asking questions about masculinity shows you're not a true man. I say, let's be brave, let's open up these questions about gender and masculinity and see what's there.

What do you think the implications of your work are for social policy? Have any of these implications been realized? Every month, it seems, a crime committed by a man becomes a touchstone for debate about gender relations: O. J. Simpson, Tailhook, Woody Allen, Mike Tyson. There's an ongoing, inadvertent debate on the politics of masculinity because we are in a moment of unbelievable transition. Think of this: My dad could attend an all-male college, serve in an all-male army, and work in an all-male workplace. It would be impossible for any student today to do that; there has been an enormous change in just one generation.

How does your research apply to students' lives? Feminists, sociologists of gender, and those in women's studies all teach us that gender is a central experience, along with race and class. Students can use this to acknowledge that constructions of masculinity didn't just fall out of the sky and can ask themselves, where did these ideas come from? Is it possible to renegotiate them? In the future I see men sharing the workplace, the home, and the family equally, and gender roles are changing to reflect this.

The recurrent problem of civilization is to define the male role satisfactorily enough—whether it be to build gardens or raise cattle, kill game or kill enemies, build bridges or handle bank-shares—so that the male may in the course of his life reach a solid sense of irreversible achievement. . . . If men are ever to be at peace, they must have, in addition to paternity, culturally elaborated forms of expression that are lasting and sure. Each culture—in its own way—has developed forms that will make men satisfied in their constructive activities without distorting their sure sense of their masculinity.[44]

Based on these conclusions, Mead would no doubt suggest that the male cadets resisted Carla's attempts to be a test pilot and degraded her when she was successful in class because her very success, in what they believed was a male domain, threatened their own identity as men. Even Eric, who was quite fond of Carla, was uncomfortable with her encroachment in areas that he regarded as a masculine preserve.

FEATURED RESEARCH STUDY
TESTING PSYCHOANALYTIC THEORY

Freud and his followers developed their theories through a process called *analysis,* in which they spent many hours listening to people talk about themselves and their childhood experiences and trying to figure out how these early experiences were related to their later lives. As you might imagine, social scientists have been fairly skeptical of this technique. Are the results reliable? Would the conclusions reached by one analyst match those of another? In addition, most psychoanalytic writing has been based on *case studies,* detailed examinations of individuals. How can we know that these individuals are representative? Can the results obtained with these individuals be generalized to others? The concepts of repression and the unconscious always provide easy "outs" for cases that don't fit the theory. One can always claim that someone's experiences and interpretations don't conform to the theory simply because the ideas were "repressed" or buried in the "unconscious." Given these criticisms, many sociologists and psychologists understandably have rejected psychoanalytic theories as untestable and therefore of little use in empirical social science.

Instead of dismissing psychoanalytic theory, however, some sociologists have seen it as a challenge. One who has taken up the gauntlet is Christine Williams, who examined the extent to which hypotheses from psychoanalytic theory could help explain gender stratification by looking at men and women in nontraditional occupations in the contemporary United States.

Two of the most gender-typed occupations in contemporary society are nursing and the Marine Corps. When you think of a nurse, you probably envision a gentle, compassionate woman who will stroke your brow when you are sick. When you think of a marine, you probably envision a tough, no-nonsense man who isn't afraid of anything. Yet some nurses are men, and some marines are women.

What, Williams wondered, are their experiences like? How do male nurses and female marines maintain their views of themselves as men and women? What happens when such strong gender-typed norms are violated? How do the Marine Corps and hospitals respond to their presence? Do the answers to these questions resemble what the tenets of psychoanalytic theory would predict?

Williams used the techniques of in-depth interviewing and field research to investigate the issue. She attended a convention of women marines and interviewed female veterans, new recruits, drill instructors, and officers. She observed the training of both male and female recruits and even participated in one of the most challenging parts of the training—rappelling off a forty-five-foot tower.[45] To find male nurses, Williams made a few initial contacts and gradually developed a more extensive sample.

Williams used the data from her interviews and observations to, as she put it, "illustrate how the conflicts that psychoanalysts associate with gender identity development may manifest themselves in adult life." In other words, by looking at the information her respondents gave her, she assessed the extent to which their experiences conformed to what psychoanalytic theory would suggest.

Williams's interviews and observations strongly supported the notion that men's gender identity, or view of themselves as men, is more tenuous than women's. As she put it,

The psychic dispositions psychoanalysts describe are practically caricatured in these two groups. Male nurses go to great lengths to carve for themselves a special niche within nursing that they then define as masculine; preserving their masculinity *requires* distancing themselves from women. Women in the Marine Corps feel they could maintain their femininity even in a foxhole alongside the male "grunts." For the women I interviewed,

The champion drag racer, Shelly Anderson, followed in the footsteps of her father, who is shown here helping her prepare for a race. What obstacles might Anderson have faced in entering this occupation, which is generally held by men? How might her father's earlier success have affected her experiences?

femininity was not "role defined" in the way that masculinity was for the male nurses. It mattered little what activity she was engaged in; a woman in the Marine Corps could be employed in *any* job specialty and still be considered a "lady" by her female peers.[46]

Men carve their "special niche" in nursing in several ways. They tend to be concentrated in certain areas, such as acute care, intensive care, and psychiatric care, and are overrepresented in the various administrative ranks. These areas are seen as more demanding and prestigious than other types of nursing. They allow for more autonomy and individual decision making, thus adding to their prestige. They also are more highly paid, which helps account for the fact that male nurses earn more than female nurses.[47]

Beyond the specialties they choose, male nurses describe how they work in a manner that helps reinforce their views of themselves as masculine. Because so much of nursing involves caring for others and being sympathetic and responsive to their needs, many of their comments centered on this theme. For instance, as one male nurse said,

> Men have the capacity to care—I think it's definitely in a different way—but they have the capacity to care for people. . . . I don't see little details, but I see the main things I need to do. I'm strong, I can lift. . . . And the women focus on different things. . . . I wouldn't put makeup on somebody. Or take care of somebody's hair.[48]

The men who are in the Marine Corps have no need to carve out a special niche to demonstrate their masculinity, because becoming a marine is synonymous with being a man. As Williams describes it,

> Marine Corps basic training promises to "make a man" out of new male recruits, who are often

called "girls" until they prove their masculine prowess by performing feats set by the training regimen. Not unlike male initiation rites in other cultures, basic training marks off the difference between masculinity and femininity; it has been culturally defined as an activity only masculine males can accomplish. It unambiguously fulfills men's largely unconscious desires to prove once and for all that they are masculine.[49]

But what happens if women can be marines? What does this say about the men and their masculinity? According to Williams the presence of female recruits diminishes the rites of basic training because it undermines their purpose—to separate masculinity from femininity:

> If a woman can do it, the value of the ritual for proving masculinity is thrown into question. In other words, whenever men witness women accomplishing tasks they regard as masculine, their own masculinity is threatened.[50]

To deal with this threat, the Corps has a whole set of regulations and practices designed to preserve female marines' femininity and thus show that they really are different from male Marines. Thus their basic training includes instruction in hair care, poise, and etiquette. All female marines are required to wear both lipstick and eye shadow on duty. Female drill instructors at Parris Island are not allowed to wear the slacks version of their uniforms unless the temperature is below freezing.[51] In addition, just as in the civilian labor force, women in the marines have traditionally been relegated to a limited range of occupational specialties, with the majority holding secretarial, data processing, nursing, and dental assistant positions. (Unlike for civilian jobs, however, men and women are paid equally within each rank.) In 1993 the secretary of defense announced that women would begin training to serve in combat, but, according to news reports at the time, "the Marines fought the change to the end!"[52]

Thus Williams suggests that the experiences of both male nurses and female marines illustrate the notions of psychoanalytic theory. Male nurses go to great lengths to demonstrate that they really are masculine, even though they are nurses; female Marines are required to demonstrate their femininity in a variety of ways just to reassure their fellow male marines that their masculinity is intact.

Based on her research, Williams suggests that many more women would join the Marine Corps if they were recruited as actively as men are. As with most blue-collar, male-typed jobs, the pay is much better than that available in female-typed jobs. Most important, female marines do not see their work as threatening their sense of themselves as feminine. The situation with nursing, however, is just the opposite. The major reason there aren't more men in nursing, Williams argues, is that

> in general, men do not want to be nurses—including the men who *are*. . . . As long as nursing is defined as "women's work," men will simply *not want to* engage in it. Not only does nursing pay less than comparable "male" occupations, it has low status by virtue of being female dominated. Not only money, but their higher social status as "masculine" men is at stake when they enter nontraditional occupations.[53]

Williams's research and her application of psychoanalytic theory can help us understand more about the recent changes in gender stratification in education and in occupations, especially the lack of men entering female-typed areas. According to psychoanalytic theory women have not been reluctant to train for traditionally masculine areas, such as test pilot, because entering these fields does not threaten their gender identity. In contrast, very few men have entered female-typed jobs precisely because if they were to do so, without elaborate safeguards and justifications such as those male nurses use, their definition of themselves as masculine

could be threatened. The theory can also help explain why women earn less than men. Rewarding women's work less than men's helps to shore up men's prestige—to show, in a very concrete manner, that women and their contributions are worth less than men's.

This work also shows us how patterns of gender segregation and stratification are related to the ways in which people develop their gender identities. According to the psychoanalytic perspective the motivations that prompt the day-to-day decisions underlying gender segregation have their roots in childhood and in early development of our self-identities. These connections then also point toward one way to lessen gender stratification. Sociologists who adopt the psychoanalytic perspective suggest that in the long run gender stratification and segregation in the occupational world won't change unless there are also changes in child-rearing practices and family life. These theorists also suggest that some of the more disturbing aspects of everyday interactions between the sex groups, such as sexual harassment, will become less common only when the motivations that underlie gender segregation are changed. (See Box 9-3, Applying Sociology to Social Issues.)

CRITICAL-THINKING QUESTIONS

1. Do you know people who have entered occupations that typically are held by members of the other sex group? How do their experiences compare with those Williams observed with male nurses and female marines?

2. Think about the theories of social interaction and socialization that were described in Chapters 4 and 5. How might the ideas of these theories be used to help understand the experiences of the male nurses and female marines?

3. What other methods, besides field research, might Williams have used in her research? What would be the advantages and disadvantages of using such methods?

Perhaps you have witnessed, or even been part of, interactions like those Carla experienced with her classmates at the Academy when the men whispered "ratings" of the female cadets' bodies just loud enough for them to hear. In recent years, the issue of sexual harassment has been in the spotlight: the subject of popular movies; the focus of rules and regulations in workplaces, colleges, and universities; and the source of lawsuits. Yet, even though the behaviors that we now call sexual harassment have been around for centuries, only recently have they been labeled and defined as deviant.[54]

Many observers have noted that sexual harassment is at root a demonstration of power—a way to concretely show both others and oneself, in daily interactions, who is more powerful. Is it possible, then, that sexual harassment would diminish if the patterns of gender stratification in our society were to change? If men and women had more equality in the occupational world, would sexual harassment lessen?

Or, as psychoanalytic theorists might argue, is the problem more deep-seated? Does sexual harassment stem from some men's need to demonstrate their masculinity and power to both themselves and others? If so, sociologists who have used psychoanalytic theory would suggest that changes in women's occupations and income might not be sufficient to alter patterns of sexual harassment.

Recall that the root of men's more tenuous gender identity is the early close ties with their mother or other females and the less strong ties with their father or other males. Psychoanalytic-oriented theorists argue that this pattern could be altered if fathers were more involved in child rearing, and especially in the nurturant caring for very young children. If, they suggest, fathers were more involved in caring for very young children, boys would grow up with a stronger gender identity and a reduced need to promote gender segregation and to devalue the contributions and activities of women. In other words, actions that affect the development of males' self-identity could alter the motivations that underlie the perpetuation of male dominance and gender stratification in other areas of society, including everyday life. Supporters of this viewpoint suggest that changes to promote greater gender equality in the long run should also involve changes in the work world that allow both mothers and fathers to spend more time with their children.[55]

Finally we can apply the contact hypothesis, which was introduced in Chapter 8, to this issue. In contrast to her interactions with some of the other cadets, Carla's experiences with Eric would probably not be interpreted as sexual harassment. Eric cared about Carla as a friend and clearly respected her abilities, but he was relaying his own fears and his perceptions of the ideas of others. In general, sexual harassment is less likely to occur with close friends than with acquaintances. Remember that the contact hypothesis suggests that individual discrimination and prejudice toward members of a minority group will diminish once members of the two groups have direct interpersonal contact within certain egalitarian conditions. So, too, might we expect sexual harassment to decrease when men and women interact as equals in a wide variety of settings, such as school, work, the family, and other areas of social life that we will examine in the next section of this book.

CRITICAL-THINKING QUESTIONS

1. What do you think sexual harassment is? Remember how a male cadet asked Carla if she was "trying to show us how tough you are" when she earned an award. Was that comment harassment? What about Carla's reply ("I'm sure you can make the list yourself next time")?

2. What is the line between teasing, flirtation, and harassment? How do you know when harassment exists? Why do people not always agree on what harassment is and when it has occurred?

3. Based on what you have read so far in this book, what do you think can be done to lessen the frequency of sexual harassment?

SUMMARY

Like social and racial-ethnic stratification, gender segregation and stratification exist across cultures and in our own society. Gender stratification is a complex issue that is intertwined with occupational roles, family lives, and personal relationships, as well as with cultural patterns of male dominance. Key chapter topics include the following:

- Gender stratification differs from racial-ethnic and social stratification largely because men and women are closely linked within families and work together to benefit the economic well-being of their families and kin groups.

- All cultures have gender segregation of tasks and, at least to some degree, value the activities of men more than women, whatever they may be.

- Within the contemporary United States women and men have, on average, very similar levels of education. In recent years women have been more likely to major in disciplines that were once dominated by men. Men have not entered female-typed fields, however, nor have vocational training programs for blue-collar jobs changed their patterns of gender segregation.

- Occupational gender segregation remains pervasive in the United States and in other countries, and the closer one looks at occupations and job titles, the more pervasive it becomes.

- Even with similar levels of education and general job titles, women workers earn much less than men. Much of this difference is due to occupa-tional gender segregation, such that women work in areas that are less highly paid than those men work in.

- Gender stratification also appears in the political arena in countries throughout the world, with women less likely than men to hold elective offices.

- Roles in the family are also segregated by gender, and women tend to do more of the household chores and to have less influence on major family decisions than men do. However, women who work outside the home tend to have more power in the family than those who do not.

- Many explanations of gender segregation and stratification focus on the concept of discrimination against women. Some macrolevel theories examine segmentation of the labor force and internal labor markets. Other, more microlevel theories, such as human capital theory, the theory of compensating differentials, and the idea of statistical discrimination, concentrate on the actions and decisions of individual employers and employees. But these theories can't explain why discrimination occurs—that is, what motivates these attitudes and behaviors.

- Psychoanalytic theory has been used to explain the motives that underlie male dominance and gender stratification, focusing on the development of children's gender identity. Christine Williams's study of male nurses and female marines provides support for this theory.

KEY TERMS

compensating differentials 231
feminism 218
feminist movement 218
gender discrimination 229
gender identity 219
gender segregation 217
gender segregation of the labor force 224

gender stratification 217
human capital 231
human capital theory 231
internal labor market 231
labor market 231
labor market segmentation theory 231
male dominance 217

occupational gender segregation 224
primary labor market 231
repression 233
secondary labor market 231
self-identity 219
statistical discrimination 231
unconscious 233

INTERNET ASSIGNMENTS

1. Locate the United Nations Web site and look for information on gender stratification in countries around the world. To what extent do women and men have different educational and occupational opportunities? How do their experiences compare with those of people in the United States? To what extent is the status of women a subject of discussion in documents of the United Nations?

2. Locate the Bureau of Labor Web site. The Current Population Survey results contain information on employment and wages for persons in census occupational categories. Obtain the most current information for women and men and compare the data to that shown in the text.

3. Using a search engine, locate Web sites and discussion groups that focus on women's studies and men's studies. What types of issues are being discussed and what types of resources are available? Try to locate research projects or research organizations that are centered around women's studies or men's studies. What are the goals of this research? Do researchers link their research to practice? That is, how might this research help us understand issues central to our lives as women and men?

TIPS FOR SEARCHING

In the United Nations Web site, information on the status of women is most often found in reports on population and economic development. Several Web sites also focus on women's issues and a growing number focus on men's issues. Although much of the material in these sites is related to current events, conferences, and activities, the sites sometimes also include a wide variety of scholarly information.

INFOTRAC COLLEGE EDITION: READINGS

Article A21204331 / Weistart, John. 1998. Equal opportunity? Title IX and intercollegiate sports. *Brookings Review, 16,* 39–44. Weistart documents the extent to which intercollegiate sports activities are still unequal for men and women several decades after the passage of Title IX.

Article A19027735 / Allan, Kenneth, and Scott Coltrane. 1996. Gender displaying television commercials: a comparative study of television commercials in the 1950s and 1980s. *Sex Roles: A Journal of Research, 35,* 185–204. Allan and Coltrane look at how the portrayal of gender has changed in television commercials over a period of three decades.

Article A20420200 / Konrad, Alison M., and Kathy Cannings. 1997. The effects of gender role congruence and statistical discrimination on managerial advancement. *Human Relations, 50,* 1305–1329. Konrad and Cannings provide an example of the use of the theory of statistical discrimination.

FURTHER READING

Suggestions for additional reading can be found in *Classic Readings in Sociology,* bundled with this book. If you purchased a used copy of this book that does not include this custom-published reader, go to www.sociology.wadsworth.com for ordering information.

RECOMMENDED SOURCES

Blau, Francine D., and Marianne A. Ferber. 1992. *The Economics of Women, Men, and Work,* 2nd ed. Englewood Cliffs, N.J.: Prentice-Hall. Two economists summarize data and theories regarding gender discrimination in the work world.

Coltrane, Scott. 1996. *Family Man: Fatherhood, Housework, and Gender Equity.* New York: Oxford University Press. Coltrane explores issues related to gender segregation in the family, especially regarding the role of fathers.

England, Paul, and George Farkas. 1986. *Households, Employment, and Gender: A Social, Economic, and Demographic View.* New York: Aldine. This book, coauthored by the sociologist profiled in Box 9-1, is a nice example of the ways in which sociologists analyze gender stratification.

Johnson, Miriam M. 1988. *Strong Mothers, Weak Wives: The Search for Gender Equality.* Berkeley and Los Angeles: University of California Press. This is a fascinating example of the use of psychoanalytic theory by a sociologist.

Kimmel, Michael S. 1996. *Manhood in America: A Cultural History.* New York: Free Press. This fascinating look at masculinity and American culture was written by the sociologist profiled in Box 9-2.

Reskin, Barbara, and Irene Padavic. 1994. *Women and Men at Work.* Thousand Oaks, Calif.: Pine Forge Press. This is a short textbook summarizing theory and data regarding gender stratification in the work world.

Risman, Barbara J. 1998. *Gender Vertigo: American Families in Transition.* New Haven, Conn.: Yale University Press. Risman describes how social actions, especially within the family, produce gendered social structures.

Spain, Daphne and Suzanne M. Bianchi. 1996. *Balancing Act: Motherhood, Marriage, and Employment Among American Women.* New York: Russell Sage Foundation. This is a detailed statistical analysis of changes in the status of women for birth cohorts throughout the twentieth century.

Stockard, Jean, and Miriam M. Johnson. 1992. *Sex and Gender in Society,* 2nd ed. Englewood Cliffs, N.J.: Prentice-Hall. This general textbook summarizes much of the literature on gender stratification both in the United States and cross-culturally.

Thomas, Sue and Clyde Wilcox (eds.) 1998. *Women and Elective Office: Past, Present, and Future.* New York: Oxford University Press. This collection of articles analyzes reasons related to women's representation in elective offices at the local and national level.

Williams, Christine L. 1989. *Gender Differences at Work: Women and Men in Nontraditional Occupations.* Berkeley and Los Angeles: University of California Press. This is a very readable and full account of the research featured in this chapter.

Williams, Christine L. 1995. *Still a Man's World: Men Who Do "Women's Work."* Berkeley and Los Angeles: University of California Press. This is an extension of Williams' 1989 book in which she looks at the experiences of men and women in the traditionally female fields of nursing, librarianship, social work, and elementary teaching.

INTERNET SOURCES

Population Reference Bureau Web Site. This site provides a wide array of links to sources of international data on women and current updates of the information.

Center for the American Woman and Politics. This center, part of the Eagleton Institute of Politics at Rutgers University, provides a wealth of information on women in political life.

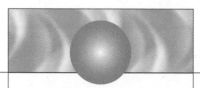

Pulling It Together

Listed here are other chapters where you can find additional discussion or examples
of some topics covered in this chapter. You can also look up topics in the index.

TOPIC	CHAPTER
Gender stratification	4: Socialization and the Life Course, pp. 87–89 15: Education, pp. 397–398 16: Population, pp. 470–472
Pre-Industrial societies	3: Culture and Ethnicity, pp. 52–57 7: Social Stratification, pp. 165–170 10: The Family, pp. 248–251 19: Technology and Social Change, pp. 513–516
Contact hypothesis and relationships between members of different status groups	3: Culture and Ethnicity, pp. 59–67 5: Social Interaction and Social Relationships, pp. 113–118 8: Racial-Ethnic Stratification, pp. 209–210 18: Communities and Urbanization, pp. 496–503

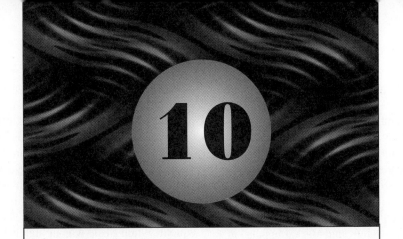

The Family

CRAIG WHITE COULDN'T DECIDE where he should go for Thanksgiving vacation. He wanted to spend the holiday with his family—but which family? His mother, Rachel, and stepfather, Phil, wanted him to come see their new house and become better acquainted with Phil's parents. His father, Joe, and stepmother, Lea, wanted him to come meet his new baby half-brother. He knew that his grandparents would also love to have him visit. In addition, his sister, Elaine, and Mark, the man she had lived with for the past five years, had invited him to their house in a nearby town.

Not surprisingly, Craig sometimes found it difficult to talk about his family with other people. What should he call his stepfather's parents? He was very fond of them, but calling them his grandparents seemed strange, especially since both his mother's and his father's parents were still living. Besides, his biological grandparents certainly wouldn't care to hear him calling anyone else Grandma or Grandpa. And then there was Mark, Elaine's boyfriend, whom he liked a lot and with whom he could spend hours shooting hoops and trading information about the latest software. But he had to be careful how he talked about Mark, too, especially with certain relatives who had never been told that Elaine and her boyfriend were living together.

Craig's family life had once been a lot simpler. When he was very young, the Whites were a "model" family—two parents and two happy children. But when Craig was four and Elaine was eight, their parents divorced. As Craig grew older, he continued to love both Joe and Rachel, but sometimes he couldn't help resenting the loss of the time when they were all together. He knew he wasn't to blame for the divorce, yet he often felt a little guilty and ashamed that he came from a "broken" family. In addition, relationships with family friends were often complicated, for they seemed to be divided into opposing camps—either Joe's friends or Rachel's friends.

After the divorce Craig and Elaine lived with Rachel but spent most weekends with their father. Joe was good about giving Rachel the child support money the court had awarded in the divorce. Although Craig was too young to know it, several of his playmates weren't so lucky; their fathers only sporadically gave any financial help, and some of them simply disappeared. Still, it was hard for Rachel to make ends meet, and, when Craig was six, she and the children moved to a less expensive but rougher neighborhood. Both Craig and Elaine continued to see Joe regularly on their weekend visits and for camping trips and other outings. Meanwhile Rachel obtained training in office skills and eventually landed a good job. Gradually life settled into a new routine.

When Craig turned twelve, his life changed again. His father married Lea, a woman he had met at work. Lea was quite a bit younger than Joe and had no children of her own. Although she tried to be nice to Craig, he was never quite sure how to relate to her, and the weekend visits with Joe became less frequent. When Craig was in high school, Lea and Joe had a baby daughter. The child support checks appeared less regularly, and Craig knew that his mother was distressed and angry.

When Craig was fifteen, Rachel married Phil. Phil had three daughters from his previous marriage, all of whom lived with their mother. Phil had little in common with Craig and Elaine, and he seemed to disapprove of their friends, their taste in music, and even the way they dressed. Family life was even more strained when Phil's daughters visited during school vacations. The house was crowded, the girls didn't seem to like their stepmother, and the tensions spilled over into the girls' relationships with Craig and Elaine.

Craig was relieved when he graduated from high school and went away to college, even though paying for school wasn't easy. Neither of his parents was in a position to help much, and Phil's money was already committed to his daughters. The only way Craig could stay in school was to work long hours during the summer and hold a part-time job during the semester. Somehow he couldn't help resenting that, too, especially when many of his friends had to work only to earn spending money.

Elaine hadn't made it to college. After high school she took a job as a clerk at a clothing store in the local mall, where she met Mark. Before long she and Mark moved in together. Although they talked vaguely about getting married someday, it wasn't at all clear whether that would happen. Sometimes Craig thought that their parents' divorce had left Elaine feeling ambivalent about committing to a marriage.[1]

As for himself, Craig often wondered what his future family life would be like. On the one hand, he thought it would be nice to be married and have kids; on the other, he was worried that he might end up divorced like his parents and estranged from his children. Things seemed so much clearer to his grandparents' generation. Are relationships today destined to be only temporary? Are traditional two-parent families a thing of the past? Have divorce, remarriage, and single-parent families become the norm?

THE EXPERIENCES OF CRAIG and his relatives reflect what sociologists call the social institution of the *family,* the area of social life that includes the relationships between kin and is responsible for nurturing young children. Craig's experiences also illustrate many of the changes that families in the United States have confronted in recent decades. In this chapter we explore how sociologists look at the family and show how their analyses can help us understand Craig's experiences, as well as our own. We focus specifically on the varieties of families, the changes in family arrangements over time, and the ways in which these changes affect our lives. We take a macrolevel view of the family, looking both at different family forms cross-culturally and at changes in the United States during the past century. We then move to a mesolevel perspective, looking at how individual families can be structured and how these structures can change over time. Finally, we move to a microlevel perspective and look at how different types of family structures can affect our individual life chances and especially the lives of children. First, however, let's examine precisely what sociologists mean when they talk about social institutions.

DEFINING SOCIAL INSTITUTIONS

Imagine that you were on a space ship that became stranded on a distant planet with no hope of rescue. Fortunately your new planet is much like Earth in its atmosphere and habitat. Gradually you and your fellow passengers realize that if you are to survive, you will have to become self-sufficient—cooperate to get food, build shelters, make clothing, and so on—to ensure the group's survival. In short, you will need to establish a society.

Your new society will require ways to settle quarrels and maintain order, as well as deal with any other inhabitants of the planet. You will need to develop rules about who can mate with whom and how children will be raised and schooled. In time you will undoubtedly develop explanations of your existence to pass on to future generations—how you came to be on the planet and why citizens of your society should follow its rules. In doing all this, you will gradually develop what sociologists call **social institutions,** complex sets of statuses, roles, organizations, norms, and beliefs that meet people's basic needs. In sociological terms, social institutions are not specific organizations, such as particular universities, churches, or hospitals. Rather, they are broad areas of social life, such as the family, the economy, political systems, and religion, involving norms, statuses, and roles.[2]

In many ways the institutions that your new society creates will probably resemble the ones you knew back on Earth. Yet they might also evolve in unique ways to suit the special circumstances of life on another planet. Whatever shape they take, these social institutions will greatly influence the way your society works. Of course the same thing is true of earthly societies: Societies everywhere meet their basic needs through social institutions. Sociologists and anthropologists have suggested that all societies have several basic social institutions including these:

1. The government or polity, which defines the legitimate use of power and the ways in which order will be maintained

2. The economy, which defines how goods will be produced, distributed, and used

3. Religion, which defines our relationship with the supernatural

4. Education, which is responsible for the training of younger generations

5. The family, which regulates adult sexual relations and the reproduction and rearing of children

6. Medicine, which promotes the health and well-being of the citizenry.[3]

As you would expect, the nature of social institutions varies from one society to another. In general, in comparison with pre-industrial societies, modern, industrialized societies are much more *differentiated.*[4] In Chapter 1 we discussed Durkheim's use of the idea of differentiation to describe general changes in societies that occur with industrialization. In this chapter and those that follow, we will use the notion of differentiation in a more specific way to describe how social institutions become distinct or differentiated from one another. For instance, our society has a highly developed system of organizations devoted to social control, with a variety of law enforcement agencies at the local, state, and national levels. In contrast, smaller pre-industrial societies, such as hunting and gathering groups, generally depend on family or kin groups to maintain social control and order. In our society religion is practiced mostly under the auspices of religious organizations, many of which have millions of members. Other societies have had vastly different arrangements, often involving the kin group as the centerpiece of religious activities. Among the ancient Greeks, the father not only served as head of the family but also assumed the role of priest in the family worship of ancestors and even performed marriage ceremonies for his sons.[5]

The family is considered the most basic social institution because it is found in all societies. How might this Thai family be similar to the one in which you grew up? How might it differ?

Other social institutions have also become more differentiated over time. For example, with industrialization, factories and other organizations developed that required large numbers of people to work outside the home, and work became more complex. As a result the economy, as a social institution, became differentiated from the family. The production of goods and services no longer occurred within the kin group but in separate organizations and groups. Separate norms, roles, and expectations specific to the economy, rather than the family, also developed. Similarly, as technical knowledge and skills expanded, the family was no longer capable of transmitting all this information. Thus education became differentiated from the family, developing its own complex set of statuses, roles, and organizations. In the same way, during the last century, medical care began to more frequently occur outside the family, so that today medicine can also be seen as a social institution.

The growing differentiation of institutions does not mean that these institutions become completely independent of one another. Within any society social institutions are connected by a complex web of interrelationships. Changes within one institution, such as the family, will affect other institutions such as education and the economy, and vice versa. In addition, changes that affect one social institution also affect others. For example, stratification based on social class, race-ethnicity, and gender affects our experiences in the family, the economy, politics, the medical world, and even religion. At the microlevel ideas and theories regarding social interaction, such as exchange theory and role theory, can help explain our experiences within each social institution.[6]

Many sociologists and anthropologists consider the family to be the most basic social institution because it can be so clearly identified in all societies. Within all societies kinship ties and rules about who should care for children are clearly understood.[7] In addition, as noted previously, tasks now embedded within other social institutions, such as the economy, education, and religion, were once included within the sphere of the family in many societies. For these reasons we'll begin our exploration of social institutions with the family. In the next chapter we look at formal organizations, such as schools, churches, and businesses, that often develop within social institutions and are deliberately created to accomplish certain goals. In subsequent chapters we turn to the institutions of religion, the political world, the economy, education, and medicine.

All social institutions can be studied at three broad levels of analysis, each of which reveals different aspects of the area we are studying. From a macrolevel perspective we can look at the general features of social institutions and the ways in which they vary from one society to another. Moving in a bit closer, we can take a mesolevel view and study organizations or groups within institutions, such as individual families. Finally, we can take a microlevel view, focusing on how institutions affect us as individuals. For example, we can explore how we see ourselves as part of our families or how the norms or stratification patterns within social institutions affect our individual life chances. In this section of the book we will use each of these levels of analysis to look at the family and other social institutions.

SOCIOLOGICAL STUDY OF THE FAMILY

Of all the topics in this book, the family is probably the one with which you are most familiar. Each of us is part of a family and an extended kin network, and we often have strong emotions related to this area of our lives. Like Craig, many of us have derived not only joy and companionship from our families, but also sadness, pain, and frustration. In addition, our thoughts and feelings about our families both reflect and evoke some of our most deeply held values and beliefs. Regardless of what kind of family we come from, we all have images of what families *should* be like. The form of a family within a society is supported by deep-seated cultural beliefs and practices, as well as by the formal norms we call laws. Children learn what families are as part of the socialization process, not just from their experiences within their own families but from the media and friends. Actually, the issue of contemporary family life in the United States is particularly complex because, as we will see, cultural images of "the family" lag behind the reality of what families actually *are*.

Craig, for example, felt self-conscious about his "broken" family. Yet the reality is that only slightly more than half of all children today live in a "traditional" nuclear family with their mother and father and, perhaps, brothers and sisters.[8] In recent years, other kinds of "nontraditional" relationships have emerged. For example, many children live with a parent and a stepparent, and many others are raised by single parents. Many more couples than in the past live together without marriage. Gay and lesbian couples establish households and maintain committed, monogamous relationships, and many of these households include natural or adopted children. These social changes, involving as they do perhaps the most basic unit in society, are profoundly important and, to many people, profoundly disturbing. It is little wonder that politicians have seized on "family values" as a campaign issue in recent years. Too often, however, such political debates are not informed by the kind of perspective involved in sociological inquiries.

Like all of us, sociologists experience the joys and sorrows of family life, and they often have strong values they try to live by. But, in their work as sociologists, they also try see the family not as the realm in which they were raised and in which they might be raising their own children, but instead as a social institution that appears in a great variety of forms both within and across societies. Whatever their own family experiences and values, sociologists' goals are to describe and understand different kinds of families as they actually are.

Inevitably this work involves sociologists in family-related social issues. Sociologists want to know, for example, what effect various family arrangements have on the life chances of the children brought up in those families. Specifically they might ask whether children who are raised by single mothers or by a parent and stepparent have different life chances than other children do and what could account for these differences. In doing so, however, researchers rely as much as possible on established methods and reliable data, rather than their preexisting beliefs, to answer such questions. As in other areas of sociology, sociologists attempt to take a dispassionate approach to an intimate and often emotional subject.

Sociology, however, does develop in a specific historical context. Although sociologists have long been fascinated by the different types of families found in societies around the world, until relatively recently most sociological research in the United States focused on the traditional nuclear family. For instance, to date only a fraction of the research on the family has involved gay and lesbian couples.[9] One future challenge for sociologists is to test the extent to which the insights they have developed regarding traditional families apply to developing and ever-changing family forms.

When thinking about the various types of families in the United States and around the world, I sometimes picture a closet full of sweaters of various patterns, materials, and colors. All the sweaters are designed to meet the basic need of providing comfort and warmth, yet each can look and feel very different from the others. Some are a solid color, and others are woven out of many different-colored yarns that you can distinguish only from up close. Some have complex and tight stitches while others are very loosely woven. Some are made of heavy material that can withstand the coldest of weather whereas others provide warmth only on mild days and need to be supplemented with other sweaters or a coat when the temperature drops. Families, too, are woven in many different ways and in a variety of designs. But, just as all sweaters are intended to provide warmth, families everywhere represent the basic way in which society provides for the care of children and close relationships between kin. We turn now to analyses of these many forms of family life.

VARIATIONS IN FAMILY FORMS: A MACROLEVEL VIEW

What constitutes a family? Do families everywhere perform the same functions for their members and for society at large? How do family forms change over time, and how do these changes relate to changes in other social institutions? To answer such questions, sociologists look at families from a macrolevel perspective. In this section we first examine varieties of families across cultures and then explore how the family has changed in the United States since pre-industrial times.

Cross-Cultural Variations in the Family

In looking at the family cross-culturally, social scientists are often interested in how families are arranged and who is seen as part of the family and kin group. Sociologists are also interested in the functions that the family plays in society. Much as you might compare different styles of sweaters to see which might work best in various situations, so have sociologists tried to understand why certain family types have developed in different environments and societies.

Nuclear Families and Kin Groups   How is "family" defined in different cultures? In Chapter 3 we described how the **nuclear family,** consisting of a mother, father, and their children, exists in virtually all societies and is the basic unit for child rearing. All societies also recognize a broader range of relatives, or what social scientists call **kin.** The definition of who is kin, however, and the obligations of various kin toward children, vary a great deal from one society to another. For instance, the traditional Navajo family is matrilineal and matrilocal; that is, kin are counted through the mother's side of the family and newly married couples live with the mother. This structure is very different from that of the traditional patrilineal and patrilocal Chinese family.

Nuclear family units can also be combined into larger kin units.[10] Sometimes they are combined through the practice of plural marriage, generally with the husband having more than one wife, a practice called **polygyny.** Scholars have investigated the incidence of polygyny in societies throughout the world by using the data compiled by George Murdock in the Human Relations Area Files (HRAF), which were first described in Chapter 3. At least three-fourths of the societies included in the HRAF are polygynous.

Even so, most of the world's people practice **monogamy,** with marriages involving only one man and one woman. (The term *monogamy* is formed from the root *gamy,* which means "marriage," and the prefix *mono,* which means "one.") This occurs because the polygynous societies are relatively small and largely disappearing in the face of global industrialization. In addition, even within societies that value polygyny, only a few men can afford the prac-

tice. After all, it's expensive to support several wives and children, and polygynous societies generally have strict norms that require husbands to treat all their wives and children similarly. Further, given that all societies have approximately equal numbers of women and men, as well as norms that support marriage for all adults, there generally aren't enough extra women to allow more than a few, usually wealthy, men to practice polygyny.[11]

Another way in which nuclear families can combine is through developing extended families, as in traditional Navajo and Chinese society. In such societies a newly married couple goes to live with the family of the husband or wife. In extended families priority or weight is given to blood ties (relationships with a person's own extended family) rather than to marriage ties, in what is called a **consanguine family system** (the term *consanguine* means "of the same blood or origin" or "descended from the same ancestor").[12] Thus, if you were part of a traditional Navajo family, you would see yourself as belonging to your mother's clan, and your relationships with your brothers and sisters would supercede the ties with your spouse.

In contrast, in modern industrialized societies such as our own, priority generally is given to marriage ties rather than to blood ties, in what is called a **conjugal family system** (the term *conjugal* comes from the Latin for "husband" or "wife").[13] Marriage vows in our society emphasize the nature of this conjugal system as we promise to "cleave only unto" our new husband or wife, and part of the usual transition to married life is the establishment of a household totally separate from the homes of either set of in-laws.[14]

Functions of the Family   What accounts for the particular forms that families take in different cultures? As explained in Chapter 3, when sociologists investigate similarities and differences across societies, they often use the ideas of structural functionalism, looking for the function, or part, that structures play in maintaining or altering a society. They also use the theory of cultural materialism to help explain why societies develop different family structures, suggesting that different family forms reflect the nature of the material resources of a society's physical environment, just as a different type of sweater might be more appropriate for one type of climate than another.

In general, sociologists have suggested that, whatever particular shape or structure the family might have within a society, the nuclear family has three major functions.[15] First, in all societies the family is responsible for the *economic support of its members,* including the provision of shelter, food, and clothing.

Although all Shi'ite Muslim men technically are allowed to practice polygyny, only a few actually do so. Why do you think the Pakistani Shi'ite man pictured here was able to have two wives?

When you were a child, you could not have survived without the food your parents gave you, the clothing they provided, and the house you shared with them. Similarly, the child support agreement Craig's parents reached after their divorce clearly specified how his parents were responsible for this function.

Throughout the world nuclear families share a common residence and cooperate in economic activities. In pre-industrial societies families often work together to produce needed goods and services. In industrial societies parents work outside the home for pay, which they then use to purchase necessary goods. In describing this change, sociologists often say that differentiation has occurred, with the family changing from a *productive* to a *consuming* economic unit. In this process the economy, as a social institution, has taken over many of the productive tasks once performed by the family.[16]

Although cultural beliefs and value systems can offer alternative explanations for the forms that families take, the various types of family structures that societies have developed seem to be highly related to their economic activities. For instance, polygyny apparently is more highly favored in pastoral, or herding, societies, where having more wives is both more economically efficient and more profitable for the family. In these societies, while the men and boys can herd numerous sheep or other animals, several women may be needed to process all the meat, hides, and milk so many animals produce, as well as keep the household going. In contrast, in hunting and gathering or industrial societies, a small family unit is just as efficient as a larger one, and so monogamous relationships seem to be favored.[17] According to structural functionalists and cultural materialists, then, the kind of family most of us have come to view as "natural" actually reflects specific material and economic conditions.

The major exception to the conjugal family pattern within our own society is also related to the economic function of families. Recall the discussion of Michael Allen's research on the corporate rich in Chapter 7 and the ways in which members of the corporate rich pass their wealth to their descendents through complex systems of trusts. Membership in the capitalist class usually depends on one's ancestors. Thus members of the capitalist class are much more likely than the rest of us to emphasize their consanguine ties—their relationship with the extended family that provides their wealth. And, in fact, cross-culturally, consanguine families are much more likely to be found in societies in which wealth and property are transferred across generations.[18] Thus the traditional Chinese family passed on its household and farm tracts from generation to generation, just as the traditional Navajo family passed on its herds and grazing areas. In contrast, most people in modern industrialized societies, as well as in hunting and gathering societies, rarely have enough wealth to worry about its preservation through the generations.

The second function of the family is related to our psychological and emotional needs. The family, especially in modern industrialized societies, is an important source of *emotional security,* for it is a primary group and the site of intimate, face-to-face, and long-lasting interactions. As you will recall from previous chapters, primary group relationships provide us with emotional support and guidance. Of course the family can also be the site of great sadness and dismay. This simply reflects the emotional significance the family has for us. Craig, the young man described at the beginning of the chapter, has experienced great joy and companionship in his relations with his family, but also sorrow and pain.

The family members who provide emotional support and security can vary from one society to another. For instance, within our own society, with its conjugal nuclear family system, much of our emotional support typically comes from our parents when we are young and from our spouse and children when

Even within a single society such as the United States, families come in many different forms. How does each of the families pictured here fulfill the functions of the family that sociologists have identified?

we are older. If, however, you were a woman in a polygynous family, you might get most of your emotional support from your co-wives. If you were a man in an extended consanguinal family, you might get most of your emotional support from your brothers and parents. In any case, what's important is that the support originates in the family. No other social institution includes role relationships that can provide this type of intimate, emotional support.[19]

Finally, the family is responsible for *regulating reproduction and child rearing.* Even though many societies allow young people a great deal of sexual freedom before marriage, sociologists have found no society that encourages childbearing and child rearing outside of the family. All societies have norms that regulate whom you may or may not marry and that determine who is responsible for raising children. The norms regarding marriage often involve what are called rules of endogamy and exogamy. **Endogamy** is a rule that requires people to select marriage partners from within their own tribe, community, social class, racial-ethnic, or other such group. **Exogamy** is a rule that requires mate selection outside the group. (The terms are formed from the root *gamy,* which means "marriage"; the prefixes *endo* and *exo* mean "inside" and "outside," respectively.) For instance, our own society is endogamous to the extent that people are more likely to marry others with similar educational, religious, racial-ethnic, and social class backgrounds than to marry people with different characteristics.[20] Yet our society is also exogamous, for both laws and mores prohibit marriages between close relatives.

No matter what specific form the norms regarding marriage take, they all result in the linkage of individuals—the formation of social networks composed of individual men, women, and children, as well as larger kin groups. In other words, the rules regarding family formation and kinship help promote social ties and bonds between people within a society. In this way the norms ensure that all people have some type of primary group membership and that this primary group involves people from both sex groups and of different ages.

These norms also ensure that children will be cared for and, especially, that young children are provided with the nurturance, or physical and emotional support, essential for survival and good health. Clearly no society could sustain itself if no one cared for its children. Further, as you will remember from previous discussions, children's bonds with their parents are an important predictor of their development of a self-identity, as well as a conscience and self-control. Many sociologists suggest that providing for the nurturance of children is the fundamental function of the family. This function of the family appears universally, and attempts to substitute other arrangements, such as caring for infants in orphanages or other formal organizations, have never been successful.[21]

Again, however, the way in which societies structure the early socialization of children can vary widely. For instance, the Nayar, a warrior society that lived on the southern coast of India, had a polyandrous family structure. (**Polyandry** is a family system whereby a woman can have multiple husbands.) The women lived in large houses with their sisters and all their children, no matter who the father was. Because the Nayar were a caste of professional soldiers, all the young men were away for most of the year fighting or training for battle. Most of the men's time at home was spent tending to responsibilities associated with their matrilineage, that is, with their mothers and sisters and their children. Each child had a designated father, who would give small gifts to the child and pay small fees associated with the birth. Nevertheless, the responsibility for child rearing fell on the mother and her eldest brother, who was responsible for the household. Observers of the Nayar structure suggest that this system was functional for that society because the men were away from home a great deal. This family arrangement allowed them contact with their wives when they returned, linked all individuals within the society into family units, and provided a stable home for the children.[22]

If we take a macrolevel view of the family in the United States today, we can see that it performs each of the three functions we have been considering. At the same time, the family is a rapidly changing institution, a fact that is reflected in contemporary debates about "family values." Exactly what do these changes mean, especially for the lives of children? To understand more about this issue, we need to take a historical overview of the family in the United States.

The Changing Family in the United States

During the past two centuries the United States has undergone dramatic transformations in the economy, in education, and in the health of its citizens. As you will recall, the United States, like many other countries, has changed from an agricultural to an industrial society. As workers left the farms and migrated to the towns and cities, they also moved into wage-paying jobs outside the home. During the first part of this transition, men were far more likely than women to be employed outside the home, except among the poor, especially immigrants and racial-ethnic minorities. In recent decades, however, many more women, especially mothers of young children, have entered the paid work force. This development appears to be a response to the greater availability of jobs seen as appropriate for women and the rising wages of women's jobs, as well as the stagnation and even falling rates of men's wages.

As the Industrial Revolution progressed, education also changed dramatically. Two centuries ago many children never attended formal schools, and only the very wealthy had a college education. By the twentieth century most states had some form of compulsory education for children, and the amount of schooling that young people received increased a great deal.

In addition, there have been enormous changes in people's health and life expectancy. In colonial times women commonly died in childbirth, many infants died before reaching adulthood, and the average life span of adults was much shorter than it is today.[23]

Collectively these changes have altered family life in several ways. Our chances of marrying and the age at which we marry, our decision whether to have children and the number of children we have, the likelihood of our marriages ending through divorce or the death of a spouse—all differ markedly from the experiences of earlier generations. And because of these developments, the environments that children experience as they grow up have also changed dramatically.

Getting Married Historically, since data were first compiled in the mid-1800s, more than 90 percent of all women and men have married at some point in their lives, although the actual percentages have varied over the years. For instance, people who came to adulthood in the 1950s were more likely than those born before or since to marry, and statistical projections indicate that people who have reached adulthood in recent years are less likely to marry than those in previous generations.[24] The age at which people marry has also varied over time, as shown in Figure 10-1. In the late 1800s the average groom was twenty-six years old and the average bride was twenty-two. These figures had dropped substantially by the 1950s but in the 1970s began to rise again, so that today the average groom is again twenty-six years of age while the average bride is twenty-four.[25]

Farm machinery such as this tractor and mower made it possible for only a few farm workers to cut and bale fields of hay. Today, farm machinery is even more complex, requiring even fewer workers to handle very large farms. These technological changes had enormous impacts on social institutions, including the family.

Underneath these general patterns we can also detect varying trends for different social groups. For example, African Americans are much more likely than Euro-Americans to delay marriage or to never marry at all. Sociologists estimate that among women born in the 1950s, about 91 percent of Euro-Americans but only 75 percent of African Americans will ever marry. This is in sharp contrast to previous years: Until about 1950 African-American women were more likely than Euro-American women to marry.[26]

What accounts for these changes? Given the importance of economic factors in helping to shape family arrangements, you might suspect that changes in the economy are a likely cause, and you would be right. Wages, after adjusting for inflation, have fallen over the past two decades. This decline has most affected men, especially African-American men. Many sociologists suspect that these economic shifts are a major influence on the decision of many people to delay marriage or simply not marry at all.[27]

Even though people are marrying at later ages, they have not delayed sexual activity. Beginning in the late 1970s, when people began to get married at older ages, the rate of **cohabitation**—the practice of men and women living together without marriage—also began to increase. In fact, the increase in cohabitation balanced out the tendency for young adults to delay marriage. Men and women in their twenties are just as likely to be sharing a household with a partner now as they were in the 1970s; they are simply less likely to be legally married.[28] Studies of people who are cohabiting indicate that the vast majority intend eventually to marry their current partner.

Thus sociologists suggest that cohabitation can be seen as a new stage in the development of families, one that simply precedes the step of legal marriage.[29]

Just as cohabitation has become more common in recent decades, so have homosexual unions (in which two men or two women live together in a committed relationship) become more open. Because homosexual marriages are not legally recognized in the United States, the government has no data on how many homosexual couples live together or how rates have changed over the years. In recent years, however, gays and lesbians clearly have been able to be more open about their relationships, and homosexual unions now represent another viable alternative to traditional marriage.[30]

In general, sociologists suggest that the rising incidence of singlehood, the increase in cohabitation, and the increased visibility of homosexual unions reflect a growing tolerance for different family forms. In comparison with previous times, especially the 1950s when the vast majority of men and women were in heterosexual marriages, adults today generally have many more options in choosing a family lifestyle.

Having Children Since 1800 the number of children women bear has declined markedly as more children have survived infancy and fewer were needed to help with work on family farms. Figure 10-2 illustrates this trend. As you can see, the **general fertility rate**—the number of births in a given year for every 1,000 women of childbearing years—fell fairly steadily throughout the nineteenth and early twentieth centuries. It dipped even lower during the 1930s and the Great Depres-

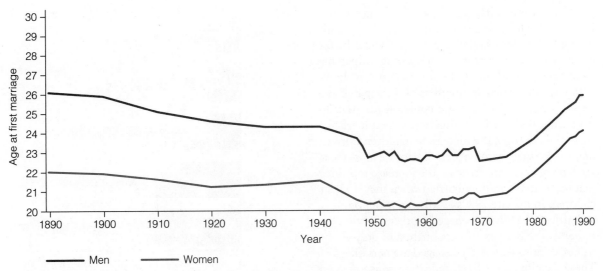

FIGURE 10-1 *Median Age at First Marriage, by Sex, 1890–1990*

Source: U.S. Bureau of the Census, 1975, p. 19; 1994, p. 103; 1997, p. 106.

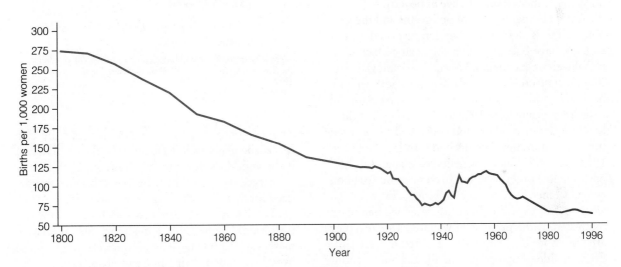

FIGURE 10-2 *General Fertility Rate for Euro-Americans, 1800–1996*

Note: The general fertility rate is the total number of births in a given year for every 1,000 women ages fifteen to forty-four, the years in which women are most likely to have children.

Source: U.S. Bureau of the Census, 1975, p. 49; 1994, p. 78; 1995, p. 76; 1998, p. 77; Ventura, Martin, Mathews, and Clarke, 1996, p. 28.

sion, rose sharply during the post–World War II years, but since the 1960s has dropped to levels that are even lower than in the 1930s.[31]

The fertility rate has declined for all racial-ethnic groups. African-American and Euro-American women now have very similar numbers of children, but in recent years they have tended to have their children at different ages. Through the 1950s Euro-American and African-American women bore children at about the same age. Beginning in the 1960s, however, Euro-American women began to delay childbirth whereas African-American women did not. By the mid 1990s, on average, Euro-American women were a little more

than two years older than African-American women when they had their first child.[32]

As noted previously, African Americans are more likely than Euro-Americans to postpone marriage. At the same time, they are less likely to postpone childbearing, so that African-American women are much more likely than Euro-American women to have children while they are unmarried. In the mid 1990s, more than two-thirds of all African-American children and about one-fourth of all Euro-American children were born to an unmarried mother.[33]

Both Euro-Americans and African Americans are more likely to have children out of wedlock now than they were thirty years ago. If Craig's sister, Elaine, were to have a baby with Mark, her grandparents might be scandalized, but she would be far less atypical than when her parents or grandparents were having children. As Figure 10-3 shows, however, births to unmarried mothers over the past three decades have been far more common for African Americans than for Euro-Americans. Despite rhetoric to the contrary, the increase in children born to unwed teenagers hasn't occurred because more teens are having children. Instead, the increase indicates that teens aren't getting married when they do bear children. In the late 1980s less than one out of every ten African-American teenaged mothers married the father of the baby before its birth, and only about half of all Euro-American teenaged mothers did so.[34] These developments have important consequences for parents and children alike. As you will recall from Chapter 8, single-parent families are much more likely than other families to live in poverty.

A popular perception of politicians and the general public is that single women, especially poor women, have babies so that they can collect additional welfare payments. However, sociologists cite three facts to refute this claim. First, both welfare payments and the incidence of single parenthood rose dramatically from 1960 to 1974. After that time, however, single motherhood continued to rise even though welfare benefits declined. In fact, between 1972 and 1992, welfare benefits fell 26 percent. Second, single motherhood has become more common among women in all social class groups. Even though we wouldn't expect college-educated women to be motivated by the promise of welfare payments, they are also more likely now than in 1960 to have children outside of marriage. Finally, European countries have much more generous welfare payments for single mothers than the United States does, but the United States has a much higher percentage of single teen mothers. (Some countries, such as Denmark, France, and Sweden, have a higher percentage of births to unmarried women than does the United States. But in these countries teen pregnancies are extremely rare. In addition, the parents are usually living together on a permanent basis, so these children are much more likely to have mothers who have completed their education and established work careers and to live with both parents.)

Thus sociologists generally suggest that the increase in births outside of marriage reflects changing economic conditions and social norms rather than rising welfare benefits. The increasing economic independence of women and the declining wages of men, a well as changing norms regarding women's independence and the importance of marriage, are probably important contributors to the rising incidence of

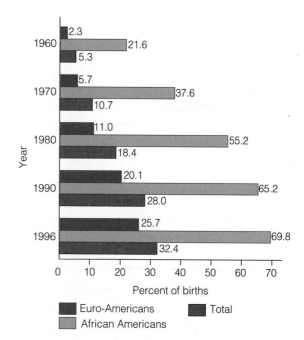

FIGURE 10-3 *Percentage of Births to Unmarried Mothers, by Race-Ethnicity, 1960–1996*

Source: For 1960–1970, Bianchi, 1990, p. 9; for 1980–1990, U.S. Bureau of the Census, 1997, p. 79; for 1996, Ventura, Peters, Martin, and Maurer, 1997, p. 3.

single parenthood.[35] If Elaine did have a baby with Mark, she might well decide to raise the child herself if she and Mark never married. A couple of generations ago, she would have been much more stigmatized for that decision than she would today.

Finally, during the past two centuries, as the fertility rate declined, life expectancy also increased markedly. Parents have had fewer children and also lived longer after their children were grown.

All these changes have resulted in contemporary women spending a much smaller proportion of their lives caring for children than did women in previous generations. In addition, couples live many more years without their children in the home. In the nineteenth century you could have expected to have your children living with you for most of your life, and you might well have died before the youngest was grown. Today you can expect to live with your own children for only a fraction of your adult years and to survive for many years after they are grown. As a result of these changes, a much larger proportion of the population today lives in homes without children. Only about a third of all households include children under age eighteen.[36]

Ending a Marriage If you had been married in the mid 1800s, you would have been very unlikely to divorce your spouse. Since then, however, your chances of divorcing would have increased every

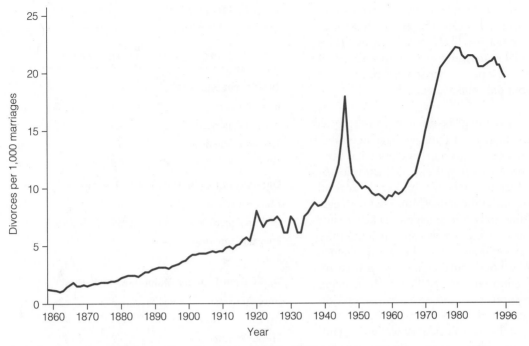

FIGURE 10-4 *Annual Divorce Rate, United States, 1860–1996*

Note: The divorce rate is calculated as the number of divorces per 1,000 existing marriages.

Source: For 1860–1920, Jacobson, 1959, p. 90; for 1920–1939, U.S. Bureau of the Census, 1975, p. 64; for 1940–1990, Clarke, 1995, p. 9; for 1991–1995, U.S. Bureau of the Census, 1998, p. 111.

year. Figure 10-4 shows how the annual **divorce rate**—the number of divorces in a year relative to every 1,000 existing marriages—in the United States has risen almost continually from 1860 to the 1990s. The spikes in the graph around 1866, 1920 and the late 1940s indicate a temporary rise after the Civil War and the two world wars; data from other countries show a similar pattern after wars. This temporary increase largely results from marriages hastily entered into during times of war (so-called wartime romances). Such marriages often dissolved once the partners had time to reflect on their actions and got to know each other better. After dipping a bit from 1950 to 1960, the divorce rate has risen sharply, reaching its highest point in 1979. Since that time there has been only a small decrease.[37] Over the years divorce has become more common for members of all racial-ethnic groups, but African Americans are more likely than either Euro-Americans or Hispanics to divorce or separate.[38]

Although the United States has always had a higher divorce rate than the European countries or Canada, recently the divorce rate also has risen there. Divorce simply seems to be more common throughout the industrialized world in recent years.[39] These changes seem to be largely related to the growing economic independence of women. As women have worked more outside the home and, especially, as their wages have grown, they have become better able to afford to live independently. Attitudes toward divorce also have changed, with people being much more accepting of divorce today than they were in earlier generations. Note, however, that this change in attitude has largely occurred after, rather than before, the sharp rise in divorce.[40]

For many years the death of a spouse was the most common way in which a marriage would end. But, at the same time that the divorce rate was rising, the mortality or death rate was falling. From 1860 to 1970 these two changes balanced each other out. In the early 1860s the *dissolution rate* for marriages (a measure of how many marriages ended by either divorce or death) was 33.2 per 1,000 marriages; in 1970 the rate was 34.5. After 1970, however, the divorce rate continued to rise while the mortality rate stayed about the same. By the mid 1970s, for the first time in U.S. history, more marriages were ended by divorce than by death.[41]

The end of a marriage, by either divorce or death, has not meant that the individuals involved will always live alone. Historically, remarriages have been common in the United States. For a long time almost all remarriages involved people who had been widowed, but the situation has changed rapidly in recent years. About one-third of all brides have been divorced, and people who have been divorced are actually more likely than those who have never been married to cohabit. Like Craig's parents, about two-thirds of divorced women and three-quarters of divorced men eventually remarry, although African Americans are much less likely than Euro-Americans to do so.[42] **Serial monogamy** refers to this pattern of marriage in which a person has several spouses over the life time, but only one at a time.

Growing Up Sociologists do not always agree about the effects of the trends we have been discussing, but in recent years much attention has been focused on how changes in the social institution of the family affect the lives of children.[43] Several trends are evident in the data. Compared with earlier eras, children today are much more likely to live in smaller families. Children also are much healthier than in previous eras and have a far greater chance of living to adulthood. Further, because people are living longer, children are very likely to have living grandparents, and four- and even five-generation families are not at all uncommon.[44] At the same time, because of the rising incidence of divorce, children are also more likely to have only one parent in the home. Even when two parents are present, both of them are more likely to work outside the home. In addition, because a high percentage of divorced parents remarry, many children, like Craig and Elaine, become enmeshed in large kin structures, with stepparents, stepsiblings, and assorted other new relatives. Government estimates indicate that about one out of every seven children currently lives with a parent and a stepparent.[45]

The type of family in which we grow up is often called the **family of orientation,** because it is the family that orients us to the world. (It is often contrasted with the **family of procreation,** the family in which we live as adults and in which we procreate, or have children.) The family of orientation has varied greatly over the years. One significant change is the reduced amount of time children spend with their parents, helping with farm and household chores. During the past two centuries, first fathers and then mothers became employed outside the home, and children began to spend more time away from home in school. In short, as the social institutions of the family, the economy, and education became more and more differentiated, children spent an ever-greater proportion of their lives away from parents.

Most recently a higher percentage of children have been growing up in homes without the father present. Table 10-1 shows the sharp increase since 1970 in the number of families with children that did not have both parents in the home. As you can see, in 1970 the majority of families with children, among both Euro-Americans and African Americans, included both a mother and a father. Since then, however, the situation has changed dramatically,

TABLE 10-1   *Family Types by Race-Ethnicity, 1970–1997*

	YEAR			
	1970	**1980**	**1990**	**1997**
Two-Parent Families				
Euro-American	90%	83%	77%	74%
African American	64	48	39	36
Hispanic American	—	74	67	64
Total	87	79	72	68
One-Parent Families Maintained by Mother				
Euro-American	9%	15%	19%	21%
African American	33	49	56	58
Hispanic American	—	24	29	31
Total	12	19	24	27
One-Parent Families Maintained by Father				
Euro-American	1%	2%	4%	5%
African American	3	3	4	7
Hispanic American	—	2	4	5
Total	1	2	4	5

Note: Data are not available for Hispanic Americans for 1970.
Source: Data for 1970 are from U.S. Bureau of the Census, 1995, p. 61; Data for 1980–97 are from U.S. Bureau of the Census, 1998, p. 65.

especially for African Americans. Today children in the United States are twice as likely as they were in 1970 to live with only their mother.[46] Even though this trend is appearing in all industrialized countries, single-parent families in the United States are much more likely than those in other societies to have a teenage parent and to live in poverty.[47]

The picture of families that we see from a macrolevel perspective suggests that during the past two centuries, and especially in the latter part of the twentieth century, families in the United States have become much more diverse. The most dramatic change, and one that appears to be unprecedented cross-culturally, is the rapid increase in the number of children living in homes without a father. Historically, a number of societies have had high divorce rates, such as rural Japan before the mid-eighteenth century and Malaysia, Indonesia, and many Arab countries before rapid industrialization. Yet these societies also had a very high rate of remarriage, thus keeping most men and women in family situations and helping ensure that children would experience economic security and day-to-day relationships with both a mother and a father.[48] We will return to these issues later in the chapter, but first we will look more closely at family interactions and the family as a social group.

Allen Family   Baker Family   Clark Family

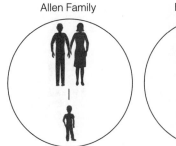 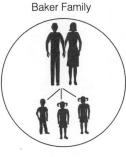 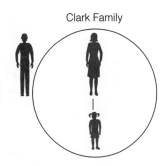

FIGURE 10-5 *Three Possible Family Structures*
Source: Inspired by Rodgers and White, 1993, p. 237.

THE FAMILY AS A SOCIAL GROUP: A MESOLEVEL VIEW

The actual structure of our families affects our interactions within the family no matter what historical period we grow up in. From a mesolevel perspective sociologists look at different types of family structures, at the roles we play within our families, and at changes in family structures and roles over time.

Family Structure: A Combination of Statuses

Using the terms we first encountered in Chapter 4, we can say that within our families each of us has a status or position. Within the nuclear family you might be a son or daughter, wife or husband, mother or father, or brother or sister. Within your extended kin group you may be a grandson or granddaughter, niece or nephew, aunt or uncle, perhaps grandmother or grandfather, or even great-grandparent or great-grandchild. The social structure of the family—the combination of statuses within a given family, particularly the number and type of statuses that are included—is sometimes referred to as a **family structure.**

Just as one might develop many different designs for sweaters, some with many different strands of yarn and some with only a few, family structures can take many different forms. Figure 10-5 illustrates three possible family structures. The Allen and Baker families represent the nuclear family of a mother, father, and children. The Clark family, like Craig's has experienced divorce. The daughter lives primarily with her mother but maintains contact with her father, who lives in a separate household.

The structure of our families is an important influence on how we fulfill family functions. Consider first the issue of economic support. Let's assume that each father depicted in Figure 10-5 earns about $30,000 a year, and each mother about $20,000 a year. The Allens can spread this $50,000 yearly income among three family members, whereas the Bakers must spread it among five people. The Clarks face a special problem, because they must support two households. Because a large proportion of a family's budget goes to pay for housing, this means that the

Clarks will have far less money to spend on their child than the Allen family does and will encounter more strains and difficulties in fulfilling this function.

Family structure also influences the ways in which families can provide emotional support and care for their children. Simply because the Allens have one child and the Bakers have three, the Allens can focus more attention on their child than the Bakers can give any of theirs. The Bakers might manage to give each of their children as much attention as the Allens give their son by the way they arrange other areas of their lives, but it is harder for them to do so simply because their family is larger. The Clarks also find it more difficult to give as much attention to their child as the Allens can give to theirs. Because they are maintaining two households rather than one, they will struggle to find as much time to devote to their child. In addition, many studies of families that have experienced divorce indicate that, over time, the father (who is usually the parent who does not have custody) tends to have much less contact with his children.[49] Thus, for the Clarks, providing emotional support and care for the child will probably be the predominant responsibility of the mother, and with only one parent present it will be harder for the Clark child to get as much attention and support as the Allen child will get.

Families that have experienced divorce, such as Craig's, often have much more complex structures than the one shown for the Clarks in Figure 10-5. Figure 10-6 summarizes Craig's complex family structure. Craig has relationships with one biological sibling, three stepsiblings, and two half-siblings; two biological parents and two stepparents; and two sets of biological grandparents and two sets of step-grandparents. Craig's mother, Rachel, has a relationship not only with her current spouse, Phil, but also with her former husband, Joe; with Joe's new wife, Lea; and with Phil's former wife, Bonnie.

These strains and difficulties that families face because of their size and composition are what sociologists call *structural* in origin. No matter how much parents in larger families love their children and how hard they work to care for them, they face a more difficult task than do parents in smaller families.

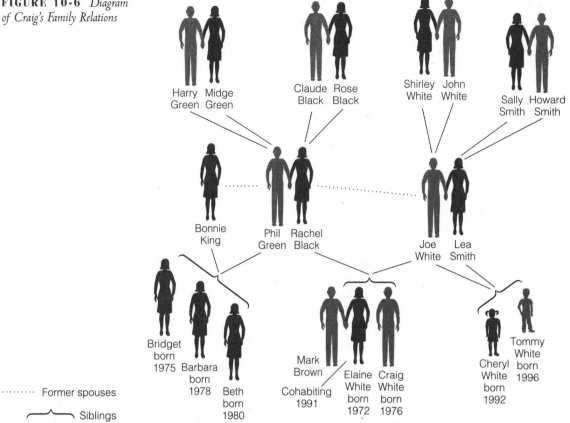

FIGURE 10-6 *Diagram of Craig's Family Relations*

Harry Green   Midge Green

Claude Black   Rose Black

Shirley White   John White

Sally Smith   Howard Smith

Bonnie King

Phil Green   Rachel Black

Joe White   Lea Smith

Bridget born 1975

Barbara born 1978

Beth born 1980

Mark Brown

Cohabiting 1991

Elaine White born 1972

Craig White born 1976

Cheryl White born 1992

Tommy White born 1996

········ Former spouses

⏜ Siblings

Similarly divorced parents have a tougher time providing and caring for their children than they would if they were not divorced. Many of the problems they face result from the structures in which they live, not from their motives, desires, or emotions.[50]

One way in which families can try to deal with these problems is to alter the family structure. In other societies with historically high divorce rates, virtually all divorced people quickly remarried. In our own society, for many years, the majority of women who were divorced or widowed moved in with other kin, often parents. Yet today, probably because they can afford to do so, separated or divorced women often maintain their own households.[51]

Family Roles: Obligations and Expectations

Each status that we hold is attached to **roles,** the obligations and expectations that define what we should do as sons or daughters, brothers or sisters, or mothers or fathers. These role expectations define how we help fulfill the functions of the family—to provide economic support, emotional security, and child care. As we've discussed in earlier chapters, social roles, especially in the family, are often differentiated by our sex and our age. Thus, for example, in the United States, parents traditionally have been ex-

pected to provide for the economic and emotional support of the family. Specifically fathers were expected to provide the bulk of the economic support while mothers were expected to provide more of the emotional security and child care.

The ways in which these roles are defined is part of the social institution of the family—that is, the roles are a societally defined aspect of our culture. These roles reflect the way we assume things should be, often unconsciously. Even though only a minority of children now live in families with an employed father and a stay-at-home mother, research suggests that we still carry the role expectations associated with a more traditional family structure in our mind.[52] In short, our beliefs and images concerning families and family roles might differ significantly from the reality of family life today.

Stereotypical images of what families are like are portrayed in classic television shows such as "Father Knows Best," "Leave It to Beaver," "The Donna Reed Show," and "The Brady Bunch." When our families don't match these stereotypes, we can feel as though we have somehow failed in our family roles. For instance, fathers, especially those who have been hard hit by the decline in real wages in recent decades, might feel guilty or inadequate because they can't provide for their families with their own salaries.[53] Mothers might feel guilty for missing a child's school performance because they have to work. Children might feel cheated that their own family doesn't resemble what they sense families "should" be like.

The lack of role expectations is especially noticeable for families that have experienced divorce. Roy Rodgers and Constance Ahrons are, respectively, a sociologist and family therapist who have written extensively on divorced and remarried families. They note that such families are, to use Durkheim's terms, in a very anomic, or normless, situation.[54] For instance, we don't even have specific terms to describe the kinship relationships of all of the people depicted in Figure 10-6. Even though both Lea and Rachel are concerned about Craig and Elaine, Rachel can describe Lea only as "my former husband's current wife" or "my children's stepmother." Even more important, we don't have clear guidelines or norms attached to the roles in these newly developing family structures. We don't know how stepparents and stepchildren should relate to each other, how stepsiblings and half-siblings should interact, what the relationships should be between step-grandparents and their step-grandchildren, or how parents should relate to the people with whom their children are cohabiting. The institutional norms surrounding the family in our culture assume that both the biological mother and father are in the home, something that is simply not the case for a growing number of families.[55]

As Durkheim would predict, this lack of norms can make life very stressful, not unlike the difficulties Craig has faced over the years. Sociologists who take a mesolevel view of the family suggest that many of the problems Craig has experienced reflect the complicated structure of his family and the lack of norms that prescribe behavior for these family statuses and associated roles.[56] This lack of clear role definitions is also another example of what sociologists call **cultural lag,** the delay between some change in society and the readjustment of other parts of the society. Even though family structures have become much more differentiated in recent years, the role expectations associated with the statuses in these emerging and more diverse structures have not yet become wholly defined. You might well have experienced this lack of clear role expectations in your own family or witnessed it in others.

When we look at family structure and family roles, we often see just a snapshot of families at one point in time. If you think about your own family of orientation, you will quickly realize, however, that the nature of the family structure in your home has changed over the years. Perhaps your siblings have grown up and left home; perhaps your parents have divorced or remarried; perhaps additional children have entered the household. All families experience these structural changes, and they affect our families and our lives within them. Another mesolevel perspective of the family focuses on these changes.

Family Development: Changes over Time

In Chapter 4 we talked about the progression through social roles that occurs as we grow older. Our analysis there focused on the individual as a unit of analysis—on what happens to each of us over the life course. But we can also use this perspective to look at changes in family structure, or the statuses and roles within a family over time, to understand how family life changes and how it affects individuals.[57] This point of view is often called **family development theory.**[58]

Family development theorists assume that different stages in a family's life correspond to different family structures and different household compositions. Figure 10-7 illustrates significant stages in Craig's family from the time Joe and Rachel established a household until Craig went off to college. Each of these stages is unique because the network of people within the household, or the family structure, changed.

Along with these structural changes come different role expectations that affect every status in the family. For instance, when Elaine was born, Rachel gained the status of mother with all its attendant role responsibilities. When she and Joe divorced, these responsibilities became even heavier, but they altered and eventually diminished as the children grew older and became more independent. Similarly Elaine acquired the new status of sister when Craig was born. At this point her life changed dramatically. Now she had to share her parents' attention with her infant brother, and in later years she assumed some of the responsibility for watching out for him.[59]

The various family roles have social norms associated with them that affect our family lives in four general ways. First, some norms are *age-graded:* They tend to change as we become older or to influence what we do at different ages. For instance, a very young child is not ordinarily expected to help care for other children in the family, but a teenager often is.

Second, some norms are *stage-graded:* they change as we enter new stages or family structures. For instance, a husband's responsibilities alter substantially after children are born, as he is expected to work with his wife in nurturing and caring for the children. They alter again after the children leave home, when these child care responsibilities diminish and disappear.[60]

Third, norms influence the *timing* of stages. In pre-industrial societies, where people often died at very young ages, couples commonly had their first

FIGURE 10-7 *Stages in the Development of Craig's Family*

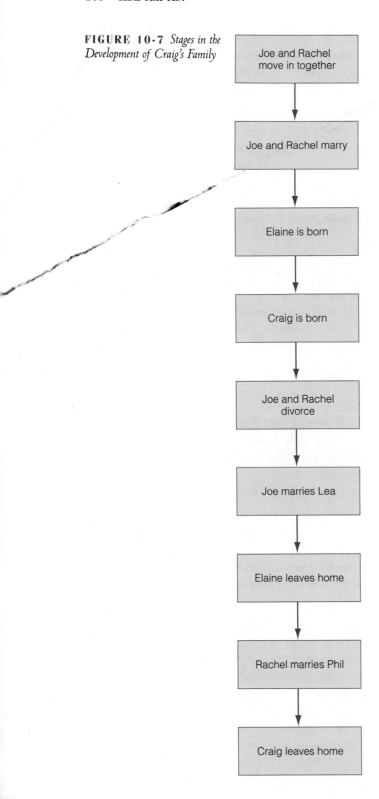

family if they have not finished their schooling, and thus early childbearing is seen as problematic.

Finally, norms influence the *sequence* in which stages occur. A popular children's rhyme illustrates this:

> Bill and Suzie sitting in a tree,
> K-I-S-S-I-N-G.
> First comes love, then comes marriage,
> Then comes Suzie with a baby carriage.[61]

We generally expect that people will fall in love, then marry, and then have children. We tend to worry about people who have children before they marry, and we may be shocked to hear of people marrying "only for money" or choosing to be in a "loveless marriage." This sequence, however, and especially the prerequisite of love, is far from universal. Many societies, such as premodern China and some contemporary Muslim countries, have arranged marriages, whereby the couples have not dated or declared their romantic love for (or even interest in) each other. In some societies the bride and groom do not even *meet* until the day of the wedding. In those societies marriage comes first; love might or might not follow.

The various stages in a family's life are marked by events such as moving in together, marrying, having a child, divorcing, or having a child leave home. Often these events are marked by **rites of passage**—ceremonies or rituals, such as funerals, baby showers and christenings, weddings, and anniversary celebrations, that help us deal with role transitions through the life course. Rites of passage represent public acknowledgment of new role identities and also serve as social events in which others can give emotional support as well as advice and gifts to help in the years ahead. Even with all the recent changes in family life, rites of passage remain very important. For example, you might expect that couples would be less likely to have large weddings since cohabitation has become more common. In fact, an extensive survey of couples married during the past half-century suggests that weddings now are more likely to be preceded by bridal showers and bachelor parties and to include elaborate receptions. Even though many couples have been openly living together, the marriage ceremony still seems to be an important ritual that formally marks a couple's commitment to each other and the beginning of their new stage of life.[62]

Notably, no developed rituals are associated with the end of a marriage through divorce. Even though a divorce can be as personally devastating as a death in the family, we have no rituals to mark the occasion for those who are experiencing it and for their friends and family. We also don't have set words, or ritual comments, that are appropriate to say to divorcing

child while quite young. In our society, in contrast, people worry a great deal about teenage marriages and teen childbearing. This change in norms regarding the timing of parenthood reflects the association between the family and other social institutions, such as education and the economy. Today young people are hard-pressed to find well-paying jobs that can support a

Arranged marriages are still common in Pakistan. In accordance with this custom, the former prime minister, Benazir Bhutto, did not meet her husband until the day of the wedding ceremony, pictured here.

couples. (Do you say, "I'm sorry" or "Congratulations"?) Nor do we have customs that provide other types of support for divorcing families. We typically send flowers, condolence notes, or memorial contributions when a family member dies, and we give presents at the birth of a new baby and at marriage. Yet we have no gift-giving customs associated with divorce, even though a family might face many new expenses in dividing one household into two. Nor do we have public or ritualized means of supporting these families as members try to develop new ways of relating to one another and new roles to go with their new family structure and statuses.[63] Just as the family structures that result from divorce and remarriage are anomic, so is the transition from a married state to a divorced state—even though this transition has become more and more common in recent years. Perhaps such rituals would have helped Craig and his family adjust to their new lives after the divorce.

It is important to realize that families develop or change in a probabilistic rather than a deterministic process—that is, family structures do not change in some type of uniform, lockstep way. For instance, if you cohabit with a partner, it is probable, but by no means certain, that you and your partner will eventually marry. If you marry, there is a significant chance that your marriage will end in divorce and also a probability that it will end in widowhood. The size of

these probabilities can be related to several variables. For instance, your chance of divorcing is somewhat higher if you marry while in your teens or if you cohabit before marrying.[64] The probability is also influenced by period factors. For instance, if you marry in the 1990s, you have a much higher probability of divorce than if you had married in the 1950s.[65]

Similarly the path of stages in your family's development is not set in stone, but some paths are more likely than others are. For instance, the probability that a young married couple will move to the stage of a family with a child is relatively high. But some couples never have children, and once a couple has been married for a long time without children being born, the probability of having a child begins to decline. And, as we have seen, today many more people are following a different path by choosing to have babies outside of marriage.[66]

The mesolevel perspective of family development complements the macrolevel analysis by helping us understand the nature and consequences of changes in specific families, including our own. If we zoom in even closer and take a microlevel perspective, we can scrutinize how our families and their structure affect us as individuals.

THE FAMILY AND INDIVIDUALS: A MICROLEVEL VIEW

In previous chapters we have encountered several examples of how our families affect our lives. In Chapter 4 we saw how family experiences help us learn about the social world and influence the development of our self-identity through the process of socialization. In Chapter 6 we discussed how our relationships with our parents influence the development of self-control and a conscience. And in Chapter 7 we saw how family background is an important determinant of social status and how social class is often seen as a characteristic of our families and not of just ourselves as individuals. To be sure, our individual efforts and decisions, and our innate talents and abilities, all help determine the direction our lives will take. Still, unquestionably, our family's structure, resources, and interactions greatly influence our life course.

Many social scientists have wondered how the recent changes in family structures are affecting the lives and life chances of children. Specifically, how does growing up in a family with only a mother or in a family with a stepparent influence children's lives?

As you might imagine, these can be politically and emotionally charged questions. In this section we explore some ways in which sociologists have examined them and the sometimes passionate reactions

BUILDING CONCEPTS

Macrolevel, Mesolevel, and Microlevel Analysis of the Family

The types of questions that sociologists seek to answer differ in scope and focus depending on the level of analysis. Below are examples of the types of questions sociologists have asked concerning the family.

Level of Analysis	Focus	Types of Questions Asked
Macrolevel	The family as an institution 	What constitutes a family? Do families everywhere perform the same functions for their members and societies at large? How have families changed over time, and how do these changes relate to changes in other social institutions?
Mesolevel	The family as a social group	What changes can occur in family structures over time? How do family role relationships change?
Microlevel	The family and individuals	How does growing up in a single-parent household or with a stepparent influence children's lives?

to their work.[67] We then focus on the recent research of Sara McLanahan and her colleagues as they adopt a more dispassionate sociological perspective on these issues in the face of political pressures.

Studying Single-Parent Families: A Politically Sensitive Issue

In the 1950s and 1960s, when divorce and single parenthood were far less common than they are today, several studies showed that children in single-parent homes exhibited more behavioral and academic problems than did those who lived with both of their biological parents. These studies generally gathered dust on library shelves. Then, in 1965, Daniel Patrick Moynihan, a former Harvard University sociologist and current United States senator from New York State, took a new approach to the issue.[68] In the process he generated political fallout that has affected work in the area to this day.

In what has come to be known as the *Moynihan Report,* Moynihan became one of the first scholars to comment on the rise in the incidence of single motherhood in poor, urban, African-American communities, linking this increase to the growing unemployment of African-American men. He feared that this change in family structure would undermine the gains of the civil rights movement and suggested that the federal government should focus more efforts on finding jobs for African-American men to help keep families together.

Thirty years later, this conclusion doesn't seem all that unwarranted. Even though the wages of individual African Americans have grown considerably over the years, the sharp increase in single-parent families (most of which occurred after Moynihan wrote his report) has offset many of these gains. As a result racial-ethnic differences in poverty are as high now as they were three decades ago. The strains faced by people in single-parent families generally arise from their family *structure,* not from individual motives or beliefs.

However, Moynihan's report outraged civil rights activists and scholars alike. Instead of focusing on Moynihan's call for increased job opportunities for African Americans, they latched onto his concerns regarding the rise in single parenthood. They suggested that, by concentrating on people who had decided to divorce or to have children outside of marriage, he was somehow implying that these people were personally responsible for the fact that they were poor. In short, his critics suggested that Moynihan was guilty of "blaming the victim."[69]

In the early 1970s, in the wake of the controversy over Moynihan's work, sociologists' conclusions from the 1950s and 1960s regarding the fate of children in single-parent homes were challenged on methodological grounds. Many of these studies had looked at children who were in psychological treatment or were wards of the court. Critics wondered if the same results would have been obtained if a probability sample, which could be generalized to the entire population of children, not just those in treatment, had been used. Because many of these studies did not control for the social class of the children, it was also impossible to know whether the problems experienced by children from divorced families occurred because of their experience with divorce or because they were poorer than other children.[70]

You might expect that sociologists would respond to these criticisms by redoing the earlier studies and correcting the problems. Interestingly, they did not. Undoubtedly they were deterred by the strong reaction to Moynihan's work and didn't want to face such criticisms themselves. Perhaps, too, they were influenced by the rapidly rising divorce rate. By the 1970s divorce was no longer rare, and many social scientists and their friends had experienced it personally. The prevailing wisdom was that children would be better off if their parents were happier living apart than together. No one wants to think that their own or their friends' children might be at risk because of their parents' decisions, and these feelings might have influenced sociologists' research.

Whatever the reason, most sociologists throughout the 1970s avoided studies of single mothers and their families. The few exceptions involved field research into the ways in which single-parent families coped with their problems by tapping networks of extended kin for social support and financial aid.[71] However, these studies tended to focus on the strengths of single-parent families and ignore the actual condition of children in the growing number of single-parent families.

Eventually, toward the end of the 1970s, some scholars began to return to the issues first examined in the 1950s. Perhaps still fearful of the political controversy that surrounded Moynihan's work, they used small samples of middle-class Euro-American children who had experienced divorce.[72] The results proved to be disturbing, for they replicated the findings obtained in the 1950s and 1960s. Children who had experienced divorce were more likely to perform poorly in school and to exhibit behavioral problems. Many of them also seemed to carry resentments, sorrow, and anxieties related to the divorce into their late adolescent and adult years. Despite the hopes of parents that their children would be better off if their parents were happy, this often didn't seem to be the case.

Even though these findings were replicated over several different samples, controversies about how changing family structures affect children still did not subside, perhaps because another political issue clouded the debate. Remember that in earlier decades most mothers who were divorced or widowed lived with extended kin, and most teenage mothers married their child's father. Women's growing ability to establish separate homes coincided with their greater economic independence. Several political commentators and scholars have feared that any criticism of single motherhood might challenge this newfound independence of women.[73]

One exception is Sara McLanahan, a sociologist who was herself once a single mother. In a book written with Gary Sandefur that summarizes much of her work, McLanahan explained why she thinks it is important to study this issue, even though it is politically sensitive:

> Well over half of the children born in 1992 will spend all or some of their childhood apart from one of their parents. If we want to develop policies to help these children, and if we want to persuade citizens that government should try to help, we must begin by acknowledging that a substantial proportion of our nation's youth is at risk.
>
> While talking about the downside of single motherhood may make some adults (and children) feel worse off in the short run, it may make everyone better off in the long run. At a minimum, parents need to be informed about the possible consequences to their children of a decision to live apart. (No one would argue that information on the potential benefit of exercise should be withheld because it stigmatizes couch potatoes.)[74]

McLanahan, Sandefur, and others who have looked at how family structures affect children suggest that we can't afford to ignore what happens to children in single-parent homes. Ignorance is far worse than knowledge; we can begin to help children only if we understand why they might face problems. The best way to understand these issues is through careful sociological research.[75]

TAKING A DISPASSIONATE APPROACH TO THE ISSUE OF SINGLE-PARENT FAMILIES

McLanahan and Sandefur decided to study the long-term consequences of living in a single-parent family. How do the family structures in which we live as children affect our life chances in adulthood? More important, why do these effects occur? Is it because economic resources are stretched tighter in single-parent households? Or is it because children get less attention and support?

McLanahan and Sandefur focused on two basic concepts that you have already encountered: (1) **social capital,** the resources that we gather from our associations and networks with other people, and (2) **family income,** our family's material or economic resources. Social capital influences the amount of support we have during emotional crises as well as the opportunities and advantages we have in tapping resources in our communities. From the discussion in Chapter 5 of the status attainment model, you will remember that family income is an important influence on our educational attainments and our subsequent occupations and income in adulthood.

McLanahan and Sandefur reasoned that both the income available for children and their social capital can be adversely affected by divorce. With the concept of family structures in mind, you can see how this happens. When divorce occurs, families must spread their economic resources over two households.[76] Because each household includes only one adult rather than two, there is less adult supervision and interaction with children. Further, after leaving the home, fathers' relationships with their children often weaken, which can result in children losing contact with the resources available through their fathers' social networks, such as friends and coworkers. In addition, if economic strains force a family to move to a different neighborhood, as happened to Craig and his family, children might have to sever ties with friends, neighbors, and teachers, thus diminishing social capital even more.

McLanahan and Sandefur wanted to test these ideas. In other words, they wanted to see whether the differences in income, parental supervision, and community resources that accompany different family structures could explain why children in single-parent and stepparent families do less well than children who live with both of their biological parents.

Collecting and Analyzing the Data

Instead of gathering their own data, McLanahan and Sandefur did a **secondary analysis** of four different sets of survey data—that is, they analyzed data that had already been gathered by other researchers. Secondary analyses allow researchers to use very high-quality data sets and provide an excellent way to replicate studies by seeing if the same results occur in each sample.

Three of the data sets used by McLanahan and Sandefur were panel studies, in which children were followed over a number of years. The fourth was a carefully designed cross-sectional, retrospective study. This means that it looked at individuals at only one point in time (the cross-sectional part) but that people were asked to recall their earlier experiences (the retrospective part). All the data sets provided information on children's family structure while they were growing up and on their well-being in young adulthood.[77] All the surveys were also very carefully designed, with probability samples that represented the United States as a whole and with questions that were known to be very reliable.

McLanahan and Sandefur focused their analysis on three general dependent variables, all related to how well children might do in adulthood: schooling, employment, and early parenthood. The researchers compared sixteen-year-old children in three different family structures: (1) those living with both biological parents, (2) those living with only one parent (in most cases the mother), and (3) those living in stepfamilies (usually the mother and a stepfather). Note that even though children in stepparent families are currently living with two parents, almost all of them have lived in single-parent families at some time. They have experienced the strains and difficulties that accompany changes in family structures, and thus their experiences are probably more similar to children with single mothers than to children living with both biological parents. As a shorthand reference term McLanahan and Sandefur sometimes called all of these families "disrupted," referring to the fact that they all experienced changes in family structure at some point.

Like the researchers in the 1950s and 1960s, McLanahan and Sandefur found striking differences between children raised in different types of family structures. These differences appeared on all the measures and in all the data sets. Children from disrupted families were more likely to drop out of high school, to score lower on achievement tests, and not to expect to go to college, let alone actually attend and graduate from college. They also tended to have lower high school grades and poorer school attendance.[78] Among those who didn't continue their education, children from single-parent families were much more likely to be unemployed, even if they had similar levels of achievement in high school.[79] Finally, young women from single-parent families had a much higher risk than others of bearing a child while still in their teens.[80]

These differences appeared even after the researchers statistically adjusted for other variables that might influence these outcomes, such as the children's race-ethnicity and sex, the parents' education, and the place of residence. The disadvantages associated with family disruption also are very similar for children whose mothers have never been married, whose parents have divorced or been widowed, and who live with stepparents.[81]

Not all children who have experienced single parenthood drop out of school, are unemployed, or have babies while in their teens. But the *probability* that they will do so is greater than for children from nondisrupted families. For example, children from one-parent families are about twice as likely as are those living with both of their biological parents to drop out of high school. McLanahan and Sandefur calculated that the educational disadvantage associated with a disrupted family structure is about the same as that associated with having a mother with less education. (How much education our mothers have is generally the best predictor of how well we will do in school.) In other words, the advantage children gain from having an educated mother is canceled out if they have a disrupted family structure. By the same token, the disadvantage children have from living with a poorly educated mother is doubled when they also live in a disrupted family. McLanahan and Sandefur also calculated that the high school dropout rate in the United States would fall from 19 percent to 13 percent if all children lived in two-parent rather than disrupted families. To put it another way, disrupted families account for about one-third of the nation's high school dropout rate.[82]

Clearly, family structure isn't the only influence on why children drop out of school, are unemployed, or have babies while in their teens. But, as McLanahan and Sandefur suggested, this does not mean that we should ignore the effects of family structure on these problems, for they can result in a society losing a great deal of potential talent and skill. Thus McLanahan and Sandefur looked for reasons that these effects occur.

Explaining the Effect of Family Disruption

For children with just one parent, family income is much lower than for children living with two biological parents or with stepparents. Among the children in McLanahan and Sandefur's study, those living with only their mother were five times as likely as those living with both their mother and father to have family incomes below the poverty level.[83] Children living with just their mothers or with stepparents also generally have less supervision and involvement with their parents, such as getting help with their homework or eating meals together. In addition, children in disrupted families, especially those in stepparent families, are more likely to have moved several times during childhood and thus experienced disruptions in their community networks. Can these differences in social capital and family income explain the differences in children's education, work experience, and early parenthood?

The simple answer, based on McLanahan and Sandefur's extensive and complicated statistical analyses, is yes. The details differ slightly, however, for children living with only a mother and for those living in stepparent families. For children who live with only a mother, either because of divorce or because she never married, a low income accounts for about half of the difference between them and other children in the measures of the dependent variables. In other words, if children in single-parent families had as much household income as those in two-parent families, their chance of dropping out of school, being unemployed, or having a baby while still a teen would be reduced by half. The rest of the difference can be explained by the fact that their mothers have lower aspirations for them and can't supervise or interact with them as much and that they have had to move more often.[84] In other words, when a single mother has been able to provide adequate supervision, hasn't had to move, and has a sufficient income, her children have life chances that are equal to those of children raised in other family structures with similar characteristics.

Children in stepparent families have family incomes equivalent to those of children living with both of their biological parents. These children also generally experience more parental supervision and involvement than children living with only their mothers. Yet these children experience the same problems with school achievement, employment, and teenage pregnancy as children living with only their mothers. Differences in family income don't account for any of these effects. But McLanahan and Sandefur found that changes in where children lived, including moving to schools where the children tended to associate with peers with lower aspirations, could account for about half of these differences. Differences in parental involvement and supervision could account for a little bit more. Thus, although small differences still remain, McLanahan and Sandefur's findings suggest that children in stepparent families who had fewer disruptions in their social networks and who received adequate parental supervision had life chances that were much more similar to children raised in nondisrupted families.[85]

In general McLanahan and Sandefur's findings refine our understanding of the effects of family structure on children's well-being by demonstrating the importance of both social capital and family income. The findings also help us see how we can use different levels of sociological analysis—macro, meso, and micro—to understand families and their influence on our lives. A macrolevel analysis reveals trends, such as the increase in divorce and single parenthood, that affect the probability that children will live in disrupted families and experience a decline in family income and social capital. A mesolevel analysis shows us how changing family structures affect the interactions and resources of families. Finally, a microlevel analysis reveals how these structures affect individuals' life chances. Scholars and policymakers who want to alleviate the problems faced by people living in single-

parent families will need to realize that solutions must deal with issues at all these levels of analysis (see Box 10-1, Applying Sociology to Social Issues). Applying our sociological imagination to each of these levels of analysis can also help us understand other issues related to family life, such as the disturbing problem of family violence (see Box 10-2, Sociologists at Work).

CRITICAL-THINKING QUESTIONS

1. Think about changes in the family during the last two centuries other than the rise in divorce and single parenthood, such as the increase in life expectancy and the change in work patterns described earlier in this chapter. How might these changes have affected family interactions? How might they have affected the lives of children?

2. How could a sociologist study the ways in which these other changes in families have affected individuals? What hypotheses could be tested? What methods could be used?

SUMMARY

The family is often considered the most basic social institution because some form of the family is found in all societies. Sociologists study the family using a variety of perspectives, and their research on the family can be influenced by historical and political factors. Key chapter topics include the following:

- In sociological terms social institutions refer to the areas of social life that meet people's basic needs—such as the family, religion, politics, the economy, education, and medicine. In comparison with pre-industrial societies, social institutions in modern industrialized societies are much more differentiated or distinct.

- At the macrolevel of analysis, sociologists have found that nuclear family units appear in virtually all societies. These nuclear units can be combined into larger kin units in a variety of ways, and the kinship system that a society develops is often related to the nature of its economy.

- In all societies the family seems to fulfill three major functions: provide for the economic support of its members, provide emotional security for its members, and regulate reproduction and child rearing.

- During the past two centuries the family in the United States has changed markedly in response to changes in the economy, education, and

health. The most dramatic change has been the rapid increase in the proportion of children living in homes without a father.

- A mesolevel view of the family reveals the significance of family structure and the ways in which the combination of statuses within a family affects the lives of family members.

- Role expectations are attached to our family statuses. Many of the role definitions and expectations associated with families in our society have not kept pace with changes in actual family structures.

- Family development theory shows how family structures, statuses, and roles change over time and affect us as individuals.

- Using a microlevel analysis, Sara McLanahan and Gary Sandefur found that children from single-parent and stepparent families tend to have lower levels of education and a greater chance of being unemployed and having a baby while in their teens than do children raised in two-parent families. These differences can be explained by the fact that children in single-parent and stepparent families have lower incomes and less social capital, particularly less parental supervision and fewer community resources.

BOX
10-1

APPLYING SOCIOLOGY
TO SOCIAL ISSUES

*Helping Children
in Disrupted
Families*

In the last chapter of their book, McLanahan and Sandefur discuss the implications of their research for social policy. Three basic principles underlie their policy recommendations. First, they believe that the *economic insecurity* of children growing up in single-parent homes must be reduced. Currently more than half of the children in single-parent homes live in households with incomes below the federal poverty level. Second, there must be *shared responsibility.* Healthy future generations will benefit the entire society, and the responsibility for raising children must be shared by mothers and fathers, by parents and nonparents, and by government bodies at the federal, state, and local levels. Third, programs and benefits should be *universal*—available to all children, whether they live in one- or two-parent homes. Currently many welfare programs are available only to single-parent families. By extending support to all children, government agencies would not convey the message to parents that they will receive benefits only if they live apart.

McLanahan and Sandefur propose several different specific policies and programs:

■ Help parents understand the risks of divorce to their children by making research findings widely known.

■ Provide help with health care and child care to both two-parent and single-parent poor families (currently most help goes only to single-parent families).

■ Replace the current child deduction on income taxes with a child allowance of $500 per child for all families, a policy similar to that found in all other Western industrialized countries, including Canada.

■ Guarantee both resident and nonresident parents a minimum wage job.

■ Increase enforcement of child support payments.

■ Provide good day care and after-school programs while parents are working.

■ Extend the school day with supervised activity and consis-

tent mentoring by responsible adults in the community.

■ Provide housing allowances and subsidies to help promote residential stability.

CRITICAL-THINKING QUESTIONS

1. How do these policy recommendations address various levels of analysis, such as macrolevel trends in divorce and single parenthood, mesolevel issues related to family structure, and microlevel issues such as family interactions and support?

2. How do these policies address the processes that account for the disadvantages that face children in single-parent families—the lack of income, the lack of adequate supervision, and the greater residential mobility?

3. Which of these policy recommendations do you think could be implemented? Why or why not? Would you make other policy recommendations?

SOCIOLOGISTS
AT WORK

Murray Straus

So much of the course of human society is found in human social arrangements, not in individual pathologies.

In Chapter 2 we described the **Conflict Tactics Scale, a series of questions that sociologist Murray Straus and his colleagues have used to ask people about their experiences with domestic violence. From his years of research in this area, Straus has found that family violence affects millions of men, women, and children each year.**

Why did you become a sociologist, and how did you come to focus on family violence? I had always been fascinated by why things are so different in different countries. As a child I spent some years in England, and I remember stories my grandfather would tell of his years in India and the Middle East. In sociology this is one of the key issues: Why are societies so different?

I got into the area of family violence through a combination of accidents and social processes. It's the old scientific saying, "If you stumble over something interesting in your research, drop everything and study it." I did research on the family for twenty years before I started focusing on violence. It came up in my studies, but I always thought it was a problem with the research, not with the families.

By the 1960s there was a growing concern with and sensitivity to violence, in light of the Vietnam War, the assassinations of Robert Kennedy and Martin Luther King, Jr., and urban and racial tensions. I was revolted by the Vietnam War, and all of these factors sensitized me to the issue of violence. Then I surveyed my intro students and found that 25 percent had been hit by their parents during their senior year of high school. From thinking about why these parents were hitting their kids, I jumped to family violence.

What surprised you most about the results you obtained in your research? There were two surprises. One was how much violence there is between partners; what parents were doing to their children, they were doing to each other. Kids reported that 15–16 percent of their parents had suffered what the law calls simple physical assault, a slap or a thrown plate of food, in the last year. The level of family violence continues to surprise: A recent study showed that 52 percent of 13- to 14-year-olds had been hit by parents, and this is with the trend away from corporal punishment.

KEY TERMS

cohabitation 252
conjugal family system 248
consanguine family system 248
cultural lag 259
divorce rate 255
endogamy 250
exogamy 250
family development theory 259

family income 264
family of orientation 256
family of procreation 256
family structure 257
general fertility rate 252
kin 248
monogamy 248
nuclear family 248

polyandry 251
polygyny 248
rite of passage 260
roles 258
secondary analysis 264
serial monogamy 256
social capital 264
social institutions 245

BOX 10-2

SOCIOLOGISTS AT WORK

Continued

The second surprise was our finding that women assault their partners at the same rate that men do and that mothers assault children, even adolescents, more than fathers do (because mothers have more child-rearing responsibility). Also, who hits first is at the same rate, which shows that women assaulting partners isn't a matter of self-defense.

It's important to note the difference between men and women in results of injury; women suffer a seven-times-greater rate of injury because of the superior strength of men. A punch to the chin from a woman will probably result in a bruise, while a punch from a man may result in a broken jaw.

What was the reaction to your research among your colleagues? The reaction has been contradictory. In the late 1960s, before the Conflict Tactics Scale was published, they said there is no scientifically valid way to study family violence because no one will give you a true report of what's going on. Now that the results are out, some say the rates are too high and people are exaggerating and over-reporting.

The truth is, people don't want to hear the bad news about the American family. It is a highly valued institution. Studies have shown that at least 90 percent of family violence is committed by people without psychological pathologies. Family violence is built into the nature of the family.

This is the point I try to get across in my intro courses: so much of the course of human society is found in human social arrangements, not in individual pathologies.

Has your research been replicated? Yes, in thirty studies so far. We've made some changes and revisions of the scale in order to take social context into account more and have had the same results. The scale has made research into family violence very practical. In 1994 it was cited in an average of ten journal articles every month.

What do you think the implications of your research are for social policy? The reception of my research shows that social research is not likely to be used if it doesn't agree with prevailing opinions and trends. There first needs to be a shift of culture and a rise of advocacy groups before research is used to buttress opinions. For example, my study showed that corporal punishment is bad for kids, but it was ignored for twenty years until, for various reasons, it was decided that spanking was bad. Now my research is widely cited!

If I could dictate social policy, I'd institute campaigns of education and prevention that focus equally on men and women who assault their partners. Services should still focus more on women, however, since they receive more serious injuries and are more economically dependent.

INTERNET ASSIGNMENTS

1. Locate Web sites pertaining to two different types of families, such as adoptive families, foster families, gay/lesbian parents, single parents, or stepparents. Try to find scholarly research as well as discussion groups and support groups. What are the most salient issues covered in these sites? How do these issues differ for the two types of families you have selected? How do they relate to concepts and issues discussed in this chapter, such as family development, social capital, and changes in the economy?

2. Locate the Bureau of the Census Web site. What are the most recent estimates of single-parent (mother only and father only) families in the United States? While at this site, also locate the estimates of the average number of children per family. How do these estimates compare with the data discussed in this chapter? What other types of data regarding the family can you find?

TIPS FOR SEARCHING

Searching with the term "family" can yield a wide variety of sites from support groups to public interest groups to scholarly information to companies providing products and services for families. To limit your search for question 1, try searching with terms related to specific types of families, such as gay/lesbian parents.

The Census Bureau Web site contains a very large amount of data. To simplify your search, try using the alphabetical subjects list (titled "Subjects A to Z") on the site and looking for such topics as "families," "children," and other key words.

InfoTrac College Edition: Readings

Article A21059915 / Reher, David Sven. 1998. Family ties in Western Europe: persistent contrasts. *Population and Development Review, 24,* 203–235. Reher examines differences among European nations in family norms and structures, such as the time at which young people leave the parental home, both currently and historically.

Article A19586664 / Ruggles, Steven. 1997. The effects of AFDC on American family structure, 1940–1990. *Journal of Family History, 22,* 307–326. Ruggles explores the relationship between changing welfare benefits and trends in nonmarital births, divorce, separation, and living arrangements of unmarried mothers.

Article A19313702 / Kattakayam, Jacob John. 1996. Marriage and family among the tribals of Kerala: a study of the Mannans of Idukky District. *Journal of Comparative Family Studies, 27,* 545–559. Kattakayam describes how the traditional family system of the Mannans, a tribal group in India, has changed in recent years.

Article A17716122 / Kumagai, Fumie. 1995. Families in Japan: beliefs and realities. *Journal of Comparative Family Studies, 26,* 135–164. Kumagai describes the ways in which contemporary Japanese families combine traditional and modern elements in their day-to-day lives.

Article A21263203 / Cherlin, Andrew J. 1998. Marriage and marital dissolution among black Americans. *Journal of Comparative Family Studies, 29,* 147–159. Cherlin describes trends in marriage, divorce, separation, and childrearing among African Americans and the relationship of these trends to economic factors.

Article A53643906 / South, Scott J.; Kyle D. Crowder; Katherine Trent. 1998. Children's residential mobility and neighborhood environment following parental divorce and remarriage. *Social Forces, 77,* 667–681. South, Crowder, and Trent look at how changes in family structure affect the probability that children will move and the types of neighborhoods that they enter.

Further Reading

Suggestions for additional reading can be found in *Classic Readings in Sociology,* bundled with this book. If you purchased a used copy of this book that does not include this custom-published reader, go to www.sociology.wadsworth.com for ordering information.

RECOMMENDED SOURCES

Ahrons, Constance R., and Roy H. Rodgers. 1987. *Divorced Families: A Multidisciplinary Developmental View.* New York: Norton. The authors show how the family development perspective can be used to help families as they go through a divorce and establish new family structures and roles.

Beeghley, Leonard. 1996. *What Does Your Wife Do? Gender and the Transformation of Family Life.* Boulder, Colo.: Westview. Beeghley uses historical and international data to examine how gender relations and family life both affect and are affected by premarital sex, abortion, divorce, and employment and income.

Cherlin, Andrew J. 1992. *Marriage, Divorce, Remarriage,* rev. and enl. ed. Cambridge, Mass.: Harvard University Press. This is an excellent discussion of changing family patterns, especially related to marriage, in the United States; includes a thoughtful discussion of social policy.

Farley, Reynolds. 1996. *The New American Reality: Who We Are, How We Got Here, Where We Are Going.* New York: Russell Sage Foundation. Farley provides an extensive discussion of changes in American families and social life during the last half century.

Gottman, John Mordechai. 1994. *What Predicts Divorce? The Relationship Between Marital Processes and Marital Outcomes.* Hillsdale, N. J.: Erlbaum. A psychologist uses many sociological insights to examine the relationship between marital interactions and the probability that a couple will divorce.

Hansen, Karen V,. and Anita Ilta Garey (eds.) 1998. *Families in the United States: Kinship and Domestic Politics.* Philadelphia: Temple University Press. This is an extensive collection of articles on family life, many of which are written from a feminist perspective.

Hernandez, Donald J. 1993. *America's Children: Resources from Family, Government, and the Economy.* New York: Russell Sage Foundation. Hernandez provides a very detailed analysis, based on census data, of the status of children in families in the United States throughout the twentieth century.

McLanahan, Sara, and Gary Sandefur. 1994. *Growing Up with a Single Parent: What Hurts, What Helps.* Cambridge, Mass.: Harvard University Press. McLanahan and Sandefur describe the research reported in this chapter in an accessible style that readers should be able to understand without having a statistics course.

Mintz, Steven, and Susan Kellogg. 1988. *Domestic Revolutions: A Social History of American Family Life.* New York: Free Press. This is an extensive and detailed history of American family life from colonial times to the present.

Staples, Robert, and Leanor Boulin Johnson. 1993. *Black Families at the Crossroads: Challenges and Prospects.* San Francisco: Jossey-Bass. Staples and Johnson provide an extensive and thorough review of research on African American families.

Stephens, William N. 1963. *The Family in Cross-Cultural Perspective.* New York: Holt, Rinehart & Winston. Williams provides numerous detailed examples of family life in other societies.

Straus, Murray A., with Denise A. Donnelly. 1994. *Beating the Devil Out of Them: Corporal Punishment in American Families.* New York: Lexington. The sociologist profiled in Box 10-2 extends his work on family violence to an extensive analysis of child abuse and corporal punishment.

White, James M. 1991. *Dynamics of Family Development: A Theoretical Approach.* New York: Guilford. This is a recent statement of the ideas underlying family development theory and a good example of how sociologists take a mesolevel view of the family.

INTERNET SOURCES

Russell Sage Foundation. The Foundation supports the application of social science research to a variety of social issues, including those involving the family. Its Web site provides descriptions of much of this work.

The Bureau of the Census Population Division has reliable longitudinal data on the family.

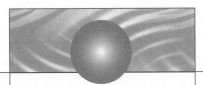

Pulling It Together

Listed below are other chapters where you can find additional discussion or examples
of some topics covered in this chapter. You can also look up topics in the index.

Formal Organizations

YOLANDA BEGAN WORKING at the local Harry's Hamburgers, a fast-food chain, while she was still in high school. By the time she entered City College, a commuter school in her hometown, she had worked her way up to assistant night manager, and when she transferred to State University as a junior, she became assistant manager at the Harry's outlet in the local Metro Mall. Certain parts of the routine at Harry's bothered Yolanda, such as the rigid rules and the stodgy uniforms, but her job provided a lot of good experience and a steady paycheck. As time went on, Yolanda began to dream of starting her own restaurant some day—one in which she could design the menu, the decor, and the ambiance. Harry's had started out as just a small cafe, and Yolanda dreamed of one day being as rich as Harry's founder is.

Yolanda was aware of the many risks involved in starting one's own business. For years her mother had run a successful bookkeeping business out of their home, so successful that she needed to hire temporary help each year at tax time. However, with the introduction of computer programs that allowed people to do their taxes at home, the business began to decline. She now had only a few clients and no employees.

Yolanda's father was a middle-level manager in the loan division of a bank. In the past few years the company that owned the bank had been bought and sold several times. His job title, his division name, and even the person to whom he reported seemed to change every few months. He usually tried to hide his frustration with work from Yolanda, but one day, after the latest change, he let his feelings slip. He told her how he felt like a little cog in a big machine, how he believed he no longer had any influence in the company, and, even worse, how he feared that his own job might disappear with all of the corporate restructuring. Yolanda had enough problems with the rules at Harry's. She didn't think she could spend her entire life taking orders from other people, especially in a big corporation. She didn't just want to manage someone else's restaurant—she wanted to run her own.

Yolanda had transferred to State University because it had a large business school and she thought that a business degree might help her actually fulfill this dream. Although the transition to a new school hadn't been easy, she now loved the fun and excitement of the large university and living away from home. She'd been elected to the Dorm Council and found that she was good at listening to students' concerns and mediating conflicts. She'd also joined the "College Sounds," an a cappella group that performed everything from gospel to jazz, and was part of a committee planning a small concert tour to nearby communities.

However, other aspects of her life at State were frustrating. She'd known that the university had strict graduation requirements, but she wasn't prepared for all the red tape and rigidity. For instance, she had to take a foreign language class to graduate, and languages had always been hard for her. She had pored over the regulations in a futile attempt to figure out some way to bypass the rule. As a last resort she had even petitioned the university's Graduation Requirements Committee, asking that she be excused. But they had refused, and now Yolanda feared that she might have to spend an extra year in school just to fulfill this requirement.

To make matters worse, on the day of the big biology midterm, two counterpeople didn't show up at Harry's and she missed the bus back to campus. Buses came only twice an hour, and when she finally got to class the exam was almost over. When she asked about a makeup exam, the instructor said, "Sorry, the biology department won't let us give makeups in large lecture classes. You'll just have to hope for really good grades on the other exams."

As if that weren't enough, there were also problems at work. Because of competition from other fast-food places at the mall, business at Harry's hadn't been that good recently, and they had been forced to cut work hours for almost all the employees. She'd even heard a rumor that Harry's corporate headquarters would close the mall outlet, and Yolanda wondered if she would have a job there much longer.

Sometimes she just got frustrated. "Why is it such a hassle?" she wondered. "I need my job at Harry's. I want my degree from the university. But life just seems so complicated. It feels as though somebody else—the university, Harry's, even the stupid transit company—is always trying to control my life. Why can't they be more flexible?

"Maybe," she thought, "I should just quit school and start my own restaurant now. But would that work? Could I ever really start a successful business myself? Would my business fail just like my mom's?"

Why are large organizations like Harry's or the university necessary when so much of the time they seem inefficient and impersonal? Why are they often so slow to change when it is clear that they need to adapt? Why do so many people who work for these organizations start out with the best of intentions and end up as inflexible bureaucrats?

GIVEN WHAT YOU HAVE LEARNED about sociology, how might you interpret the ins and outs of Yolanda's life? You might say that she is having to conform to the norms of the university culture, such as graduation requirements and rules about makeup exams. You might also suggest that she is experiencing role conflict because of the competing demands of her job and school. In addition, knowing her parents' occupations, you might point out that she comes from a middle-class background but that her mother's business problems have put a financial strain on the family.

Sociologists would suggest yet another way of looking at Yolanda's life, as well as at the daily experiences of all of us. We can look at how our daily interactions involve **formal organizations,** groups that have been deliberately created to accomplish certain goals.[1] Every day Yolanda interacts with many different formal organizations, including her dormitory, her university, the city bus system, Harry's Hamburgers, and the College Sounds. These organizations provide her with opportunities that she would not otherwise have. Yolanda's college degree can open up many job possibilities; her participation in the College Sounds provides an enjoyable leisure activity; her work at the mall provides income, as well as companionship and chances for career advancement. At the same time, these organizations often are a source of frustration, constraining her activities and limiting what she can do. For instance, Yolanda must take French to graduate from college (a requirement imposed by her college), she can catch the city bus only at certain times of the day (a limitation imposed by the transit system), she can't make up a midterm in a large lecture class (rules set by the biology department), and she might lose her job at Harry's (a decision of the restaurant corporation).

Like Yolanda, we all participate in formal organizations many times each day and throughout our lives, even though we often don't even think about it. We are born in organizations—hospitals; we are educated by organizations—schools; the vast majority of us work for organizations for most of our lives.[2] We even spend much of our leisure time interacting with organizations—from the National Park Service to the local movie theater. Organizations are as varied as the many aspects of our lives they touch, from educating us to healing us when we are sick, from comforting us when our loved ones have died to selling us cars and clothes.

Like Yolanda, most of us probably have ambivalent feelings about many organizations with which we interact. They constrain our lives in many ways, yet they provide opportunities that we couldn't possibly have without them. For instance, we can appreciate and need the financial services that banks provide yet resent their rigid credit requirements and cumbersome application procedures. We can enjoy the learning opportunities of our colleges and universities but be frustrated by the red tape involved in registering for classes, paying library fines, and so on. We can find comfort in the fellowship of our church yet disagree with a policy or position the church has taken.

Part of this ambivalence about formal organizations reflects the fact that organizations can be extremely powerful. Just as individuals wield power over others in daily interactions, organizations such as universities and business corporations make decisions and act in ways that affect not only individuals but other organizations and even entire societies as well. For instance, if the Harry's Hamburger chain decides to close the mall outlet, Yolanda and her co-workers will lose their jobs. When large corporations decide to close manufacturing plants or move their operations to other cities or even other countries, entire communities suffer the consequences. Organizations in the form of manufacturing plants pollute our air and waterways; organizations in the form of legislative bodies, such as Congress, appropriate funds for war on other nations and authorize the drafting of young people for battle. Other organizations, such as the United Nations, negotiate peace between warring countries and policies regarding the ways in which companies and nations are allowed to pollute.

No matter what form a given organization takes or how powerful it might be, sociologists look at it as having a life of its own, totally apart from the individuals who created it and who participate in it.[3] In their analyses, sociologists look at both the structure and the actions of organizations. It is also important, however, to remember that organizations are created by human beings. A recurring theme in this book is how all of us, as social actors, help create social structure—the world in which we live—even if we don't see it happening in our daily lives. After all, we really don't often directly observe ourselves creating stratification hierarchies or cultural norms or values. When you think about formal organizations, however, you can actually see how we create our social world. People devise the structures of organizations and the rules under which the organizations operate. They decide to what extent some people will have greater power than others will; they decide what statuses will be part of an organization and what roles will be attached to those positions. You probably haven't started most of the organizations in which you participate, but somebody did. Moreover, you can actively work to change or alter these organizations. Many sociologists who study organizations

Your college or university is one type of a formal organization. What other organizations do you interact with on a day-to-day basis?

hope that by gaining insight into how the organizations work, they can make the organizations more responsive and useful to those who participate in them.

To better understand sociologists' views of organizations, picture a city street with a variety of buildings, from a small delicatessen on the corner to high-rise apartments and office complexes. As varied as these structures are, all were designed and built by human beings, with doors through which we enter, hallways that regulate our movements, and rooms designed for different purposes. So, too, are organizations designed by people. They have ways for us to enter them, paths along which we interact within the organization, and different areas for different purposes. Like the delicatessen, some buildings are small and compact; like the high rises, some buildings are extremely large and serve many different purposes. Organizations also vary, from relatively simple designs like the College Sounds or a neighborhood church to complex and elaborate structures like State University or the federal government. And the design of organizations, like the design of buildings, affects our experiences within them. For instance, although you might know everyone in a group like the College Sounds or a small church, you might have no idea who works in another department of a large corporation in which you are employed.

Buildings often must adapt to their environments and undergo change as the environment changes. Perhaps, when the neighborhood was more residential, the corner deli that now serves sandwiches to office workers was a grocery store while the pricey condominiums were fleabag hotels. Organizations, too, must adapt to their environments and change as

the environment changes if they are to last. Some, like Yolanda's mother's bookkeeping business, do not survive these changes, but others can last for many years, often long after the people who designed them have left.

Finally, although architects can design houses, they can't make those houses homes. People who live and work in buildings influence both the way they look and the way they affect us. Although every structure imposes some limitations on the inhabitants, suburban tract houses or city apartments with exactly the same floor plan and even the same furnishings can be very different places with different people living in them. Similarly individuals are constrained by organizations, but they also affect and change them.

As you will recall, early sociologists were fascinated by the changes in society that came about with industrialization. One of these changes was the development of many large formal organizations to deal with the growing complexity of government and business activities. Max Weber was especially interested in these developments. Although his writings describe a very specific type of formal organization, they have provided the basis for much of sociologists' later work in this field. In this chapter we first examine Weber's ideas about organizations. We then look at how contemporary sociologists have examined organizations from the meso-, macro-, and microlevels of analysis to explain how organizations change and why some are more likely than others to survive.[4]

Our discussion of organizations in this chapter will be somewhat generic in nature, focusing on the sociological study of organizations in many different areas of social life. In the other chapters in this section, we will look at organizations within specific social institutions—the political world, religion, the economy, education, and medicine—as part of our discussion of how societies meet basic human needs.

BUREAUCRACIES AS AN IDEAL TYPE: WEBER'S CLASSICAL VIEW

In his work on formal organizations, Weber drew on his historical studies of ancient societies such as Egypt, Rome, and China, as well as his own observations of contemporary Europe. Weber was especially fascinated by **bureaucracies,** highly structured and formalized organizations that are governed by laws and rules.[5] Not all formal organizations are bureaucracies, but bureaucracies are found today in all areas of life, including businesses, voluntary organizations (groups, like the American Association of Retired People, that people choose to join), churches, and schools. When organizations become large or must deal with complex issues, a bureaucracy often is seen as the solution.[6] Each of us daily encounters many different bureaucracies, including government bureaucracies such as the post office or the local library, businesses from the phone company to the local grocery store chain, schools ranging from the elementary level to universities, and even churches and religious organizations. Each of these bureaucracies provides important services or goods, but each can also produce frustrations, much like those experienced by Yolanda.

Characteristics of Bureaucracies

Weber was fascinated by bureaucracies precisely because he saw them as becoming much more common as society became more complex. Weber believed that the mushrooming of bureaucracies was a phenomenon that demanded explanation. After all, many other kinds of organizations are possible. Why, he wondered, did bureaucracies develop as the means by which societies handle the complexities of modern life?

To answer these questions, Weber used the concept of ideal types to describe the characteristics of a bureaucracy in its purest form. Recall that **ideal type** refers not to ideal in the sense of perfect or best, but rather to a pure *conceptual* description of a phenomenon. In other words, Weber tried to describe the essence of bureaucracy, independent of any particular organization.

Weber's description of bureaucracies involved four basic characteristics. First, roles within a bureaucracy are highly specialized. Individuals are assigned only a small number of tasks, thus becoming experienced at their jobs and able to perform them well.

Second, bureaucracies are hierarchically organized, with layers of authority and elaborate lines of communication. In large organizations with many employees, no one person can possibly supervise all the workers. Instead, officials in higher positions supervise those in lower positions, who in turn supervise others below them.

Third, bureaucracies are governed by rules. These rules determine what each official in the bureaucracy can and cannot do. The rules confer the official authority to carry out tasks, but they also limit the ways in which this authority can be carried out. Reams of regulations and procedures are a common characteristic of bureaucracies.

Finally, bureaucracies are impersonal. Official business and private affairs are strictly separated. Bureaucrats don't own the resources they use in conducting business, nor do they own the offices they occupy. There are no "property rights" to jobs, and they cannot be handed down to children or friends. In addition, official business is always based on written documents and records so that others who might occupy a given position can always do the work. No one in a bureaucracy is irreplaceable.[7]

Because Weber's description is an ideal type, no formal organization matches this description perfectly. One that comes close, at least in theory, is the United States Army. As Figure 11-1 shows, roles within the army are highly specialized, and the hierarchy is rigid and clear-cut. All soldiers have a rank within the army structure, and these ranks involve responsibilities both to those below and to those above them. For example, sergeants command corporals and privates, but they must also follow the orders of lieutenants and captains, who in turn take direction from majors, colonels, and generals. Strict rules dictate what people at each rank can and cannot do relative to others in the hierarchy. For instance, a sergeant can chew out a private, but the private would be in serious trouble if he or she did the same thing to the sergeant.

In addition, strict rules govern the separation of soldiers' private lives and their roles in the army to ensure that the army's work is as impersonal as possible. Soldiers must be careful where they wear their uniforms. They are not allowed to make political pronouncements, to publicly announce their allegiance to a political party or candidate, or to reveal their sexual orientation. Nor do soldiers own their positions—that is, they can't pass along their rank to a son or daughter—and both promotions and demotions are governed by strict guidelines. Finally,

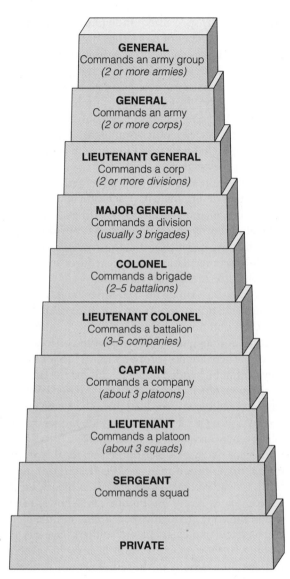

GENERAL
Commands an army group
(2 or more armies)

GENERAL
Commands an army
(2 or more corps)

LIEUTENANT GENERAL
Commands a corp
(2 or more divisions)

MAJOR GENERAL
Commands a division
(usually 3 brigades)

COLONEL
Commands a brigade
(2–5 battalions)

LIEUTENANT COLONEL
Commands a battalion
(3–5 companies)

CAPTAIN
Commands a company
(about 3 platoons)

LIEUTENANT
Commands a platoon
(about 3 squads)

SERGEANT
Commands a squad

PRIVATE

FIGURE 11-1 *The United States Army as a Bureaucracy*
Source: Ziskind, 1962, p. 611.

modern armies are flooded with paper. A large proportion of soldiers in the United States Army serve as clerical workers or are otherwise engaged in paperwork. Decisions must be documented and justified in writing so that anyone can take over a task and carry it out efficiently.

All this might seem to be in the nature of armies, but as Weber noted, armies were not always structured in this formal, hierarchical manner. For instance, feudal armies were manned by knights who purchased their own armor and horses and allied themselves with noblemen and kings of their own choice. This tradition persisted into the early days of the American Civil War, in which many companies of troops furnished their own equipment and even elected their own leaders and in which many generals were political ap-

pointees or simply rich men who could supply badly needed equipment. These less formal systems worked fine as long as wars primarily involved individual-level combat, as in medieval times. But as armies grew larger and the technology and style of combat changed, so, too, did the style of armies. Effectively using guns, cannons, and other firepower required vast numbers of people and a high degree of discipline and order. As a result, bureaucratic arrangements involving strict training and hierarchical organization developed.[8] By the end of the Civil War, for example, both the Union and Confederate armies were starting to resemble modern armies.

By the time Weber wrote on bureaucracies, in the early twentieth century, they were a fixture not just of armies but also of governments throughout Western Europe and the United States. Bureaucracies were also common in work organizations, such as large corporations, and even in universities. Weber suggested that this growing prevalence of bureaucracies reflected the increasing complexity of other areas of society.[9] As technology became more sophisticated, life became more complex and the need for organization became more obvious. No one person can run complex modern transportation systems, such as canals, roads, railroads, and airports, or modern communication systems, such as telephones, telegraphs, and post offices. According to Weber, only some type of rational, organized system could perform these tasks effectively, and bureaucracies were best suited to fill the bill.[10]

Dysfunctions of Bureaucracies

At the same time, Weber realized that, ironically enough, the attempt to structure bureaucracies to function effectively can also result in negative consequences, or what sociologists call *dysfunctions*. This "dark side" of bureaucracies includes the characteristics that many people think of when they complain about this type of organization. The specialization and predictability of bureaucracies can translate into great power and inflexibility. Because of the intense specialization of work in bureaucracies, few, if any, ordinary people can be as knowledgeable as a bureaucrat is about his or her area of expertise. Thus, for example, it is very difficult for an individual to argue successfully with an agent of the Internal Revenue Service. The IRS almost always wins. In addition, with their rigid structures and procedures, bureaucracies often aren't very responsive to public opinion or to individuals' special circumstances. Bureaucracies take on a life of their own and are often highly resistant to the opinions and desires of the people whom they serve. Finally, once they are established, it is very hard to get rid of bureaucracies. They become

How does this painting by George Tooker, entitled Government Bureau, *depict the characteristics of a bureaucracy that Weber described?*

entrenched and seemingly indispensable, and, because we generally know no other way to accomplish the tasks bureaucracies do, we cannot imagine any other type of system that would do a better job.[11]

These dysfunctions are probably the reason that the term *bureaucracy* sometimes is used like a dirty word. You might have experienced the "bureaucratic runaround" when a clerk in an office can't answer your question and directs you to another office, the clerk there directs you elsewhere, and so on. This runaround is a direct result of the specialization of tasks in a bureaucracy. Similarly, like Yolanda, you might resent the insistence by officials at your school that you fulfill each and every requirement before graduation (and produce the documentation to prove it). This insistence is part of the rule-governed nature of bureaucracies, or what you might derogatorily label "bureaucratic inflexibility." Each person who comes into contact with the bureaucracy is treated in as similar a way as possible, and exceptions to rules are granted only very infrequently. Thus, the very characteristics that enable bureaucracies to handle complex situations also make them frustrating for individuals who interact with them and, at least in some cases, can undermine overall organizational efficiency and effectiveness.

Throughout modern history bureaucracies have been responsible for both great good and great evil. For instance, Adolf Hitler's horrendous policies of extermination of European Jews, Gypsies, and homosexuals in World War II were carried out through a bureaucratic system. At the same time, the Allied forces that eventually defeated Hitler's armies were also bureaucracies. Bureaucracies themselves are probably neither good nor bad in and of themselves, but they can be put to good or evil uses. In addition, the dysfunctions associated with bureaucracies can perhaps be minimized through a better understanding of how they arise. Many sociologists who study organizations today hope that by exploring how large organizations work, we can develop structures that not only effectively carry out tasks but also best serve the interests of all people.[12]

We will return to this issue at the end of this chapter. Now, however, we turn to an examination of how contemporary sociologists have built on Weber's work to analyze formal organizations. These insights can help you apply a sociological perspective and better understand why organizations work in the way they do and why they either adapt and change or resist changing.

ORGANIZATIONS: A MESOLEVEL ANALYSIS

Based on Weber's description of a bureaucracy, you might think that all organizations look the same, much the way tract houses might all be built from the same plan. Actually formal organizations can take on a wide variety of shapes, just as houses are built to many different specifications. Further, just as the design of a house affects the lives of the people who live there and the degree to which it serves their needs, so, too, do the design and nature of organizations affect both people's lives and organizations' chances of serving their purpose and surviving changes in the environment. Many sociologists have taken a mesolevel view of individual organizations, focusing on (1) their structure, or the way they are organized, (2) their technology, or the type of work they do, and (3) their culture, or way of life.

Structure

When contemporary sociologists talk about **organizational structure,** they are referring to how the various parts of an organization are arranged, much the way an architect talks about the design, or structure, of a building. A major influence on how organizations are structured is simply their size.[13] Just as a building that is designed for only one person need not have many rooms, so can organizations that are designed for only a few people have fairly simple structures. As organizations become larger, however, they tend to develop much more complicated, formal, and centralized structures.[14] Each of these aspects can be seen in Weber's ideal model of bureaucracy, but contemporary theorists tend to focus on how they vary from one organization to another and even from one part of an organization to another.

Complexity The extent to which the work of an organization is broken up and differentiated among various units reflects its **organizational complexity.** One way of measuring complexity is to look at **horizontal differentiation,** the extent to which the work of an organization is divided up among different units or subgroups. As illustrated in Figure 11-2, some organizations, such as colleges and universities, have a great deal of horizontal differentiation, with several different departments and colleges assigned to teach different subjects. Other organizations, such as the College Sounds singing group, have little formal division of labor beyond the election of officers and assignment of committee responsibilities.

A second way of measuring complexity is to look at **vertical differentiation,** the number of supervisory levels in an organization. For instance, the United States Army, as depicted in Figure 11-1, has a great deal of vertical differentiation, whereas the College Sounds group has very little.

Finally, complexity can be measured by looking at **spatial dispersion,** the extent to which the various units of the organization are spread out in different locations.[15] For example, the Roman Catholic church is a very large organization with a great deal of spatial dispersion in the form of churches scattered throughout the world. Similarly the fast-food organization that Yolanda works for has outlets nationwide. In contrast, her university has little spatial dispersion, with all the units concentrated within the area of a few city blocks.

In general, organizations with greater complexity will have more problems with communication, coordination between the various units, and social control.[16] The more units in an organization, the more difficult it is to ensure that all the various parts communicate with one another and work toward the same goals. For this reason complex organizations, such as large universities or fast-food chains, often invest a great deal of money in communications consultants and full-time staff who create internal newsletters and other means of sharing information and passing on directives. Less complex organizations, such as the College Sounds or businesses with only a few employees, can rely more on face-to-face contact and informal patterns of communication.

Despite this need to work harder on communication and coordination, sociologists suggest, complex organizations also tend to experience more change. Because they involve more specialized roles, complex organizations also tend to have employees with more knowledge and expertise. These highly trained employees are often more willing to try innovations.[17]

Formalization The extent to which an organization uses written rules and procedures to control individuals within it is its level of **organizational formalization.** Behavior in highly formalized organizations is highly specified, and individuals have little leeway in how they choose to do their jobs. For instance, Harry's Hamburgers has extensive written rules dictating everything from the number of pickles to put on a hamburger to the words employees should use in greeting customers. In less formalized organizations individuals typically enjoy much greater freedom in carrying out their role responsibilities. Thus, as a member of the Dorm Council, Yolanda can dress as she desires and, within limits, use her own best judgment in dealing with the complaints of a dorm resident. In her role on the committee in charge of transportation for the College Sounds' concert tour, no one tells her which bus and lodging companies to call.

Formalization is probably the aspect of organizational structure that most affects how we feel about our experiences in an organization. For instance, Yolanda enjoys figuring out how to solve dorm residents' problems and how to schedule the College Sounds' concert tour, but she often tires of the routine at the restaurant.[18] At the same time, however, customers at Harry's Hamburgers might appreciate being able to count on very similar service and food at all the chain's outlets.

The amount of formalization within an organization also influences the possibility that it will be able to innovate and change. In general, the more formal-

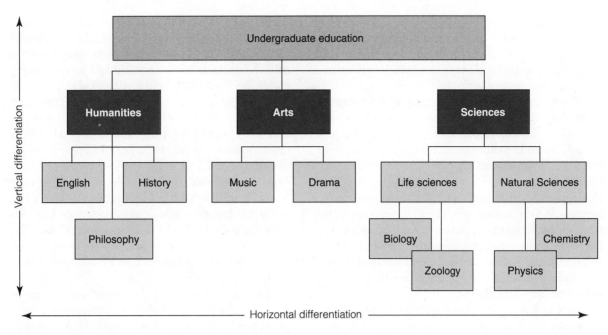

FIGURE 11-2 *A Partial View of Horizontal and Vertical Differentiation in the Organizational Structure of a College*

ized an organization, the less flexible it is. Rules within highly formalized organizations tend to set limits and discourage workers from suggesting new patterns of behavior or even from preventing disasters.[19] You no doubt remember the explosion of the space shuttle Challenger, in which several astronauts, including schoolteacher Christa McAuliffe, perished. Subsequent investigations by NASA and Morton Thiokol, the company responsible for producing the defective O-rings that caused the disaster, revealed that engineers at Thiokol had warned NASA officials that the rings might malfunction. Tragically, officials higher up in the organizational hierarchy very much wanted the launch to proceed as scheduled, and none of the workers who were concerned about the O-rings had the authority to call it off. As a result, seven lives were lost.[20]

Centralization The extent to which organizational power is centralized and decisions are made hierarchically is referred to as **organizational centralization.** An organization like the United States Army or the Roman Catholic church, in which decisions are made hierarchically (from the top down), is said to be highly centralized. Where people farther down the chain of command can make a range of decisions, the organizational structure is said to be less centralized or more decentralized. For example, even in school districts employing thousands of teachers, individual teachers typically make many significant decisions about curriculum, teaching methods, and discipline.

As you might expect, more decentralized organizations are likely to show more innovation, especially when the rank and file are highly educated or are professionals.[21] For instance, loud music and talking are a constant problem in the dorms at Yolanda's university. In a highly centralized organization, such as the army's military academy at West Point, an edict might be issued by the higher echelons prescribing quiet hours and the volume at which sound systems may be set, if they are permitted at all. In contrast, in decentralized organizations people at lower levels have greater freedom to implement their own solutions. For instance, realizing that noise is a problem in all the dorms, Yolanda and the other members of the Dorm Council might discuss the issue at one of their regular meetings and then meet with the residents in their own dorm units to try to develop rules about quiet hours and noise levels and ways that they could be enforced. Each dorm might come up with slightly different regulations, depending on the preferences of the residents. (To use the ideas introduced in Chapter 6, we could say that the residents *legitimized* the norms regarding noise control.)

Informal Structure Often the formal blueprint for a house doesn't give a full picture of how the various rooms will be used or how life will unfold once the house is occupied. Perhaps your house has a formal dining room but your family rarely eats there; or perhaps you and your guests tend to gather in the kitchen rather than the living room. Similarly the formal structures displayed in Figures 11-1 and 11-2 don't reveal all the interactions and relationships that occur in organizations because many associations skip hierarchical steps or jump boundaries from one unit to another.

These relationships reflect an **informal structure,** a social network within an organization that exists alongside the formal structure.[22] For instance, even though Yolanda's supervisor at work is the branch manager, Yolanda also happens to be friends with the district manager, whom she met while working at the Harry's outlet in her hometown. Although she is supposed to discuss any work-related concerns with her own outlet manager, she often finds herself talking to the district manager about these issues as well. In fact, through these conversations she found out about the possible closure of the Harry's mall outlet before her manager did!

These informal structures are found in all kinds of organizations, and your location in these networks can affect your relationships with others and your effectiveness in the group, no matter what your formal position is. For instance, many organizations, such as large businesses, colleges and universities, and even the United States Congress, provide gymnasium facilities and sponsor leisure activities such as golf tournaments and bowling teams for their workers, enabling individuals from many different areas of the organization to get to know one another. As one consequence, business deals, university curricula, and fine points in congressional legislation can be worked out in conversations in the locker room or on the golf links. Of course people who don't participate in these activities are excluded from these deliberations. Further, like many areas of social life, these leisure activities tend to be sex-typed. For instance, men and women have separate locker rooms, and men are much more likely than women are to play golf and to have the time to participate in tournaments. As a result, as women have entered higher-level positions in businesses and universities and been elected to Congress, they have often missed out on these "behind-the-scenes" deals.

Because of informal structures, high-level positions in an organizational hierarchy are not the only ones that allow the occupants to influence the actions and behaviors of others. Even when a formal outline of the organizational structure suggests that they were very low in the hierarchy, some positions actually can be quite powerful. The experiences of a student named Becky illustrate this phenomenon. Becky attended a small junior college as a work-study student. Because she was assigned the scheduling job in the media center of the college library, all teachers and students who wanted access to films or videos had to go through her. In addition, because she knew how to work audiovisual equipment, she was allowed to fix broken machines. Thus, even though she received extremely low wages and had no permanent status, Becky actually held a great deal of power on campus simply by virtue of her position. If someone was rude to her, she could easily lose a film order or take her sweet time fixing a machine needed for a class. Anyone who wanted to use a film or a projector soon learned that it was not wise to offend Becky. You can probably think of a few "Beckys" in your own life—people who hold positions that aren't formally recognized as powerful or important but who have the ability to control some aspect of your life, such as what type of equipment you can use, when you can get an appointment, or how long you must wait for something.

Elements of exchange theory and the concept of social stratification can help us understand how people in positions such as Becky's can acquire power. Exchange theory suggests that if you depend more on a relationship than someone else does, or if the relationship is more important to you than it is to the other person, you will have less power within that relationship. If you apply this notion to organizational structures, you can see how some positions within organizations can be more powerful than others are. The powerful positions are those that we depend on and can't do without. Sociologists suggest that positions and subunits within organizations become more powerful when they possess three characteristics: (1) They can successfully cope with uncertain situations, (2) other units can't be substituted for them, and (3) their position within the organization is central and other units are highly linked with them.[23]

According to the functionalist theory of stratification, units that are less replaceable should tend to garner more power. For instance, Becky's power didn't derive from her intellect or skills, because others could easily learn to do her job. Instead, her power depended on her position, which, despite its low level in the hierarchy, was pivotal to the work of other units in the organization. There was simply no other unit that could replace hers.

Both an organization's informal structure and the existence of formally unrecognized but powerful positions are important in understanding change in organizations. For instance, if plans for change are formulated only by those who hold formal power, those with informal power can easily sabotage those plans. Similarly informal lines of communication can actually be much more effective than formal communication lines at spreading official information, as well as rumors and gossip. The frequent failure of large organizations to carry out their

Both this man at his loom and the people in this large factory are in the textile business. The factory, however, has a complex structure and a great deal of formal organization. Why do you think this organizational form developed?

leaders' wishes is notorious. For example, during a tense confrontation between the former Soviet Union and the United States in the early 1960s, a U.S. pilot strayed into Soviet air space, leading President John F. Kennedy to comment, "There's always some S.O.B. who doesn't get the word."[24] Many chief executives, like Kennedy, recognize the importance of informal structures in facilitating or hindering their directives. Whatever the organizational chart says, these executives know that their attempts at change will be more effective if they recognize and use these informal structures. Similarly, individual workers will be more effective when they are linked into these informal networks.

Technology

For many years people who studied organizations thought that it might be possible to find "one best system" for all organizations. Much as some of Weber's writing implies, these theorists thought that one optimal type of organizational structure might exist that could be adopted by all kinds of groups. Social scientists now realize that this is not the case. The type of structure that works well for schools generally is not best for small businesses or religious organizations. The structure that works well for local charitable organizations might not work well at all for large corporations. Just as architects design the building that best suits their clients' needs, people who design organizations must develop the structure that best meets the needs of the organization.

One primary determinant in this task is the technology that an organization uses. **Organizational technology** includes the skills and knowledge of workers in an organization, as well as the characteristics of the materials they work with and the machines they use. Even these terms are used broadly, for the "materials" for some types of organizations, such as schools or hospitals, are actually people; and "machines" can refer to all the various diagnostic and ancillary materials that teachers or health professionals use, such as textbooks, testing apparatus, and laboratory equipment. Because organizations vary so widely in their technology, sociologists have tried to develop ways to divide the concept into specific dimensions. Two that have proved useful are *complexity* and *uncertainty*.[25]

The technology an organization works with can be more or less complex and diverse. For instance, a firm that manufactures commuter airplanes produces a much more complex product than does a firm that processes chickens to sell in supermarkets. Similarly an elite private school in a small upper-middle-class community will probably have a less diverse technology than will a large public school that serves students of all ability levels from a broad range of social classes.

In general, organizations that have greater technological complexity and diversity also tend to have more structural complexity.[26] Thus we could expect that an airplane manufacturing plant would be more structurally complex than a chicken-processing firm would, because more departments and divisions would be needed to handle the production of all of the various elements of the aircraft. Likewise, in contrast to the elite private school, the big public school might have multiple tracks and a wide range of specialized teachers to accommodate students with different needs and skills.

Some technologies are much less certain and predictable than others. For instance, the raw food that helpers in fast-food restaurants must prepare is carefully designed to be exactly the same, every day and in all restaurants. A Big Mac in San Diego will look exactly like a Big Mac in Boston because the McDonald's corporation controls the shape and design of the food that each restaurant receives. Cooking in McDonald's is a fairly predictable task. In contrast, elementary school teachers can be faced with thirty children of widely differing abilities, personalities, and family backgrounds. Teachers must tailor their interactions with these children in ways that will best suit each student's particular needs and learning styles.

Organizations that deal with less certain or predictable technologies also tend to have less formalized and centralized structures.[27] Thus an elementary school teacher would face far less rigid rules and have much more discretion in his work than would a counterperson or cook in a fast-food restaurant. In general, an organization's technology influences the types of structures that best suit the organization. That is, if an organization is to be successful, its structure must be suited for the technology it uses.

Organizational Culture

Just as an architect's blueprint can't reveal what a house will look like after the inhabitants have moved in, understanding an organization's structure and technology can't tell us about what it is like to participate in the organization on a daily basis. Perhaps, like Yolanda, you have attended two different colleges. Even if each of these schools had similar structures and technology, their way of life could have been vastly different.

For instance, the first college Yolanda attended, City College, had started as a night school that students could attend in the evening after a full day's work. When Yolanda was there, the school still had many students who worked during the day, and faculty and students alike were proud of the fact that classes were scheduled from early morning through late evening. A number of the students at City College liked to think of themselves as "working students." Stories (probably apocryphal) were told of students who went to evening classes after finishing work and were so tired when class was over that they slept in the empty classroom and went directly to

work the next morning. Many of the students had never thought that they would be able to go to college, and most were very serious about their schoolwork. They fretted over their grades and consulted with their teachers regularly. The majority hoped that their education would help them get better jobs than they currently had.

The culture of Yolanda's current school is strikingly different. State University is informally known as "Party U." In fact, parties are a mainstay of the school's social life, and stories are told and retold of parties that went on for days. In contrast to City College, most of the students at State started right after high school. Like Yolanda, many students work part-time in the local community, but most of them live on campus rather than commute to school. Relatively few students try to do more than the minimum schoolwork needed to get by. They rarely talk to their teachers outside of class and generally don't start worrying about what they will do after college until their senior year. Whereas alumni of City College often think about how hard they worked during their college years, alumni of State University remember how much fun they had. And whereas City College alums tend to stay in contact with their former teachers, State University alums are more likely to keep in touch with their friends.

A sociologist would say that these two schools had very different **organizational cultures,** or ways of life. Organizational cultures include all the elements of culture described in Chapter 3, such as norms, values, belief systems, ideologies, folklore, myths, and symbols. Organizational cultures help people feel part of the group. Thus, students at both City College and State University identify with their schools, much as they also come to identify with other subcultures to which they belong. This identity helps them feel connected to their school and to see its functionings, or operations, as legitimate or right.[28] Similar dynamics are at work in other kinds of organizations. For example, much has been made of the contrasting corporate cultures of "button-down" computer giant IBM and its smaller, more freewheeling rival, Apple Computer—companies in the same industry that are as distinct as a regulation blue suit and a pair of faded Levi's.

Even though identification with an organizational culture can contribute to a sense of group belongingness, it can also mitigate against change.[29] For instance, members of Congress and various state legislatures generally also belong to either the Democratic or Republican party. Because their party affiliation usually occurred long before their election, many of these legislators tend to identify more strongly with their party than with Congress or their state legislature. As members of a political party, they are expected to take strong stands on issues and display the "party line." Yet, to be effective legislators, they often need to step across party lines and collaborate with their political adversaries. Several commentators have suggested that this strong party identification is a major factor in "congressional gridlock," the failure of legislative bodies to work together, across party lines, to solve major social problems.

As individuals, we can help decide how we will structure organizations, and we all participate in developing and perpetuating organizational cultures. For instance, as Yolanda makes plans for her own restaurant, she will have to decide how it will be structured, how formalized it will be, and what technology will be used. As her plans come to fruition, she and her employees will develop an organizational culture, one that will no doubt be influenced by other work experiences they have had, but also one that can be identified with Yolanda's own business.

In addition, however, the ways in which organizations are designed affect both how we make decisions in organizations and how good these decisions are. The structure of organizations—both the recognized formal design and informal structure—influences who has the power to make decisions and how individuals communicate with one another. For instance, Yolanda's father is frustrated in his job as a loan officer because the new organizational structure excludes him from decisions he once participated in.

An organization's technology also influences the nature of decisions that it must make. When organizations have more complex and uncertain technologies, they find it more difficult to fall back on "usual" or "tried-and-true" practices and must devote more energy to solving problems.

Finally, an organization's culture and individuals' identification with the organization affect the ways in which decisions are made. For instance, an organization that prides itself on maintaining harmonious relationships among its members can become overly focused on developing consensus, to its own detriment. This process of **groupthink,** whereby group members become so oriented to maintaining the cohesiveness of the group that they ignore or suppress information that might be critical to their decisions, can sometimes have disastrous results. For instance, historians who have studied several major U.S. foreign policy fiascoes—from the lack of preparation for the attack on Pearl Harbor to the stalemate in the Korean War, the escalation of the war in Vietnam, and the Bay of Pigs invasion—suggest that in each case the decision-making group had developed such cohesion that it ignored contrary and critical information.[30]

Even though formal charts of an organization's design suggest that decisions will be rationally considered, studies of the actual decision-making process suggest that it more often reflects a **bounded rationality**—that is, the structure, technology, and culture of organizations all provide boundaries in which individuals make decisions. In this sense all of the decisions we make as individuals within organizations are bounded or constrained by the nature of the organization itself.[31] Sociologists have often studied how organizational structures and culture affect individuals. To learn about one sociologist's analyses of how organizations affect workers in Japan and the United States, see Box 11-1, Sociologists at Work.

Many decisions that organizations must make involve reactions to the organization's environment. Just as the environments in which we live influence the kinds of houses we build, the surrounding environment influences the structures and cultures that an organization develops. We turn now to work that focuses on the environments surrounding organizations.

ORGANIZATIONAL ENVIRONMENTS: A MACROLEVEL VIEW

The environments in which we live are constantly changing. If organizations—whether Harry's Hamburgers, the Internal Revenue Service, or your local church—are to serve people as well as they could and if they are to survive and prosper, they need to adapt to these changes. Several sociologists have examined how organizations adapt to environmental changes. Other sociologists have focused on why some organizations are more likely than others are to survive these changes.

Adapting to the Environment

Sociologists have described three general changes in organizational environments: (1) changes in technology and knowledge, (2) changes in other organizations and their capacity and needs, and (3) changes in norms and expectations in the institutions that surround them. The ways in which organizations adapt to these changes affect the probability that they can survive and prosper.

New Technology and Knowledge Technological advances and new knowledge appear virtually every day. A good example today is the explosive growth of the Internet. A few years ago, relatively few people beyond a small circle of engineers and scientists were even aware of the existence of this "network of networks." Today businesses, universities, and many

SOCIOLOGISTS AT WORK

James Lincoln

Motivation, commitment, and discipline are shaped by the actual work environment and organization, not so much by culture.

James Lincoln, a sociologist who teaches in the Haas School of Business at the University of California at Berkeley, has studied formal organizations for many years. Some of his work involves comparing business organizations in America and Japan.

How did you get interested in studying Japanese and American business organizations? I was an assistant professor at the University of Southern California in the 1970s. I had a Japanese graduate student, Mitsuyo Hanada, with whom I did surveys and interviews at fifty Japanese-owned companies in Southern California. I was interested in cultural effects on organizations, and not much had been done beyond James Abegglen's book *The Japanese Factory.* It seemed like an important area and subject. Also, there were many companies in the Los Angeles region available to study, and my student's father was a prominent Japanese businessman, which made access easier.

At Indiana University I did the study with Arne Kalleberg described in *Culture, Control, and Commitment* on 106 factories and 8,302 employees in Japan and Indiana, collaborating again with Hanada. This was an opportunity to do a large-scale study of cultural differences in a work organization.

What has surprised you most about the results you obtained in your research? One surprise was that levels of commitment to the company and job satisfaction were both higher among American workers than Japanese workers, despite expectations. An explanation may be the different reference points that Americans and Japanese may be using: Ameri-

cans seem to compare their job satisfaction to how they might feel if laid off or unemployed, while Japanese seem to compare job satisfaction to an abstract ideal. So if Americans' working environments are disappointing, their expectations weren't as high as the Japanese workers' to begin with.

The other surprise is the fact that welfare-corporatist organizational practices originating in Japan such as quality circles, company-sponsored courses, and secure employment have the same effects on American workers' attitudes. The benefits of these practices don't seem to be rooted in the culture and can be successfully transplanted to American organizations. For a variety of reasons this organization is further along in Japan and thus workers are generally more committed and disciplined.

What do you think the implications of your work are for social policy? Our core finding was that motivation, commitment, and discipline are shaped by the actual work environment and organization, not so much by culture. American and European managers who provide workers with secure employment, plentiful benefits, and career ladders will get a more committed and disciplined work force, leading to real payoffs for the organization.

How does your research apply to students' lives? You spend a lot of time in the workplace and that environment has a lot to do with how you feel day to day. Understanding something about work, authority, and formal rules can help you understand the workplace and your position in it better.

other organizations are compelled to establish a presence in the net to communicate with one another and with potential customers. Similarly computer technology has dramatically changed retail stores. Supermarkets, bookstores, department stores, and clothing boutiques have computerized their inventory, installed scanning devices, and developed linkages with

banks for credit card and ATM transactions. Because many of these changes aid customers, stores that haven't adopted them find themselves at a disadvantage. Recall how the introduction of tax preparation software for home computers has drastically altered Yolanda's mother's bookkeeping business, to the point where it might not survive.

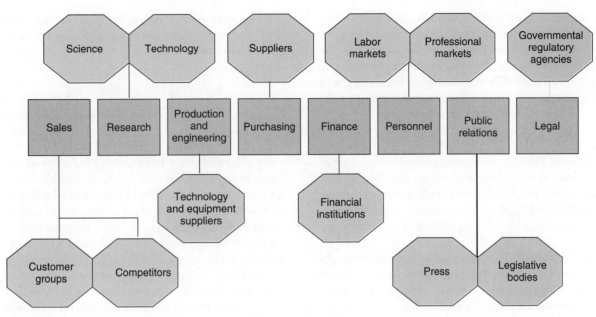

FIGURE 11-3 *Organizations in the Environment of the Amalgamated Widget Company*
Source: Lawrence and Lorsch, 1992, p. 231.

Technological advances have affected all kinds of organizations. For example, until the mid 1950s the March of Dimes organization was devoted solely to soliciting money to find a cure for polio, a crippling disease that swept the country each year. When Jonas Salk developed a vaccine that effectively wiped out the disease overnight, the March of Dimes could have claimed success and folded up shop. However, many people worked for the March of Dimes, and they didn't want to lose their jobs. In addition, the organization had a very effective fund-raising structure in place. Therefore, as a means of survival, the organization adapted to this new knowledge and turned its attention to the problems of birth defects.[32] Today the March of Dimes is still going strong, gathering large sums of money to fight this enormous health problem.

Other Organizations In addition to the knowledge and technology of the surrounding world, an organization's environment involves other organizations that interact with it. Because modern societies contain so many different and highly interrelated organizations, this aspect of organizational environments is extremely complex.

To illustrate this complexity, Figure 11-3 lists some of the organizations that form the environment of different units of the Amalgamated Widget Company. As you can see, AWC has a fair amount of horizontal differentiation, with various units responsible for sales, research, production, finance, personnel, legal issues, and so on. Besides communications that occur between units, units frequently interact with a range of other organizations. For instance, the sales unit must be concerned with the actions and concerns of customers, who might be organized in consumer protection or lobbying organizations, as well as with competitors who make similar products. The personnel department often interacts with labor unions and professional organizations, as well as with government agencies that enforce regulations regarding employees. The finance department deals with financial institutions that provide loans, sell and buy stock, and perform other services.

We can use the ideas of social networks and exchange theory to understand more about how the other organizations in the environment affect an organization's chances of success. Interorganizational relationships involve social networks—both formal and informal links between organizations. These relationships also involve ongoing exchanges between groups that are more or less rewarding or costly. If an organization is to survive, the costs of these exchanges must not exceed the rewards. For this reason organizations constantly must decide how best to respond to changes that occur within this interorganizational environment.[33]

For instance, the sales department of AWC might suddenly be faced with competition from a new firm with highly innovative widgets. AWC must then decide how to respond, whether by introducing its own new product, starting a different advertising campaign, or simply ignoring the upstart and hoping it fails. Or perhaps a group of consumers, who previously had been only writing letters of complaint about the quality of AWC's widgets, has started to air its complaints in the media. Now the sales department, along with the public relations unit, must figure out how to deal with this new problem. Meanwhile, the personnel division might face a decline in available workers when other manufacturing firms move into the area, or new leadership in the local union may demand changes in workers' contracts.

Like all organizations, AWC must figure out how to deal with each of the changes related to its associations or linkages with other organizations. Just as our lives as individuals are built on our daily interactions with many different people in our social networks, the life of an organization is significantly affected by its interactions with other organizations.

Social Norms To a large extent, organizations' decisions are influenced by ideas of what different types of organizations "should" be like. These **institutional norms** prescribe appropriate structures and behaviors of organizations and other aspects of social institutions. Organizations, like individuals, are expected to conform to these norms, and those that more readily conform are more likely to survive. For instance, the social institution of education is defined in part by norms that concern schooling. In our society most of us assume that schools should have classrooms and teachers who instruct students on a regular schedule and formally assess their work. Those who depart from these norms, such as "home schoolers" who educate their own children, can be seen as not providing "real" schooling.[34] Similarly we assume that electronics stores should display a variety of products and have clerks available to describe the merchandise and explain how to use it. We assume that a doctor's office should have an attractive reception area, a nurse who escorts us to a private examining room, and framed diplomas hanging on the wall. If stores or doctors' offices don't conform to these expectations, we might lack confidence in them. In sociological terms we could say that when organizations conform to these norms, we tend to see them as *legitimate*. As a result these organizations are more likely to be successful.[35]

A by-product of this legitimation is that we tend to ignore the behavior or performance of organizations as long as they seem to conform to institutional norms. This helps account for our surprise when we find that the doctor with the luxurious office was actually practicing without a license or that the store with the attractive displays was knowingly selling defective products.[36] The idea of legitimation also helps explain how hard it is for organizations to break new ground by being the first to introduce a new product or to provide a new service. It's not easy to be the "new kid on the block," whether as an individual or an organization. Organizations whose actions counter existing norms are more likely to fail; those that conform to institutional norms are more likely to succeed.

Surviving the Environment

Despite the various pressures on organizations to continually change and adapt, much research indicates that organizations tend to resist change, especially after they have been around for a few years. Once an organization establishes a workable structure, inertia seems to set in.[37] Instead of adapting to environmental changes, many organizations simply continue with "business as usual." Eventually they disappear or die, often to be replaced by other organizations.

Instead of looking at how organizations adapt to their environment, some sociologists have pulled back even further to examine the process by which organizations develop, die, and are replaced. Using an analogy to studies of biological systems, these theorists call their work **organizational ecology.**[38] The term *ecology* refers to the branch of science that studies the relationship between organisms and their environment. Whereas biological ecologists study living creatures, organizational ecologists study the relationship between organizations and their environments.

Organizational ecologists often study entire populations of organizations over long periods of time to see when they are founded and when they disappear or die. One of their most important findings is the discovery of what seems to be a common pattern in the *density,* or number, of organizations found within a population of organizations at any one time. In some ways these patterns tend to parallel those seen in living populations, although, of course, these patterns exist for quite different reasons.

The sociologists Michael Hannan and Glenn Carroll helped develop this perspective. In one of their studies they conducted detailed research into the history of seven different types of organizations, beginning when the very first organization of each type appeared. The seven populations of organizations were (1) national labor unions in the United States from 1836 to 1985, (2) beer brewing firms in the United States from 1633 to 1988, (3) life insurance companies in the United States from 1759 to 1937, (4) banks in the borough of Manhattan in New York City from 1791 to 1985, (5) newspapers in Argentina from 1800 to 1900, (6) newspapers in Ireland from 1800 to 1975, and (7) newspapers in the San Francisco Bay Area from 1845 to 1975.[39] Figure 11-4 depicts the density of each of these different populations over time. In years when more organizations are present, the density is high; in years when fewer organizations are present, the density is low.

Each of these populations has had many organizations, but few survived for very long. For instance, between 1845 and 1975, there were more than 2,000 newspapers in the San Francisco Bay Area, but fewer than 250 were left in 1975 (see Figure 11-4g).[40] Similarly, before 1980 almost 500 banks had been established in Manhattan, but by 1980 only 67 were still in operation (see Figure 11-4d).[41]

Just as physicists calculate the half-life of atomic particles, organizational ecologists can calculate the half-life of organizations—how long it takes for half of the organizations in a population to die. This figure differs greatly from one type of organization to another. Among the organizations Hannan and Carroll studied, labor unions and banks tended to have the greatest longevity, with half-lives of sixteen and fifteen years, respectively. Newspapers had much shorter life spans. For example, half of the nineteenth-century Argentine papers lasted less than a year, and half of the San Francisco Bay Area papers had disappeared within about seven years (see Figures 11-4e and 11-4g).

The organizations Hannan and Carroll studied differ from each other in many ways: when they existed, how long they survived, what technologies they used, and where they were located. Yet, despite these differences, the graphs in Figure 11-4 have remarkably similar forms. Initially organizational density is fairly low, but then many more organizations appear, culminating in a very high density of organizations. For instance, there were very few labor unions in the United States until the Civil War era (see Figure 11-4a). Then the numbers grew modestly until about the mid 1880s, when many more suddenly began to appear. Modest growth occurred in the early twentieth century until the density of labor unions reached a peak in the mid 1950s. Similarly the number of Manhattan banks increased rapidly from 1791 until the early 1900s, with a few temporary drops in density that coincided with economic panics in the 1840s and 1870s (see Figure 11-4d). In both cases, however, this period of very high density was quickly followed by a substantial drop in the number of organizations and then a period of relative stability.

Even the data for the American brewing firms fit this general pattern (see Figure 11-4b). During the Prohibition era of the 1920s (when the sale of alcoholic beverages was prohibited by the Eighteenth Amendment to the U.S. Constitution), no firms were allowed to operate, as indicated by the years when no organizations are noted on the graph. But as soon as Prohibition ended, there were almost as many firms operating as before, and the general trend of declining density that was established before Prohibition continued.

Similar density patterns have been observed with other sets of organizations. Organizational ecologists suggest that these patterns reflect the relationship of density to the *legitimation* of organizations and the *competition* among them. As more organizations enter the population, they tend to be seen as more legitimate, or as an acceptable part of the social world. For instance, when fast-food restaurants such as McDonald's first appeared, people found the idea of ordering "instant" food from a very limited menu to be somewhat unusual. Gradually, however, the idea began to catch on, and the notion of fast-food restaurants became much more acceptable or legitimated. Not surprisingly, when a type of organization becomes more legitimate, more of them tend to appear. As people come to see an organization as viable, they will be more likely to try to start a similar one. At this point the number of organizations starts to increase dramatically, and the density becomes much greater.

Just as biological environments can support only so many organisms, however, within any kind of organizational population only so much demand exists. People can eat only so many hamburgers or read only so many local newspapers. Just as organisms compete with one another for survival in biological environments, so do organizations compete within their environments. In high-density organizational populations competition tends to be greater. Some organizations inevitably are going to disappear, and fewer new organizations will be founded. Eventually the density will fall to a relatively stable level, one that organizational ecologists suggest can be supported by the environment. Even with this new stability, however, organizations come and go because some are more likely than others to be able to adapt to changing environmental conditions.[42]

Organizational ecologists have tested their theoretical ideas with complicated mathematical models. The results, with a large variety of different organizational types, generally support their theory about the relationship between density, legitimation, competition, and organizational births and deaths.[43] Whether the population of organizations includes fast-food restaurants, video stores, religious groups, hospitals, or colleges, organizational ecologists hypothesize that over time their density will follow the patterns shown in Figure 11-4. Harry's Hamburgers, where Yolanda works, might be falling victim to this pattern. Metro Mall contains many fast-food outlets, and Harry's doesn't seem to be changing fast enough to attract the customers necessary to make a profit. As Yolanda thinks about starting her own restaurant, her chances of having one that will survive will no doubt improve if she considers how it will fit into the organizational ecology.

Organizational ecologists take a broad macro-level view of organizations that spans many years and many different organizations. If, however, we

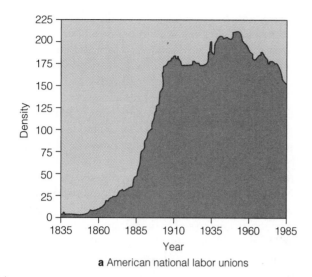

a American national labor unions

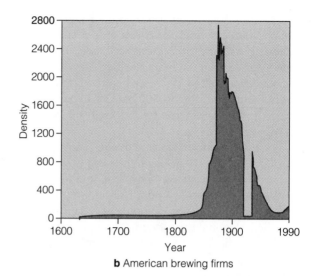

b American brewing firms

FIGURE 11-4 *Density of Organizations Studied by Hannan and Carroll*

Source: Hannan and Carroll, 1992, pp. 7, 8, 10, 12.

e Argentine newspapers

bring our sociological lens inside a single organization, we begin to get a very different picture. We can see that individuals decide to found an organization. They also decide how it will be structured, act in ways that influence the nature of its climate, decide how it should adapt to change, and, eventually, might decide that it must end. Several sociologists have analyzed organizations from this microlevel perspective. Much of this work has focused on **organizational leaders,** people within organizations who are able to influence the things others do and believe.

ORGANIZATIONS AND INDIVIDUALS: A MICROLEVEL PERSPECTIVE

Sociologists who have studied leadership emphasize that it differs from power. To put it simply, all leaders

are powerful, but not all people with power are leaders. People in powerful positions in organizations can compel members to follow directives simply because their position in the organization's structure gives them authority to do so. Using the distinctions developed by Weber and introduced in Chapter 6, we would say that their authority has a legal basis. Leaders, too, are powerful, but their power stems from our willingness to follow them. Unlike formal power, leaders' power comes from their social interactions with other people. It is not simply given to them because of the status they hold in an organization; it develops out of their interactions with others in the group and thus resembles the charismatic authority described by Weber and discussed in Chapter 6.[44] In short, leadership is attributed to people by their followers, not by the organization.

Yolanda's experience at Harry's Hamburgers illustrates the difference between leadership and

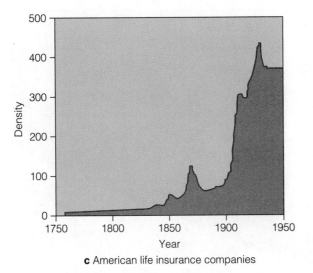

c American life insurance companies

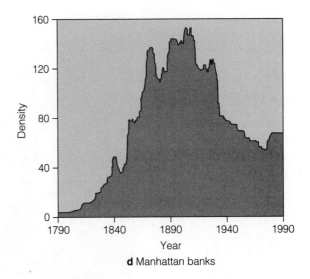

d Manhattan banks

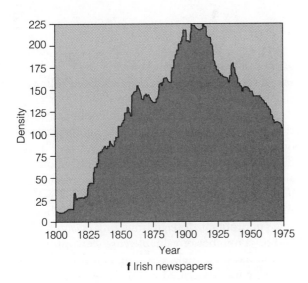

f Irish newspapers

g San Francisco Bay Area newspapers

power. In both of the Harry's outlets where Yolanda worked, the food, the decor, and the rules and regulations governing employees were exactly the same. Yet at the outlet in her hometown, no one seemed to enjoy work, employee turnover was high, and the workers habitually mocked their manager and the customers. Sometimes the workers would even dream up ways to sabotage the restaurant and "get back" at the manager for something they didn't like, such as a change in schedules. One of Yolanda's co-workers, Greg, was especially good at devising practical jokes, and the other workers delighted in helping him carry them out. In contrast, at the Harry's outlet in Metro Mall, the workers seemed proud of the company and often volunteered to work extra hours. They had fun together, but not at the expense of the manager or customers. The workers respected and truly liked the manager, and he seemed to like and respect them as well.

At both restaurants the manager held the most powerful position in the organization. But only the manager at the mall outlet was truly a leader, in the sense that his subordinates legitimated his authority and granted him power. In fact, in Yolanda's hometown outlet Greg could be seen as an organizational leader, even though he did not hold a formal position of power in the organization, because he was able to influence others to do things he wanted to do. Unfortunately, although his leadership skills helped his co-workers counter their tedium and boredom, they did nothing to help retain customers or increase profits at the restaurant.

Yolanda's experiences at Harry's also illustrate two distinct facets of leadership. **Instrumental leadership** deals with the tasks or work of an organization; **expressive leadership** deals with interpersonal relationships within a group or organization. Organizations and groups often have two leaders, one specializing in

All groups, even nonhuman ones, seem to have leaders. Sociologists are still trying to understand more about what leadership is and how it affects organizations.

task-related or instrumental activities and the other in relational or expressive activities. In both of the restaurants Yolanda worked in, the manager was the instrumental leader (although the manager at the mall outlet was much more effective than the one in her hometown was). The managers organized the work schedules and supervised the employees. In her hometown restaurant, however, Greg was the expressive leader, the one who helped the other employees feel connected to one another and part of the same group.

Sometimes the same person will emerge as the instrumental and expressive leader; sometimes, as in the two Harry's Hamburger outlets, different people will fill the two roles. Either way, both types of leadership contribute to the success of organizations. Organizations, and especially small work groups, often operate more smoothly if tasks are accomplished *and* if the people within the organization have positive feelings about the organization and their relationships with one another.[45]

Based on the discussion in this chapter, you know that the characteristics and environment of an organization can strongly influence its chances of success and its ability to change. How much difference, then, can a leader make to an organization? Leaders must operate with the resources and constraints that they find in an organization, and as the old saying goes, "You can't make a silk purse out of a sow's ear." If an organization has paltry resources, strong competition, and an inappropriate structure, any one leader will find it difficult to have a strong immediate impact on its chances for success.[46] For instance, professional and college coaches are often fired when their teams have losing seasons. Yet new coaches are faced with the same players and, often, the same or-

ganizational structures and budget. Dramatic reversals in sports teams' fortunes immediately following coaching changes are, in fact, rare.[47]

At the same time, you have no doubt heard stories in the media about leaders taking organizations on the brink of bankruptcy or otherwise "in the dumps" and turning them around. One famous example involves Lee Iacocca and Chrysler Motor Corporation. By the mid 1970s Chrysler was losing huge sums of money annually. The company had bought some unprofitable foreign firms and, because it had laid off a number of valuable engineers, produced several flawed models, resulting in massive recalls. Chrysler was described as a mess—so much of a mess that even a $1-billion cut in operating expenses left the corporation in deep trouble.

Into this situation stepped Lee Iacocca, a long-term Ford executive who headed the team that built the 1960s classic, the Mustang. Chrysler hired him as chief executive officer in a last-ditch effort to save the company. Enormously confident, Iacocca inspired other members of the company to continue to work hard. He successfully lobbied Congress to guarantee Chrysler $1.5 billion in bank loans and convinced the president of the United Auto Workers labor union to support cutbacks in auto workers' pay and to make other concessions. To improve the company's image, he appeared in television advertisements to hawk its wares. He took risks as well, such as having the company build convertibles—something American automobile firms hadn't done for years. He even cut his own salary to $1 a year. As a result, and contrary to the predictions of business experts around the country, Iacocca turned Chrysler into a highly profitable company.[48]

Most people sense intuitively that leaders like Iacocca can make a difference, and a number of sociological studies support this notion. These studies usually employ complex statistical techniques to control for the characteristics of the individuals involved in the organization and the nature of its structure so that the leader's effect can be isolated and measured. Studies of this type have shown, for example, that students in elementary schools managed by particularly skilled principals have higher achievement than students in other schools.[49] Similarly, even though putting the best manager in the league in charge of a baseball team can't make as much difference as adding a number of star players, it can increase the won-lost record substantially—enough to propel a mediocre team into the playoffs.[50]

Yet sociologists have found it virtually impossible to identify exactly what good leaders do. For a long time sociologists tried to develop lists of personality traits of effective leaders.[51] Although traits such as intelligence and self-assurance often appeared, the results varied enough from one study to another that researchers began to focus instead on leaders' behaviors. This approach also proved problematic, for again no simple description of effective behaviors appeared to apply to a wide range of situations.

Several sociologists then turned to what is called a **contingency approach,** suggesting that effective leadership traits and behaviors vary depending on a particular situation. For instance, the behaviors that make a good leader at Harry's Hamburgers might not be the same as those that make a good baseball manager or an exemplary school principal. In addition, the managerial traits that work with one group of people might not necessarily work as well with another. In other words this research suggests that effective leaders must be flexible and sensitive, able to adapt a range of behaviors to particular situations.[52] As Yolanda dreams about starting her restaurant, she will need to think about her own leadership skills. Is she better at interpersonal leadership or task leadership? When she can afford it, what kinds of people should she hire to help run the restaurant?

Common Sense versus Research Sense

COMMON SENSE: When starting a new business, it is a good idea to start small and reduce the risk of financial disaster.

RESEARCH SENSE: Josef Brüderl's study of variables related to an organization's ability to survive suggests that the more capital one invests in a business, the better chance that business has of surviving.

FEATURED RESEARCH STUDY

BUILDING SUCCESSFUL ORGANIZATIONS: THE IMPORTANCE OF STRUCTURE, ENVIRONMENT, AND LEADERSHIP

In this chapter we have looked at organizations from a meso-, macro-, and microlevel perspective. Until relatively recently, most sociological researchers tended to focus on only one of these levels when they looked at organizations, undoubtedly because multilevel analyses are difficult simply to think about, let alone carry out. However, because factors at all three levels likely affect simultaneously the way organizations operate, single-level analyses can't tell us about how these multilevel influences work.[53] Fortunately, as the statistical analyses and data sets that sociologists use have become more sophisticated, more and more researchers have been able to combine several levels of analysis of organizations. One example of this work was conducted by three German sociologists—Josef Brüderl, Peter Preisendörfer, and Rolf Ziegler—who were interested in the findings of organizational ecologists regarding the survival of organizations, especially businesses.[54]

The fortunes of the capitalist class originated from very profitable business corporations founded by family members. In fact, apart from winning the lottery or becoming a movie or sports star, about the only way to get rich in modern industrialized societies is through establishing very profitable businesses. Each year many new businesses are founded, and each year many new businesses also fail. For instance, in the United States more than 600,000 new businesses are started each year, but almost 100,000 businesses fail.[55] Each of these failures results in individuals losing their jobs and often their life savings and their dreams. Why do some new businesses thrive while others fail? Why do some last longer than others do? How can entrepreneurs start businesses that will help propel them into the ranks of the capitalist class? What can Yolanda do to increase the probability that she will build a successful business?

Brüderl and his colleagues agreed that the variables used by organizational ecologists, such as the density of other organizations and issues of legitimacy, were important. But they suspected that the characteristics

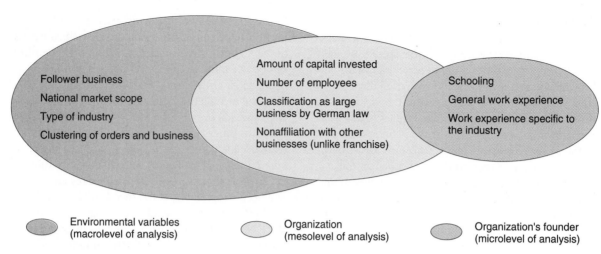

Follower business	Amount of capital invested	Schooling
National market scope	Number of employees	General work experience
Type of industry	Classification as large business by German law	Work experience specific to the industry
Clustering of orders and business	Nonaffiliation with other businesses (unlike franchise)	

Environmental variables (macrolevel of analysis)   Organization (mesolevel of analysis)   Organization's founder (microlevel of analysis)

FIGURE 11-5 *Variables Related to Organizations' Survival Time*

of specific organizations and their leaders would also be influential. Thus, applying each of the perspectives described in this chapter, they tried to determine the extent to which the individual characteristics of a business's founder (the microlevel), the organizational structure of the business (the mesolevel), and the environment in which a firm operates (the macrolevel) can affect the probability that a business will survive.

Brüderl and his colleagues gathered information about almost two thousand businesses that were founded in 1985 in Munich and Upper Bavaria, a region in southern Germany. After identifying firms that were still active in 1990, as well as those that had gone out of existence, the researchers conducted long interviews with the founders. To try to explain what can enhance the length of time an organization lasts, or what they called *survival time,* the researchers examined several independent variables related to each level of analysis. They then used complicated statistical techniques to determine which of these variables were most influential in the organizations' success. Their results are summarized in Figure 11-5.[56] As you can see, their hypothesis that variables from all three levels of analysis would be important was supported. This suggests that a budding entrepreneur like Yolanda should pay attention to variables on all three levels.

On the mesolevel, with variables related to the organization itself, Brüderl and his colleagues found that failure rates were much higher for small businesses—those that started with little capital and few employees and that were not registered with the government as large businesses. In other words, they found that it is far better to start big than to start small and try to build up.[57] One way that people without a lot of money can start a business is to buy into a franchise operation, but the researchers found that this doesn't work well either. In fact, franchised

firms, which are affiliated with larger firms, actually had lower survival rates than independent firms. At first the researchers found this result puzzling, for they had expected that the social networks of affiliated businesses would enhance the possibility of success. In their interviews with the entrepreneurs, however, Brüderl and his colleagues found that many who had signed on as affiliates had been "used" by the larger organization. As they described it,

> There are some companies in the franchise sector giving licenses with difficult or even fraudulent conditions. Founders trapped into such contracts have little chance for success. Although the founders may not know it, from the perspective of their partners or parent firms, affiliated businesses are often "experimental" or "special purpose" firms. When these experiments end or the special purposes are fulfilled, these organizations are forced out of business.[58]

Results with macrolevel variables related to the environment of the organizations, listed in the first part of Figure 11-5, support the ideas of the organizational ecologists. For instance, "follower businesses," which follow in the footsteps of already established firms and can be seen as having more legitimacy, were more likely to survive. Organizational ecologists also look at the "niche," or the place in the organizational environment that a group will fill and the amount of competition it will face. Brüderl and his colleagues found that organizations that aimed at a national market and those in certain industries (particularly manufacturing as opposed to transportation or wholesale and retail trade) tended to be more successful. The researchers also looked at the types of markets, or demands from customers, that the firms faced and found that businesses that were able to sell services or products in large chunks (such as big computer systems or large luxury boats) rather than smaller but continuous sales or business (such as a restaurant that was open each day) had better chances of making it.

For microlevel, or individual, variables, Brüderl and his colleagues found that both the founders' schooling and work experience influenced the chances that an organization would survive. Founders with more schooling and more work experience specific to their organization's industry or field had much greater chances of building successful organizations. The influence of greater work experience in general was more complex and is what sociologists term **curvilinear** in nature—that is, the relationship between the two variables is better expressed by a curved line than a straight line. The probability that an organization would survive increased fairly steadily as a founder's work experience increased, but only up until about twenty-five years. After that point having more years of experience was actually related to a greater chance of failure. Apparently, for younger workers, more experience helps, but for older workers it doesn't.[59]

Collectively these results show how an organization's probability of surviving is affected by the characteristics of its leader, its organizational structure, and its environment. Although Germany's economy is similar to that of the United States and other modern industrialized countries, Brüderl and his colleagues stress that their results might apply only to the sample that they studied—businesses within Germany in the late 1980s. Additional research will be needed to replicate their results, perhaps looking at other types of businesses as well as nonprofit groups and other organizations. Sociologists also might want to study different time periods, in different countries, as well as selecting measures of organizational success beyond mere survival.

If you don't want to wait for these replications, you might still consider the implications of this work for someone (such as Yolanda or even yourself) who is contemplating starting a business. What do these results suggest you should do if you want to be a successful entrepreneur? First, like Yolanda, you should stay in school. More highly educated entrepreneurs tend to embark on more successful ventures, even in areas with higher overall failure rates, such as smaller firms or restaurants.

Second, you should get experience within the field you want to enter. General work experience helps, but experience specific to the industry is more important, especially positions in which you can see all the various aspects of a business's operation. For instance, Yolanda's experience in Harry's should serve her well in starting her own restaurant.[60]

Third, you should make sure that the environment in which you begin your business will be hospitable. It is important that your business be seen as legitimate, but it is also important that the density of

Two Stanford University engineering graduates, William Hewlett (standing) and David Packard, developed an innovative audio oscillator in their Palo Alto, California, garage in 1939. The large and successful Hewlett Packard company grew out of this beginning. How does the history of Hewlett Packard conform with the findings of Brüderl and his colleagues?

similar organizations not be so high that intense competition will diminish your chances of success.

Finally, you shouldn't start too small. As Brüderl and his colleagues explain,

> Though smallness reduces the risk of a financial disaster, it also increases the mortality rate of a business. Depending on the branch of industry, there is evidently a certain minimum start-up size, and it does not make much sense to begin a business below this level.[61]

These results also show why it is so very hard for middle- or working-class individuals to develop businesses that will propel them into the capitalist class and why it will be extremely difficult for Yolanda to ever "make it big." Self-made millionaires are rare simply because starting a business that is large enough to actually survive, let alone thrive, takes a great deal of start-up capital. People who already have such resources at their disposal or who have the collateral to obtain financial backing from others are better able to start big. Members of the capitalist class are also much more likely to have the social networks to obtain experience in various aspects of an enterprise and to be able to afford advanced schooling. For example, even though Yolanda's father works in the loan division of a bank, she will have no special advantage in obtaining financial backing because of the bureaucratic rules of the banking business. She will have to start small, and despite her experience, her education, and her motivation, this smallness limits her probability of success.

Although the research of Brüderl and his colleagues focused on businesses (and a number of sociologists who study organizations work in business departments in universities), other research has studied how the variables discussed in this chapter affect other types of organizations. Some sociologists have focused on how organizations can be structured in ways that are not repressive to individuals and that can guard against the tendency of a small group of individuals to control the organizations' actions (see Box 11-2, "Applying Sociology to Social Issues"). Other sociologists have looked at how organizations help meet basic needs of individuals and, especially, how they are part of social institutions. The next five chapters deal directly with these issues.

CRITICAL-THINKING QUESTIONS

1. How could research like that of Brüderl and his colleagues be conducted on organizations other than businesses, such as voluntary organizations or nonprofit groups? Could other research methodologies be used? Could you look at characteristics other than survival time?

2. Think about organizations with which you are familiar, such as work organizations, your college or university, or a group to which you belong. Can you identify ways in which the activities of this organization are affected by variables at each of the levels of analysis discussed in this chapter?

SUMMARY

We all participate daily in formal organizations, groups that have been deliberately created to accomplish certain goals. Because formal organizations such as colleges and universities, banks, and businesses are so powerful, people have ambivalent feelings about them. By applying micro-, meso-, and macrolevel analyses we can better understand how organizations function and why they survive. Key chapter topics include the following:

- According to Weber's ideal type, bureaucracies have specialized roles, a hierarchical nature, and rule-governed and impersonal procedures to deal with the complexities of modern life. Although they are designed to be both efficient and effective, bureaucracies can also be dysfunctional, especially in the way individuals experience interactions in large organizations.

- A mesolevel analysis of organizations focuses on individual organizations. Organizational structure involves the elements of complexity, formalization, and centralization, all of which can vary a great deal from one organization to another. Informal structures of organizations exist alongside formal structures and can influence both power and interactions within organizations.

- The technology that organizations use also can influence their structure, especially through its complexity (or simplicity) and predictability (or uncertainty).

- Organizational culture includes elements such as norms, values, belief systems, ideologies, folklore, myths, and symbols. Individuals come to identify with the culture of their organizations, and this can work against the possibility of change.

- A macrolevel analysis focuses on how organizations are affected by the environments in which they exist. If organizations are to survive, they must adapt to new knowledge and technology, changing relationships with other organizations, and the institutional norms of the surrounding society.

- Organizational ecologists have focused on long-term patterns of birth and death of organizations. They have documented a pattern of increasing density of organizations as an organizational form becomes legitimated, followed by a decline when greater density of organizations produces more competition.

- One type of microlevel analysis of organizations focuses on organizational leaders. Leaders derive their authority from their relationships with others in the group. Two types of leadership are important in organizations, one that deals with tasks or instrumental activities and another that deals with interpersonal relationships or expressive activities.

- Although leaders' activities are restricted by an organization's structure and environment, individual leaders can and do affect the success of an organization. Sociologists have not been able to pinpoint precisely how effective leaders behave, but current theories suggest that effective leadership traits and behaviors are contingent on the particular organizational situation.

- The research of Josef Brüderl and his colleagues suggests that variables from all three levels of analysis (micro-, meso-, and macro-) are important determinants of the success of newly established businesses. Their finding that small businesses are less likely to survive helps explain why it is difficult for people from lower social classes to move into the capitalist class.

KEY TERMS

bounded rationality 285

bureaucracies 277

case study 297

contingency approach 293

curvilinear 295

democratic organization 297

expressive leadership 291

formal organizations 275

groupthink 285

horizontal differentiation 280

ideal type 277

informal structure 282

institutional norms 288

instrumental leadership 291

organizational centralization 281

organizational complexity 280

organizational culture 284

organizational ecology 288

organizational formalization 280

organizational leaders 290

organizational structure 280

organizational technology 283

spatial dispersion 280

vertical differentiation 280

INTERNET ASSIGNMENTS

1. Enter a corporation/company name into a search engine (for example, Kmart, IBM, Microsoft), and locate Web sites for these organizations. What can you learn about the structure of the organizations from the information provided? Is there information about how complex, formalized, and/or centralized the organizations are? Are there discussions about the organizational cultures? How are the organizations similar? How are they different?

2. Enter the term "bureaucracy" into a search engine. What variety of links do you obtain? How are people using this term? Is the usage and understanding of bureaucracy and bureaucratization consistent with Weber's definition of these terms? Is bureaucracy viewed mostly in a positive or negative fashion? How are these views related to the functions

and the dysfunctions of bureaucracies, such as those discussed in this chapter?

3. Enter the terms "business development" or "small business" into a search engine. Do you find guidelines and suggestions for people interested in starting a business? How do these suggestions compare to the results of the research by Josef Brüderl and his colleagues that was featured in this chapter?

TIPS FOR SEARCHING

Information provided by companies/corporations on the World Wide Web varies a great deal. Some give a lot of information about the way that they are organized. Some, however, focus more on product information only. Therefore, it may take careful searching to find the type of information you are looking for regarding businesses. Annual reports of companies that are available on-line may also be good resources.

INFOTRAC COLLEGE EDITION: READINGS

Article A19155710 / Godwin, Glen J., and William T. Markham. 1996. First encounters of the bureaucratic kind: Early freshman experiences with a campus bureaucracy. *Journal of Higher Education, 67,* 660–692. Godwin and Markham examine how college freshmen learn to cope with university bureaucracy.

Article A18939634 / Parker, Martin, and Mike Dent. 1996. Managers, doctors and culture: Changing an English Health District. *Administration & Society, 28,* 335–362. Parker and Dent use a case study of an English Health district to examine the ways in which participants in the same organization may have different views of the organization's culture.

Article A17352542 / U.S. Department of State Dispatch, May 30, 1995 v6 nSUPP-3, p1-10. *The U.S. Department of State: Structure and organiza-*

tion. This description and justification for changes in the structure of the U.S. State Department gives a first-hand look at how organizations plan and conduct changes within their structures and the reasons for doing so.

Article A16606371 / Linney, Barbara J. 1995. Leading an organization through change. *Physician Executive, 21,* 25–28. Linney interviews a successful executive in a health care organization and provides a good description of how effective leadership can promote change within organizations.

Article A20991474 / Juechter, W. Mathew, Caroline Fischer, and Randall J. Alford. 1998. Five conditions for high-performance cultures. *Training & Development, 52,* 63–68. Juechter, Fisher, and Alford present their views of the optimal corporate culture.

BOX
11-2

APPLYING SOCIOLOGY TO SOCIAL ISSUES

Creating Humane Organizations

Many sociologists have been concerned with the fact that organizations sometimes come to be controlled by a small group of people who are more interested in preserving the organization and their own leadership positions than in working for the welfare and interests of the members. Robert Michels, a German sociologist who was a contemporary of Weber, believed that this process was so common that it should be called the "Iron Law of Oligarchy." (Oligarchy means "government by the few" and refers to the situation in which only a small portion of a group controls decisions, often for their own selfish purposes.)

At the same time, organizations seem to be an essential part of modern societies. Organizations, and especially bureaucracies, appear to be the most effective means humans have devised for dealing with the complexities of modern life. Given that modern societies can't seem to exist without formal organizations, how can we create **democratic organizations,** those that are responsive to the desires of their members and do not degenerate into oligarchies?

One possible answer comes from a case study of the International Typographical Union (ITU) conducted in the early 1950s by the sociologists Seymour Martin Lipset, Martin Trow, and James S. Coleman.[62] A **case study** is an in-depth look at one case, such as one person, one group, or one organization. It can be especially useful when that case is atypical, or very different from others, as was true of the ITU. Unlike other unions the ITU had a long tradition of highly democratic governance—two political parties within the union, frequent changes in leadership, a high level of political consciousness among the members, and a very low wage differential between union officials and the rank and file. Of course, the printing business, as well as the ITU, has changed a great deal since Lipset and his colleagues did their study. Nevertheless, their results are still in-

structive in showing how organizations can be structured to help minimize the probability that an oligarchy will form.

Lipset, Trow, and Coleman found that both historical features and structural factors contributed to the democratic nature of the ITU. First, unlike other manual workers, printers have always been highly literate, for only people who could read and write could accurately set type. Lipset and his colleagues believed that the printers' education helped promote the democratic nature of the organization because highly educated individuals simply are less easily duped or manipulated.

In addition, the ITU was structured in ways that helped maintain a democratic process. There was a strong belief in the legitimacy of dissenting views, and the presence of two opposing political parties in the union helped promote an ongoing dialogue. The printers also tended to form what has been called an "occupational community." In part because they generally worked at night and found it difficult to socialize when most other people did, they tended to have strong interpersonal ties.

FURTHER READING

Suggestions for additional reading can be found in *Classic Readings in Sociology,* bundled with this book. If you purchased a used copy of this book that does not include this custom-published reader, go to www.sociology.wadsworth.com for ordering information.

APPLYING SOCIOLOGY TO SOCIAL ISSUES

Continued

This promoted a sense of solidarity and community that extended beyond the workplace. Also, although union members worked in a variety of settings, from very small print shops to quite large work sites, in all cases the workers had a great deal of autonomy and power. Indeed, at the time Lipset and his colleagues conducted their research, the foremen who supervised printers were required to belong to the ITU, and the printers, not the managers, were the only ones who could determine who would be allowed on the shop floor.[63]

Together these findings suggest how we, as individuals, can structure organizations that are more democratic—that is, how we can guard against the possibility of oligarchy. For example, we can work to establish organizations that legitimate opposition and promote a culture that encourages broad participation and freedom of expression. In other words, to develop more democratic organizations, we need to encourage free expression of all views and provide legitimate opportunities for dissent. We can also nurture and promote social ties among organization mem-

bers, especially ties that exist outside the formal organization itself. And we can foster greater autonomy among individuals within organization units.

Some contemporary writers suggest that modern colleges and universities, such as the one that Yolanda attends, include many of these elements.[64] Like the ITU, people who work at universities are very highly educated. In addition, the structure of organizations in higher education generally includes rules that encourage freedom of expression. Colleges and universities also usually give a great deal of freedom to individual units. For instance, academic departments have a great deal of control over their own curricula. Even living units, such as Yolanda's dorm, often are allowed, within limits, to develop their own rules.

Of course we participate in many organizations that are already well established or are so large and complex that it's difficult to conceive how as individu-

als we might help promote more democratic organizations. This task is especially daunting when we think about large corporations and government bodies—immensely powerful organizations with so many levels of authority that the top executives are very far removed from everyday citizens such as ourselves. In subsequent chapters we will discuss these political and economic organizations in more detail.

CRITICAL-THINKING QUESTIONS

1. How democratic are the organizations in which you participate? Is the responsiveness of these organizations related to their effectiveness?

2. The sociologist Peter Blau, who has studied organizations for many years, suggests that sociological knowledge provides power—it gives us the ability to analyze the organizations in which we participate and to help make them more responsive and democratic.[65] How might you apply your sociological understanding of organizations to help ensure that the organizations you participate in can remain or become more democratic?

INTERNET SOURCES

U.S. Department of Commerce. This Web site includes links to sites of many government agencies that regulate and serve businesses throughout the country.

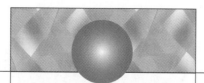

Pulling It Together

Listed below are other chapters where you can find additional discussion or examples
of some topics covered in this chapter. You can also look up topics in the index.

RECOMMENDED SOURCES

Blau, Peter, M., and Marshall W. Meyer. 1987. *Bureaucracy in Modern Society,* 3rd ed. New York: Random House. Blau and Meyer provide a readable and thorough discussion of the nature of bureaucracies and how they affect our lives.

Boden, Deirdre. 1994. *The Business of Talk: Organizations in Action.* Cambridge, U.K. Polity. A conversation analyst examines the ways in which the social actions of conversations create the social structures of organizations.

Hall, Richard H. 1998. *Organizations: Structures, Processes, and Outcomes,* 7th ed. Englewood Cliffs, N.J.: Prentice-Hall. This well-written, complete textbook is devoted to organizations.

Kalleberg, Arne L., David Knoke, Peter V. Marsden, and Joe L. Spaeth. 1996. *Organizations in America: Analyzing their Structures and Human Resource Practices.* Thousand Oaks, Calif.: Sage. The authors report the results of a large scale survey of a representative sample of work organizations throughout the United States.

Lincoln, James R., and Arne Kalleberg. 1990. *Culture, Control, and Commitment: A Study of Work Organization and Work Attitudes in the United States and Japan.* Cambridge: Cambridge University Press. This in-depth, cross-cultural look at the work environment in these two countries was written by the sociologist profiled in Box 11-1.

Scott, W. Richard. 1998. *Organizations: Rational, Natural, and Open Systems,* 4th ed. Englewood Cliffs, N.J.: Prentice-Hall. This is another well-written textbook devoted to organizations.

Shafritz, Jay M., and J. Steven Ott (eds.). 1997. *Classics of Organization Theory,* 4th ed. New York: Harcourt Brace. This thorough and varied collection of articles discusses many different aspects of organizations.

Trice, Harrison M., and Janice M. Beyer. 1993. *The Cultures of Work Organizations.* Englewood Cliffs, N.J.: Prentice-Hall. An entertaining discussion of the organizational cultures of a variety of different corporations and worksites.

12

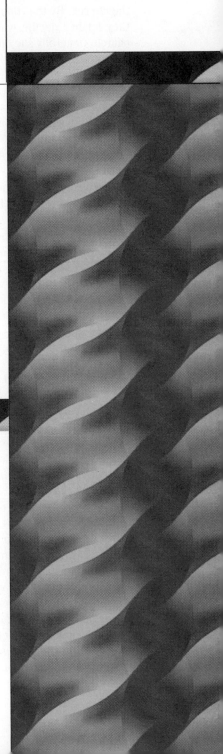

Religion

RIVERSIDE COLLEGE WAS FOUNDED by members of the Presbyterian church many years ago. There is a chapel on campus, but it is now used only on ceremonial occasions and for weddings and other special services. Today students at Riverside come from all kinds of religious backgrounds, and some, like Amber, aren't active in any church.

Amber chose Riverside College because of its academic reputation, not its religious heritage. Her grandparents had been active Presbyterians, and she attended services with them when she visited their home. Her parents sent her to Sunday school when she was young but had themselves drifted away from the church. By the time Amber was in high school, they let her decide whether she wanted to go to church. Generally she didn't.

Lately, however, Amber had begun to think more about religion. "Am I missing something?" she wondered. "What if God is watching me all the time, just waiting for me to come around? But what is God like, anyway, assuming there is one?" Try as she might, she couldn't imagine a being who could be responsible for the whole universe *and* care about her or her troubles.

At the beginning of her junior year, Amber signed up for a class entitled "Religions of the World." She wasn't really expecting to find answers, yet she had some vague hope that the class would give her some sense of where to look for them.

On the first day of class, the students were assigned to small work groups. The members were to meet each week to discuss the readings and plan a presentation to the class at the end of the semester. Amber's group decided to hold their first meeting at a pizza parlor near campus.

"How about a pitcher of beer?" asked Amber, trying to start things off on a friendly note.

"Sorry, I don't drink," said Jamaal and Roger, almost in unison.

"Why not?" chorused the other students.

"I'm Muslim," answered Jamaal matter-of-factly. "Good Muslims don't drink."

"I've given up alcohol since I was born again," said Roger.

For a moment the other students didn't know quite what to say. "Well," Amber quipped, "I guess this is a good way to start. Maybe we can talk about our own religious backgrounds and views. Meanwhile let's each order our own drinks."

The rest of the hour went more smoothly. Jamaal told how he tried to follow Islam's rules about avoiding alcohol and certain foods, such as pork and shellfish, and how he prayed several times a day. He explained how these practices helped him feel more centered and in control of his life. Cathy, who had grown up in a devout Catholic family, recounted how the church's rituals had comforted her during difficult times, such as when her grandmother suddenly died. Naomi's family was Jewish. Although they attended synagogue only on High Holy Days and didn't observe the traditional Jewish dietary laws, she was proud of her Jewish heritage. She told the group about the fasting and long services associated with Yom Kippur, the Day of Atonement, and how, after the service, she always felt inspired to live a better life.

Of all the students Roger was the most enthusiastic about his faith and the most excited about describing it to others. His parents had joined an Assembly of God church when he was young, and when he was a teenager Roger also became a committed member. He told the others about his conversion—how he felt that God spoke through him and how he then entered into a new life with Jesus.

Amber felt as if she didn't have much to share with the others. "I really don't know what I believe," she said. "I guess I'm just sort of thinking about a whole lot of things right now."

Amber and Roger stayed at the pizza parlor to talk after the other students left. Amber told him of her questions about life and God, and Roger told her more about his faith. He attended Christian Center, a very large church near campus that was affiliated with the Assembly of God denomination. "Why don't you come to a service?" he proposed.

Amber shook her head. "I'd just be out of place," she replied.

"God welcomes everyone," Roger answered. "Why not give it a try?"

After a moment's thought, Amber agreed.

The Sunday morning service at Christian Center was totally unlike the Presbyterian services she had attended. The music was contemporary and spirited, and the congregation sang joyfully and clapped along with the music. Unlike the Presbyterian church she'd attended with her grandparents, Christian Center's large sanctuary was filled to capacity. What most surprised her was how members of the congregation participated in the service and how they seemed to be transported outside of themselves. Several people prayed out loud. Some had what were called "charismatic" experiences, in which they could hear God speak through them, sometimes in tongues, or languages, that they did not know. Watching them, Amber felt a shiver. She wasn't sure whether it was excitement or fear.

Over the course of the semester, Amber returned to Christian Center with Roger several times and became friends with some other students there. They told her about times when they felt filled with the Holy Spirit, as if God were speaking directly to them, and how they were at peace with themselves afterward in the knowledge that God was with them all the time.

Although in a way she envied them, Amber knew that Roger and the other students at Christian Center had given up a lot. They no longer went to beer parties and had often broken ties with old friends. "Would I ever want to make sacrifices like these?" she

wondered. At the same time, Roger and his friends seemed very centered and more content than most people she knew did. "Could I ever feel this way?" Amber asked herself. "Could I experience God the way they do? Could I believe as they do or would I just be fooling myself? Why are there so many different religious groups anyway? Why do some people believe one thing and others believe something else?"

*I*F YOU LIVED IN A traditional nonindustrial society, you probably would never engage in conversations like those that Amber and her classmates had. Nor would you think about other religious traditions or develop new beliefs as you grew older. In many societies, religion and the supernatural are simply part of the culture—part of the way life is.

For instance, if you had grown up in a traditional Navajo family in the nineteenth and early twentieth centuries, you would have been taught from earliest childhood to behave in ways that would help maintain the goodwill of the supernatural world. You would know that you should never build a fire with wood from the woodpile of a dead person; nor should you ever step over a fire or a sleeping person. You would learn Blessing Way songs that you would sing each morning when you got up and each night before you went to bed. Throughout your life you would practice a variety of rituals that would help produce bountiful harvests and good health, and you would participate in elaborate joint rituals, or chants, to help cure a relative afflicted with a disease or a "troubled spirit."[1]

In modern industrial societies, religious ideas are far less central to many aspects of our lives than they were in pre-industrial societies. We understand that crops thrive because of weather conditions, proper planting procedures, and correct methods of caring for the soil. We know that germs and viruses cause many diseases and that many of these diseases can be cured by medical procedures rather than by religious rituals. In the United States and many other highly industrialized countries, government and public education are explicitly separated from the practice of any single religion, and many people follow standards of moral behavior that do not depend on religious faith. In fact, you have probably heard religious lead-

ers, as well as politicians, claim that the move away from religion and toward a more secular society is one of the most disturbing aspects of modern life.

Sociologists, too, have often assumed that religion was becoming less and less central to modern societies. Yet, as Amber's experiences illustrate, our "secular" society actually harbors a sometimes bewildering variety of religious beliefs and practices. Although it may take many forms, religion shows little sign of fading away.

Confronted with this reality, in recent years a number of sociologists have begun to rethink their ideas about religion. In the process they have devised fascinating new ways to explore this area of social life. They are asking new questions, and they are coming up with some surprising answers. In this chapter we examine their findings from a variety of perspectives. But first, to appreciate these recent developments, we need to put them in context by briefly considering how sociological thinking about religion has evolved over time.

THE SOCIOLOGICAL STUDY OF RELIGION

The diverse experiences of Amber and her acquaintances only begin to hint at the enormous variety of religious belief and expression. For many believers the differences between forms of religion probably matter most. These believers might feel that their particular beliefs and forms of worship are the best or truest path to what is most real and important. From a broader sociological viewpoint, however, we can see that all religions have several things in common. From a sociological perspective **religion** is the social institution that deals with the area of life people regard as holy or sacred. Religion involves the statuses, roles, organizations, norms, and beliefs that are related to humans' relationship with the supernatural. Defined in this way, religion has several aspects: Religion includes *shared beliefs* about what is divine or sacred, such as the revered Word of God in the Torah, the Bible, and the Koran, in the Jewish, Christian, and Muslim traditions, respectively; the ideas of spirits and gods in the Navajo and Chinese traditions; and the special places set apart to honor the presence or activity of the divine, such as temples and churches. Religion involves *ethical rules or codes* about how we should behave—for example, the Ten Commandments in the Jewish and Christian tradition. Religion also involves *rituals and ceremonies,* from the curing ceremonies and initiation rites of tribal cultures such as the Navajos to the holiday observances of the Jewish tradition. Finally, in all these aspects, religion involves *communities,* or groups of people, who join together in common beliefs, ethical standards, and rituals.[2]

This last aspect of religion is particularly interesting from a sociological point of view. Without attempting to judge the validity of people's religious beliefs, sociologists explore the nature and effects of religion as a *social* phenomenon. For example, in his analysis of primitive religions, such as the totemism of the Australian aborigines, Émile Durkheim suggested that religious beliefs, norms, and rituals support the collective consciousness—they help bind members of a society to one another and also promote adherence to the society's norms.[3] As you will recall from Chapter 1, Karl Marx saw religion as an important means of social control, as one way those with power in a society control those with less power.

Many of the early sociological thinkers rejected the religious faiths of their contemporaries and were sometimes openly hostile toward the established religions. Durkheim was an agnostic; Marx was an atheist. The French philosopher Auguste Comte, who first coined the term *sociology,* suggested that sociology itself would eventually become a new religion. That is, he believed that the rational science of society would ultimately replace supernatural interpretations of the world.[4]

In general, like many of their contemporaries, the early European sociologists assumed that modern society would experience a process of **secularization,** a transformation to an outlook on life based on reason and science rather than on faith and supernatural explanations.[5] Later sociologists accepted this perspective, and for many years most sociologists, in both Europe and the United States, assumed that secularization was a fact. They believed that religion was less important in contemporary society than in previous eras and that it would continue to diminish in importance as people became more educated and adopted a more naturalistic view of the world.[6]

This general theoretical view is what philosophers of science sometimes call a **paradigm,** a fundamental model or scheme that guides people's thinking about a particular subject.[7] Because the secularization paradigm was so ingrained within sociological thought, most sociologists did not question whether the modern world actually was becoming more secular.

The only problem was that the empirical evidence—the data—didn't support this assumption. Most people worldwide continue to adhere to some type of religious belief, even as greater and greater proportions of the world's population receive more education. Conflicts fueled in part by religious beliefs continue to erupt worldwide, from Ireland to the former Yugoslavia to the Middle East. As we will see later in this chapter, even while living in a highly industrialized and modern society, most people in the United States believe in God.

Eventually a number of sociologists who study religion realized that the old paradigm simply wasn't describing reality. Philosophers of science sometimes call this a **paradigm crisis,** a time when an explanatory scheme begins to break down, requiring a fundamental reorientation or rethinking of basic assumptions.[8] These periods are referred to as a crisis precisely because they are often very difficult periods. The familiar worldview no longer works, so a new paradigm must be found. Sociologists began to look for something other than the idea of secularization to help them better understand the nature of religion in society.[9]

Because of this paradigm crisis, recent years have been exciting for sociologists who specialize in the study of religion. Some of them have begun developing a new paradigm, a task that involves much rethinking of old ideas, including ideas developed by the founders of the discipline. This process of creating a new paradigm is far from complete, and not all sociologists agree on the direction it should take or even whether a new paradigm is necessary.[10] Nevertheless, this very process illustrates how sociology as a discipline continues to grow and change as social scientists create and test new ideas about the social world. It also can show us how sociologists continually use data—observable facts gathered from the world around us—to test and reshape their theories.

In this chapter we examine various aspects of the emerging paradigm, especially as it focuses on the vitality of religious groups and tries to understand why secularization hasn't occurred. Using a macrolevel perspective, sociologists ask why more people belong to religious groups in some societies than others and why religious involvement can change in a

For many years sociologists, following in the tradition of the classical theorists, believed that religion would become less important in modernized societies. Today sociologists realize that religion is a very important element of the lives of people throughout the globe, no matter how modernized a society is or how educated people may be.

society over time. Using a mesolevel perspective, sociologists ask why some religious organizations attract more adherents than others. Why do some groups prosper while others decline, and why may religious involvement vary from one part of a society to another? Finally, using a microlevel perspective, sociologists try to understand why individuals choose to affiliate or not affiliate with one religious group or another. What influences individuals' acts of faith and their decisions to join particular religious communities?

In exploring this work, we will be considering religion as a social institution, much as we considered the family in its various forms as a fundamental feature of society in Chapter 10. You may have heard people refer to religion as "food for the soul," a way in which people are spiritually nourished and renewed. We can use this metaphor to understand how religions emerge in different societies, how individuals participate in religion, and how sociologists study this area of life. All societies have religious beliefs that provide a sense of meaning and purpose in life. Yet, just as the foods people eat vary widely from one society to another, so does the nature of religious life—the ways in which people nourish their souls—vary from one group to another. Similarly, just as some of us find some types of food more satisfying than others do, we differ in our religious practices and beliefs and, like Amber's classmates, gain inspiration and solace from different religious traditions.

Sociologists who study religion understand that many people are wary of applying scientific methods to something as personal as religion. For these scholars, however, a sociological understanding of religion

is only one among several ways of looking at religious faith and commitment. Philosophers and historians of religion analyze and compare different religious belief systems, whereas theologians formulate and interpret doctrines within particular religious traditions. Clergy seek ways that religion can inspire their congregations and respond to human needs. Faithful adherents look to their religion as a way to make sense of their lives and guide their behaviors. Sociologists are interested in all these aspects of religious faith and behavior, but sociologists try to understand religion as a social phenomenon—how societies develop ways to nourish the soul.[11]

RELIGIOUS BELIEFS AND ECONOMIES: A MACROLEVEL VIEW

Many of sociologists' new ideas about religion have developed from taking a very wide-angle or macrolevel perspective of religion. In this section we first look at the religious beliefs and practices of people in different parts of the world. We then examine some sociological explanations for why people in some places and times are much more likely than are others to be involved in religious groups and activities.

Religious Beliefs Around the World

Studies of the religious preferences of people around the globe indicate that about 80 percent of the world's people profess some type of religious belief. Figure 12-1 breaks down religious adherence by denomination. Note that only about 4 percent call themselves

FIGURE 12-1
*Adherents of
Major Religious
Traditions, World-
wide Mid 1993*
Source: Famighetti,
1996, p. 646

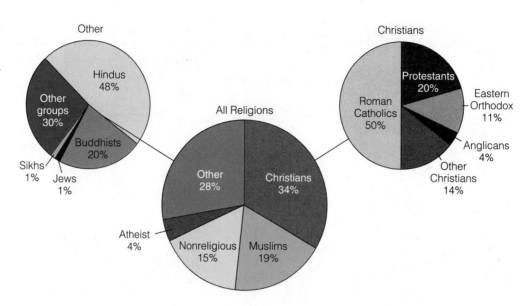

atheists, whereas another 15 percent are termed "non-religious," which means either that they are agnostic and unsure of the existence of God or that they have no interest in religion. The vast majority of the atheists and nonreligious live in Asia and the former Soviet Union.[12] This can be explained by the fact that both China, the most populous country in Asia, and the former Soviet Union have long espoused an official hostility toward all religious beliefs. Most important, although the secularization hypothesis would suggest that the most highly industrialized areas of the world would be the least religious, the data suggest just the opposite. People in Europe, Oceania, and North and South America—all modernized regions—are most likely to profess religious beliefs.

In all the continental areas of the world except Asia, Christians are the most numerous group. More than three-fourths of the population in Europe, the Americas, and Oceania profess to be Christian. Worldwide, about half of all Christians are Roman Catholic, but in the United States and Canada, the majority belong to various Protestant groups. In Asia, Muslims and those with no religion substantially outnumber Christians, whereas in Africa, Muslims are close in number to Christians.[13] These differences reflect the historical origins of various religions. Christianity took root in Europe during the Roman Empire and, centuries later, was spread throughout the Americas and Oceania by European immigrants. Islam began in the Middle East and spread to nearby Africa and Asia. And, not surprisingly, Hinduism and Buddhism, which began in Asia, are most often found in Asia and Oceania.

In addition to finding out the religious traditions with which people identify, sociologists want to know what people believe and whether they go to

church. To do so, sociologists have tapped a large data set called the World Values Survey, which includes information collected from citizens of twenty-two different countries. Some of these data are shown in Table 12-1.

As you can see, even though the majority of people in both Europe and the United States describe themselves as Christians, the United States stands out as a particularly religious country. Nearly half of all people in the United States report that they attend church weekly, while one-fifth or fewer of the citizens of many Western European countries do so. In the Scandinavian countries, only 5 percent or fewer people attend church weekly.[14]

Interestingly, individuals differ much less in their religious *beliefs* than in how often they attend church. Besides asking people how frequently they attended church, the World Values Survey also asked them whether they believed in God, whether they considered themselves to be a "religious person," and if they were a "convinced atheist." According to these measures people of the United States are also the most religious, with 96 percent saying they believe in God and over four-fifths describing themselves as "religious." Yet substantial majorities of people in most of the other countries also say they believe in God and describe themselves as religious. Very few say that they are "convinced atheists."

The old secularization paradigm suggested that as societies modernized and people became more educated, they would be less likely to engage in religious practices or hold religious beliefs. Some adherents of this view suggested that the very low church attendance in countries such as Iceland supported the thesis. Yet the data in Table 12-1 suggest that most people continue to profess religious beliefs

TABLE 12-1 *Religious Participation and Religious Beliefs in Selected European Countries and the United States, 1990–1993*

| | RESPONSE | | | |
Country	Attend Church Weekly	Believe in God	Religious Person	Convinced Atheist
Iceland	2	85	75	2
Denmark	2	64	73	4
Finland	4	76	59	3
Sweden	4	45	31	7
Norway	5	65	48	3
Netherlands	20	65	61	5
Hungary	21	65	57	4
United States	44	96	84	1

Note: The question about "Religious Person" was "Independently of whether you go to church or not, would you say that you are a religious person?"
Source: Calculated by the author from the World Values Survey.

even in societies in which relatively few individuals participate in formal religious services. What explains this disparity? Why are people who are relatively similar in their beliefs so different in their church-going behavior?

Religious Economies

The answer provided by the developing paradigm in the sociology of religion revolves around the notion of **religious economy**, the various religious options or choices available within a society. For example, Amber and her classmates can participate in a rich religious economy, with choices ranging from the Muslim and Jewish traditions to Catholicism and various branches of Protestantism.

One scholar who has helped develop ideas about the religious economy is the sociologist Rodney Stark, often in collaboration with Roger Finke, Laurence Iannaccone, and William Bainbridge. In contrast to secularization theorists, who focus on individuals' beliefs, Stark and his associates have pulled back to look at entire societies and, especially, the extent to which a society's religious economy is pluralistic. (*Religious pluralism* refers to how many different religious groups are present.) Whereas secularization theorists contend that people in modern societies are relatively indifferent to religion, Stark and his associates argue otherwise. To illustrate their ideas, think about a society with a narrow range of foods and one with a bountiful array of food choices. People in the former society might only "eat to live,"

but those in the latter might make eating a much more central part of their lives. Similarly Stark and his associates suggest that when the religious economy is more pluralistic—when there are more options available for "nourishing the soul"—more people will be attracted to religious organizations, no matter how modernized a society is. Thus, suggests Stark, instead of focusing on individuals' beliefs, sociologists should look at the social structure of organizations and religious options.[15]

To date, the theory of religious economies has been tested in two different ways. One involves looking at contemporary societies that have had little religious pluralism, or what are often called "controlled religious economies," and seeing what happens in these societies when more religious options appear. The other involves historical analyses of religious affiliation in the United States.

Changes in Controlled Religious Economies

Two types of controlled religious economies have existed in the twentieth century. One involves centrally controlled churches that are officially sanctioned by the government; the other involves attempts by the state to eliminate religious beliefs. Both systems have tended to alter over the years. Societies with officially sanctioned, or state, churches have allowed greater religious freedom, and in formerly Communist countries, more religious options have appeared as the Marxist-Leninist governments that restricted religious freedom have fallen. The experiences of both types of religious economies illustrate how the presence of more options influences the tendency of people to affiliate with a church.

If you grew up in the United States, the idea of a state church, one that is officially sanctioned by the government, is probably foreign to you. Yet many modern societies have state churches. For example, the Roman Catholic church is the official church of several European countries, including Italy, Ireland, Belgium, Portugal, and Spain; the Anglican (or Episcopalian to people in the United States) church is the state church of England.[16]

A particularly interesting example is the Evangelical Lutheran church, which is the state church of Sweden. In Sweden, the church is literally a part of the government. All Swedes are automatically members of the church unless neither parent belonged to the church or unless they request release from their membership. The government institutes church law, appoints bishops, pays pastors, and provides funds for the construction of church buildings. All Swedish citizens, whether or not they belong to the church, contribute taxes to support it. Yet, despite this official support and the fact that 90 percent of all

Swedes belong to the Church of Sweden, very few attend services. Huge cathedrals with large staffs of well-paid clergy generally have very few people in attendance at Sunday services.[17]

In recent years, other religious groups have begun to enter Sweden. Even though these groups have run into bureaucratic roadblocks in attempts to build sanctuaries and otherwise establish themselves, these religious groups have managed to attract about 10 percent of the Swedish populace. Moreover, in contrast with Church of Sweden members, an estimated three-fourths of the "free church" members attend services each Sunday. Proponents of the idea of the religious economy suggest that Swedes are more likely to attend the free churches because these churches are ones that they have chosen to match their own religious interests and needs. In other words, once more choices appear within the Swedish religious economy—ones that appeal to individual tastes—citizens will more often participate.[18]

Similar changes have been noted in various Latin American countries where for many years the Roman Catholic church has been the state religion. As in Sweden, only a small minority of the population was actively involved in the officially established church. Beginning a few decades ago, first through the efforts of missionaries from countries such as the United States and later with locally trained clergy, several Protestant groups gained a foothold. Even though Protestants are still a minority, they, rather than Catholics, are the most active churchgoers. Thus, for example, in countries such as Brazil and Chile, the majority of those who actually attend church and the majority of the clergy are not Catholic.[19]

At the opposite extreme of controlled religious economies are the societies in which the government has officially discouraged all religions. The experiences of the former Soviet Union are especially interesting. Before the Communists came to power in 1917, the Russian Orthodox church was the state religion. For centuries the Russian government had funded numerous Orthodox churches, clergy, monasteries, and convents. Following the 1917 revolution and inspired by the writings of Marx, however, the new political regime systematically tried to eliminate religious symbols and stamp out religious organizations. The persecution was probably most extreme during the Stalinist era (which lasted from the 1920s to 1953), when the number of churches dwindled to only a tenth of their previous numbers. Although restrictions loosened in subsequent years, the official atheistic position of the government remained. Anyone who wanted to gain a prestigious government position prudently either claimed to be an atheist or hid any religious beliefs. In sum, for most of the twentieth century, seminaries and churches were closed;

clergy were tightly controlled; and the government made a concerted, and historically unprecedented, effort to stamp out religion.[20]

A few years ago the government of the Soviet Union collapsed, and many former republics became independent countries. Sociologists wondered what the role of religion would be in a Russia that was no longer dominated by Marxist ideology. To what extent would belief in God survive decades of government-sanctioned atheism?

Andrew Greeley, another sociologist who is contributing to the new paradigm, used survey data gathered by Russian sociologists shortly after the fall of communism to begin framing an answer. These data indicate that, depending on how the question is worded, between about half and three-fourths of all Russians say that they believe in God, almost half believe that God cares for each human being personally, and a third believe in heaven and in hell. About a quarter of Russians say that they pray, and 10 percent pray each day. Fourteen percent report that they have had a personal religious experience, and almost a third report that they attend religious services.[21]

Greeley found that almost a quarter of the Russians in the sample said that they were once atheists but now believe in God. This change has been most notable among those younger than twenty-five. Additional data gathered in Russia seem to indicate that the number of Russians who identify themselves as religious has continued to increase.[22] Thus, Greeley suggests, in line with the concept of the religious economy, religiosity does increase once appealing opportunities are available.

The experiences in all these countries suggest that a controlled religious economy can dampen religious beliefs and practices but that religion blossoms when more options become available. But what about a country that has historically allowed religious freedom, such as the United States? Can the ideas of the religious economy explain its religious history?

The Pluralistic Religious Economy of the United States   Because the Bill of Rights (the first ten Amendments to the U.S. Constitution) guarantees freedom of religion, people in the United States generally think that there was never anything like a state church in their country. In fact, during colonial times and the early years of the new nation, this was not necessarily the case. For example, in colonial New England the Congregationalist church received tax money to support its endeavors. In colonial New York, Virginia, Maryland, North and South Carolina, and Georgia, the Episcopalian church was legally sanctioned. Its legal status disappeared with the defeat of the British in the Revolutionary War because Americans didn't want to support a church affiliated

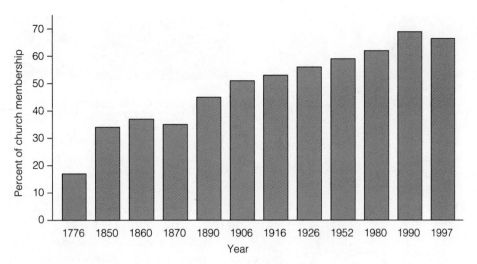

FIGURE 12-2 *Rates of Affiliation with a Church in the United States, 1776–1997*

Source: Finke and Stark, 1992, p. 16; Newport and Saad, 1997.

in any way with the Church of England. And, despite the federal guarantee of freedom of religion, the Congregationalists continued to receive tax support from state governments until the early 1800s.[23]

Even without knowing that churches were legally sanctioned by various states, most of us probably picture colonial America as highly religious. We often hear about the religious heritage of the thirteen colonies and the prayers offered by the Pilgrims at the first Thanksgiving. We might naturally assume that the early United States was a very religious country, where most people were devout churchgoers.

Amazingly, quite the opposite is true. Roger Finke and Rodney Stark combed libraries and archives to find extensive historical data on church membership in the United States from 1776 to the mid-twentieth century. They discovered that colonial America actually was not very religious at all. In 1776 only 17 percent of all citizens were affiliated with a church. Over time, however, this number grew rapidly and now includes about two-thirds of the population, as Figure 12-2 shows.

Why did this happen? Why is the proportion of church membership in the United States more than three times as large today as two centuries ago? One answer involves the notion of the religious economy and the supply of religious organizations. In 1776, according to Finke and Stark, only seventeen different denominations existed. Further, almost 80 percent of the slightly more than 3,000 churches represented only five denominations (Congregationalist, Presbyterian, Baptist, Episcopalian, and Quaker), and more than a third of the churches were actually supported by the state.[24] The United States' religious economy was not yet free and open but, rather, more closely resembled Sweden's.

Over time the situation changed dramatically. *The Encyclopedia of American Religion,* which lists and describes all religious groups in North America, doc-

uments more than 1,500 different types of religious groups that have operated at one time or another within the United States, a phenomenal increase from the seventeen groups that existed in 1776.[25] The theory of religious economies suggests that more people became affiliated with churches simply because there were more options from which to choose. When the range of choices widens, a greater proportion of people are likely to find a sufficiently attractive religious option to entice them to join.[26]

It is important to realize that this expansion of religious options resulted from social actions—concerted activities of people who were deeply moved by their faith and wanted to share it with others. For instance, throughout the first half of the nineteenth century, the Methodists and Baptists used large numbers of circuit riders (ministers who rode horses from one community to another spreading the gospel) and held big revival and camp meetings. As a result these groups, and especially the Methodists, experienced phenomenal growth.[27]

Some religious organizations were introduced and grew because of immigration. For instance, the English Puritans, German Mennonites, Russian Jews, Tibetan Buddhists, and Iranian Bahais all came to the United States to escape religious persecution in their homelands.[28] Other religious groups prospered as immigrants seeking economic opportunities brought along their religious heritage. For instance, Lutherans increased in numbers because of the arrival of Scandinavian and German immigrants, and the ranks of Catholics increased with Irish, German, and Italian immigrants.[29] Yet many of these immigrants didn't come to the shores of the United States with a firm faith. In their homelands, they generally showed the type of attachment found in Scandinavia today. Why, then, did they show greater affiliation in the New World? Andrew Greeley has stressed that in the United States the church often became a place

SOCIOLOGISTS AT WORK

Andrew Greeley

Sociology helps young men and women understand themselves and their religion.

In addition to being a well-respected sociologist affiliated with the National Opinion Research Center at the University of Chicago, Andrew Greeley is a popular novelist and a Roman Catholic priest. His writings often combine sociological insights, the elements of faith, and a clever way with words. In summarizing the results of his study of the enduring nature of religious faith in Russia, Father Greeley wrote, "God is alive and well in Moscow (and elsewhere) because She never left."

As a man of faith specializing in the sociology of religion, have you had any conflicts with the dominant sociological paradigm of secularization? I've had enormous problems with the dominant sociological paradigm of secularization. Even my closest colleagues and friends simply pay no attention to my work because it does not fit the model. I'm something of an outcast in the profession, precisely because I find evidence that shows that secularization is not a useful model.

How does the sociological study of religion apply to students' lives, especially those who believe? Sociology helps young men and women understand themselves and their religion: how it is shaped and formed, what influences the culture has on them, and how they influence the culture.

Do you think that the paradigm of secularization seems to be shifting to a more complex one encompassing ongoing or increasing religious involvement?

While the paradigm of secularization seems to be shifting, it nonetheless is fundamentally hostile to religion. It is a dogma that seeks to explain religion and explain it away.

How did you get interested in studying religious belief in Russia? I became interested in studying religious belief in Russia because the International Social Survey Program (in which I participate) did a study in 1991 of religious belief and practice in Russia.

What surprised you most about the results you obtained in your research? What was the reaction to your research among your colleagues? The biggest surprise was the enormous momentum of the religious revival in Russia. Most of my colleagues, on the other hand, have done their best to find things that will explain away the revival. I think by now it is an accepted truth, not because of my research, but because the media have also reported it.

Has the study been replicated by you or others? What were the results? Our Russian colleagues in the ISSP have replicated the study and assert that the religious revival continues and, if anything, is stronger.

What do you think the implications of your work are for social policy? One of the implications of my work is that religion is still an important factor in American cultural norms, more than it ever was, and the attempts to say it isn't important are the result not of scholarly research, but of academic dogma.

---

where immigrants could establish ties with others from their homelands—and thus an important element of their cultural and ethnic identity.[30] Many immigrants today continue to add new religious groups to the religious economy. For instance, recent immigrants from Asia and the Middle East have spurred a rapid growth in the number of adherents to Islam, Buddhism, and Hinduism.[31] To learn more about Andrew Greeley and his work, see Box 12-1.

At the end of the Civil War, African Americans attained the freedom to expand their own developing religious tradition, and this also added to the diversity of

the religious economy. The first exclusively African-American denominations were started by free blacks: the African Methodist Episcopal church in 1794 in Philadelphia and the African Methodist Episcopal Zion church in 1820 in New York.[32] Many other religious groups have also developed within the United States, including the Mormon church, the Jehovah's Witnesses, and the Pentecostal and Holiness movements.[33]

The growth of the religious economy in the United States has often involved intense conflict. At times these conflicts have paralleled other divisions within the society, particularly those between racial-

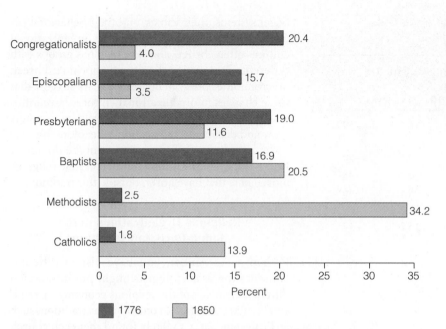

FIGURE 12-3 *Religious Adherents by Denomination, as a Percentage of All Adherents, 1776 and 1850*

Source: Finke and Stark, 1992, p. 55.

Congregationalists: 20.4 / 4.0
Episcopalians: 15.7 / 3.5
Presbyterians: 19.0 / 11.6
Baptists: 16.9 / 20.5
Methodists: 2.5 / 34.2
Catholics: 1.8 / 13.9

Percent

■ 1776   ▨ 1850

How might attending Saint Demetrios Cathedral help support the cultural and ethnic identity of adherents? Why might Greek immigrants be more likely to attend church in the United States than when they lived in Greece?

ethnic groups. For instance, animosity toward Italian and Irish immigrants was often conjoined with anti-Catholic sentiments and actions.[34]

Religious groups remain highly segregated by race and ethnicity today; churches are usually far more segregated than schools or workplaces. In fact, as religious scholars have noted, the most segregated time in American society is Sunday morning. It is estimated that more than 80 percent of all African-American Christians belong to one of the seven largest independent African-American Baptist or Methodist denominations.[35]

Even as new religious groups historically have appeared and gained many adherents, others have lost ground. Figure 12-3 displays some of the changes from 1776 to 1850. As you can see, the Baptists,

Methodists, and Catholics gained during this period while the Congregationalists, Episcopalians, and Presbyterians lost. After the mid 1800s, however, the Methodists began to lose ground relative to other groups.[36] In recent years, the decline of groups such as the Methodists has continued. As shown in Table 12-2, many religious groups that were at the forefront in the late eighteenth and early nineteenth centuries (often called "mainline religious denominations" because of their early dominance) have faced declining membership. From 1960 to 1990, groups such as the Disciples of Christ, Unitarians, Congregationalists, and Episcopalians declined by almost half, and the Methodists and Presbyterians declined by more than a third. In contrast, the Southern Baptists continued to grow, and other groups, including the Seventh-day

TABLE 12-2 *Growth and Decline of Some Religious Denominations, 1960–1990*

DENOMINATION	MEMBERS PER 1,000 US. POPULATION		
	1960	1990	% CHANGE
Christian Church (Disciples)	10.0	4.1	–59%
Unitarian-Universalist	1.0	0.5	–50
United Church of Christ	12.4	6.4	–48
Episcopalian	18.1	9.8	–46
United Methodist	58.9	35.8	–39
Presbyterian Church (U.S.A.)	23.0	15.2	–34
Evangelical Lutheran Church in America	29.3	21.1	–28
Roman Catholic	233.0	235.5	+1
Southern Baptist Convention	53.8	60.5	+12
Church of the Nazarene	1.7	2.3	+36
Seventh-day Adventist	1.8	2.9	+60
United Pentecostal Church	1.0	2.0	+100
Church of Jesus Christ of Latter-day Saints (Mormons)	8.2	17.1	+109
Jehovah's Witnesses	1.4	3.5	+150
Church of God (Cleveland, TN)	0.9	2.5	+177
Assemblies of God	2.8	8.8	+214
Church of God in Christ	2.2	22.1	+905

Adventists, Mormons, Assemblies of God, and Church of God in Christ, have shown phenomenal growth. For instance, the Assemblies of God, the denomination with which the Christian Center visited by Amber is affiliated, grew from including only 2.8 people of every thousand in the United States to almost four times that many in 1990.

Why have these changes occurred? Why have some churches gained in membership and others lost? Why did the Methodists do so well in the early 1800s and then start to lose out later on? Why have the Assemblies of God, as well as other Pentecostal and Holiness groups, done so well in more recent years? To answer these questions, sociologists have focused on religious organizations themselves.

RELIGIOUS ORGANIZATIONS: A MESOLEVEL VIEW

To understand more about the growth and decline of different religious groups, sociologists have often looked at the groups' organizational cultures: their belief systems, their values, and their behavioral expectations for members. Over time, organizational cultures often change, and new religious groups with different organizational cultures appear and break off from older ones. Many sociologists suggest that these changes in organizational cultures have influenced the way in which different religious groups grow and decline. In this section we explore this idea by looking first at religious groups in the dominant Judeo-Christian tradition and then at new religious movements that have arisen outside this tradition.

Religious Organizations in the Judeo-Christian Tradition

Perhaps, like Amber, you have visited different churches (or, as sociologists might put it, sampled different elements of the religious economy). Even if you limited your visits to one religious tradition, such as Protestantism, you likely found that each church had its own organizational culture. Perhaps in some, like the Presbyterian church of Amber's grandparents, the music was stately and majestic; in others, like Christian Center, drums accompanied the choir and the congregation clapped in time to the music. In some the pastor spoke of world problems and the importance of finding peaceful political solutions; in others the pastor dwelt on sin and the importance of personal salvation to help improve your life and your relations with others. In some the parishioners and members were expected to follow stringent behavioral standards, such as avoiding alcohol, smoking, gambling, and even dancing; in others they faced few restrictions on their personal life and behavior. Many sociologists have tried to develop ways to conceptualize or describe these differences.

The Church-Sect Dimension The data in Table 12-3 illustrate some of the differences in the organizational cultures of various religious groups as represented by the responses of Oregon clergy to a survey questionnaire. The researchers, Benton Johnson and Susie Stanley, asked these clergy how they felt about certain behaviors and how often they preached to their congregations about them. As you can see, they expressed very different views. For example, clergy in the United Church of Christ tend to be much more approving of all of the behaviors listed than do clergy in any of the other denominations. Those in the Holiness and Pentecostal groups are most disapproving. Almost all the Pentecostal and Holiness clergy are opposed to smoking cigarettes or marijuana, engaging in premarital adolescent sex and in homosexual activity, and drinking alcohol other than wine. A substantial number also disapprove of dancing. The Conservative Baptist and Southern Baptist

TABLE 12-3   *Views of Clergy in Twelve Selected Denominations, in Oregon*

			PERCENT OPPOSED		
Mainline	**Smoking Cigarettes**	**Smoking Marijuana**	**Premarital Adolescent Sex**	**Homosexual Relations**	**Drinking Alcohol**
American Baptist	76%	97%	97%	95%	32%
United Methodist	78	84	87	39	22
Presbyterian	73	86	84	53	14
United Church of Christ	53	50	68	14	3
Evangelical					
Southern Baptist	97%	98%	100%	99%	96%
Conservative Baptist	90	99	99	99	61
Holiness					
Church of God	97%	97%	100%	97%	78%
Evangelical	95	100	98	98	71
Nazarene	98	100	100	100	92
Pentecostal					
Open Bible Standard	100%	100%	100%	100%	85%
Foursquare	94	100	100	100	46
Assemblies of God	100	97	100	100	87

Note: For Holiness and Pentecostal groups, drinking measures apply to alcohol other than wine. For other groups, the measure applies to all alcohol.
Source: Personal communication, Johnson and Stanley, 1995.

clergy are closer to their Pentecostal and Holiness counterparts in their beliefs, and the Methodist, Presbyterian, and American Baptist clergy are a little more lenient.

The differences in views shown in Table 12-3 reflect a distinction sociologists make between *churches* and *sects*. Max Weber first introduced these terms to distinguish between official, large-scale religious institutions such as the state Church of Sweden, to which one belongs simply by being born (churches), and smaller religious groups that individuals join voluntarily, usually as adults, and that have strict rules regarding moral behaviors and beliefs (sects).[37] Over the years sociologists have elaborated on Weber's definition, trying to capture the subtle differences between different types of religious groups and often developing fairly elaborate and complex distinctions.[38]

One of the most widely used definitions, developed by Benton Johnson, is relatively simple yet extremely useful. It involves the notion of how much a religious organization accepts the norms of its surrounding environment. In Johnson's words, "A **church** is a religious group that accepts the social environment in which it exists. A **sect** is a religious group that rejects the social environment in which it exists."[39]

This definition allows us to conceive of religious organizations as falling along a continuum, with the ideal type of a church at one end and the ideal type of a sect at the other. Figure 12-4 illustrates this dimension for a set of denominations in the United States. The rankings were obtained from a number of religious experts, such as church historians, sociologists of religion, seminary professors, and denominational leaders. These individuals were asked to rate the extent to which each of the denominations emphasized "maintaining a separate and distinctive life style or morality in personal and family life, in such areas as dress, diet, drinking, entertainment, uses of time, marriage, sex, child rearing and the like." That the experts concurred virtually across the board in their ratings suggests that the ordering is quite reliable.[40]

As you can see, religious groups that were established in colonial times, such as the Episcopal, Congregational, Unitarian, Presbyterian, and Methodist churches, are closest to the "church" end of the continuum. These denominations accept the norms of the larger social environment and don't require other behaviors for their members. For example, if you want to be a good Episcopalian or Presbyterian, you don't have to do anything out of the ordinary, such

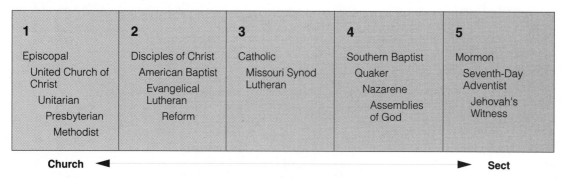

1	2	3	4	5
Episcopal	Disciples of Christ	Catholic	Southern Baptist	Mormon
United Church of Christ	American Baptist	Missouri Synod Lutheran	Quaker	Seventh-Day Adventist
Unitarian	Evangelical Lutheran		Nazarene	Jehovah's Witness
Presbyterian	Reform		Assemblies of God	
Methodist				

Church ◄——————————————————► Sect

FIGURE 12-4 *Expert Rankings of Selected Denominations on the Church-Sect Continuum*
Source: Iannaccone, 1994, p. 1191.

TABLE 12-4 *Differences Among Religious Groups in Beliefs and Behaviors*

	LOCATION ON CHURCH-SECT CONTINUUM			
Behavior/Belief	**Liberal**	**Moderate**	**Conservative**	**Sect**
Sunday attendance (services attended per year)	20.9	25.2	31.3	48.5
Weekday attendance (% attending weekday meetings)	2.7	6.3	11.5	32.3
Church contributions (% of yearly income)	1.9	1.9	2.8	3.2
Membership in church-affiliated groups (% belonging)	37.8	40.1	44.6	49.5
Secular memberships (number of memberships)	1.9	1.5	1.3	0.9
Strength of affiliation (% claiming to be "strong" members)	32.6	38.7	45.5	56.0
Biblical literalism (% believing)	23.2	40.4	57.8	68.1
Belief in afterlife (% believing)	79.5	85.1	88.9	87.8
Other				
Household income ($1,000)	38.0	31.0	31.6	27.0
Education (years)	13.8	12.5	12.1	11.3

Note: Unless otherwise noted, the figures in the table are simple averages. The data were compiled from National Opinion Research Center General Social Surveys from 1984 to 1990. Denominations included in the liberal group are Christian (Disciples of Christ), Episcopalian, Methodist, and United Church of Christ. Moderates include American Baptist, Evangelical Lutheran, Presbyterian, and Reformed. Conservative groups include Missouri Synod Lutheran and Southern Baptist. Sects include Assemblies of God, Church of Christ, Church of God, Jehovah's Witness, Nazarene, Seventh-day Adventist, and other Fundamentalists and Pentecostals.
Source: Iannaccone, 1994, p. 1193.

as dress in a particular way, give up specific foods, or avoid certain types of entertainment. Closest to the "sect" end of the continuum are the Holiness and Pentecostal groups, which began in the late nineteenth and early twentieth centuries, as well as the Mormon, Jehovah's Witness, and Seventh-day Adventist groups, which also appeared relatively recently. In contrast with the groups at the church end of the continuum, these sectlike groups have much more stringent behavioral norms and expectations. For instance, a good Mormon does not drink alcohol or caffeinated drinks, gives up at least a year of

his or her life while young for missionary work, and spends several hours each week in church-related activities. A good Seventh-day Adventist abstains from both alcohol and meat.

As you can see from Table 12-2, the groups toward the sect end of the continuum have experienced a great deal of growth in the past thirty years, while those toward the church end have declined. Groups toward the sect end of the continuum also have much more dedicated members. Table 12-4 illustrates this phenomenon with comparative data on beliefs and practices of groups across the church-sect

TABLE 12-5 *Differences Among Jewish Groups in Beliefs and Behaviors*

	GROUP		
Individual Practices	Reform	Conservative	Orthodox
Attends synagogue regularly	8.6%	19.0%	54.1%
Lights candles each Friday	5.4	15.3	56.8
Avoids money on Sabbath	6.5	13.1	57.7
Household Practices			
Buys kosher meat	3.5%	20.7%	68.5%
Separates meat and dairy dishes	3.4	18.9	66.7
Lights Hanukkah candles	53.4	66.9	76.6
Has Christmas tree	21.8	15.4	7.2
Networks			
Closest friends are all Jews	7.6%	15.8%	39.1%
Lives in Jewish neighborhood	6.6	11.5	36.0
Opposes marrying non-Jew	1.9	9.6	47.7
Household Contributions per Year			
Gives more than $1,000 to Jewish causes	5.4%	9.4%	22.5%
Gives more than $1,000 to non-Jewish causes	6.5	4.3	0.0
Other			
Volunteer for Jewish organizations (hours per week)	1.1 hrs	2.3 hrs	5.6 hrs
Household income	$64,700	$55,500	$41,700
Education (years)	15.8 yrs	15.2 yrs	14.8 yrs

Source: Iannaccone, 1994, p. 1196, based on data from the National Jewish Population Survey.

continuum. As you can see, when compared to members of the most churchlike groups, members of sects attend more than twice as many Sunday church services, are ten times more likely to attend church events during the week, donate a much larger proportion of their yearly income, belong to more church-affiliated groups and fewer secular (nonchurch-related) groups, are much more likely to say that they are "strong" members of their church, and tend to interpret the Bible much more literally.

Similar results appear when members of the three major Jewish groups within the United States—Reformed, Conservative, and Orthodox—are compared. Table 12-5 depicts differences in beliefs and practices among the three groups. Of the three, the Orthodox are the most sectlike, with the most traditional expectations regarding dress, diet, and behavior. The Orthodox members also are much more likely to be faithful in individual and family-related religious activities, such as attending synagogue regularly and following traditional dietary practices. And they are much more likely to have predominantly Jewish

friends, to give money to Jewish causes, to volunteer for Jewish organizations, and to oppose marriages between Jews and non-Jews. In contrast, the Reformed are the most churchlike, with much more lenient expectations regarding obedience to the traditional laws.

The Dynamic Church-Sect Relationship
Churches and sects are closely related to each other. Historically sects have tended to become churches, and these churches, in turn, have experienced divisions and conflicts, producing new sects. (One exception is the American Jewish tradition; the more churchlike Reformed movement actually broke off from the more sectlike Orthodox group.[41]) The result has been the dynamic development of many new religious groups, in many religious traditions, over the years. For instance, Buddhism was an outgrowth of Hinduism, and numerous different Buddhist sects developed as Buddhism spread across Asia.[42] Most sociological attention, however, has focused on the United States, where this process has resulted in many different denominations populating the religious economy.

Methodist circuit riders, energized by their faith and commitment, often would ride miles through miserable weather to reach their congregations. The zeal and enthusiasm of these early Methodist clergy no doubt helped the organization gain many adherents in the first half of the nineteenth century.

The evolution of the Methodists illustrates the dynamic relationship between churches and sects. The Methodist church grew out of the preaching and leadership of John Wesley, a clergyman in the Church of England in the 1700s. In his mid-thirties Wesley was given "saving faith" when his "heart was strangely warmed" during a church service, much like what Roger and his friends described to Amber. In contrast to traditional Anglican teachings, Wesley came to believe that salvation could come not from personal efforts but only from faith in God's mercy and grace.

Even though Wesley was an ordained Anglican minister and never officially left the Church of England, church officials disapproved of his views and prohibited him from preaching in their pulpits. Therefore he began speaking on street corners and in open fields, attracting numerous followers and training lay preachers to travel around and organize small "class meetings," groups of twelve people who met weekly for prayer, Bible study, and discussions about Christian living.

One of these lay preachers began the Methodist movement in the American colonies in 1766. In 1784 the rapidly growing movement was officially established in the United States as the "Methodist Episcopal church," with eighty-three preachers and more than 15,000 members.

Throughout the early 1800s, Methodist circuit riders journeyed from town to town, conducting camp meetings and enthusiastically preaching the gospel. The class meetings of twelve or so individuals provided the opportunity for intimate fellowship among believers, as well as a way to monitor the behavior of members. Good Methodists did not play cards, dance, or drink alcohol; nor did they attend the theater, horse races, or the circus. The local ministers received little, if any, pay and attained their position on the basis of their religious zeal rather than their education. In contrast to the Anglican and Congregational clergy, Methodist preachers talked about sin and damnation and warned of the perils of hell. Members of the congregation would spontaneously shout "Amen" or "Praise the Lord" during services when they felt moved by the spirit, and salvation was seen as a very personal experience.[43]

In sum, Methodism of this era fits the profile of a sect. The organizational culture emphasized the difference between Methodists and people who were "of the world" and required that members conform to the expectations of the church. During this time, the Methodists were also more successful than any other group in attracting new members. In 1776 less than 3 percent of all church adherents in the United States were Methodists, but by 1850 more than a third were.

By the mid-1850s, however, things began to change. Instead of relying on local, lay clergy, most Methodist churches had permanent ministers, most of whom had higher levels of education and earned regular salaries. Instead of being predominantly working-class people and farmers, more of the members were middle class and well-to-do. Church buildings became ornate and expensive edifices, and camp meetings changed from being large-scale experiences of religious fervor to something that resembled a summer lecture series, often on only semireligious subjects. Most important, by 1900 few Methodists cared about or followed the traditional restrictions on behavior. Thus, in 1924, the rules against activities like card playing, dancing, and attending "frivolous" entertainment, such as horse races, the theater, or the circus, were officially suspended.[44]

These changes illustrate the way in which the sect-like Methodist denomination of the late eighteenth and early nineteenth centuries was transformed into the more churchlike organization of today. Requirements that emphasized the differences of Methodists from others in the society were largely discarded as both clergy and adherents gained in social status. The Methodist church became more secularized.

Some people were distressed by these changes, believing that as Methodism became more worldly, it was straying from the path that Wesley had blazed. As a result, beginning in the 1820s, dissenters began to break off and form new groups, such as the "Methodist Protestant church," the "Wesleyan Methodists," and the "Free Methodists." These groups generally returned to the doctrines and teachings of the early Methodist church and required members to abstain from alcohol and tobacco and to dress in modest ways. The splintering away of these groups also represented the start of the Holiness movement. Groups that are part of this tradition emphasize Wesley's early teachings and the importance of salvation through maintaining faith and avoiding the evils of the world.

Throughout the late nineteenth and early twentieth centuries, several of these groups appeared. For instance, the Nazarene church, one of the fastest growing churches today, is part of the Holiness movement and began in the early twentieth century. Break-off groups, such as the Nazarene, have characteristics that place them near the sect end of the continuum.[45] Overall, the dynamic process of new groups emerging out of older ones has resulted in the Methodists being divided into more than thirty different denominations. Similarly the Baptists, which began at about the same time as the Methodists, have divided into more than sixty groups.[46]

Sociologists like Finke and Stark who study religious economies suggest that they contain a self-correcting dynamic. Thus many sects tend, over time, to become more churchlike as restrictions about behaviors diminish and church doctrine becomes more accepting of the surrounding environment. In other words, over time sects tend to become more secularized. Yet, as a religious group becomes less distinctive and more churchlike, some members miss the special qualities of the more sect-like atmosphere and break away from the church to develop alternative organizations. As a result, as long as the society allows pluralism, the religious economy remains dynamic and mutable.[47]

Churches and sects both represent the dominant Judeo-Christian heritage in the United States, but other groups fall outside this tradition. These groups compose the third element of the religious economy, what sociologists call cults, or new religious movements.

Cults or New Religious Movements

When sociologists use the term **cult,** they often are referring to a religion that is in its beginning stages and is new to a society. In this sense, all the major religions in the world, such as Judaism, Christianity, and Islam, were once cults, for they all were, at one time, new entrants into their society's religious economy. Unfortunately in recent years the word *cult* has been applied in a derogatory manner to groups that hold esoteric or occult ideas and that sometimes use intense and even unethical means of recruiting members. For this reason sociologists have started substituting the term **new religious movements** to describe emerging religions.[48] In their examination of new religious movements, sociologists have looked both at how they fit into the larger religious economy and at how they develop within societies.

New Religious Movements in the Religious Economy Some of the best-known new religious movements are Scientology, the Unification church (or Moonies), Hare Krishna, Wiccan (or witchcraft), the Theosophical Society, the Rosicrucian Fellowship, and various New Age groups. Altogether, according to the Institute for the Study of American Religion, more than eight hundred nonconventional religious groups have appeared in the United States over the past fifty years.[49] Most of these organizations are quite small. For instance, experts estimate that there are only 5,000 Moonies and 3,000 Hare Krishnas in the entire United States, compared with 8.8 million Methodists, 2.5 million Episcopalians, and 2.2 million members of the Assemblies of God.[50]

New religious movements are a more important part of the religious economy in some geographical areas than in others. For example, new religious movements are much more common in the western states of California, Oregon, Colorado, Nevada, and New Mexico than in other areas of the country. Why might new religious movements be more common in some places than in others?

As you might suspect, one answer involves the idea of the religious economy. New religious movements tend to crop up in areas where the established religious tradition is relatively weak and fewer people are involved in traditional religions. For instance, only 53 percent of the inhabitants of the western states report belonging to a church or synagogue, compared with 69 percent in the East, 72 percent in the Midwest, and 77 percent in the South.[51] Thus new religious movements seem to gain a foothold where the religious economy is not densely populated.[52]

These findings support the ideas of the organizational ecologists, which were introduced in Chapter 11. Organizational ecologists suggest that new organizations will form and survive as long as the demand continues and too many similar groups do not already exist. According to this view, new religious movements are much more likely to attract followers in areas where more traditional religious groups have fewer members because the current density of religious organizations is too low to satisfy the demand.

The Development of New Religious Movements
Where do new religious movements come from? Some are the result of *cultural innovation,* as people develop new religious ideas and organizations. But most result from *cultural diffusion,* the movement of elements of one culture into another. In other words many "new" religious movements aren't created out of whole cloth, but are simply new to a particular society. Just as the Western nations spread Christianity throughout the world in earlier centuries, some sociologists suggest, the religions of other countries are now showing up in the United States and Europe because the religious economy is becoming more global.

This process is illustrated by the appearance in the United States in the 1970s of more than 250 new groups associated with the Asian religions of Buddhism and Taoism.[53] Did these new groups appear because God was suddenly speaking to Americans in a more Buddhistic or Taoistic voice or because Americans were developing a lot of cultural innovations? Actually, sociologists suggest, the dramatic rise in new religious movements linked to Asia was influenced by changes in immigration laws. Until 1965 strict quotas limited the number of Asian and Middle Eastern immigrants to the United States. When immigration restrictions were eased, allowing more people from Asian countries to enter, a number of "gurus" also entered the United States and set up religious movements devoted to their own particular teachings and personalities. As a result, hundreds of new movements appeared, most of them quite small.[54]

Like splinter groups within the dominant Judeo-Christian tradition, new religious movements exhibit a dynamic process of change and spin-off into new organizations. For instance, some scholars trace today's "New Age" religions, which emphasize humans' bonds with nature and the environment, to earlier ancestors, such as the Rosicrucians, whose order was founded in the United States in 1693 (the Rosicrucians trace their origins to ancient Egypt), and the Theosophical Society, which began in the nineteenth century and often used the term "new age" in its writings. Americans have long been interested in ideas and beliefs outside the Judeo-Christian heritage, and how these beliefs are packaged has altered over the years, just as Judeo-Christian religious groups have altered.[55]

Changes in organizational culture can help to explain why groups increase or decrease in membership, but why do people choose to join a particular religious movement in the first place? How does an individual choose to become a Hare Krishna or a Methodist, a Presbyterian or a member of the Assemblies of God? What might influence Amber's decision to affiliate with Christian Center? To explore these questions, let's take a microlevel look at individuals' religious choices.

INDIVIDUALS' BELIEFS AND RELIGIOUS CHOICES: A MICROLEVEL VIEW

In recent years officials of the mainline religious denominations have fretted over their declining membership. Because several of the more sectlike groups have experienced rapid growth, officials couldn't blame this decline simply on greater secularization of society. Instead, the mainline denominations seem to be losing the "church race."

Many sociologists have also been intrigued by this phenomenon. Although sociologists haven't yet agreed on the reasons for it, they do know that the mainline denominations have not been as successful as the more sectlike groups in holding onto people who have been raised in the church.[56] That is, people like Amber's parents and Amber herself, who went to Presbyterian or Methodist or Episcopalian or Congregationalist Sunday schools when they were young, are far less likely than those who belonged to more sectlike groups to stay active when they are grown up. Three sociologists—Dean Hoge, Benton Johnson, and Donald Luidens—decided to investigate this issue by focusing on the Presbyterian church. Their results illustrate some of the explanations sociologists have developed for the decisions we make regarding our religious beliefs and affiliations.

FEATURED RESEARCH STUDY

LAPSED PRESBYTERIANS

Several people had speculated about why people raised as Presbyterians had left the church; Hoge and his associates decided to go one step further and ask them! In 1989 Hoge and his associates interviewed a sample of people who had been confirmed as Presbyterians when they were adolescents. (Confirmation is the process by which one becomes a full-fledged member of the Presbyterian church. Confirmands must complete a course of study and formally confess their faith in God and the church, typically around age thirteen or fourteen.) Because young adulthood is often a time of religious searching and wondering, as Amber's experience indicates, Hoge and his associates decided to limit their sample to people who were at least thirty-three years old in 1989. To limit the age group and thus avoid any complications that might arise from combining different birth cohorts, they also set an upper age limit of forty-two. Thus the members of their sample were at the point where many people have "settled down"—established careers and begun their families.[57]

Hoge and his associates focused on a random sample of confirmands from twenty-three churches that were carefully chosen to represent different areas of the country and different church sizes.[58] The confirmands were contacted by phone and interviewed at length about their religious experiences, beliefs, and patterns of church attendance. The researchers also conducted more extensive personal interviews with a subsample of this group and contacted some of the parents. As they expected, they found that only a minority of the confirmands were still active Presbyterians, but at the same time, few of the confirmands had explicitly rejected religion.

To better understand the experiences of the confirmands, Hoge and his colleagues used a **typology,** a classification of a group into discrete categories. In this case the typology of confirmands was based on both whether they belonged to a church and whether they actually attended church services at least six times a year. This resulted in eight different groups, as shown in Figure 12-5.

As you can see, the majority of the subjects both attend church (62 percent) and officially belong to a church (61 percent). But only 29 percent of the total group both belong to and attend a *Presbyterian* church. In other words, less than half of the confirmands who still attend church go to a Presbyterian church. Some have switched to other mainline denominations, and a few have affiliated with more conservative sectlike groups. None had joined a new religious movement. Among those who don't belong to a church, some (10 percent of the total group) do attend a church regularly but haven't joined; a large number (21 percent of the total) say they are religious but don't attend; and only 8 percent say that they are nonreligious.

Hoge, Johnson, and Luidens tried to understand why the confirmands had made their religious choices. What distinguishes those who are now active Presbyterians from those who are not? Why have some people switched to other denominations? Why have some fallen away altogether? Their explanations involve concepts and theories we have discussed in

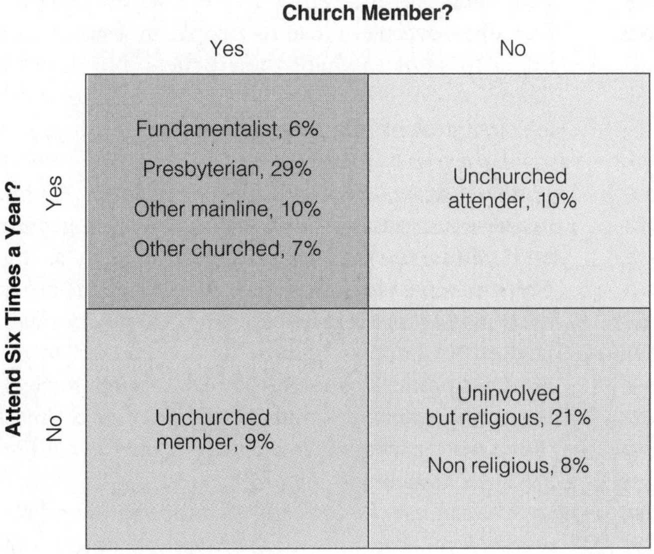

Church Member?

	Yes	No
Attend Six Times a Year? Yes	Fundamentalist, 6% Presbyterian, 29% Other mainline, 10% Other churched, 7%	Unchurched attender, 10%
No	Unchurched member, 9%	Uninvolved but religious, 21% Non religious, 8%

FIGURE 12-5 *Hoge, Johnson, and Luiden's Typology of Presbyterian Confirmands' Religious Behavior*
Source: Hoge, Johnson, and Luidens, 1994, p. 71.

earlier chapters: the ideas of social networks, subcultures, and exchange theory.

Social Networks and Religious Affiliation

Probably the most important influence on your decision to belong to a religious group is your relationship with others in that group. Perhaps, like some of the Presbyterian confirmands and several of Amber's classmates, you are active in the religious denomination that you grew up in. This probably reflects the long-standing associations you have had with others—your family and your friends in church. In fact, most people follow the general religious tradition in which they were raised.[59] The tendencies are particularly strong for Jews and Catholics, as well as for Mormons—all fairly distinctive faiths in the American religious economy. Of people raised in these faiths, 80 percent or more remain in them when they are grown.[60]

When people do switch denominations, they tend to join groups that are relatively similar to those in which they were raised.[61] Thus 10 percent of the confirmands now belong to another mainline Protestant denomination, one with churchlike characteristics similar to those of the Presbyterians. This seems to reflect what Hoge and his colleagues call a "personal comfort zone," a range of religious organizations that we understand and feel comfortable with—groups that have a familiar, identifiable organizational culture.

As you might expect, people who move away from their home community are less likely than those who stay close to home to remain active in their home church and denomination. For instance, more than half of the active Presbyterians in the Hoge study live within twenty miles of the church they attended in high school, compared with only 37 percent of the overall sample.[62] People who move away from their home community generally tend to have weaker ties with their home denomination. We simply find it easier to break old ties and establish new ones when we're farther from home.

The idea of social networks can also explain why someone would choose to convert to a different religious group. A major reason that Amber decided to attend Christian Center, a far more sectlike group than the Presbyterian church of her parents and grandparents, was that she had social ties to people in the organization. Her friend Roger introduced her to others in the church, and over time, as she became more involved in their activities, she also became attached to others within the congregation. Similarly many of the people who switched from the Presbyterian church to other denominations did so because of their ties to people involved in these other groups,

often through marriage. For instance, three-fourths of the people in the study who had become Catholics did so when they married someone who was already a Catholic.[63] In addition, many of the "unchurched attenders," people who attend churches to which they don't belong (see Figure 12-5), are attending because their spouse belongs. Often these churches are outside their personal comfort zone, such as Catholic or Christian Scientist, and so they are reluctant to take the formal step of affiliation.[64]

Studies of new religious movements have also highlighted the importance of social networks in gaining converts. For instance, the Moonies (members of the Unification church) gain almost all of their recruits through friendship networks. When trying to attract new members, they often target people who appear to be transient or alone. Once potential recruits are exposed to the group, intensive interaction and the development of strong friendships and a devotion to the group help to solidify commitment.[65] Similar processes have been documented for other groups.[66]

Subcultures, Reference Groups, and Religious Affiliation

When sociologists talk about how social ties influence our decision to join a religious group, they do not mean to diminish the importance of religious faith. Religious beliefs, experiences, and faith clearly are very important to those individuals who are active in different churches. In fact, the strength of the confirmands' religious belief was the best predictor that Hoge and his associates could find of whether they were committed to a church in adulthood.[67]

Yet studies of people who convert to different religious groups indicate that they tend to accept a group's beliefs only *after* they have established ties with the group and started to feel a sense of belonging. Once we feel close to people in a group, we begin to want to share their beliefs. No doubt a major reason the confirmands confessed their faith as Presbyterians was that Presbyterian theology was what they had been exposed to and understood. Similarly, after visiting Christian Center, Amber began to ask herself if she could believe as Roger and his friends do. Converts to groups such as the Moonies report that, once they came to see the group as an important reference group, they very much wanted to share its beliefs. As one convert to the Moonies put it, "I wanted to break through my feeling of isolation badly enough that right then it almost didn't matter what they believed—if only I could really share myself with them."[68]

We can use the concepts of subcultures and reference groups and the ideas of symbolic interaction

Norma McCorvey, the "Jane Roe" in the Roe v. Wade *court case about abortion, was a longtime symbol of the pro-choice movement. In 1995, after becoming friends with members of a church near the pro-choice office where she worked, she became a born-again Christian and was baptized by the national head of the pro-life group Operation Rescue. How might you use the theories and ideas presented in this chapter to explain McCorvey's experiences?*

theory to understand this process. When we join a new religious group, we become part of a subculture, attuned to its norms, values, ideologies, and myths. As we interact with members of the subculture, they come to be our reference group, and we interpret our interactions with them in ways that conform to the elements of the subculture. The interpretations that make sense are those that seem to reflect the subculture's own belief system. Thus two people can experience the same feelings as part of worship, but they might interpret them quite differently depending on the subculture in which they are interacting. For instance, people involved in a Holiness or Pentecostal group might believe that they have been "filled with the Holy Spirit," whereas members of a Presbyterian church might not even recognize the feelings (because they aren't looking for them) or simply interpret them as a general sense of well-being.

Even though we tend to become attached to a meaning or belief system after we have established ties to the group that espouses it, sociologists have found that such meaning systems are very important in explaining our commitment to a religious group and, thus, the reasons that some groups are more likely than others to have committed members. The

ideas of exchange theory and its close cousin, rational choice theory, can help us understand why these differences in commitment appear.

Rational Choice and Religious Affiliation

Exchange theory emphasizes the way in which we balance the costs and benefits of our actions in our interactions with other people. The costs and benefits can involve not only money and other goods but also many nonmaterial rewards such as love, esteem, approval, and affection. As an outgrowth of this perspective, rational choice theory looks at both ongoing interactions between individuals and their day-to-day decisions. Specifically, **rational choice theory** suggests that in making decisions about how we will behave, we balance the costs and benefits involved and then choose the actions that provide the lowest costs and the greatest benefits.

Using this perspective to understand religious choices directly counters secularization theory's implicit assumption that religious conversion and beliefs are illogical. Instead, this approach assumes that humans are active, thinking beings who make rational choices about ideas that can make sense of their world and lifestyles that they will follow.[69]

Costs and Benefits of Religious Affiliation
Some of the costs and benefits involved in religious membership revolve around the *behaviors* that a group might require. For instance, if you were to belong to a very strict group, such as the Hare Krishnas, you might have to break ties with your family, shave your head, wear bright-colored robes, and chant words in public places. If you became a Moonie, you would have to marry a person someone else chose for you. If you joined one of the less strict but still sectlike groups, such as the Jehovah's Witnesses, Mormons, or various Pentecostal and Holiness denominations, you would also have costs, even though they wouldn't be as great as those associated with the Hare Krishnas or Moonies. You might have to abstain from certain foods and drink, wear certain clothes, follow certain medical practices, be involved in a number of church activities, and perhaps break ties with some of your former friends.[70] For instance, neither Jamaal nor Roger, Amber's classmates, drink alcohol because of their religious beliefs, and so they have drifted apart from some of their old friends.

Clearly, the costs would be much less if you were to join a denomination that fell toward the church end of the continuum, such as the Methodists, Presbyterians, or Congregationalists. In these groups you wouldn't have to make significant lifestyle changes and you wouldn't stand out as different from the people around you.

BUILDING CONCEPTS

Social Networks, Subcultures, and Rational Choice Theory

What social processes contribute to a person's choice of religion? Below are three concepts and theories that sociologists have used to examine this question. Note how each explanation supplements the others.

Concept/Theory	Reasoning	Example
Social networks	People join or convert to a religion because they have social ties to others in that religion; for example, they might have friends, relatives, or a spouse who is a member of that religion.	A person converts to Judaism after marrying someone who is Jewish and who introduces her to Jewish beliefs.
Subcultures/ Reference groups	People join or convert to a religion because they have come to identify with members of that religion and want to strengthen their sense of belonging by conforming to the norms, values, and ideologies of that religious subculture.	A person becomes a cult member and adopts the cult's beliefs because of the sense of belonging that he feels when he is with people in that cult.
Rational choice/ Exchange theory	People may join or convert to a religion after they have weighed the costs and benefits of joining the religion and concluded that the religion will be a positive force in their lives.	A person decides to become a Buddhist after deciding that the benefits of learning Buddhist philosophy are worth the "cost" of the time needed to study Buddhist teachings.

Costs and benefits also revolve around *beliefs,* and you face clear choices among different groups. Groups at the sect end of the continuum have much stricter belief standards than do more churchlike groups. Sectlike groups have more clear-cut and compelling spiritual messages, with straightforward answers to questions about why we are here on Earth, how we should live our lives, and what will happen to us after death.[71]

Some of these differences were shown in Table 12-4. For instance, less than a quarter of the members in the churchlike Protestant groups, but more than two-thirds of the members of the sectlike groups, believe that the Bible is true word for word. Johnson and Stanley found similar differences in their study of the clergy, mentioned previously in this chapter. Almost none of the clergy in the groups that fall toward the sect end of the continuum (the Holiness, Pentecostal, and Evangelical groups) question traditional Christian beliefs. Almost all believe that Jesus was the Son of God, that there is life after death, that heaven is a real place, that Christ actually rose from the dead, that the Bible is literally true, that Christ will personally return to Earth at some point, and that belief in Christ is the only way to ensure salvation. Those at the church end of the continuum hold strikingly different beliefs. For instance, only a very few clergy in the United Church of Christ believe that heaven is a real place, that Jesus actually rose from the dead, that the Bible is literally true, or that Jesus will return. None accepts the view that believing in Christ is the only way to ensure salvation.[72] (To learn more about Johnson and his work, see Box 12-2.)

The confirmands in the Hoge study also demonstrated these differences in behaviors and beliefs. Those who currently belong to the more sectlike Fundamentalist churches were much more likely to frequently attend church services, as well as Bible study groups, spiritual growth seminars, and prayer and healing services. They were also much more likely to hold very traditional Christian beliefs and to be sure of their faith.[73]

Why do some people choose to attend the more sectlike groups, with their stricter behavior standards and belief systems, whereas others affiliate with the more churchlike groups, with their more lax behavioral and belief standards? Drawing on the ideas of rational choice theory, several sociologists have suggested that people simply balance out the costs and benefits.

Balancing Costs and Benefits of Religious Affiliation One way to think about religious choices is to focus on what people would lose from electing to join a more restrictive sect. Some people clearly would lose more than others if they made such a decision. For instance, if you had a family and a well-established and high-paying career, you would be highly unlikely to throw it all away to join the Moonies or Hare Krishnas. But if you were young, unmarried, and unemployed, you would have far less to lose.

In general, rational choice theory suggests that the people most likely to join sectlike groups are those with the least to lose, and data tend to support this hypothesis. Individuals who have fewer economic opportunities, such as women, members of racial-ethnic minorities, and young people, are more likely than are others to join sects. Similarly, when economic times are tough and job opportunities are scarce, sects tend to flourish. Historically, members of sectlike denominations have tended to earn less and to work at lower-status occupations than members of more churchlike groups are. (These patterns also appear in the data reported in Tables 12-4 and 12-5, which include information on household income and education.)[74]

Besides looking at how much someone would lose in a decision, we can also weigh the benefits. The lack of rigid belief requirements in a churchlike group might be attractive to some people, for membership does not require them to make large "leaps of faith" and hold strong beliefs. Yet, the strong belief systems of the more sectlike groups provide firm messages about the meaning of life and the reasons we are on Earth. If this knowledge is important to someone, then the benefits of such knowledge can outweigh the costs involved in joining a sect. In other words, for many people, the comfort attained from knowing God's role in one's life and the meaning of one's place on the world can outweigh the strict behavioral and belief requirements associated with these groups.

The rational choice perspective can also be used to explain why sectlike groups are growing in number and membership.[75] Precisely because these groups have stricter standards of belief and behavior, people who join sects are more committed. These dedicated people make the organization stronger and more vibrant and thus, potentially, more attractive to others. Churchlike groups, which don't make these demands, might have some members who have the strong faith and dedication of sect members. Generally, however, they aren't as dedicated, simply because the churchlike groups' organizational cultures don't require the same level of commitment. As a result, the churchlike organizations are less strong and vibrant than sects and thus less attractive to potential members.

Interestingly, the recent strong growth of some sectlike groups seems to be changing the association between social status and church membership that was noted previously. Several different studies have suggested that affluent middle-class individuals increasingly are joining more sectlike Holiness and Pentecostal groups.[76] In recent years, the benefits of belonging to more sectlike movements—such as the more vibrant organization and the stronger belief structure—might have come to outweigh the losses that result from membership in such groups. Sociologists who study religion haven't yet reached any firm conclusions about the nature of this trend, but it illustrates the way in which sociology continues to evolve, as well as the kinds of questions being addressed by the new paradigm in the sociology of religion.[77] (For another example of how sociologists are applying this new paradigm to real-life issues, see Box 12-3.)

CRITICAL-THINKING QUESTIONS

1. Would it be harder for sociologists to study religion than other areas of social life? Why or why not?

2. Do you think that Hoge and his associates would have obtained different results if they had studied a group other than Presbyterians? Why? Explain.

3. What other methods, besides survey research, can sociologists use to study individuals' religious beliefs and decisions?

SOCIOLOGISTS AT WORK

Benton Johnson

Many college students will easily be able to recognize themselves in the profiles we draw.

Benton Johnson, University of Oregon, has studied the sociology of religion for many years. His research is described throughout this chapter.

The paradigm of secularization seems to be shifting to a more complex one encompassing ongoing or increasing religious involvement. Why do you think this has happened? In my view, the shift of sociological opinion on the topic of secularization occurred in response to some big surprises of the past twenty-five or so years. First, there was the strange appeal that new religious movements had for many privileged young people during the counterculture years of the late 1960s and early 1970s. Second, there was a decline in membership in the most accommodated or "secularized" religious bodies that set in during the late 1960s and has not been reversed. Third, there was the continued growth and political mobilization of religious conservatives in the United States. And fourth, outside the United States there was an upsurge of militant fundamentalism, especially among Muslims.

What surprised you most about the results you obtained in your research? Several findings surprised me. The first was the fact that most lapsed baby boomer Presbyterians haven't rejected the Presbyter-

ian church or what they think it stands for. It's just that these things aren't very important to them. Most of them pray now and then, admire Jesus, and want their children to have a religious education, but they are reluctant to attend church regularly or put a lot of time into doing church work. And they are not firm believers of the traditional sort.

This relates to the second surprising finding, which was the existence among these baby boomers of a theological outlook we called "lay liberalism." It is "lay" because it doesn't come directly from anything being taught in seminaries or from the pulpit. It's "liberal" because its defining feature is the rejection of the orthodox notion that Christianity is the sole keeper of religious truth. Lay liberals recognize the authenticity and underlying "truth" of all religions that teach the fundamental moral precepts of loving one's neighbor and not hurting others. In fact, for many of our baby boomers, the basic value of religion is that it does teach those things.

Lay liberals do not push their own religious views on others. Indeed, they seldom talk about religion at all, and they find distasteful any religion that tries aggressively to convert other people. So although these baby boomers accept the

SUMMARY

Among the theoretical ideas sociologists have used to understand the social institution of religion is a new paradigm based on the concept of a religious economy. By focusing on actual data, sociologists have prompted a change in the theoretical approach to religion—specifically, a shift away from the long-held theory that modern society is becoming increasingly secularized. In addition, concepts drawn from social network, social exchange, and rational choice theory can help explain why individuals decide to affiliate with specific religious groups. Key chapter topics include the following:

- Religion is a social institution that revolves around the area of life people regard as sacred. From a sociological perspective, religion promotes social cohesion and provides meaning systems for societies, but it can also be used by the powerful to dominate those with less power.

- The early sociologists believed that modern societies were experiencing a long-term process of secularization, but empirical data do not support this idea, leading to a recent change in the way sociologists look at religion.

BOX
12-2

SOCIOLOGISTS
AT WORK

Continued

right of Mormons and TV evangelists to try to convert others, they themselves aren't potential converts to any "hard sell" religion or to any faith that claims to have a monopoly on the truth or requires a high level of commitment. This finding, assuming it is also widespread among baby boomers brought up in other "mainline" denominations, means that there are clear limits to the growth of evangelical Protestantism or any other religious movement with a claim to exclusive truth. It also means that there are clear limits to the growth of the religious right as a political force.

One finding that wasn't so surprising to us, but that might be surprising to others, is that baby boomers haven't abandoned the "mainline" churches in protest against the left-wing social programs promoted by the national leaders of these bodies. For the most part, our baby boom respondents have only the vaguest ideas of what national denominational leaders are up to. For them, "church" refers to local congregations, not denominations and their official programs. These respondents tend to be Republican and consider themselves politically conservative, but they tend to have liberal positions on the controversial "social" issues agitated by the religious right. Although most favor school prayer, on the whole they're pro-choice on abortion and only a minority would object to the ordination of an openly gay minister.

What research have you done since completing the study described in the chapter? Does it build on your previous research? I am currently completing the third and final phase of a long-term study of the Protestant clergy of Oregon. Phase one, which was completed in 1962, was the first research to document a strong and clear-cut relationship between theological position (liberal-conservative) and sociopolitical outlook among American Protestant clergy. Although phase three does not build directly on my study of baby boom Presbyterians, it does build on my previous work on theology and political preference among American Protestants. It should help show, among other things, how the theology-politics relationship has changed over the past thirty-five years.

How do you think the study of the sociology of religion applies to students' lives? Our study, like many other studies of personal religious involvement, shows that very young adults are the least likely of all age groups to participate in organized religious activities. Many college students will easily be able to recognize themselves in the profiles we draw of our lay liberal baby boomers. Others, however, and especially those who were raised in very strict and demanding faiths, will have their horizons broadened by getting their first inside glimpse of the perspectives of people brought up in a more liberal and relaxed religious milieu.

- Compared with other countries, people in the United States are much more likely to attend church. But, even in countries with low levels of church attendance, only a small minority usually call themselves nonbelievers or atheists.

- Sociologists who use the idea of "religious economies" suggest that when a religious economy is more pluralistic, or has more available options, more people will be involved in religious groups. Data from both controlled and free religious economies in countries around the world seem to support this theory.

- Sects and churches are religious organizations with different types of organizational cultures. Sects tend to reject the social environment in

which they exist but often develop a more churchlike culture over time. At the same time, however, churches often have splinter groups break off that have more sectlike characteristics. This dynamic relationship helps explain growth and change within religious economies.

- Cults, or new religious movements, are religions that are new to a society. New religious movements tend to be more common when a religious economy is sparsely populated. Often the appearance of such movements is due to cultural diffusion.

- In recent years in the United States, many of the churchlike, mainline religious denominations have lost members while more sectlike groups have gained members.

The media today are crammed with reports about conflicts that seem to be fueled by religious divisions—whether between Protestants and Catholics in Northern Ireland, between Jews and Muslims in the Middle East, or between Muslims, Orthodox Christians, and Roman Catholics in the former Yugoslavia. The religious identity of people who participate in such conflicts is no doubt more salient than many other aspects of their self-identity. In addition, the conflicting factions in these disputes have strong group boundaries. For example, when an Israeli Jew and an Israeli Muslim interact, they could quite literally be defined as associating with the enemy.

It is not at all uncommon for members of religious organizations, especially sectlike groups, to have strong religious identities, simply because these groups require such a high level of commitment. In addition, by definition, sects tend to have strong group boundaries—their organizational culture involves a great deal of separation from other religious organizations. Yet only a very small minority of the people in the world whose religious

identity is a salient aspect of their self-concept or who belong to sectlike groups with strong boundaries actually participate in violent sectarian conflicts. When, sociologists have asked, do religious identity and affiliation tend to be associated with violent conflict?

To answer this question, we need to back up and take a more macrolevel view of broader divisions within a society. Religious identity is often closely associated with ethnic identity. For instance, in the Middle East Jews, as a cultural and ethnic group, are pitted against Muslim Arabs. In Yugoslavia, Eastern Orthodox Serbs clash with Roman Catholic Croats and Muslim Bosnians. In Northern Ireland, Catholics of Irish descent battle Protestants of Scottish descent. As you will recall from previous chapters, ethnicity can also be associated with economic differences. For

instance, in Northern Ireland the Protestants tend to be the wealthier landowners whereas the Catholics tend to be poor laborers. Conflict between religious groups seems to be much more common when religious divisions in a society also involve differences in culture, ethnicity, and economic well-being. In other words, when religious differences are associated with racial-ethnic and social stratification, conflict between the involved groups is more likely to occur.[78]

Many scholars suggest that what often appears to be religious conflict is actually conflict based on economic stratification. Because religious identity provides an easily recognizable group boundary, as well as a system of meanings and ideological support, many people understandably interpret such conflicts as religion-based. However, sociologists who study conflicts such as those in Northern Ireland, the former Yugoslavia, and the Middle East caution against mistakenly labeling economic and ethnic conflicts as religious in nature.[79] In fact, these conflicts often have much broader causes and involve larger ques-

■ Dean Hoge and his associates gathered survey data from a sample of adults raised as Presbyterians to examine people's decisions regarding their religious beliefs and affiliations. The researchers found that an important influence on whether we choose to affiliate with a religious group is the nature of our social networks. In addition, becoming part of a religious group involves becoming part of the subculture and seeing members of the organization as a reference group. Finally, deciding to join a group can be seen through the

lens of rational choice theory as involving a balancing of the potential costs and benefits.

■ Rational choice theory suggests that sects have been more likely than churches to experience growth because their stricter behavioral and belief requirements encourage stronger commitment and produce organizations with fewer "free riders." These organizations thus appear more attractive to and produce more rewards for members.

BOX
12-3

APPLYING SOCIOLOGY
TO SOCIAL ISSUES

Continued

tions about the distribution of power within a society. (In Chapter 13 we look at these power differences by examining the political world.)

Even though many seemingly religion-based conflicts have economic and ethnic underpinnings, the emotions and allegiances generated by religious identities and group boundaries are very real. In addition, even in societies such as our own, where violent conflicts between religious groups are virtually unheard of, members of various religious groups sometimes display a good deal of open animosity and distrust. For instance, members of conservative groups accuse those with more liberal beliefs of being "secular humanists," while those with a more liberal bent label conservatives as unenlightened or ignorant. Political debates over issues such as abortion, school prayer, vouchers for parochial education, and religious displays during holidays have split many communities.

How can animosity between religious groups be lessened? How can we develop ties between disparate groups that will build connections and trust, not just in our own society but around the world? The discussions in previous chapters indicated that an important element of greater understanding is increased interaction, or social linkages, between people with different views and backgrounds. The contact hypothesis, which has been well supported by empirical research, suggests that intergroup hatreds decline substantially when members of different groups have contact with each other in circumstances that promote cooperation and the well-being of both parties.

The research reviewed in this chapter suggests that as a society's religious economy becomes more pluralistic, the number of religious groups will increase. In addition, as groups become more secularized, or churchlike, sect-like offshoots will develop, attracting adherents with their vibrancy, strong group boundaries, and sense of identity. In other words, within a free religious economy, religious differences could be expected to proliferate.

CRITICAL-THINKING QUESTIONS

1. How can societies have a free and pluralistic religious economy without the danger of fostering separatism, hatred, and conflict? Might some of the ethnic and religious hostility that has flared up in the republics of the former Soviet Union and its satellite countries, such as Yugoslavia, be a tragic by-product of the blossoming of religious freedom in those regions?

2. How can people have their spiritual needs met—perhaps including strong religious identities and membership in spiritual groups with strong group boundaries—without simultaneously developing such rigid views that they harm their neighbors?

3. Have you witnessed religious animosity? What can be done to address this issue?

KEY TERMS

church 313
cult 317
new religious movement 317
paradigm 304

paradigm crisis 304
rational choice theory 321
religion 303–304
religious economy 307

sect 313
secularization 304
typology 319

INTERNET ASSIGNMENTS

1. Find Web sites that focus on new religious movements, such as the Unification Church, Hare Krishna, or New Age religions. What kinds of information do you find? Try to locate information both in support of (such as members' home pages, information on membership, activities) and in opposition to (such as outreach organizations and reform or support groups) these groups. Can you find any evidence in these sites that relates to the discussion in this chapter about why new religious movements develop and why people choose to join them?

2. Search for information about a number of different religious organizations. If you have a religious preference, compare the doctrines and beliefs of your religious organization with those of other religious groups. Can you use the information provided on the Internet to identify groups as either more sect-like or more churchlike? Can you find out if the groups are growing or declining in size? How might the information provided on the Internet influence individuals' decisions to affiliate with these religious organizations? How might the concepts related to social networks, subcultures, reference groups, and rational choice theory be used to explain the impact (or lack of impact) of the Internet on individuals' decisions regarding religious affiliation and belief?

3. What types of "electronic" religions and religious services can you find on the Internet? How does the development of electronic religions relate to the idea of the religious economy? Who do you think might be most attracted to such electronic religions? Why? How could you test this hypothesis?

TIPS FOR SEARCHING

Searching by the name of a specific religious group is often more useful than searching under religion as a general category.

INFOTRAC COLLEGE EDITION: READINGS

Article A21206032 / Chong, Kelly H. 1998. What it means to be Christian: the role of religion in the construction of ethnic identity and boundary among second-generation Korean Americans. *Sociology of Religion, 59,* 259–287. Chong describes results of his field research into how religious affiliation influences ethnic identity.

Article A18911734 / Miller, Alan S. and Takashi Nakamura. 1996. On the stability of church attendance patterns during a time of demographic change: 1965–1988. *Journal for the Scientific Study of Religion, 35,* 275–285. Miller and Nakamura examine cohort differences in church attendance.

Article A21015907 / Iannaccone, Laurence, Rodney Stark, and Roger Finke. 1998. Rationality and the "religious mind." *Economic Inquiry, 36,* 373–390. In this article, Iannaccone, Stark, and Finke develop one of the major ideas of the new "religious economy" paradigm—the notion that religious belief and activity is just as common, if not more so, among the well educated as among those with less education.

Article A18902832 / Williams, Andrea S., and James D. Davidson. 1996. Catholic conceptions of faith: A generational analysis. *Sociology of Religion, 57,* 273–290. Williams and Davidson examine differences and similarities in the beliefs of Catholics in three different birth cohorts and discuss reasons for the patterns that they find.

FURTHER READING

Suggestions for additional reading can be found in *Classic Readings in Sociology,* bundled with this book. If you purchased a used copy of this book that does not include this custom-published reader, go to www.sociology.wadsworth.com for ordering information.

RECOMMENDED SOURCES

Bainbridge, William Sims. 1997. *The Sociology of Religious Movements.* New York and London: Routledge. This detailed and expansive book examines the development of many different religious groups during the past two centuries.

Demerath, N.J., III, Peter Dobkin Hall, Terry Schmitt, and Rhys H. Williams. 1998. *Sacred Companies: Organizational Aspects of Religion and Religious Aspects of Organizations.* New York: Oxford University Press. The authors use sociological theories and concepts related to organizations to analyze religious organizations and religious aspects of secular organizations.

Finke, Roger, and Rodney Stark. 1992. *The Churching of America, 1776–1990: Winners and Losers in Our Religious Economy.* New Brunswick, N.J.: Rutgers University Press. This book contains a lively discussion of the changing religious economy in the United States.

Greeley, Andrew. 1995. *Religion as Poetry.* New Brunswick: Transaction. The sociologist profiled in Box 12-1 includes an extensive analysis of survey data regarding religious beliefs and presents, in a very readable manner, his theoretical views regarding the development and maintenance of religious belief.

Griffith, R. Marie. 1997. *God's Daughters: Evangelical Women and the Power of Submission.* Berkeley: University of California Press. Griffith reports results of her field research among Charismatic women's prayer groups.

Hoge, Dean R., Benton Johnson, and Donald A. Luidens. 1994. *Vanishing Boundaries: The Religion of Mainline Protestant Baby Boomers.* Louisville, Ky.: Westminster/John Knox. This book gives the full report of the research featured in this chapter.

Kaufman, Debra Renee. 1991. *Rachel's Daughters: Newly Orthodox Jewish Women.* New Brunswick, N.J.: Rutgers University Press. Kaufman provides a fascinating account of why people choose to convert to a more sectlike group.

Kosmin, Barry A. and Seymour P. Lachman. 1993. *One Nation Under God: Religion in Contemporary Society.* New York: Harmony. The authors report the results of a large scale survey of Americans' religious identification and preferences.

Miller, Donald E. 1997. *Reinventing American Protestantism: Christianity in the New Millennium.* Berkeley: University of California Press. Miller describes his field research of several newly established Protestant churches and provides an inside look into how new religious bodies become established and grow.

Roberts, Keith A. 1995. *Religion in Sociological Perspective,* 3rd ed. Belmont, Calif.: Wadsworth. This extensive textbook reviews many different aspects of the sociology of religion.

Stark, Rodney. 1996. *The Rise of Christianity : A Sociologist Reconsiders History.* Princeton, N.J.: Princeton University Press. One of the major developers of the notion of a religious economy analyzes the development of early Christianity from the years 30 to 600 A.D.

Wuthnow, Robert. 1998. *After Heaven: Spirituality in America Since the 1950s.* Berkeley: University of California Press. Wuthnow examines developments and changes in Americans' spiritual lives during the last half century.

Young, Lawrence A. (ed.) 1997. *Rational Choice Theory and Religion: Summary and Assessment.* New York: Routledge. This series of essays covers rational choice theory and religious commitment.

INTERNET SOURCE

The Center for Social and Religious Research. This Web site, which is based at Hartford Seminary, includes reports of research in the sociology of religion and link to other useful sites.

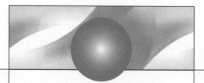

Pulling It Together

Listed below are other chapters where you can find additional discussion or examples
of some topics covered in this chapter. You can also look up topics in the index.

The Political World

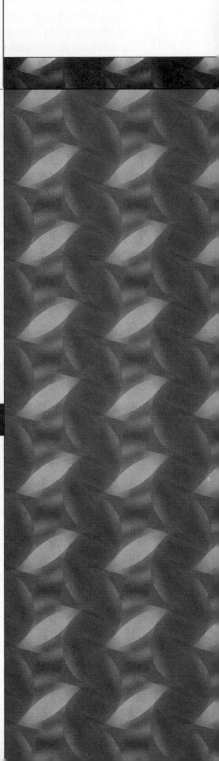

WHEN DAVID PICKED UP the morning paper and glanced at the headline, his heart sank. In big, bold letters it proclaimed, "CCC Tuition May Rise, Enrollment May Fall." The article described the deliberations of the governing board of City Community College. The board not only was planning to raise tuition and enrollment but also was considering scaling back the college transfer program and adding more courses in hotel management and food service.

Like many two-year schools, City Community College received most of its funding from local taxes. When the college needed additional money, as often happened when enrollment swelled or new programs were added, the community voted on whether to increase its contribution to the school. Traditionally, these requests for additional money had been approved by wide margins.

Recently, however, things had changed. The steel mill that formerly employed many of the town's residents had shut down, resulting in economic hardship for many families, and many former employees enrolled at CCC to train for new careers. Also, several large new retirement communities had been built and many people who were new to the town, and even the state, had moved in. The tourist industry was also becoming more important in the community, and three new resorts with large golf courses and recreation complexes were being built just outside town.

Many of the new citizens in the community vocally opposed increased payments to the college. "Let the Students Pay," proclaimed a popular sign displayed in store windows and yards. The CCC board was afraid that their next tax request would be soundly defeated, and they were trying to figure out what they would do without the sorely needed money.

In addition, a committee from the city's Chamber of Commerce had recommended that the college cut back on its college transfer program and add courses in food service and hotel management to help train people for jobs at the new resorts and retirement communities. Some CCC board members were beginning to think that these policy changes might not be a bad idea and might help generate more community support for the college.

The possibility of higher tuition and cuts in college transfer courses worried David a great deal. His family was not well off, and he was paying for school through a combination of part-time work and student loans. He couldn't afford to pay much more, and he certainly couldn't afford to go to a different school in another town. David knew that other students were in even more dire circumstances—for example, single mothers trying to get an education so they could get better jobs, and those who had lost their jobs when the steel mill closed. What could be done?

David was still pondering this question when he got to his first class. His friend Jeff informed him that many other students were also concerned and that the student government was holding a meeting at noon to discuss the problem and formulate ways to influence the vote. It would be great, he said, if David would come along. Maybe he could help too.

David had never paid much attention to any kind of politics before, but in the next few weeks he found that he was spending almost as much time thinking about the upcoming budget vote as about his studies. Following Jeff's lead, he joined the newly formed "Students for CCC" committee. Along with the other committee members, he began canvassing the campus to convince students to register to vote and to participate in the budget election. The committee also tried to convince voters throughout the community of the need for the budget increase. Some students participated in public forums, called newspaper reporters, and spoke to civic organizations. Others simply talked to people they knew who were potential voters. A subcommittee was formed to lobby CCC board members through phone calls, letters, and a petition drive on campus.

But the students were up against stiff competition. A group of citizens calling themselves Voters for Fair Taxation was working just as hard to defeat the tax measure. Not only did they speak at many of the same forums as the students did, they also bought radio, television, and newspaper ads to voice their concerns.

On election day, David and Jeff reminded all their friends to vote. That night they attended a party with other students who had worked on the campaign and anxiously awaited the election results. As David looked around the room he wondered, "Will all our hard work have been in vain? What will we do if the tax vote fails? Can 'little people' like us 'buck the system'? Or is political power inevitably concentrated in the hands of powerful organizations and individuals such as politicians, business leaders and the like?"

*D*AVID AND JEFF'S EXPERIENCES with the school budget issue are part of what sociologists call the *political world,* or the **polity,** the social institution that includes all the various ways that societies have developed to maintain social order and control. All societies have ways of making decisions about what actions a group should take or who should wield direct power, such as votes on taxes to help finance a school or votes on individuals to hold political office. Societies also have ways to protect themselves against both external threats and the actions of their fellow citizens, such as military forces and police departments. In short, like the family or religion, the political world is a universal feature of societies.

Unlike the family or religion, we might not often be aware of the political world. Certainly we hear politicians on the evening news, and we usually know when an election is occurring. But politics is much more than electing officials or voting on tax measures. If you have a job, political decisions determine how much of your income goes to taxes and how that tax money will be spent. If you want to go to college, political decisions might determine what type of school you can afford and how much scholarship and loan money is available to you. Other political decisions determine how old individuals must be to drink alcohol, drive a car, and vote in an election. Still others determine whether students can pray in school, how easy it is to get a divorce, whether a woman may get an abortion, and whether a country will send its young people to war. In times of war, political decisions determine who will serve in the armed forces and who will be exempted. Political decisions regarding other countries go far beyond issues of war and peace and determine such everyday matters as what kinds of goods will be available in our stores and how much those goods will cost. In fact, political decisions affect virtually every area of our lives, from the cost and availability of medical care and housing to the money that is spent on scientific research, parks and recreation, and prisons; from the additives in our food to the kinds of programs on television. In short, political actions continually create many aspects of social structure. In turn, the social structure of the political world influences our social actions in many different areas of life.

Given the importance of the political world to our lives, why are so many people indifferent to or even cynical about politics? As we will see in this chapter, a large number of people, especially in the United States, rarely vote in elections; even fewer participate in other ways. Perhaps, like many of these people, you are disenchanted with politics and

government. Maybe political scandals, such as accusations of sexual or financial misconduct, have left you distrustful of anything politicians say or do. Or maybe you are disgusted by government actions (or inaction) regarding health care, tax reform, the environment, military involvements, or other issues. In addition, maybe you feel disconnected from the political world because you believe that you can't really influence what happens anyway. The political world can seem so big and so distant that you feel powerless to affect it.

Sociologists realize that many people have these feelings, and at times sociologists might share the apathy. But they also believe that these attitudes raise important and troubling issues. How, many sociologists have asked, can we have a just and equitable society when so many people feel powerless within its political structure? More generally, what is the real nature of the political world? When you look beyond the hoopla, patriotic chest thumping, and mudslinging of political campaigns or the public posturings of officeholders, who really has power? How do they get it? And whose interests are being served? As the political scientist Harold Lasswell once asked, "Who gets what, when and how?"[1]

In studying the political world, sociologists often use metaphors from the physical world.[2] In the natural world people use machines and tools to increase their power and accomplish difficult tasks. For example, levers and inclined planes allow us to move much heavier loads than we ever could solely with our own muscle power. Similarly we can use various "machines" to augment our political muscle. For instance, we can band together into political parties and other organizations to try to increase our "leverage" or influence on the political process. But power can also be much more hidden—so hidden that we aren't even aware of its influence, just as we rarely think about Earth's gravitational pull. Much of the power exercised in the political world is of this hidden and unrecognized variety. Yet many sociologists believe that this hidden power is much more potent than the more visible forms and has far-reaching consequences for our everyday lives.

We begin our exploration of the political world by discussing precisely what power is and how it is exercised and distributed in the political world. At the microlevel we look at how individuals like Jeff and David can influence the political world. We then pull back to take a macrolevel look at how sociologists have described the arrangement of power within the United States. Finally, we move back in a little closer for a mesolevel view of how organizations, especially those associated with businesses and wealthy corporations, influence government operations.

In the fall of 1998, former pro-wrestler Jesse Ventura stunned the political world by being elected governor of Minnesota. What factors do you believe contributed to his election?

POWER, PUBLIC GOODS, AND THE POLITICAL WORLD

In our discussion of social interaction in Chapter 5, **power** was defined as the ability of one social element to get another to do what the first element wants. In this sense power affects all of our relationships, including those within our families and religious organizations. But power is also absolutely integral to the polity. The very nature of the political world involves the ability to coerce others to behave in certain ways. For instance, if the budget measure is defeated at the polls, City Community College will not have sufficient funds to handle additional students and classes, and it will have to drastically alter its activities. But if the budget measure passes, the citizens of the community will be compelled to pay more taxes. After all, people who refuse to pay their taxes can be arrested, have their property confiscated, and even be jailed. Similarly, if you are speeding down the freeway and hear a siren and see flashing lights behind you, you know that you have been nabbed by a police officer—a representative of the government. You also know that you must pull over and that you will probably have to pay a fine to the government for exceeding the speed limit.

None of us likes paying taxes or fines. Yet we also generally recognize that if no one paid taxes or heeded the speed limit, the social structure would crumble. This illustrates the other central aspect of the political world—it provides **public goods** (the necessities of group life that individuals cannot provide by themselves but must obtain through cooperation with others) and promotes the general welfare. Many daily necessities are provided by the government—the highways we travel, the sewer systems that carry waste from our homes, the regulations that keep traffic flowing, the transit systems that carry us to school and work, the police protection that keeps our streets and homes safe, and the parks that we play in. Large-scale government, or what is often called the **state,** first appeared as societies became larger and more complex. In modern societies daily life is affected in innumerable ways by the political world.

As you will also recall from Chapter 7, as states developed, so, too, did the extent of social stratification. Along with the growing technological and political complexity of horticultural and agrarian societies came increased domination and repression. In agrarian societies the state became a powerful agency for administering vast territories, collecting taxes, and protecting the interests of the rulers. The rulers and their close associates had much more power and wealth than others in the society did, and the bulk of the state's activities were directly oriented toward maintaining this power and wealth. For instance, it is estimated that in Dahomey, an ancient West African kingdom, the king had a thousand tax collectors and many other government officials, all devoted to maintaining his power.[3]

Because maintaining the rulers' power ultimately depends on their being able to force others to do what they want, the state represents an organized monopoly over the use of force in a society. Within modern industrialized societies, such as the United States, the government remains the area of society that controls and regulates the use of force. Police officers can openly carry weapons, whereas the rest of us usually must obtain permits to do so. And most of us accept the idea that granting police this power to use force helps promote both social order and individual well-being.

At the same time, however, there are limits on the force that the police can apply. When they overstep these boundaries, they can be prosecuted and punished. For instance, when videotape cameras documented members of the Los Angeles Police Department beating Rodney King in 1991, four officers eventually were tried, convicted, and jailed for violating King's civil rights. Of course official boundaries are not always clearly defined, and government actions sometimes result in intense controversy. For instance, federal officials who raided cult leader David Koresh's Branch Davidian compound in Waco, Texas, in 1993, killed a number of people, including many children. The officials involved faced severe criticism but no official sanctions. Similarly deadly actions of military officials, such as the bombing of Cambodia during the Vietnam War or raids by U.S. forces on Arab targets in the Middle East, may be either reasonable actions in the line of duty or morally reprehensible actions, depending on your point of view.

With industrialization, the massive inequalities found in agrarian societies began to lessen, and political power became more dispersed. There have, of course, been extremely repressive industrial societies, such as Nazi Germany and the former Soviet Union and its satellites. Compared with agrarian societies, however, citizens in industrialized societies tend to be more politically involved and to have more political freedom. For instance, in industrial societies most citizens generally have the right to vote and to organize politically. Further, where monarchies still exist in industrialized nations, as in England, the royal families are only figureheads. Mass political parties develop and enlist ordinary citizens, a **democratic ideology**—a belief that the state belongs to the people and should serve the interests of all citizens—begins to become an accepted part of life.[4] As a result, ordinary citizens can regularly participate in the political process. We can bring charges against public officials who we think have erred; we can vote for or against ballot measures and candidates for public office; we can lobby, or try to influence, our elected officials. Like Jeff and David, we can campaign for ballot measures that we feel strongly about.

To understand how power works, it helps to realize that it can take two different forms. The first is **domination,** whereby one party, such as a police officer, the court, or a king, controls people's behavior through **sanctions**—either offering or taking away some type of benefit or harm. (You might recall our discussion in Chapter 6 of domination and Weber's views about how it becomes legitimated or seen as proper.) For instance, in feudal monarchies, land owners paid taxes to the king because they were afraid that otherwise the king's army would seize their land; in modern society we obey traffic laws because we don't want to have to pay a fine. By the same token, you might vote for a candidate for city council because she promises to build a new park in your neighborhood, and you see the fulfill-

ment of this promise (a "positive sanction") as a reward for your vote.

The second form of power is **influence,** whereby simply providing information and knowledge leads others to take different actions than they would have otherwise. For instance, after listening to the arguments of the students and the Voters for Fair Taxation about the CCC budget, voters might be persuaded to vote either for or against the levy.[5]

In a democracy, influence is usually more common than actual domination, yet some members of modern societies continue to profit more than others from the collective benefits of the state. For example, the wealthy owners of the resorts near City Community College undoubtedly have much more influence on the CCC governing board than do students like Jeff and David. These owners probably contribute more tax money to the city coffers, they might employ many members of the community, and they might be linked with board members through social and business networks.

In addition, the U.S. tax code calls for people with higher incomes to pay a higher percentage of their income as taxes than people with lower incomes do, yet many loopholes and exemptions specifically help only higher-income people. For instance, most of the federal tax reductions enacted in the 1970s and 1980s benefited the wealthy much more than other segments of society.[6] Similarly police protection and support can vary a great deal from one part of a community to another. Generally, if you live in more expensive areas of town, you can expect much better police services than if you live in a poorer area. Also, members of racial-ethnic minorities often are treated very differently than Euro-Americans are by state officials, such as police officers.

Thus, a basic tension underlies democratic societies such as the United States. Virtually everyone accepts democratic ideologies, believing that the state should serve the people and that all individuals should have a voice in the way the government is run. In practice, however, some people have more influence and power than others do, and the state appears to serve the interests of some people more than others. In short, at the same time that the political world promotes the social order and provides benefits to all of us, it can also enhance the well-being of the wealthy and powerful.

How and why do these differences appear? Sociologists have examined the distribution of power from the micro-, macro-, and mesolevels. In each of these analyses, two theoretical notions are applied: the concept of *social networks* and *rational choice theory*.

Sociologists have long been interested in why people vote. Why do you think this Staten Island polling place is virtually empty, whereas Palestinians on the Gaza Strip wait in line to vote?

INDIVIDUAL INFLUENCE: A MICROLEVEL VIEW

As David and Jeff quickly discovered when they started to talk to people about voting for the CCC election, some people regularly participate in elections, but many others have never even registered to vote. Even fewer participate in other political activities, such as campaigning for candidates or around political issues. Why is this? Why do some people participate in the political process and others don't? Do patterns of participation result in some people having more influence on the political process than others do?

In trying to answer these questions, several sociologists have used **rational choice theory.** As you will recall, rational choice theory is related to the ideas of exchange theory and involves three basic assumptions. First, we all have goals in life. For example, we all have certain goods that we want to attain, from the basics of life, such as shelter and food, to love, happiness, security, or wealth. Second, because we can't attain all these goals, we face choices regarding which goals to strive for and how to go about doing so. Third, our choices regarding these actions are rational; that is, they make sense to us when we make them. In short, according to rational choice theory, we deliberately choose the actions that we think will be most useful in achieving our desired goals.[7]

This perspective suggests that when we think about whether to participate in the political process, we balance the costs and benefits involved. After all, it takes time and energy to learn about candidates and issues and to cast a ballot. Thus we ask ourselves how much difference voting will make in our lives and whether supporting a particular candidate or getting involved in a ballot issue will help us reach our personal goals. Many people feel that even if they don't vote, they will still benefit from the public goods the government provides. Traffic lights will still work, the police will still answer calls for help, and the city sewage system will still remove waste. As rational choice theorists put it, we have the ability to be **free riders,** easily benefiting from the public goods of the society even if we don't take the time and energy to participate.[8]

Many people in the United States apparently decide that the costs of political participation outweigh the benefits. Only a little more than half of all eligible citizens typically vote in presidential elections; even fewer vote in congressional elections.[9] Fewer still participate in other ways, such as working on political campaigns, donating money to causes and candidates, contacting and pressuring political officials regarding specific issues, joining with other people in political activities, or running for and holding elective office.[10] Voter turnout is generally much higher in European countries than in the United States. For example, about 90 percent of the electorate in Germany and 70 to 80 percent of the electorate in Great Britain and France participate in national elections.[11]

Several sociologists have studied why some people choose to get involved in political affairs. Because voting is the most common political activity, we will look at it first.

Choosing to Participate: Who Actually Votes?

Why do some people choose not to participate at all in politics? One theory suggests that humans must fulfill certain basic needs before they can engage in other types of activities. In general, political scientists have found that we are unlikely to have the energy or desire to participate in political activities unless we first have food, water, and shelter; a safe and orderly physical environment; and connections with other people and a sense of self-worth. Therefore individuals highly unlikely to be politically active include those who have been deprived of food, such as might occur during wartime; those who have lived in very unpredictable and difficult environments, such as repressive police states or prison camps; and social isolates, or people with few interpersonal associations.[12]

Apart from psychological motivations, our ability to vote and otherwise participate in politics is also affected by the law. In the United States, for example, women were not allowed to vote in presidential elections until 1920. Although African-American men were granted the legal right to vote after the Civil War, in practice they were deterred from voting by several regulations specifically targeted at the former slave population: complex and tricky literacy tests, "grandfather" clauses (which restricted voting to people whose ancestors had also voted) and "poll taxes" and property ownership requirements, which restricted voting to the well-to-do. Not surprisingly, such laws kept the actual numbers of African-American voters very low for many years.[13] Today voting is still restricted in many ways. For instance, you must have reached a minimum age, generally eighteen, and you must have lived in your community for a specified period of time. You must also be a citizen, a requirement that limits the political participation of temporary residents and of immigrants, both legal and illegal, who are nevertheless greatly affected by political decisions. Finally, you must be registered to vote.[14]

A few years ago registering to vote usually required visiting a particular office during specified hours and filling out lengthy forms. In 1993, however, Congress passed laws requiring the states to allow voter registration through the mail and in readily accessible offices, such as the Department of Motor Vehicles. As Congress hoped, these changes resulted in an increased proportion of the voting age population registering. By 1996, almost 73 percent of all eligible adults were registered to vote. This increased registration rate, however, did not translate into an increase in voter participation. In fact, in 1996, the first Presidential election after the law was enacted, the number of Americans who voted declined by 5 percentage points from the last Presidential election in 1992.[15]

This finding suggests that voter apathy is influenced by more than simply the convenience of voter

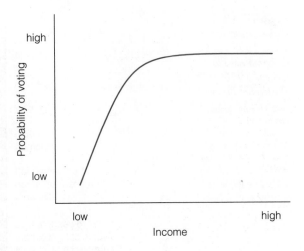

FIGURE 13-1 *Relationship of Probability of Voting to Family Income*
Source: Wolfinger and Rosenstone, 1980, p. 26.

registration. Many studies have documented that people in higher social classes are more likely than others are to participate in politics. For instance, people who have higher incomes are more likely to vote than are those who have less money. However, the pattern of this relationship is what sociologists call **curvilinear**—that is, the graph of the relationship is a curve rather than a straight line. As shown in Figure 13-1, voting increases with income level, but only to a point. Once a person is earning enough to satisfy basic needs, having more money doesn't increase the chance that he or she will vote.[16]

A much more important influence on whether we will vote is how much education we have. Within all income levels, people who have more education are more likely to vote than are those with less education.[17] This probably reflects the fact that education makes it easier for us to learn about politics and to understand the issues involved. Education also helps us deal with the bureaucratic arrangements needed to register to vote and complete a ballot. For instance, simply by negotiating the ins and outs of attending college, you have learned a great deal about how to work within bureaucracies. Thus voting is less costly for people who have more education. Further, our occupations are highly related to our education, which explains why people in higher-status occupations are more likely than are those with lower-status jobs to vote.

As Figure 13-2 shows, characteristics besides level of education generally have a much smaller influence on voting patterns. For instance, today women and men are about equally likely to vote, reflecting the fact that they have, on average, about the

FIGURE 13-2

Characteristics of Voters, 1994 Congressional Election and 1996 Presidential Election

Note: Hispanic group may include either Euro-Americans or African Americans. NA means the data were not available.

Source: U.S. Bureau of the Census, 1997, p. 288.

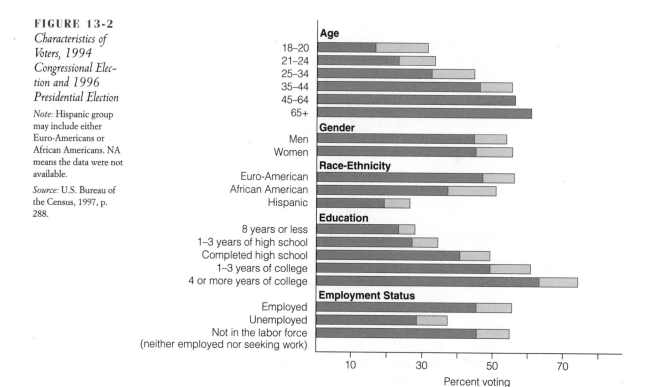

same levels of education. This wasn't always the case, however. Women who were born and grew up before the passage of the Nineteenth Amendment in 1920 were less likely than men to vote, no matter how much education they had. But most of the women who were of voting age when women were first allowed to vote are no longer living, so this pattern no longer influences who votes.[18] In fact, women today are slightly more likely than men to vote. This characteristic, coupled with the fact that women tend to live longer than men, results in women constituting more than half of the voting public.[19] That is why you sometimes hear about politicians soliciting the "woman's vote."

Even though, on average, younger people have more education than older people, older people are much more likely to vote. David and Jeff were wise to encourage their fellow college students to go to the polls, at least the younger ones. People age eighteen to twenty-five are typically less likely than people in any other age group to vote, although students in this age category are more likely than nonstudents are to vote. That young people vote less often than older people do has traditionally been explained by their increased mobility and thus their need to change their voting registration. It also probably reflects the nature of their social networks because older people are more likely to be involved in roles that link them with others in the community and

with people who participate in politics. As a result, older people might feel that they have more to gain or lose by the outcome of an election and thus might be more likely to participate.[20]

Finally, for the overall population, members of racial-ethnic minorities are somewhat less likely than Euro-Americans are to vote. Members of most racial-ethnic minority groups typically have lower levels of education than the Euro-American majority, as well as having the restricted social networks and the lower incomes and occupational statuses that generally are associated with low education. Among people with equal amounts of education, however, racial-ethnic differences in voting generally disappear. In fact, when both age and education are controlled, African Americans and Mexican Americans are actually more likely than other people in the general population to cast ballots.[21]

These studies indicate that the outcome of an election, as in the vote on a tax increase for City Community College, is usually determined by only a subset of the people who live within a society. People who choose to vote have at least an adequate standard of living, they have met certain legal and bureaucratic requirements, they are more likely to be older than other people, and, most important, they tend to have higher levels of education. From the rational choice perspective, we could say that people with these characteristics find that voting represents less

Why are some people more likely than others to be politically active?

of a cost than it does for others and, thus, they tend to believe that the benefits of voting outweigh the costs. Therefore, this subset of people is much more powerful than others are in influencing issues that are determined by public votes.

Although about half of the adult population votes in national elections, far fewer Americans participate in other, more costly, political activities. We turn now to an exploration of who these politically active citizens are.

Beyond Voting: Who Is Politically Active?

In addition to voting, political activity involves organizing voters, campaigning, participating in party caucuses, lobbying elected officials, providing money and other services to political candidates and issue-advocacy groups, and running for elective office. All these activities are crucial parts of the political process. Just as the use of machines enhances our ability to perform physical tasks, these activities, which generally involve contacts or networks with other people and organizations, magnify our ability to affect the political world. Actually, they are often more influential than voting because they help define the issues on which candidates run and on which elections are held. For instance, the CCC budget ballot proposal was first developed and approved by the elected board that governs City Community College. The deliberations of the CCC board about the possibility of scaling back the college transfer program and increasing classes in food service and hotel management resulted from the suggestions of representatives of local businesses. Likewise city councils,

state legislatures, and the U.S. Congress routinely approve and implement policies that were first developed by lobbyists and political activists.[22]

But *why* are some people more likely than others to get involved in politics? Probably the most important influence is the social networks to which we belong.[23] People whose social networks include others who are already involved in political activities are much more likely to become involved themselves. For instance, David became involved with the campaign in support of CCC's budget because he talked to Jeff, who in turn knew the members of student government who were active in the campaign. Similarly, you might feel quite strongly that Jane Smith, who is running for your local legislative seat, is a far better candidate than John Brown. Chances are, however, you won't contribute money to Smith's campaign, put up a lawn poster for her, or engage in other activities unless someone solicits your help or you already know political activists and tap these networks to find out what you can do.[24]

Using a rational choice perspective, we can also see that the costs and benefits associated with campaigning, lobbying, running for office, and so on differ from those associated with voting. David and Jeff and other members of the Students for CCC committee invested a lot of time in talking to people and even donated some of their own money to help pay for posters and other campaign supplies. Members of Voters for Fair Taxation also devoted time to working on their own campaign and donated money for advertising for their side of the issue. Both college students and members of the Voters for Fair Taxation would benefit if their side prevailed in the election, but the costs they experienced were far different, for the well-to-do Voters for Fair Taxation group could much more easily afford the campaign expenses.

In general, higher-income people can better afford many of the costs associated with political activity than the rest of the population.[25] For instance, a $100 campaign contribution represents a much larger proportion of the assets of someone who earns $20,000 a year than of someone who earns $100,000 annually. If you are a member of the upper-middle or capitalist class, you are also much more likely to be involved in social networks that link you to other people with money.

As a result, not surprisingly, elected officials and appointees to high offices, especially at the national level, are much better off financially than the average U.S. citizen. Although people of all income levels are equally likely to vote once their level of education is considered, income is highly associated with whether people participate in other ways, especially whether they run for or are appointed to government positions.

This influence is most noticeable at the national level. For example, members of Congress are required to disclose information about their income and wealth each year. Even though these disclosures don't reveal the full extent of politicians' wealth, they provide enough information to show that the average congressperson has many more assets than the average American does. For instance, in 1994 at least 28 percent of members of the Senate and 14 percent of members of the House of Representatives had assets worth more than $1 million, a level of wealth reached by less than 1 percent of the total population.[26]

Even though few of the recent U.S. Presidents have come from the upper or capitalist class, they are very likely to appoint people from the wealthiest segments of the class structure to key positions in their cabinets. This practice has been common throughout the last century. For instance, a study of people appointed to cabinet positions from 1897 through 1973 found that more than three-fourths had linkages to large corporations, either as corporate officials or partners in law firms that served these corporations. Studies of more recent cabinet members have found similar patterns. Cabinet officials, and appointees to major diplomatic posts, are far more likely than the general population to have access to large family fortunes, to have worked at high levels in the corporate or banking field, and to have attended exclusive private schools and belong to exclusive private clubs. The majority of these top political appointees would be classified as members of the corporate rich or capitalist class.[27]

The fact that members of governing bodies are so unlike the population they represent has often been the subject of criticism. U.S. Senator Daniel Patrick

Moynihan, a sociologist who once taught at Harvard University, described the situation in this way: "At least half of the members of the Senate today are millionaires. That has changed the body. We've become a plutocracy.... The Senate was meant to represent the states; instead it represents the interests of a class."[28] (**Plutocracy** refers to government by the wealthy; **democracy** refers to government by the people and, especially, majority rule.)

Members of the capitalist class certainly have different interests and concerns than members of the underclass or working poor do and often have different interests than the working or middle classes. For instance, members of the capitalist class are much more likely than the rest of us are to have to pay capital gains taxes and to want laws that allow them to pass on their estates to their heirs with as few taxes as possible. In addition, they are much more likely to be able to afford quality health care and elaborate insurance policies, are extremely unlikely to have ever been on welfare, and may well have never attended public schools. In contrast, members of the middle and working classes generally live from paycheck to paycheck rather than off their investments, must worry about the cost of health care and housing, and must rely on public education for their schooling. As long as these people are not directly represented within governmental bodies, sociologists like Moynihan argue, their concerns and needs will not be as influential as those of the rich.

These concerns are reflected in sociologists' descriptions of the **power structure,** the way that power is distributed in society. To understand the power structure, sociologists often step back to analyze the political world from the macrolevel.

PLURALIST OR ELITIST: TWO MACROLEVEL VIEWS OF THE POWER STRUCTURE

Although *democracy* means "rule by the people," a power structure that is a true *direct* or *participatory democracy,* in which all citizens contribute to all decisions, is quite impractical in large groups. Even communities in New England, which historically have held town meetings in which all citizens can vote on important issues, actually are governed by only part of the community. Many individuals choose not to participate because the benefits don't outweigh the costs. In addition, genuine participation by everyone simply becomes too complex and time consuming. If each town citizen spent only two or three minutes expressing a view about an issue, the meetings might last for hours or even days.

As the pluralists suggest, we all have power in the form of the ballot. As a result, to get elected, politicians must appeal to a wide range of voters. In an effort to attract younger voters, Bill Clinton appeared on MTV before the presidential election in 1992 and often played his saxophone in other settings.

As a result, modern democracies use a system of **representative democracy,** in which elected officials represent the people and make important governmental decisions. In the United States we elect fellow citizens to serve as members of legislative bodies, such as city councils, state legislatures, and the U.S. Congress. We also elect people to executive offices as mayors, governors, and president of the United States. In turn, these executives appoint (usually with the consent of the legislative bodies) people to various administrative posts, such as cabinet positions and agency administrators.

As discussed previously, however, these elected and appointed officials are hardly truly representative of the citizenry. Like the heads of large corporations and charitable organizations such as the Red Cross or United Way, they are far more likely to come from the capitalist and upper-middle classes than from the other 85 percent of the population.

Although sociologists agree that these powerful officials represent only a minority of the population, over the years these sociologists have developed two quite different pictures of which social networks constitute the power structure and how it actually works.

Power as Competing Interests: The Pluralist View

The pluralist view tends to be relatively optimistic about the nature of the power structure and the ability of a wide variety of groups to influence political decisions. Although the tenets of this perspective were first developed through a close study of the decision-making process in the city of New Haven, Connecticut, other scholars have expanded on these ideas to describe the power structure of the nation as a whole.[29]

Pluralism refers to the view that the political power structure involves a number of powerful groups and individuals, all of which can potentially influence the decision-making process. According to pluralists, there are many different sources of power. For example, bankers may control money and credit; newspaper editors and television executives may control the information citizens receive; and business owners may control jobs. Likewise, famous economists or other scholars can have a great deal of knowledge and experience, and revered religious figures might command a high degree of esteem. Some groups of people are powerful because of their solidarity, as when members of a racial-ethnic group vote in a bloc or when people with common interests join together to lobby, as older Americans do when they join the American Association of Retired Persons (AARP).

Pluralists note that these sources of power aren't concentrated in just a few people. The famous economist usually doesn't own a bank or edit a newspaper. The esteemed religious leader probably doesn't own a large company employing thousands of people. The wealthy business owner might not be held in especially high regard. Most members of the AARP don't hold other sources of power. To the extent that different people hold these different sources of power, pluralists stress that power is dispersed rather than concentrated.

In addition, pluralists emphasize that different types of power are important in influencing different types of issues. For instance, bankers who control money and credit and business owners who control jobs might be much more influential in decisions that involve economic issues. Voters, and especially those who campaign to get like-minded people to go to the polls, can be especially influential in the outcome of elections, such as the CCC budget vote. **Interest groups,** organizations that concentrate their activities on specific policy issues, also are common players in the American political scene. For instance, the AARP keeps close tabs on legislation and policies related to older Americans, and the National Rifle Association (NRA) is heavily involved in fighting gun control laws.[30] Not all powerful groups are active in all political decisions, but pluralists generally suggest that virtually all citizens, or at least all groups of citizens, have the capability of influencing some political decisions, even if only through their votes, their association with particular interest groups, or their lobbying of government officials.[31]

Pluralists tend to recognize two types of power that influence political decisions. The first reflects the characterization of power presented in previous chapters—getting others to do what you want them to do, as when the business interests in David's community tried to persuade the CCC board to drop the college transfer program and substitute food service and hotel management classes. The second type of power represents the flip side of this coin—stopping others from doing something you don't want them to do, as when the CCC students tried to persuade the CCC board not to adopt this course of action.[32] In other words, pluralists see power as two-dimensional, promoting decisions that are favorable to a group's interests and opposing actions that might be unfavorable.[33]

In general pluralists suggest that the power structure involves several **elites,** powerful people who can influence the political process. Individual elites might have different bases for their power and tend to focus on different areas of social life. Some might specialize in issues related to economic policy, some in education and schooling, some in international relations, and some in religious life and organizations.

Pluralists generally recognize that some interest groups have more influence than others do, just as some machines are more powerful than others.[34] For instance, the AARP, with 30 million members, is a far more powerful entity than are groups that lobby for the interests of children, who, of course, can't vote. Legislators almost never cut programs or alter policies that benefit the elderly, such as Social Security, Medicare, or special tax breaks. In contrast, programs designed to benefit children, such as school lunches, educational programs, and child care subsidies, are often the target of cuts.

Yet pluralists tend to imply that the playing field is relatively level. They see the power structure as a very loose and large social network, one that can incorporate almost everyone, in at least some way. Pluralists emphasize that political elites include many different elements and that most of the population has ties to at least some of them, whether through voting or through more direct ties with special interest groups.[35] In general, pluralists believe that the ideal of democracy is reasonably well served, at least indirectly, by these competing interests. Pluralists also suggest that the exertion of power is usually visible, that we can see the mechanisms that various interest groups use to exert their influence.

In contrast to this relatively optimistic view of a power structure that involves a fairly level playing field, other sociologists emphasize the extent to which only a very small minority of the population actually influences political decisions and how the views of this small minority come to dominate the political scene and, actually, the entire culture. These sociologists also argue that political power is often much more hidden and concentrated than the pluralists suggest.

The Power of the Few: The Elitist View

The basic notions of the elitist view were developed by C. Wright Mills in his 1956 book *The Power Elite.* (Mills was also the sociologist who coined the term "sociological imagination.") In contrast to pluralism, **elitism** suggests that political power and influence in the United States are dominated by a handful of people who are relatively unified and form a comparatively small, tight-knit social network. Even though members of the elite might have different views on particular issues, these differences pale in comparison with their similarities and shared interests. Figure 13-3 illustrates the key differences between the pluralist and elitist models of the political power structure.

Even though the pluralist view conforms with popular ideas about how the political world works, in recent years most sociologists have come to believe that the preponderance of the data supports the elitist perspective.[36] As we have already discussed, high government officials are much more likely to be from the capitalist and upper-middle classes than other class groups are. These officials have similar educational and occupational backgrounds and often have shared experiences in social clubs, recreational activities, and friendship networks. Mills and other proponents of the elitist view believe that these commonalities, and especially the common class

FIGURE 13-3 *The pluralist view of the political world suggests that the power structure is loose and sizable and that most of us have the potential to exert influence on political activities. In contrast, the elitist view suggests that the real power is concentrated among only a small proportion of the population who have strong social ties with each other and who disproportionately represent the capitalist class.*

background, result in elites having interests and concerns that are much more similar to one another's than to those of the rest of the population.[37]

Besides their similar class background, elites also develop strong networks through the various roles they play in powerful organizations. Corporate officials may move into high-level government positions for a while and then return to their corporate positions. After leaving government service, former government officials often enter the corporate world. Likewise high-ranking members of the military sometimes fill important corporate or government positions when they retire from the service. For instance, Robert McNamara went from president of Ford Motor Company to secretary of defense under Presidents Kennedy and Johnson and then became president of the World Bank. General Alexander Haig temporarily abandoned his army career to serve as President Nixon's chief of staff, returned to serve as commander of the U.S. armed forces in Europe, retired from the army to become president of the United Technologies Corporation, returned to government as secretary of state for President Reagan, and then left government to become president of Worldwide Associates.[38]

Elites sometimes also hold influential positions in more than one area at the same time. Individuals commonly serve on the boards of directors of several large corporations while maintaining linkages to other powerful groups. For instance, Wendy Lee Gramm, the wife of Texas senator and one-time presidential candidate Phil Gramm, is a former economics professor at Texas A&M and a former director of the Commodities Futures Trading Commission. She also serves on the board of several corporations, including State Farm, the Chicago Mercantile Exchange, and Iowa Beef Processors.[39] Similarly Vernon Jordan, a close friend of President Clinton, is a prominent lawyer and lobbyist in Washington, D.C., overseeing the work of about 100 other registered lobbyists. Jordan also sits on the board of eleven major corporations, including American Express, Bankers Trust, Dow Jones and Co., JC Penney, and Revlon; and his wife serves on the board of six other companies.[40]

Of course, when someone like McNamara, Haig, Gramm, or Jordan has powerful ties to various areas of the society, his or her individual influence is enhanced. But elite theorists stress that these linkages have a more important social consequence: They provide a social network that helps increase cohesion among powerful business and government organizations. The fact that one person, such as Wendy Lee Gramm, serves on the boards of directors of several major corporations, is married to a U.S. senator, and also was once a top executive in a federal agency means that these organizations are tied to one another. Just as your linkages with other people in your own social networks can be advantageous to you, so too are these linkages advantageous to organizations. These organizational linkages help organizations reduce uncertainties or ambiguities about other organizations in their environment. After all, the stronger the ties among organizations, the more their members will know about one another's activities. As a result, these organizations should be able to make better business decisions and be more effective in the political world.[41]

Sometimes, of course, people from modest origins gain high political office and other powerful positions. Ross Perot, who now heads multimillion-dollar companies and has run for president as an independent candidate, began as a salesman in his native Texas. President Bill Clinton grew up without a father in modest circumstances in Arkansas. Similarly, someone from the middle or upper-middle class can become the leader of a large organization, such as the AARP or the United Way, and a union member with a working-class background can rise through the ranks to become the leader of the Teamsters or the AFL-CIO. Might we expect that, because of their modest origins, such individuals would think and act differently than elites who came from higher-status backgrounds?

To some extent they might, but elite theorists suggest that, most often, they do not. Instead, such leaders are "coopted." They begin to adopt the worldview of and to identify with the elite circles in which they currently operate, rather than the networks from which they originated. As a result, these leaders no longer truly represent their original constituencies. Instead, they are more likely to consider the interests of other elites and try to convey, or sell, these elite interests to the working- or middle-class people they purportedly represent.[42]

Elite theorists see power as being more complex than do pluralists. Elite theorists accept the pluralists' idea that power involves both getting others to do what one wants and being able to stop unwanted events and policies. More important, though, they see power as both pervasive and hidden, going beyond interactions between groups or individuals, and instead influencing people's worldviews. To the elite theorists, power more closely resembles an invisible magnetic or gravitational field than a readily observed machine. Gravitational fields operate on everything in our environment; they are the reason that we don't float off into space and that it's harder to go uphill than down. Yet we don't think about

these fields—they are simply part of life. Similarly, elite theorists suggest, political power can affect our lives without our even being aware of it. Precisely because we aren't aware of this type of power, it is much stronger than it might otherwise be.[43]

The Italian political theorist Antonio Gramsci used the term **hegemony** to refer to this special kind of hidden but pervasive power which dominates our social life so thoroughly that we seldom recognize the power or question its legitimacy.[44] For instance, as noted in Chapter 10, sociologists didn't even think about the idea of gender stratification until the late 1960s and the beginning of the current feminist movement. Gender stratification and inequality were such an ingrained part of everyday life (or so hegemonic) that they weren't even recognized.[45]

Many elite theorists believe that in the United States the power of corporations, and especially of the capitalist class, approximates a hegemony. Political decisions, they suggest, reflect the power of business, and especially large corporations, much more than they reflect the needs of individual people. For instance, in the late 1980s and early 1990s there were exhaustive debates over passage of a law that would guarantee all working mothers the right to an unpaid leave of absence from their jobs after giving birth. (Most European nations provide paid maternity leaves and even "child allowances," cash grants of money to families to help pay for the costs of child rearing.[46]) But the proposal for an unpaid maternity leave met fierce opposition in Congress, fueled particularly by the Chamber of Commerce and other groups that represent business interests. Eventually a bill was passed, but its application was limited to large companies, thus excluding many mothers who might need such a leave. In contrast, at about the same time and with minimal debate, Congress committed more than $500 billion of taxpayers' money to bail out the nation's savings and loans associations, which had gotten themselves in deep financial trouble through misguided and even illegal loan practices.[47]

In general, elite theorists suggest that the hegemony of business limits the way we look at and define important issues. As Charles Wilson, the former president of General Motors and someone who was certainly part of the power elite, once put it, "What is good for the country is good for General Motors and vice versa!"[48] This perspective of examining social policies in the light of how they affect business permeates many different aspects of our lives—so much so that we often find it difficult to imagine alternative ways in which our world could be structured. For instance, virtually every industrialized country and many developing countries provide

BUILDING CONCEPTS

Microlevel, Mesolevel, and Macrolevel Analyses of the Political World

As we have seen throughout this text, sociologists often use microlevel, mesolevel, and macrolevels of analysis. Below are some of the sociological questions explored in this chapter, classified by level of analysis. If you were a sociologist, what other kinds of questions would you ask about the political world? Try to come up with one question for each level of analysis.

Level of Analysis	Example of Question Posed
Microlevel	What motivates individuals to be politically active?
Mesolevel	How do political action committees (PACs) influence legislation?
Macrolevel	Who controls the power structure: an elite few, or the mass citizenry?

government funding for health care for all citizens. Yet the prospect of government-funded health care for all American citizens has provoked extreme opposition, spearheaded by the business community, since the 1950s. The result has been that health care policies that are routinely accepted by people in most other countries are seen as radical in the United States.

The power and influence of wealthy business interests can be seen on the local level as well. For instance, even though David and Jeff and the other students mobilized political support for the passage of the budget levy, the city's business interests, as represented by the Chamber of Commerce, actually wielded much more power. In fact, the original proposal to cut the college transfer program came from business interests. These business interests are able to more strongly influence decisions—both those made in public at board meetings and those made behind the scenes—because they share social networks with the CCC board members and others who have authority to make decisions regarding the school. Students like Jeff and David simply don't have those social resources.

There are, of course, limits on the power of the corporate world and the capitalist class. Over the years many interest groups, such as labor unions, en-

vironmental organizations, and consumer groups, have challenged the power of business, lobbying successfully for laws and regulations that affect the wages and hours of employees, the safety and environmental impact of manufacturing processes, the design of products such as toys and automobiles, and the way businesses treat customers. Even though the interests of business are often seen as coinciding with the interests of the country as a whole, business does not automatically run the show, especially when other interests are obviously affected by its practices. Consequently business interests must work to maintain their power and influence, an effort that is greatly facilitated by the dominance of the capitalist class in the political arena.[49]

Only a minority of the capitalist class ever directly participates in political decisions through appointments to cabinet posts or elections to public office. But a large proportion of the capitalist class influences the political process through campaign contributions. In recent years, in an attempt to curtail such efforts to purchase political favor, the amount of money that individuals may contribute to political candidates has been greatly restricted. Undeterred, the elite capitalist class and large business corporations have used organizations as a way to try to influence the political process.

INTERACTIONS BETWEEN ORGANIZATIONS: A MESOLEVEL VIEW OF THE POLITICAL WORLD

The concept of social action might lead you to picture individuals and their everyday decisions and interactions with others. Yet social action occurs on other levels of analysis as well. Just as individuals wield power over others in daily interactions, so, too, do organizations and nations wield power and influence over other groups and individuals and thus help to shape the social structure. When such larger entities engage in social actions, sociologists call them **corporate actors,** a term that refers to any unified group (not just businesses).[50]

One of the most powerful types of corporate actors in the political world in the United States is the political action committee, or PAC. PACs are organizations that represent certain groups or interests and serve as a means by which these groups gather money and distribute it to candidates.[51] In this section we look at how some sociologists have studied PACs, and other corporate actors, so we can understand power and influence at the mesolevel—how organizations wield power and influence, and how these processes support elite theorists' views of the power structure.

The Nature of PACs

From a pluralist perspective we can see PACs as a means by which individuals who are interested in specific issues can join forces and present a stronger front than any of them could individually. Many PACs not only give money directly to candidates but also employ professional lobbyists who track legislation, represent the PACs' interests to policymakers, and engage in public relations and advertising campaigns. With the financial contributions of large numbers of interested parties, a PAC can have much more influence than can the same number of citizens acting as individuals, much as a lever or pulley can transform our energy into a greater force than we can supply by ourselves. In fact, PACs were started by labor unions in the 1930s as a way for workers, who could afford to donate only small sums of money individually, to pool their resources and let the union use it for maximum advantage.[52]

PACs represent many different issues and groups of people. For instance, professors and administrators of the state-supported colleges and universities in my area have formed a PAC to represent their interests in the state legislature and to the governor. Recently the students at these schools have also formed a PAC, funded in part by student fees. Both

TABLE 13-1 *Examples of Federal Political Action Committees in the United States*

Air Line Pilots Association International (AFL-CIO)

Aircraft Owners and Pilots

American Federation of Teachers

American Medical Association

Associated Milk Producers

Coca-Cola Company

Dow Chemical Company

Hollywood Women's Political Committee

Maytag Company

National Association of Letter Carriers of the U.S.A.

National Association of Social Workers

Sierra Club

United Mine Workers of America

Source: Almanac of Federal PACs, 1990.

PACs employ lobbyists who keep track of potential legislation regarding colleges and universities, recommend which candidates should receive campaign money, and testify in legislative hearings.

Today there are more than four thousand PACs in the United States. A sample of these is shown in Table 13-1. Some of them are associated with citizen interests, such as the AARP, which represents older Americans; others are associated with environmental groups, such as the Sierra Club. More than three-fifths of PACs, however, are associated with business corporations or various trade groups,[53] and, not surprisingly, most of the influence from PACs involves these groups. For instance, in 1995–1996, PACs donated more than $215 million to candidates for the United States House and Senate, and almost two-thirds of this money came from business and trade PACs.[54]

Elite theorists emphasize that business PACs are fundamentally different from PACs associated with voluntary groups such as the Sierra Club or the AARP. As an individual you are generally free to decide whether to contribute money when the Sierra Club or AARP asks you for a donation. You can toss the solicitation in the trash can or you can write a check. If, however, you are a high-level manager in a large business and someone asks you to contribute to your company's PAC, you will probably be much more reluctant to ignore the request. Some large corporations aggressively solicit employee contributions and reward those who donate funds with, for example, lunch with the company president, eligibility for drawings for prizes, or even trips to Washington, D.C. Most important, businesses always have the ability to pressure employees, however subtly, to contribute to the PAC based on the notion that those who decline aren't really "team players" and don't really care about the company. As a result, business PACs have the potential to gather much more money than PACs associated with various voluntary interest groups do.[55]

Many sociologists believe that the large number of corporate PACs, the huge amount of corporate giving, and the potential to raise even more money reflect the hegemony of business interests and the capitalist class. Sociologists have looked at the actions of business corporations and their PACs to try to understand more about how this hegemony operates and how it affects both national legislation and policy (at the macrolevel of analysis) and our lives as individual citizens (at the microlevel). Two sociolo-

gists who have done a great deal of this work are Dan Clawson and Alan Neustadtl. We turn now to an examination of their work.

Common Sense versus Research Sense

COMMON SENSE: Political action committees (PACs) make campaign contributions to influence the way politicians vote on certain issues.

RESEARCH SENSE: Clawson and Neustadtl found that although business PACs do make campaign contributions to gain access to legislators, they generally use that access to *shape the way legislation is crafted* rather than attempt to influence the final votes on legislation.

FEATURED RESEARCH STUDY

THE POLITICAL POWER OF BUSINESS

Clawson and Neustadtl focused most of their initial work on contributions to people running for Congress. As you probably know, political campaigns can be quite costly. For instance, in the two-year cycle that ended with the 1996 congressional elections, candidates for Congress spent more than $735 million on their campaigns.[56] Even though these candidates are almost always from the upper-middle or capitalist class, few can afford to finance their campaigns totally on their own. Instead, they must rely on contributions from others. In recent years, in part because of the newly developed limits on individual contributions to candidates, PACs have become a very important source of campaign funds.

All PACs must officially register with the Federal Elections Commission and report the amount of money they receive and distribute. Clawson and Neustadtl began their research into PACs by using these official reports to find out how much money PACs gave to candidates, who PACs gave the money to, and how these contributions might be related to the outcome of elections.

Clawson and Neustadtl found that the social actions of business PACs generally take two forms. Some PACs tend to focus almost all their contributions on incumbents, candidates who already hold office. Because incumbents are highly likely to be reelected, Clawson and Neustadtl call these *pragmatic donations* and suggest that they represent efforts of the PAC to ensure future access to a senator or representative. Other PACs tend to give more money to challengers or to people running for open seats. Clawson and Neustadtl call these *ideological dona-*

tions and suggest that they are designed to influence who actually gets elected and thus the ideological composition of Congress.

In reviewing the Federal Election Commission's records of donations from corporate PACs from 1976 to 1996, Clawson and Neustadtl concluded that the vast majority of PAC donations, especially in more recent years, are pragmatic in nature. In other words they are aimed at gaining access to the political world, rather than changing the types of candidates who are elected.[57]

Although their review of FEC reports told Clawson and Neustadtl how much money was given and to whom, these reports did not indicate *why* the PACs, as rational corporate actors, chose to give money to candidates and why they chose either a pragmatic or an ideological strategy. To help answer these questions, Clawson and Neustadtl decided to actually talk to the people who make the decisions about corporate PAC donations. With the help of Denise Scott, then a graduate student at the University of Massachusetts, they conducted in-depth interviews with the corporate officials in charge of PAC donations for thirty different corporations.

The PACs were carefully selected to include both larger and smaller companies and those that made both pragmatic and ideological donations.[58] The researchers guaranteed all the respondents total confidentiality, both for themselves and their companies.[59] Clawson and Neustadtl began the interviews by showing the officials a set of graphs and charts derived from their analysis of the FEC data that illustrated the PAC's pattern of political contributions

The movie actor, Charlton Heston, often testified before Congress in his role as President of the National Rifle Association, which has a powerful political action committee.

and compared it with those of other large PACs. This tactic broke the ice and helped the interview move beyond vague generalities about PAC activities to more detailed questions about donation strategies and decisions.

Gifts, Networks of Obligation, and Access

When you hear about a business PAC giving money to a congressional candidate, you might think that the business is trying to "bribe" the candidate to vote a certain way on an issue. As the famous humorist Will Rogers once commented, we have "the best Congress money can buy." Despite some striking and well-publicized exceptions, however, scholars have generally found that legislators rarely change their minds on issues after they have received a donation.[60]

Clawson and his associates suggest that an analysis of campaign contributions as bribes is too simplistic. Instead, they suggest that we can see contributions as gifts that create *social ties* and, as a result, a sense of obligation. For example, though you or I might send a check through the mail to a campaign fund, Clawson and his associates found that PAC officials almost always deliver their company's contributions to members of Congress in person. They visit the senator's or representative's office, chat with him or her, and then present the contribution.

Thus, over time, the politicians and PAC officials build relationships with one another. These relationships give the PAC officials *access* to legislators—the ability to meet and talk with them, informally, off the record, and without the knowledge of the public, on issues that are important to them. PACs also use a variety of other means to establish and maintain these relationships. Businesses invite legislators to speak at company gatherings, often at attractive vacation sites; they finance trips to "investigate" issues; and they give personal gifts, such as dinners, tickets to ball games, or works of art. However, by far the most common type of gift is a contribution from a business PAC to a candidate's campaign fund.[61]

It is important to realize that PACs are often *invited* to contribute to a legislator's campaign fund. Most often a legislator's request is low-key, such as an invitation in the mail. But the PAC officials Clawson and his associates interviewed reported other requests that more closely resembled the Godfather's "I'll make 'em an offer they can't refuse." For instance, a member of Congress might refuse to meet with a PAC official until a donation has been made. If a PAC hasn't contributed recently, its officials may be allowed to speak only to the most junior members of the legislator's staff.[62]

When a PAC responds to a request from a legislator's office for help, this reaction enhances social ties. For one thing, Clawson and his associates suggest that politicians feel indebted to people who contribute to their campaign. In addition, the fact that the gifts are almost always hand-delivered helps personalize the ties and make them more meaningful.

Interestingly Clawson and his associates found that business PACs usually don't use this access to try to influence final votes on legislation. Instead, they use it to shape the way that legislation is constructed—for example, by focusing on creating loopholes that exempt their companies from costly or inconvenient laws and regulations. As Clawson, Neustadtl, and Scott put it,

> If tax law is going to be changed, the aim of the company's government relations unit, and its associated PAC, is to be sure that the law has built-in loopholes that protect the company. The law may say that corporate tax rates are increased, and that's what the media and the public think, but section 739, subsection J, paragraph iii, contains a hard-to-decipher phrase. No ordinary mortal can figure out what it means or to whom it applies, but the consequence is that the company doesn't pay the taxes you'd think it would. For example, the 1986 Tax "Reform" Act contained a provision limited to a single company, identified as a "corporation incorporated on June 13, 1917, which has its principal place of business in

Bartlesville, Oklahoma." With that provision in the bill, Phillips Petroleum didn't mind at all if Congress wanted to "reform" the tax laws.[63]

Special tax considerations are one of the most common objectives of business PACs. When Clawson and his associates asked the PAC officials for an example of what their office tries to do, 90 percent gave an example of a tax loophole.[64] Such special considerations appear in other areas as well. For instance, the Clean Air Act, which was passed during the Bush administration, contains *forty pages* of exemptions and qualifications![65] The minimum wage bill, which was passed in 1996 and raised wages for low-paid workers for the first time in over a decade, includes tax breaks for corporations that totaled $16.2 billion![66]

These successes for business PACs don't automatically happen because members of Congress consciously try to craft legislation that benefits business. Rather, they come about because of *social actions.* As the pluralists suggest, many different interest groups can push for new laws and regulations. Environmentalists lobby for controls on the emissions from factories into the air and water; labor unions lobby for higher minimum wages and more favorable vacation and leave policies. When businesses find that legislation, such as the Clean Air Act or a new tax proposal might affect them in some way, they act to counter these efforts. Not only do businesses try to develop alternative legislation and devise loopholes that will protect their interests, they also work to convince legislators to accept these changes.[67]

At the same time, these successes reflect the enormous power and vast economic resources that businesses hold within our society. Few individuals have the resources to match the monetary power of corporations, let alone the time and skill needed to understand legislation well enough to draft alternatives that will be personally beneficial. But this power also reflects the hegemony of business—the unquestioned assumption that what is good for business must be good for the rest of us. This assumption is clearly reflected in the actions of ideological PACs.

Ideological Giving and Rational Choice

As we mentioned previously, the vast majority of PACs are pragmatic in their giving. To help build social ties, create feelings of obligation, and gain access, they donate to both liberals and conservatives, Democrats and Republicans. They also give primarily to incumbents—people who are likely to remain in office and who will be in positions to help them down the road. But some PACs, and their associated officials, see this strategy as less than honorable and

detrimental to the long-term interests of business. It would be far better in the long run, these people suggest, if PACs spent their money on candidates who are sympathetic to business interests rather than merely trying to buy access to whoever is in office. According to rational choice theory, such ideological donations represent a rational, deliberate decision to gain benefits by changing the nature of legislative bodies rather than by gaining access to legislators.

Interestingly, Clawson and his associates found that the only time business PACs generally behaved in an ideological rather than a pragmatic manner was in the 1980 election in which Ronald Reagan won the presidency and several other similarly conservative candidates gained various offices. In 1980 almost ninety PACs donated at least 30 percent of their funds to Republican challengers in congressional races. In addition, in 1980 the amount of money given to Democrats declined sharply. In 1972 incumbent Democrats who were running for reelection received four times as much as their Republican challengers, but in 1980 they actually received less than their challengers.[68] After 1980 and continuing through the 1990s, PACs returned to their more pragmatic behavior, giving about equally to Democrats and Republicans and giving the majority of their money to incumbents.[69]

Clawson and his associates suggest that the ideological strategy used in 1980 was successful. More people who were openly sympathetic to business interests were elected, the Republicans gained control of the Senate, and enough Republicans were elected to the House of Representatives to substantially increase their influence on legislation. Subsequent policy changes included tax cuts for the wealthy, cutbacks in programs for the poor, and the elimination of many regulations on business.[70] After 1980 the income of the capitalist class increased substantially, and overall inequality in incomes (the gap between the rich and the poor) also increased.

Why, given these successes, would PACs return to the pragmatic pattern and donate as much money to Democrats as Republicans? Part of the answer, according to the officials interviewed by Clawson and his associates, has to do with changes in the actions of the Democratic party and reminds us that PAC giving involves interactions between donors and recipients (see Box 13-1). After the 1980 election Democratic party officials tried to persuade corporate PAC officials that the Democrats deserved support as well. These efforts were successful partly because, based on the results of the 1980 election, the Democrats may have modified their views and decided that their chances for funding would be enhanced if they were more supportive of business interests. One PAC director who formerly worked for a senator described the impact he thought the 1980 election had on Democrats: "[This] is a perfect example of that. Here's a guy who as a representative was conceived of as quite liberal and now is conceived of and is a genuine moderate."[71] Or, as Clawson and his associates described the change,

> After 1980 and 1981 business and the Democratic party moved toward each other. The Democratic party changed to be more accommodating to business, and corporate PACs became more willing to give to moderate Democratic incumbents. Most Democratic members of Congress are now housebroken and do not challenge business. Corporate PACs, in their turn, are willing to make contributions to most of these members, providing they will grant access and do "minor" favors for the corporation.[72]

As a result of these realignments, most PACs, as rational corporate actors, have decided that it simply doesn't make sense to try to change the political makeup of Congress. Because members of both political parties are responsive to the needs of business, business can already get pretty much what it wants, most of the time.[73]

Soft Money: Big Donors and Access to High Levels of Politics

Although most PAC donations go to candidates for congress, high level political players, such as Presidential candidates and the leaders of political parties, are much more likely to benefit from "soft money." Even though federal laws limit the amount of money that individuals and PACs can contribute to individual candidates, there are literally no limits on the amount of money that can be contributed to political parties as long as the money will be used for nonspecific purposes such as "party building" or "get-out-the-vote drives." This has generally been interpreted to include "issue ads," political advertising that doesn't directly support individual candidates but a party as a whole. In addition, soft money contributions do not need to be made through Political Action Committees but can come straight from corporate coffers.[74] After regulations regarding soft money were relaxed in the late 1970s, the amount contributed to these funds dramatically increased— from a little over $19 million in 1980 to more than $250 million by 1996![75]

Who donates soft money? From analyzing information about contributions in the 1990s (the first period in which records were made public) Clawson and his associates found that 90 percent of the donors gave less than $10,000, and most of these people and groups donated less than $1,000. The contributions from these "small" donors amounted to only about 12 percent of all the money received. In contrast, contributions from slightly fewer than 500 donors who gave $100,000 or more each accounted for almost half of all the receipts. Most of these large money contributions came from corporations or from groups or individuals associated with business corporations. Some companies contributed vast amounts: Tobacco company Philip Morris gave $3.3 million; and Seagram (a liquor company), R. J Reynolds (tobacco), Atlantic Richfield (oil), and Nations Bank Corporation each contributed more than $1 million. Almost all the largest individual contributors were either business owners or top executives.[76]

Just as PAC donations to congressional campaigns create social ties and a sense of obligation, so do soft money donations. Some large individual contributors are "rewarded" with ambassadorships to foreign countries. For instance, Felix Rohayton, who gave $362,500 to the Democratic Party in 1995–1996, was appointed as ambassador to France six months after the election. Many other large donors to the Democratic party have been overnight guests at the White House or invited for exclusive "Whitehouse Coffees." Similarly, large donors to the Republican party have been given photographs and meals with Congressional leaders. Whether one just gets a photo or actually gets to sit beside a congressional leader at the head table depends on the size of the donation.[77]

Despite these perks, Clawson and his associates stress that soft money contributions, like PAC contributions, don't allow business interests to directly buy votes or legislation. Instead, they offer access and the opportunity for donors to draw attention to their concerns and to influence the positions of the highest political officials. Thus, Clawson and his associates suggest that the huge amounts of money contributed by tobacco interests are no accident. Over the past few years tobacco companies have been faced with legal actions and potential regulations that could cost billions of dollars. If politicians could be persuaded to look even a bit more favorably on these companies, the savings could be enormous.[78]

PACs, a Unified Elite, and the Hegemony of Big Business

Pluralist theorists emphasize that members of the elite often work at cross-purposes and independently of one another. For instance, financial and banking corporations might support certain legislative changes regarding loans, but manufacturing industries might oppose them. According to the pluralists, even though each industry has a great deal of power

BOX
13-1

SOCIOLOGISTS AT WORK

Dan Clawson

Real change requires the commitment and mobilization of millions of people.

Dan Clawson teaches at the University of Massachusetts at Amherst. Most of his research has focused on power and the political world.

Where did you grow up and first become interested in politics? I grew up in the suburbs of Washington, D.C., and the summer after my junior year in high school worked in the basement of the U.S. Senate for a senator from Alaska. That was 1965, before the days of computers, but we had cutting edge equipment that could generate personalized form letters explaining how much the senator valued the voter's advice. We even had a machine that held a real fountain pen and signed the senator's name.

How did you get interested in studying political action committees (PACs)? I've been a radical for as long as I can remember, committed to creating a world where opportunities in life, and the ability to influence government actions, do not depend on whether a person is rich or poor. When Ronald Reagan was elected president, and the 1981 tax and budget cuts took money from the poor to give it to the rich, it seemed to me that the most pressing political problem was how and why this had happened. One key part of the answer appeared to be what big business had done and was doing, so we set out to study that, by looking at corporate PAC donations to congressional candidates.

We started by doing careful quantitative studies, using computers to analyze 20,000 corporate PAC donations. Those established that in 1980 a major fraction of business had unified to support conservatives, a break from the normal pattern of support for incumbents, even liberal ones. Then, in order to

learn why corporations acted as they did, we interviewed corporate executives.

What was the reaction to your research among your colleagues? Our quantitative studies have been highly praised by other sociologists, replicated by some, and published in top journals. The book based on interviews, however, has frequently been criticized as too political, too hostile to corporations, and not sufficiently academic. Our critics would like us to use more quantitative evidence, but the main objection is that we are too partisan, that we are constantly trying to discredit corporate PACs. I say: darn right we are, and that's the right thing to do.

Do you feel your work has had some effect on campaign finance reform? When the book came out, I had hopes it would be picked up by the media and help make reform a major political issue. No such luck. I appeared on a couple of dozen major talk radio shows, with combined audiences of several million people, and had pieces on the editorial pages of several major newspapers, but that had no significant effect on the issue. Except at election time, politicians no longer seem interested in real reform, and somehow the interest expressed during the campaign always disappears after the candidate is elected.

Academic analyses can contribute to social change, but they don't accomplish much in isolation; real change requires the commitment and mobilization of millions of people, and for whatever reason, there is lots of discontent about the influence of money on politics, but there has not yet been a powerful and effective movement.

and money, they will not have a great deal of cumulative or total power if they work at cross-purposes. However, Clawson and his associates found little support for the notion that businesses will work against one another. Instead, they found a great deal of business unity.[79] For example, their research into

donations in a large number of congressional races revealed that business PACs almost always contribute to the same candidate. Even though the researchers' interviews suggested that PAC officers often disagree, they rarely oppose one another in public or testify on opposite sides of a bill.[80]

These agreements reflect the extent to which businesses are in common social networks. The PAC officials Clawson and his associates interviewed reported that they read the same newsletters and magazines; often hire the same lobbyists, lawyers, and consultants; talk to the same members of Congress and congressional staff people; and attend the same fundraisers. Most important, the PAC officers know one another and socialize through golf, lunches and dinners, parties, the theater, and charity events. They also help one another out by passing on information about bills and other news, as well as attending fundraisers or contributing money to campaigns that other PAC officers chair.

In contrast to the implications from pluralist theory, these relationships are *most* likely to appear between competitors. As one PAC official put it,

> In our X division, we are major competitors with a company called Y. You couldn't get more competitive. We're suing each other all the time. We hate each other, but I have a very good relationship with them down here, and often we work together on the Hill for certain bills.... [W]hen we're petitioning the government, our interests are very similar . . . [81]

Clawson, Neustadtl, and other sociologists who subscribe to the elitist view of the power structure see these common actions, as well as the successes of PACs, as reflecting the hegemonic power of big business. Business is united in trying to develop and maintain an atmosphere that is supportive to business—all types of business.[82]

These sociologists also remind us that this hegemony is not total. As we have seen throughout this book, significant social change does occur. Much of this change involves the political world and results from the concerted efforts of organizations and interest groups. Racial-ethnic minority groups earned many protections and legal guarantees against discriminatory treatment through the political system. The ability of workers to unionize, the guarantee of a forty-hour week, and the minimum wage all came about through political actions. Consumer protection laws and environmental regulations have forced businesses to change many of their practices, to our benefit. For instance, the improvements in automobile safety in the late 1960s and 1970s, such as the requirements of seat belts and stronger body frames, happened only over the fierce objections of manufacturing firms. Yet these changes are estimated to have saved tens of thousands of lives each year.[83]

Clawson, Neustadtl, and Scott use the analogy of football to try to explain these relationships:

> Business's vast resources and its influence on the economy may be equivalent to a powerful offensive line that is able to clear out the opposition and create a huge opening, but someone then has to take the ball and run through that opening. The PAC and the government relations operation are, in this analogy, like a football running back. When they carry the ball they have to move quickly, dodge attempts to tackle them, and if necessary fight off an opponent and keep going. The analogy breaks down, however, because it implies a contest between two evenly matched opponents. Most of the time the situation approximates a contest between an NFL team and high school opponents. The opponents just don't have the same muscle. Often they are simply intimidated or have learned through past experience the best thing to do is get out of the way. Occasionally, however, the outclassed opponents will have so much courage and determination that they will be at least able to score, if not to win.[84]

In short, even though business interests wield a disproportionate amount of influence in the United States, they are not all-powerful. Sociologists who study the political world often do so with the hope that understanding more about how it operates will suggest how to level the playing field so that all social actors can have more equal opportunities and access. Sociologists who study the political world also hope that by developing a greater understanding of the polity, more people will be able to participate in it (see Box 13-2).

CRITICAL-THINKING QUESTIONS

1. Given what you have read about power and the political world, think about David and Jeff's involvement in the budget election at City Community College. What do you think the outcome of the election is likely to be? More important, what factors—both obvious and hidden—will influence the outcome?

2. What must each side do, based on what you have read in this chapter, to be successful? How do the ideas of rational choice and exchange theory influence your suggestions?

3. How do you suppose David and Jeff will feel about the political world, and their own involvement, if they win? If they lose? Might they be more determined than ever to influence the decisions that affect their lives? Or will they become either discouraged or complacent? Explain your answers.

4. Finally, think about the ideas of the classical theorists discussed in Chapter 1. Which of these sets of ideas most closely resembles the conclusions reached by Clawson and his associates? Explain your answer.

SUMMARY

The theoretical ideas of social networks and rational choice are useful tools for exploring the distribution and use of power in the political world. The role of political action committees in influencing legislation at the national level is of particular interest. These ideas and research findings can help you understand not only the national political scene but also how the political world operates at the state and local level and how you and others can influence the decisions that affect your life. Key chapter topics include the following:

- The political world both controls the behavior of people within the society and provides common or public goods to members of society.

- Power is integral to the political world and reflects the ability both to coerce citizens and to provide public goods for all citizens. Power can involve both domination and influence.

- At the microlevel rational choice theory and the concept of social networks can be used to explain why some people are more likely than others to vote and engage in other types of political activity.

- At the macrolevel the pluralist perspective suggests that a large number of different interest groups influence the political process, thus serving the interests of democracy by representing most citizens, at least indirectly. Pluralists see the power structure in U.S. society as a loose and sizable social network.

- In contrast, the elitist view asserts that real power is concentrated in a relatively small number of people with common interests who have strong social ties with one another. According to this perspective, the political playing field is decidedly uneven and is dominated by members of the capitalist class. Most sociologists believe that the available evidence supports the elitist perspective.

- At the mesolevel sociologists have examined organizations such as political action committees (PACs) and their interactions and influence as corporate actors. Specifically Dan Clawson and his associates found that the contributions of business PACs to congressional candidates help establish social ties and avenues of access that allow businesses to influence legislation. Most PAC donations are pragmatic rather than ideological, although contributions are sometimes used to influence the outcome of elections in order to create a Congress that is more supportive of business. In general, this line of research helps to illustrate the nature of the hegemonic power of business in the United States.

KEY TERMS

corporate actors 346

curvilinear 337

democracy 340

democratic ideology 335

domination 335

elites 342

elitism 342

free riders 336

hegemony 344

influence 335

interest groups 342

pluralism 341

plutocracy 340

polity 333

power 334

power structure 340

public goods 334

rational choice theory 336

representative democracy 341

sanctions 335

state 334

FURTHER READING

Suggestions for additional reading can be found in *Classic Readings in Sociology,* bundled with this book. If you purchased a used copy of this book that does not include this custom-published reader, go to www.sociology.wadsworth.com for ordering information.

BOX 13-2

APPLYING SOCIOLOGY TO SOCIAL ISSUES

Voting in the United States and in Other Countries

In the United States one may vote on everything from various local referendums to candidates for county sheriffs, judges, city councils, state legislatures, the U.S. Congress, and the presidency. In contrast, citizens of many other countries typically have fewer elections in which to participate, especially local-level elections, and they do not have a system that includes a primary election as in the United States.[85] In the United States fewer people vote in the various local elections than in nationwide, highly publicized presidential elections. Still, when just national-level elections are compared, U.S. citizens are far less likely to vote than those in many other countries, including virtually all of the European nations, Canada, Japan, Turkey, and India.[86] Why do these differences appear? Why are people in some countries much more likely than those in others to vote? Could the answer to these questions suggest ways to encourage more U.S. citizens to participate in the political system?

Part of the reason that U.S. citizens vote less than in other countries may involve the mechanics associated with voting. Many other nations simply make it easier to cast a ballot. In many countries elections are held on more than one day, polls are kept open for longer periods of time, and special provisions are made to enable people to get to the polling place. For instance, Italy keeps voting booths open from 6:00 A.M. to 10:00 P.M. on Sundays and again from 7:00 A.M. to 2:00 P.M. on Mondays. Both days are declared legal holidays, and transit fares are reduced to help people get to the polls. Further, many countries automatically register citizens to vote based on their place of residence; voters need not sign up to vote as they must in the United States. Some countries, such as Belgium, even require voting and hold out the possibility of arrest or fines if a citizen doesn't participate (although such sanctions apparently are rarely actually applied). Finally, the simple fact that many European countries have fewer elections than does the United States may make voting more of a "special" event and encourage participation.[87]

To make voter registration and participation easier, the United States has started distributing voter registration materials through post offices and departments of motor vehicles, offices that people often visit when they have recently moved. However, these procedures still require potential voters to indicate their willingness to participate, and, as noted earlier, participation rates have actually declined since these policies were instituted.[88] Estimates suggest that the rate of participation in elections in the United States would increase by about 10 percent if a registration system like that used in Europe, where citizens are automatically included on a roster of eligible voters, were adopted.[89] Some states also have experimented with conducting elections with

INTERNET ASSIGNMENTS

1. Use a search engine to find information on political action committees (PACs). What types of organizations are represented on the World Wide Web? What are the goals of these groups? Can you find information on contributions to political candidates? How does this information provide support for the pluralist and/or the elitist views of the power structure?

2. Locate information on elections. Search for statements by candidates, by political parties, and by supporters and opponents of the candidates. What kinds of information are provided? How

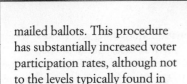

BOX 13-2

APPLYING SOCIOLOGY TO SOCIAL ISSUES

Continued

mailed ballots. This procedure has substantially increased voter participation rates, although not to the levels typically found in European countries.

Besides the simple mechanics of voting, low voter turnout in the United States may be related to the nature of political choices and debate—the structure of the political system. In many European countries elections offer choices among a range of parties with sharply different ideological positions, from extremely conservative to quite radical. A number of European countries also have systems of proportional representation, whereby seats are assigned to political parties in proportion to their share of votes in the election. In contrast, the United States has only two major political parties. In addition, only one candidate may win a given election—a "winner-takes-all" system. As a result differences between political parties tend to be much smaller than in countries with a number of different parties and a different electoral system. As the research of Claw-

son, Neustadtl, and their associates indicates, both the Democratic and Republican parties are highly attuned to the needs and desires of business and the capitalist class. This both reduces the choices available to voters and makes election debates less broad-ranging.[90]

Among the range of political parties in most European countries is a Socialist party that explicitly aligns itself with working-class interests. From their analysis of survey data from a variety of sources and years, sociologists Reeve Vanneman and Lynn Weber Cannon concluded that many nonvoters in the United States are similar, especially in their social class characteristics and political attitudes, to Europeans who vote Socialist. Thus, the researchers suggest, the absence of a range of viable political choices, particularly on the left of the political

continuum, can help account for the lower voting rate in the United States.[91]

CRITICAL-THINKING QUESTIONS

1. Why do you think voter participation is low in the United States? Do you know people who neither vote nor participate in politics in other ways?

2. Do you think that changing the mechanics of voting in the United States might increase participation?

3. What about the structure of the American political system? Should it be changed? Could it be changed? Would changes increase participation? Why or why not?

4. Given the strong role of business interests in American politics, as documented by Clawson, Neustadtl, and their associates, could a leftist political party, like those found in European countries, ever become established in the United States?

might this information affect the probability that someone would participate in the political process? Can you use the ideas of social networks and rational choice theory to explain your answer?

TIPS FOR SEARCHING

For question 1, if you are not successful searching under "political action committees," try searching under "interest groups." For question 2, try using the terms "elections" and "political candidates."

INFOTRAC COLLEGE EDITION: READINGS

Article A21099884 / Rothman, Stanley, and Amy E. Black. 1998. "Who rules now? American elites in the 1990s." *Society, 35,* 17–21. Rothman and Black examine characteristics of elite and powerful groups in America, including ways in which they are similar and ways in which they differ from each other.

Article A20791115 / Dodson, Debra L. 1997. "Change and continuity in the relationship between private responsibilities and public officeholding: The more things change, the more they stay the same." *Policy Studies Journal. 25,* 569–585. Dodson analyzes factors related to the representation of women in elective offices.

Article A20567328 / Southwell, Priscilla Lewis, and Marcy Jean Everest. 1998. "The electoral consequences of alienation: nonvoting and protest voting in the 1992 presidential race." *Social Science Journal, 35,* 43–52. Southwell and Everest examine data from the 1992 national elections to determine why some voters opt for candidates that are not affiliated with a major political party.

Article A20560468 / Gerber, Elisabeth R., Rebecca B. Morton; Thomas A. Rietz. 1998. "Minority representation in multimember districts." *American Political Science Review, 92,* 127–145. Gerber, Morton and Rietz examine the ways in which different systems of voting (for example, winner take all systems vs. choosing several candidates at large) influence the possibility that minority candidates will be elected.

RECOMMENDED SOURCES

Clawson, Dan, Alan Neustadtl, and Denise Scott. 1992. *Money Talks: Corporate PACs and Political Influence.* New York: Basic Books; Clawson, Dan, Alan Neustadtl, and Mark Weller. 1998. *Dollars and Votes: How Business Campaign Contributions Subvert Democracy.* Philadelphia: Temple University Press. These books provide entertaining reports of the research featured in this chapter.

Dalton, Russell J. 1988. *Citizen Politics in Western Democracies: Public Opinion and Political Parties in the United States, Great Britain, West Germany, and France.* Chatham, N.J.: Chatham House. Dalton's short, readable book compares political involvement in four countries.

Domhoff, G. William. 1970. *The Higher Circles: The Governing Class in America.* New York: Random House; Domhoff, G. William. 1978. *Who Really Rules?* New Brunswick, N.J.: Transaction. Domhoff provides easy-to-read and entertaining discussions of the nature of political elites and their power in the United States.

Garcia, H. Chris (ed.). 1997. *Pursuing Power: Latinos and the Political System.* Notre Dame, Ind.: University of Notre Dame Press. Garcia provides an extensive overview and analysis of the participation of Hispanic Americans in the political process.

Mitchell, Neil J. 1997. *The Conspicuous Corporation: Business, Public Policy, and Representative Democracy.* Ann Arbor: The University of Michigan Press; Vogel, David. 1996. *Kindred Strangers: The Uneasy Relationship between Politics and Business in America.* Princeton, N.J.: Princeton University Press. These books provide a contemporary development and defense of the pluralist perspective.

Nie, Norman H., Jane Junn, and Kenneth Stehlik-Barry. 1996. *Education and Democratic Citizenship in America.* Chicago: University of Chicago Press. The authors examine both how and why education is so strongly related to participation in the political process.

Phillips, Kevin. 1994. *Arrogant Capital: Washington, Wall Street, and the Frustration of American Politics.* Boston: Little, Brown and Company. Phillips analyzes the relationship of the low level of political involvement of citizens in the United States to the political role of business.

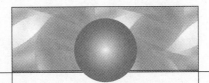

Pulling It Together

Listed below are other chapters where you can find additional discussion or examples
of some topics covered in this chapter. You can also look up topics in the index.

INTERNET SOURCES

Center for Public Integrity. The Web site for this non-profit, nonpartisan research organization provides a wealth of information on political issues, including campaign financing.

The International Foundation for Election Systems (IFES). This private, nonprofit organization monitors and reports on political activities in more than 90 different countries throughout the world.

Project Vote Smart. This nonprofit, nonpartisan group provides extensive information about political candidates, office holders, and legislation at both the national and state level.

14

The Economy

SEAN'S LIFE REVOLVED around music. As far back as he could remember, he had loved listening to music and picking out tunes on the piano. As he got older, he explored other instruments. He took drum lessons, played around with the trumpet and violin, and became a skilled guitar player. With the encouragement of his piano teacher, he began to write down the tunes that came into his head. In high school he played guitar in a band he formed with some friends, performing at school dances and at a local teen club. Occasionally the band would play some of Sean's own compositions. Sean wanted to attend a conservatory after graduation and become a professional musician, but his parents knew how hard it was to make a living as a musician and insisted that he go to college.

In his sophomore year Sean took a class called "Music in Film." By the end of the first week he knew what he wanted to do when he finished college. He had always loved movies, and the more he thought about creating music for films, the more excited he became. He shared his hopes and dreams with his teacher, Professor Sherman, showing him his compositions and playing tapes of his music.

Professor Sherman was impressed with Sean's potential as a composer and called some former students who were employed in the film industry to see whether they might be able to help Sean break in. Because Sean already belonged to the musicians' union through his work with his band, they suggested that he try to pick up work playing background music for commercials and television shows.

The following summer Sean went to Los Angeles and managed to get a few small jobs playing for commercial jingles. More important, he began to get a feel for the Hollywood music world. He met a number of people involved with the music business and talked to them about the ins and outs of their jobs and the nature of their careers. Most were fairly encouraging, but many of them emphasized that breaking in wasn't easy. Sean would have to knock on a lot of doors and take whatever work he could find. "It's just like that old saying," one of his contacts told him, "it's not *what* you know—it's *who* you know."

During his junior year Sean stayed in touch with several of his contacts, and the following summer he returned to Hollywood determined to make many more. This time he was even more professional in his preparation, bringing high-quality audition tapes with him and systematically working every lead he uncovered. In July he landed a job helping to score a television movie for an independent production company. Although the movie was never completed, the experience left him even more excited about working in films. He dreamed of composing his own music to enhance the terror of a movie like *Jurassic Park*, or the playfulness of one like *Grumpy Old Men*, or the tenderness of one like *Titanic*. In his senior year Sean took as many classes as he could in music composition and theory, as well as film and literature. Meanwhile he continued to play with his band at local clubs, sprinkling in more and more of his own compositions. By now his parents had reluctantly accepted that he wasn't about to be deterred from pursuing a career in music.

After graduation Sean packed his bags and moved to Hollywood, hoping to fulfill his dream of a career as a film composer. He knew that his summer jobs and his contacts would give him a head start, but he also recognized that the competition would be rough and really good jobs would be difficult to find. Probably he wouldn't find work scoring movies or even television shows right away, but he was willing to take on whatever jobs he had to as long as they were connected with music. "I'll give it at least a few years," Sean told his parents. "I've invested so much in this already that I'm not about to give up too soon."

Despite the brave front he showed his parents, inwardly Sean worried a great deal about his future. All around him he could see less talented composers who seemed to have all the work they wanted, apparently because they knew the right people. Often he winced at the quality of music they produced, although a few thrilled him with scores that were consistently original and appropriate to the film or television show. For the most part, the few composers he admired were already well established. Maybe, he mused, it wasn't just a lack of talent that explained mediocre scores. Perhaps the successful composers had more freedom to make the kind of music they really wanted to make. But in that case, how did someone attain success in the first place?

Despite the advice he had been given during his previous visits to Hollywood, Sean was convinced that it wasn't just *who* he knew that would matter. *What* he knew, and what kind of music he knew he could create, were important, too. But would he get the chance to show what he could do? How could he break into the magic circle of composers, musicians, producers, and directors who all seemed to know one another? Would any of them be willing to lend him a hand? Or would they view him as simply another starry-eyed dreamer—or even an unwelcome competitor? And how much would this opportunity to create be constrained by dictates from producers and others who wanted a certain kind of music? Could he really make a life out of his dream?

*V*ERY FEW OF Us have Sean's talent to compose music, much less earn a living from such work. But each of us possesses other skills and talents that can provide a living for our families and ourselves and goods and services for other people. However, no matter what type of work we want to pursue when we finish school, we also face uncertainties about the future. What kind of job will we have? Will it provide us with stable employment and enough income to buy the things we need and want? Will it offer opportunities for advancement?

These concerns involve much more than our abilities and our willingness to work. They also relate to the **economy,** the social institution involving the production, distribution, and consumption of goods and services. As a social institution the economy includes all the norms, organizations, roles, and activities related to making, exchanging, and using everything from necessities—such as food, shelter, and clothing—to luxuries—such as records, movies, and books. We all participate in the economy each day, whether as producers or consumers. In addition, the economic world influences our daily lives and the opportunities and constraints that we experience. One of its most important influences is in stratification. The jobs we hold and the amount of money we earn are prime determinants of our place on the stratification mountain, and the constraints that women and members of minority groups face in the economy are one of the defining features of racial-ethnic and gender stratification.

In studying the economy, sociologists have often been interested in how its workings as a social institution relate to opportunities and constraints—not just for individuals, but for communities and entire societies. Economic success—as reflected in the ability to distribute, exchange, and consume goods and services—has a direct effect on the power and well-being of individuals and societies. How do these advantages appear? Much as Sean asks why certain composers, producers, and directors have been able to penetrate the magic inner circle of filmmaking, sociologists have asked why some individuals and some societies have enjoyed greater economic success than others have. As with their analyses of other areas of the social world, many sociologists hope that by understanding more about the economy as a social institution, they can help create a world where inequality is lessened and more people have the opportunity to use their talents and skills to their fullest capacity.

Sociologists have been interested in the economy since the beginning of the discipline. Karl Marx suggested that economic relationships were the most fundamental feature of a society, greatly influencing the social institutions of the family, religion, and the political world. He linked the inequalities in society of his time to capitalism, the economic system that characterizes the United States and many other modern industrialized countries and in which private individuals or corporations develop, own, and control business enterprises. Other early sociologists also wrote about the economy and the development of capitalism. For instance, much of Max Weber's work was devoted both to understanding the economic life of ancient societies and to explaining why capitalism first appeared in Europe and not elsewhere.

When his work was first published, Marx was seen as an economist, not a sociologist. Until the mid-to-late 1800s there wasn't a sharp separation between people who wrote on the economy and those who wrote about other areas of society, such as the family and the political world. Besides Marx, early economists such as Adam Smith and John Stuart Mill often wrote about issues and ideas that we would today consider to be related to both economics and sociology, and Weber was trained in this tradition. Unfortunately, by the late 1800s, bitter controversies split the field of economics, and the outcomes of these controversies affected both economics and sociology for many years.[1]

We begin this chapter by briefly describing the relationship between economics and sociology and the ways in which sociologists describe the economy. We then apply macro-, meso-, and microlevel analyses to see how the workings of the economy translate into opportunities and constraints for societies and individuals.

SOCIOLOGICAL STUDY OF THE ECONOMY

The fight among economists began in Germany and Austria, where it was called the "battle of the methods." The more sociologically oriented economic historians, like Weber, were on one side, and the more mathematically oriented economists were on the other. The dispute soon spread to Great Britain and the United States. Within a few years the more mathematical economists began to predominate, so that economics focused increasingly on abstract, deductive models of how the economy works. Economic historians, who were much more likely to consider the relationship of the economy to other areas of society, in essence abandoned the field of economics and developed their own academic specialty.[2] Even today most colleges and universities teach courses on economic history in departments of history, not in departments of economics.

A busy restaurant can be the scene of many different economic relationships, perhaps involving customers, waiters, owners, and suppliers. How many different possible economic relationships can you identify in this picture?

In time, therefore, economics became characterized by analyses that used equations and graphs and mathematical formulations to study the world, rather than historical or sociological analyses, such as those Marx and Weber used. Meanwhile sociology emerged as a distinct area of study and research. Although their fields of interest overlapped, economists and sociologists maintained an uneasy truce. Economists tended to study money and **markets** (defined as the process of buying and selling goods and services and the values that are set in this process); sociologists studied workers, corporations, and industries. Economists developed complex mathematical models of how markets, finance, and world economies work; sociologists tended to do more descriptive studies.[3]

An old joke about the two disciplines summarizes the historical differences between them: "Economics is all about how people make choices; sociology is all about how they don't have any choices to make."[4] In general, economists focused on how actors, such as individuals, firms, or nations, make their economic choices. In contrast, sociologists often looked at the constraints that people face in making these choices and the ways in which social structures limit those choices. But this quip really only caricatures the two disciplines, and many contemporary sociologists and economists cringe when they hear it. As emphasized throughout this book, most sociologists today believe that we need to look at both social action and social structure. We need to understand how we make choices in our daily lives and how the social situations in which we live constrain these choices. We need to understand both opportunities and constraints.

Fortunately for social science, the century-old rift between economics and sociology is beginning to narrow. These changes began with the efforts of some economists, such as Nobel Prize winner Gary Becker, to apply economic theories to areas of social life such as the family, religion, and the political world. A number of sociologists joined these efforts, often using the insights of exchange theory, and have helped develop the tradition of rational choice theory that was discussed in Chapters 12 and 13.[5] At the same time, sociologists began to apply concepts developed within their own field to areas that economists traditionally studied, such as economic transactions and markets.[6] One of the central ideas in this developing tradition is the notion that economic transactions, or what economists often call **economic exchanges** (the exchange or transfer of goods and services in return for some other type of valued item, such as money), involve **social networks**—linkages between individuals, organizations, and entire societies.

Three elements are integral to the development of these interactions and economic relationships. The first of these is *trust*.[7] We go to certain mechanics, hairdressers, or physicians because we trust their work, and they in turn trust that we will pay them what they are due. Trust also underlies economic relationships at higher levels of analysis, such as the decisions of nations to trade with one another or the decision of one organization to contract work with another. In other words, economic relationships become stable to the extent that the units that are involved in the relationship trust each other. The second important element is *cooperation*. Movie

Sociologists emphasize that market relationships and economic exchanges involve elements of trust, cooperation, and competition. How might the relationships that people have in this Mexican market be similar to your own market relationships? How might they differ?

directors can't complete a film without the cooperation of many different people. Frozen food can't get to the market without the cooperation of several different organizations. Economic relationships and the market itself can't operate without the cooperation of the involved units. Finally, although it may not be so obvious, *competition* is also influenced by our relationships. In particular, social units tend to have a competitive advantage, to be more successful in their economic endeavors, when they have a stronger social network. Salespeople who know more people will have a greater chance of making sales; nations that have treaties with numerous other countries will have more opportunities to sell their goods; corporations with more contacts in Congress will have a better chance of influencing legislation.

Contemporary sociologists who study the economy stress that all aspects of the economic world—production, distribution, and consumption of goods and services—can be seen as involving social networks, or relationships between social units. For instance, we usually work with other people to produce things or provide services, whether teaching students, building cars, or making movies. Also, goods and services are distributed through linkages between people, organizations, and nations. For example, farmers ship raw produce to processing plants, where the produce is frozen, canned, or packaged fresh and then shipped to local grocery stores. The consumption or use of goods and services also involves exchanges between social units, as when we purchase the services of a doctor, mechanic, or hairdresser or buy food from the supermarket. As we will

see, the economic success of societies and individuals is related to their linkage to and control of extensive networks.

This key role of social networks in economic exchange can be seen in all societies. For instance, in hunting and gathering societies, hunters commonly shared their game with others in the group. This sharing helped build relationships among group members because they all would, at some time, exchange food with one another. Through these exchanges people became obligated to one another, reinforcing their social ties.[8]

Economic relationships became more complicated in horticultural societies, in which people grew crops for food. Almost everyone in hunting and gathering societies was involved in procuring food for the group. In contrast, horticultural societies were able to produce a surplus of food, enabling some group members to fill other roles. For instance, some became shamans whereas others specialized in making weapons or tools. Of course these specialists still needed to eat, and so they would trade the services or goods that they produced for food. Just as in hunting and gathering societies, these exchanges established social ties. If you were satisfied with the new tool, you would be likely to return to the toolmaker the next time you needed one; if the toolmaker was satisfied with the payment, she or he would be willing to make another tool for you. However, whereas hunting and gathering societies were relatively egalitarian, horticultural societies had more inequality. The greater division of labor, combined with the growing surplus of food, enabled some people to accumulate more goods than others. In particular, individuals who provided or produced more highly valued or rare services or goods began to develop more extensive social networks and to wield greater control over these networks, allowing them to obtain more than others do.

These inequalities became even more pronounced in agricultural societies, which were much more productive than horticultural societies were and which had a much more complex division of labor. Because members of agricultural societies were engaged in many different occupations, the kinds of goods and services that were exchanged and the geographic area in which the exchanges occurred grew substantially. Cities prospered, and elites amassed large fortunes.[9]

Today's economy is even more complex than that of agrarian societies, for it involves a vast mix of goods and services, as well as transactions that literally span the globe and involve large corporate bodies that have roots in countries throughout the world. On any given day you might wear clothes manufactured in China, Mexico, and the United States; eat food grown and produced in Canada, the United States, and several South American countries; drive a car manufactured in Japan or Germany; and watch television on a set made in Japan or Korea—without ever thinking about where these goods originated.[10] Manufactured goods that might once have been made in your own neighborhood or in nearby communities are now manufactured on the other side of the world. And whereas people in earlier eras might have bartered for goods and services or, later, used gold and silver coins, you routinely conduct electronic economic exchanges, using credit and debit cards virtually anywhere in the world. Today economic exchanges and markets can involve not just local communities, but the entire world—what is often called a **global economy.**

Much as Georg Simmel described society as a "web of social relations," so can we see the economy as a web of relationships between social units. Economic sociologists see social networks as the *social structure* that underlies the economy. And just as our personal social networks develop from social interactions, economic networks develop from economic relationships or exchanges.[11] Much as spiders spin webs, societies and individuals create webs of economic relationships through *social actions*—their interactions, or social relationships, with other social units. In the sections that follow, we will see how societies and individuals construct their webs of economic relationships and how these relationships provide greater opportunities and power for some than for others.

THE ECONOMY: A MACROLEVEL VIEW

When we step back and take a macrolevel view, we can see how linkages between societies and communities produce economic relationships. Although some societies have been largely isolated and self-sufficient, most have had at least some links with other groups. In today's increasingly global economy, many societies have extensive relationships with one another. For instance, television movies such as the one Sean worked on often are aired not just in the United States, but in Europe, Oceania, and Asia as well. Your community might advertise its tourist attractions to potential visitors from other nations, and an international conglomerate might own the company you work for. Such linkages exist at all levels of economic activity, from the interactions between businesses and customers in the same community to the complex arrangements that allow a Japanese-owned firm to manufacture and sell cars in the United States that burn fuel derived from Middle Eastern oil.

From a macrolevel perspective we can also see that some countries have had more economic success than others have. One way in which sociologists and economists measure the economic success of a country is through its **gross national product,** or **GNP,** the value of all goods and services that a nation produces. This is often expressed in U.S. dollars and adjusted for the number of people who live in the country (this adjusted figure is referred to as the **per capita GNP**). As shown in Map 14-1, countries around the world vary enormously in the size of their per capita GNPs—from $90 in Mozambique to almost $29,000 in the United States to more than $45,000 in Luxembourg.[12] Sociologists often describe these differences between nations as a **global stratification system,** whereby some societies have more power, prestige, and property than others.

Although the differences usually aren't as great, areas within a country can also have varying economic fortunes, in what is sometimes called **regional stratification.** For instance, the average income of residents of the different states varies from a high of more than $36,000 per person in Connecticut to a low of a little more than $18,000 per person in Mississippi.[13] Even within a single metropolitan area, such as Los Angeles, where Sean moved, communities vary tremendously in their economic fortunes. There are very poor inner-city neighborhoods such as Watts and Echo Park and luxurious communities such as Beverly Hills and Malibu.

Why do these differences exist? Why do some nations, states, and communities have more opportunities and fewer constraints than others do? How

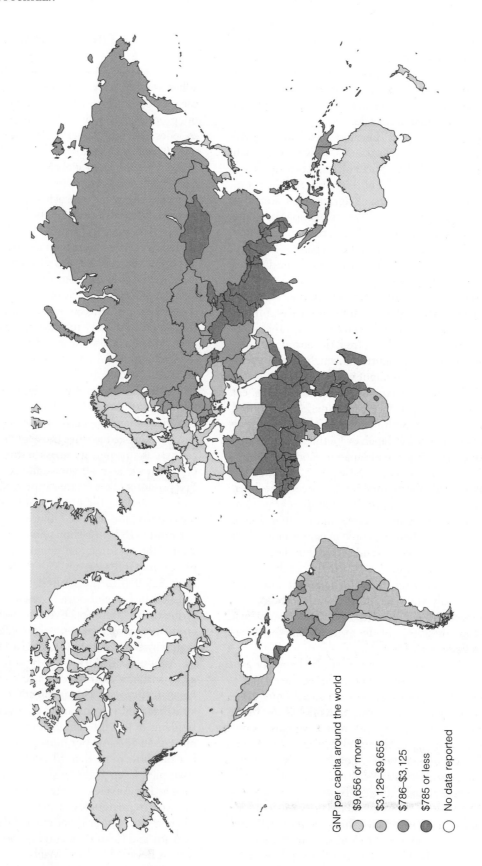

GNP per capita around the world

⬤ $9,656 or more
⬤ $3,126–$9,655
⬤ $786–$3,125
⬤ $785 or less
◯ No data reported

MAP 14-1 *Gross National Product of Countries Around the World, 1997*
Source: World Bank, 1998.

is global and regional stratification related to the economy as a social institution? At the macrolevel sociologists have sought answers by looking at a society's natural resources and geographic characteristics, its level of technological development, and the relationship between the political world and the economy. Each of these aspects influences the web of economic relationships that a society or a community can develop. The characteristics of this economic web, in turn, influence societies' economic power and success.[14]

Resources, Geography, and Economic Opportunities

In our discussion of cultural materialism in Chapter 3, we described how the physical environment in which people live influences their way of life. These physical characteristics, however, influence not only a society's culture but also its economy.

The effect of natural resources on economic activity is apparent in nonindustrial societies, but we can see it in industrialized societies as well. For example, the inhabitants of what is now Alaska and northern Canada obtained their living by fishing, hunting, and foraging a few native plants; only in the more temperate climates much farther to the south could the land be farmed. In more modern times, automobiles and other manufacturing firms flourished near sources of iron ore and steel in the midwestern United States rather than on either the east or west coasts, which lacked these natural resources. The movie industry flourished in Hollywood in the early twentieth century, before the development of extensive indoor lighting capacities, because the dry, sunny climate allowed filming in natural light. Countries in the Middle East with bountiful supplies of oil have become enormously wealthy because of their relative monopoly of this valuable natural resource.

Geographical areas differ not only in their natural resources and climate but also in their accessibility. Access to other groups facilitates the establishment of extensive social networks and trading relationships. For instance, advanced horticultural societies in Africa developed several sizable cities to facilitate trade with other groups. A sixteenth-century Dutch visitor to one of these cities (Benin, in what is now Nigeria) reported that it was "very great," with streets that were much broader than those of Amsterdam. Notably, however, none of these cities developed in the heart of the rain forests, where the thick vegetation hindered communication and transportation between neighboring villages.[15]

The first large cities to develop in the United States, such as New York, Boston, Philadelphia, Charleston, and Baltimore, were all located at ocean ports. This allowed them to become focal points for gathering goods to ship to other countries and for receiving imports in return. In later years Chicago, on Lake Michigan, and St. Louis and New Orleans, on the Mississippi River, used lake and river transportation to help bolster their economic activities. And Atlanta became a hub of commerce and rail transport in the southern United States because it lay in a flat area in central Georgia.[16]

A society's economic activities also are influenced by the resources and characteristics of the neighboring societies with which it interacts. For instance, wealthy state societies developed in Africa because people were able to trade with their neighbors for goods they could not produce themselves. The auto manufacturing plants in the midwestern United States received raw materials and component parts from several neighboring communities and then shipped the finished products throughout the country and to other countries. In the contemporary United States, the success of Las Vegas, Nevada, as a tourist center is aided by its being the closest city with legalized gambling to the populous Los Angeles metropolitan area.

As a result of these connections, the economies of societies and large geographical areas become networks of interdependent activities.[17] In small, isolated societies, which have limited communication with other groups, these networks are simple and sparse. In contrast, in populous societies with easy access to transportation, networks can be highly complex and dense. Because of these links, the economy can be much more productive and provide many more opportunities for people.

Even today, our geographic environment influences our work opportunities. Part of the reason that Connecticut is so much more prosperous than is Mississippi is its close proximity to the major urban centers of the northeastern United States. In fact, the four states with the highest average personal income (Connecticut, New Jersey, Massachusetts, New York) are in the Northeast; and three of the five with the lowest average incomes (Mississippi, Arkansas, West Virginia) are in the South.[18] Similarly people who live in poverty-ridden areas of large cities often have few job opportunities nearby because manufacturing firms have relocated to suburban areas or other countries, and young people who grow up in isolated rural areas often must leave their hometowns for more populated areas to find work. Further, specific industries are largely concentrated in certain areas. Thus, when Sean wished to pursue a career in the movies, his best hope was to go to Hollywood.

To some extent the poverty of the countries at the very bottom of the global stratification mountain can also be linked to their limited natural resources and their less-than-ideal geographic location. As we will describe in more detail in Chapter 17, many countries in sub-Saharan Africa, almost all of which are very poor, face severe problems with **desertification** (the increase in desert land) and **deforestation** (the disappearance of forested land). The amount of arable land in this region is disappearing, and the area has few other natural resources. In addition, because of the geography of the region and the extreme poverty of its inhabitants, building roads, trains, and other transportation systems that can effectively link these nations with other countries is difficult. Moreover, as in other regions around the globe, primarily in Africa and Asia, the neighboring countries and potential partners in national economic exchanges are also quite poor—placing still more constraints on the possibility of future growth.

Other countries with more favorable natural resources and geographic locations have more economic opportunities and brighter prospects. For instance, even though it is still considerably poorer than its neighbors to the north are, Mexico has experienced considerable economic growth in recent years, and experts expect this trend to continue. Some of Mexico's growth has been fueled by the discovery of substantial natural oil reserves, perhaps the largest in the world.[19] Growth also has been spurred by the North American Free Trade Agreement (NAFTA), which was ratified in 1993. NAFTA provides for the gradual end of all tariffs (a form of tax) on products that travel between Mexico, the United States, and Canada, which will create a much larger market—a much more extensive web of economic relationships—for Mexico and is expected to prompt even more economic growth in the future.

Technology and Economic Opportunities

Throughout history humans have developed **technology** applying knowledge to solve problems such as those associated with a lack of natural resources or a given climate or geographic location. Technological innovations have allowed many societies to develop economic activities in areas that once would have been considered almost uninhabitable. For example, water is piped many miles to cities in the arid southwestern United States. Wealthy Middle Eastern nations fund massive desalinization projects to convert seawater to drinking water. Technological inventions such as air conditioners, heat pumps, and furnaces keep us cool on the hottest days and warm on the coldest nights, and electric lights allow us to work and play at all hours. Throughout history technological innovations related to transportation, from wheeled carts to sailing ships, have allowed people to trade farther and farther away from their homes. Today telephones, satellites, and computer technology allow us to communicate instantaneously with people all over the world. The wonders of technology allow us to build much more extensive social networks and much broader economic relationships than would have been possible even a few years ago. The global economy could not exist without advanced technology.

But not all nations have equal access to this often-expensive modern technology. The United Nations and other governmental bodies often distinguish among countries based on how developed their economies are. You can see these differences by looking again at Map 14-1. **Industrialized countries**—including the United States, Canada, the European nations, Australia, Japan, and New Zealand—have highly industrialized economies. The **least developed countries,** most of which are in sub-Saharan Africa and Southeast Asia, are the poorest and least industrialized.[20] These least developed nations are at the bottom of the global stratification mountain, with an average per capita GNP of only $350. In addition, they are much less likely to be able to afford the basic technology needed for clean water and sanitary facilities, to have modern communication techniques such as television and radio accessible to their citizens, and to provide secondary levels of education for their children.[21] In between are the **developing countries**—including many nations in Central and South America. These nations have started the process of development, as indicated by gross national products that are higher than those of the least developed countries. Nevertheless, the developing countries are still far poorer than the highly industrialized nations.

In a sense the gap between the developed and less developed countries represents a vicious cycle. Technological advances allow societies to develop broader webs of economic activity and thus enhance their economic fortunes, but technological advances also require money. The least developed countries, which most need to broaden their economic webs, are least likely to be able to afford technological advances that can help them do so. Just as the possession of technology can provide economic advantages and opportunities, the lack of technology produces economic constraints. The plight of very poor countries is, in many ways, also associated with their political history, and we turn now to the relationship between the political world and the economy.

The Political World and Economic Activity

In our analysis of the U.S. political system in Chapter 13, we saw how political action committees (PACs) linked to business corporations disproportionately influence the political process through PACs' involvement in campaign financing and the development of legislation. Many sociologists suggest that the power of business interests reflects the interrelationship of the political world and the economy, often referred to as the **political economy.** We can look at the political economy both within societies and among nations.

National Political Economies Sociologists and economists typically distinguish between two general types of economic systems in contemporary societies: capitalism and socialism. As an ideal type in the Weberian sense, **capitalism** generally is considered to have four major characteristics. First, individuals own the means of production—the ways products and goods are produced, such as farmland, factories, and machines. Second, a profit motive—a desire to make money by selling goods and services for more than production cost—is the major incentive underlying economic relationships. Third, profits from economic exchanges are reinvested to expand business opportunities and generate greater future profits. Finally, capitalism involves free competition in the market: Buyers and sellers are able to seek each other willingly and freely.[22]

In contrast, as an ideal type, **socialism** involves public, not private, ownership of the means of production. The government owns farmland, factories, railroads, and all natural resources such as oil, timber, and minerals. Socialism also involves the elimination of economic competition and the profit motive through centralized government control of the economy. Government agencies plan how much should be produced, where the products will be made, how much will be charged for them, and where they will be distributed. The goal of economic exchange is not profit but providing for the needs of the people.[23]

With pure-form capitalism the government has no control or influence over business operations. This is sometimes called **laissez-faire** (literally, "let do" or "let things take their own course") **capitalism** and implies that the social institutions of the economy and the polity are unrelated—that, in fact, there is no political economy. In contrast, with pure-form socialism, exactly the opposite is true: The web of economic relationships is determined by political decisions, and the polity and the economy are merged.

Of course neither capitalism nor socialism actually exists in a pure form. For instance, the United States is often said to have a system of **welfare capitalism,** whereby a broad system of laws has been developed to protect workers and consumers. These regulations emerged gradually in response to a wide range of activities that often seriously harmed people physically and economically. For example, only a century ago you could have been exposed to dangerous, untested medicines; worked in a factory with no regulations regarding the safety of your environment, your work hours, or your wages; and been forced to buy products whose prices had been set by strong **monopolies** (where one company controls a single market).[24] Today new drugs must pass stringent testing by the Food and Drug Administration; laws regulate working hours, wages, and occupational safety; and federal statutes protect consumers and businesses against monopolies. In addition, many government actions in the United States are taken to protect businesses and enhance their profits. For instance, the price of many products not only is set by competition, as a pure capitalist model would suggest, but also is strongly influenced by price supports, such as those given farm products ranging from wheat to milk to peanuts. When natural disasters such as earthquakes, hurricanes, or floods occur, the government provides grants to individuals and businesses to help rebuild a region's economy.

Just as the pure form of capitalism is extremely rare, so is the pure form of socialism. Even within the former Soviet Union and its satellite nations, some workers earned more than others did. For instance, as the functionalist theory of stratification would predict, factory managers were paid more than factory workers and professionals more than laborers to induce people to take on added responsibilities. Even nations that continue to practice socialism, such as the People's Republic of China, have incorporated elements of capitalism within their economy, allowing some individuals to start their own businesses.

The most common form of socialism today is probably **democratic socialism,** sometimes called **welfare socialism,** which contains elements of both capitalism and socialism. This system has emerged in a number of European countries, such as Sweden and Denmark, through the democratic election of Socialist parties. Democratic socialism involves government ownership and operation of some aspects of the economy, such as mining, forestry, telephones, and airlines, and private ownership and operation of other aspects, such as farms, factories, and retail stores.[25]

All political economies involve social structures—webs of economic relationships—that tend to provide greater power to some members of the society than to others. In the United States, as we saw in Chapter 13, immense wealth enhances the attempts of business in-

BUILDING CONCEPTS

Capitalism and Socialism

Sociologists and economists generally distinguish two general types of political economies: capitalism and socialism. Below is a summary of the characteristics of each, as an ideal type.

Capitalism	Socialism
Individuals privately own the means of production (the ways products and goods are produced, such as farmland, factories, and machines).	The means of production are publicly owned by the government.
Making a profit is the major motivation underlying economic relationships. Profits are reinvested to expand business opportunities and make more profit.	Centralized government control of the economy eliminates a profit motive. The major motivation is providing for the needs of the people.
The market is characterized by free competition. Buyers and sellers are able to seek each other willingly and freely.	Centralized government control of the economy eliminates economic competition.

terests to influence political decisions, and many tax breaks and legislative proposals are designed to bolster business interests. In the former Soviet Union the greatest power tended to accrue to high-ranked government officials, who also controlled economic decisions. In short, the structure of the national political economy influences who in a society is more likely to have greater economic power and success.

It is also important to remember that changes in political economies come about through social actions. In the United States, the protections that characterize welfare capitalism resulted from the efforts of concerned citizens, just as the benefits that the government provides business often resulted from the lobbying efforts of business organizations such as PACs. The overthrow of oppressive political regimes in the former Soviet Union and the Eastern European nations occurred through social actions—the concerted efforts of thousands of citizens. Even though political economies, like all social structures, strongly influence the way that power and economic success are distributed within a population, these structures are created and changed through social actions.

In short, the polity is closely entwined with the economy in all areas of the world, thereby affecting the web of economic relationships. The United States, whose economy more closely resembles pure-form capitalism than does that of many countries, both supports businesses and protects consumers and workers. Some enterprises, such as services for the poor and elderly, are largely financed and run by government agencies. In other countries, even those whose governments advocate rigorous socialism, elements of capitalism and free enterprise are apparent. Some scholars suggest that political economies around the world are becoming more similar. **Convergence theory** suggests that a **mixed economic system,** one blending elements of socialism and capitalism, is beginning to characterize most countries around the world.[26]

In large part these changes might be occurring because the various economies of the world are not isolated but, rather, are all part of a global economy. This global economy involves an enormous number of economic exchanges each day—exchanges that involve material goods, such as food, clothing, and appliances; cultural artifacts and ideas; financial credit and capital; technological innovations and developments; and even arms and weapons. For example, television shows and movies, such as those Sean hopes to work on, are shown around the world; books are translated and distributed across the globe; and clothing styles, food, music, and art are exchanged worldwide. Financial capital for movie production comes from countries around the world. Technical aspects of editing, sound, and animation can occur any place on the globe. Conflicts, even within nations, are powered by weapons and ammunition purchased on worldwide markets. Because of the immensity of the global economy, it is important to also look at political economies from an international perspective.

International Political Economies Political power and relationships between societies have always been intertwined with economic needs and opportunities. Frequently these relationships have involved nations with more advanced technology enhancing their wealth by conquering and dominating other societies. For instance, historically large state societies, such as the Incan and Mayan empires in the Americas and the Chinese empire in Asia, developed as rulers aggressively expanded their territories to incorporate more land and resources.[27] Beginning in the sixteenth century, European countries such as England, Portugal, and Spain created vast colonial empires by forcibly acquiring territories in other parts of the world, a practice often termed **imperialism**. These nations extracted large amounts of raw materials, such as gold, cotton, sugar, rubber, and tobacco, from their colonies and shipped them home, where they were manufactured into various products and sold for high prices.[28] As a result, these European nations became the hub of highly lucrative social networks. Their web of economic relationships gave them a competitive advantage, not only over their colonies but also over other industrialized countries that did not have such empires.

Once a group of colonies itself, the United States never owned vast reaches of other continents, as the European countries did, but instead controlled small, strategically placed areas throughout the world such as the Philippines, Hawaii, Guam, Puerto Rico, and the Panama Canal Zone. Moreover, the United States has actively engaged in what has been called **neo-imperialism,** especially after World War II ended and the European colonial empires collapsed. Although the United States then had the largest armed forces in the world, its dominance was based less on direct military strength than on economic might. The United States was the only large nation to emerge from World War II with its industrial plant unscathed, and many countries owed it money for purchases made during the war. Even

though it did not directly rule less developed countries in the way that colonial nations did, the United States imported raw materials from the less developed countries, such as rubber, a wide variety of metals and minerals, and many different plant and food crops. These raw materials were then used to manufacture goods, from tires, airplanes, and cars to household appliances and furnishings. Like the imperialistic countries, the United States gained great competitive advantage from its dominance of the network of nations. Thus it experienced huge economic growth in the 1950s and became the world's leading producer of all types of products, from automobiles to electronic goods.[29]

A perspective called **world systems theory,** which was developed primarily by the sociologist Immanuel Wallerstein and inspired largely by the writings of Karl Marx, suggests that this history of imperialism and neo-imperialism can explain why the system of global stratification persists. Why are some societies so much richer than others are, and why are so many of the societies that were once colonies now among the least developed nations? World systems theory looks at the network of economic relations between nations and suggests that, much as the capitalist class dominates the social class hierarchy, a powerful group of nations, called the **core nations,** dominates and exploits other nations around the world. Core nations were the first to industrialize and develop capitalistic systems, and thus they have many competitive advantages over other nations.

Most of the power of core nations today comes from the strength of **multinational corporations,** business enterprises that have outlets around the globe. Many of these multinational corporations (sometimes called MNCs) are huge. Multinational corporations based in the United States employ almost 26 million people around the world and have assets of more than $10 *trillion.*[30] These corporations span the globe, producing a global economy that is strongly influenced by interests based in the United States and other industrialized countries. Familiar examples of multinational corporations include Colgate-Palmolive, which manufactures and distributes household and personal care products and which maintains branches in thirty-nine other countries; Dow Chemical, which manufactures chemicals, plastics, metals, pharmaceuticals, and household products and which has branches in thirty-eight other countries; and Hewlett-Packard, which manufactures and markets computers, electronic instruments, and medical products and which has branches in a whopping eighty-nine other countries.[31]

In contrast to the core nations, the **peripheral nations,** are at the edges of the world system, primarily because their role is selling raw materials to the core nations. Traditionally, the peripheral nations have produced agricultural goods and raw materials used in manufacturing. In recent years they have also provided a cheap source of labor as many multinational corporations have established factories in these countries to manufacture everything from shoes and clothes to computers.

Besides these two major elements of the world system, some nations belong to what is called the **semiperiphery,** a group of nations that falls between the core and periphery. Semiperipheral nations include countries like Mexico, Turkey, and Malaysia that are moving up in the world system hierarchy as they develop their own industries. It also includes countries like Portugal and Greece that are part of the European Union (EU) and have extensive trading relationships with core nations but that are still much poorer than the core countries. Other nations belong to the **external area,** which includes countries that have few economic ties with the core nations and were not involved in the development of capitalism. This group includes the least developed and most impoverished nations, mainly in Africa and Asia.[32]

It is important to realize that imperialism and the world system of economic relationships would not have been possible without the political and military support of powerful governments. Colonial governments, such as the British in India, built extensive railroads to facilitate the movement of raw materials to ocean ports, and large and powerful armies and navies ensured safe transport of materials from the colonies to the home country.[33] Today the industrialized nations no longer directly control the political structures of other countries, because virtually all the former colonies have achieved political independence. Yet the industrialized nations still influence political events in other, less powerful nations. For instance, repressive dictators, such as Ferdinand Marcos in the Philippines, often have been supported by governments such as the United States because they were thought to be supportive of U.S. business interests.[34] Similarly the United States and other core nations readily participated in the Gulf War in the early 1990s because the loss of Kuwaiti oil was seen as a potential threat to their economies. By contrast, many of these same countries have been much more reluctant to join in other bloody conflicts around the globe because their economic interests were less affected.

World systems theory suggests that the global stratification system exists because it benefits the core nations; core nations are wealthy largely because they have been able to exploit much poorer societies. At the same time, Wallerstein and other proponents of world systems theory stress that the global stratification system is dynamic—it changes as the fortunes of individual nations change. For instance, Spain once ruled the high seas but lost that position to Great Britain several centuries ago. In turn, Great Britain eventually lost its vast empire and its dominance of world trade and industry. Japan rose from the ravages of World War II to become a dominant member of the group of core nations and one of the wealthiest countries in the world. As Wallerstein once put it, there can be a game of "musical chairs" among the nations as some gain dominance and some lose dominance, especially at the top.[35]

A number of sociologists and economists suggest that the global economic playing field recently has become somewhat more level or, at least, less dominated by only a few politically powerful countries.[36] Although there are still enormous differences in the economic wealth of the world's nations, a number of Asian countries, such as Korea, Singapore, and Taiwan, have made rapid economic progress.[37] Similarly many oil-rich nations in the Middle East have experienced a great deal of economic success.

Some of these changes have occurred because nations such as Korea have rapidly industrialized and modernized, as we will discuss more thoroughly in Chapter 19. Some of this greater equality among nations also reflects changes in the relationships among the less powerful nations. Countries that produce raw materials, such as oil or coffee beans, that are in wide demand throughout the world have joined together in **cartels,** organizations that seek to limit competition among the members and set the most profitable prices for their products. For instance, OPEC (the Organization of Petroleum Exporting Countries) includes members from countries as far-flung as Saudi Arabia, Kuwait, Nigeria, and Venezuela.[38] This cartel has been widely successful in controlling the price of oil and greatly enhancing the economic fortunes of its member countries. By acting as a joint body—by forming tight-knit webs of economic relationships—these countries wield much more power collectively than any one of them would individually.

Many other international organizations connect nations with the purpose of promoting economic development and cooperation. (See Table 14-1 for a list of these groups.) For example, in addition to OPEC and the EU, major international organizations include the Commonwealth of Independent States (CIS), composed of twelve former republics of the U.S.S.R.; the Commonwealth, composed of fifty-one nations that once were part of the British colonial empire; the Organization of American States (OAS), which has thirty-five member nations from the Americas and the Caribbean; and the Group of Seven, which includes the United States and six other highly industrialized nations.[39] Most of these groups link countries from specific areas of the world, although some, such as the Commonwealth and the CIS, unite countries that once were part of a much larger political entity. Because these groups develop economic policies and regulations, they help to establish expectations about how different countries will conduct economic activity. These groups set guidelines for such varied activities as the nets that will be used in fishing, the wiring that will be used in electric appliances, and the inspection standards that will be followed for fruits and vegetables. The policy determinations of these groups are the social actions by which nations build their webs of economic relationships.

Two specialized agencies of the United Nations also promote international economic networks. The International Bank for Reconstruction and Development, more commonly known as the World Bank, grants loans to member nations for the development of specific projects. When the Bank was first established after World War II, loans were mainly given to European nations to help reconstruct areas that had been damaged in the war. In recent years most money has been lent to the poorest nations in the world to finance the development of facilities that the developed world takes for granted such as safe water, sanitation, health care, education, and housing. The International Monetary Fund (IMF), governs the level of exchange rates, essentially determining how much one country's currency will be worth in the world market and thus the prices that can be obtained for its products.[40] Although organizations such as OPEC, the EU, and most of the other groups listed in Table 14-1 comprise nations with relatively equal power relations, the World Bank and IMF are structured in a way that mirrors the global stratification system. Wealthier countries have more authority than poorer countries in decisions, thus allowing these wealthier nations significant influence on the development decisions and economic well-being of poorer nations.

Even though we rarely think about these large-scale webs of economic relationships, they are important determinants of the well-being of people around the world. The extent of a society's economic relationships is a prime determinant of its economic fortunes and the health and well-being of its citizens. Each of us is affected in our daily lives by these relationships—in the goods and services we can purchase, the prices we pay for these products, and the jobs that are available to us. But the nature of these large-scale economic webs provides only part of the picture of societies' different economic opportunities and constraints. To understand more about this, we can move in closer and look at the ways in which economies are organized.

TABLE 14-1 *Major International Organizations Concerned with Economic Cooperation and Development*

Asia-Pacific Economic Cooperation Group (APEC)—includes 19 nations spanning Asia and the Pacific region such as Australia, Canada, Chile, Japan, Mexico, Singapore, and the United States. Headquarters are in Singapore.

Association of Southeast Asian Nations (ASEAN)—includes 9 nations in Southeast Asia, such as Brunei, Indonesia, Philippines, and Vietnam. Headquarters are in Jakarta.

Caribbean Community and Common Market (CARICOM)—includes 14 nations in the Caribbean, such as Barbados, Belize, Guyana, Jamaica, and Suriname.

Commonwealth of Independent States (CIS)—includes 12 of the former republics of the USSR, such as Armenia, Belarus, Georgia, Ukraine, and Uzbekistan. The capital is in Minsk, Belarus.

The Commonwealth—includes 52 nations, all of which are former members of the British Empire, as well as various colonies and protectorates, such as Australia, Canada, New Zealand, Papua New Guinea, Brunei, Pakistan, Swaziland, Western Samoa, and Zimbabwe.

European Union (EU)—15 full members, including such nations as Austria, Belgium, Denmark, Netherlands, Spain, Sweden, and the United Kingdom.

European Free Trade Association (EFTA)—includes 4 northern European countries: Iceland, Liechtenstein, Norway, and Switzerland. Many former members now belong to the European Union.

Group of Eight (G-8)—includes 7 major industrial democracies: Canada, France, Germany, Italy, Japan, the United Kingdom, and the United States. Russia was invited to join in 1997 in recognition of its political importance.

League of Arab States (Arab League)—22 member nations including such nations as Algeria, Bahrain, Iraq, Jordan, Kuwait, and Palestine. Headquarters are in Cairo.

Organization of African Unity (OAU)—53 member nations. Headquarters are in Addis Ababa, Ethiopia.

Organization of American States (OAS)—35 member nations from North, Central, and South America and the Caribbean. Headquarters are in Washington, D.C.

Organization for Economic Cooperation and Development (OECD)—29 countries, mainly from Europe, plus Japan, Australia, New Zealand, Canada, and the United States. Headquarters are in Paris.

Organization of Petroleum Exporting Countries (OPEC)—12 member nations including countries such as Algeria, Indonesia, Iran, Kuwait, Libya, Saudi Arabia, and Venezuela. Headquarters are in Vienna.

Note: Membership is given as of mid 1996.
Source: Famighetti, 1996, pp. 842–843; Wright, 1998, pp. 496–505, 707.

How has Internet technology affected economic opportunities and relationships?

INDUSTRIES AND ORGANIZATIONS: A MESOLEVEL VIEW

Sociologists who analyze the economy from the mesolevel often examine how the economy is organized. Much as we can study the different structures and patterns within a spider's web, sociologists look at the different divisions and patterns of relationships within a society's economy and the ways in which these change over time. These analyses can help us see how the characteristics of the industries and the organizations in which people work provide both economic opportunities and constraints.

Industries

Sociologists and economists use the term **industry** to refer to branches or areas of economic activity. In very simple societies, where everyone does similar work, such distinctions aren't really needed. But in societies such as our own, where people are engaged in many different kinds of activities, we can better understand how an economy functions by looking at the types of work that people do. More important, the types of available work within a society significantly influence individuals' economic fortunes. We can see these effects by looking at what types of industries exist in the United States and how the size of these industries has changed over time.

Industry Divisions and Opportunities Industries can be distinguished between those that produce goods (material things that we can consume) and those that provide services (often nonmaterial or perishable items or things such as restaurant food service, health care, and professional labor and consultation). Figure 14-1 uses this distinction to show how the U.S. Bureau of Labor Statistics (an agency of the federal government) categorizes industries. As you can see, there are three basic types of **goods-producing industries** in the United States. **Extractive industries,** such as agriculture, fishing, forestry, and mining, remove (extract) raw materials used in manufacturing from the environment (we will distinguish between mining and the other extractive industries). **Manufacturing industries** process raw materials into more usable forms. The **construction industry** uses raw materials to build homes, offices, roads, and other structures. Slightly more than one-fourth of Americans are employed in these goods-producing industries.

The rest of the country's population works in industries that provide services, although the term *service industry* often is applied in several ways. The Bureau of Labor Statistics uses the term to refer to the broad group of industries listed in Figure 14-1. However, it also uses the term to describe a smaller and more specific set of industries within this larger group. These more specific **service industries** can take many different forms, but essentially they provide various nonmaterial things people want or need, such as medical care, education, social welfare, and entertainment. This is the sector Sean hopes to work in, and it now employs more people than any other industry—more than one-third of the population. The next-largest industrial sector is **wholesale and retail trade,** involving the sale of goods to stores and directly to individual consumers. About one-fifth of Americans are employed in this sector. Slightly less than 20 percent of the labor force are involved in other service-producing sectors, including finance, insurance, and real estate, transportation and communication, and public administration.

The industry in which we are employed affects our life chances and opportunities. As you can see in Figure 14-2, unemployment rates vary from one industry to another. Your chances of being laid off or unemployed are much greater if you work in construction or agriculture than if you work in government, finance, insurance, or real estate. Average salaries also vary from one sector to another, as Figure 14-3 shows. Workers in mining, communication, and electric, gas, and sanitary services earn much

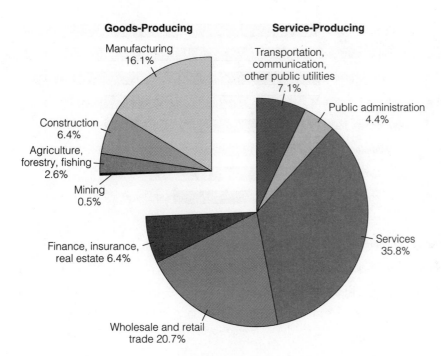

Goods-Producing

Manufacturing 16.1%

Construction 6.4%

Agriculture, forestry, fishing 2.6%

Mining 0.5%

Finance, insurance, real estate 6.4%

Wholesale and retail trade 20.7%

Service-Producing

Transportation, communication, other public utilities 7.1%

Public administration 4.4%

Services 35.8%

FIGURE 14-1 *Employment by Industrial Sector, 1997*

Source: U.S. Bureau of the Census, 1998, p. 421.

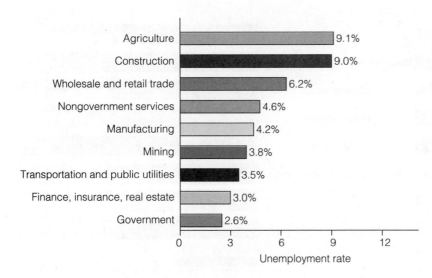

Agriculture 9.1%

Construction 9.0%

Wholesale and retail trade 6.2%

Nongovernment services 4.6%

Manufacturing 4.2%

Mining 3.8%

Transportation and public utilities 3.5%

Finance, insurance, real estate 3.0%

Government 2.6%

Unemployment rate

FIGURE 14-2 *Unemployment by Industry, 1994*

Note: The unemployment rate is calculated as the percentage of the labor force in each industry that is out of work.

Source: U.S. Bureau of the Census, 1998, p. 424.

more than the average worker does; those in agriculture, forestry, fishing, and retail trades earn much less.

Some of these differences reflect the fact that workers in some industries are more highly skilled, on average, than workers in other industries. But part of the difference reflects the fact that workers in some industries are more likely to belong to **labor unions,** organizations of workers who have joined together to collectively promote their interests. As shown in Figure 14-4, approximately 37 percent of government workers and 26 percent of workers in utilities and transportation belong to unions, compared with less than 10 percent of workers in wholesale and retail trade, nongovernment services, and finance, insurance, and real estate. More highly unionized indus-

tries tend to have higher wages for all workers, and within almost all industries workers who belong to unions earn more than those who do not.

It is important to realize that the advantages unionized workers now enjoy are a result of social relationships and social actions. Historically unions have been much more successful in organizing workers who share a common workplace and thus have more opportunities to share their concerns and views—in other words to establish strong social ties with one another. Unions typically have been relatively uncommon in service areas, where companies assign clerks to different parts of a store or office and sales representatives to different geographical areas. These workers therefore have had more trouble

FIGURE 14-3 *Average Annual Wages and Salaries by Industry, 1996*

Source: U.S. Bureau of the Census, 1998, p. 434.

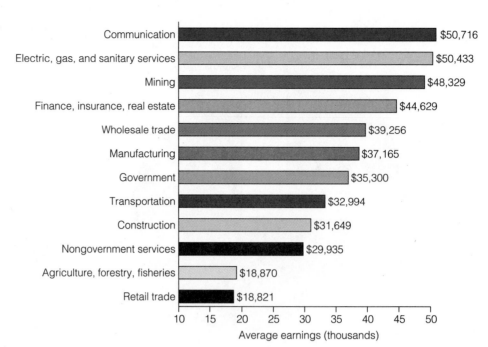

Average earnings (thousands)

- Communication $50,716
- Electric, gas, and sanitary services $50,433
- Mining $48,329
- Finance, insurance, real estate $44,629
- Wholesale trade $39,256
- Manufacturing $37,165
- Government $35,300
- Transportation $32,994
- Construction $31,649
- Nongovernment services $29,935
- Agriculture, forestry, fisheries $18,870
- Retail trade $18,821

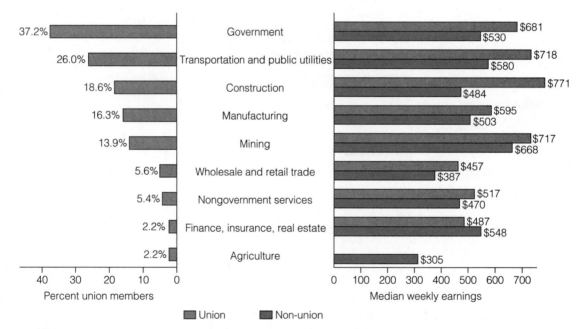

FIGURE 14-4 *Union Membership and Workers' Earnings, by Industry, 1994*

Note: For agriculture, union membership excludes owners and applies to wage and salary workers only; median weekly earnings for union members are not available for this group because data aren't provided by the government when a category has fewer than 50,000 people.

Source: U.S. Bureau of the Census, 1998, p. 444.

forming the social networks that can serve as the basis of union organizations.

Today we take for granted many of the improved working conditions spurred by the social actions of unions. These conditions have become *institutionalized,* part of the way in which we assume, and the state requires, that employers and work organiza-

tions will operate. The aspects of welfare capitalism that protect workers—the forty-hour/five-day work week, minimum wage laws, worker's compensation programs, worker safety legislation, and standardized grievance procedures—all came about through the efforts of unions. Sometimes the struggle to obtain these benefits resulted in protracted and bloody

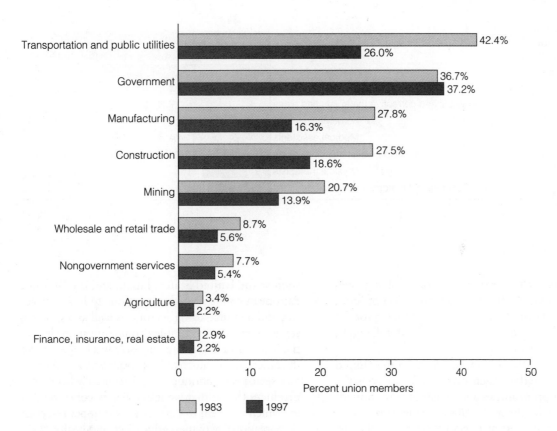

FIGURE 14-5 *Shifts in Union Membership, by Industry*
Source: U.S. Bureau of the Census, 1995, p. 445; 1998, p. 444.

conflicts with management, because wealthy capitalists did not want to give up the advantages that they accrued from paying low wages and ignoring costly safety provisions. Nevertheless, largely because of the power of collective action, the unions won many of their demands. Today virtually all workers enjoy the benefits derived from unions' efforts, whether or not they belong to unions themselves.[41]

In recent years the proportion of workers who belong to unions has declined—from 20 percent in 1983 to 14 percent in 1997. As shown in Figure 14-5, this decline has occurred in almost all industries. Some sociologists suggest that it reflects the fact that virtually all workers now enjoy the benefits that were originally gained by unions, and union membership is not seen as necessary to maintain them.[42] In addition, some industries have tried to discourage unionization and have moved plants to areas where unions are less common, such as the South and Southwest. Similarly, with the globalization of the economy and the growth of multinational corporations, manufacturers of products ranging from steel to computers to running shoes have moved out of the United States, establishing plants in countries where labor costs are much lower and labor unions are still quite rare.[43]

Part of the decline in union membership also can be traced to changes in the patterns of industrial employment. That is, there has been a decline in the proportion of the labor force working in industries that typically are unionized. Over time, the mix of industries in the United States, as well as in other industrialized countries, has changed considerably, in large part because of changing technology and increased productivity. Let's take a brief look at these changes and their effects on unions and workers.

Changes in Industries and Economic Opportunities As you will remember from previous chapters, in the mid 1800s more than half of all workers in the United States were employed in agriculture. Today less than 3 percent of the population works in this industry because machinery performs the work that humans and animals once did.[44] Technology has also reduced the need for workers in other extractive industries, with machines now efficiently extracting ore, hauling in fishing nets, and felling and processing timber.

As the proportion of individuals employed in agriculture and other extractive industries declined, other areas expanded. For instance, the proportion of workers employed in manufacturing grew from only 9 percent in 1840 to 27 percent in 1920.[45] Other industrial sectors, such as those associated with banking, transportation, trade, and education, also

FIGURE 14-6
The Projected Ten Fastest-Growing Occupations in the United States

Note: Data represent estimated growth of occupations from 1996 to 2006.

Source: U.S. Bureau of the Census, 1998, p. 420.

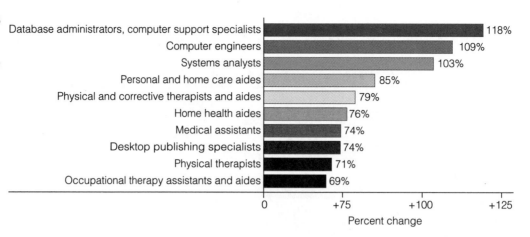

Database administrators, computer support specialists — 118%
Computer engineers — 109%
Systems analysts — 103%
Personal and home care aides — 85%
Physical and corrective therapists and aides — 79%
Home health aides — 76%
Medical assistants — 74%
Desktop publishing specialists — 74%
Physical therapists — 71%
Occupational therapy assistants and aides — 69%

Percent change

grew substantially in response to the evolving need for such things as efficient transportation of food, workers, and manufactured goods; financing for the building of factories and homes; places to sell goods; and training for workers.

Since 1970, the mix of industries in the United States has altered in other ways as well, affecting the economic opportunities and constraints of individual workers. By the mid 1990s a smaller proportion of American workers were employed in manufacturing industries than at any time since 1870.[46] Much of this change has resulted from the massive increase in productivity caused by changing technology and from the movement of many manufacturing jobs to other countries. The sharp decrease in the proportion of workers in the highly unionized manufacturing sector has been an important factor underlying the decline in numbers of unionized workers. In addition, because the expanding retail trade and service sectors often paid substantially less, workers displaced from manufacturing had difficulty finding jobs that paid nearly as well as those they had once held, with substantial numbers able to find only part-time work. This trend has been a key contributor to the decline in wages, adjusted for inflation, which was described in Chapter 7.[47]

The mix of economic activity continues to change today. You may have heard the term *postindustrial society* applied to the United States, suggesting that the nation somehow has gone "beyond" an industrial society. You might have also heard the term *information age* applied to contemporary society, referring to the amazing array of ways in which people can virtually instantaneously communicate, through computers, telephones, faxes, and so on. Although manufacturing is indeed a smaller part of the mix of industries than previously was the case and communications has become more important, most sociologists who study the economy use the term **advanced industrial societies** to describe societies—

such as the United States, Japan, and the Western European countries—that continue to have extractive and manufacturing sectors, as well as a growing service sector and a complex communications industry. The communications industry has helped make the other sectors much more productive, and the service sector can continue to grow precisely because it produces things that we immediately consume and for which we seem to have insatiable appetites, such as recreational activities, education, and health.[48]

As industries expand and contract, new jobs are created and others disappear. Figure 14-6 shows the U.S. government's predictions of the ten fastest-growing occupations through the beginning years of the twenty-first century. As you can see, the fast-growing occupations are in either the computer and communications technology industries or health care services. Fast-declining occupations involve either manufacturing industries or areas of the communications industry where work will largely be done by machines rather than by humans.[49] Much of the uncertainty that workers face today stems from such changes in industries worldwide. But the uncertainty, opportunities and constraints that workers face also come from the organizations in which they work.

Work Organizations

Sociologists often differentiate between **work establishments,** the actual place where someone works, such as a Gap store or McDonald's outlet, and **firms,** the parent company or organization, such as the entire Gap or McDonald's chain. More than half of all employees in the United States, especially those in the service sector, work in establishments that have fewer than one hundred employees.[50] But far fewer individuals, especially in manufacturing industries, work in small *firms.*[51] Some industrial corporations in the United States employ tens of thousands of workers. In fact, about a third of all workers are em-

TABLE 14-2 *Revenues and Number of Employees of the top 10 Fortune 500 U.S. Companies*

COMPANY	TOTAL REVENUES (IN MILLIONS)	NUMBER OF EMPLOYEES
General Motors	$178,174	608,000
Ford Motor Company	153,627	363,892
Exxon	122,379	80,000
Wal-Mart Stores	119,299	825,000
General Electric	90,840	276,000
IBM	78,508	269,465
Chrysler	61,147	121,000
Mobil	59,978	42,700
Philip Morris	56,114	152,000
AT&T	53,261	128,000

Source: http://www.pathfinder.com/fortune/fortune500/500list.html and http://cgi.pathfinder.com/fortune/global500/index.html. (December 21, 1998).

ployed in what are called the Fortune 500 companies, a listing of the largest industrial and service companies that *Fortune* magazine compiles each year. Table 14-2 lists the total assets and number of employees of the ten largest Fortune 500 companies. As you can see more than half a million people work for General Motors, and more than three-quarters of a million people work for Wal-Mart.

These large corporations are extremely powerful and can wield enormous influence on the political world. Much of their power stems from their immense size and vast wealth. This corporate size and power have increased over the years, primarily through the merger of different companies. In analyzing work organizations, sociologists have studied how mergers have changed the structure of business organizations and how these changes have affected businesses, individual workers, and entire societies.

Corporate Mergers and Changing Webs of Economic Relationships Almost weekly, the newspaper's business pages report the possibility of mergers of businesses around the country, often involving multimillion-dollar transactions. Mergers have been a part of the U.S. economy for more than a century. Over time the nature of these mergers has varied, reflecting different types of social relationships between organizations and different institutional norms regarding the shape and structure of business organizations. Yet, no matter what form mergers have taken, they have altered the web of economic relationships between businesses, consumers, and workers.[52]

In the 1890s and early 1900s, many corporate mergers resulted from technological breakthroughs that made large-scale manufacturing much more efficient and profitable than smaller operations. Larger firms bought smaller companies in the metallurgy, chemical, electrical machinery, tobacco, and food processing industries. These acquisitions removed a number of firms from the network of organizations and resulted in such large corporations as Westinghouse, International Harvester, Standard Oil, American Can, American Tobacco, and United States Steel.[53] These mergers also resulted in monopolies, in which one company controls an entire market. For instance, when John D. Rockefeller either bought out or crippled all his competitors in 1910, his Standard Oil Company controlled more than 80 percent of the oil refining capacity in the United States. As a result, Standard Oil dominated the entire industry and could set virtually whatever oil prices it desired.[54]

The abuses of monopolistic companies such as Standard Oil eventually led to the passage of anti-monopoly legislation. But in the 1920s a new type of merger began to become more common. In this type of structural change, called **vertical integration,** firms acquired other companies that supplied raw materials for manufacture, and the firms either did further processing or sold their products. These mergers produced companies, such as Bethlehem Steel and Republic Steel, that could control the entire extraction, manufacturing, and distribution process.[55]

From the time of the Great Depression through the early 1950s, a number of companies pursued a strategy of **diversification,** expanding their product lines and buying companies that produced related products.[56] These changes produced the first corporate giants, such as du Pont, which began as a family partnership that made munitions and explosives. During World War I, the company's profits soared, and over the next few decades Pierre du Pont, the head of the company, used these profits to buy other companies that produced paints, dyes, and chemicals, creating a large corporation that specialized in many different aspects of the chemical industry.[57]

In the late 1960s and 1970s, the nature of corporate networks altered as industrial **conglomerates** began to appear through the merger of firms with entirely different types of products. Instead of pursuing growth within a single industry, as was characteristic of earlier mergers, conglomerates involved the merger of firms in a variety of industries. For instance, in the early 1970s, U.S. oil companies earned record profits as a result of sharp increases in oil prices. They used these profits to purchase firms in totally different areas of the economy. For instance, Gulf Oil bought the Ringling Brothers and Barnum and Bailey circuses; Mobil Oil purchased Marcor

Corporation, which owned the Montgomery Ward stores and Container Corporation of America.[58] The result of this trend was even greater diversification with firms spread across many different industries

After 1980 institutionalized norms regarding business forms began to change and many conglomerates sold off firms that were perceived to be less central to their mission. The government was showing little interest in enforcing laws that restricted mergers and the development of monopolies, and corporations began to acquire and merge with firms in the same, or closely related, industries. The rate of acquisitions and the amount of money involved was higher than at any previous time, resulting in what has been called a period of **megamergers,** with the consolidation of some of the largest companies in the economy. By the late 1990s, corporate mergers and acquisitions involved more than 5,000 firms a year and transactions of more than $1 trillion, over five times the figures from a decade earlier. Many of these mergers have involved "hostile takeovers," in which a company was bought without the seller's consent.[59]

The result of such megamergers can be seen in all types of industries, from banking and finance to publishing, automobile manufacturing, and the entertainment world. The trend has resulted in corporations that control vast elements of just one industry. For instance, the merger of Capital Cities/ABC and the Walt Disney Company produced a huge corporate structure that includes the ABC television network; individual television stations, which reach 25 percent of the nation's households; a radio network; more than two dozen radio stations; several syndicated television shows; cable channels such as ESPN, the Disney Channel, A & E, the History Channel, and Lifetime; and a number of magazines.[60] Citicorp, a very large financial firm, includes a vast array of banks, brokerage services, and insurance and finance companies.[61]

Many of these recent structural changes in business corporations have also involved the expansion of multinational corporations. Disney has branches in Canada, France, Germany, Italy, Great Britain, and Japan.[62] The merger of Chrysler Motors and Daimler-Benz, the German company that manufactures Mercedes-Benz automobiles, joined very large auto manufacturing firms in two continents. Even the firm that published this book is part of a huge corporation that owns newspapers throughout North America and many publishing firms in Great Britain, Canada, the United States, and other areas of the world. But how do these changing corporate structures affect businesses, individuals, and societies?

Changing Corporate Structures and Opportunities and Constraints Corporate mergers have almost always helped businesses gain a competitive advantage in the web of economic relationships, either by removing competitors or providing control of other organizations crucial to a firm's economic success. In addition, the more diversified a company is, the better protected it is from temporary setbacks in the economy. Mergers and overseas investments can provide tax advantages and other benefits as well. For instance, the company making the purchase in a merger can receive tax write-offs, or it can acquire a company with a steady stream of income that provides needed cash.[63] Multinational companies often can take advantage of cheaper labor and raw material costs in developing countries, as well as tax advantages associated with an overseas location. Further, the immense wealth and size of conglomerates and multinational corporations greatly increase the probability that they can influence government decisions—not just in the United States but, because of their linkages around the world, in many other countries as well.[64]

Although corporate mergers are almost always profitable for companies and stockholders, they do not necessarily benefit workers, consumers, or even society. Large corporations clearly are an important source of employment for people around the world. Larger firms also tend to pay their employees more than smaller firms do, and they usually have more formalized hiring and grievance procedures, thus helping to equalize individuals' opportunities. But sociologists who study these large firms also point to dysfunctions associated with their sheer size. On the individual level, the highly bureaucratic nature can diminish workers' identification with their enterprise, contributing to the process of **alienation.** This term, first used by Karl Marx, refers to the separation of workers from the product or result of their work, which can result in feelings of powerlessness.[65]

In addition, the development of conglomerates frequently has resulted in local establishments and whole firms closing, causing severe unemployment, changes in workers' benefit packages, and loss of seniority. In recent years a number of corporations in all kinds of industries, including computers, banking, communications, and manufacturing, have attempted to ward off possible takeovers by "downsizing," or dramatically cutting the number of their employees. This sudden loss of jobs can, of course, be devastating to individual workers, their families, and to the communities in which they live.[66] Ironically, however, while many employees lose their jobs because of downsizing, in some cases the value of the company's stock has risen dramatically when

Film composer Henry Mancini is shown here in the early days of his career when he was employed in the Hollywood studio system. Like other composers, when the studio system disappeared, Mancini turned to freelance work and, through his earlier successes and strong social networks, built a successful career.

it has laid people off, yielding large profits for company owners and stockholders, many of whom belong to the capitalist class.

Even though multinational corporations introduce employment opportunities to the countries in which they invest, they have been criticized for contributing to uneven development in countries that are still industrializing. For instance, many multinational corporations obtain local financing for expansion and local operations, thus reducing the amount of money available to local corporations. Similarly, large multinational agricultural corporations encourage growing crops for export rather than for domestic use, thus contributing to a scarcity of food and higher food prices for the local population.[67]

Finally, political commentators and scholars have expressed concern that the growth of massive conglomerates will lead to a decline in competition and to increased hegemony of big business. For instance, the merger of ABC/Capital Cities and Disney joined two very large media companies, putting the ultimate control and power of large portions of the media into the hands of just one corporation. This conglomer-

ate potentially can control not only the work lives of its employees but also the content of much of the country's entertainment and news reports. The merger of ABC/Capital Cities and Disney greatly increased the corporation's potential power to act as a **cultural gatekeeper.**

The movie industry that Sean wants to enter has experienced many of the structural changes that have characterized other industries in the United States. In the Golden Age of Hollywood, which lasted into the 1950s, the industry was characterized by vertical integration. Large movie production companies controlled all phases of the industry, from the production of films to their distribution in corporate-owned local movie theaters. Composers, directors, musicians, and all the other people who worked on movies were under contract to one of the big studios, such as MGM, Paramount, or Twentieth-Century Fox. The movies were then sent to local theaters, which were owned by these companies. In short, all facets of the movie industry—production, distribution, and consumption—were controlled by the big studios.

In 1948 the Supreme Court ruled that this total control was illegal and forced the movie companies to sell the local theaters. Around the same time, with the emergence of television, fewer and fewer people went to the movies. Eventually the vertically integrated companies and the studio system disappeared, and in recent years huge conglomerates gradually have absorbed the large movie companies, such as Paramount, Universal Studios, and Twentieth-Century Fox. For instance, Time Warner owns the Warner Brothers studio, as well as several magazine and book publishing firms, and television cable channels such as HBO, CNN and TNT.[68] Paramount is part of a huge conglomerate that includes publishing giants such as Simon and Schuster and Prentice-Hall.

Today, even though movie companies are owned by huge conglomerates, they generally don't make films themselves, but instead publicize and distribute films made by independent producers. These independent producers contract with actors, directors, screenwriters, composers, and all the other people needed to make a movie. As with all workers, employees in the movie industry have been significantly affected by changes in the industry and the studio system. Yet an individual's experiences within the economy—the possibility that someone like Sean will succeed or fail—also are influenced by his or her own individual web of economic relationships. We turn now to research that has used a microlevel analysis of the development of work careers.

FEATURED RESEARCH STUDY

WORK AND CAREERS AS SOCIAL RELATIONSHIPS: A MICROLEVEL VIEW

When we use a microlevel analysis to look at the web of economic relationships, we can see that our everyday economic transactions—such as going to work or school, filling the car with gas, buying milk and bread at the store, or getting a haircut—involve social relationships. Through our actions and decisions we establish economic and social ties with some people and not with others. For example, you may decide that you prefer one mechanic to another or one hairstylist over another. Over time these actions and decisions produce the social relationships that provide the basis for the economy.

Our occupations and work careers are probably the most important influence on how each of us will fare in the economy. In our discussion of the life course in Chapter 4, we defined a **career** as a sequence of roles that we enact during our lifetimes. In our working lives, most of us move through a number of jobs and even careers. With the recent trend toward corporate downsizing, some analysts believe that the frequency of career moves will increase. More and more, individuals could find it unusual to remain with one employer throughout their work career; instead, they will make a number of transitions.

When we think about our work careers, we might assume that they reflect our own hard work, skills, and talents. We also probably assume that they involve luck—being in the right place at the right time, talking to the right person, happening to have the skills that an employer needs, and making certain decisions along the way. For instance, Sean managed to get his summer jobs in Hollywood through Professor Sherman's former students. He got to know Professor Sherman by taking the "Music in Film" class and doing well in the course. He continued getting jobs for his band because he happened to be a highly skilled musician.

As you might imagine, sociologists tend to look at work careers in a slightly different way. They suggest that if we examine the careers of a large number of people, and not just one individual's work career, we can begin to identify general patterns and develop understandings about ways in which all work careers tend to develop. In particular, we can see that they involve a series of social relationships and that these relationships have a strong effect on the kinds of jobs we manage to get.[69]

Robert Faulkner used these ideas to study the careers of film composers.[70] His analysis shows how our work careers can be understood through examining the social structures that we live in and the social networks that we construct and maintain with our friends and colleagues. His work also shows how our social actions help to shape and maintain our careers, including our interpretations of our interactions with others, our choices, and the ways in which our social networks and environments influence these choices. In turn, these choices and networks help create the social structure by helping to maintain a highly stratified occupational system.

Very few of us will pursue careers as composers, but the insights Faulkner developed can also be applied to our own work experiences. His analysis can help us see how individual webs of economic relationships provide both opportunities and constraints on our work careers and possibilities for economic success.

Studying Composers' Careers

The entertainment business is booming, and hundreds of movies are being produced each year. Yet an aspiring composer like Sean faces a great deal of uncertainty. This generally wasn't the case during the Golden Age of Hollywood, when the studio system reigned and composers were under contract. Once hired, they would be assigned to various projects. Not only would a studio contract provide budding musicians with dependable incomes, but studios also provided convenient settings in which to interact with and learn from seasoned professionals. Aspiring composers could learn their craft by helping to arrange and orchestrate music from a piano score. They might also be assigned to work as part of a composing team for one of the studio's movies. Once the studio music director became aware of a composer's talents and skills, aspirants would be given more complex assignments and, eventually, a chance to score a film by themselves. Sean's career is more uncertain because musicians and composers today generally work as freelancers. They contract with individual film producers to compose a film score and, usually, to hire and conduct the orchestra.

Although many people think of their careers as involving moves from one job to another, perhaps as a teacher in one school and then a teacher in another, freelance composers' careers involve a series of credits, or film projects. If Sean has a "big career" in the movie business, he will have a long line of credits; if not, his line of credits will be much shorter.

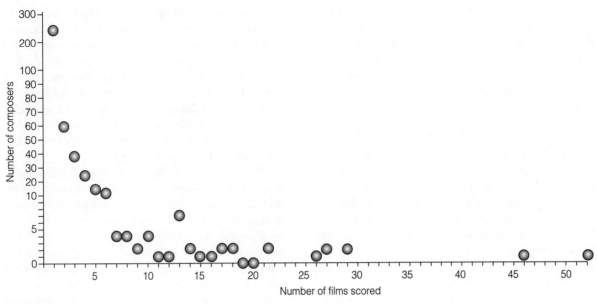

FIGURE 14-7 *Scatterplot of the Number of Films and Film Composers, 1964–1976*
Source: Faulkner, 1983, p. 28.

Faulkner conducted his research on the film industry in the late 1970s and early 1980s, long after the Golden Age of Hollywood had ended, so the freelance system had been around for many years. A musician as well as a sociologist, Faulkner used a variety of techniques to study composers' careers. At first he was interested in the ins and outs of working as a composer in the movies and the problems composers faced in developing their creative talents within a business community. To explore these issues, he spent two summers doing field research in Hollywood, interviewing composers and following them around to learn more about how they did their work.

From these interviews Faulkner became interested in the way the composers' careers were linked to the projects that they worked on and the other people involved with these projects. He combed issues of *Daily Variety,* the newspaper that chronicles events in the movie industry, and books such as the *International Motion Picture Almanac* to amass information on all the films produced from 1964 to 1975. He noted the producers, directors, and composers associated with each of these 1,355 movies, as well as any awards the movies had received.[71]

These films were scored by 442 different composers, but more than half of these composers (252 to be exact) scored only one film (see Figure 14-7).[72] Faulkner described his results:

> I found that most people who work in feature films work very little and that a small number of visible freelancers are responsible for composing far more than their proportionate share of scores for Hollywood's films. Ten percent of the population is re-

sponsible for one of every two industry projects, and only a small percent of the composers in Hollywood consists of highly productive, central figures. A few names have enormous credit lists; a great many names have only one, two, or three credits.[73]

A similar pattern appeared with Faulkner's data on producers and directors. Half of the people who make a film in Hollywood never make another one, but a small number of people make a great many.[74]

Highly successful producers, directors, and composers are linked with one another, frequently working together on the same projects. You can see this yourself when you watch movies. Directors like Steven Spielberg and George Lucas or composers like Quincy Jones and John Williams appear again and again. The result is an industry that is highly stratified: A few people become famous multimillionaires, but most do not.

Based on his earlier interviews and the data in Figure 14-7, Faulkner decided to focus on learning more about the careers of composers in different parts of this stratification system. How, he asked, do composers develop social networks that allow them to become successful in the film industry? How do these processes affect the way in which the industry is stratified? He compiled a "filmography," or list of the film credits, of every composer who had scored a film since 1964 and returned to Hollywood for a third summer of fieldwork. Armed with the filmographies, he systematically interviewed composers with different numbers of credits. Some had only one film credit and were on the *periphery,* or edge, of the industry; some worked continuously in films and thus were at the *core,* or center, of the industry, and some were in between. In these interviews Faulkner used the filmographies to get details from the composers about each step in their careers—how they got the jobs, whom they worked with, and also how they felt about their work.[75]

Entering and Moving in Career Networks

Because so few composers actually have sustained careers composing for films, they must make a living in other ways. Just as aspiring movie actors work in television, so aspiring film composers often work in the production of television movies and series, hoping that they will somehow break into films. Such famous movie composers as Henry Mancini, Quincy Jones, John Williams, and Marvin Hamlisch all worked in television at one time or another.[76]

As in the movies, television composers are freelancers participating in a highly stratified system. Savvy television producers don't want to take chances with unknown talents, so they usually turn to well-known composers to write the theme songs for new shows and to score the initial episodes of new series. But these leading composers also tend to be busy people who receive more job offers than they can handle. When this happens, they seek help from trusted colleagues who are less in demand. After writing the theme song for a television series, for example, a leading composer might ask an up-and-coming colleague to score the weekly episodes; the newcomer will then build on the composer's basic themes and style to match the specific drama and action of each episode.

Social networks are vitally important in this process. The more relationships that fledgling composers have with established composers, and the better their musical reputation, the greater the chance that they will be asked to help on a project. These relationships sometimes evolve into "sponsorships," whereby leading composers help newcomers learn the ropes, pass work on to them, and introduce them to colleagues. This sponsorship process always goes from the core to the periphery. Those who have made it can pick and choose from those on the fringes, who in turn may begin to gain access to contacts that will help them penetrate the network. But this process is slow and uncertain. Fledgling composers may spend years—even their entire careers— on the fringes, trying to make the contacts that will lead to scoring big films. Like all economic relationships sponsorship is based on trust. Sponsors trust that freelancers will not embarrass them by doing inferior work. At the same time, newcomers must trust their sponsors to pay them as promised and to acknowledge their contributions and help them move up in the field.

Gradually fledgling composers learn the ins and outs of the business, in what sociologists call **occupational socialization,** the process of learning and identifying with the norms and roles associated with a particular occupation. Just as socialization in other areas of our lives involves learning the roles associated with the statuses we hold in our social networks, occupational socialization involves coming to understand the roles associated with the web of economic relationships in our work careers.[77]

Part of this socialization process for film composers involves learning the technical aspects of writing movie music—how to match the timing and style of music to visual segments, how much time to spend on a project, what kind of orchestra to use in different types of work, and how much to charge for the work.[78] A second aspect of socialization is learning about the entertainment industry and the composer's role within that industry—learning that commercial success is the most important goal of filmmaking, that a film's music is only one part of the entire enterprise, and how one can communicate with movie producers and directors, many of whom know very little about music.[79] Finally, occupational socialization involves developing a self-identity as a film composer and reconciling the demands of the work with one's creative desires as a musician. Fledgling composers must learn to integrate the aesthetic and commercial aspects of their work. They might be capable of writing beautiful, complex fugues and symphonies, but they also realize that if they want to be film composers, they will have to write music that pleases producers and fits the movie.

Penetrating Hollywood's Social Networks

Like all webs of economic relationships, the network of Hollywood composers is competitive. There are far more aspiring composers than movies to be made, and only a few make it to the top. Based on his interviews, Faulkner concluded that the musical abilities of composers weren't as important in their ultimate career success as their social networks—whether or not they managed to hook up with successful projects, effective sponsors, and thriving producers.

The experiences of one of Faulkner's respondents and his colleague illustrate this phenomenon. The respondent was once hired by a producer to whom he had been introduced by a friend to score the pilot for a new, high-status television series. At the same time, another composer had been hired to score the pilot for another series just purchased by the network. The respondent described what happened:

> We started on two television shows with two new composers. My office was rather large, with wood

paneling—but his office didn't even have his name on the door. It had a sign on the door that said, "Sprinkler Drain"—inside were the sprinkler hose and the drain. So for six months or so I had the wood-paneled office. Then my show folded and he moved into the paneled office. He's been there ever since—very successful. His show was a hit and from that assignment he got a big-budget film and a lot of other things. This is not to say his music for the TV show was not as good as the music I did for my show, but it turned out to be music for an enormously successful show—it's still going strong—and that's the kind of thing Hollywood understands.[80]

Freelancers have little control over the success of a series. Rarely does the music make a difference in whether a show is a winner or a loser. And freelancers realize that their work is a gamble—they can't tell whether a show will hit big or bomb. They can do two things, however, to help minimize the risk, both of which involve the nature of their social networks.

First, they can cultivate diverse and broad-ranging relationships. The more contacts a composer has, the greater the chance of being called on to help someone out or to be viewed as a good risk when an opportunity opens up. Diversity of contacts also helps keep composers from being typecast, or regarded as suitable for only one kind of music or film. For example, a composer might have been very successful in scoring rock-and-roll, beach-blanket movies. But if producers only identified the composer with this genre, he or she could have difficulty finding work when such films began to fade from the scene.[81]

Second, successful freelancers gradually move toward the center of the social network. Because most of the movies are made by only a few of the producers and directors working in the industry, it is far better to be linked to one of these successful moguls than to one on the fringe. Of course, producers are the ones who have the money and who decide whom they will hire. When possible, they prefer to hire people who have already scored a money-making film or whose work they know. As a result, fledgling composers tend to work with fledgling producers, and superstar composers work with superstar producers. To make it big, composers in the middle segment of the hierarchy—those who might have scored one or two films but haven't yet penetrated the core—somehow need to hook up with a producer who is on the way up and, as one composer put it, "get into a position where they want you."[82]

John Williams is one of the most successful composers working in Hollywood today; his career illustrates the importance of both creative diversity and network penetrations. Williams began his career in the 1950s, working for television series. In the late 1950s and 1960s, he began orchestrating movies such as *Gid-*

get, Diamond Head, and *Valley of the Dolls,* which was the first of his movies to be nominated for an Academy Award. In the 1970s Williams began to hit the big time, scoring films such as *Jaws, Close Encounters of the Third Kind, Star Wars, Superman, Raiders of the Lost Ark,* and *E.T.* and receiving many more Oscar nominations and awards. In 1980 Williams became the director of the Boston Pops orchestra but continued his work in film, scoring such movies as *Jurassic Park, Schindler's List, Amistad,* and *Saving Private Ryan.* His career exemplifies the two elements that Faulkner found were crucial for success. First, his credits are diverse, ranging from adventures such as *Return of the Jedi* and *Jurassic Park,* to dramas such as *JFK* and *The Big Chill,* to comedies such as *Hook* and *Home Alone.*[83] Second, he moved gradually from the periphery of the industry in the early years of his career to the core. In recent years he has consistently worked with some of the best-known directors and producers, such as Steven Spielberg and Oliver Stone.

Based on Faulkner's research, we could expect that Sean's opportunities in Hollywood—the possibility that he will succeed as a film composer—will be related not just to his talent but also to the nature of his social networks. The more diverse his networks are and the more they can become oriented to the core of the industry, the greater the probability of his success.

Of course many people do not work in areas that are as risk-laden or as dependent on freelancers as the movie industry. But no matter what career we choose, our economic relationships and our career success will depend largely on the nature of our social networks. Even though job seekers routinely comb the want ads and visit employment agencies, many studies have revealed that over half of all workers get their jobs through personal contacts. For instance, you may hear about a job through a friend of a friend and decide to apply. Or a relative may tell her boss about your skills and your suitability for a position; based on that recommendation, you have an inside track to the job. Just as with film composers, the more diverse your social networks, and the more people who know that you are looking for a job, the better your chances of finding one. In addition, as with the film composers, having a better job is related to the success of other members of your social network. If your friends hold high-status positions, you are likely to hear about high-status jobs; but if your friends primarily hold menial positions, chances are they won't be able to help you find positions that are much better.[84] One of the first sociologists to analyze the importance of social networks in promoting economic opportunities was Mark Granovetter. To learn more about Granovetter and his work see Box 14-1.

SOCIOLOGISTS AT WORK

Mark Granovetter

I found myself reading sociology for pleasure and decided to make that pleasure a lifelong preoccupation.

Mark Granovetter, Stanford University, is the author of *Getting a Job: A Study of Contacts and Careers,* **one of the first books to document the central role of social networks in developing careers.**

Where did you grow up and go to school? I grew up in Jersey City, New Jersey, and attended a large, urban, multiethnic, predominantly working-to-lower-middle-class school. It was certainly a sociological education growing up in this area, though I didn't know it at the time! To me, it was just the place I was growing up, and like any other such place it seemed to be the most normal place in the world. Outside observers like to refer to places like Jersey City as "gritty industrial towns," and it wasn't long after going to college that I became aware that my hometown was indeed one of the most famous in this category. But that's never the reality for those living in such a place; it just seemed like home, benign and welcoming.

Why did you become a sociologist? I went from Jersey City to be an undergraduate at Princeton, which might seem like a culture shock, but somehow I didn't notice. After all, they were both in New Jersey! Although I majored in history, I found that what I enjoyed most was thinking about the patterns of behavior and institutions that were common across times and places. Rather than spend my life studying the French Revolution, for example, I thought it would be more interesting to study the causes of revolutions in general. At some point I became aware that the study of such general patterns is called sociology. It was neat to find out that the thing I liked was actually a known field of study. I found myself reading sociology for pleasure and decided to make that pleasure a lifelong preoccupation.

What has surprised you most about the results you have obtained in your research? Perhaps the biggest surprise was how tenuous the ties often were that led people to jobs. I realized that your close friends, even though they might want to help, tend to know the same people you know, and so they are giving you the same information you already have. Your acquaintances, on the other hand, are more likely to move in different circles from your own, and thus to have different information. They are really your "window on the world." Partly as the result of this

---

Common Sense versus Research Sense

COMMON SENSE: The best way to "network" and find out about new job opportunities is to go through your immediate circle of close friends.

RESEARCH SENSE: Granovetter's research suggests that your *acquaintances,* rather than your close friends, are more likely to know about opportunities that can lead you to a new job. Your close friends tend to know the same people you do and aren't as able to provide information you don't already know, whereas acquaintances move in different social circles and are thus privy to different information.

The sociological analyses we have reviewed in this chapter suggest that the economy, as a social institution, affects individuals and societies in many different ways. From a macrolevel perspective we have seen how webs of economic relationships between nations and communities influence their relative wealth and power. From a mesolevel perspective we have seen how changes in the composition of industries within a society and in the form of work organizations affect the welfare of communities and societies, as well as individuals' economic opportunities and constraints. And from a microlevel perspective we have seen how our individual social networks also influence the probability of our own economic success. Whatever level

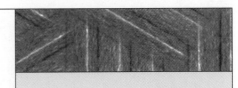

BOX
14-1

SOCIOLOGISTS
AT WORK

Continued

insight I published a paper in 1973 called "The Strength of Weak Ties." I am told this is now one of the most-cited papers in sociology, which reflects how interesting and important processes in social networks are.

What do you think the implications of your work are for social policy? One thing my work shows is that even though the movement of jobs from the inner city to the suburbs is a problem for the poor, the presence of jobs in the right place doesn't in itself guarantee that every group will have access to them. Since most jobs are filled through networks, groups with the wrong networks will have trouble even if the jobs are right next door. This is especially so since ethnic groups within workplaces often try to reserve jobs for their own group. And since employers like to hire through their current employees, because they trust the information about new prospects that they are getting this way, some groups can be excluded even without any conscious attempt at discrimination. One positive consequence of affirmative action rarely mentioned is that once a previously excluded group gets a foothold in a workplace, it can reproduce itself there without further assistance because it will now bring in new workers through referrals. Another lesson from my work is that job training programs that don't connect well to networks of employers

will probably never make much impact. If you just train people and set them loose, they still lack the contacts that will get them into the workplace.

How does your research apply to students' lives? Do you have specific advice for them on entering the job world? Because people build up networks over long periods in a career, personal contacts, at least in jobs above the blue-collar level, are probably going to be more useful later than at the very beginning. And because the contacts that are useful are often acquaintances that you may not even have seen for years, all the advice about "networking" that you may hear is probably of little or no value. "Investing" in contacts in this style—for example, some of the seminars tell you not to let a day go by without meeting five new people at your workplace—is too artificial. What are all these people going to think?

They will know you are just trying to "network," and people are always on the lookout for others who only want to use them. The most useful contacts are ones you come by naturally. But since even weak ties can be so useful, good general advice would be to keep up as many of your ties as you can, but without being crass about it, which is just counterproductive. Just be friendly and sociable, which is probably a good way to enjoy life anyway.

---

of analysis we have taken, however, the analyses suggest that social units—whether nations, communities, business organizations, or individuals—tend to have greater economic success and power when they have more extensive networks and are more centrally placed within these networks. Some sociologists also have applied these general ideas to examine the economic opportunities and constraints that have faced members of racial-ethnic minorities (see Box 14-2, "Applying Sociology to Social Issues").

CRITICAL-THINKING QUESTIONS

1. Can you think of any economic areas where social networks would not be an important contributor to economic success?

2. What methods, besides those Faulkner used, might a researcher use to study how social networks influence the relative success of film composers?

3. How could you apply the techniques that Faulkner used to study other occupations and other industries?

The sociologist Alejandro Portes and his colleagues have specialized in the study of the economic activities of immigrants to the United States. One of his important findings involves the documentation of **ethnic enclaves,** areas in which there is a great deal of entrepreneurial business activity by members of particular racial-ethnic groups. For instance, Koreatown, near downtown Los Angeles, is a well-developed area of grocery stores, restaurants, banks, import-export houses, industries, and real estate offices, all operated by Korean immigrants. Similarly, near downtown Miami, Little Havana is a five-mile stretch of almost 30,000 different Cuban-owned firms, including enterprises ranging from manufacturing and construction to finance and insurance.

Such ethnic enclaves appear to provide important benefits to ethnic communities. They provide employment opportunities for community members, and as profits are reinvested, the immigrant groups begin to gain

greater economic freedom. Although ethnic enclaves generally "start small," using savings or pooling funds from family members to provide start-up money and tapping family members and fellow immigrants for the required labor, many such enclaves have become quite successful. For instance, estimates are that the Cuban business community in Miami does close to two-thirds of the residential construction in the area, and the profits of Cuban-owned manufacturing firms increased over 1,000 percent during a recent ten-year period. Ethnic enclaves also can promote ethnic solidarity by enabling members of some racial-ethnic groups to conduct many of their economic transactions within their own ethnic community.[85]

However, there may be dysfunctions associated with ethnic enclaves as well. For instance, some scholars have found that immigrants can earn higher wages when they work outside the ethnic enclave and that employment outside an enclave can promote wider social networks and thus greater career opportunities.[86] To the extent that ethnic enclaves promote separation of racial-ethnic groups, they also may diminish contact between different groups, a key element in promoting better intergroup relationships.

**CRITICAL-THINKING
QUESTIONS**

1. Are there ethnic enclaves in your community? Do you know people who work in ethnic enclaves, or have you or members of your family worked in one? What were the advantages? The disadvantages?

2. Might there be ways of retaining the benefits of ethnic enclaves while minimizing their dysfunctions?

SUMMARY

The economy, as a social institution, involves all the norms, organizations, roles, and activities related to the production, distribution, and consumption of goods and services. It is also an important influence on the nature of social stratification, providing opportunities and constraints for individuals and entire societies. Key chapter topics include the following:

- Although sociologists and economists once tended to study different areas of the economy, these differences have diminished in recent years. Contemporary sociologists emphasize that eco-

nomic exchanges and markets involve social networks and social relationships.

- The nations of the world differ markedly in their per capita GNP, resulting in a system of global stratification. Differences can also be seen in the economic fortunes of states and local communities, a system of regional stratification.

- Natural resources and geographic location influence the economic well-being of societies and communities through their effect on the ability to develop wide-ranging economic relationships.

Although technology can be used to overcome problems associated with the lack of natural resources and adverse geographic locations, very poor societies often are unable to afford modern technological advances.

- The political economy, or relationship between the political and economic world, affects economic relationships both within and between societies. Some scholars suggest that many countries are moving toward a mixed political economy, which incorporates elements of both socialism and capitalism.

- The greater power of the richer, industrialized nations has been traced to earlier eras of imperialism and neo-imperialism. World systems theory suggests that core nations' wealth is maintained through their domination and exploitation of peripheral nations, from which they purchase raw materials for use in manufacturing. In recent years multinational corporations have become more common, establishing manufacturing plants in countries where labor costs are relatively less expensive.

- About one-fourth of workers in the United States are employed in goods-producing industries; the rest are employed in service-producing industries. Industries differ in their rates of unemployment, average salaries, and extent of unionization. In recent years the proportion of the work force that is unionized has declined. The mix of industries also has changed, with a decline in manufacturing and a rise in service industries.

- Over the years business organizations have merged and expanded in several ways: through the creation of monopolies, vertical integration, diversification, the development of conglomerates and, most recently, megamergers. Although these mergers have generally increased business profits, they have not necessarily always benefited individual workers, communities, or society as a whole.

- In his study of the careers of film composers, Robert Faulkner showed how individuals' social networks influence their economic success. Composers who have more diverse networks and who are able to move to the center of the networks of filmmakers are more successful.

KEY TERMS

advanced industrial societies 376
alienation 378
capitalism 367
career 380
cartel 370
conglomerate 377
construction industry 372
convergence theory 368
core nations 369
cultural gatekeepers 379
deforestation 366
democratic socialism 367
desertification 377
developing countries 366
diversification 366
economic exchanges 361
economy 360
ethnic enclave 386
external area 370

extractive industries 372
firm 376
global economy 363
global stratification system 363
goods-producing industries 372
gross national product (GNP) 363
imperialism 369
industrialized countries 366
industry 372
labor unions 373
laissez-faire capitalism 367
least developed countries 366
manufacturing industries 372
market 361
megamergers 378
mixed economic system 368
monopoly 367
multinational corporations 369

neo-imperialism 369
occupational socialization 382
per capita GNP 363
peripheral nations 369
political economy 367
regional stratification 363
semiperiphery 370
service industries 372
social networks 361
socialism 367
technology 366
vertical integration 377
welfare capitalism 367
welfare socialism 367
wholesale and retail trade industry 372
work establishment 376
world systems theory 369

INTERNET ASSIGNMENTS

1. Locate the Web site of the World Bank and browse the press releases and data included about conditions and activities in countries around the world. How does the information relate to the concept of a global stratification system? Can you find examples of the way in which a country's natural resources, geographic characteristics, technological development, and its government affect its economic success? Can you find evidence for the importance of relationships between countries in the development of economic success? Can you find support for or against the ideas of world systems theory? What is the nature of this support?

2. Use terms such as "corporate merger" and "multinational corporations" to find information about changes in economic organizations in recent years. Choose one or two large companies to focus on and look for their home pages. To what extent do different sources (such as news reports and documents produced by a company) discuss either the functions or dysfunctions of mergers and the development of conglomerates?

3. Search the net with terms such as "careers" and "employment." What types of advice and help for job seekers do you find? Based on the discussion in this chapter and the results of Faulkner's research on film composers, how much do you think these Internet resources will help job seekers? Explain.

TIPS FOR SEARCHING

The sites for the World Bank, the United Nations and the Population Reference Bureau can all provide links to information about the economic well-being of countries.

INFOTRAC COLLEGE EDITION: READINGS

Article A20243379 / Davern, Michael. 1997. "Social networks and economic sociology: A proposed research agenda for more complete social science." *American Journal of Economics and Sociology, 56,* 287–303. In this article Davern explains to economists how the sociological concept of "social networks" would enhance economic theories and research.

Article A20418499 / Cohen, Joel E. 1997. "Why should more United States tax money be used to pay for development assistance in poor countries?" *Population and Development Review, 23,* 579–585. Cohen reviews benefits that accrue to more developed nations when they provide aid to the poorer countries of the world.

FURTHER READING

Suggestions for additional reading can be found in *Classic Readings in Sociology,* bundled with this book. If you purchased a used copy of this book that does not include this custom-published reader, go to www.sociology.wadsworth.com for ordering information.

RECOMMENDED SOURCES

Bridges, William P,. and Wayne J. Villemez. 1994. *The Employment Relationship: Causes and Consequences of Modern Personnel Administration.* New York: Plenum. Bridge and Villemez analyze how characteristics of work organizations affect individuals' work careers.

Brinton, Mary C. and Victor Nee. (eds.). 1998. *The New Institutionalism in Sociology.* New York: Russell Sage Foundation. This edited volume contains a number of articles that illustrate the way in which contemporary sociologists analyze the economy as a social institution.

Danziger, Sheldon, and Peter Gottschalk. 1995. *America Unequal.* Cambridge, Mass.: Harvard University Press. Danziger and Gottschalk present a readable analysis of how changing relations within the economy have contributed to growing inequality and gaps between the wealthy and poor.

Faulkner, Robert R. 1983. *Music on Demand: Composers and Careers in the Hollywood Film Industry.* New Brunswick, N.J.: Transaction. Faulkner's book includes many more details than were reported here on the careers and experiences of film composers.

Granovetter, Mark. 1995. *Getting a Job: A Study of Contacts and Careers,* 2nd ed. Chicago: University of Chicago Press. This is a short, informative book written by the sociologist profiled in Box 14-1 that shows how our contacts and connections are a prime influence on the kinds of jobs we get.

Hodson, Randy, and Teresa A. Sullivan. 1995. *The Social Organization of Work,* 2nd ed. Belmont, Calif.: Wadsworth. This text covers many aspects of work life in the United States and other countries.

So, Alvin Y., and Stephen W. K. Chiu. 1995. *East Asia and the World Economy.* Thousand Oaks, Calif.: Sage. This book uses world systems theory to analyze the growth and development of the economy in East Asia.

Stinchcombe, Arthur L. 1983. *Economic Sociology.* New York: Academic. This book is a good example of how a sociologist can take a very macrolevel view of the economy.

Swedberg, Richard. 1990. *Economics and Sociology: Redefining Their Boundaries: Conversations with Economists and Sociologists.* Princeton, N.J.: Princeton University Press. Swedberg provides a fascinating collection of interviews with both economists and sociologists whose work tries to bridge the gap between the two disciplines.

Swedberg, Richard. 1998. *Max Weber and the Idea of Economic Sociology.* Princeton, N.J.: Princeton University Press. This book gives an extensive analysis of Weber's analysis of economic sociology.

INTERNET SOURCES

World Bank. This Web site includes a great deal of information about the economic conditions of countries around the world.

Fortune Magazine. The Web site for this periodical provides a great deal of information on large corporations in the United States and internationally.

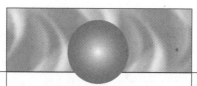

Pulling It Together

Listed below are other chapters where you can find additional discussion or examples
of some topics covered in this chapter. You can also look up topics in the index.

Education

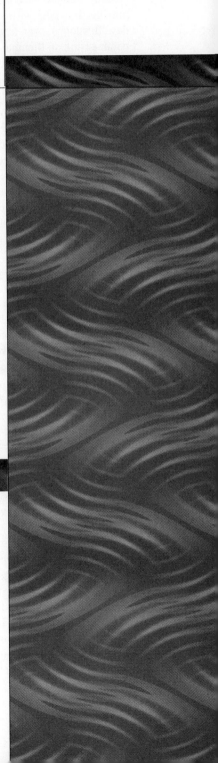

**The Development of Mass Schooling:
A Macrolevel View**

Mass Schooling as a Worldwide Phenomenon

*Explanations for the Development of Mass Schooling:
Societal Functions versus
Schools as Organizations*

**Promoting School Achievement:
Microlevel Analyses**

Family Interactions and Educational Achievement

Peer Interactions and Educational Effort

*Teacher–Student Interactions
and Educational Achievement*

**Increasing the Effectiveness of Schools:
Mesolevel Analyses**

The Effectiveness of Catholic Schools

FEATURED RESEARCH STUDY

Effective Schools: Multiple Approaches

School Structure and Student Commitment

Effective School Cultures

OR AS LONG AS SHE could remember, Abby had wanted to be a schoolteacher. She didn't know exactly why, but she suspected it had something to do with a couple of special teachers she had while growing up. For instance, there was her third-grade teacher, Mrs. Alvarez, who had taken an interest in Abby around the time her parents were going through a divorce. Abby had been an angry child then, an indifferent pupil and something of a troublemaker on the playgrounds, where she often got into fights. Then Mrs. Alvarez gently took her under her wing and encouraged her to show her talents in a constructive way. Before long, Abby was playing "teacher" with younger playmates on Saturday afternoons, even assigning them fun projects and commenting on them just as Mrs. Alvarez did.

Now Abby was a senior at City College and close to getting her bachelor's degree in education. As part of the degree requirements, she would be visiting different schools, observing teachers at work, helping in the classrooms, and, eventually, doing student teaching, taking over the class by herself.

During the fall semester Abby was assigned to Oak Street Elementary, an inner-city school separated from the surrounding busy streets by a tall chain-link fence. Inside the fence was a concrete playground and an aging brick school building. The sterile exterior of the school, however, hid a beehive of activity and warmth. On Abby's first day Mrs. Washington, the principal, greeted her warmly. Abby soon realized that Mrs. Washington was that way every morning, giving a cheery "hello" to each student and staff member she encountered and addressing many of them by name. Mrs. Washington seemed to sense almost intuitively what teachers and students needed, from books and audiovisual equipment to a receptive ear and a warm smile.

Mrs. Washington's vitality and concern seemed to permeate the school. Many of the students were from poor families and were provided both breakfast and lunch at the school. These were served in a brightly painted cafeteria that was monitored by friendly staff, who would often sit and talk with the students. Many of the students' parents were bilingual and seemed to have only a limited amount of schooling themselves. Yet almost every day, there were several parents in the school, helping in the classrooms, in the library, and on the playground. The staff was encouraged to welcome the parents' involvement and to treat them as partners in their children's education. Many of the children whose parents were able to help out seemed proud of their parents' involvement and eager to introduce them to Abby. Sometimes the teachers complained to one another that the parents could be a distrac-

tion, but all agreed that overall their presence was well worth it.

Throughout the fall semester Abby worked in Ms. Johnson's third-grade classroom. After a few days of observation, Abby could see that the children had a wide range of skills. Some were still struggling to learn to read, whereas others could read books far above their grade level and write surprisingly complex essays. Yet the children didn't seem to pay much attention to these differences, often working together on projects in cooperative learning groups.

A casual visitor might have perceived the classroom as occasionally chaotic, but there was an underlying order to all the activity. Over time, Ms. Johnson showed Abby how she carefully planned group activities so that all children could contribute and how she arranged the class schedule to give extra attention to the students who were having more difficulty with reading and math. She also explained techniques she had developed to keep the students involved with their work and to halt arguments and disputes. By the end of the semester, Abby was trying out these techniques for herself as she developed group activities, worked with students on their reading and math, and handled the class when Ms. Johnson was out of the room. Although she didn't always feel secure about her skills, overall she found herself loving the time she spent at Oak Street. The children seemed to be learning, and so was she.

In the spring semester Abby was assigned to Elm Avenue Grammar School. Because this was her second semester of field experience, she would be expected to handle even more responsibilities than at Oak Street, and initially she looked forward to the challenge. Most of the students were from middle-class families, and their parents tended to be better educated than the parents at Oak Street Elementary were. The building was situated in a pleasant, quiet neighborhood with tall trees and a grassy playground surrounding the well-maintained buildings. But Abby immediately sensed that Elm Avenue had a different feeling than Oak Street did.

Mrs. Martin, the principal at Elm Avenue, was pleasant enough, but she constantly seemed preoccupied. Unlike Mrs. Washington she tended to stay in her office and met with students only when there were discipline problems or special assemblies. It seemed as if the only students Mrs. Martin knew by name were those who regularly were detained for discipline problems or were on the honor roll.

Relationships among the faculty and staff were also very different from those at Oak Street. Abby had become used to the idea that teachers spent time with one another, sharing assignments and materials or just talking about their work. But the staff at Elm Avenue seemed rather cool and distant with one another. Many of them shut the doors to their classrooms before and after school, and those who did gather in the faculty lounge were something of a clique. Parents rarely visited Elm Avenue during the day, and when they did a number of the teachers regarded them as simply a nuisance to have around. And there were often groans in the faculty lounge when an upcoming parent–teacher conference night was announced.

At Elm Avenue Grammar Abby was assigned to work with Ms. Nelson's third-grade class. The students were being taught the same kinds of things as the students at Oak Street Elementary. They were instructed in how to write in cursive and how to use the library to write reports. They were given longer books to read by themselves and practice in learning the multiplication tables. Yet whereas the students in Ms. Johnson's class seemed to enjoy school, most of the ones in Ms. Nelson's class seemed to detest everything about it. They talked back to their teacher, moved freely about the room throughout the day, and continually talked and giggled among themselves. Nearly every day Ms. Nelson would retaliate by sending the most troublesome students to Mrs. Martin's office.

When Abby taught the class, she tried to use some of the techniques she'd picked up from Ms. Johnson. She assigned the students to cooperative group projects and led discussions of class rules and the reasons for them. But her approach was completely unfamiliar to the students, and it didn't seem to take hold. Abby thought that perhaps she might make headway, given enough time and support, but Ms. Nelson's frowns when she observed her at work told Abby that her attempts to change things weren't especially welcome. As soon as Ms. Nelson took over, the class reverted to its usual patterns.

As the semester wore on, Abby felt more and more discouraged. She liked the kids at Elm Avenue, but she often became angry and frustrated when they were disruptive or failed to apply themselves. As a group they were no less intelligent and capable than the students at Oak Street were, but they were steadily falling behind. Abby wasn't sure who or what to blame—the school, the principal, the parents, the students—or herself. "Maybe I'm not cut out to be a teacher after all," she thought on bad days. Yet she knew that wasn't the answer. And she couldn't just write off the school either, and especially the students. Somehow teachers could teach and students could learn just as well at Elm Avenue as at Oak Street, if only the conditions were right. But which conditions? Where should she—or anyone—begin in order to turn things around?

ABBY'S EXPERIENCES INVOLVE the social institution of **education**—the groups, organizations, norms, roles, and statuses associated with a society's transmission of knowledge and skills to its members. In our society, as well as in other industrialized and developing countries, this involves **schooling,** formal instruction by trained teachers that usually takes place within schools, like Oak Street Elementary and Elm Avenue Grammar. Schools are one type of formal organization, as described in Chapter 11.

Education, and especially schooling, is often the focus of national and local political debates. For instance, local communities and state legislatures argue fiercely over school budgets and the dispersal of funds. The media, policymakers, and the general public debate the purpose of education and the content of school curricula. Should schools teach only "the basics," or should they also offer music, art, and drama? Should there be more vocational education or less? Should sex education be offered? Should creationism be taught in science classes? Some debates involve the extent to which schools can help solve social problems, such as juvenile delinquency, teenage pregnancy, and racial-ethnic and gender discrimination. Issues related to religion in schools also are prominent, such as whether to allow prayer in public schools or to provide government funding for religious education. Other debates involve questions regarding who should be schooled. What programs should be available for children who do not speak English or who have disabilities? Should government funds support higher education?

Each of these concerns relates to the broad nature of education as a social institution. How can knowledge and skills best be transmitted to the members of a society, and what kinds of knowledge and skills should be taught? And how will decisions on these matters be reached? These concerns also involve the relationship of education to other social institutions. To what extent should the government support schooling as a public good? What distinguishes the role of the family, religion, and the economy? Should parents, and not schools, teach children about sex? Should employers, and not schools, prepare students for jobs? Should churches, and not schools, transmit moral values? In addition, these concerns reflect the "American Dream," which we discussed in Chapter 7—the cultural belief that each of us has a right to "success" and can be upwardly mobile if only we try. They reflect the hope that through education children can achieve to their fullest potential, whatever the circumstances of their birth.

As you will recall, the American Dream is far from a reality; the system of social stratification is such that few people ever make it to the top of the stratification mountain, no matter how much schooling they have. Even though education can't magically bestow wealth and power, it still is central to our well-being, both as a society and as individuals. From a macrolevel perspective we have seen how education is related to large-scale changes in social stratification. As knowledge increases, technology becomes more complex, individuals become better educated, economic development increases, and extensive social stratification and oppressive political domination tend to diminish. From a microlevel perspective we tend as individuals to have better life chances if we have more education: to have higher-paying jobs, to participate more in the political world, and even to be accorded more respect in interpersonal interactions. In sum, education is a key determinant of individuals' well-being and their social status, as well as a key element of stratification systems within societies.

For these reasons, many sociologists are concerned about disparities in educational achievement among individuals and across groups and seek ways to reduce these inequalities. Sociologists often share the concerns of policymakers and parents with issues surrounding schools, but they try to approach these issues dispassionately—with a scientific perspective. They try both to understand how the world works and to develop knowledge that might help make it a better place.

In this chapter we explore this work. We begin by stepping back to look at education and schooling from a macrolevel perspective. Why has mass schooling expanded worldwide over the past two centuries? How have schools developed, and why have they evolved in fairly similar ways in very dissimilar cultures? We then zoom in to take a microlevel view of students' experiences and achievements in the classroom. What are some of the factors that affect individual students' chances of success? Finally, because

schools obviously differ in their effectiveness, we use a mesolevel analysis to examine schools as organizations and explore sociologists' analyses of why some schools, like Oak Street Elementary, tend to do a better job than others do. From this work we can understand more about how the behaviors of teachers and parents and the organizational structure of schools can help students learn to their fullest potential.

Of course, hard work, intelligence, and motivation to learn explain some of the differences in students' achievement in schools. But such individual factors are intertwined with social influences operating at several levels. In thinking about education from a sociological point of view, I like to use the image of a river journey that I described in Chapter 4 to illustrate the process of socialization and the development of the self. In that chapter we talked about how some boats travel down gentle, smoothly flowing rivers, but others must navigate swift, dangerous currents. Further, some of the boats have crew members who are capable and strong, whereas others have fewer crew members or those who are less capable.

Education as a social institution provides part of this metaphorical environment and crew. Some societies provide a lush educational environment, but some can provide only the bare minimum. Similar differences can be seen from one community to another and even from one school to another. Some school staffs, like those at Oak Street Elementary, are more adept at creating effective organizations, and some individual teachers and principals, such as Mrs. Washington and Ms. Johnson, are more skilled than others are in promoting learning. Although such differences do not explain all disparities in educational achievement, more effective organizations and individuals do make a significant difference in students' lives, subtly or dramatically affecting their life journeys. Understanding what makes these individuals and groups more effective is one way to begin addressing the question of how to improve one of society's most important institutions.

THE DEVELOPMENT OF MASS SCHOOLING: A MACROLEVEL VIEW

In simple hunting and gathering societies, and even in many agricultural societies, you would be hard-pressed to locate anything that resembled education as we know it in our own society. In these societies most children learned the skills needed to survive as adults from working alongside their parents or other grown-ups. In medieval Europe, tutors sometimes taught the children of the very wealthy, and religious organizations kept the art of reading and writing

As you read this chapter, think about all of the variables that might influence the experiences these students have had at school. How might you use sociological theories and research to account for their experiences?

alive. In the Renaissance, schools developed that provided a "liberal education," emphasizing studies in disciplines such as philosophy and classical literature. But such schooling was available mostly to children from privileged family backgrounds. Most people completed their life's journey without any formal schooling at all.

In the past few centuries, however, this situation has changed dramatically. Education has become not a privilege, but a necessity and even a right. Today, throughout the world, most children attend schools that are, in many ways, quite similar to one another. Schools, rather than parents or the church, have become the most common way to train children for their lives as adults. In this section we look first at the development of schooling around the world and then at explanations sociologists have provided for these developments.

Mass Schooling as a Worldwide Phenomenon

Mass schooling, elementary-level schooling for all children, developed throughout Europe and North America in the eighteenth and nineteenth centuries. For instance, King Frederick II of Prussia decreed compulsory schooling (a law requiring all children to attend school) in 1763; Norway followed suit in 1800, and Denmark in 1814.[1] By the end of the nineteenth century, compulsory elementary schooling was found throughout Europe, as well as in other modernizing countries, such as the United States and Canada, and most children in these countries were enrolled in school.[2] Because these schools are available to all children, they are also sometimes referred

to as **common schools,** schools that are held "in common" by all. After World War II, mass schooling spread throughout the rest of the world, as virtually all countries developed schools to educate children in the elementary grades.[3]

Map 15-1 displays variations in countries in the extent to which adults are *illiterate,* or unable to read and write. You can see that the countries with the highest levels of illiteracy are in the least developed regions of the world, and the lowest levels are in the richest, industrialized countries. Still, rapid progress has occurred during the last few decades. By the late 1990s, more than 75 percent of all males and 60 percent of all females older than fifteen years in developing countries (those that are in the process of industrializing) were literate. In the least developed countries (the very poorest and least industrialized nations), 60 percent of the men and more than a third of the women could read and write. For both sets of countries, these rates are substantially higher than those from two decades earlier. Even though there are still large gaps between the poorest and richest countries, literacy rates have been increasing continuously throughout the world during the past few decades.[4]

This increased literacy can be attributed to the greater likelihood that younger cohorts have attended school, and current patterns of school enrollment suggest that literacy rates should continue to increase in the future. Because children in developing and least developed countries often must skip school to help their parents with various work tasks, it is not uncommon for them to fall behind in school. Yet, because schooling is seen as so important, large

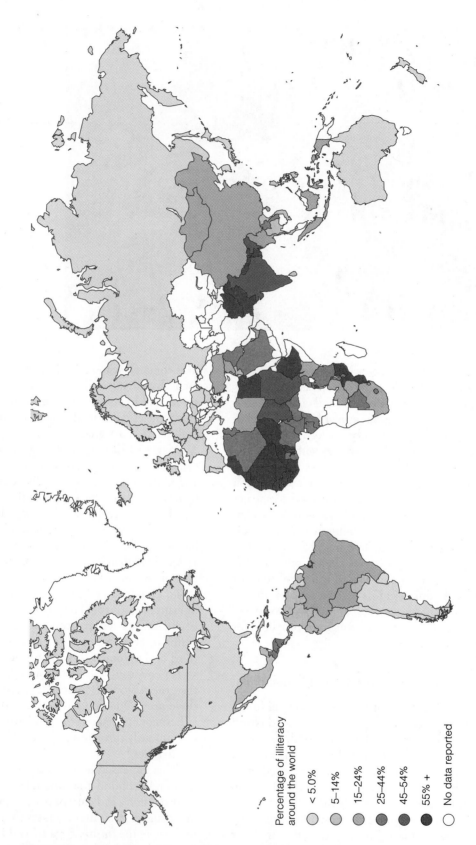

MAP 15-1 *Adult Illiteracy Rates, 1995*

Source: UNESCO, 1998, Selected Indicators (http://unescostat.unesco.org//Yearbook/Indicator_I.html).

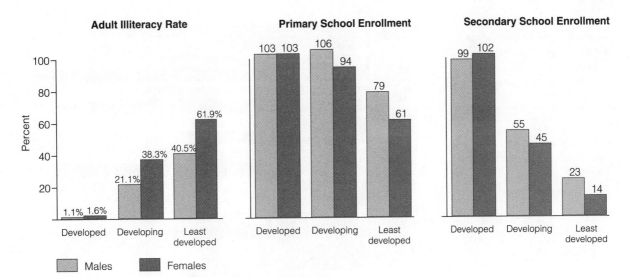

FIGURE 15-1 *Literacy Rates and School Enrollment, Worldwide*

Note: The enrollment figure reports the number of students, regardless of age, enrolled in a level of schooling as a percentage of the total number of children in the appropriate age group for that level. Because some children are held back and are thus "overage" for their grade level, this figure can be higher than 100.

Source: UNESCO, 1998, Tables 2–2 and 2–10.

numbers of these children persist in school even though they are older than the usual age. As shown in Figure 15-1, most children of elementary school age in developing countries and significantly more than half of all children in the least developed countries are enrolled in elementary level schooling.

Although all countries now offer free elementary education, they have differed significantly in access to secondary and higher education. Historically, the United States provided much greater access than other countries have, with students able to attend free, public high schools since the mid-nineteenth century. In contrast, until well into the twentieth century, students in many European countries had to pay for schooling beyond the primary years, thus severely limiting the opportunities of children from working-class families.[5] However, just as free primary education swept the globe, free access to secondary education has become very common in the industrialized world. Throughout Europe after World War II, secondary schools became free and open to all children, and as shown in Figure 15-1, almost all children in industrialized countries now attend secondary schools. However, far fewer children are enrolled in secondary schools in developing and least developed countries. In fact, only 14 percent of girls in the least developed countries attend secondary schools.

The disparity between the literacy rates and school enrollment levels of males and females in developing and least developed countries reflects gender stratification, which we discussed in Chapter 9. Both historically in the United States and cross-culturally, educational advantages have tended to accrue to males before females. This pattern is especially likely to occur when education is costly or in short supply. As a result, the differences in school enrollment levels are larger in the poorer, least developed countries than in the developing countries. These differences also are larger at the more expensive secondary level than at the primary level. There is virtually no difference in the literacy rates or in primary and secondary school enrollment rates of males and females in industrialized countries because these nations are so wealthy and have such well-established school systems that they can afford to provide education for all children.

The United States also has had the broadest system of higher education, with many more colleges and universities than other nations have, as well as the world's first extensive system of junior colleges.[6] Today in the United States more than half of all young people between the ages of eighteen and twenty-two are enrolled in some type of college, more than double the rate in any Western European country.[7] But in recent years, many other industrialized countries have expanded their university systems to accommodate more students and developed extensive scholarship programs to finance the education of young people who are unable to afford tuition. Even though the United States still has many more students in higher education than other countries do, all countries appear to be moving toward providing a much more extensive system of education at all levels.[8]

FIGURE 15-2 *Percentage of Bachelor's Degree Recipients of the Theoretical Age of Graduation in Selected Countries, 1994*

Note: Data represent the number of bachelor's degrees awarded in a year divided by the number of people in the population who are the theoretical, or usual, age at which people receive the degree—ranging from twenty-one to twenty-five.

Source: Snyder, Hoffman, and Geddes, 1997, p. 436.

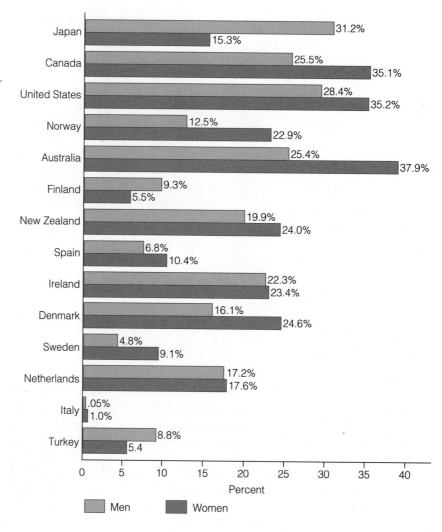

Note, however, that the high college enrollment rate of young people in the United States does not necessarily translate into the highest rates of college completion, at least in a timely manner. As shown in Figure 15-2, when only people in the usual age group of college graduates (ages twenty-one to twenty-five) are considered, students in both Canada and Australia are about as likely to be college graduates as those in the United States are. In general, the extent to which young people complete college varies a fair amount among industrialized countries. In contrast with the relatively high completion rates of Canada, Australia, and the United States, fewer than 10 percent of individuals in their early twenties in Finland, Turkey, Spain, Sweden, and Italy have a bachelor's degree.

Finally, you might have noticed that the traditional pattern of male dominance in educational attainment has largely disappeared, or even reversed, in most of the countries portrayed in Figure 15-2. In many highly industrialized countries, including the United States, Canada, Australia, New Zealand, Denmark, Norway, Spain, and Sweden, women are more likely than men to complete college in a timely fashion. In others, most notably Japan, the traditional pattern of male dominance remains, with men about twice as likely as women are to graduate with a bachelor's degree.

In general the data on school enrollment, from the primary level through higher education, suggest that mass schooling now appears throughout the world. As countries can afford to do so, they gradually expand their educational systems so that more and more children and more levels of education are included.

Explanations for the Development of Mass Schooling: Societal Functions versus Schools as Organizations

Why did mass schooling appear when it did, and why did it spread so quickly and so thoroughly? Sociologists disagree about the answers to these questions. One explanation emphasizes the societal functions performed by schools and education; the other focuses on the development of schools as organizations.

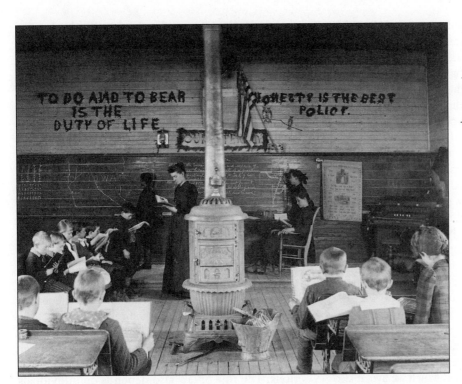

How might this one-room school in Bear Creek Township, Iowa, in 1893 be similar to schools that you attended? How is it different? What do these similarities and differences tell you about the functions of schooling?

The Functions of Schooling As you will remember from previous chapters, sociologists often talk about the function of an element of society when they want to explain why certain structures, such as schools or education, exist. In other words they ask what role a structure plays in maintaining or altering the society. Some sociologists have argued that schools developed to meet the changing demands of industrializing societies. Some scholars, inspired by the writings of Émile Durkheim, have tended to stress the ways in which schooling promotes social cohesion. Others, inspired more by the writings of Karl Marx, have stressed the ways in which schools promote the welfare of the powerful at the expense of those with fewer resources.

In the spirit of Durkheim, some sociologists have suggested that schools developed to meet the need of industrializing societies to teach children necessary cognitive skills and behaviors for future employment.[9] They have asserted as well that schools promote social solidarity, common social norms, and a sense of moral unity by giving children a uniform introduction to the nature of society.[10] In contrast, those with a more Marxian perspective have suggested that these same schools were developed to produce a compliant and efficient adult labor force and to prepare workers for low-level positions within the economy.[11]

Both groups cite the history of schooling in the United States to support their claims. Those with a more optimistic outlook point to the development of mass schooling in the United States as the avenue through which immigrant children learned skills to participate in the economy of their new country. Those with a more cynical orientation suggest that this process also indoctrinated and controlled the children of newly arrived immigrants, eradicating the remnants of their cultural heritage and preparing them to be compliant workers in the factories of wealthy industrialists.[12]

The idea that changes in education are associated with changes in the economy certainly sounds plausible, whether one sees this association as benign or as serving the interests of the wealthy and powerful. Most of us have attended school to help prepare ourselves for future employment, and in job interviews we typically are asked about our educational training and credentials. Interestingly, however, little historical evidence actually supports the idea that schools developed in previous eras to meet the needs of industrializing societies. In other words, even though we commonly assert that education functions to prepare individuals for the world of work, it is unclear whether schools actually developed for this specific purpose.

For one thing, compulsory schooling appeared in many countries before industrialization. For example, two-thirds of all children in Sweden were enrolled in schools by 1860, but only about 10 percent of the population lived in cities, and even fewer worked in industry. Similarly, in places such as Norway, Prussia, Denmark, Japan, and the northern United States, mass schooling was a part of life long before workers were needed in industry. Moreover, within specific countries such as England, France, Italy, and Sweden, schooling appeared later in areas that were highly industrialized than in those that didn't have industry, often because children in industrialized areas worked in factories rather than attending school.[13]

As for promoting social cohesion, scholars generally agree that schools do provide a means of establishing social control and teaching all children common ideas and behaviors. Many proponents of the development of mass schooling in the United States saw it as a way to help educate and assimilate the hordes of impoverished immigrant children.[14] Yet schools are not necessarily the best or most efficient way to promote social control and cohesion. For centuries the powerful have used means other than schools to control the masses, from religious prohibitions to naked force. Why would schools be seen as a better means of control, especially in *all* societies?[15]

In addition, education often has prompted revolts against established authority, at least indirectly. Although schooling conveys the requisite attitudes and behaviors of the communities in which we live, it also empowers us as individuals by promoting literacy and the examination of ideas. Historians frequently have noted the role of literacy and popular literature in fomenting social change and building revolutionary fervor. In colonial America, for instance, as many as a half-million people bought Thomas Paine's pamphlet, *Common Sense,* which called for independence from England. Pamphlets such as this, as well as quickly printed handbills and, especially, newspapers, are seen as having played a powerful role in spurring the American Revolution.[16]

In short, functionalist arguments suggesting that schools developed to prepare workers for roles in industrial societies and to serve as instruments of social control have a superficial plausibility when applied to the modern world. Yet it is hard to see how they fit the historical picture and, especially, how they can explain why mass schooling developed worldwide. Therefore, in recent years sociologists have developed another line of explanation for the growth of mass schooling.

The Legitimation of Schools as Organizations in Modern Societies The alternative explanation for the worldwide development of mass schooling draws on ideas first presented in Chapter 11 in our discussion of formal organizations and their environments. In that chapter we talked about how organizations conform to institutional norms; that is, organizations reflect cultural ideas about which organizations should exist and what forms they should take. Similarly some sociologists have suggested that

mass schooling became institutionalized as part of the definition of what modern nations or societies should be. In other words, people in all nations—rich and poor alike—believe that an educational system open to all people is simply part of what a modern political system provides its citizens. In short, modern nations believe that their legitimacy is very closely linked to the development and expansion of their educational systems.[17]

These changes are related to large-scale shifts worldwide in the arrangement of social institutions and the roles of individuals. As societies modernize, the importance of the kin group and religious life begins to decline relative to that of the political world. For instance, whereas small hunting and gathering societies maintained social control through the possibility of sanctions by family and kin and through religious prohibitions, complex modern societies depend on police officers and the rule of law. The rule of law, in turn, depends on the development of complex **nation-states,** political entities that unite large groups of people under agreed-on laws and regulations. The rule of law also depends on individuals adopting the social role of "citizen," a person who both owes allegiance to a government and is entitled to protection from it.

Modern nation-states began appearing in Europe several hundred years ago and were firmly established by the end of the nineteenth century in many countries. They now appear throughout the world.[18] No doubt you think of yourself as a citizen of the United States or another country. In fact, it is virtually impossible today not to be a citizen of some nation. Several sociologists suggest that mass schooling came to be seen as the way to develop citizens for modern nations. Governments throughout the world have established school systems to prepare children to be citizens of the national political world. In this sense schooling has become a modern "rite of passage" and the primary way that children can legitimately enter adulthood.

John Boli, one of the sociologists who developed this view, describes the importance of schooling as marking this transition:

Apart from school, the only activity routinely demanded of young people is military service, but the latter citizenship ritual is more limited in scope (it rarely includes girls) and is altogether lacking in a number of countries, not least the USA. In addition, schooling is by definition a sufficient ritual. Children must go to school, but nothing else is required of them to participate in economic, political and religious life; there are no polity-wide examinations to evaluate the competence of individuals with respect to starting a business,

Which of the four ideological goals of schooling does this photo demonstrate? How did you reach this conclusion?

voting, or joining a church. In the modern model, one must go to school to become a citizen, but once one has survived the common school there is no other ritual that the individual must endure before assuming the status of an adult.[19]

If you think about your own schooling, you can see how it prepared you to be a citizen of your own country and community. You learned the history of your country and your state or province. You also learned patriotic songs and studied the legal requirements for voting and the nature of the political world.

The legitimation of education as a social institution has brought with it some basic beliefs about what should constitute an education. Sociologists who have studied mass schooling throughout the world suggest that it always seems to exhibit an underlying ideology involving four basic elements. These elements appear in the writings of advocates of mass schooling as far back as the nineteenth century and earlier, and they continue to reflect the views of educators and the general public around the world today, especially regarding schooling at the elementary level.

First among these ideological tenets is the idea of *universality:* Schooling should be available to all children, rich and poor, male and female, urban and rural. Second is the notion of *egalitarianism:* Schooling should provide equal opportunities, and inequalities in children's social background and home environment should not hinder their life chances. The third element results from the second and involves the idea of *standardization* of the educational system: All schools should offer the same curriculum and opportunities to all students, no matter who they are or where they come from. Finally, mass schooling involves the idea of *individualism:* All children should be encouraged to develop their own potential, no matter what their individual background is.[20]

Of course few, if any, nations actually have developed school systems that fully attain these ideological goals. Even within advanced industrialized societies, which provide virtually universal access to education, educational outcomes differ greatly, which means that the goal of egalitarianism has not been met. Studies in many different countries—from the United States to Great Britain, Ireland, Israel, and Russia—have shown that children from less privileged backgrounds do not fare as well as their more privileged classmates in schools; less privileged children tend to have lower achievement and to be far less likely to go on to advanced levels of schooling. These differences appear even in countries, such as the former Soviet Union, that have explicit ideologies calling for the elimination of privileges associated with social class.[21]

Even within countries that are very wealthy by international standards and that have highly developed, universalistic systems of schooling, such as the United States, there can be wide disparities in the resources devoted to schools, making it impossible to meet the goal of standardization. For instance, the amount of money spent per child per year on education varies in the United States from a low of about $4,350 in Mississippi to more than $10,000 in Alaska and New Jersey.[22] Within states, where school funding is often the primary responsibility of local communities, the amount of money spent on education also can vary widely from one area to another. In general, communities with larger populations of poor people tend to be able to provide fewer resources for their schools than richer communities do. Therefore children from poor families are much more likely than are children from wealthier families to attend poorly funded schools, simply by virtue of where they live.

In recent years other issues related to the ideological tenets of common schools have come to the fore. For instance, educational specialists have made remarkable strides in individualizing curricula and instructional techniques to more effectively teach children who have special conditions such as mental retardation, learning disabilities, visual and hearing impairments, and physical and emotional disabilities. To help these children experience as much of the same curriculum and school experience as others (meeting the goals of universality, egalitarianism, and standardization), many schools have tried to integrate special education students into the general curriculum and school day, a practice called **mainstreaming.** Yet mainstreaming can be very costly and thus may be more likely to occur for some students than others.

In addition, however, much as Abby experienced, some schools and teachers just don't do as well as others in providing their students with a good education, even when they have similar amounts of funding and children with similar social class backgrounds. Many sociologists have been disturbed by these facts and have tried to understand more about how teachers, parents, peers, and schools can enhance children's learning and achievement. Given the strong association between social class and educational achievement, sociologists have been especially interested in how their insights can help mitigate the disadvantages facing children from less privileged backgrounds.

PROMOTING SCHOOL ACHIEVEMENT: MICROLEVEL ANALYSES

How do families, peers, and teachers promote or retard learning? Sociologists have explored this question at the microlevel of interactions that occur within families, with friends, and in classrooms with teachers. Many sociologists have been especially interested in identifying how these interactions can help overcome the disadvantages that children from lower-status families typically face. In other words, even though large-scale aspects of society influence children's opportunities, many sociologists also remain committed to examining how individuals and their actions can make a difference in children's lives.

Family Interactions and Educational Achievement

Several sociologists have examined the relationship between our family backgrounds and our level of success in school. Using a wide variety of statistics, they've estimated that between one-half and two-thirds of the differences among children in school achievement can be explained by differences in their families.[23] For instance, as you will remember from our discussion of the status attainment model in Chapter 5, children

from higher-status families tend to go farther in school than children from working- and lower-class families do, even when they are equally motivated or have the same levels of individual ability. Similarly, as you will recall from Chapter 10, children from two-parent families tend to do better in school than do children from one-parent families.

Statistical models, by themselves, cannot tell us *why* these results occur. They can tell us how large the difference in achievement is between middle-class and working-class children or between children who grow up in single-parent and two-parent families, as well as general ways in which the families differ. But these models usually can't tell us exactly what happens within families or how a family's interactions produce these differences. In addition, sociological explanations are always probabilistic, not deterministic. In other words, given their backgrounds, many children from more advantaged homes do less well in school than we would expect, and many children from disadvantaged backgrounds do much better than we would expect. To explain the difference, we need to look more closely at individual children, families, and teachers.

Sociologist Reginald Clark decided to find out more about how family interactions affect children's school success. Specifically, he wondered, what can low-income and single parents do to enhance their children's academic success? And how do disadvantaged families with high-achieving children differ from similar families with low-achieving children?

To answer these questions, Clark used field research, combining extensive observations, interviews, and questionnaires to study the lives of ten African-American families in detail. All the families lived in low-income communities in Chicago and had children in the twelfth grade. Half of the children were classified by their teachers as being high achievers in school, and half were classified as low achievers; half of the students in each group also came from one-parent families. All the families had incomes that fell below the official poverty line.[24] As a result, all the children would have been likely to be low achievers based on their family income and, for some, the number of parents in their family. But half of them defied the odds. Clark wanted to find out what might be happening in their families that helped make this success possible.

Clark observed and talked with all the families for more than forty-eight hours each. From his observations and interviews, he developed an extensive list of ways in which the families of high and low achievers differed from each other. These results are summarized in Table 15-1.

As you can see, the families differed in three important ways: (1) the nature of their social networks,

TABLE 15-1 *Comparison of the Homes of High Achievers and Low Achievers*

HIGH ACHIEVERS	LOW ACHIEVERS
Social Networks	
Parents initiate frequent school contact.	Parents initiate school contact infrequently.
Students have had some stimulating, supportive schoolteachers.	Students have had no stimulating, supportive schoolteachers.
Siblings interact as an organized subgroup.	Siblings are a less structured, less interactive subgroup.
Role Expectations and Norms	
Parents expect to play a major role in children's schooling.	Parents have lower expectations of playing a role in children's schooling.
Parents expect children to play a major role in their schooling.	Parents have lower expectations of children playing a major role in their schooling.
Parents frequently engage in deliberate achievement-training activities.	Parents seldom engage in deliberate achievement-training activities.
Parents frequently engage in implicit achievement-training activities.	Parents engage less frequently in implicit achievement-training activities.
Parents expect children to get postsecondary training.	Parents have lower expectations that children will get postsecondary training.
Parents have explicit achievement-centered rules and norms.	Parents have less explicit achievement-centered rules and norms.
Students show long-term acceptance of the norms.	Students have less long-term acceptance of the norms.
Parents establish clear, specific role boundaries and status structures with parents as dominant authority.	Parents establish more blurred role boundaries and 9 status structures.
Parents exercise firm, consistent monitoring and rules enforcement.	Parents have inconsistent standards and exercise less monitoring of children's time and space.
Quality of Parent-Child Interactions	
Parents are psychologically and emotionally calm with children.	Parents are in psychological and emotional upheaval with children.
Students are psychologically and emotionally calm with parents.	Students are less psychologically and emotionally calm with parents.
Conflict between family members is infrequent.	Conflict between some family members is frequent.
Parents provide liberal nurturance and support.	Parents are less free with nurturance and support.
Parents defer to children's knowledge in intellectual matters.	Parents do not defer to children's knowledge in intellectual matters.

Source: Clark, 1983, p. 200.

(2) the role expectations and norms established within the family, and (3) the quality of their interactions. First, high achievers' families have social networks that include supportive ties with teachers and siblings. Second, parents in these families expect that both they and their child will be actively involved in the child's schooling and that their child will continue with his or her education beyond high school. These parents have firm expectations about how their child should behave in school and at home. They talk about and help the child with school-related activities and consistently monitor the child's behavior. In turn, the high-achieving children accept these roles and norms as legitimate. Finally, high achievers' families have more emotionally calm interactions, have far less conflict between family members, offer a lot of nurturance and support, and accept the child's intellectual competence.

Clark used the idea of "sponsored independence" to describe the way in which the high achievers' families functioned:

> The interpersonal communication patterns in these homes tended to be marked by frequent parent–child dialogue, strong parental encouragement in academic pursuits, clear and consistent limits set for the young, warm and nurturing interactions, and consistent monitoring of how they used their time.[25]

In this way the parents "sponsored" or groomed their children for success at school by helping them develop the necessary behaviors and attitudes.

It is important to remember that Clark found this sponsored-independence communication style among families with low incomes and with only one parent present. The incredible strain that most low-income and single-parent families face can make it very hard to maintain the supportive communication style that is embodied in sponsored independence. But when families are able to have these interactions, their children can do quite well in school.

Other scholars have replicated Clark's results using a variety of methods.[26] But they have also found that parents aren't the only influence on students' achievement and attitudes toward schools. As you might expect, peers can also exert a large influence, especially on adolescents.

Peer Interactions and Educational Effort

One of the most extensive analyses of how peers influence students was conducted by Laurence Steinberg and his associates in a large number of high schools in Wisconsin and California. They observed and talked to 12,000 students and many of their parents over a period of three years. From their analyses, the researchers concluded that, in many respects, peers can be a more powerful influence on students' achievement than family members can—especially on how much effort students devote to their schooling.

These influential peers include (1) students' *best friends,* the one or two people with whom they spend most of their free time; (2) their *clique,* six or ten mutual friends who tend to "hang out" together; and (3) their *crowd,* a larger group of students with similar characteristics and attitudes who might share features in common but who aren't necessarily friends with each other. Close friends and cliques influence adolescents fairly directly through providing models of behavior and open pressures to behave in certain ways. The influence of crowds is more indirect and seems to involve the establishment of norms and standards that adolescents feel they must adhere to—*subcultures* to which students belong. This crowd influence appears to be most important between about sixth and tenth grade, a time when young people are often trying to establish a sense of their own identity, the way that they think about themselves.[27]

As you might recall from your own days in high school, most schools have several different crowds. From their observations and interviews Steinberg and his associates concluded that most schools have one or two socially elite crowds, such as the "populars" and the "jocks." They also tend to have one or

more alienated crowds, such as "druggies," "burnouts," or "greasers," whose members tend to be involved in delinquent activities and openly hostile to teachers and school personnel. A large proportion of students are seen as "average" or "normal," not distinguishing themselves in any particular area. Less than 5 percent of all students in a typical school identify with an academically oriented crowd, which focuses on high achievement, such as the "intellectuals" or "brains."[28]

To understand how much difference friends make, Steinberg and his associates compared students who were doing equally well at the beginning of their high school career but belonged to different crowds. Consistently they found that students whose friends were more academically oriented tended to do better in high school than similar students in a less academic crowd did. Similarly, as we would expect from our discussion of *differential association theory* in Chapter 6, students who identified with a crowd that was more inclined toward delinquent behavior tended to become more involved in delinquency and exhibit low academic performance themselves.[29]

Even though peers exert a powerful influence on students, Steinberg and his associates found that parents can influence this process. By monitoring their children's friendships, influencing where they spend their leisure time, and choosing neighborhoods and schools that have a greater proportion of academically oriented students, parents enhance the probability that their children will not become involved in crowds that promote destructive behavior.[30] Such behaviors parallel the "sponsored independence" style that was displayed by the parents of the high achieving students that Reginald Clark studied. Of course, many families don't develop these supportive patterns. For this reason other studies have focused on how schools and, especially, teachers can help students from disadvantaged homes learn to the best of their ability.

Teacher-Student Interactions and Educational Achievement

Just as some families do a better job than others of preparing their children for school and some peer groups are more supportive of academic achievement, some teachers are more effective than others in promoting learning and school achievement. Much as Abby observed differences in Ms. Johnson's and Ms. Nelson's classrooms, so have sociologists documented extensive differences in the teaching styles and effectiveness of teachers.

One long-term study documented these differences for a group of children who attended a school, fictitiously termed "Ray School," in a northeastern city of North America. Ray School was the type of school where you would expect many students to do quite poorly. As the researchers described it,

> Situated in one of the poorest areas of the city, the fifty-year-old building that housed its students stood out like a fortress in the streets . . . Across the street from the front entrance, the buildings of a brothel, thinly disguised as residences, blocked the view of a junkyard. Crowded tenement houses were interspersed with an automobile repair shop, a dry-cleaning plant, and an armature-wiring factory. The asphalt schoolyard was enclosed by a chain-link fence and the ground-floor windows were protected with vertical iron bars . . . The children were frequently unruly and often fought with one another . . . Ray had the reputation among teachers as being the most difficult school under its particular board. Not surprisingly, teachers reluctantly accepted assignments there, and a rapid and continuous turnover of novice principals and newly trained teachers was commonplace. Nonetheless, a solid core of experienced teachers continued to teach at the Ray School since the Depression.[31]

Eigel Pedersen, along with Thérèse Annette Faucher, a former minister of education of Quebec, and William Eaton, looked at the experiences of children who had attended Ray School approximately twenty-five years after they entered first grade. Using data gathered from school records and interviews with the former students, they found that the students of one particular first-grade teacher, whom they called Miss A, had done much better in adult life than the other students had. Specifically two-thirds of Miss A's former students, in contrast with less than one-third of the other students in the sample, had relatively higher educational and occupational status. Using a variety of statistical methods, the researchers determined that these results couldn't be explained by differences in the family background of the students. Amazingly, the unusual success of Miss A's students seemed to come from the special qualities and behavior of their first-grade teacher.[32]

In trying to understand why some teachers, like Miss A, are more effective than others are, sociologists have used a variety of observational methods. Sometimes they have started with fairly open-ended observations of classroom interactions, in the tradition of field research and participant observation. They then develop systems of recording classroom observations that are often quite extensive and complex. These systems generally involve classifying the types of activities that teachers and students are engaged in and timing the duration of each.

The results from these studies give us a fairly good idea of what qualities can help make a teacher more effective. One way to understand these results is to use the ideas regarding organizational leaders that were introduced in Chapter 11. There we described two different facets of leadership: (1) **instrumental leadership** deals with tasks, and (2) **expressive leadership** focuses on relationships between people. The most effective classroom teachers seem to be those who do a good job in both of these areas.

Like Ms. Johnson, the first teacher whom Abby worked with, effective teachers are very good at organizing the way that they teach material. They structure their classrooms so that children spend as much time on learning as possible. They also carefully choose the materials the children will study to ensure that they are neither too difficult nor too easy. Even more important, they present the materials at a pace and in a manner that keeps the children engaged, challenging them to move further and also making the material exciting and interesting. In this way the effective teachers are good instrumental leaders, structuring the classroom and the students' activities in ways that enhance and enliven the learning process.

At the same time, effective teachers are warm and responsive to their students. They are generous with praise and positive rewards while maintaining control of their classrooms, usually without the students even realizing that the teachers are doing this. In this way they are good expressive leaders. They create an environment in which the children feel safe and secure, connected to others in the class, good about themselves, and good about learning.[33]

Interviews conducted with Miss A's students twenty-five years after their experiences in her classroom illustrate these characteristics. Every single student who had Miss A for a teacher remembered her, whereas less than half of the other students could identify their first-grade teacher. From interviews with Miss A's students and some of her colleagues and principals, the following picture of her work appears:

> "She kept control by sheer force of personality and her obvious affection for the children, never needing to lose her temper or resort to physical restraint." It was said of Miss A's teaching that "it did not matter what background or abilities the beginning pupil had; there was no way that the pupil was not going to read by the end of grade one." One informant reported that Miss A left her pupils with a "profound impression of the importance of schooling, and how one should stick to it" and that "she gave extra hours to the children who were slow learners." When children forgot their lunches, she would give them some of her own, and she invariably stayed after hours to help children. Not only did her pupils remember her, but she apparently could remember each former pupil

by name even after an interval of twenty years. She adjusted to new math and reading methods, but her secret for success was summarized by a former colleague this way: "How did she teach? With a lot of love!" One would add, with a lot of confidence and hard work.[34]

In short, by combining instrumental and expressive leadership, Miss A managed to do an extraordinary job in a school where many other teachers had failed. Just as some skillful crews can guide boats through the most dangerous of rapids, so Miss A, and other especially skilled teachers, as well as the dedicated parents that Clark studied, can help students learn in the most difficult of circumstances.

Very few teachers can be expected to have the formidable skills and dedication of Miss A, let alone succeed in putting them into practice in such trying circumstances. Moreover, the environment of the school itself clearly affects teachers and students alike. Consequently, sociologists also have studied how schools, as social organizations, can be structured to help teachers be as effective as possible and students to learn as much as they can.

INCREASING THE EFFECTIVENESS OF SCHOOLS: MESOLEVEL ANALYSES

Sociologists have confirmed Abby's intuitive feeling about the Oak Street Elementary and Elm Avenue Grammar schools: Schools themselves do make a difference in students' learning. Most important, the influence of schools is independent of the types of families we come from and even the type of teachers we have. Some schools are much more effective than others are in providing an organizational structure and culture in which all children can learn, no matter what their social class background. Similarly some schools do much better than others at providing situations in which teachers can be effective instructors. Sociologists often call these organizations "effective schools."[35]

This finding has important implications for individuals concerned with enhancing children's educational experiences and with meeting the ideological goals that underlie mass schooling. After all, it can be quite difficult to alter interactions within families or peer groups or to inspire teachers to be as dedicated as Miss A, but it is somewhat easier to modify the organization of the schools in which students learn.

The evidence that has accrued in this area comes from two different sources of research. One involves extensive field studies of schools that have been identified as doing particularly good jobs, especially among disadvantaged youngsters. Researchers have spent many hours observing classrooms and school activities and the interactions of students, teachers, and parents. The other source involves statistical analyses of very large data sets that contain information about students and their families, teachers, and schools. Though field studies can provide details about the day-to-day operations of effective schools, statistical analyses of nationally representative data sets can reassure us that the results generalize to the entire population of students. Statistical analyses also can help researchers ensure that variables related to the school as an organization—rather than the individual characteristics of students—are really influencing changes that the researchers observe.

As often happens with new sociological research, the results of this work have been controversial. Nevertheless, sociologists today generally agree about what characterizes more effective and less effective schools. The studies that inform their judgments have been conducted with both elementary schools and high schools, and the underlying dynamics of effective schools seem to apply at both levels. In this section we describe work done on the high school level to illustrate this research.

The Effectiveness of Catholic Schools

One of the most consistent, and initially controversial, findings of this body of research involved the discovery that Catholic schools generally were more effective than public schools.[36] In the context of this research, "effective" means that students in Catholic high schools tended to have higher levels of academic achievement and higher future aspirations than those attending public schools. They also were more involved in their schools, less disruptive in class, and far less likely to drop out of school.

Because Catholic schools are private organizations that require students to pay tuition, you might suspect that these results appear because their students typically have higher social status than most students in public schools and that their parents are better educated themselves and care more about education. However, sociologists have tested this hypothesis carefully and found that it is *not* supported by the data. Some of the comparatively greater effectiveness of Catholic schools does stem from the fact that students' families are, on average, more dedicated to schooling than families in the public sector are. But extensive analyses show that the disparities in school effectiveness persist even when family differences are considered. Quite simply, students in Catholic schools tend to fare better academically than those in public schools, no matter what their family background.[37]

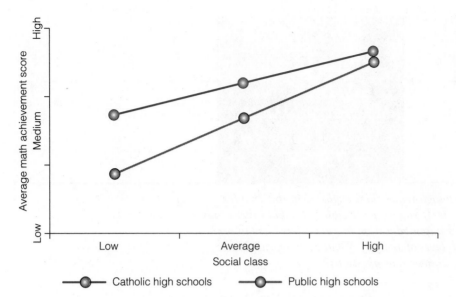

FIGURE 15-3 *Math Achievement and Social Class Composition of Seniors in Catholic versus Public Schools*

Source: Bryk, Lee, and Holland, 1993, p. 264.

Moreover, the differences between types of school do not reflect any greater racial-ethnic or economic segregation in Catholic schools or the possibility that Catholic schools serve few students from lower-income families or racial-ethnic minorities. In fact, just the opposite is true. Catholic schools typically are more diverse than public schools in family income and student race-ethnicity. Public schools usually draw their students from specific residential areas, which tend to be highly segregated by income and race-ethnicity, while Catholic schools serve students from much wider geographic areas. In addition, many Catholic schools have actively recruited students from disadvantaged backgrounds. As a result Catholic schools actually have more diverse student bodies in income and race-ethnicity than public schools, yet they still produce students with higher achievement and aspirations.[38]

Most important, Catholic schools seem to be far more effective than public schools are in teaching students from the most disadvantaged backgrounds. Although family background influences the achievement of students in all schools, this influence is weaker in Catholic schools than in public schools.

Figure 15-3 illustrates this pattern. The horizontal axis in the graph measures the average social class background of students in each type of school. The vertical axis measures the average math achievement score of seniors in the schools, which is one of many possible ways to assess student achievement. As you can see, in both types of schools, math achievement is higher in schools with more advantaged children (that is, both lines in the graph slope upward to the right). In addition, however, no matter what the social composition of the school, scores are higher for students in Catholic schools than for those in public schools. Moreover, the gap between the two kinds of schools is greatest among schools that serve primarily working-class and lower-class students and smallest among those that serve higher-class students. In other words, the effect of social class background on student achievement is greater in public schools than in Catholic schools. Ironically Catholic schools seem to come closer than public schools to meeting the ideological tenets that underlie common schools—that is, providing equal learning opportunities and results for all students, no matter what their family background might be.

Why does that happen? Why are Catholic schools, on average, better than public schools at meeting these ideals? Sociologist Anthony Bryk became fascinated with this question and, over a span of about ten years, worked with several graduate students and colleagues to try to answer it. In 1993, along with Valerie Lee and Peter Holland, he published a book that summarizes their findings, *Catholic Schools and the Common Good.*[39] Let's take a closer look at what these researchers discovered about the characteristics of effective schools.

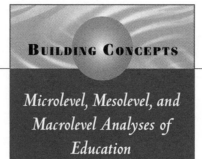

BUILDING CONCEPTS

Microlevel, Mesolevel, and Macrolevel Analyses of Education

Below are some of the sociological questions explored in this chapter, classified by level of analysis. If you were a sociologist, what other kinds of questions would you ask about education in our society? Try to come up with one question for each level of analysis. What sociological theories would you employ to help answer your questions?

Level of analysis	Example of Question Posed
Microlevel	How can parents promote student success in school?
Mesolevel	How can schools be more effective organizations?
Macrolevel	Why and how did mass schooling develop through the world?

FEATURED RESEARCH STUDY

EFFECTIVE SCHOOLS: MULTIPLE APPROACHES

In exploring school effectiveness, Bryk and his colleagues used both of the methods traditionally used by researchers in this area and added a third technique as well. First, they used the large "High School and Beyond" data set, which other researchers had compiled to examine student achievement and school effectiveness, to compare students in Catholic high schools with a representative group of students in public high schools. Second, they did extensive fieldwork in seven Catholic high schools selected from a group of schools described by superintendents in Catholic dioceses as "good." Bryk and his colleagues explained the selection process:

> In choosing the final set of schools, we . . . looked for affluent schools and schools with a weak resource base; schools ranging in size from 130 to over 1,500; coeducational and single-sex schools; schools with racial composition varying from all white to racially mixed to all black; and schools in which student enrollment was almost exclusively Catholic and schools in which half the students were non-Catholic. We rejected some schools because of their academically elite reputations. We

wanted good schools, but not necessarily the best schools. The sample of seven reflects the diversity that is Catholic secondary schooling in America.[40]

During a single school year, two-person teams visited each of these schools, once in the fall and again in the spring. In the first visit they interviewed students, staff, and parents; observed classrooms and school activities; and gathered as many documents about the school as they could find, focusing on the school's philosophy, curriculum, structure, and student culture. The second visit allowed them to focus their research observations on more specific hypotheses and research questions that they wanted to explore based on the data gathered in their first visit.[41]

Finally, Bryk and his colleagues employed a third technique, using historical research to look at the philosophical roots and history of Catholic schooling.[42] This research allowed them to examine the philosophy and values underlying Catholic schools and the ways in which these might have changed over time.

The results that Bryk and his colleagues obtained highlight some of the basic characteristics that they believe may influence the effectiveness of schools. Many of these results parallel those that have been obtained in other studies of school effectiveness and can help us understand why Oak Street Elementary, for example, is likely to produce more learning and higher achievement than Elm Avenue Grammar. The findings concern two general areas: the schools' structures and their organizational cultures.

School Structure and Student Commitment

Bryk and his colleagues, as well as others who have studied Catholic and public schools, have found that the schools are similar in many ways, including the age and condition of their facilities, the training of their staffs, and resources such as libraries and science labs. There are, however, two differences in the organizational structure of the schools that appear to influence their relative effectiveness: (1) their size and (2) the amount of structural differentiation. Together these structural characteristics contribute to what Bryk and his colleagues call the "constrained academic structure" of Catholic schools. They suggest that this type of structure helps promote students' academic achievement.

In Chapter 11 we saw how sociologists often describe an organization's structure by its complexity—the extent to which the work of an organization is broken up and differentiated among various units. In general Catholic schools are less differentiated, or less structurally complex, than public schools are. Bryk and his associates suggest that this limited differentiation is an important influence on the common school orientation of Catholic education. This can be seen in both the curriculum available to students and the roles assumed by staff.

Many public high schools are highly differentiated, with different academic departments and different "tracks" of courses that students follow. For example, public high schools often have a college prep track, a general track, and a vocational track. In addition, highly differentiated schools typically offer a wide variety of classes, ranging from advanced college-level courses in areas such as calculus, literature, and foreign languages to remedial courses in reading and basic mathematics. Many schools have introduced a number of elective courses, some designed to replace traditional courses, such as "Checkbook Mathematics" in place of algebra or "Contemporary Mysteries" to replace literature, as well as courses on personal development, such as "Personal Grooming" or "Leisure Activities."

In Catholic schools, all students generally take the same courses, and these tend to involve a traditional academic orientation. Even when students in public and Catholic schools are matched on variables such as their prior achievement and social class background, those in Catholic schools are more likely to take rigorous academic courses and to participate in college preparatory tracks. As a result, there is much less differentiation among students in Catholic schools than in public schools—students are more likely to share classes and a common academic experience. Bryk and his colleagues suggest that this emphasis on a strong academic core, combined with the limited choices given to students, is one of the major factors that promotes the common school element of Catholic education—its ability to provide equal learning opportunities and obtain similar results for all students, no matter what their family background may be.[43] The structure of Catholic school organization ensures that virtually all students in Catholic schools encounter the same academic experience.

The staffs in Catholic schools also are less differentiated. Public schools tend to be highly departmentalized. For instance, counselors deal with vocational and college counseling, vice principals handle discipline problems, teachers work in their classrooms, and coaches work in the gyms and playing fields. In contrast, teachers in Catholic schools tend to hold multiple roles, which permits them to interact with students in many ways. As Bryk and his colleagues describe it,

> Catholic school faculty typically take on multiple responsibilities: classroom teacher, coach, counselor, and adult role model. This broadly defined role creates many opportunities for faculty and student encounters. Through these social interactions, teachers convey an "intrusive interest" in students' personal lives that extends beyond the classroom door into virtually every facet of school life. In some cases it extends even to students' homes and families. In these interactions with teachers, students encounter a full person, not just a subject-matter specialist, a guidance specialist, a discipline specialist, or some other technical expert. The interaction is personal rather than bureaucratic.[44]

Thus, just as the smaller amount of structural differentiation in curricular offerings promotes a common academic experience for students, the smaller amount of differentiation in the roles of staff members promotes closer ties between students and faculty.

Part of the reason that Catholic schools are less differentiated than public schools is simply that they are smaller. Only 15 percent of all Catholic high schools but 40 percent of all public high schools have more than 900 students, and 5 percent of all public schools have more than 2,000 students.

The size of a school places clear limits on what it can do. When there are fewer students and teachers, there can't be as many course offerings or as many

In what ways are Catholic schools less differentiated than public schools?

specialized resources. Thus larger schools are more likely to offer a wider variety of courses and to be able to afford special facilities such as science labs or language resource centers. In short, larger schools—both Catholic and public—can offer a wider range of courses than smaller schools can.[45] Even so, when they do expand the curriculum, Catholic schools are much more likely to add academic courses; in contrast, public schools tend to add vocational courses.[46] Even the larger Catholic schools are far less likely than the larger public schools to offer many vocational or nonacademic courses.

Besides limiting the options that a school may offer, its size influences the roles that students and staff may play. Within any organization certain things always need to be done. For instance, staff is needed to teach, counsel, plan schedules, administer discipline, and supervise extracurricular activities. Students are needed to fill a wide variety of extracurricular roles, such as participating in school clubs, plays, and athletic events, and working on the newspaper and yearbook. Because smaller schools have fewer people to fill these roles, the chances that any one student or staff member will participate in various activities are greatly increased. Just as smaller schools have less differentiation in the academic cur-

riculum, they also have less differentiation in the roles that people play. As a consequence the individuals within a school are more connected with and less differentiated from one another.[47] The greater connections among students and staff and the stronger involvement in school life are reflected in the organizational culture that Bryk and his associates found to be typical of the Catholic high schools they studied.

Effective School Cultures

Throughout their field observations Bryk and his associates were struck by the almost total lack of student discipline problems, by the students' involvement in their schoolwork, and by the long hours teachers devoted to their work, at much lower salaries than those earned by teachers in the public sector. They also found an organizational culture that differed significantly from what they had observed in public schools:

> Whether sitting in an English class of twenty-five students, walking school corridors during class breaks, sitting in crowded lunchrooms while students were eating, or attending a sporting event after school hours, we were struck by the pervasive warmth and caring that characterized the thousands of routine social interactions in each school day. Coupled with this, we heard the claim "we are a community" repeated often. For adults, especially principals, the idea of building and nurturing a school community was a major concern.[48]

As social scientists Bryk and his colleagues report that their "initial reaction to such rhetoric was skepticism. What does it really mean to talk about a school as a community?"[49] But, as they thought more about their observations, they concluded that "this was more than a rhetorical exercise. We sensed something special in the organization of these schools, above and beyond the constrained academic structure, that was central to their operations."[50] This "something special" involves what we have described in previous chapters as organizational culture—the norms and values that characterize the schools and the ways in which the students, staff, and parents form a cohesive and caring social network. Bryk and his associates referred to this culture as a "communal organization."

Even though the field observers had been carefully trained and the schools carefully chosen, Bryk and his associates realized that their field observations might be unique to the schools they had visited or perhaps biased in some other way. To check out this possibility, they turned again to the large "High School and Beyond" data set.

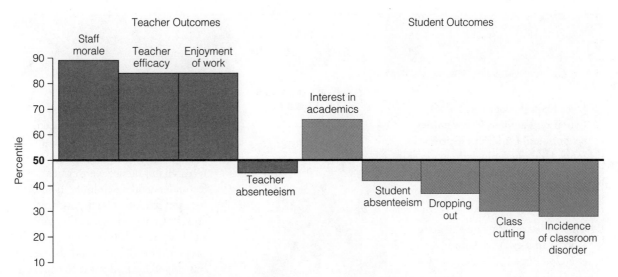

FIGURE 15-4 *Estimated Improvements in Teacher and Student Outcomes If the Average Public School Had a Communal Organization Like the Average Catholic School*

Note: The data represent the percentile score a public school would receive on each measure if it resembled the typical Catholic school on all of the measures of communal organization. The base figure (the way that public schools are now) is the 50-percentile mark. So, for example, if public schools had a communal organization similar to that of Catholic schools, they would be predicted to move from the 50th percentile on "teacher efficacy" to the 84th percentile.

Source: Bryk, Lee, and Holland, 1993, p. 288.

Using data that had been gathered from students, teachers, and administrators, Bryk and his associates developed several measures that were related to their idea of a communal school organization. These measures fell into three categories: (1) the extent to which administrators, teachers, students, and parents shared values and beliefs about the school's purpose and goals, the material that students should and could learn, and the conduct of teachers and students; (2) the extent to which all members of the school participated in schoolwide events, sharing experiences and building common traditions; and (3) the extent to which the relationships among staff and students indicated an "ethic of caring," especially collegial relationships among the adults and a tendency to see students as whole persons rather than just "problems to be solved."[51]

The hypotheses based on their field observations received strong support. Bryk and his colleagues found large differences between public and Catholic schools on each of these measures. That is, the Catholic schools consistently had broader agreement on the goals of schooling and expected behaviors, greater participation in schoolwide activities, and much more collegial and satisfying interpersonal relationships.

In addition, and perhaps even more important, these differences help explain the disparity in achievement between the two types of schools. Using complex statistical manipulations, the researchers estimated how public high schools might differ as a group if their organizational cultures had the cultural characteristics typical of the Catholic schools. The results of this analysis are summarized in Figure 15-4. As you can see, if public schools were to have more "communal organizations," the lives of teachers and students would change markedly. Teachers would have a stronger sense of efficacy, enjoy their work more, and have higher morale. Students would be far less likely to cut class, be disorderly in class, be absent, or drop out of school, and they would be more interested in academics. It's important to remember that these results all represent averages: *on average,* Catholic schools do better than public schools; however, public schools that have a constrained academic structure and a communal organizational culture do just as well as Catholic schools.

Many other less extensive studies of both high schools and elementary schools have produced results similar to those of Bryk and his associates. In many ways these findings parallel those regarding effective teachers that were described earlier in the chapter. Effective schools are organized in ways that promote both instrumental and expressive activities. In terms of instrumental activities effective schools emphasize norms that promote academic excellence and expectations of high academic achievement from all their students. In terms of expressive activities,

BOX
15-1

SOCIOLOGISTS AT WORK

Joyce Epstein

If families are so important, how can schools help all families increase students' positive attitudes and success in school?

Joyce Epstein, codirector of the Center on Families, Communities, Schools, and Children's Learning at Johns Hopkins University, has spent many years helping schools develop more effective partnerships with families and communities.

Where did you grow up and go to school? In New York City, I attended P.S. 23, P.S. 20, and Flushing High School where, year after year, I found that school was the best place to be! There was much to learn from mostly excellent teachers, and many challenging and engaging activities.

In addition to working very hard and loving to learn, I sang in the chorus, edited the newspaper, joined a "political party" for student government elections, and participated in many other activities. My spirited teachers inspired me to want to be a teacher—one of the professions that welcomed women at the time. After obtaining a bachelor's degree from Lesley College, I earned a master's degree at Harvard University's Graduate School of Education in a program that included sociology, anthropology, psychology, and human development. I taught school for several years because it seemed to me that education researchers should know about the schools they study. After a taste of the "real world" of work with hundreds of students, families, teachers, principals, lesson plans, and innovations for school improvement, I went back to graduate school.

Why did you become a sociologist? For my Ph.D. program, I selected Johns Hopkins University's sociology department (then called "social relations") where Professors Coleman, McPartland,

McDill, Holland, Entwisle, and others helped me see that, indeed, sociology of education, quantitative research methods, and the use of research in practice were critical for understanding children, families, and schools. The department and the Center for Social Organization of Schools (CSOS) not only deepened my thinking and abilities but helped me find a way to combine all of my earlier experiences in a new and exciting profession.

How did you get interested in studying the relationships among families, schools, and communities? In the 1970s, many sociologists were arguing about which is more important for students' success—families or schools. Studies that I conducted with James M. McPartland showed that both contexts—families and schools—were important for students across the grades. In my next set of studies, I changed the research question from "Are families important?" to "If families are so important, how can schools help all families do those things that increase students' positive attitudes and success in school?" This redirection required attention to whether and how schools reach out to work in partnership with families and communities in order to facilitate all children's learning. This approach also helped redefine "parent involvement" as "school, family, and community partnerships" in order to recognize that all three contexts share responsibilities for children through the school years.

Has your research been replicated by you or others? What were the results? Based on many studies, I developed a theory of "overlapping spheres of influence" that recognizes the shared goals and

effective schools provide structures that maintain pleasant and orderly environments for students and staff, promote cooperation among students, and bolster the morale and self-confidence of students and teachers. School principals often have been found to play a key role in developing these effective schools. Like Mrs. Washington at Oak Street Elementary,

principals of effective schools manage to structure an environment in which students are expected to learn and succeed, as well as one in which individuals feel supported by and connected to other people.[52]

These findings also involve the familiar ideas of social capital and role expectations. Students do better when they have more social capital associated with

missions of families, schools, and communities. This contrasts with traditional theories that emphasize the importance of separate goals and missions for effective organizational functioning. My work also has produced a framework of six major types of involvement that helps integrate social research and school improvement, so that results of studies of school-family-community partnerships can be used by educators to improve their programs and practices. My colleagues and I developed surveys that help researchers study and educators use data from teachers, parents, and students to plan and implement better programs of partnerships. The theory, framework, surveys, and studies provide a base on which others can build research, policies, and practices. Many studies have confirmed that if schools implement practices to involve families, then parents respond by conducting those practices, including single parents, working mothers, poor parents, and others who might not become involved on their own.

What has been the reaction to your research among your colleagues? Has it surprised you in any way?
When I started this line of studies, school-family-community connections were "peripheral" in the sociology of education, and schools were rarely studied by sociologists of the family. Over the past few years, it is surprising and gratifying to see many master's and doctoral dissertations in colleges and universities in sociology and several other disciplines, both in this and other countries, that build on my work with new and deeper questions of the organization, structure, processes, policies, and effects of school, family, and community partnerships.

What do you think the implications of your work are for social policy? Have any of these implications been

realized? Research on school-family-community partnerships has had direct effects on federal, state, and local policies of parent involvement. One early example is California's state policy of parent involvement, approved in 1989, which specifies the importance of the six major types of involvement that were identified in our studies. Utah established a Center for Families in Education, and other states including Wisconsin, Maryland, Alaska, and several others are moving forward with policies and leadership to assist districts and schools to implement and maintain comprehensive programs of partnership. In 1994, the National PTA and others influenced Congress to approve an eighth national goal for school programs of partnerships with parents. Thousands of schools and school districts across the country are working on this agenda. In 1995, I initiated the National Network of Partnership–2000 Schools to assist policymakers and educational leaders at the state, district, and school levels to use research as the basis for their plans and programs to improve school, family, and community connections.

How does your research apply to students' lives?
Research on school-family-community partnerships is important for any student who is or will become a parent, teacher, counselor, administrator, or social worker or otherwise work with children, families, schools, or communities. Everyone needs to understand how to work together to help all students succeed. My work is just one example of sociological research that examines both what is and what might be. Sociologists need to be able to describe, analyze, and understand existing social conditions and relationships and design, implement, and study interventions that will improve social conditions and relationships.

their role as students—when they have stronger connections with adults who care about them and their future. This involves not just school staff but also parents. Effective schools, like Oak Street Elementary, involve parents and teachers together in promoting children's education. Students also learn more when they are in situations in which the expectations sur-

rounding the student role are those of high performance. Thus teachers at Oak Street Elementary expect the children to do well, and they structure the classroom and school to help achieve this result.

One way to promote the development of social capital and expectations of high academic achievement is to strengthen the bonds between schools, families, and communities. Sociologist Joyce Epstein has studied these connections for many years. Based on this research, she has developed programs that help schools and families work closer together. To learn more about Epstein's work, see Box 15-1.

At the beginning of this chapter, using a macrolevel analysis, we saw how schooling has become a feature of societies around the globe as modern nation-states have established school systems as the means by which young people learn the role of citizen. Although worldwide there appear to be common ideological tenets that underlie the development of these schools, the reality is that few schools actually meet the goals of providing equal opportunities to all children. Using a microlevel analysis, we saw how some parents and some teachers help to attain this goal, enabling children to learn to their full potential, no matter what their individual backgrounds. Mesolevel analyses show how teachers, administrators, and parents can structure schools into organizations that help all students attain these goals. To use the metaphor of a river journey, these analyses have shown how crews can alter the design of vessels that are used in educational journeys to make these trips more successful for all children. Many sociologists who study education hope that these findings can help educators, such as Abby and the staff at the schools where she worked, develop more effective schools for all children and thus approach the ideals that have been associated with mass schooling in modern societies (see Box 15-2, "Applying Sociology to Social Issues").

CRITICAL-THINKING QUESTIONS

1. Do you think that the concept of "effective schools" can be applied to higher education, as well as at the elementary and secondary levels?

2. How might you define effective colleges and universities?

3. How could a researcher study characteristics that differentiate less effective and more effective higher education organizations and classrooms?

SUMMARY

Education as a social institution involves the groups, organizations, norms, roles, and statuses associated with a society's transmission of knowledge and skills. Schools are the formal organizations developed to carry out this task. Education and schooling are the focus of debate, often revolving around the relationship of education to other social institutions. Key chapter topics include the following:

- Mass schooling developed in Europe and North America in the eighteenth and nineteenth centuries and has now spread throughout the world. Literacy and school enrollment levels have increased in recent years in the developing and least developed countries.

- Functionalist explanations for the emergence of mass schooling suggest that schools developed to meet the needs of industrialized societies. However, empirical evidence suggests that schools developed independently of industrialization as the expansion of an educational system became associated with the definition of a modern nation-state. Schools became the way to socialize children into the role of citizen.

- The ideological tenets underlying common schools—universality, egalitarianism, standardization, and individualism—seem to appear in all countries, but few, if any, nations have developed schools that actually meet these goals.

- Reginald Clark's research showed that families' social networks, role expectations, and interaction patterns can promote high school achievement, even among students from families whose income levels and family structures might indicate academic difficulties.

- The research of Laurence Steinberg and his colleagues showed how peers can be an important influence on the effort that students devote to their schooling.

- The research of Eigel Pedersen and his associates demonstrated how a highly effective teacher could have a life-long impact on the success of her students. This and other research indicate that the characteristics of effective teachers parallel those of effective organizational leaders, as described in Chapter 11.

- The research of Anthony Bryk and his associates showed that Catholic schools come closer to the goal of egalitarianism than do public schools because their structure is less differentiated and promotes more academic involvement of all students; their smaller size promotes greater student involvement and commitment; and their school cultures promote learning and cohesive social networks. These findings support those of researchers who have studied effective schools in many other settings.

In addition to the variables related to school structure, size, curriculum, and culture, Bryk and his associates theorize, Catholic schools are more effective than public schools because they form a "voluntary community," one that students, staff, and parents choose to participate in. They suggest that part of the strength of Catholic schools comes from the moral authority of this community and from the post-Vatican II Catholic social ethic that explicitly deals with the relationship between moral issues and social action. Both the way in which many inner-city Catholic schools have reached out to enroll non-Catholic residents of their communities and their greater economic and racial-ethnic diversity reflect this social ethic.

Bryk and his associates realize that it would be very difficult for public schools to emulate the moral and spiritual structure of Catholic schools or to develop

BOX 15-2

APPLYING SOCIOLOGY TO SOCIAL ISSUES

Building Better Schools

schools that embody moral authority. Yet they also suggest that schools need to deal with moral issues. As they put it,

A democratic education demands a melding of the technical knowledge and skill to negotiate a complex secular world, a moral vision toward which that skill should be pointed, and a voice of conscience that encourages students to pursue it . . . In educating for postmodern life, schools must simultaneously help students develop the necessary technical competencies to function in the twenty-first century while also

encouraging a sense of hopefulness in confronting the unknown and a sense of belonging in a large, complex, and highly specialized society.[53]

CRITICAL-THINKING QUESTIONS

1. Should such a voluntary community and moral commitment be developed in public schools such as those Abby taught in? Do you think changes such as these could help enhance the effectiveness of schools? Why or why not?

2. Could such communities be developed in large schools with diverse student bodies?

3. How could school officials ensure that the input of all parents and families would be respected? How could schools today encourage a sense of hope and belongingness? What difference would it make if they did?

KEY TERMS

common schools 395
education 393
expressive leadership 405

instrumental leadership 405
mainstreaming 402
mass schooling 395

nation-state 400
schooling 393

INTERNET ASSIGNMENTS

1. The U.S. Department of Education publishes several newsletters and research reports on its Web site. Find the site and a report that discusses student achievement. Compare the findings in the report to those discussed in this chapter. How are they similar? How are they different?

2. The Internet has numerous resources that can be used by teachers and parents who want to help their children. Explore the Internet to examine some of these resources and think about how they might be used by teachers in different settings. Can the Internet help teachers and schools better meet the goals that underlie common schools? Explain your answer.

INFOTRAC COLLEGE EDITION: READINGS

Article A20505588 / Jencks, Christopher, and Meredith Phillips. 1998. "The black-white test score gap: Why it persists and what can be done." *Brookings Review, 16*, 24–28. Jencks and Phillips review data on racial-ethnic differences in educational achievement and ways to address this issue.

Article A20505589 / Darling-Hammond, Linda. 1998. "Unequal opportunity: Race and education." *Brookings Review, 16*, 28–33. Darling-Hammond describes how schools can be structured to enhance student achievement.

Article A20411709 / Alexander, Karl L. 1997. "Public schools and the public good." *Social Forces 76*, 1–30. Sociologist Karl Alexander eloquently reviews a great deal of evidence to show how schools promote the well being of individual students and the public good despite often daunting problems in families and communities.

FURTHER READING

Suggestions for additional reading can be found in *Classic Readings in Sociology*, bundled with this book. If you purchased a used copy of this book that does not include this custom-published reader, go to www.sociology.wadsworth.com for ordering information.

RECOMMENDED SOURCES

Bidwell, Charles E., and Noah E. Friedkin. 1988. "The Sociology of Education." Pp. 449–471 in Neil Smelser (ed.), *Handbook of Sociology*. Newbury Park, Calif.: Sage. The authors provide a historical overview of the field of sociology of education.

Boli, John. 1989. *New Citizens for a New Society: The Institutional Origins of Mass Schooling in Sweden*. Oxford, England: Pergamon. Boli discusses the development of mass schooling in Sweden, linking these experiences to the worldwide phenomenon of the development of common schooling.

Brown, David K. 1995. *Degrees of Control: A Sociology of Educational Expansion and Occupational Credentialism*. New York: Teachers' College Press. Brown examines why the expansion of higher education was so much greater in the United States than in other countries.

Bryk, Anthony S., Valerie E. Lee, and Peter B. Holland. 1993. *Catholic Schools and the Common Good*. Cambridge, Mass.: Harvard University Press. This is an in-depth discussion of the research featured in this chapter.

Clark, Reginald. 1983. *Family Life and School Achievement: Why Poor Black Children Succeed or Fail*. Chicago: University of Chicago Press. The results of this study are discussed in the chapter and summarized in Table 15-1; the book gives many in-depth examples of how families can facilitate or hinder school achievement.

Gregory, T. B., and G. R. Smith. 1987. *High Schools as Communities: The Small School Reconsidered*. Bloomington, Ind.: Phi Delta Kappa Educational Foundation. The authors describe how the structure of smaller schools promotes more effective learning climates.

Hurn, Christopher J. 1993. *The Limits and Possibilities of Schooling: An Introduction to the Sociology of Education*. Boston: Allyn & Bacon. This sociology of education textbook focuses on many of the issues discussed in this chapter.

Steinberg, Laurence. 1996. *Beyond the Classroom: Why School Reform has Failed and What Parents Need to Do*. New York: Simon and Schuster. This is an extensive and readable discussion of the influence of families and peers on students' achievement.

Stockard, Jean, and Maralee Mayberry. 1992. *Effective Educational Environments*. Newbury Park, Calif.: Corwin/Sage. This is an extensive overview of studies regarding effective schools that links their results to sociological theory.

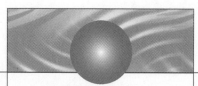

Pulling It Together

Listed below are other chapters where you can find additional discussion or examples
of some topics covered in this chapter. You can also look up topics in the index.

TOPIC	CHAPTER
Subcultures	3: Culture and Ethnicity, pp. 53, 58–67
	6: Deviance and Social Control, p. 140
	11: Formal Organizations, pp. 284–285
	12: Religion, pp. 320–321
Structural functionalism	3: Culture and Ethnicity, pp. 55–56
	7: Social Stratification, pp. 163–165
	10: The Family, pp. 248–251
Social networks and social capital	4: Socialization and the Life Course, pp. 77–79
	5: Social Interaction and Social Relationships, pp. 101–104
	6: Deviance and Social Control, pp. 140–144
	7: Social Stratification, pp. 179–180
	8: Racial-Ethnic Stratification, pp. 200, 203–204
	10: The Family, pp. 257–258, 265–266
	11: Formal Organizations, pp. 282–283, 287–288
	12: Religion, pp. 320–321
	13: The Political World, pp. 339–352
	14: The Economy, pp. 361–371, 376–385
	18: Communities and Urbanization, pp. 487–492, 495–503
	20: Collective Behavior and Social Movements, p. 535–536, 544–545
Family interactions	4: Socialization and the Life Course, pp. 77–83
	10: The Family, pp. 257–261
	17: Population, pp. 470–472
Stratification and social class	5: Social Interaction and Social Relationships, pp. 113–120
	7: Social Stratification, pp. 154–181
	8: Racial-Ethnic Stratification, pp. 187–211
	9: Gender Stratification, pp. 217–238
	13: The Political World, pp. 334–352
	14: The Economy, pp. 363–385
	16: Health and Society, pp. 422–443
	18: Communities and Urbanization, pp. 496–503
Global stratification and differences among industrialized, developing, and least developed countries	14: The Economy, pp. 363–371
	16: Health and Society, pp. 433–437
	17: Population, pp. 449–475
	18: Communities and Urbanization, pp. 484–487
	19: Technology and Social Change, pp. 524–527

Pulling It Together

Continued

INTERNET SOURCES

Education Place, Discovery Channel, NASA, and PBS. Each of these Web sites has a large collection of educational resources for teachers, parents, and students.

The Web site for the Center on School, Family, and Community Partnerships describes much of the current work done by Joyce Epstein, the sociologist profiled in Box 15-1.

Health and Society

The Health of Individuals: A Microlevel Perspective

Placement on the Health Hierarchy

Stress and Health

Health-Related Behavior

The Organization of Health Care: Mesolevel Analyses

The "Fee-for-Service" System of the United States

Systems of Guaranteed Health Care

Health and Global Stratification: A Macrolevel Perspective

Historical Changes in Health and Disease

Health and Global Stratification

FEATURED RESEARCH STUDY

Inequality and the Health of Nations: The Work of Richard G. Wilkinson

FROM THE TIME she was a little girl, Marisa knew that she wanted to work in the field of medicine. When she was accepted as a premed student at State University she felt like her dream was coming true. Some of the courses were hard, but she was doing well. And, until her time at the University Clinic, she had really enjoyed her field experience classes. These courses were required of all premed students at State and involved actual work experience in medical settings.

Marisa's first field experience was in her family doctor's office, where she had gone all of her life. When she was young, Dr. Reed had a private practice that he shared with another doctor. Now he was part of a clinic with a number of other physicians in a managed care program. In her time there Marisa got to work with many of the people from her own neighborhood. She weighed mothers when they came in for prenatal visits and played with babies as she prepared them for their well-baby checkups. She talked with older men and women as they came to have their blood pressure checked and listened to their concerns about their declining health.

Marisa knew that even though she was having a good time in her field experience there were tensions in the office. Lots of things had changed over the last few years—especially on the financial and business end. Several employees worked all the time filling out forms for the managed care system, and she often heard the doctors talking among themselves about their frustrations with "the system." Sometimes, when she ushered older patients into the examining rooms, they would ask, "Do you think Medicare will cover this? I really don't understand how all this HMO stuff works." Some of the other patients would tell her, often in great detail, about the hassles they'd had with the insurance company over their payments.

Marisa's second field experience took place in the outpatient clinic at University Hospital, and she sometimes felt like she had moved to a totally different country. Many of the people who came to the clinic were on Medicaid, a government program that pays for the health care of poor people. Others didn't have any type of health insurance. In general, Marisa sensed that the people who came to the clinic were sicker than were those who came to her doctor's office. There were only stiff hard chairs in the waiting room, and some of the people looked like they could barely sit up in them. The coughing and wheezing in the waiting room was almost more than she could handle, and she felt like she was constantly washing her hands to get rid of germs. Occasionally people would come in with diseases that she thought had pretty much disappeared—like tuberculosis and whooping cough. Others seemed to come in when they were really sick with diseases that were pretty ordinary, like the flu or bronchitis. Expectant mothers often didn't come in for their first visit until they were several months along, and many of them didn't seem to be taking the extra vitamins that Marisa knew they needed.

The staff at University Clinic was terribly busy and seemed just as worried and frustrated with "the system" as the staff at her neighborhood doctor's office. Still, all the people who worked at the clinic were really nice to Marisa, especially Dr. Jefferson, a young resident who had grown up in the neighborhood around the hospital. She would always find time to explain to Marisa what was happening, even if she had to stay a little bit late to do so. Marisa also enjoyed watching Dr. Jefferson interact with the patients. While many of the doctors seemed to just quickly tell the patients what they needed to do and didn't even explain why, Dr. Jefferson took the time to talk to each patient carefully and not in a way that made them feel ignorant or like they were taking too much time.

But then Marisa had an experience that made her begin to wonder if she had chosen the right career. Dr. Jefferson had let Marisa sit in on the classes she held for expectant mothers, and Marisa had become especially fond of Tessa, a young woman who was a little younger than Marisa and expecting her first baby. Like so many other expectant mothers in the neighborhood, Tessa had started her prenatal care several months after she knew that she was pregnant. Still, once she started to come to the clinic she had been very faithful in taking her vitamins and attending the classes. Marisa had helped her learn about childbirth and how to take care of the baby. As Tessa's due date approached, Marisa became more and more excited because Tessa had asked her to be at the birth.

Early one morning, a few weeks before Tessa's due date, Marisa's phone rang. It was Dr. Jefferson, who told her that Tessa had entered labor and was at University Hospital. Marisa pulled her clothes on and ran from her dorm to the delivery area, where she donned sanitary robes and went in to be with Tessa. She held Tessa's hand and soothed her throughout labor, anxiously awaiting the moment when they could hold a beautiful new baby in their arms.

But that moment never came. When Tessa's baby boy was born he was beautiful, but very small— much smaller than he should have been. Over the next few days he had many health problems, problems that Marisa had never seen or heard of in all of her months at her neighborhood doctor's office. Dr. Jefferson quietly told Marisa that the baby would probably not live and, together, they tried to prepare Tessa. A few days later the baby died, and Marisa cried with Tessa like she had never cried before.

As she walked back to her apartment after the baby's funeral Marisa couldn't stop thinking about the contrast between Dr. Reed's practice and University Clinic and about her own future in medicine. "What kind of doctor will I be?" she wondered. "Could I really ever help people like Tessa? Could her baby have lived if we'd done something different? And why is University clinic so different from Dr. Reed's office? Why do people seem so much sicker at the clinic?" Thinking about her biology courses, Marisa asked herself, "Is it just genetic—are some of us just genetically prone to be healthier than others? Are some people just exposed to more germs? Or is there something else going on as well?"

*H*EALTH IS ONE of the most important aspects of our lives. Because most of us are healthy most of the time it is often hard to imagine what it would be like if we weren't. But even a few days of sickness or disability, or, like Marisa, seeing the devastation that poor health brings to those who are close to us, can make us realize how precious health is. When we aren't healthy it is harder, and sometimes impossible, to do most of the things we enjoy in life—from playing to working; and our ill health can affect others who are close to us. The death of Tessa's baby destroyed many of her hopes and dreams and severely affected Marisa as well.

Health is also important on a societal level. When societies are healthier, they have a greater proportion of their citizenry that can contribute to productivity. Societies with larger proportions of their population who are infirm in one way or another face severe problems—not just in caring for those who are ill but in conducting the economic activities that are part of everyday life. In many ways, a healthy population is a key element to both a happy and productive life for individuals and a prosperous economy for a society.

We often think of health as simply "not being sick." But, in fact, health is much more. The World Health Organization, a branch of the United Nations that monitors the status of people around the world, defines **health** as "a state of complete physical, mental, and social well-being." From this perspective, health goes beyond just simply being disease free and includes general well-being—not just physically, but also mentally and socially—feeling good about yourself and others with whom you interact.[1] Thus, although Americans often tend to think of health as simply not being physically sick, this much broader definition encompasses views of health that are common to many other cultures—a fit mind and body, a harmony of body and spirit, and congenial and helpful relationships with others.

The early sociologists were concerned about issues related to health. Durkheim's classic study of suicide discussed in Chapter 1 is the first example of sociological research on health, and much of the early work of the Chicago school, described in Chapter 3, focused on the well-being of immigrant groups and racial-ethnic minorities. Only in the last few decades, however, have sociologists begun to specialize in specific studies of medicine as a social institution, the complex set of statuses, roles, organizations, norms and beliefs that revolve around meeting the health care needs of a society. In part, the relatively late attention to this area reflects the fact that medicine is much more differentiated and complex now than when sociology first began. As we will see later in this chapter, today a relatively large proportion of our workforce is dedicated to meeting health care needs, and health care organizations are large and complex.

Some sociologists who study medicine work directly with health practitioners, often in medical schools, public health agencies, and government organizations, looking for social variables that are related to particular health problems. These sociologists often specialize in **epidemiology,** which literally means the study of epidemics, widespread outbreaks of communicable disease (those that can be transmitted from one person to another, such as smallpox, polio, and measles). Over time epidemiologists have expanded their analyses to include not just these "catching diseases," but many other conditions that threaten health, such as cancer, alcoholism, malnutrition and heart disease. Specifically, *epidemiologists* study patterns and trends in the incidence of disease, injuries, and other health-related areas. They try to find out who is more likely to have certain diseases or conditions and why these patterns occur. Epidemiologists can be sociologists, but they can also be people with degrees in epidemiology and public health, medicine,

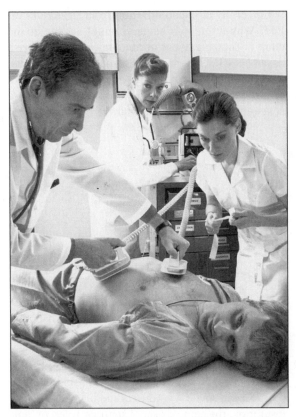

Why are some people healthier than others are? Is there something about the way medical care is organized that contributes to these differences? Or does it also involve other areas of social life?

statistics, biology, geography, anthropology, and veterinary medicine.[2]

Other sociologists who study medicine focus not just on specific health problems and issues but look more at medicine as a social institution both in our society and in other societies. These sociologists are interested in how social conditions help to produce health and illness and how health and illness affect other aspects of our lives. Just as sociologists look at social institutions of the economy, education, religion, the family or the political world, they also examine relationships and organizations within the medical world. Most sociologists who take this perspective work in departments of sociology in universities and colleges.[3]

Whatever particular perspective medical sociologists use or wherever they work, much of their research tries to answer questions such as those posed by Marisa. Why are some people healthier than others are? Is it biological—exposure to germs or inheriting a healthy genetic makeup? Or is it something more? Is there something about the way people live that causes these differences? Is there something about how medical care is organized—the social organization of medicine—that contributes to these

differences? In this chapter we explore their answers to these questions.

As we explore these analyses it might be helpful to remember the metaphor of mountain ranges that was used in our discussions of stratification. Just as individuals rank relatively higher or lower on stratification hierarchies, they also rank relatively higher or lower in health and well being. Similarly, just as some nations are wealthier and more powerful than others are, some are healthier than others are. As you will see, there are close linkages between economic stratification and hierarchies of health. In general, people and nations that are poorer in economic terms tend to be poorer in health. The correspondence to the stratification mountain isn't perfect. Some very rich people are sickly and die at young ages; some very poor people live to be quite old and are very healthy throughout their lives. Some of the richest nations are not the healthiest. Yet, in general, there appears to be a connection—and a fairly strong one. People and groups that are higher on the hierarchy of health also tend to be higher on the stratification mountain.

In this chapter we explore sociologists' analyses of why this occurs. We begin by taking a microlevel perspective and look at why some individuals are healthier than others are. We use a mesolevel view to look at the way that health care is organized, both in our society and in other societies. We then move to a macrolevel perspective and examine variations in health around the world, asking to what extent people in some societies are healthier than people in others, and why these differences occur.

THE HEALTH OF INDIVIDUALS: A MICROLEVEL PERSPECTIVE

How healthy are people in the United States? Who tends to fall at different places on the health hierarchy? Why do these differences occur? Medical sociologists have looked at these questions from a variety of perspectives. One view examines the relationship of social environments, and particularly the influence of stress, on health. Another, not unrelated, view looks at individuals' behaviors—their life styles and their interactions with the medical world. We first examine the health hierarchy to see who is more likely to fall near the top and who is not and then look at explanations of these differences.

Placement on the Health Hierarchy

Sociologists can measure health and well being in many different ways, such as asking people how healthy they believe that they are, listing diseases and

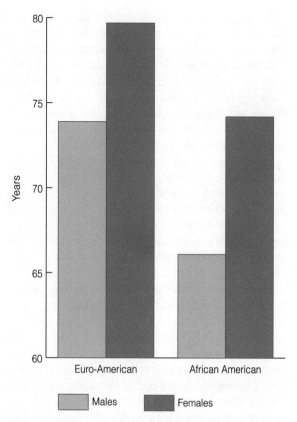

FIGURE 16-1 *Life Expectancy by Gender and Race-Ethnicity, 1996*

Source: National Center for Health Statistics, 1998, p. 200.

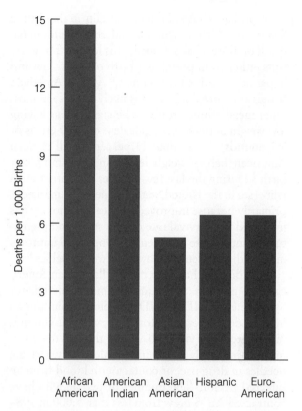

FIGURE 16-2 *Infant Mortality Rates by Race-Ethnicity, 1995*

Source: National Center for Health Statistics, 1998, p. 190

chronic conditions that individuals have experienced, and obtaining measurements of blood pressure, cholesterol levels, and other physical criteria.[4] Although these indicators are useful for describing health at a particular time or in relation to a particular disease, two other measures are commonly used when describing the health of groups within a society or when comparing one society with another. Both of these measures involve the ultimate outcome of health—whether people live or die.

The first measure is **life expectancy,** the number of years that a baby is expected to live when it is born, given the current rates of death within a society. For instance, a baby boy born in the United States today can expect to live until the age of 73; a baby girl can expect to live until she is 79. These data represent the population as a whole. As shown in Figure 16-1, the expectations are strikingly different for African Americans and Euro-Americans. Even though Euro-American females can expect to live almost until their eightieth birthdays, African American females can only expect to live until they are 74. Although Euro-American males can expect to live until age 74, African American males can only expect

to live until they are 66—fourteen years less than a Euro-American female!

These racial-ethnic differences in life expectancy occur throughout the life span,[5] and the second general measure of health, the **infant mortality rate,** captures these differences at a very young age. (The term mortality simply means death.) This rate is calculated like other types of rates, such as the crime rates described in Chapter 6, and tells the number of babies in 1,000 who are expected to die before the age of one in a given year. The infant mortality rate in the United States in 1995 was 7.6, but this varied a great deal from one racial-ethnic group to another, as shown in Figure 16-2. Asian Americans had the lowest infant mortality rate—5.3. Those for Euro-Americans and Hispanics were somewhat higher at 6.3. Rates for American Indians were substantially higher (9.0), and those for African Americans were the highest for any racial-ethnic group for which data are available at 14.6. These data indicate that whereas only 5 to 7 of every 1,000 Asian American, Hispanic, and Euro-American babies may die before their first birthday, more than twice as many African American babies will.

A primary cause of infant mortality in the United States is low birth weight, the underlying factor in the death of Tessa's baby. Usually this low birth rate results either from premature birth or from slow and impaired growth while in utero.[6] African American babies are more than twice as likely as those in most other racial-ethnic groups to be classified as having low weight at birth (the equivalent of 2500 grams or 5.5 pounds). More than 13 percent of all African American babies weigh less than this amount at birth.[7] During the last few decades, the infant mortality rate in the United States has declined markedly, primarily because improved medical technology has increased the survival rate of very small and premature infants. There has been no comparable improvement in the incidence of low birth weight babies.[8]

The disease that most often kills African Americans at mid-life is Acquired Immunodeficiency Syndrome, or AIDS.[9] AIDS develops from human immunodeficiency virus (or HIV), which is communicated from one victim to another through the transmission of bodily fluids in sexual activity, sharing needles in drug use, or contact with blood from an infected person. The number of people who have contracted AIDS grew from less than 5,000 in 1984 to more than 500,000 in 1995, and from 1990 through 1996 AIDS was one of the top 10 causes of death in the United States.[10] In contrast with conditions such as cancer and heart disease, AIDS strikes young people. Almost two-thirds of the deaths attributed to AIDS in the mid 1990s were to people less than 40 years of age. In addition, more than half of all AIDS deaths involve members of minority groups.[11]

Epidemiologists and medical sociologists have done a great deal of research trying to figure out why racial-ethnic differences in disease and death occur. Although some conditions appear to have a genetic basis, such as the greater prevalence of hypertension, diabetes, and sickle cell anemia among African Americans, most researchers now believe that the major reason underlying the large differences in illness and death is socioeconomic status. Because members of minority groups, and especially African Americans, are less likely than Euro-Americans to belong to the middle, upper-middle, and upper classes, they tend to have lower levels of health, lower life expectancies, and higher rates of infant mortality.[12] Marisa's observations of the clientele of the places where she worked was correct—poor people do tend to be sicker and in generally poorer health than their fellow citizens.

Epidemiologists have found that the relationship of social class to health is very strong. In the United States and throughout the world, people at the bottom of the socioeconomic ladder tend to be less healthy than those toward the top are. These differences occur at each stage of the stratification hierarchy. The working poor are healthier than the underclass, people in the upper middle class are healthier than those in the middle class, and those in the upper class are healthier than those in the upper middle class.[13] Health is also related to social mobility. Within any particular class group, the healthiest are most likely to have upward mobility and the least healthy are most likely to have downward mobility.[14]

Sociologists have explored various factors that might help explain why the health hierarchy is so strongly related to social class. One major factor appears to involve stress and how we are able to deal with this stress. People at lower ends of the stratification ladder experience more stress, and they have fewer resources with which to combat this stress.[15] For instance, they are more likely to live in areas that are crowded and have poor sanitation, inadequate heating and cooling, and greater exposure to various environmental hazards. They are also less likely to be able to afford healthy diets and tend to have jobs that are more stressful and dangerous. Another major factor appears to involve health-related behaviors. Those at the lower end of the stratification ladder are often less able to adopt healthy life styles or to have relationships with health care providers that promote the best health outcomes.

Stress and Health

It is hard to pick up a newspaper or magazine without seeing an article on "the harmful effects of stress," or "how to cope with stress in your life." Students often talk about the pressures of school such as deadlines, exams and just too much to do. Parents talk about the stresses involved with raising families. Workers lament the strains and pressures of their jobs.[16]

Some stressful events are catastrophic—such as losing a job, having a loved one die, or living through a natural disaster. Much more often, however, stress takes the form of daily "hassles"—inconveniences and persistent problems that can just seem to pile up in life. Although both major life changes and daily hassles appear to be associated with changes in health, research in the United States and cross-culturally indicates that the day-to-day hassles of living, the buildup of too much daily stress, is more strongly related to the incidence of illness.[17]

Physiological Reactions to Stress The way in which daily hassles and the buildup of stressful events affect the probability that we will get sick can be understood by knowing how our bodies work. You have no doubt noticed that when you start to get nervous or scared your palms will get sweaty and your heart will beat faster. This is a natural reaction—one that your body has been primed to perform when you are in threatening situations. When you sense that you are in a dangerous situation your body automatically responds by increasing the flow of blood to organs and muscles that you might need to defend yourself. Glands secrete extra doses of hormones that increase your heartrate and blood pressure and mobilize fatty acids to provide an extra burst of energy. All of this occurs outside the conscious or voluntary control of the central nervous system in what is called the autonomic nervous system.[18]

These involuntary responses no doubt evolved as a way for humans to protect themselves against dangerous predators through engaging in physical battles. Today we rarely, if ever, face the physical dangers that our distant ancestors might have encountered. Most of the threats we face are symbolic in nature, and social norms strongly disapprove of a physical response and often even of a verbal response, such as a sharp retort or an insult. The result is that our bodies become prepared for physical action, even combat, but this event never comes. Eventually, repeated exposure to the physical consequences of stress can damage our health. Scientists have not totally figured out what all of these effects may be, but a great deal of evidence indicates that, over the long term, stress is related to the incidence of a large number of conditions—from cardiovascular disease and the related heart attacks and strokes, to asthma, muscular pain, compulsive vomiting and other gastric problems, headaches, eye conditions, and a weakened immune system, which, in turn, increases the probability of contracting infections and even cancer and AIDS.[19]

Even though we know that stressful events produce these reactions in the bodies of all humans, and other animals as well, particular events are not equally stressful to all people. You might get sweaty palmed and feel your heart race during a scary movie, while your friends may simply laugh at the screen. In other words, what is quite stressful to one person might not be stressful to others. These differences are influenced by *social* factors—both our interpretations of events and the support that we receive from others.

Interpreting Stressful Life Events One way to understand why some events are more stressful to some people than to others is to use the ideas involved in symbolic interaction theory, described in Chapter 5. According to symbolic interactionism, social interaction involves a constant process of presenting and interpreting symbols through thinking about what another person is trying to communicate. Symbols can involve words, gestures, pictures, colors, emotions—any thing that we can think about

and interpret. Each of us might interpret the symbols that we encounter differently, and how we interpret these symbols determines whether or not they will be stressful to us.[20]

The way we interpret and respond to stressful events is influenced both by our emotional temperament, which appears to be highly influenced by our genetic makeup, and by socialization, the process by which we develop the ability to relate to others and learn about the social world.[21] Through the socialization process we develop views about ourselves, about others, and about the world around us. Part of this involves learning about what is dangerous and what is safe. For instance, suppose you were hiking on a trail that had a sign posted, "Look out for rattlesnakes." If you grew up in an area infested by rattlesnakes and had learned about when and where they would be dangerous to you, you might proceed on the trail without worry. If, on the other hand, you or others you knew had had bad experiences with rattlesnakes or been taught that snakes were always terribly dangerous, you might simply decide to cancel your trip. The extent to which the sign induced stress would depend upon your prior socialization.

Socialization also involves learning ways of interacting and responding to stressful events and ways of managing our personality characteristics, such as a tendency to be excitable, irritable, or placid. This learning process can affect the way that we respond to potentially stressful events. For instance, you might have once been very afraid of heights, and the thought of leaving the ground in an airplane or on a chairlift made your palms sweat and your heart race—a classic stress reaction. Over time, however, you might have learned that your fears were not based on fact and you discovered how to manage your anxieties. Similarly, when you first started college you might have felt overwhelmed with the demands of different classes and instructors and the new environment of your campus. Gradually, however, even though the difficulty of your classes did not change, you learned to handle the situation and now find it hard to even recall how stressful your first few weeks were.

In general, stress, like beauty, appears to be in the eye of the beholder. Whether or not we interpret something as stressful depends largely on how we interpret it. These interpretations, in turn, are highly influenced by our socialization experiences. They are also influenced by the support we receive from others—our *social networks.*

Combatting Stress In his classic study of suicide, Durkheim found that societies and groups with greater integration and a stronger collective consciousness tended to have lower suicide rates. On a more individual level, numerous contemporary studies have found

People who have strong levels of social support tend to deal better with stress and have fewer health problems in general.

that social support—strong social networks—can buffer the effects of stressful events. People who have strong levels of social support tend to have fewer health problems in general. In addition, when difficult events do occur, such as deaths or serious illness, those with strong webs of social support tend to have fewer negative reactions.[22]

The exact way in which social support counters the physiological effects of stress is not yet known. It appears, however, that as good relationships with others promote feelings of belonging and being accepted and needed, we tend to develop a higher general sense of well-being. This, in turn, helps relieve the symptoms of tension that can accompany stressful events.[23] You can see how this happens by considering how you and friends or family members might talk about a stressful event after it occurs. Suppose your boss had just scolded you and your co-workers about a mishap at work. You would probably feel nervous and sweaty palmed during the encounter with your boss. After he leaves the room, however, you and your co-workers might talk about it and perhaps even joke a little. You then probably start to calm down and begin to feel better. If you didn't have co-workers or friends to talk with, you might be feeling the stress of the encounter for much longer. In general, people who have strong links with others— through marriage, strong friendships, or membership in religious and other groups—tend to be better able to ward off the unhealthy effects of stress.[24]

Health-Related Behavior

Warding off stress is only part of the explanation of a healthy life. Another major factor involves our own health-related behaviors. These behaviors fall into two broad categories. Some, such as going to doctors when we are sick or for regular physical exams to check for the possibility of illness, involve contact with the health care profession. Others—such as eating a balanced diet, getting enough sleep, exercising regularly, and avoiding harmful substances—have little to do directly with health care workers and organizations and instead involve our own life styles and choices. People tend to vary in both types of behaviors, and sociologists have tried to understand why this happens.

Healthy Living and Status Groups Through the nineteenth century in the United States and other developed nations and even today in developing countries, improvements in health tended to come from providing clean water supplies, adequate sewage systems, and nutritious diets. From about 1900 to the 1960s, improvements in health generally resulted from medical advances such as vaccinations against disease and the development of new drugs. In the last few decades in rich, highly industrialized countries such as the United States, most advances in health have resulted from various social and environmental factors. Those of us who choose healthy lifestyles—as in not smoking, adhering to low-fat diets, and getting adequate exercise and preventive medicine—tend to live longer and healthier lives.[25] In fact, physicians estimate that more than a third of all "premature deaths," deaths that aren't caused by either old age or by some genetic disease, are caused by tobacco use, improper diet, lack of exercise, and alcohol abuse.[26] In other words, unhealthy lifestyles are a major factor contributing to early death in the contemporary United States.

People with more economic resources are better able to support healthy lifestyles, such as joining a health club.

Healthy lifestyles have been valued in American society for many years. In the eighteenth century Ben Franklin's motto, "Early to bed, early to rise, makes a man healthy, wealthy, and wise," became popular. Throughout the nation's history, religious groups, particularly the many Protestant groups that developed in the United States, have promoted healthy life styles including exercise, abstaining from tobacco and alcohol use, and good diets. In recent years presidential commissions and awards promote fitness and healthy living, and many large corporations support "wellness programs," all aimed at promoting better health.[27]

Even though evidence suggests that people at all levels of the stratification ladder try to lead healthy lives,[28] those who are poor are far less likely than other people are to use preventive care, health services that are designed to prevent disease or look for illness in its very early stages.[29] For instance, doctors recommend that pregnant women begin having regular checkups and larger doses of vitamins as soon as they know they are pregnant. Yet, even though more than 80 percent of all Euro-Americans, Filipino-Americans, Japanese-Americans, Chinese Americans, and Cuban Americans receive prenatal care in their first trimester (the first 3 months of pregnancy), only a little more than two-thirds of all African Americans, American Indians, and Mexican Americans see doctors at that time.[30]

Max Weber's analysis of social stratification, which was discussed in Chapter 7, can help us understand these differences. Weber suggested that one dimension of stratification was status, or *prestige* —communities or social networks of people with similar lifestyles and viewpoints. For instance, people in one status group, perhaps like those who go to

Marisa's family doctor, might shop at upscale organic grocery stores, exercise at the neighborhood health club, vacation at ski resorts in the winter, take bike tours through Europe in the summer, and have regular exams from their doctor and dentist. People in another status group, perhaps like those who go to the University Hospital Clinic, might depend on foodstamps for their food, shop at the corner convenience store because they don't have transportation to go elsewhere, cannot begin to afford membership in a health club, and get their health care through the local public health clinic or from whatever doctor they can find who will take a Medicaid card.

Weber stressed that status or prestige groups arise from economic distinctions and can also reinforce them. In other words, our economic circumstances influence the kinds of life styles that we can choose. The French sociologist Pierre Bourdieu, who has written extensively on the multiple ways stratification affects our lives, notes that we are able to make choices about our life styles, but that the autonomy needed to make choices is determined by the social and economic conditions in which we live. Even if clients of the University Clinic wanted to belong to a health club, have a personal trainer, and eat only organic foods, the circumstances of their daily lives— their positions in the social class hierarchy—would not allow them to do so. In short, people who have more resources are much more able to support healthy life styles—from eating a wide variety of fresh foods to having time and money to exercise.[31] These circumstances and choices have cumulative effects. Better health not only helps individuals maintain their social position, it also facilitates social mobility, reinforcing and maintaining economic differences.[32]

Interactions with Health Care Providers In addition to getting different levels of preventive health care, social class groups tend to differ in their interactions with health care providers. In general, recent research indicates that people from middle and upper-middle class backgrounds are more likely than others are to take a consumer oriented approach to health care. More highly educated and more affluent people tend to make decisions on their own about when and what type of health care is needed. They also tend to be much more active participants in the physician-patient encounter and, at least in non-emergency situations, to be much more likely to question their doctor's decisions. They tend to see interactions with health care providers as a mutual participation in treating health by actively presenting their ideas to their doctors and receiving more personalized service in return. In contrast, those with less education tend to be more accepting of doctors' authority, to exercise less personal control over medical decisions, and to have more impersonal interactions.[33]

These differences probably reflect social distance. When patients and physicians share social class backgrounds, they tend to have a more effective and equitable communication style. Most physicians come from middle and upper-middle class families and feel more comfortable interacting with those from similar backgrounds. Similarly, people from those backgrounds feel less intimidated by physicians. Dr. Jefferson, who worked with Marisa at the University Clinic, was, unfortunately, rather exceptional in her ability to communicate well with people from a variety of backgrounds.[34]

Even though the poor are less likely to seek medical help for preventive care and tend to have less personal relationships with their care providers, they actually are more likely than other people to seek help from physicians. This reflects differences in overall health: Poor people tend to be less healthy than those with more resources and are more likely to need medical attention. In fact, when the actual needs of patients are considered, low-income people tend to use *fewer* services relative to their needs. Extensive analyses of the reasons that low-income people do not use medical services as often as they might suggest that the major factor involves the type of care that they can obtain given what they can afford.[35]

There are strong social class differences in the type of care that people receive. In general, the poor are more likely to receive services from emergency rooms and hospital out-patient clinics. Those with higher incomes more often receive care in private doctors' offices or group practices or even over the telephone.[36] William Cockerham, one of the most noted medical sociologists in the United States, suggests that this reflects a "dual health care system." The "private system" of doctors' offices is used by higher income people, whereas the "public system" of hospitals and clinics is used by the poor, and thus, more often, by racial-ethnic minorities.[37]

These differences are related to the organization of medical care in the United States, or "the system" that the medical staff Marisa worked with referred to. Sociologists who take a mesolevel analysis have examined the organization of health care both in the United States and in other countries.

THE ORGANIZATION OF HEALTH CARE: MESOLEVEL ANALYSES

The United States spends a tremendous amount of money each year on health care, especially when compared with other industrialized countries. In 1996, more than one trillion dollars was spent on various aspects of health care from research to the building and maintenance of hospitals to the care of individual patients.[38] Throughout the last few decades the amount of money spent on health care has risen at rates far higher than the cost of living. Currently almost 14 percent of the United States' gross domestic product is spent on health care, a figure that is substantially higher than that of other nations. For instance, the next highest spending countries are Germany, Switzerland, and France, which spend about 10 percent of their gross domestic product on health. Japan and the United Kingdom spend only about 7 percent.[39]

In addition, much of the United States work force is dedicated to health care. Almost 10 million people are employed in health service industries—representing 8 percent of all employed people in the country.[40] More people are now employed in health services in the United States than in a wide variety of other economic sectors including mining, construction work, the manufacture of nondurable goods such as clothing and food, the finance and real estate industries, and federal and state governments.

Societies throughout the world generally have the same array of health care organizations and personnel—clinics and offices where doctors, dentists, and other practitioners meet patients; hospitals for the very seriously ill and injured; and nursing and care facilities for people who need rehabilitative or long-term care. At the same time societies tend to organize the delivery of health care in different ways and, specifically, to have different degrees of political involvement in the day-to-day operations and the financing of medical organizations. In general, societies vary in the amount of government regulation applied to medical practices, the way in which health care providers are paid, the ownership of health facilities, the extent to which access to health care is guaranteed to all citizens, and the extent to which private care, rather than government sponsored care, is available.[41] We can use a mesolevel perspective to examine the different systems of health care delivery. We begin by looking at the organization of health care in the United States and then move to examining how health care is organized in other societies.

The "Fee-for-Service" System of the United States

The health care system of the United States is best described as a "fee-for-service" system where doctors and hospitals charge patients a certain fee for all services that are rendered, such as office visits, lab tests, medications, surgeries, or other treatments. Over the years, although this basic nature of the system has remained, the way in which the fees are paid and managed has changed. These changes have been accompanied by changes in the power of physicians within the health care system.

The Pure-Form Capitalism Model For many years most physicians in the United States were self-employed practitioners who worked by themselves or in relatively small partnerships under the fee for service system. In the pure form of laissez-faire capitalism, where the government has no control or influence, the principles of the open market allow consumers to choose services, such as health care, at prices they can afford. Theoretically, the workings of the free market would drive out doctors who overcharged or who were somehow incompetent, leaving the most effective and reasonably priced physicians and hospitals.[42]

When you think about it, however, you realize that this type of fee-for-service system is actually much more attractive for health care providers than for patients. The free market system works on the notion of "supply and demand." When the supply of a product, such as gas, food, or medical care, is larger than the demand, the price of the product should drop. But, in the case of medical care, the providers—doctors and hospitals—determine the demand. When you are sick, your health care providers, not you, generally decide what kinds of medical services are needed and then provide the services at the prices that they set. In other words, health care providers, and especially physicians, have traditionally had a great deal of power in the health delivery system and over the extent to which their services are in demand.

This incredible power of physicians developed only within the last 100 years or so. Even though most of us now simply assume that we will seek the advice of a medical doctor when we are sick, this has not always been the case.[43] In the mid 1800s Americans could choose to receive health care from a variety of medical practitioners including patent medicine makers; botanic doctors, who offered herbal remedies; bonesetters, who specialized in the setting of broken bones; and midwives, who specialized in delivering babies. Most doctors trained through apprenticeships with other doctors; neither doctors nor medical schools were licensed; and potentially dangerous practices, such as blood-letting and large doses of purgatives, were still in common use. As one observer put it, "cholera victims were given an even chance of being done in by the disease or by the doctor."[44] It is no wonder that doctors weren't held in particularly high regard by their fellow citizens.[45]

Gradually, however, this situation changed. Today, even though they constitute less than 10 percent of the medical workforce, physicians hold authority over almost all other workers in the field, including the relatively highly paid nurses, pharmacists, therapists, and dieticians as well as auxiliary workers such as aides and orderlies. Doctors also are at the highest rungs of the stratification ladder within the society as a whole. As shown in Figure 7–2, physicians are lower than only U.S. Supreme Court justices in typical measures of occupational prestige. This high status of physicians has been documented throughout North and South America, Europe, Asia, and Australia.[46]

One explanation of the high status of physicians comes from the functionalist theory of stratification discussed in Chapter 7. Physicians perform tasks that are important to society and require skills that come from extensive specialized training. According to the functionalist position, because the role of physician is so important and because it is relatively more difficult than other work roles, higher pay and status are needed to motivate people to fill it. In the mid 1800s physicians didn't have much more knowledge—or success in curing patients—than other health professionals and they didn't have the prestige that they do today. Gradually, beginning in Europe and then in the United States, physicians began to apply the scientific method to understanding disease. By the beginning of the twentieth century the discovery of bacillus that caused anthrax, tuberculosis, and cholera and the development of the stethoscope, anesthetics, and x-rays increased the effectiveness of medical practice and helped alter the public's view regarding physicians.[47]

Though the functionalist position can explain why physicians and other health diagnostic workers are paid more than other health professionals, it cannot explain why medical doctors and not other health care workers attained this high status position, nor can it explain the vast power and pay differences between workers in the field. To understand these differences sociologists have examined the nature of medical work as a profession and the way in which physicians were able to consolidate their power as a professional group.

The Hegemony of the Medical Profession

Sociologists use the term **profession** to refer to a work group that has a highly specialized body of knowledge and a service orientation—a culture that is oriented toward providing services to others. There are many professional groups including, for example, accountants, insurance agents, clinical psychologists, and college professors. Yet few have the power and prestige of physicians. Physicians, and in many respects dentists, are the preeminent professional groups in that they have a great deal of control over the way in which they do their work, the training process and licensure requirements for their work, and their relationships with others with whom they interact, including clients, colleagues, and agencies and organizations that are outside the profession. In addition, they tend to be very strongly identified with their profession. For instance, sociologists, and other academics, rarely introduce themselves as "doctor" when in a social situation, even though most academics have doctoral degrees. In contrast, medical professionals almost always include this appellation—not just in introductions but on other items of identification such as checks and credit cards. Even though some might suggest that this strong identity reflects personality characteristics or the relative status of the two fields, it also probably reflects the strong identification with the profession that is fostered by the medical world.

The development of physicians' extraordinary power can be traced, in large part, to the efforts of the American Medical Association, a formal organization of physicians in the United States. Although the AMA was formed in 1847, it only gradually developed the authority and influence it would have in later years. One important step was the founding in 1883 of the *Journal of the American Medical Association,* which published reports of the latest medical findings and also helped practitioners develop a *professional identity,* a sense of belonging to the medical profession. Another important step was the organization of local chapters of the AMA, which determined who could and could not join. The threat of exclusion from membership could be a powerful sanction, for the association could provide network ties that would bring patient referrals and social contacts that would be especially important for the development of a young physician's career.[48]

While the development of *JAMA* and the local chapters affected the careers and identities of individual doctors, other activities of the AMA greatly increased the control of the profession over training, licensure and relationships with other formal organizations. During the 1800s medical training in the United States was largely unregulated and conducted in profit making "proprietary" schools that usually admitted anyone who could pay the tuition. In 1904 this situation began to change with the establishment of the AMA's "Council on Medical Education," which was designed to develop suggestions for improving medical education. This council's power was greatly enhanced by a report, sponsored by the Carnegie Foundation for the Advancement of Teaching (a philanthropic foundation started with funds from the industrialist Andrew Carnegie), that documented the dismal state of medical education in the United States and called for drastic changes in medical education.[49]

Because the AMA's Council on Medical Education was the only group that was available to rate medical schools, it became responsible for determining both the content and nature of medical education. The council could determine not only what should be taught in medical school, but also criteria for admitting students, and the nature of licensure requirements. By the mid 1920s the medical profession had more control over its worklife than any other work group: It determined what entrants should learn, it regulated whether or not they could join the profession, and it controlled sanctioning of deviant members by controlling the membership of review boards. Its lofty status allowed the profession to set the fees for services that it deemed necessary with virtually no control or formal evaluation from the lay public. In addition, the AMA served as a powerful political force, with a very active lobbying group that tried to shape legislation to serve the interests of its members.[50] It would perhaps be fair to describe the power of the AMA as *hegemonic,* hidden but pervasive.

The Development of Insurance Plans This hegemonic power of physicians continued unabated for many years, but the way in which doctors and hospitals got their fees began to change, especially during the Great Depression of the 1930s. Within the traditional fee-for-service system, if you didn't have the money to pay for treatment, you technically could not receive services; alternatively, if doctors or hospitals treated you, they would not get paid. When the depression hit and large proportions of people were unemployed or living in greatly reduced circumstances, doctors and hospitals naturally became quite worried that they might not get paid for their services.

The solution to this problem was insurance plans. In the 1930s, the American Hospital Association founded Blue Cross and the AMA founded Blue Shield. Both companies were nonprofit, but allowed individuals, families, or companies to buy insurance premiums that would cover a substantial proportion of hospital and medical bills. By the 1940s, commercial insurance plans entered the picture. Because the companies that sponsored these plans wanted to make a profit, they either refused to sell insurance to people with expensive diseases or charged these people very high prices.[51] The important point for our purposes is that both the nonprofit and for-profit insurance companies did not challenge the fee-for-service system nor the hegemony of the medical profession. They simply provided a way to ensure that hospitals and doctors would be paid their fees if patients were too sick to work or if the fees were substantially more than they could afford.

A different type of insurance program, which also began in the 1930s and 1940s, did challenge the fee-for-service system and tried to change the focus of medical care from one of treating illnesses after they occurred to preventive medicine. These health maintenance organizations, or HMOs, pay doctors a flat salary rather than a fee for each service that they perform. HMOs also pay the full cost of preventive care and discourage the overuse of expensive specialists by requiring that patients be referred to specialists by their primary care physician.[52]

None of these insurance programs did anything to help the poor gain access to medical treatment. Insurance policies are purchased either by individuals or by employers for their employees, in what is called a group plan. If individuals cannot afford to buy a policy or if they do not work for a company that has medical insurance as a benefit, then they have no health insurance and more difficulties accessing health care.

Increasing Regulation of Health Delivery Systems and Physicians For many years the fee-for-service system applied to most medical transactions. Until well into the 1980s HMOs enrolled fewer than 5 percent of the American public.[53] The only other exceptions were charity hospitals designed for the very poor and government hospitals for military personnel and high level federal employees such as members of Congress and the president. Over the years some policy makers and politicians noted the inequities of this system and proposed changes that would guarantee all Americans access to health care. The first major proposal was by Theodore Roosevelt in his losing bid for the presidency in 1912. The second was by President Franklin D. Roosevelt during the 1930s; his successor, Harry Truman, tried again in the late 1940s. None of these proposals received sufficient political support to pass, largely because of

very strong AMA opposition. Physicians objected to the possibility of government control over their actions and fees and feared potential loss of income that new policies might create.[54]

The situation had changed by the mid 1960s when the inequities of the fee-for-service system, and the special problems faced by the elderly and the poor, had become increasingly clear to the American public. In fact, much of the writing in medical sociology in the 1950s and early 1960s documented these problems. In 1965, under the administration of Lyndon Johnson, the United States Congress passed the bills that authorized the Medicare and Medicaid programs. Medicare is a federally funded program that provides hospital insurance and supplemental medical insurance (for visits to doctors) for all people older than 65 and for disabled people younger than 65 who receive social security benefits. Medicaid provides federal matching funds to the individual states to pay for medical services to poor people. Both of these programs were also strongly opposed by the AMA and its Political Action Committee, yet they now have widespread support from the American public.[55]

After the passage of the Medicare and Medicaid legislation, other politicians, including Presidents Nixon, Carter, and, most recently, Clinton, tried to alter the fee-for-service system even more by expanding federal benefits to cover a much larger proportion of the population.[56] The only program to pass Congress was the Children's Health Insurance Program (CHIP), which reimburses states that offer affordable health insurance to children in families that make too much money to qualify for Medicaid but too little to afford private insurance.[57] None of the larger scale proposals received sufficient support to pass through Congress.

Even though none of these attempts was successful, several observers suggest that the hegemony of the medical profession has been weakened. Although physicians are still extremely powerful, and certainly the most powerful group among health care workers, this political and organizational power was successfully challenged by the successes of the Medicare legislation. For many years the AMA and its political action committee have opposed programs that were highly supported by the American public, including workmen's compensation laws, social security, and the creation of groups that would review doctor's professional standards and practices. The public appears to have become more aware of these different views and more skeptical of the medical profession. Polls regarding patients' attitudes toward their doctors indicate that the public is generally more dissatisfied with their medical care and tend to have less respect for their doctors than in earlier years.[58]

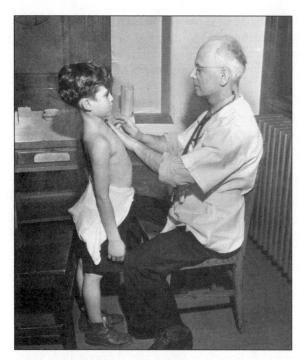

Do you think that the type of doctor-patient relationship that was common in the early years of the twentieth century could occur now? Why or why not?

Some of the decreasing respect for the medical profession appears to be related to the enormous increase in the costs of medical care during the past few decades. Not only does the United States spend a larger proportion of its gross domestic product on medical services than other countries do, but the cost of medical services has been rising much faster in the United States than elsewhere and more quickly than the general cost of living. Sociologists have tried to explain why health care in the United States costs so much more than in other countries. The high costs can not be attributed to Americans visiting doctors more, for, in fact, we see our doctors less than people in other industrialized countries do. Nor can the costs be explained by a large number of older people, for most industrialized countries actually have a larger proportion of elderly than the United States does. The costs also can't be explained by either a high incidence of malpractice suits or very advanced technology. Both of these, although more common in the United States than elsewhere, account for only a very small proportion of health care costs. Instead, most of the difference appears to arise from higher administrative costs, doctors fees, and hospital costs in the United States than elsewhere.[59]

To combat the very high and rising costs of medical care there have been increasing restrictions and strong challenges to the fee-for-service system in re-

cent years. Instead of allowing hospitals and doctors to set the fees they would charge for specific services under Medicare and Medicaid, Congress voted to have these fees set by a government agency. If doctors or hospitals want to charge more they have to directly bill their patients. Limits were also placed on these additional fees.[60] In addition, many more insurance companies have begun to use HMOs or some other form of managed care. By the late 1990s more than 80 million Americans were covered by some type of managed care plan, a figure that is more than twice as high as in the beginning of the decade .[61]

Managed care refers to health care delivery systems in which patients are allowed only to see certain preselected health care providers, including specialists and hospitals, with prearranged agreements and restrictions on the costs that will be paid for services. Both HMOs and other kinds of managed care programs directly challenge the fee-for-service system by setting limits on costs and strict reviews of recommended treatments. Utilization management and utilization review boards provide oversight to physicians' recommendations. Insurance programs can require that patients receive approval from the health plans before receiving treatment and can refuse to pay physicians or hospitals if they decide that the treatment was not appropriate or unneeded.[62]

Managed care has faced strong opposition from much of the medical establishment as well as many consumers. Although it has been successful in slowing the increase in medical costs, medical practitioners have objected to the decline in their authority, for under managed care they not only cannot set fees for their services, they can be restricted in the types of services they can give or recommend. Consumers have opposed managed care because they believe that it puts undue restrictions on their access to medical help. As I write this chapter, it is not at all clear what the future health care system in the United States will look like, and it is likely that the dust will not settle for several years to come. It is probably safe to say, however, that certain elements will remain important in a mesolevel analysis of health care delivery in this country: the power of physicians and challenges to this power, the need to control medical costs, consumers' commitment to receiving quality care, and the role of insurance companies and the government.

An important issue that has not been solved by the current U.S. system is the issue of universal access to health care. Even though Medicare guarantees health care to the aged and many disabled people, Medicaid provides health care payments for the very poor, and CHIP helps provide health insurance to children in some states, large numbers of individuals have no health insurance or only very limited health insurance coverage. Throughout the 1990s, about 15 percent of all Americans had *no* private or government health insurance. Young people between the ages of 18 and 34, people with incomes close to or below the poverty line, African Americans, and Hispanics were more likely than others to not be covered.[63] None of the attempts to contain health care costs or to alter the fee-for-service system has succeeded in addressing the lack of health care for the poor and near-poor who cannot obtain health insurance. Thus, although it can be argued that the present system adequately promotes the health of those who have health insurance, it does virtually nothing to help the millions of Americans with no insurance.

Systems of Guaranteed Health Care

The United States is the *only* industrialized nation that doesn't have an extensive system of government guaranteed health services.[64] Although the details of their systems vary, all industrialized countries other than the United States have a system of health care that guarantees benefits to all their residents. Because these countries have political economies that are similar to the United States, understanding their health care systems can provide a useful comparison.

Health reformers in the United States have often focused on Canada, our northern neighbor. Doctors in Canada are private, self-employed, fee-for-service practitioners. But, instead of being paid directly by consumers or individual insurance companies as in the United States, they are paid by a national health insurance system on a common fee schedule. The fee schedule is negotiated in each province between provincial officials and the medical association. Hospitals also operate on budgets provided by the provincial government. Although Canadians pay higher taxes than U. S. citizens do, they pay a smaller proportion of their gross domestic product for medical services.[65]

Great Britain began its National Health Service, or NHS, in 1948, becoming the first Western society to offer free medical care to the entire population. General practitioners, who work out of a solo office or in a group practice, are the major health care providers in Great Britain and are paid a certain amount each year for each patient who is on their "patient list," the size of which is restricted by legislation. Patients may choose their own doctors. When a patient needs special treatment or hospitalization, the general practitioner refers the patient to a specialist, who is paid a yearly salary by the government. Physicians may also take private patients, who tend

to receive faster treatment and more luxurious facilities. Again, even though the British pay higher taxes than Americans, a smaller proportion of their GDP is devoted to medical expenses.[66] In addition, survey data indicate that Britons overwhelmingly consider the health service to be a success.[67]

Like their American counterparts, physicians in both Canada and Great Britain initially opposed these systems. For instance, the Canadian system was not fully instituted until 1971, largely because of doctors' opposition.[68] Doctors in Great Britain have been allowed to have private patients because of their initial objections to the founding of the NHS. But, as appears to be happening in the United States, the power of physicians in these countries is no longer hegemonic, and, in contrast to the United States, quality health care is available to all citizens.

Even though they are much poorer than the United States, most developing countries also provide free health care to their citizens. For instance, Kenya is a very poor country toward the bottom of the global stratification ladder. Yet it has a national health service that owns hospitals and employs doctors and other health workers and guarantees health care to all citizens. Similar systems are found in other poor countries in sub-Saharan Africa and throughout the world. Unfortunately, because these countries are very poor, and also because of the sharp differences between cities and the rural areas, only relatively small proportions of the population receive high quality medical care. The best equipped hospitals and clinics are in the cities, where the most educated and wealthiest citizens live. In addition, there are far from enough medical personnel. Kenya has only one doctor for every 20,000 citizens.[69] In contrast, the United States has one doctor for about every 400 citizens.[70]

In short, the United States stands alone among the rich industrialized nations, and almost alone within the world, in its system of health care. It is the most expensive, yet does not try to serve all people within the country. Yet, no matter what system of health care a country may have, the final question involves the outcome: How healthy are the citizens? We turn now to a macrolevel analysis to examine health and wellbeing of people around the world.

HEALTH AND GLOBAL STRATIFICATION: A MACROLEVEL PERSPECTIVE

No matter where you might have grown up, chances are that you are healthier now than you would be if you had been born 100 years ago. In comparison

FIGURE 16-3 *Life Expectancy in the United States, for Males and Females, 1990–1995*

Source: U.S. Bureau of the Census, 1975, p. 55; U.S. Bureau of the Census, 1997, p. 88; National Center for Health Statistics, 1998, p. 200.

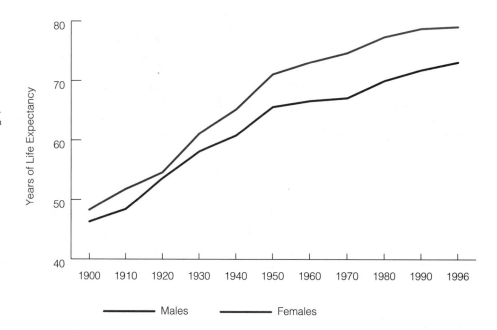

Historical Changes in Health and Disease

with just a few decades ago people in most areas of the world live longer and are more likely to survive infancy. Yet, there are still wide variations between societies. How large are these variations? Why do they exist? We begin by taking an historical look at changes in health in the United States and other developed countries over the last century and then look more closely at the health hierarchy of nations. Where do countries fall on this hierarchy and how is the health hierarchy of nations related to global economic stratification?

Historical Changes in Health and Disease

Figure 16-3 shows changes in life expectancy of males and females during the last century in the United States. In 1900, a baby boy could expect to live to the age of 46 and a baby girl to 48. By 1940, a man's life expectancy was about 64 years, and a woman's was 68. Today the figures of 73 for men and 79 for women are more than one and one-half times the figures for 1900.[71] You might have noticed that the gap between men's and women's life expectancy has grown over the years. This reflects the fact that women are now much less likely to die in childbirth than they were in earlier eras. In 1920 the rate of maternal deaths was 799 per 100,000 births. But this rate dropped rapidly—to 376 in 1940, 22 in 1970, and to only 8 by 1996.[72] Similar changes have occurred in highly developed countries throughout the world.

As life expectancy has increased people have tended to die from different causes. For most of human history, people could expect to die from a communicable disease, such as cholera, pneumonia, or tu-

berculosis, often at a fairly young age. Perhaps the most virulent epidemic in human history was the bubonic plague, which swept Europe periodically between 1340 and 1750. It is estimated to have killed 20 million people, one quarter of the total population.[73] Often underlying these deaths was poor nutrition, resulting from either a lifelong history of inadequate diets or periodic food shortages and famines that swept entire nations. For instance, the Potato Famine in Ireland resulted in the death of hundreds of thousands of people in the mid 1800s and encouraged many more to move to other countries. Another major factor was poor sanitation. The availability of clean water and adequate sewage systems has been a major factor in stemming deaths from communicable disease.

The United States has not escaped widespread death from communicable disease. As shown in Table 16-1, a century ago, the most common cause of death in the United States was some type of infectious disease, such as pneumonia, tuberculosis, and diarrhea. Gradually, however, as agricultural practices became more efficient and adequate diets and housing became available to many more people throughout the developed world, deaths from communicable disease became much less common. Today most people in the United States, as well as in other highly developed countries, tend to die from chronic and degenerative diseases, such as cancer or heart disease, which generally strike older people, or accidents, suicide, and liver disease, which come from our own actions (including the excessive use of alcohol). Although we often think that modern medicines are primarily responsible for the growth in life expectancy, epidemiologists and historians who have

TABLE 16-1 *Top Ten Causes of Death in the United States, 1900 and 1997*

RANK	CAUSE OF DEATH	PERCENT OF ALL DEATHS
1900		
1	Pneumonia	12
2	Tuberculosis	11
3	Diarrhea and enteritis	8
4	Heart disease	8
5	Chronic nephritis (Bright's disease)	5
6	Accidents	4
7	Stroke	4
8	Diseases of early infancy	4
9	Cancer	4
10	Diphtheria	2
1997		
1	Heart disease	31
2	Cancer	23
3	Stroke	7
4	Lung diseases	5
5	Accidents	4
6	Pneumonia and influenza	4
7	Diabetes	3
8	Suicide	1
9	Nephritis	1
10	Liver disease	1

Source: Rockett, 1994, p. 7; Ventura, Anderson, Martin, and Smith, 1998, p. 7.

studied these changes have found that they often occurred before the discovery of powerful drugs, such as antibiotics or immunizations. Improved nutrition and living standards gave individuals the strength to fight off infections that people in earlier generations could not withstand.[74]

Even though the general health of the populace has improved greatly over the last century, these improvements have been more concentrated among the upper and middle classes than among the working class and poor. For instance, the incidence of coronary heart disease (the forerunner of heart attacks and strokes) has declined dramatically in the last quarter of a century in the United States, but the decline has primarily occurred among those at higher points on the stratification ladder. As a result, coronary heart disease is now more concentrated among the poor than it was in earlier generations. The poor are also more likely than those with greater means to suffer from communicable diseases, such as tubercu-

losis, influenza, and the newest communicable disease to ravage the earth, AIDS.[75] The concentration of ill health among the poor becomes especially apparent when we look at nations at different points on the global economic stratification hierarchy.

Health and Global Stratification

Those of us who live in the United States can now expect to live to enjoy our retirement years and to see our grandchildren grow up, but this expectation is not common in many parts of the world, for life expectancy varies a great deal from one nation to another. As shown in Map 16-1, people in sub-Saharan Africa generally have the lowest life expectancies, and those in Europe, the United States, Canada, Japan, and Oceania have the highest. For instance, children born in Sierra Leone, Mali, Zambia, and Swaziland—all countries in Africa—cannot expect to live to their fortieth birthdays, a life expectancy that is lower than that of the United States at the beginning of the century. In contrast, children born in Hong Kong, Japan, Macao, Sweden, Switzerland, and Andorra (a very small country in Europe) can expect to live more than twice as long—until they are 79 or 80.[76]

Differences in infant mortality are just as stark. Rates vary from more than 150 deaths for every 1,000 births in Guinea and Sierra Leone in Western Africa and Afghanistan in Southeastern Asia to 4 or fewer in Singapore, Hong Kong, Japan, Finland, Norway, and Sweden. A baby born in Sierra Leone, with an infant mortality rate of 195, has a chance of dying before its first birthday that is more than 50 times as high as that for a baby born in Finland, with its infant mortality rate of only 3.5.[77]

These national differences in life expectancy and infant mortality are linked to patterns of global stratification, which were discussed in Chapter 14. Countries with the lowest life expectancies and highest infant mortality rates are also those with the lowest rates of economic growth and the smallest gross national products. (Turn back to Map 14-1 and compare it with the patterns shown in Map 16-1.) Just as our health as individuals is highly related to where we fall on the social stratification ladder, the health of nations is related to their position on the global stratification ladder. People in the rich industrialized countries are much healthier, on average, than those in the poorer developing and least developed countries are.

People in the poor, least industrialized countries are much more likely than are those in the richer and highly industrialized countries to die from the infectious diseases that ravaged countries such as the United States and those in Europe a century and more ago. Today, diseases including malaria, diarrhea, tuberculosis, measles, hepatitis, tetanus, and cholera,

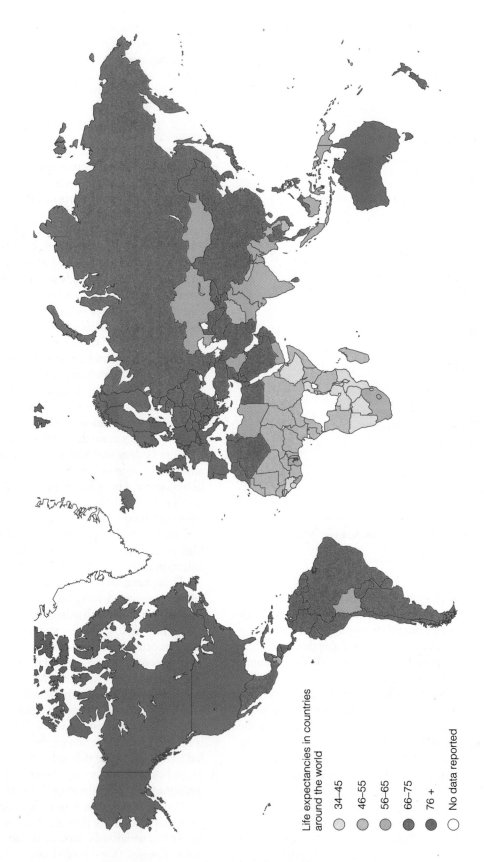

Life expectancies in countries
around the world

○ 34–45

○ 46–55

○ 56–65

● 66–75

● 76 +

○ No data reported

MAP 16-1 *Life Expectancies in Countries Around the World*

Source: Population Reference Bureau, 1998.

In some African countries, life expectancy is only 40 years of age—lower than that in the U.S. at the beginning of the century. What patterns of global stratification might account for this difference?

which are all largely preventable and extremely rare in the rich industrialized nations, kill literally millions of people in the poorer countries of the world. Infectious and parasitic diseases, including pneumonia and influenza, account for only about 5 percent of the deaths each year in the industrialized countries but for more than half of all deaths in sub-Saharan Africa (the part of Africa that lies south of the Sahara desert and is the poorest part of the continent).[78]

Even though you may think of AIDS as a disease that primarily affects homosexual men in the United States, it is much more common in other countries. In fact, the United Nations estimates that two-thirds of the people in the world who have AIDS live in sub-Saharan Africa. Seven percent of all adults there are affected, compared with only about 0.6 percent in North America and 1.0 percent throughout the world. In some countries, such as Botswana and Zimbabwe more than 20 percent of the adults are infected.[79] The potential impact of the AIDS epidemic on these poor nations is staggering. For instance, if the AIDS epidemic had not occurred experts esti-

mate that by the year 2010 people in Zimbabwe would have had a life expectancy of almost 70 years, close to that of many much more industrialized nations. However, with the spread of AIDS, it is estimated that life expectancy will fall to only 33 years—a level that was common two centuries ago.[80]

The effects of global stratification can be seen very clearly when we consider treatments for HIV and AIDS. In 1998 the standard drug treatment for one person infected with HIV cost ten thousand dollars or more each year. The widespread availability of these drugs has allowed many people in industrialized countries who are infected with HIV to live relatively normal lives and has drastically reduced the death rate from this condition.[81] Yet only countries where a small proportion of the population is affected and there is a strong economic base can absorb the extreme costs of treatment. The world's poor countries find meeting these costs virtually impossible. In sub-Saharan Africa, less than half of the population has access to adequate sanitation and barely half have access to safe water. Almost one-third of the children are moderately or severely underweight.[82] Until these countries have the resources to provide adequate nutrition and shelter, it is simply impossible for them to afford the extensive drug regimen needed to combat AIDS.

Global stratification can account for many of the variations in health status between countries, but it cannot explain all of them. For instance, you might have noticed that, even though the United States is one of the wealthiest countries in the world and spends a much larger proportion of its gross domestic product on health than any other country does, it is not the healthiest. A number of industrialized nations, all of which spend less on health care than the United States does, have lower levels of infant mortality and longer average life expectancies than the United States. Richard Wilkinson has examined this issue and has tried to understand more about why some industrialized countries are healthier than others are.

Common Sense versus Research Sense

COMMON SENSE: The wealthiest nations in the world have the healthiest citizens.

RESEARCH SENSE: The most important influence on a society's health is not how rich the society is, but how its riches are distributed. The countries that have a more equal distribution of income tend to have citizens with overall better health. Research shows that citizens of many countries can expect to live longer than Americans can, even though the United States is a wealthier nation.

INEQUALITY AND THE HEALTH OF NATIONS—THE WORK OF RICHARD G. WILKINSON

Richard Wilkinson is a sociologist who works in a medical setting, as a Senior Research Fellow at the Trafford Center for Medical Research at the University of Sussex in England. For more than twenty years, he has been interested in how social and economic characteristics of a society influence where it falls on the health hierarchy. Much of his work has involved **secondary analysis,** analyzing data that have been gathered by other researchers, including statistics from governments throughout the world.

Based on his analysis of data from countries around the globe and covering a number of years, Wilkinson has concluded that the relationship between economic wealth of a nation and health tends to be *curvilinear,* a relationship that cannot be represented by a straight line, but is instead better represented by a curved line (see Figure 16-4). When nations are relatively poor there tends to be a fairly straight, one-to-one correspondence between average income and health. Among these less developed countries, an increase in average wealth seems to translate fairly directly to an increase in health and longevity. Among the wealthier nations, however, increases in per capita income have very small, and sometimes nonexistent, influences on health. It appears that once a certain threshold is reached, improving the economic health of a nation doesn't have much effect on the physical well-being of its citizens.[83] The result is that some of the wealthiest countries in the world do not have the healthiest citizens. People who live in at least twelve different countries (Greece, Italy, Spain, the United Kingdom, Iceland, Finland, France, Belgium, Netherlands, Sweden, Canada, and Australia) can all expect to live longer than Americans do, even though their nations are poorer. For instance, residents of Greece can expect to live two years more than the average American can, even though the per capita gross national product of Greece is far less than half that of the United States.[84]

Wilkinson realized that this situation appears to be a paradox. On the microlevel of analysis, within societies, as discussed earlier in this chapter, people who are in higher social classes tend to live longer and healthier lives than those in lower social class groups. This relationship is strong and straightforward and occurs at all levels of the stratification hierarchy. Yet, on a macrolevel, when we look at developed nations and compare one society with another there is no apparent relationship between a nation's wealth and its health.

To sort out this paradox, Wilkinson looked more closely at the data about health and income of societies. He found that the most important influence on a society's health is not how rich it is but, rather, how it distributes its riches. Remember our discussion of social stratification and the idea that some societies are more egalitarian than others are. The social stratification mountain can be peaked and narrow, with only relatively few people being able to get to the top, or it can be more rounded and less steep, with fewer differences between those at the bottom and those at the top. From his analysis of data from many different countries, Wilkinson concluded that countries that have a more equal distribution of income—a more rounded and shorter stratification mountain—tend to have better overall health.

Several other researchers have replicated Wilkinson's findings. Even more important, these results continue to appear even when the researchers consider other possible influences, such as the actual income level of the country, levels of education, amount of smoking, and how much money is spent on health care. It appears fairly safe to conclude that countries that have less inequality, where the poor aren't as poor and the rich aren't as rich, tend to have better levels of health, even though they are not necessarily any wealthier. Their health hierarchy is not as steep, and both the rich and the poor are more likely to enjoy good health.[85]

To know that a relationship occurs does not explain *why* it happens. To understand why inequality is so strongly related to health, Wilkinson used **case studies,** in-depth analyses of the experiences of particular societies. From these analyses he concluded that more egalitarian societies are healthier because they are more cohesive and have stronger social relationships. This cohesion provides the social support that is crucial in combating the negative effects of stress, as described earlier in this chapter.[86]

His case analysis of Japan is especially noteworthy, for Japan is now the healthiest society in the world. In 1965 the Japanese had lower life expectancies than Britons. Twenty years later, however, life expectancy in Japan was about four years longer than life expectancy in Britain. There is no indication that during this time the Japanese had better diets, health care, or preventive health policies, or that Britons developed worse habits or experiences in these areas. Instead, these changes appear to be related to changes in the pattern of income inequality. Until about 1970 Japan and Britain had sim-

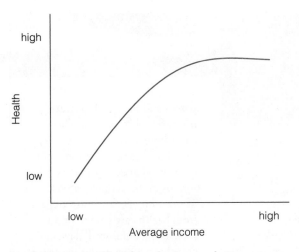

FIGURE 16-4 *The Relationship Between the Economic Wealth of Nations and Health*

TABLE 16-2 *Increases in Life Expectancy in England and Wales Each Decade: 1901–1991*

DECADE	MEN	WOMEN
1901–1911	4.1	4.0
1911–1921	**6.6**	**6.5**
1921–1931	2.3	2.4
1931–1941	1.2	1.5
1941–1951	**6.5**	**7.0**
1951–1961	2.4	3.2
1961–1971	0.9	1.2
1971–1981	2.0	1.8
1981–1991	2.4	2.0

Note: The numbers given show the additional years in life expectancy that were added during each decade. The decades that included World War I and World War II are highlighted.
Source: Wilkinson, 1996, p. 114.

ilar levels of inequality, but by the end of the 1980s Japan had the lowest level of income inequality of any country that reports data to the World Bank. In other words, the change in Japan's health appears to be related to its changing patterns of inequality.[87]

Wilkinson suggests that the extent of inequality in a society affects many different aspects of its culture and daily life and that these daily experiences of living affect health. Drawing on the work of a number of other social scientists, Wilkinson suggests that the Japanese culture is less likely than that of other societies, such as Great Britain or the United States, to emphasize social class differences and more likely to promote feelings of connectedness and group loyalty. For instance, Japanese corporations try to minimize notions of rivalry between workers and promote the importance of collective performance. Japanese police see themselves as not just law enforcers, but also as moral teachers who help bring offenders into line with community norms and obligations. Even though the Japanese economy is similar to the British and American economic systems, Wilkinson suggests that the more supportive and integrated structure of Japanese society helps combat the stress of everyday life in ways that British and American societies do not.

A second case study that can help explain why inequality is related to health involves Wilkinson's home country of England. Even though the data suggest that England is currently much less egalitarian, integrative, and cohesive than Japan is, this was not always the case. In fact, two particular historical eras stand out as periods that were far more egalitarian and in which life expectancy also increased substantially. As shown in Table 16-2, from both 1911 to 1921 and 1940 to 1951, the periods which included World War I and World War II, life expectancy in Great Britain increased by more than 6 years for

both men and women. These increases were more than twice as large as the average improvement in other decades of the century.

These wartime decades were also characterized by much more economic and social equality. Poverty declined markedly, there was little unemployment, and income differences among citizens became much smaller. At the same time, there was a sense of social cohesion and camaraderie, of people "pulling together." For instance, during World War II Princess Elizabeth, who later became Queen Elizabeth II, drove a jeep and helped in the local war effort. The King and Queen refused to leave London during dreadful bombing sieges, and the Prime Minister often spoke to the nation encouraging all citizens to devote themselves to the war effort. Even though the population experienced extreme stress including nightly bombings, the death of loved ones, and stringent food rationing, Wilkinson suggests that the increased economic equality and social integration helped mitigate the harmful effects of stress and led to sharper increases in life expectancy. When the wars ended, even though the dreadful stresses of war were no longer present, the social and economic distinctions between class groups returned and increases in life expectancy became much more modest.

A case study from the United States illustrates similar processes. For many years epidemiologists have been fascinated by the experiences of people in Roseta, Pennsylvania, a small town of about 1600 in the eastern part of the state. For a large part of its history people in Roseta had death rates, especially from heart disease, that were as much as forty percent lower than people in nearby communities. These differences

APPLYING SOCIOLOGY TO SOCIAL ISSUES

Changing the Health Care System in the United States: The Polity at Work

William C. Cockerham, who is profiled in Box 16-2, suggests that controversies over reform of the health care system in the United States revolve around the issue of whether medical care is a *right* or a *privilege*.[90] The traditional fee-for-service system of medicine assumes that medical care is a privilege, something that we can purchase, much as we buy other goods and services, such as food, housing, and clothing. Those who advocate some type of government funded health care system, such as those found in other countries, assume that medical care is a right, a benefit that all people should be entitled to receive no matter how rich or poor they happen to be.

At present, the United States health care system embodies both of these viewpoints. The Medicare and Medicaid systems ensure health care for the elderly and the very poor, thus suggesting that it is a right for those who are deemed most in need. Most employers provide health insurance for their employees, indicating that these employers believe that health care

is important for their workers' well-being. Yet, there are literally millions of Americans for whom health care is a privilege. Some work for companies that provide no health insurance; others may be unemployed and ineligible for Medicaid programs. Those who are self-employed often have trouble finding affordable insurance.

As noted earlier in this chapter, reformers and politicians in the United States have advocated changes in the health care system for much of this century, but these efforts have been strongly opposed by powerful corporate actors in the health care system. In the early part of the century the American Medical Association spearheaded the opposition. Beginning in the 1940s the AMA was joined by private insurance companies. Both insurance com-

panies and the AMA are very powerful lobby groups. They contribute huge sums of money to political campaigns and maintain very active lobbying organizations in the nation's capital. Those who have the most to gain from health care system reforms, such as the uninsured, have neither the money nor the resources to begin to match the efforts of these powerful political players.

The most recent large-scale attempt to reform the United States health care system came in the first term of the Clinton administration in 1993 and 1994. The plan would have provided basic health insurance benefits to all legal residents of the country, and employers and employees would have paid the cost of the insurance. The federal government would provide subsidies to small businesses, workers in low-paid jobs, and the unemployed to allow them to obtain coverage, and there would be a limit on the amount of money that any individual would have to pay for health care. The government funds would come from proposed

were especially striking because Rosetans were no different from their neighbors in what they ate, how much they smoked, or how much exercise they received—all factors that typically influence both the incidence of heart disease and longevity.

Roseta did differ from neighboring communities, however, in the nature of social relationships and community culture. The town had been primarily settled by people who immigrated from Roseta, Italy, in the 1880s and the residents continued to maintain close knit and cohesive relationships. They also downplayed economic differences and promoted a norm of egalitarianism. As one set of researchers described the community:

[I]t was difficult to distinguish, on the basis of dress or behavior, the wealthy from the impecunious in Roseta. Living arrangements—houses and cars—were simple and strikingly similar. Despite the affluence of many, there was no atmosphere of "keeping up with the Joneses" . . . From the beginning the sense of common purpose and the camaraderie among the Italians precluded ostentation or embarrassment to the less affluent, and the concern for neighbors ensured that no one was ever abandoned.[88]

Just as the egalitarian nature of wartime England promoted more integrative and supportive social relations, so the egalitarian culture of Roseta provided support and protection that could ward

BOX
16-1

APPLYING SOCIOLOGY
TO SOCIAL ISSUES

Continued

savings in the Medicare and Medicaid programs as well as new taxes on unhealthy items such as cigarettes. Insurance companies would administer the system, presumably through managed care or HMOs. These companies would compete with one another to obtain subscribers, but would use standard claim forms to cut down on the extremely high administrative costs that are common in the current system. A federal commission would monitor the system to ensure that all elements worked as planned.[91]

When the plan was presented to Congress in the fall of 1993, many different interest groups lobbied legislators to change or oppose the plan. Much of the opposition involved financial self-interest. The AMA, as in earlier decades, opposed the idea of government control of doctors and any potential loss in income. Hospitals, drug companies, and insurance companies opposed the possibility of limits on health care spending and their profits. Labor unions and the elderly opposed the plan for fear that it would

result in losses in health coverage that they already had. Consumer groups opposed the powerful role of the insurance companies and a lack of immediate coverage for the unemployed.[92]

As a result of this intense opposition, the plan was not passed by Congress and has not been introduced again. Cockerham, however, notes two major effects of this latest large reform effort. First, it stimulated the development of the very large managed care system as the dominant form of insurance coverage. Second, it prompted the passage, in 1996, of the "Health Insurance Reform Act," which guarantees that workers who have health plans from their employers will not lose these plans when they change or lose their jobs and which also bars insurance companies from refusing to insure people who have medical conditions that can be very expensive to treat.

As I write this chapter, political debates about medical care continue to appear. Given the very large amounts of money that our country spends on health care and that substantial numbers of the population have no regular access to medical care, this issue will not disappear in the near future. Debates about the extent to which medical care is a *right*—a public good that societies should provide their citizens—or a *privilege*—a commodity to be purchased—will no doubt continue. And the role of powerful corporate actors in this debate will no doubt not diminish.

CRITICAL-THINKING QUESTIONS

1. Is medical care a right or a privilege?

2. Should the health care system of the United States be changed? If so, what kinds of changes do you think would be most important?

3. How, given our political system, might these changes be attained?

off disease. Yet, just as equality lessened and health improvements declined in England when the wars ended, the egalitarian nature and unusual health status of Roseta has not endured over the years. By the mid 1960s the researchers who studied Roseta noted the appearance of more "status symbols," such as expensive cars and large new homes, marking the decline of the culture of egalitarianism. Soon the health advantages of Roseta also declined. These advantages were not linked to any genetic advantage or difference in diet or exercise, but to the supportive social networks that arose from an egalitarian culture.[89]

CRITICAL-THINKING QUESTIONS

1. Given what you have read about social class, inequality, and health, think about Marisa's experiences. How could these research findings account for the differences between University Clinic and her family doctor's office?

2. Can you think of ways to study the relationship between inequality and health in your community?

3. What do you think could be done to attenuate the relationship between social class, inequality, and health within your own community? Within the nation as a whole?

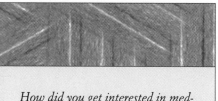

William C. Cockerham

*Health care delivery systems are
acts of political philosophy.*

**William C. Cockerham, University
of Alabama, Birmingham, conducted
the research discussed in Box 16-1.**

Why did you become a sociologist? I
did not anticipate that I would ever
become a sociologist. I grew up in
Oklahoma and graduated from the
University of Oklahoma with a B.A.
in history simply because I found
history interesting and made good
grades in it. Getting a job after college in those days (the 1960s) was
no problem because of the military
draft: A tour in the Army as an
infantry officer was waiting. Afterwards, I went to graduate school at the University
of California at Berkeley in journalism because
there was a family connection, it was a good school,
I liked the San Francisco Bay area, and I was interested in writing. Berkeley, at that time, was a major
center of social unrest and various countercultures
were active, and it was also a great place to go to
school. I finished the masters program in journalism
(which turned out to be good training for a sociologist) and worked part-time as a sports writer for the
San Francisco Chronicle, but decided against being
a journalist. While looking for an alternative, I took
some sociology courses at Berkeley from professors
like Herbert Blumer and Norman Denzin and became fascinated by the subject matter. This was a
time in sociology of great debates between symbolic
interaction (Blumer) and structural-functionalism
(Parsons), as well as an important period of change
in American society.

*How did you get interested in med-
ical sociology?* As I was finishing
my doctoral dissertation, I became
friends with Anselm Strauss who
was on the faculty at the UC Medical Center in San Francisco.
Strauss was a major figure in medical sociology and he and his colleagues occupied an old Victorian
house on a steep San Francisco
street (the kind you see in car chase
movies) near the medical center.
They were warm and friendly,
dressed and acted casually, had fun,
researched interesting topics, and
produced some of the most impor-
tant works in sociology during the late 1960s and
early 1970s. It was a very attractive way of life, and
it was through this connection that I became interested in the field. My first job was at the University
of Wyoming, and they had a course on medical
sociology which I volunteered to teach; next came a
textbook on medical sociology which I published
later at the University of Illinois in 1978 and will be
coming out in an 8th edition in 2001. Although my
training was in social psychology, this book, more
than anything else, marked me as a medical sociologist and led to many opportunities to do research
and meet people around the world.

*What has surprised you most about the results you
have obtained in your research?* Most of my
research today is cross-national and focused on
health lifestyles. Given the shift in modern societies
from communicable to chronic diseases, the way in

At the beginning of this chapter we noted that
the World Health Organization defines health as
not just being free of disease, but as "a state of
complete physical, mental, and social well-being."
We can now see how this definition applies in real
life. At a macrolevel, nations are healthier when
they are more egalitarian and supportive to all
members. At a microlevel, individuals are healthier when they have the social resources and sup-

port to counter the stresses and strains of life.
Marisa's observations of the health of poor people
in University Clinic are supported by research.
Both nations and individuals with fewer financial
resources are less likely to have the resources to
combat diseases and germs. The biological sources
of illness are, of course, real, but our social environments strongly influence whether we will suc-
cumb to the effects of these factors.

BOX 16-2

SOCIOLOGISTS AT WORK

Continued

which we live with respect to alcohol and cigarette consumption, diet, exercise, and the like play perhaps the major role in how long a person lives. Along with some German colleagues, I compared the health lifestyles of Americans and Germans in order to determine in 1985 whether a population (Germans) with state-sponsored health insurance lives as healthy a lifestyle as a population (Americans) who are generally on their own as individuals in obtaining health insurance. We found little or no differences between the two countries and, surprisingly, observed that efforts at leading a healthy life were spreading across class boundaries. Consequently, we suggested that if the United States were to provide national insurance to all Americans, people would not stop taking care of their own health and invest responsibility for their health in the government. However, the U.S. remains the only major nation in the world without government health insurance and some 37 million people still lack such insurance. The opposition to health insurance in the U.S. is grounded in our society's historical emphasis upon individual freedom and responsibility which in many ways is admirable, but other countries take greater collective responsibility for their citizens. Health care delivery systems are acts of political philosophy and are based upon a nation's historical experience, traditions, values, norms, culture, economy, political ideology, social organization, and attitudes toward the welfare state. This complex mixture of variables helps to make medical sociology an interesting subject area.

What do you think are the implications of your studies of health care for social policy in the United States? My most recent research in the late 1990s concerns a group of nations in which the people did place responsibility for their health in the hands of their government, but the government failed to deliver quality health care and the population did not practice a healthy lifestyle. These countries are members of the former Soviet bloc which are the only modern nations with a stable government to experience a decline in life expectancy in peacetime. Mortality began rising in this region in the mid-1960s, and it was not clear why this was happening. I traveled to Russia and throughout Eastern Europe visiting clinics, hospitals, ministries of health, and collecting data from numerous sources. I also interviewed experts in medical sociology, medicine, and public health and learned a great deal by seeing how the population lived. Environmental pollution, infectious illnesses, and genetics were not responsible. Stress and poor health policy decisions that failed to contain the rise in heart disease were important, but I found that the primary social determinant for the massive rise in mortality was the unhealthy lifestyles of middle-age, working-class males dying prematurely from heart disease. Heavy drinking and smoking, high fat diets, and no exercise characterized the lifestyles of these men. The social, political, and economic structure of their society limited their choices in life and unhealthy practices were curtailing their lifespan. Uncovering the sources of health problems like this is one way in which medical sociology can also help people.

Our mesolevel analysis focused on health care systems, the ways in which citizens receive health care and help ensure that their good health will continue. The health care system within the United States is unique to the world: All other industrialized countries and many poor countries guarantee that their citizens will receive health care. No other country spends as much on health care as does the United States, yet many other countries have longer life expectancies. As described in Box 16-1, Applying Sociology to Social Issues, the controversies over health care systems in the United States involve other social institutions, especially the political world. They also involve issues regarding social change, the focus of the last four chapters of this book.

SUMMARY

Health involves physical, mental, and social well-being and is influenced by not just biological variables such as exposure to germs or genetic susceptibilities, but by social environments and experiences. Understanding these relationships can help you understand variations in health around the world as well as in your community and among people that you know. The health hierarchy, at both the microlevel and the macrolevel, appears to be strongly related to economic stratification, although the nature of this relationship at the macro level is not always straightforward. A mesolevel analysis highlights the ways in which countries have different health care systems and the unique approach taken by the United States. Key chapter topics include the following:

- Throughout the world individuals in the most deprived social classes tend to have the most health problems, lowest levels of life expectancy, and highest rates of infant mortality. The poorer health of members of minority groups in the United States can be largely explained by their lower socioeconomic status.

- Stressful events produce physiological reactions that can lower resistance to disease, but both how we interpret stressful events and the extent of our supportive social networks can buffer the effects of stress. People in lower social class groups are more likely to experience stress and less likely to have support to counteract it.

- Members of higher social class groups are more likely to have the economic resources to be able to adopt healthy life styles and the social resources to take a more consumer-oriented approach in their interactions with health care providers.

- The United States health care system is a "fee-for-service" system that has traditionally allowed doctors a great deal of power over medical transactions. This power was promoted by actions of the American Medical Association, but has recently been challenged by consumers, government actions, and developments in insurance coverage.

- All industrialized countries except the United States and most less developed countries guarantee health care to all of their citizens and spend less of their gross national product on health care.

- During the last century in the United States, life expectancy has increased and infant mortality has declined. Deaths are now more likely to occur from chronic diseases than from infectious diseases, but improvements have been concentrated less among the poor than among the more affluent segments of society.

- Cross-national comparisons reveal large differences in health between rich and poor countries, with the poorest countries of the world having the lowest levels of life expectancy and highest levels of infant mortality. AIDS is especially virulent in the very poor African countries that do not have the resources to provide basic sanitary services, let alone adequate medical treatment.

- Among developed countries, economic wealth is not related to levels of health, and at least twelve countries have higher levels of life expectancy than the United States even though they are poorer.

- Richard Wilkinson's research indicates that a society's health is related to the level of inequality within the country. When societies and communities have greater economic and social equality they tend to have stronger social networks and better overall health for all citizens.

KEY TERMS

case study 438

epidemiology 421

health 421

infant mortality rate 423

life expectancy 423

profession 430

secondary analysis 438

INTERNET ASSIGNMENTS

1. Search for Web sites that provide material related to stress and health behavior. To what extent are the remedies presented in these Web pages available for people in different social classes? Choose three or four different Web sites and analyze how the information presented could or could not be used by people in different social class groups.

2. Find Web sites that deal with HMOs, managed care, and other types of health delivery systems. What types of issues are discussed on these sites? How do the discussions reflect issues regarding the professional identity and power of physicians, the nature of the "fee for service" system, and consumers' desires and needs for health care. Choose three different sites and compare and contrast the views that are given in each.

3. Find the Web site for the World Health Organization and examine documents discussing health issues that face the developing and least developed nations of the world. How do these issues compare with the health issues discussed in your daily newspaper? How are these differences influenced by the nature of global stratification?

TIPS FOR SEARCHING

When searching for information on health delivery systems try using a variety of terms such as "HMOs," "Health Maintenance Organizations," "health insurance," and "managed health care."

INFOTRAC COLLEGE EDITION: READINGS

Article A19475064 / Brunner, Eric. 1997. Stress and the biology of inequality: Socioeconomic Determinants of Health, part 7. *British Medical Journal, 314,* 1472–1477. Brunner explores how social class influences the levels of stress people encounter, which, in turn, influence physical health.

Article A19410925 / Bartley, Mel, David Blane, and Scott Montgomery. 1997. Health and the life course: Why safety nets matter: Socioeconomic determinants of health, part 4. *British Medical Journal, 314,* 1194–1197. Bartley, Blane, and Montgomery describe how stress resulting from social inequality can have a long-term effect on health and how pro-viding medical care to all people, regardless of their income, can alter this relationship.

Article A20824165 / Reynolds, John R., and Catherine E. Ross. 1998. Social stratification and health: Education's benefit beyond economic status and social origins. *Social Problems, 45,* 221–248. Reynolds and Ross study the ways in which higher levels of education promote psychological and physical health.

Article A17474787 / Williams, David R., and Chiquita Collins. 1995. U.S. socioeconomic and racial differences in health: Patterns and explanations. *Annual Review of Sociology, 21,* 349–387. Williams and Collins look at changes over time in class and racial-ethnic differences in health.

FURTHER READING

Suggestions for additional reading can be found in *Classic Readings in Sociology,* bundled with this book. If you purchased a used copy of this book that does not include this custom-published reader, go to www.sociology.wadsworth.com for ordering information.

RECOMMENDED SOURCES

Cockerham, William C. 1998. *Medical Sociology,* 7th ed. Upper Saddle River, N.J.: Prentice Hall. This readable text, written by the sociologist profiled in Box 16-2, covers all aspects of medical sociology.

Rice, Phillip L. 1998. *Stress and Health,* 3rd edition. Pacific Grove, Calif.: Brooks/Cole. This is an extensive and readable overview of the relationship between stress and health.

Starr, Paul. 1982. *The Social Transformation of American Medicine.* New York: Basic. This is a classic discussion of the history of medicine in the United States.

Pulling It Together

Listed below are other chapters where you can find additional discussion or examples of some topics covered in this chapter. You can also look up topics in the index.

Weiss, Lawrence D. 1997. *Private Medicine and Public Health: Profit, Politics, and Prejudice in the American Health Care Enterprise.* Boulder, Colo.: Westview. Weiss provides an extensive and biting critique of the inequalities within the American health care system.

Weitz, Rose. 1996. *The Sociology of Health, Illness, and Health Care: A Critical Approach.* Belmont, Calif.: Wadsworth. This textbook reviews many major areas within the sociology of medicine with special focus on inequalities within the United States.

Wilkinson, Richard G. 1996. *Unhealthy Societies: The Afflictions of Inequality.* London: Routledge. This book provides many more details from the research that is featured in this chapter.

INTERNET SOURCES

The World Health Organization is a branch of the United Nations that gathers data on the health status of people around the world. The site includes an extensive body of reports and data.

The International Health Program at the University of Washington and Health Alliance International maintain a Web site that provides up-to-date information on scholarly data regarding inequality and health.

The National Center for Health Statistics Web site contains a wide array of data and summaries of research related to health in the United States.

Population

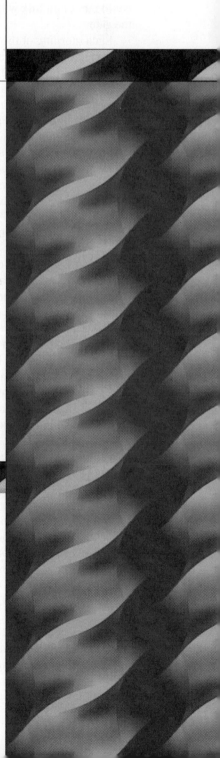

Looking at Populations: Demographic Data

Gathering Demographic Data

Using Demographic Data

**Population Growth: Balancing
Fertility and Mortality**

*Tracing Patterns in Population Growth:
Pre-Industrial versus Industrializing Societies*

*Modeling Population Changes:
Demographic Transition Theory*

Explaining Declines in Fertility

The Effect of Population Changes

Age Structures and Dependency

Economic Opportunities and Constraints

Environmental Costs

The Wild Card of Migration

FEATURED RESEARCH STUDY

Social Action and Population Change

*Studying Women's Education
and Fertility Rates in Mexico*

Explaining the Effect of Education

LUPE WAS THRILLED TO BE chosen as her university's representative to the International Conference of Students. She'd been interested in the nations of the world since she had moved from Mexico to the United States as a young girl and had quickly noticed how different the two neighboring countries sometimes seemed. In college she majored in international relations, thinking that she might someday be able to use her facility with Spanish and English to get a job with the State Department or the Diplomatic Corps. She hoped that the International Conference could be a stepping stone to this career, for each day would be filled with seminars where she could meet and talk with students from all around the globe.

Each morning of the conference, Lupe met with other members of a seminar on "Poverty, Population, and the Environment." The group was charged with developing policy recommendations to bring before the large general meeting at the end of the gathering. The first morning she began chatting with Winnie, who was from Malawi, a small country about the size of Pennsylvania in southeastern Africa. Winnie spoke perfect English with a slight British accent. "Years ago," she said, "my country was a British protectorate, but now we are independent. I'm lucky to be here," she confided, "for my country is very poor. I had a scholarship to university in Britain, and they paid my way to come."

Just as Lupe was about to tell Winnie of her own background, the seminar was called to order. The group's members introduced themselves, and then the leader showed pictures and presented a dismaying array of statistics, gathered by the United Nations, that summarized information on the lives of people in different parts of the globe. Even though Lupe knew from her trips to Mexico that many people outside the United States lived in poverty, she was shocked to learn how desperate conditions were in some parts of the world. For instance, people in sub-Saharan Africa, where Winnie was from, had an average life span of only fifty-one years, whereas people in the United States, Canada, Europe, Japan, and Mexico could expect to live until their seventies. Of every 1,000 babies born in sub-Saharan Africa, almost 100 died before their first birthday, compared with only 10 in Europe. The leader also talked about how impoverished people in many regions of the world were and how this poverty was exacerbated by extreme population growth. For example, the population of sub-Saharan Africa would double in only 27 years if the current growth rate continued; in contrast, the population of Europe has actually ceased growing! At the same time, the leader noted, scientists worried that the environment might not be able to support growing populations in the least developed areas, especially with extensive overgrazing and the cutting of forests leading to the loss of land on which crops could be grown.

As the morning wore on and the leader finished her presentation, the group began to discuss the data and the pictures they had seen. Lupe sensed that the members of the group seemed to be divided. On one side were the Europeans and North Americans, who lived in the richer, industrialized countries; on the other side were individuals from other areas of the world, particularly Latin America, Africa, and much of Asia. Christa, a smartly dressed young woman from Germany, seemed to summarize the views of the Europeans and North Americans when she said, "This is all very sad, but I just don't get it. Why can't people in these countries learn to control their population? Haven't they even heard of birth control? And, for heaven's sake, why do they keep destroying their forests and grazing lands? Don't they know that these practices just make erosion worse and make it even harder to grow crops? We've learned our lessons. Why can't they?"

As Christa went on and others around the room started to nod their heads in agreement, Lupe watched Winnie growing more and more agitated. Finally, Winnie burst out, "Wait, you don't understand. It's much more complicated than the way you've portrayed it. Let me tell you about how these people really live." In a quiet, clear voice Winnie related how in her country most of the people lived as subsistence farmers, working small plots of land and occasionally toiling on large tobacco, sugarcane, tea, or coffee plantations for meager wages. She told of how poor most of the residents were and of how lucky she was to have gone to a university. "Yes," she said, "Malawians have large families. But their babies often die, and no one wants to be childless. And people must cut the forest if they are to grow corn. Would you rather have them starve? Besides," she ended, with more than a trace of bitterness in her voice, "we don't pollute the world's atmosphere with cars or the burning of oil and gas, and we don't dump mounds of garbage into the oceans and land each year like your countries do." Winnie looked defiantly around the room and then sat down.

Lupe felt torn. She could understand the points made by Christa and others from the industrialized countries. Shouldn't countries with high birth rates be responsible for curbing their population growth? The group's leader had noted that China's policy of limiting families to one child seemed to be succeeding. Lupe was a devout Catholic and knew that at least some of China's success came from forcing women to have abortions, a policy that made her more than a little uneasy. But surely residents of these poor countries could see that having large families wasn't helping matters.

At the same time, Winnie's comments also made sense to Lupe. She suspected that the situation was much more complex than Christa had implied. Even though her own relatives in Mexico were far from poor, Lupe knew that many Mexicans lived in extreme poverty and that, until recently, many infants had died before their first birthday and few adults had lived into their seventies and eighties. In fact, Mexico was one of the countries the group leader had listed as having high population growth. Was this growth somehow linked to its history of poverty? But if so, how? People lived much longer now in Mexico than they once had, and the country was getting richer. Why then was Mexico's population still grow-

ing? Would this population growth eventually stop as Mexico continued to industrialize? And what about Winnie's comments about the way that industrialized countries were polluting the environment? She knew that she and her friends and family had many more possessions than most Mexicans or Malawians had. But could they really be harming the environment as badly as the inhabitants of the poor countries trying to scratch out a living could?

Why do countries have such different patterns of population growth and change? And is population growth and accompanying poverty as big a deal as the seminar leader implied? Or, as Christa and some others suggested, is it just something that people in Malawi and other poor countries should handle by themselves?

*T*HESE QUESTIONS REFLECT FORCES and trends in population, the number and types of people who live in a society. Even though we might not think about it, changes in population significantly influence our day-to-day lives, providing opportunities and constraints for us as individuals, for the communities in which we live, for nations, and for the world as a whole. For instance, the poverty, malnourishment, and unsanitary conditions that people in Malawi must deal with are exacerbated by Malawi's rapidly growing population. In contrast, in growing up in Germany, Christa has enjoyed a life of prosperity largely because her country has not had to deal with large population growth.

Figure 17-1 presents basic data on four countries that differ markedly in size, population, economy, education, and so on—Malawi, Mexico, Germany, and the United States. Malawi is typical of the least developed countries in the world, those that are extremely poor and have little industrialization. Germany and the United States are typical of advanced industrialized countries, the richest and most powerful nations. Mexico is a developing country, one that is in the process of industrialization; it is richer than countries such as Malawi but far poorer than countries such as the United States and Germany.

People in many different areas of social life are concerned with the size and composition of populations, not just in the very poor and developing countries but in the most developed and industrialized

countries as well. For example, government officials need to know how many children will be needing schooling and how many young people will be entering the work force in the years to come. Business leaders want to know who lives in a community and how old they are to help plan what kinds of stores and businesses they might open. Insurance companies need to know the probability of someone dying to figure out how much to charge to insure a life. Transportation departments need to know how many people live in different areas to plan the most effective highway and public transit systems. The number of people who move in or out of your own community can affect how crowded public facilities will be and what kinds of economic opportunities you have.

The scientific study of populations and their effects is called **demography.** (The term derives from the Greek root, *demos,* meaning "the people," and *graphy,* which refers to writing or describing, as well as to a branch of learning.) Scholars who study populations are called *demographers.* They can be found in many fields, such as mathematics, biology, economics, and public health, in addition to sociology. You will sometimes hear the term *social demography* in reference to the use of demographic data by sociologists as they study areas we have looked at throughout this book, such as stratification, the family, the economy, and even religion.

The study of population is more than simply the dry counting of numbers—it involves analyzing

FIGURE 17-1 *Basic Data on Malawi, Germany, Mexico, and the United States*

Note: Per capita GNP is in U.S. dollars.

Source: Population Reference Bureau, 1998; Bellamy, 1998, pp. 107–108; Famighetti, 1996, pp. 767, 795, 798, 831; U.S. Bureau of the Census, 1998, p. 421.

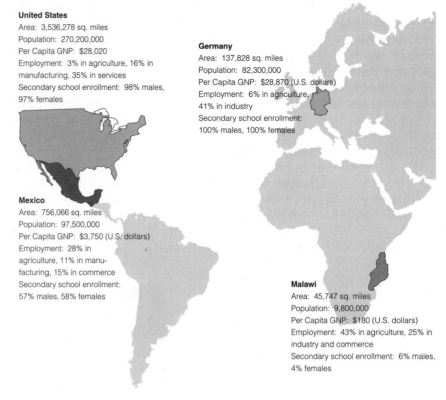

United States
Area: 3,536,278 sq. miles
Population: 270,200,000
Per Capita GNP: $28,020
Employment: 3% in agriculture, 16% in manufacturing, 35% in services
Secondary school enrollment: 98% males, 97% females

Germany
Area: 137,828 sq. miles
Population: 82,300,000
Per Capita GNP: $28,870 (U.S. dollars)
Employment: 6% in agriculture, 41% in industry
Secondary school enrollment: 100% males, 100% females

Mexico
Area: 756,066 sq. miles
Population: 97,500,000
Per Capita GNP: $3,750 (U.S. dollars)
Employment: 28% in agriculture, 11% in manufacturing, 15% in commerce
Secondary school enrollment: 57% males, 58% females

Malawi
Area: 45,747 sq. miles
Population: 9,800,000
Per Capita GNP: $180 (U.S. dollars)
Employment: 43% in agriculture, 25% in industry and commerce
Secondary school enrollment: 6% males, 4% females

processes and structures that affect many different aspects of lives. As you will see later in this chapter, some of the issues related to population involve serious problems regarding the perpetuation of poverty and environmental degradation. These problems are especially pressing in developing countries such as Malawi, but they also affect advanced industrialized countries such as the United States and Germany. We all inhabit the same globe, and population pressures, poverty, and environmental degradation in one part of the world affect other parts as well. In addition, as we will see, the economic advantages that have accompanied low population growth in highly industrialized countries have generated their own problems, particularly in the environment, much as Winnie described. Many sociologists who study populations are concerned with these problems. But, just as they approach other areas of the social world in a scientific manner, they examine population and population change in a dispassionate way, often hoping that careful, well-substantiated analyses eventually will help to ease the problems associated with population growth and change.

By now you've had a lot of practice taking a sociological perspective. You've looked for the social structures that underlie our daily lives and seen how our social actions both are affected by these structures and help to create them. In this chapter we use our sociological lens to take a *demographic perspec-*

tive, focusing on issues related to the sizes of populations, the age and sex of their members, and the ways in which people are distributed in individual societies and throughout the world. We look at how populations change in their size and composition, or what are often called **demographic** or **population processes,** and we discuss the social implications of these changes. We begin by looking at the data that demographers use and the methods they employ to analyze it. Then, at the macrolevel, we explore how large-scale shifts in populations affect individuals, communities, and societies. Finally, at the microlevel, we trace how social actions affect demographic processes by examining social research on the individual decisions that underlie population changes.

Population processes have attracted the attention of not only social scientists but also natural scientists, especially ecologists (biologists who study all forms of living life). One of the most prolific writers in the area of ecology, Paul Ehrlich, has written extensively about the relationships between different forms of life and the environment. He once described these relationships as follows:

On a sunlit summer day in a suburban backyard, or in a flower-strewn mountain meadow with butterflies flying and birds singing, a multitude of interactions is taking place between organisms and their physical environment. Energy is being captured from the sunlight by the plants, water is constantly flowing through them, and all the plants and animals are exchanging gases with the atmosphere. . . . If the physical environment changes, though, the relationship between the organisms and their surroundings is often thrown into sharp focus. . . .

[I]f a freak snowstorm hits a mountain meadow late in June, or a homeowner forgets to water the backyard for a few rainless weeks, the effects are dramatic. The flowers in the meadow are destroyed; the lawn turns brown.[1]

As we will see, just as populations of birds, butterflies, and flowers can change over time, so can the size and composition of human populations. As with changes in animal and plant populations, changes in human populations are related to the natural environment. They are also strongly related to technology, the development and application of knowledge. Most important, especially from a sociological perspective, changes in human population, like changes in the natural environment, can have far-reaching implications—for individuals, communities, and entire nations. Just as fluctuations in the weather can affect the foliage in a mountain meadow, changes in a population can influence many aspects of a society and the individuals who live in it.

LOOKING AT POPULATIONS: DEMOGRAPHIC DATA

A society's population can change in only one of three ways: (1) people are born, (2) people die, and (3) people move either in or out. Demographers often refer to these as the three basic demographic processes: (1) *fertility,* or the birth of children, (2) *mortality,* or death, and (3) *migration,* or the movement from one geographic area to another. Sociologists who specialize in demography analyze changes in these three processes and the ways in which these changes affect entire nations, societies, communities, and individuals.

Clearly, to develop an accurate understanding of population changes, demographers must have accurate data, as well as ways of representing the data so that the information is useful to citizens and policymakers. In this section we look at how demographers gather and analyze their data.

Gathering Demographic Data

Demographers typically tap three different sources for data, most of which are gathered by governmental agencies. One primary source of information is a country's **census,** such as the one conducted in the United States every ten years. Countries have been conducting censuses since biblical times, often as a way to determine who should be paying taxes and who might be eligible to serve in the military. More recently, governments have used censuses as a means of obtaining other information about their citizens—not just the number of people and their addresses but also their age, marital status, race-ethnicity, occupation, education, and housing type.[2]

The second major source of demographic data is official records of **vital statistics,** such as births, marriages, divorces, deaths, and official immigration into a country from other nations. In past centuries in Europe, the church kept these records; today governments generally keep them. Laws require that a birth certificate be issued when you are born, a marriage certificate if you marry, a divorce certificate if you divorce, a death certificate when you die, and a visa and passport when you move from one country to another. Governments keep track of these events and regularly publish summaries of the information.[3] The data that Émile Durkheim used in his classic study of suicide, described in Chapter 1, came from vital statistics that had been gathered throughout Europe.

Finally, demographers can use data obtained through sample surveys. Rather than enumerating all the citizens of a society through a census, government agencies often gather data from carefully selected samples of citizens. Because studying a sample is much less expensive than studying an entire population, social scientists throughout the world increasingly have used samples to obtain data like that gathered in censuses, as well as much more detailed information about topics such as wealth, income, disability, migration, and fertility.[4]

As in all kinds of research, sociologists have been concerned about the accuracy of demographic data. Generally the most developed countries tend to have the most accurate data, largely because they can afford the large and highly trained staffs needed to gather and check the information. In addition, staff members at the United Nations continually work to gather and check data throughout the world, making sure that it is as accurate as possible. Because of these efforts, scholars generally agree that demographic data usually do a good job of describing the nature of the world's population.[5]

Using Demographic Data

If you were to look up the results of censuses, vital statistics records, and sample surveys in your library or on the Internet, you would find page after page of numbers. Simply looking at these numbers would not tell you very much about the nature of or changes in the population. For this reason demographers often use diagrams and graphs, as well as various statistics, that summarize data so that we can see, in a graph or

The changes in population since 1790 have transformed New York City from a quiet, modest town to a bustling, crowded metropolis. Have similar changes occurred in your community?

just a few numbers, population characteristics and trends. Demographers generally look at the data in three different ways: by graphing population trends, by constructing population pyramids, and by calculating various rates and estimates of life expectancy.

Population Trends   First, demographers often use graphs to look at how the size of a population changes over time. For instance, Figure 17-2 traces the size of the U.S. population since the first census was conducted in 1790. In that year the United States had slightly less than 4 million residents, about half the current size of New York City. In 1990 the census counted almost 250 million people, more than sixty times as many people as two centuries earlier.

As you can see, the line in Figure 17-2 extends all the way to 2050. The figures shown in the graph for the years 2000–2050 represent **population projections,** demographers' estimates of the future size of the population. They base these calculations on the number of people currently living and on estimates of future birth rates, death rates, and amount of migration; they then apply various mathematical formulas to these numbers. Of course it is always hard to foretell the future, so demographers tend to hedge their bets by using different estimates. The data shown in Figure 17-2 represent the middle series of estimates, which assumes levels of fertility, mortality, and migration that are midway between the lowest and highest estimates.[6] Demographers in the Census Bureau project that there will be almost 275 million people counted in the United States in the Census conducted in the year 2000.[7]

Population Pyramids   Although graphs like the one in Figure 17-2 are helpful for displaying long-term trends in population size, other types of graphics give a better picture of the composition of a population at one time point. One of the most useful is the **population pyramid,** a graph that illustrates the population of a society by age and sex—that is, its **population structure** (or **age/sex structure**). The population pyramid for the United States is shown in Figure 17-3.

Even though they display information for just one time point, population pyramids are very useful for revealing both a society's past and its future composition. For instance, the large bulge in the middle of Figure 17-3 represents people who were born between 1946 and 1964, a period commonly referred to as the "baby boom." As you can see, birth cohorts in this period were much larger than previous or subsequent cohorts were.

This bulge in the population pyramid is associated with a number of social implications. The appearance of the "baby boomers" put a tremendous strain on society. For example, when the boomers began school, there often weren't enough classrooms to hold them, so classes were held in church basements while new schools were constructed. When they reached college age, college enrollments skyrocketed, so new buildings were constructed and new professors hired. As the baby boomers approach retirement age, there might be an entirely new set of problems, ranging from providing adequate housing and medical care to finding sufficient income for the retirement years. (We will return to these concerns in the "Applying Sociology to Social Issues" feature later in this chapter.)

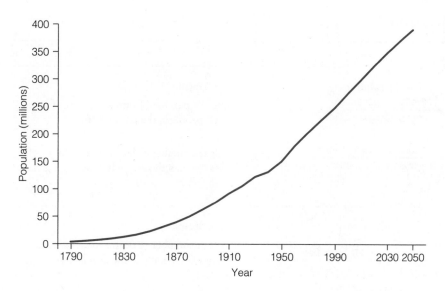

FIGURE 17-2 *Population of the United States, 1790–2050*

Source: U.S. Bureau of the Census, 1995, pp. 8–9; 1998c, p. 9.

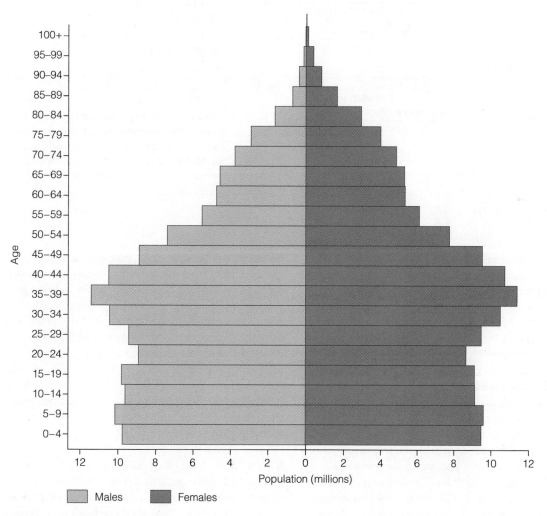

FIGURE 17-3 *U.S. Population Pyramid, 1997*

Source: U.S. Bureau of the Census, 1998c, p. 16.

TABLE 17-1 *Calculating Demographic Rates*

MEASURES OF FERTILITY

Measure	Calculation	Description
Crude birth rate	$\dfrac{\text{no. of live births per year}}{\text{population}} \times 1000$	Tells the number of births in an area for every 1,000 people in the population.
General fertility rate	$\dfrac{\text{no. of live births per year}}{\text{no. of women in childbearing years}} \times 1000$	Tells the number of births in an area for every 1,000 women theoretically capable of giving birth (generally ages 15–44).

MEASURES OF MORTALITY

Measure	Calculation	Description
Crude death rate	$\dfrac{\text{no. of deaths per year}}{\text{population}} \times 1000$	Tells the number of deaths in an area for every 1,000 people in the population.
Age-specific death rate	$\dfrac{\text{no. of deaths per year per given age range}}{\text{no. of people in given age range}} \times 1000$	Tells the number of deaths in an area for every 1,000 people of a certain age in the population.
Infant mortality rate	$\dfrac{\text{no. of infant deaths per year}}{\text{no. of infants in population}} \times 1000$	Involves children less than one year of age; the most common age-specific death rate.

MEASURES OF MIGRATION

Measure	Calculation	Description
Crude net migration rate	$\dfrac{\text{in-migrants} - \text{out-migrants}}{\text{population}} \times 1000$	Tells the number of people per 1,000 population who have migrated into an area; a positive number indicates more people entering an area (in-migrants) and a negative number indicates more people leaving an area (out-migrants).

Source: Weeks, 1994, pp. 114–115, 170–171, 195.

Being a member of such a "boom cohort" can also affect individuals' lives. As you may remember from Chapter 6, members of boom cohorts are slightly more likely to get into trouble with the law and to commit crimes than are people who grew up in smaller cohorts. Sociologists have speculated that, simply because there are so many people in these cohorts, they find it harder to maintain the conventional ties to family, community, and work that people in the smaller and so-called bust cohorts take for granted.

Besides indicating how the lives of individuals might unfold, the shape of the age/sex structure also can help demographers anticipate the future growth of societies. At the very bottom of the pyramid in Figure 17-3, you can see that the population appears to be expanding again. These are the children of the baby boomers, a group that demographers have sometimes called a "baby boomlet." Although the baby boomers have smaller families than their parents did, the population will continue to grow simply because there are so many baby boomers. This phenomenon involves what demographers call **demographic momentum,** the growth of a society that occurs simply because of the people, or demographic characteristics, that are already present. Once a society has experienced high birth rates in just one generation, it is almost certain to experience continued population growth in the next generation, much as a car gains momentum once it starts down a steep hill. As we will see throughout this chapter, demographic momentum is a problem that worries officials and citizens in many parts of the world and is part of the reason that the developing and least developed countries, such as Mexico and Malawi, are experiencing such rapid population growth.

Rates and Life Expectancy Finally, besides graphs and diagrams, demographers calculate many different types of statistics that help describe populations. For instance, they often use **birth** and **fertility rates** to measure how new births are changing the size of a population; they use **death** and **mortality rates** to indicate how deaths affect the population; and they use **migration rates** to tell how many people have moved into or out of an area. Table 17-1 gives details on how some of these statistics are calculated.

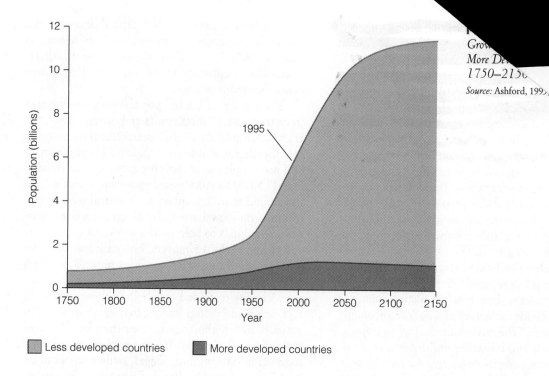

Population (billions)

12
10
8
6
4
2
0

1995

1750 1800 1850 1900 1950 2000 2050 2100 2150

Year

■ Less developed countries   ■ More developed countries

Gro...
More De...
1750–215...
Source: Ashford, 199...

Sociologists use this type of information in several ways. For instance, in our discussion of the family in Chapter 10, we looked at how changes in fertility and mortality affected family structures and members' lives as families became smaller and marriages were less likely to end in the death of a spouse. In our discussion of the life course in Chapter 4, we examined the different experiences of various birth cohorts in terms of life expectancy and marriage and how these variations can result in different role expectations and attitudes. And, in our discussion of health in Chapter 16 we examined variations in life expectancy and mortality of people in different social class groups and in different societies.

We will examine the many implications of changes in a society's population, its age/sex structure, and rates of fertility, mortality, and migration in more detail later in this chapter. But first, it is important to understand how populations grow and change. Why did the United States experience the tremendous growth shown in Figure 17-2? Have other societies experienced this growth? What do demographers expect will happen in the future?

POPULATION GROWTH: BALANCING FERTILITY AND MORTALITY

Scientists estimate that humans have been around for at least a million years. For 99 percent of this vast time span, the size of the world's population remained fairly stable. However, beginning around 10,000 years ago—barely a minute of time in evolu-

tionary terms—this began to change. At first the increases were small and gradual, but during the past 250 years the world's population has exploded. In 1750, shortly before the United States became an independent nation, fewer than a billion people (about 800 million) inhabited the planet; by 1987 this figure had grown to 5 billion inhabitants. Demographers estimate that there will be 6 billion people by the year 2000, more than seven times as many people as there were only 250 years ago.[8] And by the year 2028 the planet will be home to 8 billion people![9]

These numbers are so large that it's difficult to grasp what they really mean. One way to appreciate how many people the figure 8 billion represents is to calculate how long it would take a tireless census taker to count them one by one. If you started today and counted one person per second, nonstop, twenty-four hours per day, 365 days per year, it would take you more than 250 years—well into the twenty-third century—to count to 8 billion! (And who knows how many people will be in the world by then.)

As shown in Figure 17-4, most of the increase in the world's population is occurring in the less developed countries, those that are in the process of industrialization and modernization. These nonindustrialized countries currently contain about 80 percent of the world's population, but given their high rate of population increase, demographers estimate that they will contain an even larger proportion of the world's population in the future.[10] By the middle of the twenty-first century (the year 2050) it is expected that these countries will be home to 88 percent of the world's people.[11]

...ment of-
...oncerned
...ulations.
...people the
...vers agree that
...ace heavy demands
...ch of us as individuals.
...ve will have to find enough
...d schooling for everyone. Be-
...growth is occurring in the poorest
...the world, whose budgets are already
...thin, this trend is even more problematic.
...ally all scholars agree that the best long-term
...lution is slowing and eventually stopping the
world's population growth.[12]

But to slow population growth, we need to understand why it has occurred. Sociologists' explanations of this process show how population change is closely linked with the nature of a society's economy and the nature of the environment. Just as populations of flowers and butterflies influence and are affected by their surroundings, so, too, are human population changes linked with their environments. In addition, unlike animals and plants, humans create technology and inventions that alter their environment, and these technological changes affect the nature of the economy and population growth.

Tracing Patterns in Population Growth: Pre-Industrial versus Industrializing Societies

For most of the million years of human life, people lived in small hunting and gathering societies. The average group contained an estimated fifty people, and the largest groups, which lived in lush and favorable environments, had only a few hundred.[13]

Life was precarious in these societies, and **life expectancy**—the number of years a person is expected to live—was low. The average person could expect to live only to the age of thirty, and many infants and children died before reaching adulthood. Because women worked constantly to help gather food for their families, they generally had about four years between each birth, so that it would be it easier for them to carry their youngest child while they were working. Women's short life span, combined with the spacing between childbirths, meant that they rarely lived long enough to have more than four or five babies. Given the high rate of infant and child mortality, this number was what was needed, on average, to ensure that at least two children survived to adulthood.[14]

As a result, during the lengthy era when hunting and gathering societies populated the world, the human population remained fairly stable. In demographic terms fertility balanced out mortality. Death rates tended to vary somewhat more than birth rates

as a result of famines or other natural disasters. But overall approximately as many people were born as died, so that the number of people on Earth remained fairly constant, increasing only slightly over many thousands of years.

The pattern of limited population growth began to change about 10,000 years ago, when a few societies developed advanced horticultural techniques, specifically the ability to irrigate and fertilize crops and use simple metal tools for cultivation. Approximately 5,000 to 6,000 years ago some societies also developed more advanced agricultural techniques, forging iron plowshares to break up the soil and harnessing animals to help do the work. As you will recall from previous chapters, horticultural societies and, especially, agrarian societies tend to be much larger than hunting and gathering societies.

Mortality remained very high in these groups. In fact, some authorities believe that agrarian societies actually had higher death rates than hunting and gathering societies did. When people were grouped together into towns and villages, sanitation became a problem, and the likelihood of catching communicable disease increased. However, even though death rates were higher, women also had higher fertility than in hunting and gathering societies. Because they no longer necessarily had to carry their children with them wherever they went, they could space childbirths more closely together. In addition, the ability to grow crops led to better nutrition for mothers and babies alike. Both the improved nutrition of mothers and the decreased time that they had to nurse their infants increased the chances that a woman could become pregnant. As women's increased fertility overbalanced the high mortality rate, populations began to grow.[15]

Until about 250 years ago, this population growth was quite gradual. Between 8000 B.C., when the first agricultural societies appeared, and A.D. 1750, around the start of the Industrial Revolution, the world's population grew by only about 67,000 people per year. After 1750, however, population growth was literally explosive. By the 1990s, some 67,000 more people were added to the world's population every seven hours![16] Why did this happen?

Again, the answer involves the nature of the economy and its influence on fertility and mortality. After the mid-eighteenth century, as the Industrial Revolution progressed, both sanitation and nutrition improved. People had better diets, drank cleaner water, wore warmer clothes, and took more baths. As described in Chapter 16, all these changes meant that people could fight off disease more readily and that death rates gradually declined. And since 1900 medical technology has progressed very rapidly. Vaccinations prevent many communicable diseases, and

China has the largest population of any country in the world. What do you think of China's efforts to control population growth by establishing birth control policies such as the one-child-per-family campaign? Could these policies be applied in other countries?

a wide array of medical treatments can cure once-fatal illnesses. As a result, death rates have declined dramatically in this century, life spans have increased, and infants are much more likely to survive to adulthood than they were previously. Initially these changes affected the more developed countries of Europe and North America, but since World War II, improved sanitation and medical technology have led to declines in death rates in virtually all countries.

Population growth always results when fertility is higher than mortality, but this difference can result from changes in either the mortality or the fertility rate. In industrialized societies mortality has decreased very dramatically while fertility has remained fairly constant or has decreased less quickly than mortality has. In other words, in the past two centuries, death rates fell without an accompanying change in birth rates. Many more people were born than died, and the population grew at an exponential rate.

But why is the rate of growth in the world's population so high? Why does the population increase as much in seven hours today as it did in a year in previous times? The answer involves simple mathematics. When mothers have more children, and all of them survive to adulthood, there are more people within a society who can themselves have children. Over a few generations this can lead to a very rapid increase in the population. In contrast, if each mother only has two children, the population will remain stable (because, on average, only one of the two children will be female).

You might have heard the term **zero population growth.** This refers to the idea that the world population will stop growing if, on average, each woman only has two children. When mortality is high, as in pre-industrial societies and in some developing societies today, a mother can expect only two of her four children to live to adulthood. Today, with much lower mortality rates, chances are much greater that all four children will survive and have children themselves.

The implications of this phenomenon are staggering simply because the world's population is already so large. One way to understand how such growth occurred is by thinking about what demographers call the **growth rate,** a measure of how many people are added to a population each year divided by the number of people who are already there. The current growth rate of the world's population is estimated at 1.4 percent, somewhat lower than the 1950 rate of 1.7. But even this slightly lowered growth rate over the last half century cannot stem a phenomenal amount of population growth. In 1950 there were only about 2.5 billion people on Earth, so adding another 1.7 percent to the population that year translated into approximately 43 million more people. But today, because there are almost 6 billion people on Earth a growth rate of 1.4 percent results in almost twice as many people added to the population as in 1950![17] Lowering the growth rate to 1.0 percent would mean a drop of over 20 million in the number of people added to the world's population each year.

The rate of growth can be used to estimate how long it would take for a population to double in size. Map 17-1 shows the number of years in which populations are expected to double if they continue to grow at their current rate. The world as a whole is expected to double its population in about 49 years. But, as you can see from looking at Map 17-1, some countries, all

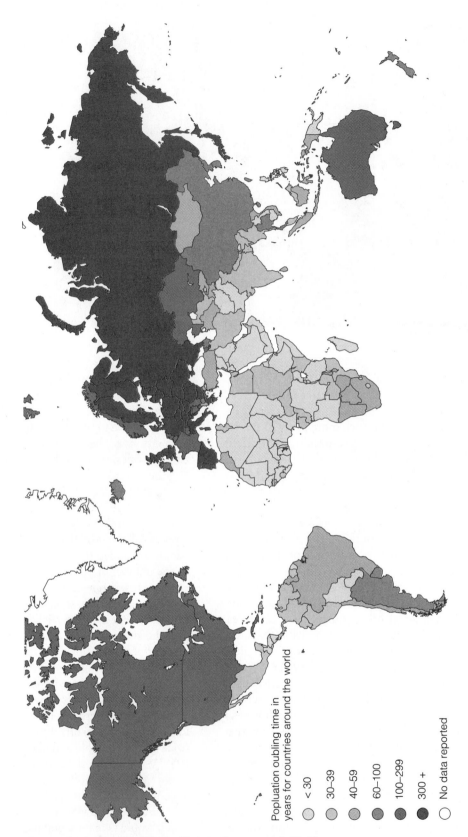

MAP 17-1 *"Doubling Time" in Years for Countries of the World*

Note: — means that the current growth rate is zero or negative. If this trend continues the population will never double in size.

Source: Population Reference Bureau, 1998.

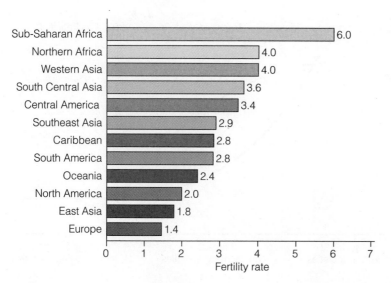

FIGURE 17-5 *Fertility Rates in Regions of the World, 1998*

Source: Population Reference Bureau, 1998

of which are among the less developed countries of the world, will double their population in a far shorter period of time, whereas others, if they continue their present rates of growth, will never double in size or will take literally hundreds of years to do so.

This low growth rate occurs because in recent years the industrialized countries of the world have reached the point at which mothers have an average of only two or fewer children. Figure 17-5 shows fertility rates in various areas of the world. As you can see, Europe, where Christa lives, as well as several East Asian countries, such as Japan, China, Korea, and Taiwan and Canada in North America, all have fertility rates below 2.0, a rate that is actually lower than the level needed for a population simply to reproduce itself (this level is sometimes referred to as the **replacement rate**). The United States has a fertility rate of 2.0, a point at which fertility approximately balances out mortality.

Note that China is among the countries with fertility rates at the replacement level even though it is not yet highly industrialized. Its low fertility rate appears to have resulted from strict government policies related to birth control. China has the largest population of any country in the world—more than a billion people, or more than one-fifth of the world's population. Given the realities of demographic momentum, even a replacement-level fertility rate could result in substantial population growth. Recognizing that such growth would place enormous economic pressures on the nation, the Chinese government has established a variety of birth control policies over the years. The most restrictive policies were instituted in 1979 when the "One-Child Campaign" began, designed to encourage families, especially those in cities, to have only one child. Although the policy is often criticized as overly coercive, both within China

(to the extent that such criticism is tolerated) and internationally, and although nontotalitarian states are not likely to emulate it, the policy has succeeded in lowering China's fertility rate significantly.[18]

Apart from China, developing countries, which contain the majority of the world's population, have fertility rates that are substantially higher than the replacement level, ranging from 2.8 in South America to 6.0 in sub-Saharan Africa, the area where Winnie lives. Even though the rates have fallen in most of these areas in recent years, they are still far higher than is required for replacement. This high fertility rate, combined with the dynamics of demographic momentum and lowered levels of mortality, is the force that underlies the rapid and extreme population growth of recent years.[19]

Modeling Population Changes: Demographic Transition Theory

In trying to develop a broad understanding of population trends in different societies, demographers have combined the historical changes we have discussed into a single model called **demographic transition theory,** depicted in Figure 17-6. It summarizes the changes in population that societies experience as they move from the high fertility and mortality typical of pre-industrial societies to the low fertility and mortality characteristic of highly industrialized modern societies.[20]

The theory includes three stages. The first stage, sometimes called the *pretransitional* stage, involves high fertility, high mortality, and a stable population size, as in pre-industrial societies. In the second, or *transitional,* stage the birth rate is higher than the death rate, and societies experience rapid population growth, much as in Malawi and other developing

FIGURE 17-6
The Demographic Transition

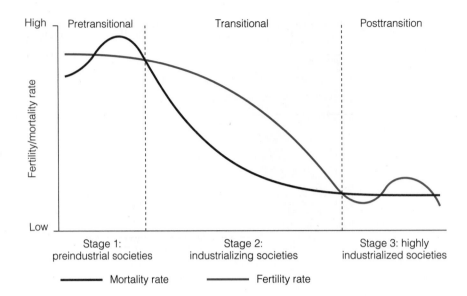

countries. Over time, however, the birth rate declines to a level that is close to the death rate, and countries move to the third, or *posttransition,* stage. This stage describes the situation in modern industrialized countries, which have both low fertility and low mortality, again producing a stable population (though one that is much larger than in the earlier stages).

Demographers believe that the demographic transition is underway worldwide, with all countries experiencing at least some drop in fertility and a much more substantial drop in mortality. The highly industrialized nations, such as the European countries, Japan, the United States, and Canada, have experienced the full transition and now have both low fertility and mortality. Developing countries, which still have higher fertility than mortality, are at various points in the transition.[21]

The major problem that demographers and government officials now face is how quickly the demographic transition will occur. Some countries, such as China, Thailand, and Korea, have experienced a very rapid decline in their fertility rates; others, such as most countries in sub-Saharan Africa, including Malawi, have not experienced as quick a drop, even though the rates are declining.[22] Given the problems posed by the growing population in these countries, both demographers and government officials are interested in understanding more about why fertility declines—thus prompting the demographic transition.

Explaining Declines in Fertility

Demographers have developed two hypotheses to explain why fertility differs from one part of the world to another. The first involves the relationship between a society's death rate and its birth rate and suggests that people adjust the number of children they have depending on the likelihood that their offspring will live to adulthood. In other words, once a society's death rate decreases, people come to believe that their children are more likely to survive and thus will have fewer children. For instance, much as Winnie expressed to the other seminar members, if a woman has lost a child or fears that such a loss is a genuine possibility, she might reasonably decide to have another one.

Empirical data support this hypothesis. Figure 17-7 summarizes data on life expectancies, or the average number of years an infant can expect to live, in the various areas of the world. (These data are also displayed in Map 16-1 if you want to see a more specific country-by-country comparison.) If you compare this figure with Figure 17-5, you can see that the areas with the lowest fertility rates are also those with the longest life expectancies—the industrialized areas of the world. In contrast, people who live in less developed areas have lower life expectancies, as well as fertility rates that can be more than three times as high as those in the developed countries.

Changes in mortality and fertility over time provide additional support for this hypothesis. Both rates generally are declining in less developed areas, even though the birth rate remains significantly higher than the death rate. Within all societies the data seem to suggest that as people live longer and children become more likely to reach adulthood, the number of births declines.

The second explanation for differences in fertility involves the processes of modernization in general and the development of mass education, which we discussed in Chapter 15. As you will remember, all societies strive to provide a basic education for

Given what you have read in this chapter, how might you expect the lives of these young Sudanese women to differ from the lives of their peers who were unable to attend school?

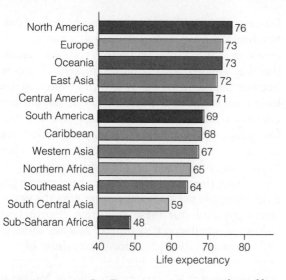

FIGURE 17-7 *Life Expectancies in Regions of the World, 1998*

Source: Population Reference Bureau, 1998.

their citizens as part of becoming a modern nation-state. Both declining death rates and declining birth rates appear to be associated with the development of mass education. More extensive analyses indicate that the education of women is much more important than the education of men in producing these effects. Countries where women have more education, such as those in Europe and North America, generally have much lower birth rates and death rates than do countries where women have less education. Once a large proportion of women within a society are exposed to education, the birth rate tends to drop dramatically.[23] Thus, although the first explanation suggests that parents reduce the number of children they have once they see that the infants are more likely to survive to adulthood, the second

explanation suggests that declining mortality and fertility can occur almost simultaneously in response to the process of modernization and mass schooling.

It is very difficult to sort out any one cause of declining fertility as most important, because modernization, mass education, industrialization, urbanization, and declining mortality and fertility all tend to appear at about the same time in most societies.[24] Nevertheless, this relationship between education, modernization, and declining fertility is an important one for citizens and government officials worldwide. Later in the chapter we will examine the implications and dynamics underlying this relationship more closely. Now, however, we will focus on how population changes affect the opportunities and constraints that face all societies and their citizens, especially in Malawi, Mexico, Germany, and the United States.

THE EFFECT OF POPULATION CHANGES

Population changes affect our societies, our communities, and our individual lives. These effects are related to several different areas, including the age of people in a population, the effects of population on the economy and the environment, and the ways in which people migrate in and out of a society.

Age Structures and Dependency

Table 17-2 lists comparative data on some key demographic characteristics for Malawi, Mexico, Germany, and the United States. As you can see, Malawi, where Winnie is from, has a very high birth rate, more than four times that of Germany, where Christa is from. It also has a high death rate, more than twice that of

TABLE 17-2 *Demographic Characteristics of Malawi, Mexico, Germany, and the United States*

	MALAWI	MEXICO	GERMANY	UNITED STATES
Birth rate	42	27	10	15
Death rate	24	5	10	9
Infant mortality rate	140	28	5	7
Life expectancy at birth	36 yrs	72 yrs	77 yrs	76 yrs
Rate of natural increase	1.7%	2.2%	–0.1%	0.6%
Doubling time	41 yrs	32 yrs	—	116 yrs
Population below age 15	48%	36%	16%	22%
Population above age 65	3%	4%	15%	13%
Dependency ratio	1.04	0.67	0.45	0.54

Source: Population Reference Bureau, 1998

Germany and the United States and almost five times that of Mexico. This high death rate results partly from the staggering numbers of infants who die each year—in Malawi 140 babies out of every 1,000 die before their first birthday. But death is more common at all ages in Malawi than in the other countries. The average Malawian can expect to live only to age thirty-six, but people in the other countries can expect, on average, to live into their seventies.

In recent years the AIDS epidemic has swept Malawi and other sub-Saharan African countries, contributing to the deaths of many people in the prime of life, as well as in childhood, and lowering life expectancy throughout the region. This tragedy has led to a lower rate of natural increase than in earlier years, but, because the birth rate remains high, the population will continue to grow rapidly.[25]

Although Mexico has a birth rate that is much smaller than Malawi's, it has a higher growth rate than Malawi and is projected to double its population in only thirty-two years if current trends continue. This incredible growth stems from the fact that mortality has declined much more rapidly in Mexico than has fertility. Mexico is right in the middle of the second, or transitional, stage of the demographic transition, with fertility rates much higher than mortality rates. This rapid decline in mortality also helps explain why the death rate in Mexico is so much lower than in Germany or the United States. Because, until recently, it had a much lower life expectancy than those two countries, Mexico now has substantially fewer people in the oldest age groups and thus fewer people who are most likely to die.

The U.S. population also is expected to grow during the next few decades, although at a slower pace than in Malawi or Mexico. In contrast, Germany actually has a declining population because more people die each year than are born. Christa and other

German women will tend to have much smaller families than Malawian women do, and they can be fairly confident, barring some unforeseen and rare tragedy, that their children will live into adulthood. However, because people in Germany and the United States have such long life expectancies, Christa and her counterparts in the United States can expect to care for aging relatives, often when they themselves are getting up in years, something Winnie and other Malawians probably will never experience.

These different experiences can be viewed from a demographic perspective by examining the population pyramids for these countries. Figure 17-8 shows the population structures of Malawi, Mexico, and Germany; the population structure of the United States was shown in Figure 17-3. The graphic representation of the age/sex structure of Malawi is what gave population pyramids their name, for it is shaped much like an actual pyramid (see Figure 17-8a). Population pyramids with this type of shape are sometimes called *expansive pyramids.* The wide base represents the society's extremely high birth rate, and the gradual narrowing of each side represents its high mortality. There are very few old people in Malawi and other countries in sub-Saharan Africa simply because most people don't live that long.[26] For this reason countries such as Malawi often are said to have a *young population.*

The age/sex structure for Germany is what demographers call a *constricted population pyramid,* so named because the bottom part of the pyramid looks as if someone had drawn a belt tight around it (see Figure 17-8b). This constriction in the lower regions represents the fact that Germany, like other European countries, has had low birth rates for a long time. As you can see, there are also many older people in Germany, reflecting the very low death rate and the long life expectancy. Because of this age

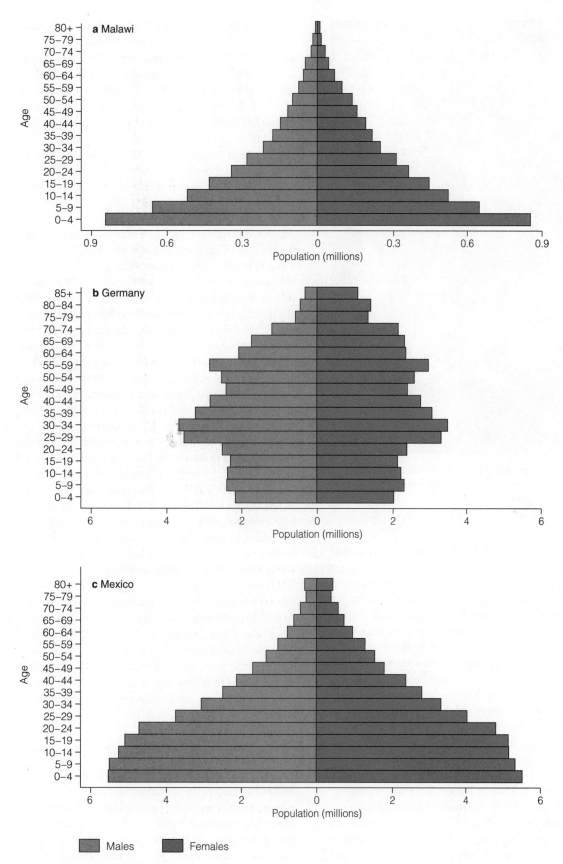

FIGURE 17-8 *Population Pyramids for Malawi, Mexico, and Germany*

Note: Data for Germany and Mexico are from 1995; data for Malawi are from 1991.

Source: United Nations, 1998, pp. 182–183, 190–191, 208–209.

This young boy from Mali is selling goods to help support his family. How might his experiences be related to the fact that Mali has a very high dependency ratio?

structure, countries such as Germany are often said to have an *old population*.[27]

The age/sex structures of Mexico and the United States fall between these two extremes. The upper portion of Mexico's population pyramid is shaped like Malawi's, indicating a rapidly expanding population (see Figure 17-8c). However, the base of the pyramid, which represents the most recent cohorts, shows the beginnings of a constrictive pyramid—smaller population increases in the younger age brackets. The constriction of the U.S. population appears across several cohorts, reflecting the "baby bust" that began in the late 1960s. The United States has had lower birth rates for several decades.

From looking at the population pyramids, you can also see that Malawi has a much smaller proportion of its people in the traditional working ages than do the other three countries. Demographers sometimes use a measure called a **dependency ratio** to understand how this aspect of the age/sex structure affects a society. The dependency ratio is the ratio of the dependent-age population, considered to be those younger than age 15 or older than age 64, to those of the traditional working ages (ages 15–64). It is thus a measure of how many people the average working person has to support. As you can see in Table 17-2, working-age people in Malawi have to support about twice as many people, on average, as do those in Germany or the United States. Note that Mexico, even with the potential of rapid population growth, has a dependency ratio that is much closer to that of Germany and the United States than to Malawi. This reflects the fact that Mexico has had lowered fertility for several years and experienced an increase in life expectancy only relatively recently.

The dependency ratio has enormous implications for both individual citizens and societies. Malawi has a much smaller proportion of its population available to support its dependent citizens. This can result in children leaving school at early ages to help support their families and can explain why Winnie is one of the very few women in Malawi who has completed secondary school, let alone gone on to higher education. It also means that money that could be funneled to economic development must instead go to help care for the dependent population. As we will see, both of these actions only perpetuate the poverty that, in turn, reinforces population growth.

It is also important to note the differences in the dependent populations in Germany and the United States compared with Malawi and Mexico. Most of the dependents in developing and least developed countries are children, whereas almost half of those in Germany and other countries in Western Europe are older people. Older residents are much more likely than the young are to have their own financial resources, such as savings and work pensions. As a result, the discrepancy between the actual impact of the dependency ratio of the countries is even greater than the numbers might suggest.

These contrasting population structures have other wide-ranging implications for individual lives. They define many economic constraints and opportunities, and they also play a key role in shaping the environmental issues facing countries.[28] We turn now to these issues.

Economic Opportunities and Constraints

Even though Winnie tried to explain conditions in Malawi to her fellow seminar participants, the people of her country face problems that Christa and other Europeans and North Americans can't imagine—from a lack of sanitation and health care, to inadequate schooling, to poverty and unemployment. For example, the United Nations estimates that slightly more than one-third of Malawians have access to safe water and less than 10 percent have access to adequate sanitation. Only slightly more than forty percent of adult women in Malawi know how to read, compared with almost three-fourths of the men. To some extent this disparity is disappearing among younger cohorts, but a fifth of all Malawian children still don't attend even primary school, and only a little more than a third of those who do attend school reach the fifth grade.[29]

In addition, people in Malawi are very poor. The United Nations estimates that 85 percent of the people in Malawi's rural areas (where the vast majority of Malawians live) don't earn enough money to have a minimally adequate diet. Thus malnutrition is common, with almost half of the children under age five suffering from either moderate or severe stunting of their growth because of the lack of nutrients. The average Malawian gets only about 88 percent of the calories needed to survive. As shown in Figure 17-1, Malawi's per capita GNP (the market value of all goods and services produced by a country, divided by the number of inhabitants) is less than 1 percent that of Germany or the United States.[30]

Of course, Malawi historically has been a poor country. Most important, however, Malawi's chances of solving its economic problems are not good as long as it continues to experience rapid population growth, which exacerbates the problems of poverty and makes economic development even more difficult to attain. For instance, as more and more children are born, it will be harder to build enough schools to educate them. As more and more people enter the labor market, it will be harder to provide jobs for them, and the land available for growing crops will have to be divided into ever smaller portions. Further, money and effort must be expended feeding the growing numbers of children rather than investing in industry and otherwise improving the economy. But, as noted previously, lower rather than higher birth rates seem to accompany industrialization and education. If the experiences of other countries are any indicator, Malawi will lower its birth rate only when more Malawians can attend school and when they have more economic opportunities.

In contrast, the economic fortunes of Germany, the United States, and even Mexico are enhanced by their lower birth rates. Because of their lower dependency ratios, larger proportions of their populations are engaged in economic activities, and a smaller proportion of each country's resources goes to care for the young. Even though Mexico still is expected to experience significant population growth in the future, the fact that it has already lowered its birth rate substantially and has a much lower dependency ratio than Malawi means that it is in a much better position to continue to expand its economy in the future.

As you can see, Malawi and a number of other countries, especially in sub-Saharan Africa, are caught in a vicious cycle. Their rapidly growing populations increase poverty by diluting employment opportunities and overstretching school systems, social services, and water and sanitation services. At the same time, the poverty of these countries increases the chances of rapid population growth. When infant mortality rates are high, parents compensate by having more children; and when families are poor, they might conceive more children to help bring in additional income or to work in the fields and homes. The ravages of the AIDS epidemic make these problems more vexing: As death rates have risen, fewer adults are available to provide for children and participate in the economy. The difficulty of this situation becomes even more apparent when we consider the relationship between population changes and the environment.[31]

Environmental Costs

Like many other countries, Malawi has experienced a great deal of deforestation (the removal of trees once indigenous to the region) in recent years. Thousands of acres of trees have been cut down to provide fuel for cooking and to clear ground for more planting of corn.[32] Of course, as the population grows, more and more land must be cleared to provide food and fuel. This pattern of deforestation has occurred not just in Africa but in South America and Southeast Asia as well. Further, in some areas more and more animals are grazing the land, leading to a problem of overgrazing. Both deforestation and overgrazing can cause the eventual destruction of the land. Without trees and their roots to help hold water in the soil, rain water can pour down mountains and valleys, carrying the fertile topsoil away; the wind can have a similar effect. Overgrazing, especially when complicated by drought, can convert once fertile farmland into desert.[33]

This environmental degradation increases the chance that a society will experience poverty. Soil erosion and desertification make it even harder to grow crops and raise animals, which, in turn, makes malnutrition, health problems, and poverty more likely. Sociologists and demographers who work for the United Nations have coined the term **PPE spiral** to refer to the intertwining relationship of poverty, population growth, and environmental degradation.[34] As shown in Figure 17-9, growing populations and poverty are both related to each other and to environmental degradation, which in turn reinforces poverty.

Highly developed countries, such as Germany and the United States, do not have the problems

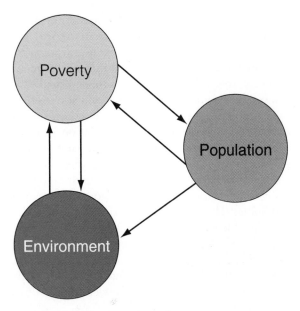

FIGURE 17-9 *The PPE Spiral*
Source: Grant, 1994, p. 25.

with population growth that lead to widespread deforestation and desertification. But, as Winnie tried to explain, these countries do contribute disproportionately to other aspects of environmental degradation, largely because their economic wealth, which has been enhanced by their low population growth, has provided the capacity to produce and consume large quantities of goods. For instance, automobiles have a large impact on the environment. The heavy industry that is required to produce them pollutes the environment; the oil needed to power them depletes natural resources and pollutes the air; the roads that must be built to carry them also use many nonrenewable resources; and the patterns of suburban and rural living that automobiles make possible increase the use of resources in transportation, as well as consuming millions of acres of once arable land. Of all the private automobiles in the world, 80 percent are in the more developed countries. There are seventy-five people for every passenger car in Africa, but only two to five people for every car in North America, Central America, Oceania, and Europe.[35]

Figure 17-10 shows the extent to which the United States and other developed countries both consume resources and produce waste in comparison with the less developed nations of the world. As you can see, even though the United States comprises only 5 percent of the world's population, it consumes 25 percent of all fossil fuels and produces 72 percent of all the hazardous waste that is emitted into the atmosphere. All the developed nations combined make up only about 22 percent of the world's population, but they consume more than three-fourths of all the raw metal and paper products that are produced.

In recent years many ecologists have expressed concerns regarding the changing climate of the world and the possibility of "global warming," as the average temperature of the world rises, the ozone layer diminishes, and ice caps begin to melt. A major contributor to this process appears to be emissions of carbon dioxide as a result of industrial processes and the burning of fuels such as oil and gas. Industrialized nations are far more likely than less developed nations are to emit such gases. Although most countries in Africa emit less than one metric ton of carbon dioxide per person and Mexico emits only 4, Germany emits 11 metric tons per capita and the United States leads the world by spewing more than 19 metric tons of carbon dioxide gas into the atmosphere for each person who lives there.[36] Thus, as Winnie suggested, from a global perspective the environmental degradation that stems from Malawi is minuscule compared with that of Germany, the United States, and the other highly developed nations.

The environmental degradation that results from the consumption by the inhabitants of highly industrialized nations makes the implications of the PPE spiral even more vexing. Even if the less developed countries are able to alter the course of the PPE spiral by slowing population growth and decreasing poverty, the environment will not benefit if their newfound economic wealth translates into the high consumption patterns of the already industrialized world. Yet, as the less developed areas of the world make economic progress, how can they be denied the consumable goods that industrialized nations now enjoy?

Many experts now are suggesting that the linkage between the environment, the economy, and population growth is such that all nations, more developed and less developed, will need to alter their behavior patterns if the quality of the environment is to be sustained. The global environment cannot sustain the energy consumption typical of the industrialized world indefinitely, nor can the least developed countries sustain extremely large populations and continue to grow at their current levels. Clearly, as Christa and others in the seminar suggested, the less developed countries will need to slow population growth. At the same time, as Winnie pointed out, the more developed countries will need to alter their patterns of consumption. Much as the demographic transition results in a stable population size within societies, so we can hope that nations develop economies based on sustainable resources—in which humans consume no more resources than can be produced.[37] We will return to these issues again in the remaining chapters of this book.

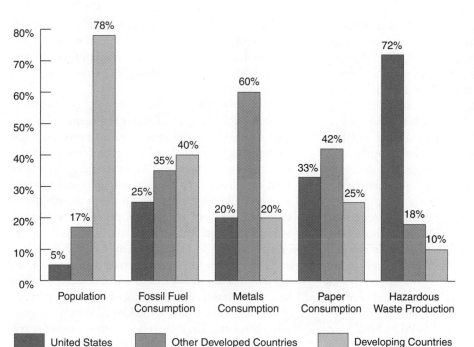

FIGURE 17-10 *Share of Population, Resource Consumption, and Waste Production*
Source: Crews and Stauffer, 1997.

The Wild Card of Migration

Until now, our discussion of the effects of population changes has focused on fertility and mortality, changes that produce what demographers call **natural increase** in population.[38] Population changes also occur within and across societies because of migration. Like fertility and mortality, migration is affected by economic and environmental changes. It is also highly affected by political events.

When migration occurs across national borders, it is called **international migration.** These movements affect both the countries that people leave (the *sender countries*) and those to which they move (the *receiver countries*).[39] Demographers have looked at international migration within the world as a whole and specifically as it affects the United States. (We will look at **internal migration,** migration within national borders, in Chapter 18.)

Migration Around the World People have migrated from the dawn of human life, initially from Africa to Asia and Europe and subsequently from the Asian landmass to North America and from Southeast Asia into Oceania, New Zealand, Australia, and the neighboring islands. These large-scale migrations usually have involved people moving from densely to sparsely populated areas. For instance, massive numbers of Europeans moved to other parts of the world, including North America, during the nineteenth century when European populations began to burgeon. Demographers estimate that by the mid 1990s 125 million people lived in a country other than the one in which they were born. Even though this is a large number, it actually constitutes less than 2 percent of the total world population.[40]

One of the major reasons that people choose to move from one country to another is political. Until the early 1990s, Malawi was inundated with refugees from Mozambique, which experienced violent warfare. As you might imagine, this influx of refugees stretched Malawi's resources even further.[41] Many other countries have also experienced the sudden impact of refugees fleeing violence, conflict, or repression in their homelands. In the late 1990s, officials of the United Nations estimated that there were about 4 million political refugees in Africa, 5 million in Asia, and 3 million in Europe. Fewer than 1 million were in the United States.[42] Most of the world's political refugees are in the least developed and developing nations because these are the areas of the world where most people live and refugees tend to go to the nearest safe country that they can find. More than half of all refugees in most developing countries are women and young children.[43]

Another important reason that people migrate is economic—to find better jobs and attain a better standard of living.[44] About half of the world's migrants live in developing nations that have many natural resources and need workers. For instance, foreign workers make up the majority of the labor force in several Middle Eastern countries that produce a great deal of oil. Migrants also constitute about one-seventh of the labor force in South Africa, which needs many workers in its diamond mines. Another one-third of the world's migrants are in just seven wealthy countries in the industrial world (Germany, France, the United Kingdom, United States, Italy, Japan, and Canada), even though these countries have less than one-eighth of the total world's population. Salaries in these countries are generally much higher than immigrants could earn in their homelands. In addition, some countries, such as Germany and Italy, have had low birth rates for so many years that they now have a shortage of labor and actively recruit immigrants to fill the gap.[45]

When immigrants work for wages and in conditions that a country's natives disdain, or when the native workers perceive the immigrants as competitors, a "split labor market" can appear. This economic competition often can result in intergroup hostilities, not unlike those exhibited toward some immigrant populations in Europe and the United States today. In recent years, several violent confrontations, as well as discriminatory actions, have occurred between the so-called guest workers in Germany (primarily Turks, Africans, Pakistanis, and Persians) and German nationals, most often when economic conditions have taken a downturn.[46]

Migration can produce difficulties in *receiver countries,* those to which immigrants move, but it also produces hardship in the *sender countries,* those that people leave. Migrants typically are workers in their economically productive years. In addition, because it takes money and knowledge to move elsewhere, migrants tend to be better educated and to have higher social class status than people who stay in a country. Thus, the dependency ratio of sender countries becomes higher, and the people who remain often have fewer skills and resources to cope with the situation.

Migration is a "wild card" in the sense that demographers have found migration much harder to predict than fertility and mortality. It is difficult, if not impossible, to predict wars and political turmoil such as the conflicts that sent Iraqi or Afghani refugees to Iran, Rwandan refugees to Zaire and Tanzania, Mozambican refugees to Malawi, and ethnic Albanians from Kosovo to Albania, Macedonia, and other European countries. Similarly it would have been difficult to foretell the economic ups and downs that led to Germany gaining more than 300,000 new residents per year through immigration in the early 1970s but only 3,000 per year a decade later.[47]

In addition, government policies of both sender and receiver countries can affect migration patterns. For example, for many years the former Soviet Union, as a sender country, severely restricted the ability of Jews to move to Israel. In the late 1990s, the Serbs as a sender country, expelled thousands of ethnic Albanians from Kosovo and other areas of the former Yugoslavia. In the past few years Germany, a receiver country, has severely tightened its regulations regarding the admission of immigrants; and although the United States continues to admit substantial numbers of immigrants each year, immigration to this country has often been politically controversial.[48]

Migration to the United States   As we discussed in Chapter 3, the United States is a nation of immigrants. The number of immigrants was very low from 1928 through 1965, when laws severely restricted the countries from which immigrants could come, but in 1965 these laws were changed and immigration again increased. Throughout the mid 1990s slightly fewer than one million immigrants entered the United States each year. About 70 percent of today's immigrants come from Asia, Mexico, Central America, and the Caribbean, almost all from nations that have higher rates of growth than the United States. In the late 1990s immigrants constituted almost 10 percent of the total population of the United States. The majority lived in states relatively near their port of entry, such as California, New York, Florida, and Texas.[49]

People choose to migrate to the United States for the same reasons that they move elsewhere in the world. Some, about 14 percent of all immigrants entering the country in the late 1990s, are political refugees, fleeing repressive regimes in areas such as the Ukraine, Viet Nam, and Cuba. Many others come for economic reasons. About 13 percent of all immigrants are professional-level and skilled workers who are given preference for immigration based on their ability to provide needed work in the United States.[50] For instance, Lupe's parents immigrated to the United States under these conditions: her mother, a physician, and her father, a computer engineer, believed that working conditions in the states were more favorable than those in Mexico. Yet, most immigrants to the United States come to join family members already in this country. Just as in so many other areas of social life, social networks are a strong influence on patterns of migration.

The largest number of immigrants to the United States comes from Mexico, our closest neighbor to the south. Over the last few decades U.S. immigration policies have, at times, encouraged migration and, at other times, tried to stem and restrict this process through means such as stringent border patrols or policies that would halt government benefits to immigrants, both legal and illegal.[51] Interestingly enough, however, these official policies appear to have relatively little effect on whether or not people from Mexico choose to move to the United States or, when they are here, to move back to Mexico. Instead, extensive interviews with immigrants suggest that these decisions are influenced by economic factors, especially the ability to obtain jobs in this country and, even more important, by social ties with people

BUILDING CONCEPTS

*Gathering and
Analyzing
Demographic Data*

*Below is a recap of the demographic processes studied by sociologists, the levels
of analysis used, the sources of data for these studies, and how data are analyzed.*

Demographic processes studied	• Fertility (the birth of children) • Mortality (death) • Migration (movement from one geographic area to another)
Levels of analysis	• At a macro level, sociologists ask how changes in demographic processes affect entire nations and societies. • At a micro level, sociologists ask how changes in demographic processes affect individuals and how individuals actively create demographic trends
Sources of demographic data	• Census data • Official government records of vital statistics, for example, births, marriages, divorces, deaths • Sample surveys
How data are analyzed	• By graphing population trends • By constructing population pyramids • By calculating birth and fertility rates, death and mortality rates, and migration rates

already in the United States. About half of all adults who live in Mexico are related to someone who currently lives in the United States. Given how important these social relationships are in the decision to migrate, some demographers suggest that policies that attempt to stem immigration through restrictive legislation have not been successful in the past and will probably not be very effective in the future.[52]

It is important to reemphasize that immigrants constitute only a small part of the total population. In addition, extensive analyses of census and other data indicate that individual immigrants rarely represent a drain on government resources in the United States and that, overall, their input enhances the economy. The few citizens who are most likely to face economic competition from immigrants are those with the lowest levels of skill, but even these effects are quite small. Of course, many immigrants to the United States, especially from Mexico, are undocumented and enter the country illegally. But studies of these undocumented workers, suggest that they generally do work that U.S. residents prefer not to do; thus they add to rather than detract from the economy's productivity.[53]

Individuals like Lupe's parents decide to migrate after balancing out the pros and cons of such an action. In general, analyses of why people choose to migrate suggest that they weigh the benefits of moving to a new location against the costs of such a move and, as rational choice theory would suggest, decide to migrate when the benefits outweigh the costs.[54] In fact, such individual actions underlie all demographic processes. We can see the influence of social actions on population change from a microlevel of analysis.

FEATURED RESEARCH STUDY

SOCIAL ACTION AND
POPULATION CHANGE

Demographic processes don't just happen on their own. Like all social patterns, fertility, mortality, and migration rates are created through social actions— decisions made by countless individuals and by societies and governments. Societies decide to invest in public sanitation and public health measures that can counteract high infant mortality. Some governments, as in China, actively discourage citizens from having large families. Others, as in some Eastern European countries a few years ago, encourage fertility by outlawing birth control. At the individual level, people decide to move from one country to another, and couples decide that they will have another child or seek out contraceptives to prevent conception. How do people make these decisions?

Many sociologists, concerned with the problems of population growth, poverty, and environmental degradation, have tried to understand how the PPE spiral might be broken. As you will recall, a key factor in fertility is greater education, especially for women. Studies in many countries have shown that when mothers have higher levels of schooling (usually beyond about the sixth grade), they tend to have fewer children, and the children they do have are healthier and less likely to die.[55] What these studies haven't explained is *why* these associations appear. Are more highly educated mothers more knowledgeable about birth control and sanitation? Or does having more education change the way that mothers interact with and raise their children? Robert and Sarah LeVine and their associates decided to explore these questions.[56]

Studying Women's Education and Fertility Rates in Mexico

A complete answer to these questions would involve looking at several different cohorts of women in many different countries, each at different stages of the demographic transition, to try to see how changes in education affect practices within a variety of different cultures. Because such a massive study would be impractical and incredibly expensive, the LeVines decided to focus on just one society—Mexico. Mexico was a good choice for this study because, as we have seen, it is in the middle of the demographic transition. After 1940 its economy began to industrialize, many people moved from the country to the city, and health and educational services expanded greatly. As a result, mortality declined, but fertility remained high and the population grew rapidly. After 1970 the birth rate began to decline markedly, and this decline

has continued to the present. As we saw in Table 17-2, however, the birth rate still has not reached the level of countries such as the United States and Germany. With the current rate of natural increase, Mexico faces the possibility of doubling its population in thirty-four years, almost as quickly as Malawi.

Within Mexico, as in other countries, women who have had more education have fewer and healthier children. This association appears in both urban and rural settings and apparently is not affected by the husband's education or occupation. Even in families in which the husband is highly educated or has a high-status occupation, both fertility and child mortality remain strongly associated with the mother's education. The LeVines and their team of researchers wanted to find out why that occurs.

They began their work in the mid 1980s in Cuernavaca, a city of 200,000 people about fifty miles south of Mexico City. To rule out the possibility that differences in social class might affect their results, they limited the study to low-income neighborhoods and to women who had completed between one and nine years of school. (Mexican law requires all children to attend primary school, usually for six years. Attendance at secundaria, or junior high schools, has not been mandatory and was the highest level usually attained by women in low-income areas at the time of the LeVines' study.) About four years later, they replicated their study in Tilzapotla, a town of 4,500 people about thirty miles south of Cuernavaca. Children in Tilzapotla had been able to attend school for several generations, so women in the town had the same range of schooling as those in Cuernavaca. In addition, the town included some women who had migrated from small villages and had received no schooling at all. The two communities had similar cultures, with the husband traditionally expected to dominate the wife in the home and a high value placed on large families. By the time the study was conducted, however, government policies that encouraged more egalitarian husband–wife relationships and smaller families had begun to change these attitudes.[57]

The LeVines and their team of researchers conducted extensive interviews with a sample of several hundred women in the two towns. Each subject had a child who was less than four years old. The interviews lasted for as long as two hours and focused on the mothers' knowledge, attitudes, and behavior surrounding reproduction and child care and development; additional questions addressed the women's views of their marriage and their aspirations for themselves and their

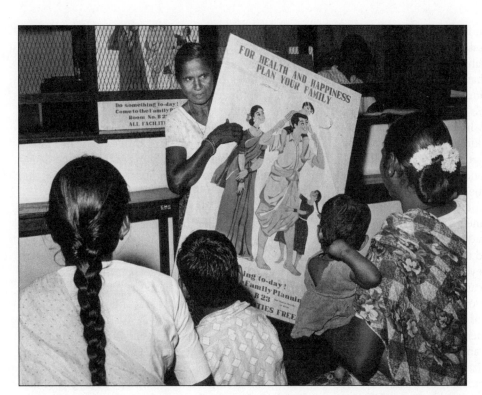

These women in India are attending a family planning class. Why might women who have gone to school be more likely to attend these classes than women who have little education might? How might their lives and those of their children be affected by their attendance?

children. The researchers also observed a subsample of mothers in Cuernavaca interacting with their children when they were about five to ten months old and again five months later, when the children were between ten and fifteen months of age.

As expected, given the results of other studies, the LeVines and their colleagues found that mothers who had more education tended to have both fewer children and healthier children than other mothers, even when controlling for variables such as the education of their husbands and their social class. Mothers who had gone beyond primary schooling were more likely to receive prenatal care while pregnant (thus helping to ensure a healthy infant at birth), to give birth in a hospital, to take a child to a clinic in an emergency, and to have children who were less often ill with ailments such as diarrhea (a significant health problem for children in many developing countries).[58] But why, the LeVines asked, do these relationships appear? Virtually all the women had access to prenatal care, hospitals, medical clinics, and contraception. They had similar incomes and access to transportation. Why were more highly educated women more likely to use these services? What is it about going to school that influences them to choose different health and reproductive actions in adulthood?

Explaining the Effect of Education

One part of the answer seems to involve the attitudes women develop as they attend school. The LeVines

and their research team found that mothers who had more schooling had attitudes and beliefs that were similar to those of people in modern industrialized countries. For instance, women who had more schooling were more likely to want more education for themselves, to believe that both husbands and wives should be involved in family decisions, to want their children to aspire to professional-level jobs, and to pay more attention to media such as television and radio. The better-educated mothers also tended to give lower estimates of the age at which infants can recognize their mothers' voices and be startled by loud noises.

These different views about the capabilities and responsiveness of infants seem to be related to differences in the ways that the mothers cared for their babies. The observations of interactions between mothers and infants showed that mothers who had more education were much more likely to speak to their babies after they made sounds or moved or when the baby looked at them. These mothers also were much more likely to respond to the baby's gaze by gazing back. In addition, by the time the infants were fifteen months old, the more highly educated mothers were less likely to spend a great deal of time holding their babies. In general, they tended to adopt a care-giving and socialization style that involved more verbal interaction.

The LeVines and their associates note that this care-giving style requires more energy than the more traditional approaches and cannot be done while engaged in other tasks. When mothers routinely soothe their babies through comforting and holding them, they can carry on other tasks at the same time. In contrast, if they use conversation as a way to soothe and amuse them, it is much harder to do anything else at the same time. The LeVines and their associates give three explanations for this association. First, most of women's tasks, such as housework, cooking, and shopping, require some kind of visual attention, and it is very hard to concentrate on these tasks while talking to a young child. Second, soothing a baby through holding tends to quiet the child, making it easier to perform other tasks; in contrast, talking with a child tends to generate excitement. Third, babies who are engaged in conversations during their first year of life become more active and talkative during their second year, thus needing even more attention as they grow older.[59] In short, a care-giving style that involves a great deal of conversation complicates the process of caring for infants, and child care is much more time consuming for mothers who adopt this style.

When the LeVines and their colleagues interviewed the women in Tilzapotla, they also examined the mothers' verbal interactions, and specifically their ability to use "decontextualized language." This is a term used to describe the style of conversation that typically is used in classrooms, in books, and in bureaucratic settings in which the participants do not share everyday experiences. The researchers found that the more highly educated mothers were much more skilled in the use of decontextualized language. Because this is the style of language used in health clinics and hospitals, women who have more schooling would be more confident in these settings and more likely to use health care services and to follow the instructions of health care personnel, thus promoting their own and their children's health.

From their analysis, the LeVines and their associates concluded that formal education changes the way in which people see the world around them and interact with other people. In attending school, through reciprocal role interactions, people learn not only the role of student but also the role of teacher— they learn how to talk to children, often in a formal language that differs markedly from what they have heard at home. As they develop literacy and competence in this more formal language, they learn how to communicate in bureaucratic settings and to use varied sources of information. And, very important, the ability to play the teacher's role allows them to engage in relationships with their children that resemble the teacher-student model and involve a great deal of verbal interaction—so much verbal interaction that child rearing becomes much more time consuming. As a result, the costs of having more children become greater, and so more educated women are more likely to limit their fertility.

Besides this impact on how many children educated women choose to have, the LeVines and their associates suggest that these changes in mother-child interactions have long-term implications for the children. Specifically, the children are better prepared to succeed in school, learn a more formal style of language from their very early years, and receive better health care than they otherwise would. Each of these changes, in turn, helps to further the demographic transition by increasing the chance that these children will also have lower fertility.[60] Although the LeVines' results await replication in other settings, they provide hypotheses that can help us understand the social actions that underlie the demographic transition. These results show us how the social interactions and socialization patterns that occur in families and schools can have far-reaching effects on population patterns and the well-being of entire societies.

Of course, education is only one part of the solution to the problems that growing populations cause for societies. But in many ways it may be one of the most important. Unlike populations of animals and plants, humans have the capacity to observe and understand their population and its changes. Many sociologists who study population hope that through developing a better understanding of population processes and their effects, human societies can alter demographic processes in ways that can make a better world. Less developed countries, such as Winnie's homeland of Malawi, can apply this knowledge to help break the cycle of poverty, population increase, and environmental degradation. More developed countries, such as the homelands of Christa and Lupe, can help develop ways to diminish the impact that their very wealthy populations have on the environment. No matter where we live, understanding demographic processes can help us deal with the wide array of political issues that involve changes in the population structure of our society (see Box 17-1, Applying Sociology to Social Issues).

CRITICAL-THINKING QUESTIONS

1. Could the LeVines' study be replicated in the United States or other advanced industrialized societies? Why or why not?

2. What other types of methods might the LeVines have used to examine how women's education affects fertility?

3. What implications does the finding that mothers' education is more important than fathers' education in explaining declining fertility have for public policies regarding education in developing countries?

BOX
17-1

APPLYING SOCIOLOGY
TO SOCIAL ISSUES

*Baby Boomers
and Retirement*

In recent years in the United States, issues such as how to balance the budget and whether to reduce funding for Medicare and Social Security have been the focus of political and public debate. Underlying these concerns are certain basic facts about the population structure of the United States and how it will change in coming years.

If you are employed in the United States, you know that a portion of your paycheck is designated for Social Security taxes. Perhaps you assumed that this money was placed in an annuity or savings account where it would generate interest to be used when you retire. In fact, the money that employees currently pay in Social Security taxes is used to fund the Medicare and Social Security benefits that current retirees receive. In other words, retirement benefits for current retirees are financed by current workers in the society.

This system works fine as long as the dependency ratio is relatively low—as long as the number of elderly citizens is not large relative to the number of people of working age. Currently, as we saw in Table 17-2, about 13 percent of the population of the United States is sixty-five or older. In 1940, shortly after the Social Security system was instituted, only 7 percent of the population was in the retirement age range. Further, in 1940 the average person in the United States could expect to live only to age sixty-three, dying before being eligible to retire. Today the average person can expect to live to age seventy-six, or more than a full decade after retirement. This incredible increase in life expectancy has placed a very large

burden on the Social Security and Medicare systems, so that expenditures for these programs alone constitute a very large portion of the federal budget.

This increase in the proportion of older workers during the past fifty years was large, but the numbers will swell even more as the baby boomers age. As you can tell from looking at the population pyramid for the United States in Figure 17-3, the baby boomers, people born between 1946 and 1964, represent the largest single age group in the population, making up almost 30 percent of the total population. Population projections calculated by the Census Bureau suggest that by the year 2030, when all these baby boomers will have retired, people over age sixty-five will compose 20 percent of the population, close to equaling the number of children under the age of eighteen. The dependency ratio for the United States will have risen to a level similar to that of today's developing countries, such as Mexico.[61]

The sociologist Judith Treas, who is profiled in Box 17-2, has looked at the changing age structure in the United States and the implications of growing longevity for individuals and for the society as a whole. From this work she has concluded,

The older years can be both the best and worst of times in an individual's life. With some exceptions, most

people now enter their older years with the health and resources to pursue full and independent lives. But the aging process does take a toll. By age 80, many older individuals are troubled by poor health, difficulty accomplishing simple tasks, and dependency on others.

The transition from active, independent living to a period when greater assistance is required can be a painful time for older individuals and their families. Addressing this need will be one of the greatest challenges that individuals and society will face in the 21st century.[62]

CRITICAL-THINKING QUESTIONS

1. What are the implications of these changes in the population structure? How will they affect you and people you know?

2. What age will you, your parents, and your children be in 2030? Which of you will be in the working-age population? Who will be retired? And how will these retirees find financial support for their retirement years?

3. What are the implications for society as a whole? Can the Social Security and Medicare systems be retained? Will they need to be altered? If so, how?

4. What are the implications for other segments of society of such a large group of older Americans? What might happen to services for children? What will be the burdens on the working-age population? How will these changes affect you?

SOCIOLOGISTS AT WORK

Judith Treas

When I signed up for my first sociology class, I was amazed. Here was a subject that explained everything.

Judith Treas teaches at the University of California at Irvine. Her research on aging is described in the "Applying Sociology to Social Issues" box in this chapter.

Where did you grow up and go to school? I grew up in a small town in Arizona that isn't there anymore. I don't mean a ghost town—Phoenix was just a small city when I was a kid, but it's grown into one of the ten biggest in the United States. Talk about social change! When the time came for college, I headed for Pitzer College in Claremont, California, and then I went to UCLA to study for a Ph.D.

Why did you become a sociologist? I had never even heard of sociology in high school. When I signed up for my first sociology class at Pitzer, I was amazed. Here was a subject that explained everything: why some small groups work so well together and other folks can't even get along, why we're shocked by some behavior and take other things for granted, why riots and revolutions happen. My sociology classes were a real revelation.

How did you get interested in studying aging? I didn't set out to study aging. At one time few sociologists had a background in the field. I took a job teaching gerontology without ever having had a class or read a book about the topic. If I had realized how little I knew, I might never have jumped into the field. Luckily sociology had given me the theoretical perspectives and basic research skills I needed to make sense of a new area like aging.

What has surprised you most about the results you have obtained in your research? When I started studying the elderly population, I just assumed most older people were like my own grandparents. Wrong! Older Americans are the most diverse population in the United States. I've worked beside affluent, active sixty-somethings who qualify as senior citizens, but I know other older Americans who are too old, too sick, and too poor to even leave the house alone. Also, since so many people have immigrated to the United States over the past couple of decades, the older population is becoming more racially diverse. If current trends continue, one-third of Americans age sixty-five and older

SUMMARY

Demography is the scientific study of population and the implications of changes in population for societies, individuals, and the world at large. Demographic transition theory attempts to explain the changes that societies experience as they move from the high fertility and high mortality characteristic of less developed countries to the low fertility and low mortality characteristic of highly industrialized countries. Key chapter topics include the following:

- Populations and population changes significantly affect the quality of life in all societies.

- Demographers use data gathered through censuses, records of vital statistics, and sample surveys. They study population trends, examine the age and sex structure of societies, and analyze

various summary statistics of population composition and change.

- The world's human population was relatively stable until about 10,000 years ago, when it began to gradually increase. Approximately 250 years ago, population growth became explosive. Most of the current growth is occurring in developing countries, which also are the world's poorest countries. This immense population growth strains national budgets that are already stretched very thin.

- Demographic transition theory describes how preindustrial societies had stable populations as a result of high fertility levels balanced by high mortality levels. As societies start to industrialize, mortality drops but birth rates remain high. Eventually fertil-

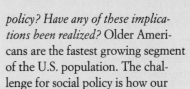

BOX
17-2

**SOCIOLOGISTS
AT WORK**

Continued

will be Latinos, African Americans, Asians, or Native Americans by the middle of the twenty-first century.

What has been the reaction to your research among your colleagues? I think the sociology of aging was once dismissed as a sleepy, backwater field of research carried out by do-gooders. That's not the case anymore. Sure, there continues to be an applied interest in improving the lives of older Americans, but the research is characterized by sophisticated methods and has made significant contributions to other fields of study, such as the sociology of the family.

Have you or others replicated your research? What were the results? Some of my earliest research showed that older Americans are unlikely to marry, despite the benefits of companionship in old age. A more recent study found that little has changed. Apparently a shortage of eligible bachelors in the older population, coupled with some social disapproval of late-life sex and romance, continues to discourage older people from tying the knot. A recent book by Rebecca Gronveld Hatch shows that a few older couples, like many of their juniors, are opting to live together outside of marriage.

What do you think the implications of your work on the growing number of older people are for social policy? Have any of these implications been realized? Older Americans are the fastest growing segment of the U.S. population. The challenge for social policy is how our limited resources can meet the real needs of a diverse population that keeps getting bigger and bigger. Without government help, some older Americans will not be able to live out their lives in dignity. Others can manage pretty well on their own or with the assistance of their families. At a congressional briefing last year, I made the point that we need to direct public funds to the neediest seniors— for example, those who are poor, disabled, alone.

How can your research help students understand events in their own lives? Our own families offer our first lessons in growing older. We get a chance to see our grandparents making big plans for retirement or welcoming their first great-grandchild. We take it for granted that the lives of grandparents and grandchildren will overlap. In earlier eras when mortality rates were much higher, fewer grandparents lived to see all their grandchildren grown. Demographic trends affect us intimately. Our holiday gatherings are apt to find more elderly relations crowding around the dining room table. Also, the fertility trend to smaller families means there are fewer youngsters to fight over the turkey wishbone!

---

ity also begins to drop, and by the end of the transition, as has occurred in modern industrialized countries, both fertility and mortality are at low levels.

- Declines in fertility rates can be explained by declines in the rate of mortality and by expanding educational opportunities, industrialization, and urbanization.

- High birth rates affect the age structure of a society and contribute to a high dependency ratio, placing great strains on a society's resources and contributing to the perpetuation of poverty and environmental degradation. In turn, environmental degradation increases the chance of poverty, and both poverty and environmental degradation increase the chances of population growth, producing a cycle of mutual influences that is hard to alter. Despite their smaller levels of population

growth, highly industrialized countries contribute much more to global environmental degradation than do less industrialized countries.

- Migration influences the size of populations apart from the influences of fertility and mortality. People migrate for both political and economic reasons, and migration affects sender countries by increasing the dependency ratio.

- Robert and Sarah LeVine and their associates explored why increased education for women seems to be related to lowered fertility. They concluded that education provides mothers with tools to interact in modern bureaucratic settings and alters the way that they interact with their children in a manner that makes child rearing more labor intensive and time consuming, thus increasing the cost of additional childbearing.

KEY TERMS

age/sex structure 452

birth rates 454

census 451

death rates 452

demographic momentum 452

demographic (population)
    processes 450

demographic transition theory
    459

demography 449

dependency ratio 464

fertility rates 452

growth rate 457

internal migration 467

international migration 467

life expectancy 456

migration rates 452

mortality rates 452

natural increase 467

population projections 452

population pyramid 452

population structure 452

PPE spiral 465

replacement rate 459

vital statistics 451

zero population growth 457

INTERNET ASSIGNMENTS

1. Find the Population Reference Bureau home page and read some of the reports provided. How do the issues discussed in this site relate to topics covered in the chapter?

2. Find the United Nations Web site and read the latest reports on world population and trends in mortality, fertility, and migration. (Look for the United Nations Population Information Network.) How does the information reported there compare with the information given in the text?

3. Find the home page for of the U.S. Bureau of the Census and look at the U.S. and World population

clocks. How quickly are the two clocks changing? Then search for the most recent population projections for the United States. Are population projections similar for different racial-ethnic groups? What are the social implications of these differences?

TIPS FOR SEARCHING
You can use links from the Population Reference Bureau Web site to get to both the United Nations and the U.S. Census Bureau and many other Web sites with demographic information.

InfoTrac COLLEGE EDITION:
READINGS

Article A16958087 / Kontuly, Thomas, Ken R. Smith, and Tim B. Heaton. 1995. Culture as a determinant of reasons for migration. *Social Science Journal, 32,* 179(15). Kontuly, Smith, and Heaton examine the role of culture and family ties in prompting decisions to move within the United States

Article A20526960 / Ehrlich, Paul R., and Anne H. Ehrlich. 1997. The population explosion: why we should care and what we should do about it. *Environmental Law, 27* 4, 1187–1208. Ehrlich and

Ehrlich eloquently describe issues regarding population growth and how they affect everyone on earth.

Article A20574466 / Eberstadt, Nicholas. 1998. World population implosion? *Human Life Review, 24,* 15(16.) Eberstadt discusses the possibility that the world could reach a state of zero population growth by the mid twenty-first century and the impact that this situation would have on social life.

FURTHER READING

Suggestions for additional reading can be found in *Classic Readings in Sociology,* bundled with this book. If you purchased a used copy of this book that does not include this custom-published reader, go to www.sociology.wadsworth.com for ordering information.

RECOMMENDED SOURCES

Bellamy, Carol. *The State of the World's Children.* New York: Oxford University Press, published for UNICEF. This small book is updated each year by the staff of UNICEF, part of the United Nations; it contains a wide range of statistical information and gives up-to-date reports on the current status of population issues around the world, focusing especially on the lives of children.

McFalls, Joseph A., Jr. 1998. "Population: A Lively Introduction." *Population Bulletin, 53*(3). Washington, D.C.: Population Reference Bureau. As implied by the title, this is a very readable, and short, introduction to the study of populations.

Pedraza, Silvia, and Rubén G. Rumbaut. 1996. *Origins and Destinies: Immigration, Race, and Ethnicity in America.* Belmont, Calif.: Wadsworth. This wide-ranging collection of essays discusses immigration in the United States.

The Population Reference Bureau, 1875 Connecticut Avenue, N.W., Suite 520, Washington, D.C. 20009–5728. This nonprofit organization publishes a wide range of materials related to population issues, including small pamphlets, entitled *Population Bulletin,* and a monthly newsletter, called *Population Today,* that contains the very latest demographic information.

Portés, Alejandro, and Rubén G. Rumbaut. 1996. *Immigrant America: A Portrait,* 2nd ed. Berkeley: University of California Press. The authors provide an extensive discussion of the nature of immigration to and settlement in the United States and the characteristics of immigrants.

Poston, Dudley L., Jr., and David Yaukey. 1992. *The Population of Modern China.* New York: Plenum. This edited volume provides a wide range of readable articles by demographers about the most populous country in the world.

Stack, Carol. 1996. *Call to Home: African Americans Reclaim the Rural South.* New York: Basic Books. Stack provides a very readable examination of the motivations and experiences of African Americans who have migrated from the northern to southern regions of the United States in recent years.

Treas, Judith. 1995. "Older Americans in the 1990s and Beyond." *Population Bulletin* 50(2). Washington, D.C.: Population Reference Bureau. This short, easy-to-read pamphlet, written by the sociologist who is profiled in Box 17-2, succinctly describes recent changes in the elderly population in the United States, the health and economic well-being of the elderly, and future implications of changes in this population.

Weeks, John R. 1998. *Population: An Introduction to Concepts and Issues,* 7th ed. Belmont, Calif.: Wadsworth. This well-written textbook gives a complete introduction to the concepts and issues addressed by sociologists who specialize in demography.

INTERNET SOURCE

The Population Reference Bureau, 1875 Connecticut Avenue, N.W., Suite 520, Washington, D.C. 20009–5728, is a nonprofit organization that publishes a wide range of materials related to population issues, including a small pamphlet titled *Population Bulletin* and a monthly newsletter called *Population Today.* Its Web site includes on-line versions of *Population Today,* current press releases, and links to many other useful sources of demographic information.

Pulling It Together

Listed below are other chapters where you can find additional discussion or examples
of some topics covered in this chapter. You can also look up topics in the index.

Communities and Urbanization

Urbanization and Changes in Communities
Urbanization in the United States
Urbanization in Developing Countries

Sustaining Communities in Modern Cities
Theories of Urbanization and Community
Research on Urbanization and Community

Urban Structures and Opportunities and Constraints
Social Ecology and the Structure of Cities
Neighborhood Segregation

FEATURED RESEARCH STUDY

Urban Segregation

BRIAN WASN'T SURE WHAT he wanted to do when he finished college. The previous two summers he had worked as an assistant in the advertising firm where his father worked and once thought that he would also enter advertising when he finished school. Lately, however, Brian had been thinking a lot about social work or some other type of job in which he might be able to help people on a day-to-day basis. He knew that he wouldn't make a lot of money, but he sometimes felt a little guilty about all the advantages he'd enjoyed, especially compared with other people who weren't so lucky. When he saw an ad in the campus newspaper recruiting counselors for an inner-city summer camp, Brian decided to apply.

The camp director, Jim, came to campus to recruit and interview applicants for counselor positions. In a slide presentation that described the camp's activities, Jim explained that the camp was housed in a community center in a poor neighborhood. Each counselor would be responsible for about eight youngsters—taking them on field trips, supervising recreational activities, and helping with crafts and games. "I don't want to hide anything from you," Jim told the prospective counselors. "Some of these kids have had pretty rough lives. But they deserve to have a good time in the summer. We'll provide a week's training before camp begins to help you learn how to work with them. During this time you can also help out in our after-school program and start to get to know the kids."

A few weeks after the interview session, Jim phoned Brian to offer him a job as a counselor. He told Brian he would be sending him a camp T-shirt and a map that would show him how to get to the community center from the train stop. "You shouldn't drive your car here," said Jim. "And be sure to wear your T-shirt. That way people on the street will know that you're working with us."

As Brian walked down the stairs from the train platform, he was grateful for Jim's advice. He tried not to look shocked, but the sight of decaying buildings along the streets and residents standing around the street corners, seemingly with nothing to do, unnerved him.

Brian was a little embarrassed about being nervous during his walk to the community center, but Jim and the five other new counselors gathered around a conference table in Jim's office soon put him at ease. Like Brian, the other counselors had grown up on the outskirts of the city and were now college students on summer break, considering a career in social service professions. Like most of the children who would be in the camp, they were African Americans, but, like Brian, none of them had ever lived in neighborhoods like those that surrounded the community center. Jim had anticipated their discomfort. "The first thing we're going to do," he announced, "is get to know the neighborhood and the surrounding area. You need to understand where the kids live if you are going to work with them."

"I've worked here now for twenty years," said Jim, as he drove the community center's van slowly through the streets, "and I've learned that there are lots of separate neighborhoods all over the city, some of which are like their own little towns." Jim pointed out corner restaurants and churches that served as gathering points. As he maneuvered the van along the route that Brian had walked that morning, Jim waved at people along the street. "I know most of the people right around the center," he said. "You'll find out that even though this neighborhood may not look like the ones you grew up in, it is a community—people usually know and help each other. I'm proud that we at the center have helped make this happen."

As they continued the tour, Jim reeled off the names of different neighborhoods and pointed out landmarks that separated them, such as busy streets, small parks, or large buildings. He also characterized each neighborhood by who lived there. Most of the neighborhoods near the community center were African-American, but eventually Jim drove through an area that was mainly Hispanic and pointed across the street to another that he said was predominantly Irish and Italian. Some areas seemed to have older residents; others had younger residents. Some were unbelievably dilapidated—far worse than the area around the community center; others were trim and tidy and indicated, at least to Brian, that the residents might have a bit more money. Jim pointed out areas that were a haven for drug dealers and plagued by a lot of crime and other areas that were much safer.

The afternoon provided Brian's first opportunity to meet some of the children in the community center's after-school program. Jim had already told the counselors more about the children—many of them came from families near the center, often had no father in the home, and that many of them were quite poor. He also offered the counselors some tips about how to get to know the children better and introduced them to the after-school staff and the neighborhood residents who volunteered at the program. Before Brian had time to get too worried the children came running in the door, shouting greetings to Jim, the staff, and the volunteers. Soon Jim's advice paid off. Within a short time Brian was playing ball and trading jokes with some of the youngsters and helping with craft projects.

As the summer wore on, Brian became more comfortable working with the children and less nervous during his walk from the train station. But he was becoming more and more aware of the characteristics of the neighborhoods around the community center and more and more concerned about how these neighborhoods might be affecting the children in his group at camp. When he took the children to the swimming pool in the nearby park, he realized that it was only used by children from the Hispanic and African-American neighborhoods. Jim explained that the children in the nearby Irish and Italian areas went to a different park.

The intense poverty that surrounded the lives of many of the children shocked and depressed Brian. Some of the kids seemed to get a full meal only when they were at the community center, and he soon discovered that only a few of the children had parents who were employed. Brian also came to realize how dangerous some neighborhoods around the community center actually were when one of the boys in his group told him that he had heard gunshots the night before and he and his family had huddled together in the bathtub, terrified that a bullet might pierce the walls of their apartment. And Brian began to sense how rarely the children got out of their own neighborhoods. When he took his group downtown to the art museum, only one of the eight children recognized the downtown area, even though it was only a few miles from their homes. It almost felt as if invisible barriers kept the children confined to certain areas of the city—sometimes very dangerous ones.

As he rode the train home at night and watched the passengers getting on and off, Brian often thought about the variety of neighborhoods scattered throughout the city. He noticed that Euro-Americans tended to get off at some stations, Hispanics at others, and African Americans at still others. He realized that, even though he lived in a neighborhood that was wealthier than those that surrounded the community center were, it was almost as homogeneous as those near where he worked. He and his parents enjoyed the community and its churches and stores, but most of the residents were middle-class African Americans like his parents and the parents of the other counselors; there were far fewer residents from other racial-ethnic backgrounds.

Brian began to wonder if there was some type of invisible barrier around his neighborhood. Why are people so separated anyway? Can people really know and help one another in large cities and in neighborhoods like those around the community center? And what does all this mean for the kids at the camp? So many of them are so poor and live in such dangerous circumstances. How might their experiences in their neighborhood affect them and their future?

*B*RIAN'S THOUGHTS INVOLVE WHAT sociologists call community. You might think of community as simply involving people who live in the same town or neighborhood, much as Jim used the word. However, to sociologists **community** describes a group of people who have frequent face-to-face interactions, but who also have common values and interests, relatively enduring ties, and a sense of personal closeness to one another.[1] You might be part of a community in the neighborhood in which you live, and much of Jim's work at the community center has involved attempts to build this type of neighborhood community. You might also feel that you are part of a community with your friends at school. You might see yourself as part of a racial-ethnic community, a community of fellow workers, or a religious community. Through our social networks each of us belongs to one or more communities.

Brian's thoughts and experiences also relate to **urbanization,** the process of societal change that involves large numbers of people leaving their rural or small-town homes to move to cities and nearby metropolitan areas. In the past few centuries, societies throughout the world, including the United States, have experienced a great deal of urbanization. Small towns and rural areas, especially in highly industrialized societies, have seen their populations decline, while large metropolitan areas, such as the city in which the community center is located, have rapidly expanded. Urbanization has altered individuals' social networks, created new conditions in which these networks develop, and challenged communities and entire nations to adjust to changes in the size and age structures of both urban and rural populations.

Think for a moment of a huge city like New York or Chicago—dominated by skyscrapers and filled with noisy traffic and scurrying pedestrians. Then compare this image with that of a small, rural town—one where the residents all know each other, the main street contains only a few stores, and social lives revolve around school and church activities. Intuitively we might expect that city residents would feel rootless, friendless, and afraid of their environment and that they would be overwhelmed by the vast impersonality of the metropolis. In contrast, we might expect people in the small town to feel more connected to others and more comfortable in their familiar world. Many early sociologists had exactly these expectations. Since the beginning of the discipline, sociologists have been concerned about the relationship between urbanization and community. As they looked at the rapid industrialization and urbanization of their societies, they wondered how communities could endure. What would happen as communities became larger? Could people maintain the social ties that bound them together?

In this chapter we explore the relationship between urbanization and community. First, from a macrolevel perspective, we examine how the process of urbanization has changed communities in both the United States and developing countries. As we will see, urbanization affects all communities—cities, small towns, and suburbs. Citizens and policymakers worldwide have been concerned with the problems facing rural areas as populations decline and with those facing urban areas as populations grow and sometimes shift from inner cities to suburbs.

We then move in closer to examine how urbanization affects individuals' community relationships. Does urbanization destroy the close relationships that typify small towns, as several early sociologists expected? Are the problems that appear in some urban neighborhoods, such as those that surround the community center, a result of urbanization? Or can people in urban areas develop community ties no matter how large a city they live in? As we will see, contemporary sociological studies have not supported the fears of the early sociologists. Rather, humans develop communities even in the most urban of settings. Although a city might look huge and impersonal, it is actually composed of millions of connections and thousands of communities, some of them based on physical proximity and others based on shared backgrounds and interests.

Finally, we step back to look at how cities are structured—the different ways in which boundaries between city neighborhoods and communities are created—and how these structures affect individuals' opportunities and constraints. We will see how city structures and patterns of growth promote not only the development of neighborhoods and com-

How might you use the metaphor of a spiderweb to describe the communities in which you participate?

munities but also divide and separate groups. In addition, we will examine patterns of segregation in large metropolitan areas, much like those Brian witnessed, and the ways in which these patterns help perpetuate racial-ethnic and social stratification.

When thinking about communities and urbanization, I often reflect on the metaphor of the spiderweb that we used in discussing social interaction in Chapter 5. When we look at cities from afar, they seem to resemble a mass of unconnected, independent beings much as, from a distance, a large spiderweb in a dirty basement can simply look like a big cloud of dust. Yet when we closely examine even the largest of cities, we can see that what looked like an amorphous mass is actually a complex web of connections. Sometimes these interrelationships involve several different networks, each spun in relative isolation from the others and having only tenuous links; other times the interrelationships involve overlapping networks with multiple connections. Some people in cities have dense, rich social networks; others have far sparser networks.

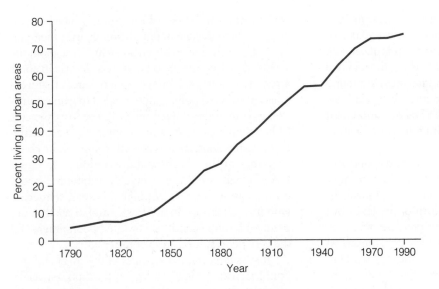

FIGURE 18-1 *Proportion of the U.S. Population Living in Urban Areas, 1790–1990*

Note: The Census Bureau defines urban areas as those that have 2,500 or more residents.

Source: U.S. Bureau of the Census, 1975, pp. 11–12; 1994, p. 43.

You might have noticed that spiderwebs are incredibly strong, capable of supporting a great deal of weight. In the same way, webs of relationships provide support and connections for individual humans, whether in urban areas, suburban areas, or rural communities. You might also have noticed that, no matter what the circumstances, spiders keep spinning webs. In the same way, humans seem to develop webs of social relationships in all circumstances, no matter where they live. Yet, much as physical objects such as trees or posts can alter the shape of a spider's web, the boundaries and structures of neighborhoods can limit the interactions city residents have with one another. As we will see, the way in which cities grow and the structure of neighborhoods affect the ways that communities can be formed, much as Brian sensed that invisible barriers separated the neighborhoods around both the community center and his home community.

URBANIZATION AND CHANGES IN COMMUNITIES

Although cities have existed since ancient times, the concentration of people in urban areas occurred relatively recently. As you will recall from Chapter 17, the Industrial Revolution led to a decline in the high mortality rate that had been characteristic of agrarian societies, and the world's population began to grow exponentially. Much of this growth occurred in cities. In fact, in Europe and North America industrialization and urbanization have been highly interrelated processes, occurring during the same time period and seemingly dependent on each other. The mechanization of agriculture made urbanization possible by reducing the number of people who were needed to grow crops. This allowed many people to move to cities and work in business and industry. Factories required large numbers of workers, and cities provided this labor pool.

Because of these large-scale societal changes, millions of individual workers and families gradually found that the circumstances of their lives had altered. Farmers began leaving their farms and rural communities to work as laborers in large, noisy cities. In the United States, millions of European immigrants created ethnic enclaves as they sought work in rapidly growing urban areas. These individual actions collectively produced the process of urbanization. Today a greater proportion of the world's population lives in urban areas than at any previous time in history. This relatively sudden shift in the pattern of human interaction has brought enormous changes for communities and individual lives. We turn now to examinations of how urbanization occurred in the United States and how it is occurring in developing countries.

Urbanization in the United States

Since the first census was taken in 1790, the U.S. Census Bureau has gathered information about where people live. Figure 18-1 shows changes that have occurred since the first census. As you can see, in 1790 only about 5 percent of the nation's population lived in urban areas. But the country gradually became more urbanized over the years, so that today about 75 percent of the population lives in towns and cities.[2]

The Census Bureau also designates some areas of the country as **metropolitan areas**, or **MAs.** MAs include both a populous central city or core area and nearby communities that are closely related to the core city. To qualify as an MA, an area must include either one city with a population of at least 50,000 or several adjacent cities with a total population of 100,000 (except in New England, where the minimal population is 75,000).[3] Metropolitan areas are found in every state of the union, ranging from relatively small cities like Bismarck, North Dakota, and Great Falls, Montana, to the sprawling areas of Boston, New York, and Los Angeles.

MAs cover only about 19 percent of the land area of the country, but they include almost 80 percent of the country's population. In other words, four-fifths of the population live on one-fifth of the land. But when we look still closer at census data, we can see an even greater concentration of people within the largest metropolitan areas. Three-fifths of the metropolitan population live in areas with more than a million inhabitants.[4] Many of these people live in what is called a **megalopolis** (literally, "great city"), a vast area of land with a dense population. Megalopolises also are found throughout the country. For instance, in California dense residential and commercial areas extend well over a hundred miles from San Diego to north of Los Angeles. On the East Coast there is virtually unbroken development from southern Maine through Boston, New York City, Philadelphia, and Baltimore to south of Washington, D.C. This eastern megalopolis includes one-fifth of the nation's population.[5]

Many metropolitan areas also include **suburbs,** cities and towns outside a city's boundaries but either adjacent to the city or within commuting distance. In recent years populations of suburbs have risen substantially, while those of areas inside city boundaries often have declined. As a result, more people actually live and work in suburbs than within the boundaries of central cities. Although residents of suburbs have access to the resources of the central city, the pattern of workers commuting from their homes in the suburbs to jobs in the city appears to be diminishing as more and more offices and industries relocate to suburban areas.[6] These movements have many consequences for city residents, including the loss of jobs, the deterioration of neighborhoods as wealthier residents leave, a decline in business opportunities, and even the loss of professional sports teams.

Population growth brings both opportunities and constraints to metropolitan areas and suburbs. Although growth often fuels economic prosperity and provides jobs, it also can spur demands for services and facilities, such as expanded schools, new highways, and new housing. In addition, communities face growing problems with environmental pollution and the crowding and overuse of existing facilities.

The growth of the country's urban population has been accompanied by a sharp decline in rural population as farming has become more mechanized. As a

result some rural areas of the country have experienced a sharp decline in population density, the number of people who live within an area. Throughout the 1990s, however, this trend reversed slightly in some areas of the country. Nonmetropolitan areas that were adjacent to metropolitan areas tended to gain population as commuters moved further away from the city. Other rural areas that gained population were retirement destinations, such as coastal regions of the country, areas in the South, and areas along the Great Lakes. Finally, rural areas where the economy is based on recreation, such as popular ski areas, tended to gain in population. In contrast, rural areas that depend on farming continued to lose population as did the very poorest rural areas of the country.[7]

Just as important as changes in the sheer size of a population are changes in its composition, especially its age structure. As with international migration, **internal migration** (permanent changes in residence that occur within a nation's boundaries) tends to involve young people in the prime working ages. When young people cannot find jobs in a local farming economy and move to cities, the age structure of rural areas changes dramatically. The average age rises, and there are fewer children. As you will remember from Chapter 17, changes in the age structure affect the opportunities and constraints that communities face. For instance, the relatively large number of older residents can place severe demands on medical facilities, yet the shrinking overall population might make it difficult to attract health care professionals. The decrease in the number of children can force schools to merge with those in nearby communities. And the decline in the number of potential customers can prompt businesses to close down, which can result in even fewer economic opportunities and even more people leaving.[8]

Highly industrialized countries throughout the world have faced these changes. For instance, like the United States, Japan has experienced a dwindling population in its rural areas and more and more crowding within cities. As in the United States, rural Japanese communities often struggle to find new jobs for their citizens as young people move away and many fewer workers are needed in traditional industries such as agriculture.[9] Similar patterns have occurred throughout the industrialized world as urbanization has accompanied industrialization and growing development.

Urbanization in Developing Countries

Map 18-1 portrays the variations in urbanization among countries throughout the world. More than three-fourths of the population in industrialized countries live in urban areas.[10] The situation is different,

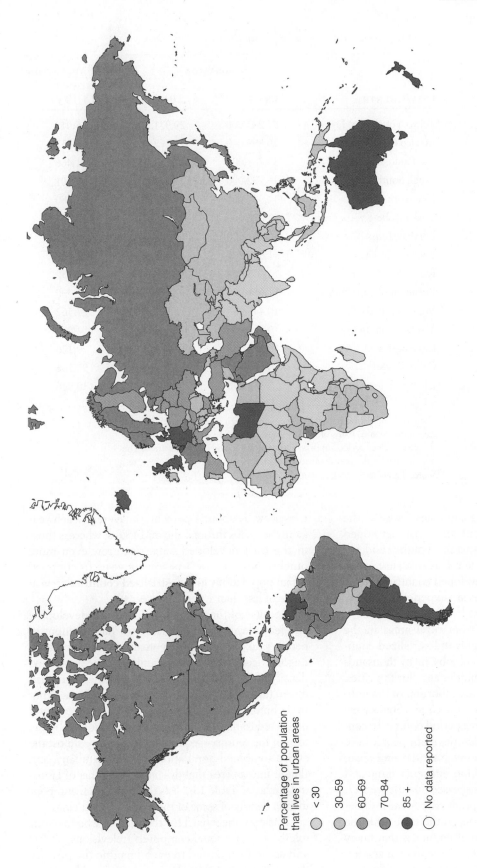

MAP 18-1 *Percentage of Population That Lives in Urban Areas*

Source: Population Reference Bureau, 1998.

TABLE 18-1 *Population and Population Density of Selected Cities*

CITY/COUNTRY	POPULATION 1991	POPULATION 2000	POPULATION DENSITY (1991)
Tokyo-Yokohama, Japan	27,245,000	29,971,000	25,019
Mexico City, Mexico	20,899,000	27,872,000	40,037
São Paulo, Brazil	18,701,000	25,354,000	41,466
Seoul, South Korea	16,792,000	21,976,000	49,101
New York, U.S.A.	14,625,000	14,648,000	11,480
Osaka-Kobe-Kyoto, Japan	13,872,000	14,287,000	28,025
Bombay, India	12,101,000	15,357,000	127,461
Calcutta, India	11,898,000	14,088,000	56,927
Rio de Janeiro, Brazil	11,688,000	14,169,000	44,952
Buenos Aires, Argentina	11,657,000	12,911,000	21,790
Moscow, Russia	10,446,000	11,121,000	27,562
Manila, Philippines	10,156,000	12,846,000	54,024
Los Angeles, U.S.A.	10,130,000	10,714,000	9,126
Cairo, Egypt	10,099,000	12,512,000	97,106
Jakarta, Indonesia	9,882,000	12,804,000	130,026

Note: The population density is per square mile; it was calculated by consulting detailed maps of each city and taking into account nonresidential areas such as parks, airports, bodies of water, and industrial areas. The populations for the year 2000 are estimates.
Source: Famighetti, 1994, p. 840.

however, in the developing countries (those in the process of industrialization) and least developed countries. Slightly more than a third of the residents of developing countries live in urban areas; just over a fifth of those in the least developed countries are urbanized. Yet these countries are experiencing a very rapid process of urbanization.[11]

Just as young people from rural areas in the United States and other highly industrialized countries move to cities in search of jobs, many thousands of citizens in developing countries also flock to cities. As shown in Figure 18-1, the proportion of the population living in urban areas in the United States grew gradually over a relatively long period of time. In contrast, in many developing countries many people have migrated to urban areas in a much shorter time span, largely because of population and environmental pressures, such as those discussed in Chapter 17. Much land has become less productive because of deforestation and desertification. At the same time, the increasing rural population often means that many residents have no access to land of their own and experience intense poverty. It is no wonder, then, that many rural residents move to cities in search of economic opportunities. Data from the United Nations estimate that urban populations in developing countries grew by about 4 percent, on average, each year from the 1980s through the mid 1990s, whereas those in the least developed countries grew even more quickly, at a rate of 5 percent a year. In contrast, urban populations in industrialized countries grew at a rate of less than 1 percent a year.[12]

Because of these trends, cities in the developing and least developed countries now have millions of people packed into relatively small areas. For instance, Lagos, Nigeria, has an almost unbelievable population density of 143,000 people per square mile, more than fifteen times that of Los Angeles. Bombay, India, is not far behind, with a density of 127,000. Mexico City is projected to have almost 28 million residents by the turn of the century—equivalent to packing almost the entire combined populations of Ohio, Indiana, and Illinois into an area slightly smaller than that of Houston, Texas![13] Table 18-1 lists the population and population density of some of the world's largest cities.

It is extremely hard for those of us who take our microwaves, personal computers, televisions, VCRs, and so on for granted to even imagine the poverty that exists in many developing countries. Many people live in makeshift shacks, often constructed of cardboard. They lack sanitary facilities and must get water from a common pump. The streets in their

Even though this family in India is trying hard to have clean clothes and healthy food, it is immeasurably harder for these family members to complete these tasks than it is for people in industrialized societies.

neighborhoods are full of filth and raw sewage, and it is virtually impossible for residents to ensure that their food, let alone their houses and bodies, is clean. Disease and malnutrition are rampant. Jobs are scarce, and whole families are reduced to begging for a little money to scrape by. Even so, remarkably enough, life in urban areas is often better than that in the country, where problems of sanitation, malnutrition, and poverty are even worse. Thus many people continue to choose to migrate to the cities.[14]

Because the urban population is growing so quickly in these areas, governments cannot meet all the demands for housing, roads, sewage, water, or schools. As if the sheer numbers weren't enough, however, most developing countries that are facing this rapid urbanization also are quite poor. They simply don't have the resources to meet the needs of the burgeoning cities.

At the same time, a small minority of residents of cities in developing countries is extraordinarily wealthy. In fact, as discussed in Chapter 7, many de-

veloping countries have a level of economic inequality that is substantially higher than that in the United States and other industrialized nations. For instance, in the United States, the United Kingdom, and Canada the wealthiest 10 percent of the population earns about 25 percent of the total income generated by the country. But in many poorer countries, including Brazil, Chile, Guatemala, Kenya, Paraguay, South Africa, and Zimbabwe, the wealthiest 10 percent receives almost one-half of the total income.[15] This concentration of wealth among a relatively small proportion of the population makes it even more difficult for developing countries to deal with all the demands that face them.

In the previous chapter we described the PPE spiral, referring to the intertwining relationship of poverty, population growth, and environmental degradation. The phenomenal growth in the urban areas of developing countries makes the PPE spiral especially problematic.[16] We will return to these issues in Chapter 19 when we discuss how societies change. Now, however, we will focus more closely on how the process of urbanization affects individuals. How do people in urban settings, wherever they might be in the world, manage to build and maintain communities?

SUSTAINING COMMUNITIES IN MODERN CITIES

Early sociologists like Émile Durkheim and Georg Simmel were concerned by the rapid social changes brought about by growing industrialization and urbanization. Among many other issues these scholars wondered about the implications of these changes for human communities. How would the shift from small rural villages where everyone knew everyone else to large, seemingly impersonal urban centers affect individuals and the social structure, the fabric of social life? Could communities exist in large urban centers, or would urbanization destroy community? Would it be possible to spin strong webs of social relationships in these new settings? As noted previously, the classical theorists and some early American sociologists gave relatively pessimistic answers to these questions. But, as we will see, empirical data and more recent theories suggest that communities are remarkably resilient and can take many different forms. Humans create connections no matter what type of area they may inhabit.

Theories of Urbanization and Community

As you may recall from Chapter 1, Durkheim's theory of society and social relationships included a distinction between mechanical solidarity and organic solidarity. Durkheim suggested that with industrialization and urbanization societies were moving from a society based on mechanical solidarity, in which individuals had relatively similar responsibilities, tasks, and behaviors, to a society based on organic solidarity, with a more complex division of labor. Durkheim didn't see mechanical solidarity as better or worse than organic solidarity; rather, it simply provided a different way in which the collective consciousness developed.

Another early sociologist, however, took a different view of these changes. Ferdinand Tönnies (1855–1936), a German sociologist who lived at about the same time as Durkheim, distinguished between two types of community relationships. He termed the first one **Gemeinschaft,** a German word that means something close to "community." Tönnies suggested that a *Gemeinschaft* involves relationships that might appear in small, close-knit communities where people are involved in social networks with relatives and long-time friends and neighbors, much like those that appear in primary groups. The second, which he called **Gesellschaft** and which can be roughly translated as "association," involves relationships that come about through formal organizations and economic activities rather than through kinship or friendship; thus these relationships are similar to those that appear in secondary groups. Tönnies asserted that relationships in small towns epitomize a *Gemeinschaft* while those in large cities are typical of a *Gesellschaft.* Unlike Durkheim, Tönnies worried that the shift from *Gemeinschaft* to *Gesellschaft* that was occurring in the Europe of his day would prove to be destructive. He thought that the changes would result in a loss of community, a loss of meaningful relationships with other people.[17]

To some extent Georg Simmel's thoughts coincided with those of Tönnies. Simmel used his microlevel perspective to look at interactions in urban settings. He hypothesized that because cities are often noisy and hectic, we try to block out some of these external stimuli. As a result we could become less sensitive or caring toward other people than we might be if we lived in smaller communities. At the same time, Simmel suggested, the wide web of social relationships that we can maintain in cities allows us to feel more anonymous in our actions and to develop more individuality.[18]

A member of the Chicago School of sociology, Louis Wirth, writing in the 1930s, advanced ideas similar to those of Tönnies and Simmel. Wirth argued that simply because cities include more people, they will tend to have more diverse citizens than would be found in a smaller community. In addition, the increased numbers make it very unlikely that inhabitants of a city can know each other personally. Therefore contacts with other people will much more often involve secondary, rather than primary, relationships. Wirth hypothesized that these relationships would be "more complicated, fragile, and volatile" than those in smaller communities are. Further, individuals would have much less control over many aspects of their social interactions.[19] Wirth suggested that these conditions could lead to a number of serious problems, including "the weakening of bonds of kinship, and the declining social significance of the family, the disappearance of the neighborhood, and the undermining of the traditional basis of social solidarity."[20] At the end of his essay, Wirth called for testing this "theory of urbanism."[21] Like contemporary sociologists he believed that the science of society can advance only when we test our theories and, if necessary, revise them to more closely match the empirical world. We turn now to the attempts by contemporary sociologists to test the ideas of Wirth and the earlier theorists.

Research on Urbanization and Community

The empirical data suggest that the reality of urban life is much more complex and far less dire than these early theorists claimed. To indicate this complexity, we will review research on urban neighborhoods, social networks, and individuals' attitudes toward and tolerance of other people.

The Importance of Neighborhoods When you ask people from New York City where they live, they almost always will respond with something like "the Upper West Side," "Columbus Circle," "Queens," or the "South Bronx." Rarely will they just say "New York." Even though they might all root for the Yankees or Mets in baseball and the Giants or Jets in football, they define where they live by their neighborhood, not the larger city. By shopping at the local market, eating in nearby restaurants, and getting to know people who live nearby, they acquire a sense of belonging to their local neighborhood.[22]

For many years urban neighborhoods, like the area surrounding the community center where Jim and Brian work, were identified with different ethnic groups. For instance, in Philadelphia Italians tended to live in South Philly while the Irish lived in Fishtown and Kensington. As we will describe more fully later in this chapter, today in cities throughout the country, African Americans and, to a lesser extent, other people of color live in separate areas of town, apart from Euro-Americans. For example, in Chicago, African Americans more often live on the South Side than in other areas. In New York African

How might their scout troop help these boys identify with their neighborhood?

Americans often live in Harlem, the South Bronx, and some parts of Brooklyn. San Francisco, New York, and other large cities also have areas known as Chinatown, where large numbers of Chinese immigrants and their descendants live. As you might imagine, this separation of ethnic groups into distinct neighborhoods limits the possibility that members of different groups will have contacts with one another and can promote both a sense of ethnic identity and ethnocentrism.

At the same time, however, the fact that people inhabit and identify with their neighborhoods can counter the isolation and weakened ties that early sociologists such as Tönnies and Wirth associated with urban life. Urban neighborhoods can have patterns of social interaction and social support that resemble those in the small rural towns idealized by Tönnies and Wirth. One of the most famous demonstrations of this phenomenon developed from a study Herbert Gans conducted in Boston's West End in 1957 and described in his book, *The Urban Villagers*.[23]

Gans used the method of participant observation to study this low-income, working-class area; he

rented an apartment and lived there for an extended period. The residents were primarily second- and third-generation Italian Americans, although there were also small enclaves of other white ethnics such as Poles, Jews, Greeks, and Ukrainians. The area was seen by the government as a decaying slum and was, in fact, scheduled to be razed within a few years. Yet Gans found that it housed a thriving community, or what he called an "urban village." Even though the West End had 7,000 residents and they clearly couldn't all be acquainted with one another, residents did know their immediate neighbors. The West Enders routinely conversed in their hallways, on their front stoops, in the streets, and in neighborhood shops. Several times each week small, more intimate groups would gather in individual residences to talk for hours on end. Peer groups were a very important part of the West Enders' lives from early childhood through adolescence and into adulthood. In this way West Enders formed a community—long-lasting, supportive ties—based on their associations with others in their immediate neighborhood.

But only some urban residents live in ethnic neighborhoods like the West End. What about other city dwellers? When you walk around a large city or tour it in a vehicle as Brian and the other counselors did with Jim, you can easily see the many different types of neighborhoods and many different kinds of people. Do they all experience the sense of community that the West Enders enjoyed?

Gans used Max Weber's technique of "ideal types" to distinguish five different types of urban residents he observed in a number of urban areas. First, there are "ethnic villagers," people like the West Enders who live in tightly knit communities centered on a common heritage. Second, there are "cosmopolites," people such as students, professionals, artists, and entertainers who are drawn to the city because of its cultural advantages and conveniences. Third, there are the "singles," unmarried and childless people who come to the city to work and live before moving on when they begin a family, as well as those who never begin a family and might stay in the city for the rest of their lives. Fourth, there are the "deprived," the poorest people in the city, who generally belong to the underclass and who might live in extremely dangerous and desolate neighborhoods. Finally, there are the "trapped" and downwardly mobile, who really don't want to live in the city but are unable to move elsewhere, often because they can't afford to.[24]

Gans suggested that people in the trapped and deprived groups are much less likely than people in the other groups are to have a sense of community. Wirth's depiction of the extreme problems that city dwellers face probably applies best to these people, for they are

Sociologists have found that disasters such as the San Francisco earthquake of 1906 can often, ironically enough, promote community ties. Why do you think such tragedies might bring neighbors together?

Common Sense versus Research Sense

COMMON SENSE: Urbanization contributes to the decline of community ties.

RESEARCH SENSE: Contemporary research suggests that city dwellers are no more socially isolated than rural residents are. Diverse subcultures within urban areas have patterns of social interaction and support similar to those in small rural towns.

Communities as Social Networks   Gans looked at neighborhoods as the basis of community, but sociologist Claude Fischer chose a different approach, studying the ways in which individuals create their own communities through their social networks. Do people in cities primarily have secondary relationships and thus more psychological problems than people in smaller communities do, as Wirth suggested? Or are people in large cities as likely to have close, primary relationships as people in smaller communities are?

To answer these questions, Fischer used survey research. In the late 1970s he and his student assistants interviewed more than a thousand people in fifty communities in northern California. The communities ranged from high-rise neighborhoods in San Francisco to small rural towns several hundred miles to the north.[27] They asked each of these people about their social networks—with whom they regularly interacted and what kinds of support they received. For instance, the researchers asked the respondents who would care for their home if they went out of town, with whom they had recently engaged in social activities, whose advice they would seek in making an important decision, and from whom they might borrow money in an emergency.[28] They also asked the respondents several questions designed to measure their psychological mood, including how upset and angry they generally felt or how pleased they were with life.[29] Based on this information, they tested the notion that people living in urban communities have less supportive social networks and more psychological difficulties than those who live in more rural settings.

In contrast to the ideas of the early sociologists, Fischer found no indication that urban residents were more psychologically distressed than people who live elsewhere were. To further check this finding, he controlled for a wide variety of variables that might affect the amount of stress people experience, such as the presence of children, their marital status, any recent stressful life events (such as the death of a close relative), and their education, income, and occupation. Even with all of these controls, the same basic finding appeared: Urban life does not seem to cause the psychological distress predicted by the early theorists.[30]

much more likely than other urbanites to experience crime, despair, and loneliness. But do these negative experiences result from living in a city? Gans thought not. Instead, he asserted that these problems result from the fact that people in the deprived and trapped groups are poor, might be victims of racial-ethnic discrimination, and might have other problems that have forced them to live in these situations.[25]

Unlike the deprived and the trapped, Gans hypothesized, the cosmopolites and singles do experience community in the midst of the urban environment. To some extent this can reflect the neighborhood in which they live. For instance, singles may live in apartment complexes with other singles, cosmopolites might choose to live in neighborhoods near cultural attractions and with residents who have interests similar to their own. Cosmopolites and singles also have the opportunity and means to develop connections with other people through their work and leisure activities. In other words they are much freer to develop their social networks—to create their own communities.[26]

TABLE 18-2   *A Sample of Community Activities Listed in the Newspapers of Three Northern California Communities*

GROWING MOUNTAIN TOWN	SMALL WINE COUNTRY CITY	MEDIUM-SIZED BAY AREA CITY
Two swimming activities	Two bridge clubs	Cross-cultural couples meeting
Three bus trips (to zoo, a fort, and Lake Tahoe)	Model railroaders	African-American women
Girls' club	Five branch meetings of Alcoholics Anonymous	African-American co-eds
Teenagers' dance	Two square dances	Middle-years groups
Request for people to show travel slides	Senior citizens' group	Lesbian parents
Reflexology class	The Grange	Transvestites/transsexuals rap
	The Masons	Drop-in group for parents and children
		Round dancing

Source: Fischer, 1982, p. 198.

Again contrary to the theoretical ideas of the early sociologists, Fischer did not find that urbanization produces social isolation. Urban residents' social relationships were as supportive and personal as were those of people who lived in smaller communities. However, urban residents were more likely to have friends who lived farther away. They also were more likely to get support from friends they met at work or in voluntary organizations, such as clubs or civic organizations. In contrast, people in more rural areas were more likely to have relationships with relatives and people who belonged to the same church. In other words, people in all types of areas—from small towns to large cities—create social networks of individuals with whom they have meaningful relationships. In urban settings, however, there might be more contexts from which people can draw their relationships, and so their support network comes from a broader range of places. As Fischer put it, "Urbanism tends to expand people's opportunities for building social ties beyond the family and the neighborhood."[31] The vastness of the city increases the ways in which individuals can spin their webs of social relationships.

In contrast to the early theorists, Fischer and several other sociologists suggested that urban life doesn't destroy community. Instead, it creates *plural communities,* or a variety of **subcultures** much like those we discussed in Chapter 3. These subcultures can have many different bases, such as ethnicity, occupation, lifestyle, or leisure interests. In fact, the more urban a place is, the more subcultures it seems to have and the more varied these subcultures are.[32]

Table 18-2 illustrates the wide variety of subcultures that can appear in cities by listing notices taken from newspapers in some of the communities Fischer studied. As you can see, the activities listed in the Bay Area paper are much more varied than those available in the two smaller communities. With their large and diverse populations, cities allow people to develop social networks based not only on proximity or family ties, but also through shared interests and worldviews.[33]

Just as Gans found a grain of truth in Wirth's characterization of the city in his observations of the deprived and the trapped urban dwellers, so did Fischer's analysis provide support for Simmel's observation that city dwellers tend to block out many stimuli and thus to seem more aloof, distant, and uncaring than people in smaller communities. In addition, Fischer found support for the notion that urban dwellers are more distrustful of their fellow citizens than are people in smaller cities and towns. In public places, city dwellers tend to be careful, reserved, and watchful, even though their private relationships are generally just as satisfying as are those of people who live in smaller towns.[34]

Fischer suggested that this urban reserve has two sources. First, it reflects a realistic fear of crime, which is more common in large cities. Second, the presence of the very subcultures that give cities a sense of community and vitality also produces tensions and the possibility of friction between groups. As he described it, "As urbanism strengthens the little worlds of the urban mosaic, it also increases the strain between pieces of that mosaic."[35] Simply because urban areas contain many more people, the chances of encountering people from different subcultures are much greater than in a small town. As a result, casual observations of the public scenes in cities emphasize the differences and the lack of connection between city residents.

Finally, even though each of us can create our social networks, the extent to which we can build lasting ties and connections to our community depends not just on our individual actions but also the larger,

Why do you suppose no one on this city street is paying much attention to the man who is so skimpily dressed? Can you imagine seeing this scene in a small town? Why or why not?

or more macrolevel, context in which we live—much as the shape and nature of spiderwebs varies from one location to another. As you might expect, the longer people live in a community, the more likely they are to develop friendships, participate in local affairs, and feel connected to the community. But the extent to which this attachment can occur is also affected by the community's overall stability. If you live in a place where people frequently move in and out, your ability to form lasting friendships and participate in the community in other ways is lessened simply because there are fewer opportunities for lasting contacts. In other words, the social actions that create social networks occur within a broader context, and the characteristics of that larger context significantly influence our opportunities and constraints.[36] We will return to this interaction of social structure and social action in the last part of the chapter.

Urbanism and Tolerance The early theorists worried that city life, by destroying community, would also eliminate the social ties that create social control and thus destroy traditional values and promote deviance. In apparent support of these claims, Fischer did find that residents of the most urban areas were much more likely than were residents of smaller cities and towns to report behaviors that could be considered "deviant"—that is, that departed from prevailing norms. For instance, individuals in the most urban areas were 1½ times as likely to be cohabiting with someone, 2½ times as likely to have no religious identity, and 12 times as likely to be openly homosexual.[37]

Are these differences produced by the lack of social networks, as the early theorists suggested? As you might expect from the preceding discussion, the answer is no. People who engaged in less traditional lifestyles did not have social networks that were any weaker than those of more conforming individuals.

Might the differences result from people who want to pursue alternative lifestyles moving to large cities so that they can do so in relative anonymity? Again, this doesn't seem to be the case. To some extent "selective migration" can explain these patterns, but Fischer suggests that the differences are so great that migration can't account for all the variation.

Instead, Fischer argues that city dwellers are simply more tolerant of behaviors that others see as deviant. This greater tolerance develops because urbanism promotes subcultures that nourish new ideas and styles and that promote nontraditional values. These new views are then transmitted gradually to other subcultures within the urban environment.[38] For instance, new fashions or hairstyles typically originate in small subcultures in large cities and then spread to other areas through the process of *cultural diffusion,* first introduced in Chapter 3.[39] The diversity of lifestyles found within cities seems to affect the attitudes of all city dwellers, whether or not they participate in various subcultures themselves. In short, the greater tolerance of urbanites seems to be because they are much more likely to be exposed to and to have social relationships, however fleeting, with people who are part of a variety of subcultures.[40]

In sum, the presence of many subcultures and neighborhood groupings in cities enables individuals to spin new social webs that give them a sense of community, a process that was not foreseen by the early sociologists. Paradoxically, however, the very diversity that permits urban communities to develop and that fosters tolerance also can produce divisions and strains between groups. If we look at cities from a more macrolevel perspective, we can see how these divisions are exacerbated when cities and metropolitan areas develop structures that reinforce group distinctions and create very different opportunities and constraints for individuals of different groups.

URBAN STRUCTURES AND OPPORTUNITIES AND CONSTRAINTS

As you will remember from our discussion in Chapter 3, the Chicago School of sociology developed at the same time that the city of Chicago was experiencing a great deal of growth and change. As a result, much of the work of the Chicago School concerned the large immigrant and growing African-American populations of the city. Wirth's analysis of city life focused on how city life affects individuals; other members of the Chicago School stepped back to take a broader macrolevel view, developing models of how urban structures change as a city's population expands.

In this section we look at how the structure of cities affects the ways in which individuals spin their webs of relationships and how these webs affect their individual lives. We first examine models sociologists have developed to describe how cities grow and change and then look at how the growth of metropolitan areas results in divisions among neighborhoods and patterns of segregation. Finally, the featured research explores how city structures affect patterns of racial-ethnic stratification and what consequences these patterns have for the life chances of individual citizens.

Social Ecology and the Structure of Cities

In our discussion of formal organizations in Chapter 11, we introduced the theory of organizational ecology. As you may remember, ecology is the branch of science that studies the relationship between organisms and their environment. Several social scientists, beginning with members of the Chicago School, have used similar ideas to study **social ecology,** or the ways in which people interact with their physical and geographic environment. With the tremendous growth of urban areas during the past two centuries, sociologists interested in social ecology have given a great deal of attention to the patterns of this growth and the implications for individuals and communities. Over the years these sociologists have created three models to describe how cities expand to deal with changes in their economy and the size of their population and with the nature of their particular geographic and physical characteristics (see Figure 18-2). Of particular interest to many sociologists has been the way that city expansion is related to where citizens live and who lives in these areas.

The first model of urban growth, called the **concentric-zone model,** was developed by Ernest W. Burgess at the University of Chicago in the 1920s. Based on his observations of how Chicago had changed over the years, Burgess suggested that cities expand outward from their central business district in a series of concentric zones (circles within circles) (see Figure 18-2a). The development of strong structural steel and the invention of mechanical elevators led to the construction of tall buildings, allowing thousands of people to work in a relatively small vicinity. Further, the development of urban transportation systems such as rail lines allowed people to live outside this core business district and commute to work. Therefore the city could expand outward, with residential areas springing up around the central business district.

Burgess's concentric-zone model builds on these observations. Radiating out from the central business district are a series of residential zones, with each successive zone housing people of a higher social class status and containing newer houses that are less crowded together. Many cities in the United States in the early twentieth century, and especially Chicago, fit this model. The core business district remained in the center of town, and new construction occurred first around this core area and then, in successive waves, farther and farther out into the surrounding countryside. The more luxurious houses were located farthest from the city center, whereas ethnic groups tended to be crowded together closer to the city center in various neighborhood enclaves.[41] Over time neighborhoods in the central city have often become extremely run-down and home to the city's poorest residents.[42]

A second model, often called the **sector model,** was proposed in the late 1930s. This model modified Burgess's ideas by suggesting that cities often are composed of sectors or segments (see Figure 18-2b). These segments can cut across the circular zones and extend outward from the central business district, resembling slices of a pie. Each of these slices or sectors represents an area where people of specific social classes or racial-ethnic groups tend to live. As the population of the city expands, the members of these groups gradually move into adjoining neighborhoods.[43] For example, while African Americans occupied only part of Chicago's South Side in the early part of the century, African-American communities expanded throughout the area and farther from the city center as more and more African Americans migrated to the city.

The third model, usually called the **multiple-nuclei model,** directly challenges Burgess's key idea that cities develop around a core business district. Instead, this model suggests that cities can have a number of distinct core areas (or nuclei) influenced by the geographic and physical characteristics of an area (see Figure 18-2c). For instance, communications

BUILDING CONCEPTS

Three Models of City Growth

Over the years, sociologists have created three models of city growth that describe the way cities have expanded to deal with changes in their economy, the size of their population, and the nature of their particular geographic and physical characteristics.

Model	Theory of City Expansion
Concentric zone model	Cities expand outward from their central business districts in a series of concentric zones, with each successive zone housing people of higher social class status.
Sector model	Cities, in addition to expanding in a series of concentric zones, are composed of sectors or segments that can cut across the circular zones. Each of these sectors represents an area where specific social classes or racial-ethnic groups tend to live.
Multiple-nuclei model	Instead of expanding outward from a central business district, cities have a number of distinct core areas (nuclei) influenced by the geographic and physical characteristics of an area.

FIGURE 18-2 *Three Models of Urban Growth*

Note: CBD stands for "central business district."

Source: White, 1987, p. 119.

  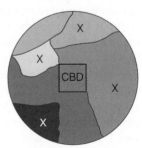

a Concentric-zone model   **b** Sector model   **c** Multiple-nuclei model

and financial enterprises might be located in the center of town; transportation industries might lie along major freeways and water routes; and industrial plants might be located in suburban areas where land is cheaper. Many metropolitan areas in the South and West, such as Los Angeles, Dallas, and Seattle, appear to fit this model, with a great deal of industry and development occurring in suburban areas all across the metro area.[44]

Each of these models of urban growth tries to describe how zones or sectors of a city differ in terms of inhabitants or activities. To a large extent the models also describe historical changes in the ways that cities have expanded. The changing structures of older cities in the United States and Europe were highly influenced by the development of mass transit, but the changes in newer cities have been more strongly influenced by the use of the automobile and the development of freeway systems. Thus, although one model of city growth may accurately reflect patterns in some cities, another model might provide a more accurate picture of other metropolitan areas.

No matter how a city has grown and changed, however, its structure affects the lives of its residents because its structure influences the kinds of networks to which they are exposed. We now examine exactly how the areas of a city can differ and what consequences these differences have for individual lives. How do the divisions between neighborhoods, like those Brian saw around the community center and throughout his city, affect patterns of stratification and individuals' life chances?

Neighborhood Segregation

If you think about your own neighborhood, you will realize that many of the people who live near you share characteristics with you. For instance, if you are a young, unmarried college student, you likely live near other young, unmarried students. If you have children, you probably live near other people who also have children. If you are retired, you are more likely to have other retirees as neighbors. If you are wealthy, your neighbors probably are affluent as well, and if you are poor, your neighbors probably are too. From a sociological perspective, what you are observing is **segregation,** the division or separation of neighborhoods in ways that lead to the inclusion of some groups and the exclusion of others.

Sociologists often study segregation by using data gathered in the decennial census conducted by the U.S. Census Bureau about census tracts. **Census tracts** are small, relatively permanent areas within cities and towns that are designed by the Census Bureau to be fairly homogeneous in population, economic situation, and living conditions and that represent, as much as possible, a city's neighborhood boundaries. On average, census tracts contain about 4,000 people, and boundaries between tracts typically consist of a major thoroughfare, a large shopping center, or a park. The Census Bureau publishes summary information about the characteristics of people in census tracts, such as their age, occupation, income, and race-ethnicity. Sociologists can use these data to understand more about the characteristics of a neighborhood's residents and differences among neighborhoods within a metropolitan area.

Based on their analyses of census tracts in several metropolitan areas, sociologists have identified three variables that define the structure of urban segregation: (1) the social class status of the residents, (2) their age and family status, and (3) their race-ethnicity. In some neighborhoods, residents have very high incomes, whereas in others virtually all the residents live below the poverty line. In some neighborhoods most people live alone, and in others the majority live in families with children. And in some neighborhoods there are no African-American or Hispanic residents, but in others virtually all or a substantial majority are members of a racial-ethnic minority group. These differences consistently appear in metropolitan areas throughout the country.

Segregation based on race-ethnicity has been characteristic of U.S. cities for years. Many European immigrants to the United States settled in areas that came to be known as "Little Italy" or "Germantown." Such ethnic enclaves form the basis of the neighborhood studied by Gans that was described earlier in this chapter. Yet in the United States, these neighborhoods of white ethnics were actually quite diverse. For instance, the U.S. Department of Labor prepared a color-coded map of Chicago's West Side in 1893 that used eighteen colors to show the location of distinct European ethnic groups. No block in the entire area included just one group, and four-fifths of each *lot* within the blocks included a mixture of different ethnic groups. Even in the areas that were called "Little Italys" there were no blocks that had only Italian residents.[45] In addition, as time went on and residents of these neighborhoods experienced social mobility, they were able to translate their new economic well-being into residential mobility, moving into areas of their own choice. Today descendants of European immigrant groups experience virtually no residential segregation.[46]

The experiences of people of color have been very different. For example, Hispanics experience residential segregation, although the extent of segregation seems to vary from one part of the country to another.[47] Many urban areas have neighborhoods called "Chinatown," which house a large number of Asian immigrants. But African Americans and, in some instances, Puerto Ricans have experienced the most segregation.

Residential segregation based on race and ethnicity has been illegal for several decades. Why, then, does it persist? How does it affect individuals in the society? How are patterns of residential segregation related to issues of racial-ethnic and social stratification? Two sociologists, Douglas Massey and Nancy Denton, have tried to answer these questions.

URBAN SEGREGATION

Like other researchers who have studied segregation, Massey and Denton used census data. They combed through information collected on census tracts over several decades in metropolitan areas throughout the United States. They analyzed the extent of segregation, its shifting patterns during the past century, and its effects on individuals and the larger society. They also used computer simulations to try to understand more about how the structure of cities contributes to the development of impoverished inner-city ghettos, such as those described in our discussion of racial-ethnic stratification in Chapter 8.

The Extent and Persistence of Segregation

Sociologists have developed several ways to measure how segregated communities are. The measures are based on the general idea that an unsegregated community would be one in which the characteristics of an individual neighborhood were very similar to those of the city as a whole. One frequently used measure is an **index of dissimilarity** (or **segregation index**), which is calculated from data on the characteristics of people living in different census tracts. For example, when used to measure the extent of segregation that affects African Americans, the index gives the percentage of African Americans in a city who would have to move to create an "even" residential pattern, one in which every neighborhood has the same proportion of African Americans as the city as a whole. The closer the index is to 100, the more segregated a city is. Sociologists often use a figure of 60 or greater to indicate a high level of segregation, between 30 and 60 to indicate a moderate level, and less than 30 to indicate a low level.[48]

Figure 18-3 shows the index of dissimilarity for the thirty metropolitan areas in the United States with the largest population of African Americans. As you can see, all but one of these metropolitan areas (Norfolk–Virginia Beach) have a high level of segregation, and not one has a low level. On average, more than three-fourths of African Americans in northern cities and two-thirds of those in southern cities would have to move to create unsegregated communities.[49]

As you will recall from Chapter 8, the residential segregation of African Americans became an entrenched pattern in northern cities during the early twentieth century as more and more African Americans moved from the rural South to work in factories in the North. Economic competition exacerbated conflicts between African Americans and Euro-Americans, and African Americans who tried to live away from "black" neighborhoods sometimes had

their houses ransacked and burned. This tide of violence, or its potential, drove African Americans in virtually every northern city into homogeneous neighborhood enclaves.[50] Since that time, residential areas have remained extremely segregated despite extensive migration within the country. Many African Americans have moved from the South to northern cities and from cities to suburban areas, and many Euro-Americans also have moved to urban and to suburban areas. Segregation declined slightly from the 1970s to the 1980s and 1990s, but it remains extremely high in virtually every metropolitan area in the United States.[51]

Massey and Denton found that residential segregation affects all African Americans, no matter what their income level. Using data from the 1980 census, they calculated the average dissimilarity index for African Americans earning over $50,000 a year in northern cities to be 83.2 and the average figure for those who earned less than $2,500 a year to be 85.8, only a few percentage points higher. Even with substantial incomes, African Americans like Brian's family can't buy their way out of segregated neighborhoods.[52]

An analysis of data from the city of Philadelphia illustrates this phenomenon. As shown in Table 18-3, both low-income African Americans and low-income Euro-Americans in Philadelphia live in bleak circumstances—neighborhoods where more than a third of the births are to unwed mothers, home values are low, and substantial numbers of students have level levels of school achievement. More affluent Euro-Americans are able to escape these circumstances, living in neighborhoods where few teens are unwed mothers, house values are higher, and students achieve more. In contrast, high-income African Americans remain in disadvantaged neighborhoods, with house values only slightly higher than those in the neighborhoods where the poorest African Americans live.[53]

The experiences of well-to-do African Americans contrast sharply with those of other people of color. Massey and Denton's results from their analysis of Los Angeles were typical. They found that the dissimilarity index for the poorest Hispanics was 64 while the index for the wealthiest Hispanics was 50, considered to be moderate. In contrast, the index for the most affluent African Americans in Los Angeles was 79. In other words the wealthiest African Americans in Los Angeles are more segregated than the poorest Hispanics.

Why are wealthy African Americans more likely to experience segregated housing than poor Hispanics are? Massey and Denton found that the answer seems to involve skin color. The only racial-ethnic group that experiences as much residential segregation as African Americans is Puerto Ricans. Many Puerto Ricans, like

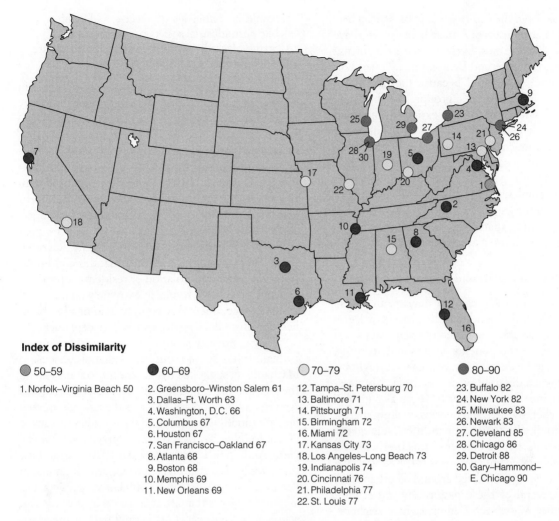

Index of Dissimilarity

○ 50–59   ● 60–69   ○ 70–79   ● 80–90

1. Norfolk–Virginia Beach 50

2. Greensboro–Winston Salem 61
3. Dallas–Ft. Worth 63
4. Washington, D.C. 66
5. Columbus 67
6. Houston 67
7. San Francisco–Oakland 67
8. Atlanta 68
9. Boston 68
10. Memphis 69
11. New Orleans 69

12. Tampa–St. Petersburg 70
13. Baltimore 71
14. Pittsburgh 71
15. Birmingham 72
16. Miami 72
17. Kansas City 73
18. Los Angeles–Long Beach 73
19. Indianapolis 74
20. Cincinnati 76
21. Philadelphia 77
22. St. Louis 77

23. Buffalo 82
24. New York 82
25. Milwaukee 83
26. Newark 83
27. Cleveland 85
28. Chicago 86
29. Detroit 88
30. Gary–Hammond–
    E. Chicago 90

FIGURE 18-3 *Index of Dissimilarity for the Thirty Metropolitan Areas with the Largest African–American Populations, 1990*

Source: Harrison and Weinberg, 1992; Massey and Denton, 1993, p. 222.

TABLE 18-3 *Characteristics of Neighborhoods Inhabited by African Americans and Euro-Americans at Different Income Levels in Philadelphia, 1980*

Neighborhood Characteristic	LOW-INCOME		MIDDLE-INCOME		HIGH-INCOME	
	Euro-American	African-American	Euro-American	African-American	Euro-American	African-American
Births to unwed mothers (%)	40.7%	37.6%	10.3%	25.8%	1.9%	16.7%
Median value of homes (in thousands)	$19.4	$27.1	$38.0	$29.5	$56.6	$31.9
Students scoring below 15th percentile on high school achievement test (%)	39.3%	35.5%	16.5%	26.6%	5.7%	19.2%

Source: Massey et al., 1987, pp. 46–47, 50; Massey and Denton, 1993, p. 152.

other people from the Caribbean, are of African heritage. And Puerto Ricans of African heritage, who have the darkest skin, are most likely to experience intense segregation. As Massey and Denton put it, "Puerto Ricans are more segregated because they are more African than other Hispanic groups." In short, the color of one's skin seems to be an extremely important indicator of where one lives—much more important even than how much money one makes.[54] This segregation is so intense, so extensive, and so persistent that Massey and Denton gave their study the title *American Apartheid* to emphasize the similarity of this pattern of segregation to the legally constituted system of racial segregation, called apartheid, that was recently dismantled in the country of South Africa.

American Apartheid and Hypersegregation

How, exactly, does segregation develop in cities? Massey and Denton found that the process of segregation can involve several different components and that, when these components occur simultaneously, the results can be devastating. Segregation might involve an uneven distribution across the city, as measured by the index of dissimilarity. It can also involve isolation, whereby African Americans rarely share a neighborhood with whites. In addition, segregated neighborhoods can be clustered together within a city to form a large enclave, concentrated within a very small area, or centralized around an urban core.

When several of these patterns of segregation occur together, Massey and Denton assert, a neighborhood experiences **hypersegregation,** extreme isolation from other areas of the community. In their examination of census data, they found such hypersegregation in a wide variety of cities, including Buffalo, Chicago, Cleveland, New York, Los Angeles, Philadelphia, St. Louis, Atlanta, Baltimore, and Dallas-Fort Worth, among others. Massey and Denton describe the results of this residential pattern:

> Thus one-third of all African Americans in the United States live under conditions of intense racial segregation. They are unambiguously among the nation's most spatially isolated and geographically secluded people, suffering extreme segregation across multiple dimensions simultaneously. Black Americans in these metropolitan areas live within large, contiguous settlements of densely inhabited neighborhoods that are packed tightly around the urban core. In plain terms, they live in ghettos.
>
> Typical inhabitants of one of these ghettos are not only unlikely to come into contact with whites within the particular neighborhood where they live; even if they traveled to the adjacent neighborhood they would still be unlikely to see a white face; and if they went to the next neighborhood beyond that, no whites would be there either. . .

> Ironically, within a large, diverse, and highly mobile postindustrial society such as the United States, blacks living in the heart of the ghetto are among the most isolated people on earth.[55]

In our discussion of racial-ethnic stratification in Chapter 8, we noted that these very poor inner-city neighborhoods, like some of those surrounding the community center, experience extreme social problems. Because poverty is so concentrated in these areas, almost all the inhabitants are poor and unemployed. This means that there are few employed people who can provide linkages into social networks leading to jobs or who can help children learn job skills. High unemployment also means that few men can afford to marry or to support a family. Research discussed in Chapter 8 suggested that this concentration of poverty and isolation of residents is part of a "vicious cycle" that results in extremely high rates of violent crime, unemployment, female-headed families, out-of-wedlock births, and welfare dependency.[56]

Why is poverty so concentrated in these areas? William Julius Wilson's historical analysis, discussed in Chapter 8, suggested that it reflects, to some extent, the fact that many middle- and working-class African Americans abandoned these extremely depressed neighborhoods when they had the economic means to do so. In other words it reflects segregation by social class. And, in fact, analyses of the 1990 census show that middle-class African Americans are continuing to flee inner-city areas for more affluent suburban areas, albeit often ones that are segregated by color.[57] But segregation by social class is only part of the story. Using computer simulations, Massey and Denton demonstrated how city structures that promote racial-ethnic segregation foster the development of the underclass and of impoverished ghetto communities.

Segregation and the Development of the Underclass

You might have played computer games such as "Sim City" that allow you to construct imaginary towns and see what happens to these towns as a series of events unfolds. Such games are based on the notion of **computer simulations,** in which computers are programmed to simulate, or model, selected characteristics of real life. Because sociologists often study events and behaviors that they can't manipulate in the laboratory or field, they sometimes use such computer simulations to conduct *hypothetical experiments.* Massey and Denton did such a computer simulation to understand how racial-ethnic segregation contributes to the development of the underclass and the concentration of poverty in urban ghettos.

The results of one of these simulations are depicted in Figure 18-4. In this simulation they constructed four

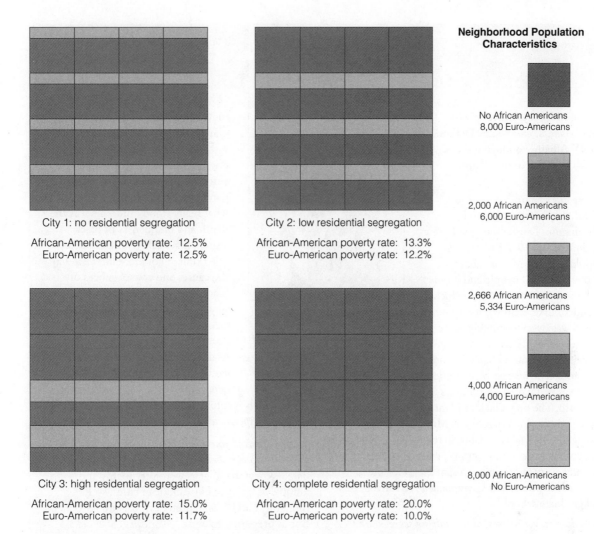

City 1: no residential segregation
African-American poverty rate: 12.5%
Euro-American poverty rate: 12.5%

City 2: low residential segregation
African-American poverty rate: 13.3%
Euro-American poverty rate: 12.2%

City 3: high residential segregation
African-American poverty rate: 15.0%
Euro-American poverty rate: 11.7%

City 4: complete residential segregation
African-American poverty rate: 20.0%
Euro-American poverty rate: 10.0%

Neighborhood Population Characteristics

No African Americans
8,000 Euro-Americans

2,000 African Americans
6,000 Euro-Americans

2,666 African Americans
5,334 Euro-Americans

4,000 African Americans
4,000 Euro-Americans

8,000 African-Americans
No Euro-Americans

FIGURE 18-4 *Denton and Massey's Computer Simulation of the Effects of Different Levels of Racial-Ethnic Segregation on the Neighborhood Poverty Ratio*
Source: Massey and Denton, 1993, pp. 120–121.

imaginary cities, all of the same size and each with sixteen neighborhoods. To mirror reality, they assumed that each city was one-quarter African American and that the rate of poverty among African Americans was 20 percent, compared with 10 percent among Euro-Americans. The only difference between the four cities was in the amount of residential segregation based on race-ethnicity. City 1 had no residential segregation, and everyone lived in a neighborhood where 25 percent of the residents were African Americans. City 2 had a low level of segregation, such that one-fourth of the neighborhoods had no African Americans, but all of the other neighborhoods had an equal number. City 3 had a high level of segregation. City 4 had total segregation, with all of the African Americans concentrated in one-fourth of the neighborhoods and all of the Euro-Americans concentrated in the other three-fourths.

Denton and Massey then used computer simulations to determine the characteristics of neighbor-

hoods that the average African American and Euro-American would experience in each of these cities. As shown in Figure 18-4, in City 1, where there was no segregation, both the average African American and the average Euro-American family would live in a neighborhood where the poverty rate was 12.5 percent. But as segregation becomes more intense, the average African-American family would tend to live in a neighborhood with more poverty, whereas the average Euro-American family would tend to live in a neighborhood with less poverty. In City 3, which actually has less segregation than most major metropolitan areas today, the average African-American family would live in a neighborhood with a poverty rate of 15 percent, and the average Euro-American family would live in a neighborhood with a poverty rate of 11.7 percent. And in the totally segregated City 4, the average African-American family would inhabit a neighborhood with a poverty rate that was

BOX
18-1

SOCIOLOGISTS
AT WORK

Douglas Massey

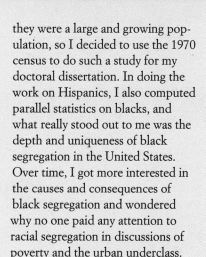

Racial segregation is one of the most important forces shaping the nature of American cities and the perpetuation of urban poverty.

Douglas Massey, University of Pennsylvania, and Nancy Denton, SUNY Albany, conducted the research featured in this chapter.

Where did you grow up and go to school? I grew up in Olympia, Washington. During the time I grew up there, from 1952 to 1970, its population ranged from about 13,000 to 25,000 people and it only had one public high school, which I attended. Like the town itself, the school was overwhelmingly white and European. There were a smattering of Asians and Hispanics, but no blacks. The nearest black settlement was in Tacoma, thirty miles to the north. The only black resident of Olympia until I was in junior high school was Mr. Grace, a blind man who tuned all the pianos in town. The school was diverse in economic terms, however, since virtually every family in town sent their kids there, from rich to poor. I left Olympia in 1970 for college and graduate school.

How did you get interested in studying segregation? As a graduate and undergraduate I had done a lot of work studying the spatial structure of cities. I was particularly interested in the Chicago School and its studies of segregation, and I thought I wanted to do work in this tradition. One day I realized that no one had done a systematic study of Hispanic segregation in the United States, despite the fact that

they were a large and growing population, so I decided to use the 1970 census to do such a study for my doctoral dissertation. In doing the work on Hispanics, I also computed parallel statistics on blacks, and what really stood out to me was the depth and uniqueness of black segregation in the United States. Over time, I got more interested in the causes and consequences of black segregation and wondered why no one paid any attention to racial segregation in discussions of poverty and the urban underclass. As a young faculty member, I studied the patterns, causes, and consequences of black segregation in U.S. cities, which eventually led to the publication of *American Apartheid.*

What surprised you most about the results you obtained in your research? After working with census data, I knew that blacks were very highly and uniquely segregated in American cities, so I thought I had become pretty cynical and hardened. But in preparing to write *American Apartheid* I read many historical studies of how the black ghetto was created, and I was shocked at the incredible record of violence, racism, and prejudice and how deeply segregation had been institutionalized at virtually all levels of American life for so many years. It's a pretty depressing record that left me shaken.

twice that of the neighborhood of the average Euro-American family.

The simulation shown in Figure 18-4 did not take into account social stratification. In reality, neighborhoods, even within the African-American community, are segregated by social class. When class segregation is factored in, the concentration of poverty in the very poorest African-American neighborhoods becomes even more extreme. Thus, when faced with both class and racial-ethnic segregation, the average poor African American moves from a neighborhood with an expected poverty rate of 20 percent to one that is twice as high.[58] In other words, the simple structural

facts of segregation by social class and race-ethnicity combine to produce the incredible concentration of poverty that is found in urban ghettos.

Using computer simulations, Denton and Massey also showed how economic downturns that affect highly segregated cities increase the relative poverty of ghetto areas. Because segregation results in poverty being concentrated in relatively small areas of a city, it also concentrates changes in the economic fortunes of African Americans. For instance, during the Great Depression of the 1930s, African-American areas of cities were especially hard hit. Similarly, when economic hard times struck many northern cities in the

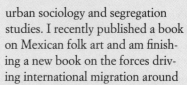

BOX
18-1

SOCIOLOGISTS
AT WORK

Continued

What was the reaction to your research among your colleagues? Did it surprise you in any way? I was surprised at the intense interest in my research by the press and the field in general. As my articles and then *American Apartheid* were published, they received a great deal of attention and numerous reviews. The book won two prizes, one being the Distinguished Publication Award of the American Sociological Association, and it continues to be widely read and frequently used in college classes around the country. I am always surprised by the number and variety of people who have read *American Apartheid,* from Cornel West to Ted Koppel to Bill Bradley.

What research have you done since completing the study described in the chapter? Does it build on your previous research? After publishing *American Apartheid,* I decided not to do a major study of segregation using the 1990 census. If the same person does the same research for a long time, it is easy to fall back into the same routinized patterns, and I thought that sociology would benefit from the involvement of new investigators. I have done some work since then on the consequences of segregation, specifically the consequences of growing up and living in poor, racially isolated neighborhoods, but most of my work has been on international migration, specifically on Mexican migration to the United States. Even while I was working on segregation, I had other projects ongoing about immigration, and I am as well known in the fields of immigration and Latin American studies as I am in

urban sociology and segregation studies. I recently published a book on Mexican folk art and am finishing a new book on the forces driving international migration around the world, written in conjunction with a committee of the International Union for the Scientific Study of Population.

What do you think the implications of your work are for social policy? Have any of these implications been realized? One of the people who read *American Apartheid* was Henry Cisneros, when he was President Clinton's Secretary of Housing and Urban Development. In fact, Cisneros not only read it, he assigned it to all his assistant secretaries and senior staff members and invited me and Nancy Denton to spend a day at HUD giving a talk and meeting with his people just as they were formulating plans to launch major fair housing initiatives. The book ended up providing the intellectual and theoretical justification for Cisneros's open housing agenda.

How can your research help students understand events in their own lives? In my view, racial segregation is one of the most important forces shaping the nature of American cities and the perpetuation of urban poverty. Drive through any large American city and see the abandonment or watch the evening news on TV and you will see very directly the consequences of segregation. Not as easy to see, but no less important, are the decline in American competitiveness and living standards and the withering of public life, all of which makes us poorer as a nation and as a people.

1970s, these areas again experienced an increased concentration of poverty. The computer simulation showed that simply increasing the average poverty rate of African Americans from the 20 percent assumed in the original model to 30 percent would increase the percentage of poor people in the neighborhood of the average African-American family from 40 percent to 60 percent.[59]

In general, Massey and Denton's computer simulations demonstrated how the social structure of cities—the simple fact of segregated neighborhoods—vastly increases the chance that African Americans will live in areas with higher levels of poverty and de-

creases the chance that Euro-Americans will. In short, racial-ethnic segregation of cities results in advantages for Euro-Americans and disadvantages for African Americans. (To learn more about Douglas Massey and his work with Denton, see Box 18-1.)

Denton and Massey, as well as other sociologists who have studied contemporary cities, suggest that this extreme segregation has several effects and can exacerbate the problems that Brian observed in the area where he worked. Because of their isolation, individuals who live in urban ghettos have far fewer chances than do other people of establishing the social networks that can help them find jobs and live the

BOX
18-2

**APPLYING SOCIOLOGY
TO SOCIAL ISSUES**

*How Can Residential
Segregation
Be Reduced?*

Racial-ethnic discrimination in housing has been illegal for a quarter of a century and, as we noted in Chapter 8, Euro-Americans have developed increasingly more liberal attitudes toward members of minority groups. Why, then, is racial-ethnic segregation in housing so pervasive? Two explanations often are given: (1) individual African Americans and Euro-Americans prefer to live in separate areas of the city and (2) real estate brokers and mortgage lenders actively discriminate against African Americans in their day-to-day business practices.

A team of sociologists headed by Reynolds Farley examined these two hypotheses using data from large-scale surveys conducted in the Detroit, Michigan, area in 1976 and 1992.[60] Detroit is a particularly apt choice. With a dissimilarity index of 89 in 1990, it is the most highly segregated metropolitan area in the United States that has a million or more residents, and one of the areas that Massey and Denton identified as being hypersegregated.

To measure the attitudes of respondents, Farley and his colleagues showed them a series of cards, each of which represented a neighborhood with fifteen homes and a different amount of racial-ethnic segregation. One of the questions the respondents were asked was whether they would be willing to move into any of these neighborhoods "should they find an attractive home they could afford."[61] As shown in the accompanying figure, African Americans were much more likely to be willing to live in integrated neighborhoods than Euro-Americans were. Almost a third of the African-American respondents were willing to be the only minority resident in an all-white community. In addition, however, substantial numbers of Euro-Americans were willing to enter neighborhoods with African-American residents. For example, by 1992 over two-thirds of the Euro-Americans were willing to move into a neighborhood where 20 percent of the residents were African American, a number that was significantly higher than that obtained in 1976. In general, Farley and his associates conclude that Detroit's extensive segregation does not simply reflect the attitudes of the area's residents, for a substantial proportion would be willing to live in integrated neighborhoods. Younger and more highly educated residents were especially likely to express more liberal attitudes.

Why, then, does segregation persist? In part, Farley and his colleagues suggest, this persistence reflects the belief of real estate brokers and mortgage lenders that Detroit residents want to live in segregated neighborhoods. As a result these housing professionals market homes and apartments in ways that perpetuate these presumed values.[62]

What can be done to diminish the amount of segregation in Detroit? Because younger cohorts are more liberal than older cohorts are, one option would be to wait for the gradual changes that would occur as these younger cohorts replace older ones. But, as Farley and his associates point out, this process will take a very long time, especially in a city that is as divided as Detroit.

"American Dream." The concentration of poverty within such confined areas might also support the development of what Denton and Massey call an "oppositional culture," one in which joblessness, crime, and other negative features of ghetto life seem normal. In the political sphere, racial-ethnic segregation can virtually ensure the election of African Americans to political positions that are based on area representation, such as city councils or state legislatures. At the same time, however, segregation can work against political alliances with other groups. For example, attempts to build housing in a ghetto area can become perceived as an "African-American project" rather than a project for the entire community, simply because only African Americans live in the ghetto area.

As noted previously in this chapter, the fear of early sociologists that community life would be destroyed by urbanization has not been substantiated. But the research of Massey and Denton raises other troubling issues regarding the nature of community in metropolitan areas. Their research shows how, much as the structure of an area

BOX
18-2

APPLYING SOCIOLOGY TO SOCIAL ISSUES

Continued

Instead, Farley and his colleagues suggest that the city could initiate a number of strategies that would actively promote the development of integrated housing, including the following:

- Make sure that real estate brokers are aware that both Euro-Americans and African Americans are willing to live in mixed neighborhoods.

- Assure home owners that neighborhood services and loans for home purchases will be maintained when neighborhoods are integrated.

- Mount efforts to prevent Euro-American residents from moving from neighborhoods when they are integrated and to welcome minority residents to newly integrated areas.
- Develop programs to protect property values against declines when integration occurs.
- Provide slight discounts in interest rates for first-time

home buyers in areas where they will be in the racial-ethnic minority.[63]

CRITICAL-THINKING QUESTIONS

1. Could these policies help promote racial-ethnic integration in communities you have lived in? What other policies do you think might be effective?

2. How can social actions in your community help alter the structures of segregation?

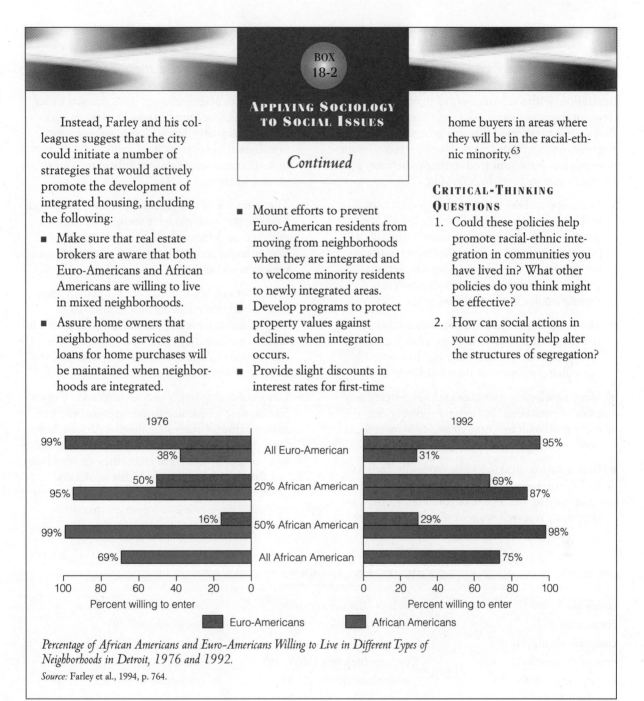

Percentage of African Americans and Euro-Americans Willing to Live in Different Types of Neighborhoods in Detroit, 1976 and 1992.

Source: Farley et al., 1994, p. 764.

determines how spiders spin their webs, the structures of cities limit the extent to which individual social actions can create inclusive webs of human relationships. How can communities that include a full cross-section of an area's residents be created in highly segregated cities? How can residents establish social networks that include people of different racial-ethnic backgrounds if their social environment is homogeneous? (See Box 18-2 for potential answers.)

CRITICAL-THINKING QUESTIONS

1. Can you think of data sources other than the census that could be used to study segregation within cities? What are the relative advantages and disadvantages of these sources?

2. As computer technology becomes more sophisticated, it will undoubtedly be used more and more by sociologists. Think of some other issues from this book. Are there any for which computer simulations might provide a useful research tool?

SUMMARY

During the past several centuries, societies throughout the world have experienced a great deal of urbanization, with vast numbers of people migrating from rural areas to cities. Although many of the classical sociologists feared that this process would have negative consequences for communities, more recent research has not supported these predictions. Key chapter topics include the following:

- In Europe and the United States, industrialization and urbanization occurred simultaneously and seem to have been interdependent. Originally a mostly agricultural country, the United States has become highly urbanized during the past two centuries. Today more than three-quarters of the population live in metropolitan areas.

- The sharp decline in rural populations that accompanied the growth of urban areas has resulted in special problems for rural areas in maintaining services for their inhabitants.

- Many developing countries are experiencing very rapid urbanization, but extreme poverty can make it difficult for these countries to meet the needs of their burgeoning populations.

- Earlier sociological theorists, especially Tönnies and Wirth, suggested that urbanization would contribute to the decline of community ties. Contemporary research suggests that these fears are largely unfounded and that city dwellers are no more socially isolated than rural residents are. Some urban neighborhoods have characteristics not unlike those found in small towns. In addition, urban areas allow people a wide range of possible social contacts and facilitate the development of numerous diverse subcultures and greater tolerance.

- Three models of city growth have been proposed: a concentric-zone model, a sector model, and a multiple-nuclei model. All three models stress that different areas of a city have distinctive compositions and activities.

- Studies of neighborhoods of metropolitan areas have found that residents differ markedly in social class, in their stage in the life course, and in their race-ethnicity. People of color, and especially those of African descent, are most likely to experience severe segregation.

- Douglas Massey and Nancy Denton used Census Bureau data to document the extensive and persistent nature of the segregation of African Americans in U.S. cities. Using computer simulations to construct hypothetical cities, they showed how city structures that include highly segregated ghetto communities make economic down-turns especially devastating for these communities.

KEY TERMS

census tracts 495

community 481

computer simulation 498

concentric-zone model 493

Gemeinschaft 488

Gesellschaft 488

hypersegregation 498

index of dissimilarity 496

internal migration 484

megalopolis 484

metropolitan area (MA) 483

multiple-nuclei model 493

sector model 493

segregation 495

segregation index 496

social ecology 493

subculture 491

suburb 484

urbanization 481

INTERNET ASSIGNMENTS

1. Using a search engine, find the home page of three or four cities in the United States. What does the information provided by the cities suggest about the nature of the communities and their local cultures? Then go to the U.S. Census Bureau home page and locate information on the demographic characteristics of these cities, such as the size of their population, the average income of residents, and their racial-ethnic diversity. Are these demographic characteristics related in any way to the observations you made about the cultural characteristics of the cities or the ways they were portrayed on their home pages? What hypotheses about the relationship between demographic characteristics and the nature of communities might you develop based on these observations? How could you test these hypotheses?

2. Cities in the United States face many different issues, from economic development to affordable housing. Find the Web site for the U.S. Department of Housing and Urban Development and read one or two of its reports on contemporary urban problems or community programs. How do the descriptions provided there compare with the issues and concepts discussed in this chapter?

TIPS FOR SEARCHING:
When you are looking for home pages of cities, it can help to specify "city of" in the search engine, as in "City of Atlanta."

INFOTRAC COLLEGE EDITION: READINGS

Article A20791186 / Brockerhoff, Martin, and Ellen Brennan. 1998. The poverty of cities in developing regions. *Population and Development Review, 24,* 75–115. Brockerhoff and Brennan examine how the very rapid urbanization of developing countries has exacerbated problems of poverty in cities.

Article A19136822 / Krivo, Lauren J., and Ruth D. Peterson. 1996. Extremely disadvantaged neighborhoods and urban crime. *Social Forces, 75,* 619–649. Krivo and Peterson examine the extent to which segregation based on both race-ethnicity and social class influence neighborhood differences in crime.

Article A20572908 / Van Kempen, Eva T. 1997. Poverty pockets and life chances: On the role of place in shaping social inequality. *American Behavioral Scientist, 41,* 430–450. Van Kempen examines how the neighborhood in which one lives affects the chance that one will be poor.

FURTHER READING

Suggestions for additional reading can be found in *Classic Readings in Sociology,* bundled with this book. If you purchased a used copy of this book that does not include this custom-published reader, go to www.sociology.wadsworth.com for ordering information.

Recommended Sources

Brooks-Gunn, Jeanne, Greg J. Duncan, and J. Lawrence Aber (eds.). 1997. *Neighborhood Poverty, Volume I: Context and Consequences for Children,* and *Volume II: Policy Implications in Studying Neighborhoods.* New York: Russell Sage Foundation. This extensive collection of articles explores how living in a poor neighborhood affects children and policies that can address these issues.

Fischer, Claude S. 1982. *To Dwell Among Friends: Personal Networks in Town and City.* Chicago: University of Chicago Press. In this highly readable book Fischer analyzes how people construct their communities in cities and towns of varying sizes.

Fong, Timothy P. 1994. *The First Suburban Chinatown: The Remaking of Monterey Park, California.* Philadelphia: Temple University Press. Fong analyzes political, economic, and racial-ethnic conflicts that arose from the migration of large numbers of Chinese to a southern California community.

Gans, Herbert J. 1962. *The Urban Villagers.* New York: Free Press. Despite being more than thirty years old, this book is a good example of participant observation research in urban neighborhoods.

Gugler, Josef (ed.). 1997. *Cities in the Developing World: Issues, Theory, and Policy.* Oxford, U.K.: Oxford University Press. This collection of essays discusses the rapid urbanization that is now occurring in developing countries.

Jargowsky, Paul A. 1997. *Poverty and Place: Ghettos, Barrios, and the American City.* New York: Russell Sage Foundation. Jargowsky examines neighborhood segregation based on social class among Euro-Americans, Hispanic Americans, and African Americans.

Kleniewski, Nancy. 1997. *Cities, Change, and Conflict: A Political Economy of Urban Life.* Belmont, Calif.: Wadsworth. This text examines issues related to the changing shape of cities and the relationship of these changes to opportunities and constraints of citizens.

Massey, Douglas S., and Nancy A. Denton. 1993. *American Apartheid: Segregation and the Making of the Underclass.* Cambridge, Mass.: Harvard University Press. The authors describe the research featured in this chapter.

Mumford, Lewis. 1961. *The City in History: Its Origins, Its Transformations, and Its Prospects.* Mumford provides a classic historical analysis of the process of urbanization from human's earliest days.

Orum, Anthony M. 1995. *City-Building in America.* Boulder, Colo.: Westview Press; Sugrue, Thomas J. 1996. *The Origins of the Urban Crisis: Race and Inequality in Postwar Detroit.* Princeton, N.J.: Princeton University Press. Both of these books examine the history of cities that have experienced a decaying core area and thriving suburbs.

Scott, Alan J., and Edward W. Soja (eds.). 1996. *The City: Los Angeles and Urban Theory at the End of the Twentieth Century.* Berkeley: University of California Press; Waldinger, Roger, and Mehdi Bozorgmehr (eds.) 1996. *Ethnic Los Angeles.* New York: Russell Sage Foundation. These books both include extensive collections of articles that describe the development of Los Angeles as a multiethnic metropolitan area.

Vergara, Camilo José. 1995. *The New American Ghetto.* New Brunswick, N.J.: Rutgers University Press. This fascinating collection of photographs of poor city neighborhoods throughout the United States is accompanied by insightful commentary.

Yinger, John. 1995. *Closed Doors, Opportunities Lost: The Continuing Costs of Housing Discrimination.* New York: Russell Sage Foundation. Yinger documents the extent of housing discrimination in the United States, its impact, and policies that can combat it.

Internet Source

The Department of Housing and Urban Development. This Web site contains an abundance of information on housing, cities, and neighborhoods.

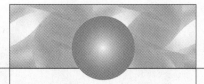

Pulling It Together

Listed below are other chapters where you can find additional discussion or examples of some topics covered in this chapter. You can also look up topics in the index.

Technology
and Social Change

Technology, the Environment,
and Social Institutions

Changing Societies: A Long-Range View

*The First Revolution: Domestication
of Plants and Animals*

The Second Revolution: Invention of the Plow

The Third Revolution: The Steam Engine

A Fourth Revolution? The Global Information Age

FEATURED RESEARCH STUDY

The Telephone, Social Action, and Social
Change—The Work of Claude Fischer

Choosing to Use the Telephone

A Technology of Sociability

Social Change and Technology
in a Global Perspective

Modernization

*Rural Development, Land Reform,
and Sustainable Development*

Joel's family had lived in Oak Valley since the town was founded in the mid 1800s. His great-great-grandparents had once had a dairy farm a few miles from the town center on land that was now part of a large shopping mall and the local community college. When Joel was very young, his Great-Grandmother Nora would tell him stories of the fun she had on the farm as a child—jumping into piles of hay, playing in the creek, and picking blackberries during lazy summer days. She also told Joel of many "firsts" in her life—the first car she ever saw, the first time she heard a radio, the first talking movie she went to, and even the first time she rode in an airplane! Joel loved to listen to her stories and felt lucky that he had been able to know her for just a little while before she died.

Nora had lived with Joel's Grandpa Steven and Grandma Bessie. Grandpa Steven worked for many years in the local lumber mill. By the time he was grown there wasn't enough farm land for all of his brothers and sisters to continue to work on the farm, and the mill provided a good living. Joel's father Matt also got a job at the mill when he finished school, and Joel always assumed that he would eventually work at the mill, too.

All this changed when Joel was beginning high school. The supply of lumber began to dwindle. For many years, the company had been cutting more trees than could be replaced, and there just wasn't enough timber left in the mountains to keep the mill working at full production level. In addition, as if the overcutting weren't bad enough, many of the trees that were left were affected by insect–borne disease. In the process of cutting trees and replanting, many of the species that were native to the area but weren't useful to the mill had disappeared, along with the birds that lived in these trees. Unfortunately, these birds fed on insects, and without these other tree species and the birds, the insect–borne diseases began to overtake the forest.

As a result of these changes, the lumber mill announced that it would slow down production. Along with several hundred other workers, Matt was laid off of work, and, within a few weeks, the mill announced that it was closing permanently. Even though Matt always tried to hide his worries from his family, Joel knew that Matt was devastated. He'd always been so proud of how he could provide for his family, and now he couldn't. Joel's mother, Anne, tried to get more hours at her job at the local beauty shop, but with so many people out of work there weren't that many customers. And, besides, she couldn't earn nearly as much as Matt had in the mill.

After the mill closed, Matt looked hard for other work. He applied at all the other mills in town, but they were also laying off workers. He was good at fixing things and tried to pick up jobs with the local mechanics, but hours were short and they just couldn't pay enough to support the family. When some new computer chip factories moved into town Matt thought that his problems would be solved, but his hopes were dashed again. The personnel officer told him that he didn't have the right qualifications—his experience at the mill wasn't relevant.

Joel had always considered his father one of the bravest people he knew. One day, when Joel was a sophomore in high school, Matt confirmed this belief when he announced, "I guess I just have to face facts. There aren't going to be any more mill jobs here for me. Oak Valley has changed, and I've got to change with it. I'm going to go back to school." The next day he drove out to the community college and signed up for computer technology courses. Matt hadn't cracked a textbook in more than 20 years, but he soon discovered that he could do just fine in his classes. It took two years, but by the time Joel finished high school, Matt was finishing his course work at the community college and working part time at the chip factory, with a full-time job waiting for him when his course work was done.

In the fall after high school, Joel followed in his father's footsteps, not to the mill as he envisioned when he was young, but instead to the computer tech program at Oak Valley Community College. Sometimes when he parked his car and walked toward class he looked around and tried to imagine what the area must have been like when his great grandmother was young. "How much life has changed," he thought. "Nora had so many new experiences in her life. My father's whole career has changed. Will I face so many changes as I get older? Why are things always changing anyway?

The experiences of Joel and his family reflect **social change,** the way in which societies and cultures alter over time. Many of these changes involve alterations in *social institutions,* the complex sets of statuses, roles, organizations, norms, and beliefs that meet people's basic needs within a society. For instance, during the time Joel's family has lived in Oak Valley, its economy has changed from a dependence on agriculture to much more dependence on the production of lumber to a reliance on manufacturing computer components and writing software. Families have become smaller, the roles of women and men have changed both in the home and in the economy, and schooling opportunities have increased.

Other changes have involved the introduction of **technological innovations,** the development of new material elements of culture that help us deal with the problems of day-to-day life. Great-Grandmother Nora remembered the changes brought about by cars, radios, movies, the electric light, and many other technological innovations. Matt is now working with computer technology that was only a dream when he was young.

Many of these changes have also involved alterations in the *environment,* the physical, biological, and natural world in which we live. The development of industry and urbanization in Oak Valley led to the destruction of farm land. The overcutting of the forests and depletion of natural species led to changes in timber production that affected the timber industry. Each of these changes has affected not just Matt and Joel and their family but all of the people who live in Oak Valley.

Throughout this book we have seen many examples of how changes in society affect individuals. For instance, in Chapter 7 we saw how changes in the occupational structure made it much less likely that individuals will experience upward mobility. In Chapter 10 we saw how the employment of women outside the home and the increased likelihood of divorce and single-parent families have changed the environments children experience as they grow up. In Chapter 18 we saw how urbanization has led to many more people living in cities and suburbs rather than in rural, agricultural areas. Chances are, your great-grandparents and grandparents were more likely than you are to experience upward social mobility, families without divorce, and life in a small rural community.

Although these social changes have affected our lives as individuals, it is important to remember that they result from social actions—the activities of individuals and social groups. Even though we might not often think about it, large-scale changes such as urbanization, alterations in the occupational structure,

and changes in the divorce rate result from the accumulation of individual decisions to move, to take or create new jobs, and to divorce rather than to stay married. Individuals also develop *cultural innovations.* Individuals create new technology, or material cultural elements, as well as nonmaterial cultural elements, such as ideas and beliefs, that alter how we experience the world. New jobs in the occupational structure, the laws that govern divorce, and the means humans have developed to cope with urban life, from sewer systems to public transit to shopping malls, are all cultural innovations.

C. Wright Mills suggested that the sociological imagination allows us to link biography and history, to see how our own lives and the individual lives of others occur within the historical context of changing societies.[1] Joel's Great-Grandmother Nora grew up in a world in which most people worked on farms. She experienced the introduction of cars, radios, and talking movies, and by the end of her life people had flown to the moon and routinely took planes to go around the country and even the world. Joel's father Matt grew up and entered adulthood at a time when manufacturing jobs, such as those in the timber industry, were still common. When he reached middle age, these jobs began to disappear and he found himself out of work and training for a new career. Joel's biography will be different still. As he enters the work force, computers and related inventions we probably can't even imagine will shape opportunities and constraints in his life. The personal troubles that we each experience in our own biographies relate to public issues that are grounded in an historical context.

In our discussion of socialization and the life course in Chapter 4, we looked at how members of different birth cohorts, such as Nora, Matt, and Joel, have different socialization experiences—how historical periods influence the patterns of our lives. In this chapter we step back and look not just at what happens to individuals in different historical periods but at these historical changes themselves. How have societies and social structures changed over time? How have these changes affected individuals' opportunities and constraints, and how have social actions contributed to the nature of these changes? Why have different areas of the world experienced different rates of change? How might the poorer nations of the world gain greater economic independence and security?

In trying to understand social change, I sometimes think about gardens and natural landscapes, such as the farm where Great-Grandmother Nora lived, and how landscapers and gardeners can affect these areas. All gardens and natural areas change

This Yemenite man can communicate with virtually any corner of the world using modern communication techniques. How does this reflect the process of globalization?

over time as plants get larger, grow and multiply, and die. Sometimes these changes result from the maturation of plants, the introduction of new species, or destruction wrought by wind, floods, and droughts. Other times the changes come from human interventions. Landscapers might trim trees, introduce new shrubs and bushes, and move plants around to create a more pleasing setting. Segments of a park might be removed to allow construction of a road or a house. Similarly, societies might change gradually over time—through the development of cultural innovations that are adopted over a span of years—or they might alter because of the deliberate and concerted efforts of human beings over just a few years.

Sometimes the introduction of new species in a natural environment will have totally unintended consequences. For example, a bamboo plant meant to add decorative appeal to a garden corner eventually might overwhelm the surrounding plants and grass. In the same way, technological innovations can have totally unanticipated consequences. For

example, few, if any, people in the nineteenth century could have anticipated the long-term effects of the telephone and electricity. In addition, however, based on their knowledge and experience with many different settings, skilled gardeners often can anticipate what will happen to a tree after it is pruned and know how plants should be arranged in a new garden plot to create a pleasing arrangement in the months and years to come. As you know by now, sociology is still a young discipline, and sociologists' understandings of many aspects of the world are just beginning to develop. Yet many sociologists hope that through a careful study of social change, they can eventually help all of us understand how societies change and modify and, perhaps, then be able to help the social world become a better place for all people.

In the sections that follow we will take a closer look at social change and how it involves the intertwining developments of changing technology, environments, and social institutions. We use a broad macrolevel approach to examine social change over the long span of human history, tracing the gradual shifts in the conditions of life and, especially, the development of technological innovations that have remade societies over time. We then focus on how these changes result from social actions by looking at how one technological development, the telephone, has affected social life. Finally, because social change has not occurred at the same rate in different parts of the world and because the varying rates of change are highly related to global stratification, we examine ideas about social change in the less developed countries of the world.

TECHNOLOGY, THE ENVIRONMENT, AND SOCIAL INSTITUTIONS

Perhaps, like Joel, you have been lucky enough to have older relatives and friends who loved to tell stories about their lives and experiences. From these conversations you know that life has changed a great deal in just a few decades as many new technological innovations have been developed. We now have household appliances such as microwave ovens, means of communication such as e-mail, and entertainment devices such as CD players and video games that would have belonged in the realm of science fiction a few years ago. We consider pursuing jobs that did not even exist in previous decades, and we think nothing about traveling distances that once involved a journey of several days. When we take an even longer view and compare our lives with those of our ancestors who lived in hunting and gathering societies, the differences become even more extreme. It is almost impossible for us to imagine life without written communication and wheeled vehicles, let alone electricity and the telephone.

Technological innovations involve the development of *material* elements of culture, new ways of applying knowledge to solve problems such as a lack of natural resources or the nature of a society's climate or geographic location. People have developed technological innovations that span the whole of human life. Some involve new methods of getting food, from the invention of better bows and arrows to learning how to preserve food through drying and salting to the development of advanced fertilizers and farming methods. Others involve the invention of new ways to build shelters (from simple huts to complex skyscrapers) and new methods of constructing clothing (from cloaks made of skin to Gortex rain parkas). People have devised new ways of traveling from place to place, from the invention of wheels to the use of animals in transport to mechanized vehicles such as cars, planes, and rockets. New methods of communication have also developed, including calendars, numbers, written languages, newspapers, books, and electronic mail.[2]

In Chapter 14 we described how such technological innovations produce changes in economic opportunities and allow the countries and regions that have more advanced technology to accrue economic gains. This happens because technology provides its owners with a great deal of power; technology defines what is possible for a society to accomplish. Thus, throughout history, societies with more advanced technology have been able to subdue and conquer those with less advanced resources. The bloody history of the Americas is an excellent example; here the more technologically advanced colonial Europeans, and later the United States and other newly founded countries, systematically annihilated or subdued the native peoples.[3]

Social change also involves alterations in social institutions and changes in the *nonmaterial* aspects of culture—new ways of thinking, new organizational forms, and new norms and values. For example, the recent development of computer technology and the Internet has altered the roles of students and teachers, influenced beliefs about what should be taught in school, and fostered the development of new organizations that guide and aid people in the use of computers. These changes have involved not just one social institution, such as the economy or education, but several at once. In fact, it appears virtually impossible for just one part of a social institution, or just one social institution, to change by itself. Rather, changes in one part of life seem to spur changes in others. For instance, the introduction of

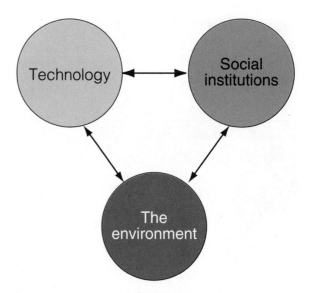

FIGURE 19-1 *The TIE Connection: Social change involves the intertwining of changes in technology (T), social institutions (I), and the environment (E).*

the computer has altered how people spend their money, how families spend their leisure time, and how students learn in school.

Changes in both social institutions and technology are related to how individuals interact with their environment and to alterations in the environment. For instance, innovations related to transportation during the past 150 years, such as trains, cars, and planes, have drastically altered organizations and norms, as well as how individuals relate to their environment. We no longer have to walk from place to place or have animals carry us. Instead, we can ride in cushioned comfort at a much faster pace. At the same time, however, these innovations have changed the nature of the environment. The vehicles that transport us require fossil fuels, which are extracted from the environment and add pollutants to the atmosphere, altering the planet for future generations and likely compelling our descendants to develop new means of transportation.

At the same time environments also influence the types of technology that are developed within a society and how social institutions are structured. For instance, societies that lack abundant and regular sources of water and tillable soil would find it very difficult to develop the irrigation systems needed for advanced agriculture, whereas groups that live near rivers with fertile soils and regular supplies of moisture could more easily make this transition. Societies that live near oceans and lakes tend to develop technologies related to fishing and boating; societies that live in mountainous areas develop technologies that facilitate hunting and gathering of food.[4]

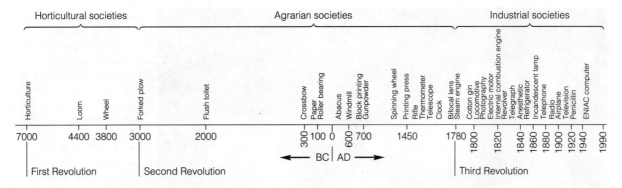

FIGURE 19-2 *Significant Technological Innovations in Human History*
Source: Johnson, 1995, pp. 550–557.

In a very broad sense, social change can be seen as the result of the intertwining of changes in social institutions, technology, and the environment, much as the developing shape and contours of a garden reflect the interaction of growing seeds and plants, the efforts of the landscaper, and the soil and climate in which the garden is located. As depicted in Figure 19-1, changes might first appear in any one of these areas, but change in one aspect tends to affect other areas as well. However, just as the form and changes in a garden or a natural area become apparent only over a long span of time, so do the results of changes in social institutions, technology, and the environment appear full-blown only after they have largely occurred. By examining changes in human societies over time and variations among contemporary societies, sociologists have learned more about how societies tend to change and how these changes have affected individuals and their life chances. We turn now to this very long-range view of human history.

CHANGING SOCIETIES: A LONG-RANGE VIEW

Changes have always occurred within societies, but the nature of these changes and the pace at which they have taken place have varied over the centuries. If you had been born into a hunting and gathering society thousands of years ago, your life probably would have resembled that of your parents, grandparents, and great-grandparents. Today, however, your life, like Joel's, no doubt differs markedly from that which your parents led at your age, let alone your grandparents and great-grandparents. In general the pace of social change has quickened throughout human history, simply because one technological innovation can spur two or even three more, leading to an ever more rapid rate of innovation and related changes in social institutions and the environment.[5] This quickening

pace of innovation can be seen in the time line in Figure 19-2, which marks the development of a number of technological innovations through the last 9,000 years.

Although the potential of these innovations probably wasn't apparent to people living at the time they appeared, in hindsight we can see how important certain technological innovations have been. The sociologist Gerhard Lenski and his colleagues have identified three major technological innovations, pinpointed in Figure 19-2, as most important—so important that they have used the term "social revolutions" to describe the changes in social institutions that they produced.[6] As we will see, these technological and social changes were also associated with changes in the environment, changes that have often been apparent only from a very macrolevel, historical view.

The First Revolution: Domestication of Plants and Animals

The first human groups were hunting and gathering societies that obtained food from hunting game and gathering wild plants. Over the course of thousands of years, hunters and gatherers developed a number of technological innovations that dramatically changed their lives. For instance, roughly 35,000 years ago, a number of groups began gradually to modify and alter the spear, originally just a long, sharp stick. The eventual result was the bow and arrow, a much more powerful and accurate weapon that enabled hunters to obtain more food, especially large game animals such as deer and elk. This increased availability of food also allowed them to support a larger population. In turn, however, this larger population needed to be fed, thus prompting the need to obtain even more game.[7]

Eventually this improved hunting technology, combined with a gradual warming of the earth's

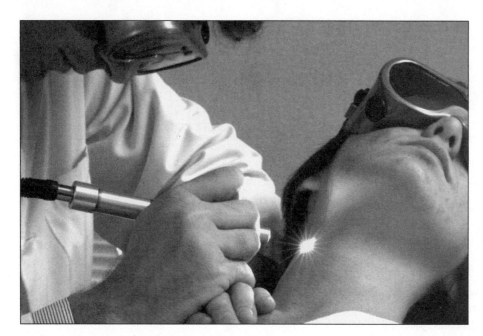

In the 1950s, when scientists at Bell Labs developed the laser, they were just following their intellectual curiosity, seeing if they could solve some interesting problems. Since then the laser has spawned many other inventions such as CDs, laser printers, supermarket checkout scanners, laser-guided missiles, and various surgical procedures. What does this series of inventions indicate about the pace of social change and technological innovation?

temperature, changed the environment. Over a long span of time, as habitats changed and as more and more large animals were hunted, large animals weren't able to reproduce fast enough to match the rate at which they were being killed. Thus, the fossil record indicates that between 13,000 and 7000 B.C. (the period in which many innovations in hunting technology as well as global warming appeared), numerous species disappeared from different regions of the earth. Horses, giant bison, oxen, elephants, camels, antelopes, pigs, ground sloths, and giant rodents became extinct in North America, and the woolly mammoth, woolly rhinoceros, steppe bison, and giant elk disappeared from Europe.[8] The decline of available game, in conjunction with the slight increase in the size of populations, resulted in a food shortage.

Beginning about 7000 B.C., several societies turned to horticulture, rather than to hunting and gathering, as a means of subsistence.[9] Horticultural societies appeared slowly—in fact, so slowly that if you lived in one of the societies that experienced this transition, you probably would not have been aware of it. Over many generations, people gradually moved from gathering food to planting food and to becoming more dependent on the harvest from these cultivated plants. Yet the overall effect of this transition on the nature of social institutions was enormous—so enormous that Lenski and his colleagues refer to it as the "first great social revolution."[10]

As you recall from our discussion of social stratification, when compared with hunting and gathering societies, horticultural societies tended to have a more extensive division of labor, more social stratification, larger populations, more permanent resi-

dences, and a greater incidence of war, often over access to land.[11] Thus, if you had lived in a horticultural society, you would probably have had access to a more stable food supply and more cultural innovations than if you had lived in a hunting and gathering society. But you also would have been more likely to experience political domination and the possibility of war and slavery. Moreover, you probably would have accepted the idea that political domination and social inequality were simply part of the way that life should be. In other words, the changes in social institutions that accompanied the development of horticulture involved alterations not just in the food supply and the use of environmental resources but also in individuals' interactions, ideologies, worldviews, and daily lives.[12]

The Second Revolution: Invention of the Plow

Of course people in horticultural groups also developed their own technological advances, such as weaving techniques to make cloth and clay-shaping techniques to make pottery. The most important invention was the discovery of metallurgy (the ability to use fire to make tools and weapons out of metal), specifically the use of metallurgy to make the plow. Again, this innovation would have been virtually unnoticeable to members of these societies because it evolved over thousands of years. Further, only some horticultural societies developed this technology, and among those that did, the process of innovation often differed.[13]

But the social consequences of developing the plow were enormous. The plow, together with the innovation of harnessing animals to this new tool, enabled farmers to dramatically increase the productivity of the land—to extract much more food from the environment. The result was what Lenski and his colleagues term the "second revolution"—the emergence of agricultural or agrarian societies, which first began to appear in about 3000 B.C. Although that might seem like ancient history, this type of society remained dominant until about two centuries ago, around the time the United States became an independent country.[14]

The change from horticultural to agrarian societies appears revolutionary only in retrospect, but the development of agriculture unquestionably led to massive changes in social institutions. The availability of more food allowed populations to increase because there was more food to go around. In addition, because the plow was more efficient, fewer people needed to be involved in food production, and many different occupations appeared. Cities developed, and trade with other societies became much more common and extensive. All these changes affected the environment because as more land was tilled, the range of plant and animal species that could survive changed.

Agrarian societies were much more unequal than horticultural societies were. Along with the first cities came large political systems. Powerful governments that instituted extensive taxation, and military systems designed to preserve the rulers' power. Slavery and warfare were commonplace. Many agrarian societies built large empires by conquering nearby societies that had less well-developed technology. For instance, the Roman Empire stretched from Rome to present-day England in the north, to Spain in the west, through the Middle East, and to Northern Africa in the south; it included 70 million people.[15] If you had lived in an agrarian society, you probably would have led a dreary existence, for most of the population worked as peasants, toiling on the lord's land in return for his protection.

This economic system, in which the political elite determines how goods and services will be produced, distributed, and consumed and in which workers' labor is coerced rather than freely hired, is called a **command economy.** Clearly such an economic system would affect not only individual lives but also the pace of social change. In contrast to earlier eras, the pace of change actually slowed in agrarian societies, even though populations were larger and societies had more contact. Sociologists generally attribute this slower pace of change to the extreme social stratification associated with a command economy. Most of the population that was involved in day-to-day labor (the peasants) had little motivation to develop new inventions because the governing elite would simply have appropriated any resulting benefits. However, because these elites were so powerful and wealthy, they did little of the day-to-day work themselves and lacked the knowledge or experience to innovate. Thus the elites, who were in a position to gain by innovations, didn't have the technical expertise to develop them; and the peasants, who might have had the expertise, didn't have the motivation to do so.[16] In short, although innovations continued to crop up, the pace at which they appeared did not quicken as in earlier eras and in different types of societies, and so life varied relatively little from generation to generation.

The Third Revolution: The Steam Engine

Even though the pace of change in agrarian societies was somewhat slower than in horticultural and hunting and gathering societies, technological innovations continued to appear. Some of the most important innovations related to sailing ships—the invention of the compass and the rudder and refinements in the design of sails and ships. Eventually these alterations allowed explorers to find routes to Africa, the Middle East, and Asia and to discover the American continents.

The invention of the printing press in the fifteenth century opened up a new intellectual world to much of the population by making written materials much more accessible than they had been. The distribution of Bibles and religious literature contributed to the spread of the ideas of the Protestant Reformation throughout Europe, leading to sweeping changes in beliefs and ideologies.[17] As described in Chapter 1, Max Weber traced the development of capitalism in Europe at least partly to the ideology that characterized some Protestant groups. These beliefs also helped inspire the exploration of new lands and the search for greater scientific understanding—the basis of new inventions and improved technology.[18] As the power of merchants increased (the result of capitalism and increased trade with newly discovered and colonized lands) and a money economy began to replace barter and command economies, the pace of innovations increased. For instance, crop rotation and selective breeding of livestock made farms much more efficient and productive. Machines such as the spinning jenny and the shuttle loom made it possible to produce cloth more quickly. But by far the most important technological innovation was the steam engine, which vastly increased the energy available to produce goods. No longer was the energy of humans, animals, wind, or running water needed to power machines. Instead, the steam engine could turn gears and drive equipment much more rapidly and efficiently than previously imagined.

The changes in social institutions that were wrought by the development of the steam engine are so massive that Lenski and his colleagues have termed it the "third revolution"—the Industrial Revolution. Within a few years, machine-based industry replaced agriculture as the dominant feature of the economies of the European nations and the United States. With the increased mechanization of agriculture, many poor rural families moved to cities and provided a source of cheap labor for burgeoning factories. In the late nineteenth and early twentieth centuries, the pace of change quickened even more with the invention of the electric light, the internal combustion engine, and, after World War II, large-scale aircraft. New occupations appeared, and families changed as fertility declined, both women and men began to work outside the home, and schooling became available to virtually all children. Further, the development of new means of communication by telegraphs and telephones linked people in widely separated geographic areas in unprecedented ways. At the same time, political and social inequalities diminished. As described in our discussion of stratification in Chapter 7, industrialized societies have smaller gaps between the rich and the poor than do agrarian societies, and workers in industrialized countries have more political rights. In sum, our lives in industrialized countries today, just like those of Joel and his family, differ tremendously from those of our great-grandparents and even our grandparents. We are much more likely to live in cities, to have smaller families, to enter occupations our grandparents never heard of, to communicate instantaneously with people around the world, to travel much more widely than they ever dreamed possible, and to have greater political rights.[19]

Despite these enormous changes, some basic elements of society have endured. Just as the essential shape and content of a garden might still be evident after years of growth, so, too, many basic elements of a society remain intact. For instance, the nuclear family, which was the focal point of the organization of hunting and gathering societies, remains the central and normative form of the family in industrialized societies. Similarly, as described in Chapter 12, religious beliefs remain strong across industrialized societies, and in societies where a variety of religious organizations have been allowed to proliferate, religious involvement is high.

In addition, just as with previous social changes, the social changes associated with the Industrial Revolution have affected the environment. Most machines used in the industrial era are powered by resources extracted from the earth, such as coal, gas, and oil. Because these resources can't be replaced, they are referred to as nonrenewable resources. Even when water power is used, as in the production of electricity, huge dams must first be constructed, which involves the use of heavy machinery and, thus, nonrenewable resources. Industrialization also has prompted the emergence of an **industrial worldview,** or ideology, suggesting that the earth's resources are valuable because they provide for human needs but that human ingenuity, inventiveness, and technology can continually expand to cope with or offset any depletions in these resources. As with most elements of culture, we are seldom aware of these ideas. But they are nonmaterial elements of culture that emerged as part of the changes in social institutions that accompanied the Industrial Revolution, and the industrial worldview helps provide the underlying rationale for our everyday lives.

A Fourth Revolution? The Global Information Age

What about the age we are living in now? Did the invention of the computer herald the start of a new era? Did it mark yet another technological revolution? Will the development of so many new and different forms of transferring and using information that have developed in the past few decades—from Xerox copies, faxes, and cell phones to personal computers, e-mail, and the World Wide Web—fundamentally change the ways in which humans live, much as earlier technological innovations have? Might these technological innovations also be related to vast changes that have occurred in relationships among people throughout the world and in the political systems in many different countries?

New Technology and Individuals' Lives Many people, like Joel's father Matt, have lost their jobs in traditional industries, such as timber, and have had to find new ways to make a living. The manufacturing sector of the economy has shrunk in recent decades as the increasing efficiency and mechanization of manufacturing has resulted in greater productivity for each laborer, and a decreased demand for large numbers of workers. At the same time, industries devoted to finance, personal services, and other nonmanufacturing areas have grown. As a result, several popular and scholarly authors have suggested that we are in an "Information Age," in which our economy is based more on ideas and data than on the actual production of material goods.[20] The changes could affect not just what kinds of jobs we have but also where we work, as communications technology allows workers to "telecommute" and link with their offices and co-workers through modems, fax machines, and telephone conferencing.

This "information age" also appears to be affecting how we communicate with each other and form relationships. Jokes, some of which only make sense when seen on the computer, quickly circulate around the country via e-mail. Virtual greeting cards can replace traditional birthday and holiday greetings sent through "snail mail." Support groups provide comfort to individuals who can live literally thousands of miles from each other, and couples meet and even develop intimate relationships over the Internet, as in the hit movie "You've Got Mail."[21]

Because the possibility of Internet relationships is so new, research in this area is just beginning to appear. The few studies that are available indicate that, just as people are able to build meaningful and strong social relationships within large, seemingly impersonal urban settings, so are they able to build rewarding and supportive relationships in the new medium of cyberspace. Within these new interaction formats such as e-mail, bulletin boards, and chat rooms, people appear to engage in a wide array of activities. As one observer put it,

> People in virtual communities use words on screens to exchange pleasantries and argue, engage in intellectual discourse, conduct commerce, exchange knowledge, share emotional support, make plans, brainstorm, gossip, feud, fall in love, find friends and lose them, play games, flirt, create a little high art, a lot of idle talk. People in virtual communities do just about everything people do in real life . . . you can't kiss anybody and nobody can punch you in the nose, but a lot can happen within those boundaries.[22]

It appears that on-line relationships are another *frame* of interaction, as described in Chapter 5, and we can understand these new interactions in the same way that we understand other types of relationships. When we communicate on line we play social roles and engage in impression management, as role theory and Goffman's dramaturgical approach would predict. Our responses to others are guided by our interpretations of what others have communicated to us, as symbolic interaction theory suggests. And our decisions regarding what we will say are based on balancing relative costs and benefits, as predicted by exchange theory. In addition, in line with the work of conversational analysts, we can see that our interactions are guided by norms and rules, such as turn-taking and the order in which topics occur. Even though we don't see others when we send or receive e-mail, symbols called "emoticons" have developed to portray emotions that can generally be seen only through our nonverbal communications (see Figure 5-2).

Even though on-line communications share many characteristics of in-person communications, they are obviously different. Characteristics that affect interpersonal relationships, such as age, perceptions of physical attractiveness and personality traits such as shyness, are not apparent to the participants. Researchers suggest that electronic communication also allows individuals to get to know each other more quickly, partly because they are not influenced by these other characteristics. In addition, computer relationships allow people to communicate over great distances and allow those who find it difficult to leave home, perhaps because of illness, disabilities, or other obligations, to interact with others.[23]

It will, of course, be several years before sociologists fully understand how the vast array of new technology will affect individuals' lives. One area that is receiving a great deal of attention, however, is how new communication technologies cross societal borders. Some scholars and authors refer to the present time as a "Global Age," whereby all areas of the world are becoming interdependent and linked with one another in a process known as **globalization.**[24]

Globalization and Changing Political and Economic Systems   As discussed in Chapter 14, the economy is now global in scope with multinational corporations and economic transactions spanning the world. Each day we use products that might have been made on any continent. You might work for firms that are based in another part of the world. By just pressing a few buttons on a phone you can talk to someone in Australia or China or, with a few strokes on a computer, send an e-mail message to someone in Iceland, Russia, or even Antarctica. Cable television stations, such as CNN, are seen throughout the world and instantly transmit news from one part of the globe to another. Partly because of this global communications system we realize how environmental change in one part of the world affects people in others. "El Nino" and its counterpart, "La Nina," bring excessive rain or drought, heat or cold, to all parts of the world in a complex but interrelated pattern.[25] This global communications system also allows political news to travel almost instantaneously throughout the world and might be connected to the changes in governments and political systems that have swept the world in recent years.

One of the most dramatic changes in the last few decades has been the collapse of dictatorships in countries throughout the globe. In the 1970s, dictators in Portugal, Greece, and Spain were replaced with democratic governments, and in the 1980s similar fates befell dictators throughout Latin America and in the Philippines. Most recently, communist governments throughout Eastern Europe and the former U.S.S.R have collapsed. In the beginning of 1989, twenty-three countries could be defined as having communist governments, but by 1994 only five (China, Cuba, Laos, North Korea, and Vietnam) could.[26] This incredible change largely took sociologists by surprise.[27] Few social scientists accurately predicted these massive changes, and researchers are still sifting through data trying to understand why these changes occurred when they did and how they are related to other aspects of social change.

So far these scholars have concluded that the downfall of communist governments reflects a change in ideology and beliefs, or the nonmaterial culture of the society, among leaders and activists, as well as many ordinary citizens. To use the terminology developed by Max Weber and introduced in Chapter 6, these changes represented a crisis or challenge to the *legitimation of authority,* the mechanism of social control within the society. The prevailing form of government was no longer seen as the legitimate legal authority. In the once–communist countries these changes could be seen in several ways. Throughout the countries where communist governments fell, other political parties arose, the communist party either changed its name or split into factions, competitive elections were held, and new or highly modified constitutions were adopted.[28]

Some scholars speculate that these trends represent a general, global transition that involves the gradual movement of all industrialized societies to more democratic and inclusive political structures.[29] Much as people and nations around the world support the ideological tenets that underlie common schooling, as discussed in Chapter 15, they also share norms and ideologies that support notions of equality, democracy, free expression of ideas, and development of more humane societies.[30]

Most likely, if the transition toward greater democracy is to occur, the changes will not always be simple, easy, or quick. Studies conducted in Russia after the collapse of the Soviet Union have found declining incomes, rising unemployment, and greater income inequality.[31] Soviet block countries throughout Eastern Europe have also experienced economic difficulties and, in many cases, intense ethnic hostilities and even warfare.

Speculations about the future political course of nations throughout the world are, of course, hypotheses that can only be tested with the passing of time. Most important, deciding whether or not the present era involves a large-scale technological revolution and understanding how such a revolution will alter social institutions will depend on the hindsight of many years, for the picture that reveals such large-scale social changes can only be seen from a very broad macrolevel perspective. Yet such a broad historical view is not the only way to analyze social change. In fact, such a perspective obscures a detailed view of the social actions by which technological changes are adopted and affect social institutions. We turn now to a much more detailed analysis of the adoption of one technological innovation and how the decisions of individuals and groups regarding this innovation affected social life.

FEATURED RESEARCH STUDY

THE TELEPHONE, SOCIAL ACTION, AND SOCIAL CHANGE—THE WORK OF CLAUDE FISCHER

Until the late 1800s, if you wanted to communicate with someone, you had to physically go see that person, unless you sent a letter through the post office or a personal messenger or used the new technology of telegraphy. In 1876 Alexander Graham Bell's invention of the telephone removed this limitation from our lives. With the phone, we could almost instantly communicate with people on the other side of town, on the other side of the country, and even on the other side of the world! Today, with cordless phones, cell phones, voice mail, and live video and audio transmissions, phone technology has expanded into realms that Bell probably never even dreamed of. We can now talk with people who are almost anywhere from virtually any place in the world.

The telephone is just one of a number of "space-transcending technologies," inventions that changed peoples' ability to travel and speak across space, that were generally developed between 1850 and 1950. Trains, street cars, bicycles, and then automobiles and planes allowed people to more easily travel long distances; the telegraph and telephone allowed people to communicate with others in far-away places without having to travel. Although even the idea of some of these inventions seemed fanciful in 1850, a century later people just accepted them as part of daily life. At the same time, these innovations changed the nature of actual and potential social networks. No longer were people likely to converse and visit with only people who lived within a distance of a walk or horse ride. Instead, people could talk with others literally around the world and could travel, in a span of a few hours, to places that would previously have taken days to reach. Sociologist Claude Fischer wanted to know how the development of these space-transcending technologies affected human lives. Why and how did people use this new technology? How did it alter their daily lives? And how did people's use of the telephone change culture and social structure?[32] Though we can speculate about how the current space-transcending technologies of e-mail and other forms of virtual communication are changing our lives, Fischer was able to use the vantage point of history to examine the effects of an earlier form of these technologies.

Sociologists often become interested in rather general issues of research and pursue *research agendas,* a series of studies all directed toward understanding more about a general phenomenon. Fischer's study of the telephone was, in many ways, an extension of his earlier work on how urbanization affects social networks and relationships in communities, described in Chapter 18. As part of his general research agenda focused on social relationships and social change, Fischer wanted to learn more about how the new technology of the telephone affected individuals' relationships with each other, as well as culture as a whole.

Fischer was especially interested in two elements related to social change. The first involved understanding how social actions produce social change. Many analyses of social change use what Fischer calls a "billiard-ball model," suggesting that technological developments roll in from outside and then impact various elements of a society, which in turn then impact other elements.[33] For instance, the introduction of the automobile reduced the demand for horses. This, in turn, reduced the demand for feed grain, which then allowed farmers to grow more food for humans, which then reduced the price of food, and so on. Although this very broad, macrolevel analysis helps us understand the long-term outcome of social change, it doesn't help us understand much about how this change came about—how individual members of a society were involved in the adoption of new technologies.

Fischer's second interest involved the ways in which social change affects individuals' lives and the society as a whole. Predictions about the effects of new technologies seem to appear hand-in-hand with innovations. A century ago, it was predicted that the telephone, as well as the automobile, would make "one neighborhood of the whole country" and that use of the telephone would rid the country of regional speech dialects. It was hoped that the telephone would make many daily chores simpler, such as allowing women to shop over the phone. It was also feared that the telephone could lead to a "rootlessness" or "placelessness," by freeing people from geographical constraints.[34] Such speculations are similar to those made today about innovations such as cell phones and the World Wide Web, and, because they are speculations, we do not know the extent to which they can be supported by empirical evidence. Fischer wanted to see to what extent predictions might resemble reality by looking back in time.

To explore these issues, Fischer used *historical research,* focusing on the United States from the invention of the telephone in the late nineteenth century until 1940, when telephones were in wide use throughout the country. In this process he used a variety of sources and techniques: He closely examined publications, reports, and correspondence of the telephone industry and government agencies that were involved with the industry. He examined statistics relating to the use of the telephone in the various states. He conducted extensive case studies of three different communities and changes in their use of the telephone from 1890 to 1940 and interviewed elderly people who had lived in these towns during that period.[35] Through these efforts he hoped to get beyond the simple "billiard ball" explanation of social change and understand more about the social actions that are involved in people using technological innovations and how these innovations come to affect social life.

Choosing to Use the Telephone

Like so many technological innovations, Bell's invention of the telephone occurred as he was trying to improve on an earlier innovation, the telegraph. For the first year or so after patenting his invention, Bell and his associate Thomas Watson gave demonstrations of the new device around the country, entertaining people at exhibitions, fairs, and town meetings. Although these demonstrations awed the observers and certainly created interest in the new device, the major organizational impetus to change was the development of the Bell Telephone Company in 1877. The company grew rapidly, setting up its first switchboard in early January, 1878. Two years later the company had 60,000 subscribers and Americans' access to telephones continued to rise.[36] By 1920, there were more than 9 million telephones in residences and more than 4 million in businesses,[37] and the telephone appeared to be simply accepted as a part of middle-class life.

Reviewing all the information he gathered, Fischer gradually concluded that the decision to have a telephone reflected both what people felt they needed as well as what they wanted—their tastes and preferences. For instance, some of the first phones were in local businesses, such as lumberyards, drugstores, and newspaper offices, and in doctors' homes and offices. Fischer concluded that these people generally wanted phones for business purposes, and, in fact, drugstores often advertised the use of their phones as a way to attract customers. As phones became more common in homes, their social value became more noticeable. From studying which households in communities were more likely to have phones, Fischer found that those with younger people and more women were more likely than households with older people and fewer women to have phones. In other words, some people, especially women and young people, were more likely to *want* to have a telephone than others were . This desire helped fuel the diffusion of this new technological innovation.

Of course, not everyone who might want to have a telephone could have one. The most obvious restriction was income, for telephones were not cheap. Thus, the first people in communities to have residential telephones were the wealthiest, and those who were least likely to have phones were the poorest. Whether or not people could have a phone was also influenced by the decisions of telephone companies. A community could obtain phone service only if a company believed that it would be profitable to offer phone service, and individuals were subject to the type of service and rates that the company charged.

An indication, however, of how much some people wanted the new technology of telephones was the development in many rural communities of independent telephone cooperatives. When the commercial telephone companies wouldn't take phone lines into rural areas because they believed that the effort would not be profitable, many smaller communities took the matter into their own hands and developed telephone cooperatives. These were nonprofit organizations established under the leadership of small town merchants, doctors, or farmers. The so-called "farmer lines" usually involved only a few dozen farmers at most, whose homes were connected by wire (sometimes barbed wire fencing) and had switchboards in a home or local business. In 1902, the U.S. Census counted about 6000 mutual telephone companies throughout all parts of the nation; by 1907 there were 18,000 such companies![38]

A Technology of Sociability

As more and more people got telephones in their homes, concerns about the ultimate impact of these changes were openly discussed. For instance, in 1926 the Knights of Columbus (a fraternal organization associated with the Catholic Church) proposed that its members discuss the influence of modern inventions in their regular group meetings. Among the discussion questions were the following: "Does the telephone make men more active or more lazy?" and "Does the telephone break up home life and the old practice of visiting friends?"[39] In many ways, these concerns echoed the thoughts and worries of early sociologists, such as Ferdinand Tönnies, Georg Simmel, and Louis Wirth. As described in Chapter 18, each of these scholars worried that changes associated with modern life might lead to social disintegration, a loss of community and of meaningful relationships with other people.

As we saw in our discussion of urbanization, these worries have been largely unsupported by the empirical evidence. People continue to develop and maintain close social ties no matter where they may live. Similarly, the introduction of new technology such as the telephone appears to have led to very few changes in basic social structures. Most institutional patterns and social norms appear to be quite impervious to technological innovations. As Fischer put it, "basic social patterns are not easily altered by new technologies, . . . they are resilient even to widespread innovations."[40]

In fact, based on his analysis of the historical data, Fischer concluded that the telephone helped to reinforce and deepen existing social ties. Even though it was technically possible for Americans to call people in far away places and create new extensions to their social networks, this didn't happen. Instead, the telephone helped people more easily maintain their current relationships.

> Americans—notably women—used the telephone to chat more often with neighbors, friends, and relatives; to save a walk when a call might do; to stay in touch more easily with people who lived an inconvenient distance away. The telephone resulted in a reinforcement, a deepening, a widening, of existing lifestyles. . . .[41]

Before acquiring a telephone, you might be able to visit with friends and kin once a week or so, but with a telephone you could call and chat several times a week.

Even though Fischer concluded that the telephone did not undermine local community ties, he did come to believe that it had a role in increasing "privatism." He used this term to refer to an increasing valuation of and participation in private social networks, such as the primary groups of family and close friends, rather than larger public communities,

SOCIOLOGISTS AT WORK

Claude Fischer

We, individually and as a society, make decisions about how technology will be used.

Claude Fischer is a professor of sociology at the University of California at Berkeley.

Where did you grow up and go to school? I was born in Paris, France, and grew up in Paterson, New Jersey (also hometown of Morris Janowitz) but graduated from high school in Los Angeles. I was an undergrad at UCLA (graduated 1968, in Sociology) and got my Ph.D. in Sociology at Harvard (1972).

Why did you become a sociologist? I was always interested in social studies, especially history. As a high school junior, I was part of a summer-at-college program sponsored by the National Science Foundation. I went to Arizona State in the summer of 1963 and took a sociology course from John Kunkel. That did it. I entered college with the goal of being a sociologist.

How did you get interested in studying social change? One of my key interests has been social change in this sense: A major concern of sociologists has been to understand the major changes in western society we call "modernization." It is embedded in most sociological theory. It suggests looking closely at the key changes of the nineteenth and early twentieth centuries.

What has surprised you most about the results you have obtained in your research? There are a few modest surprises from the America Calling project. One is the degree and speed with which telephones

faded into the woodwork of American society, becoming noncontroversial and almost unnoticed. Another is the degree to which people bent the telephone to their interests (rather than vice-versa).

What has been the reaction to your research among your colleagues? The book has gained a good reputation among historians of technology. (It won the Society for the History of Technology's annual book prize, the Dexter Award.) It has been well-reviewed by historians, as well as sociologists, which I appreciated. For most sociologists, of course, the topic — telephones 50-100 years ago — seems far outside their interests and so it is hardly a central study in the discipline. But those who study technology have been positive. The book has been translated into Italian and a Japanese translation is scheduled.

How can your research help students understand events in their own lives? Perhaps the research can help students think more critically about technology—in particular, that technology doesn't drive us, but that we, individually and as a society, make decisions about how technology will be used. More generally, a good social science study should make students more analytical about the world around them: Don't take things for granted, question, see complications, ask for evidence, and examine the evidence.

such as the secondary groups that might include neighborhood gatherings and civic groups. Fischer found little evidence that the telephone led to new social connections; instead people tended to visit more frequently with people they already knew. As a result, he concluded that society might have become more "compartmentalized," with people becoming more active in primary group relationships rather than in the wider, public realm of secondary relationships.[42] (To learn more about Fischer and his work, see Box 19-1.)

CRITICAL-THINKING QUESTIONS

1. How might Fischer's research be replicated with more recent technological innovations such as the Internet, e-mail, and the World Wide Web?

2. What hypotheses would you develop regarding the ways in which these technologies have affected social relationships? Why?

3. What sociological methods could you use to test these hypotheses?

BOX
19-2

APPLYING SOCIOLOGY
TO SOCIAL ISSUES

*How Can Change
Efforts Be
Effective?*

The residents of the least developed countries—those at the bottom of the global stratification mountain—are much more likely to suffer severe health problems than residents of other countries are. As we discussed in Chapters 16 and 17, people in the least developed countries have much shorter life spans, babies are more likely to die during infancy, mothers are more likely to die during childbirth, children's physical growth and intellectual development are often stunted because of malnutrition, and inadequate diets and communicable diseases often lead to permanent disabilities. For example, the United Nations estimates that the lack of adequate iodine places millions of children and adults at risk of a variety of ailments ranging from goiter to brain damage and severe retardation.[58]

Many of these problems can be corrected by adequate diets, basic health care services, such as medical care for mothers and children, including immunizations, and sanitary water supplies and sewage disposal systems. For instance, iodine can be added to the salt supply (as it is in most industrialized societies). Vitamins can be administered to children and mothers. Vaccines are available to combat diseases such as polio, measles, and tetanus, which still ravage the least devel-

oped countries, and simple antibiotic treatments can effectively halt the onslaught of pneumonia. The United Nations, under the leadership of its Children's Fund (UNICEF), has spearheaded efforts to improve children's health around the world. The story of these attempts illustrates a number of the sociological concepts introduced in this book and shows how concerted actions can help solve even the most serious and difficult social problems.

In 1990, the United Nations sponsored a "World Summit for Children," which was attended by representatives of more than 150 governments, including 71 heads of states. The representatives formally adopted a series of goals to be attained by their countries by the year 2000. Among these goals were "a one-third reduction in child deaths, a halving of child malnutrition, immunization levels of 90 percent, control of the major childhood diseases, the eradication of polio, the elimination of micronutrient deficiencies [such as iodine deficiency], . . . [and] the

provision of clean water and safe sanitation to all communities."[59] Intermediate goals were set for the year 1995.

The participating nations have met with enormous success. Although some countries, particularly those in sub-Saharan Africa, have had more problems in meeting the goals than others have, a majority of the countries appear to be meeting a majority of the goals. In describing the results, the United Nations reports,

> Such progress means that approximately 2.5 million fewer children died in 1996 than in 1990. It also means that tens of millions will be spared the insidious sabotage wrought on their development by malnutrition. And it means that at least three-quarters of a million fewer children each year will be disabled, blinded, crippled, or mentally retarded.[60]

The success of the United Nations and the participating countries can provide lessons for all groups and individuals interested in social change. How did they manage to turn these words into deeds? How did they mobilize thousands of organizations and millions of people into social actions that could have such far-reaching results?

Officials of the United Nations point to several important strate-

In general, Fischer's research helps us see how social actions produce changing social structures. When we look at social change with a long-range macrolevel perspective, we can see, in retrospect, how certain technological innovations produced revolutionary changes in a society. Yet, if we move closer and examine particular innovations and historical periods much more carefully, we can see how social change results from many different social actions. People choose whether or not to adopt innovations and how to use them; thus, the direction and speed of cultural diffusion is determined by human decisions. People's choices and decisions might not be anticipated or even promoted by those who devel-

BOX
19-2

APPLYING SOCIOLOGY
TO SOCIAL ISSUES

Continued

gies. First, they broke down the broad goals that were developed in 1990 into doable, and measurable, propositions. They carefully evaluated the broader goals to find the aspects that were most achievable with the available resources and then targeted these activities. They also selected goals for which progress could be monitored and emphasized the importance of social statistics, such as those we have used throughout this book, in letting people know how well the goals are being met.[61]

Second, UN officials mobilized political commitment at the highest levels in each country, as well as at lower levels of political administration. The heads of UNICEF and the World Health Organization (WHO) personally met with the majority of the presidents and prime ministers of developing countries, seeking their personal support for the goals. UN officials followed up by maintaining communication with these heads of state and by contacting leaders at all levels of government. In several countries, including China, India, and several Latin American nations, plans for improvement were developed at the national level and again within provinces or regions. In other words, UN officials mobilized leaders within each country and built strong networks that could keep this mobilization active.[62]

Third, they mobilized a broad range of resources. This process has been aided by rising literacy in many developing and least developed countries and the radios and television sets that are found throughout the world. Thus printed materials and media broadcasts have been used to spread health messages and advertise immunization services. Paraprofessionals and voluntary organizations have been enlisted to help provide health care services and spread information. In addition, the necessary technology and knowledge have been simplified so that people with all levels of education can use them.

An unfortunate footnote to this campaign is that the United States has not made similar progress in promoting the health of its children, even though it is one of the richest countries in the world. For instance, less than three-quarters of all one-year-olds in the United States are immunized against polio. Although the rate of infant mortality in the United States is not as high as that in many less developed countries, it is, as we saw in Chapter 16, almost twice as high as the rates in Japan, Singapore, Sweden, and

Finland.[63] Moreover, the infant mortality rate for African-American children is more than twice the rate for Euro-American children and resembles the rate found in many Latin American countries.[64]

Notably the United States is one of very few countries in the world that does not guarantee medical care for children, even though it does do so for senior citizens. In addition, African Americans and Hispanic Americans are far less likely than Euro-Americans are to have medical insurance, either through a private plan or a government plan such as Medicaid. It is estimated that one-fifth of all African Americans and one-third of all Hispanic Americans have no medical coverage.[65]

CRITICAL-THINKING QUESTIONS

1. What might the experiences of other countries and your knowledge of social change suggest about the possibility of improving the health of all Americans?

2. What would need to happen for the United States to reach the levels of health enjoyed in other highly industrialized countries?

3. What do you think the prospects are for such a far-reaching social change? Why?

oped the innovation. For instance, the Bell company did not expect farmers to be one of their prime users, nor did they expect social calls, especially by women, to be such an important aspect of telephone usage.

Just as snapshots of a garden taken over a period of years can show large-scale changes in the size and placement of plants, if we were to look at a garden on a daily basis and much more closely we could see how these larger changes result from an accumulation of many smaller alterations. Box 19-2 describes how such social actions have brought about social changes in some of the world's poorest nations, and we now turn to a more detailed discussion of such nations.

SOCIAL CHANGE AND TECHNOLOGY IN A GLOBAL PERSPECTIVE

Gardens and natural areas around the world don't look alike, and neither do societies. It is virtually impossible to grow tropical flowers in an arid desert, and cacti and other plants that are suited to the desert would soon perish in a tropical rainforest. Similarly, even though people throughout the world have the same biological and genetic heritage, the same social needs and desires, and the same basic social institutions, societies have changed and developed in different ways.

As we noted in our discussion of global stratification in Chapter 14, there is an incredible amount of variation in the extent to which societies around the world have industrialized and in the extent to which they have advanced communication. Some societies, such as those in Europe, North America, Oceania, and Japan, are very highly industrialized, have very high gross national products, and rely almost exclusively on energy sources developed during the past two centuries, such as hydroelectric power, coal, petroleum, gas, and nuclear energy. Other societies, especially in Asia and Africa, are very poor and rely much more on traditional energy sources, such as humans, animals, wind, water, and wood fuel and are very poor. These societies are more characteristic of agrarian societies than of industrialized societies and their citizens are far less likely to have ready access to modern technological innovations.

Map 19-1 illustrates some of these differences by portraying the number of telephones in countries around the world. In highly developed countries such as the United States, Sweden, Denmark, and Switzerland there are more than 60 main telephone lines for every 100 people. In India there is only one. Similar differences appear with other technologies. For instance, in the United States there are almost 800 television sets and more than 2,000 radios for every 1000 people, but in Pakistan there are only 22 televisions and 88 radios for every 1000 people.[43] Differences in access to the most advanced communications technology such as e-mail and the Internet are even more extreme, both because the very poor countries cannot afford computer equipment and because many poor countries do not have phone lines that are capable of handling the information load.[44]

Even though people in the poorer countries of the world are far less likely to have as many radios, televisions, and telephones in their homes as Americans, Canadians, and Western Europeans, they are still exposed to western media. With the development of extensive telecommunications systems as part of the process of globalization, no area of the world is to-

How might this steel factory change the lives of these three women? How might it affect their families, their chances of education, their role in the economy, their political world, and their religious lives?

tally isolated. There are now some 2 billion radios and 900 million television sets around the world.[45] With satellites, even the inhabitants of remote areas of South America, Asia, and Africa can pick up the same television signals that Europeans and North Americans receive. Although there might be only one television or radio in each village in many areas of the world, compared with several in each house in industrialized societies, people in the most remote and poverty stricken areas of the world have seen how the far richer Americans and Europeans, who produce most of the media, live. Even a half century ago many people would not have been aware of how different their economic situation was from that of other people in the world. But with globalization such ignorance is no longer possible. Reference groups may no longer be just people in one's own community but those portrayed on media produced literally on the other side of the world. Television shows such as "Bay Watch" and "Dallas" are shown around the globe, and as a result people who might once have been unaware of their poverty relative to the rest of the world now know that other ways of life exist.

Many sociologists, citizens, and government officials have been concerned with the possibilities for social change in less developed societies such as

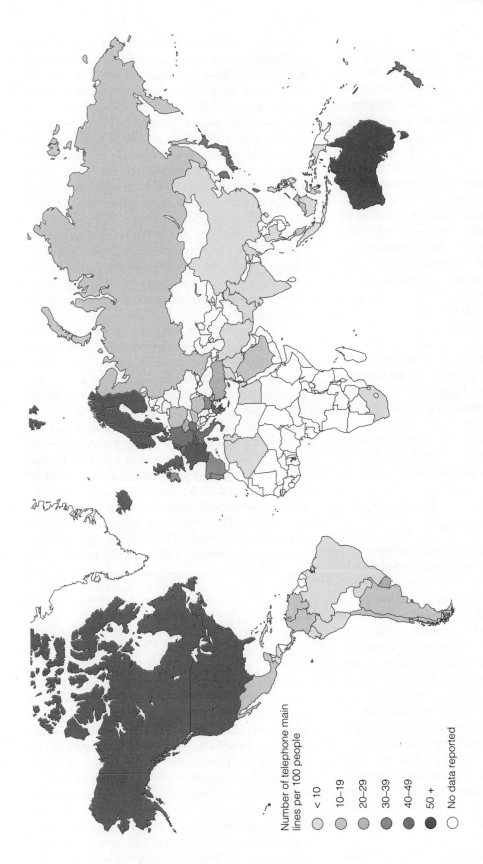

MAP 19-1 *Number of Telephone Main Lines per 100 people*

Source: U.S. Bureau of the Census, 1998c, p. 840.

India and Pakistan. The poverty and inequality within less developed societies is often excruciating, with many people not even having enough food to sustain a healthy life. The experiences of the industrialized world suggest that poverty and inequality decrease as industrialization progresses. But can the processes of social change that occurred in the industrialized countries also be effective in the less developed nations? Can the technology used to modernize industrialized nations also be used to help the less developed nations? How will such attempts affect the environment and social institutions of these developing countries? Much as gardeners might ponder the best way to landscape a garden, sociologists and government officials have asked, and sometimes fiercely argued about, what social policies could be implemented to help the less developed countries combat poverty and inequality.

Two different proposals have been considered. One suggests that the solution involves a process of modernization and industrialization similar to that experienced by the already industrialized nations. The other, which has gained more advocates in recent years, involves programs of rural development and land reform that take into account the ways in which the less developed nations differ from their more industrialized counterparts and the relationship between changing social institutions and the environment.

Modernization

For many years many sociologists and economists accepted the tenets of **modernization theory,** which suggested that the majority of less developed countries would eventually industrialize in the manner of countries such as England and the United States.[46] Government organizations, such as the United Nations, based many of their policies and programs on this theory. Two ingredients were believed to be crucial to this process: (1) the development of cultural beliefs and values that supported the growth of capitalism, such as the "Protestant ethic" identified by Weber, and (2) the development of the reliable physical infrastructure of education, transportation, and communication needed to support the growth of industry. In other words, modernization theory suggests that countries will industrialize as they acquire the necessary physical resources and cultural attributes—the technology and the social institutions—that support industrialization.[47]

Several countries have followed the modernization model, especially those in Asia. For example, South Korea, Taiwan, Hong Kong, and Singapore have industrialized rapidly and now have gross national products similar to those of Australia, New Zealand, and some members of the European Community.[48] Yet modernization has not occurred in much of the world, especially the least developed countries that are excluded from the world economy, such as many nations in Asia and sub-Saharan Africa.[49] Many sociologists now assert that, much as people who belong to the underclass find it difficult to experience social mobility, many of the least developed nations find it difficult to move up the global stratification ladder. *World systems theory,* an approach that was inspired by the work of Karl Marx and introduced in Chapter 14, suggests that highly industrialized nations are wealthy largely because they have been able to exploit much poorer societies. These poorer countries face an arduous uphill climb in combating poverty and other national problems. Given their extreme poverty, they may not be able to follow the modernization route that other nations have taken.

As described in the previous chapter, urbanization is occurring in many least developed nations at a much more rapid pace than that experienced by countries such as the United States but without an accompanying rapid increase in industrialization. Instead, urbanization appears to be a response to the extreme population and environmental pressures in rural areas. The population growth in urban areas is so intense that most governments cannot provide enough schools, housing, transportation, and jobs to keep up with the demand. As a result, economists and sociologists who work with international organizations such as the United Nations are now suggesting that social change—and especially a break in the cycle of population growth, poverty, and environmental degradation—might best be accomplished by systems of land reform and the development of rural areas.[50]

Rural Development, Land Reform, and Sustainable Development

As described in earlier chapters (see especially Maps 7-1 and 7-2), many less developed nations have extremely strong internal systems of social stratification and a high degree of inequality, often related to their history of colonization. Much of the land and raw materials are owned and controlled by a very small segment of the population. For instance, in Latin America the Spanish conquistadors and their descendants settled and farmed the best land. The native Indians were pushed into other areas, and African slaves were imported to provide labor. Later, when slavery was outlawed, a system of peonage emerged whereby the landlords gave laborers loans to buy necessities such as food and clothing but never paid them sufficient wages to pay off these loans. The result, in Latin America and other less developed nations, is a very small, wealthy upper class and a large, very poor group of peasants, most of whom own no land of their own—a system not unlike that found in the agrarian societies of medieval Europe.

Much of the food that is produced in less developed nations, such as tea, coffee, sugar, and various fruits and vegetables, is exported to the industrialized nations. Thus many of these countries must import food to feed their own people, while exporting food to the industrialized nations. However, the very poorest people in the least developed countries frequently cannot afford to buy this imported food.[51] As James Grant, who was once the executive director of the United Nations Children's Fund (UNICEF), described the problem,

> In Latin America today, fewer than 10% of landowners own almost 90% of the land. In the Philippines, the proportion of rural workers who are landless has risen from 10% in the 1950s to 50% in the 1990s. . . . In Africa . . . it is increasingly the case that most productive lands are devoted to export agriculture while the lands of the poor majority are of lesser quality, receive less investment, and are rapidly becoming degraded and depleted.[52]

Although land reform once was advocated only by revolutionaries, today officials of government bodies and international agencies such as the United Nations, as well as noted mainstream economists such as John Kenneth Galbraith, promote it as a means of combating the problems of intense poverty and environmental degradation in much of the world.[53] **Land reform** involves breaking up and redistributing very large parcels of land, many of which are now owned by extremely wealthy landowners or multinational corporations, to landless peasants. When peasants own their own land, they can grow food for their own consumption and for direct sale. Thus land reform would give them the opportunity to escape the command economy and enter a free economy. The hope is that this transition could help reduce the intense poverty in rural areas.

Experts also suggest that the very poorest countries would benefit from programs of **rural development,** industrialization in rural areas. Instead of linking industrialization with urbanization, they suggest that small-scale industries can be established in rural areas and employ the large populations already living in these areas—thus helping to stem the problems associated with rapid urbanization and to combat rural poverty.[54]

Although scholars and government officials hope that land reform and rural development will directly address the issue of poverty in the least developed and developing countries, another issue that affects all areas of the world is environmental degradation. In Chapter 17 we examined environmental issues that face developed countries, such as heavy patterns of consumption and pollution, and those that face less developed countries, such as deforestation and desertification. In addition, many environmental issues, such as ozone layer depletion, pollution, and endangered species, potentially affect people worldwide.[55] Many scholars and political commentators suggest that the *industrial worldview,* the belief that the earth's resources are valuable because they provide for human's needs, has provided the ideological basis for behaviors that contribute to these environmental problems. For instance, the extensive cutting of timber and the use of only one species in replanting the forests near Joel's home were justified because the logs formed the basis of the timber industry. Yet, over time, these practices depleted the forest resources and promoted insect-borne diseases.

In recent years, largely because of the development of such environmental problems, another worldview has been promulgated that takes a different view of the relationship between technology and the environment. This alternative perspective, which can be called an **ecological worldview,** suggests that new technology can create as many problems as it solves, that the earth's resources are limited, that the world's population has reached the earth's carrying capacity, and that population growth must be curbed. Instead of using nonreplaceable resources from the earth, the ecological world view encourages policies of **sustainable development,** economic development that does not deplete or degrade natural resources. In short, whereas the industrial worldview suggests that humans can control nature, the ecological worldview suggests that humans must live within nature as part of a total ecosystem.[56] Business people and economists who advocate this approach suggest that countries throughout the world could move to a full adoption of this mode of development within 40 to 50 years, the span of time in which the economies of most of the industrialized nations today were rapidly transformed.[57]

The emergence of the ecological worldview shows how social change is still occurring. This worldview simply didn't appear out of thin air. Instead, it has been widely promoted by activists in the environmental movement, a social movement that is attempting to improve the quality of the world's air, water, and land. Sociologists have long been interested in social movements because, unlike the long-term patterns of social change that were described earlier in this chapter, social movements involve concerted social actions that are explicitly designed to change the social structure. Much as gardeners and landscapers can alter the appearance of a garden, people can work together to change the nature of society. In the next, and final, chapter of this book we examine social movements and other forms of collective behavior.

SUMMARY

Social change reflects the way societies alter over the long span of human history. By studying social change, sociologists hope to make the world a better place. Key chapter topics include the following:

- Social change involves alterations in social institutions, including statuses, roles, organizations, norms, and ideologies; technological innovations; and the environment and the way that individuals interact with the environment.

- In general, the pace of social change has increased throughout human history because one technological innovation typically spurs several others.

- The first great social revolution, or massive social change, was spurred by the domestication of plants and animals, which produced horticultural societies. The second involved the development of the plow and agrarian societies; the third was the development of industrialized societies, prompted by the invention of the steam engine. Some suggest that we might be in the midst of another large scale change with the development of global communications technology.

- Claude Fischer's historical analysis of the effect of the telephone on social life showed how the

adoption of technological innovations depends on individuals' perceptions of their needs and desires and how the telephone actually reinforced and deepened existing social ties.

- Societies around the world differ greatly in the number of modern technological innovations such as telephones, televisions, and radios that are available to their citizens. Globalization and the development of worldwide telecommunications systems, however, have allowed people throughout the world to be exposed to western media.

- Two proposals for promoting social change in less developed countries have been promulgated. Modernization theory suggests that less developed countries can follow the model of already industrialized countries by developing the needed physical infrastructure and beliefs and values that support the development of capitalism. This model appears to describe the changes that have occurred in South Korea, Taiwan, Hong Kong, and Singapore. The second proposal involves land reform and rural development and is often seen as more appropriate for the extremely poor countries that have strong internal systems of social stratification, a high degree of inequality, and histories of colonization.

KEY TERMS

command economy 515

cultural innovations 510

ecological worldview 527

globalization 517

industrial worldview 516

land reform 527

modernization theory 526

rural development 527

social change 510

sustainable development 527

technological innovations 510

INTERNET ASSIGNMENTS

1. Locate information on the usage of the Internet and the World Wide Web over time. Recall the finding discussed in this chapter about how the pace of social change tends to increase as new innovations appear. Does the rate of usage of the Internet and World Wide Web support this finding? Why or why not?

2. Recall how technological changes often are related to changes in social institutions and the environment. What impact might the development of the Internet

and World Wide Web have on social institutions? On the environment? Cite specific examples of these impacts, using observations from the Internet.

3. Using a search engine, locate Web sites that discuss issues related to sustainable development. Try to find two sites that focus on industrialized societies and two that focus on developing societies. Compare and contrast the issues discussed in these sites. Based on what you have read in this chapter, why do you think these differences appear?

InfoTrac College Edition: Readings

Article A53436506 / Badaracco, Claire Hoertz. 1998. The transparent corporation and organized community. *Public Relations Review, 24,* 265–273. This article is part of a special issue on "Technology and the Corporate Citizen."

Article A18730579 / Duchin, Faye. 1996. Population change, lifestyle, and technology: How much difference can they make? *Population and Development Review, 22,* 321–333.(10). Duchin discusses the relationship of technological develop-ment, affluent lifestyles, population growth, and environmental issues.

Article A20151724 / Minnegal, Monica, and Peter D. Dwyer. 1997. Women, pigs, God and evolution: Social and economic change among Kubo people of Papua New Guinea. *Oceania, 68,* 47–61. Minnegal and Dwyer's description of recent changes in a tra-ditional village in the interior of New Guinea illus-trates how changes in technology, the environment, and social institutions are interrelated.

Further Reading

Suggestions for additional reading can be found in *Classic Readings in Sociology,* bundled with this book. If you purchased a used copy of this book that does not include this custom-published reader, go to www.sociology.wadsworth.com for ordering information.

Recommended Sources

Bellamy, Carol. 1998. *The State of the World's Children 1998.* New York: Oxford University Press. Written by the executive director of UNICEF and reissued each year with complete, readable updates regarding the status of children throughout the world. This series often includes reports of change efforts direct-ed at improving the health and status of children.

Chirot, Daniel. 1994. *How Societies Change.* Thousand Oaks, Calif.: Pine Forge. Chirot's short, readable book describes long-term changes in human societies.

Couch, Carl J. 1996. *Information Technologies and Social Orders.* New York: Aldine. Edited and with an introduction by David R. Maines and Shing-Ling Chen, Couch describes the ways in which means of communication have changed over the centuries.

Dertouzos, Michael L. 1997. *What Will Be: How the New World of Information Will Change our Lives.* New York: HarperEdge. Written by one of the founders of the Internet, this book describes ways in which computer technology could change our lives within the next few decades.

Diamond, Jared. 1997. *Guns, Germs, and Steel: The Fates of Human Societies.* New York: Norton; and Landes, David S. 1998. *The Wealth and Poverty of Nations: Why Some are So Rich and Some So Poor.* New York: Norton . Both of these books take a very broad view of history and explore why some nations have become so much richer and more powerful than others have.

Farley, Reynolds. 1996. *The New American Reality: Who We Are, How We Got Here, Where We Are Going.* New York: Russell Sage Foundation. This is a comprehensive and extensive analysis, based on census data, of social change in the United States.

Fischer, Claude S. 1992. *America Calling: A Social History of the Telephone to 1940.* Berkeley: University of California Press. The research featured in the chapter is presented in full detail in this book.

Ireson, Carol J. 1996. *Field, Forest, and Family: Women's Work and Power in Rural Laos.* Boulder, Colo.: Westview. Ireson explores how changes since the late 1960s have affected the lives of rural women in Laos.

Lenski, Gerhard, Patrick Nolan, and Jean Lenski. 1995. *Human Societies: An Introduction to Macrosociology,* 7th ed. New York: McGraw Hill. This comprehensive textbook discusses the broad scope of social change from hunting and gathering to modern advanced industrialized societies.

Pulling It Together

Listed below are other chapters where you can find additional discussion or examples
of some topics covered in this chapter. You can also look up topics in the index.

INTERNET SOURCE

The United Nations Environment Programme
provides a great deal of information on programs
used to promote sustainable development throughout
the world.

20

Collective Behavior and Social Movements

Collective Behavior: People Acting Together
Collective Behavior and Other Group Behavior
Explaining Collective Behavior

Social Movements: Creating Social Change

FEATURED RESEARCH STUDY

The Environmental Movement as a Social Movement
Resource Mobilization: A Mesolevel Analysis
Worldviews and Political Environments: A Macrolevel Analysis
Individual Involvement in Social Movements: A Microlevel Analysis

A Final Word

Amy and her good friends Cathy and Jill loved to walk in Forest Park. It was the last big stand of tall trees within fifty miles of Central City, and the deep shadows, the rich scent of pine and cedar, and the songs of the birds always made Amy feel calmer and a little more contented with life. Sometimes Amy would bring her beloved retriever, Sam, and let him run through the trees, delighting in the variety of smells and in the vast area in which to roam.

Yet the area around Forest Park was changing. New housing developments had been built right up to the edge of the park, and a large shopping mall was under construction. Traffic in the area was becoming more and more congested, and city officials and the mall owners had decided that a new freeway was needed to help get shoppers to the mall. The logical place for the freeway, they declared, was right through the middle of Forest Park.

When Amy and her friends heard of the plan for the freeway, they were shocked. "We can't lose Forest Park," Amy told them at lunch in the college cafeteria. "It's my favorite place on Earth." Cathy and Jill shared her concerns. None of them wanted Forest Park to be broken up by a freeway. They all knew that traffic was awful in that area, but surely there was another solution. Couldn't something be done to save the park?

On their way out of the cafeteria, Amy saw a notice that read, "Let's Save Forest Park—Join the Committee for Alternative Transportation." The notice invited students to attend a meeting at the student union that evening to discuss the proposed freeway. Amy and her friends looked at each other and, almost in unison, said "Let's go." At least they could see what it was all about.

When they arrived at the meeting, the three friends were pleased to see other people they knew. Some they had met in classes; others they knew from hikes they'd taken together in the park. The first hour or so of the meeting was consumed by speaker after speaker denouncing the freeway plans. Some of the speakers were very impassioned, and their remarks drew loud rounds of applause and cheers. At one point the crowd started to chant, "Stop the freeway! Stop the freeway!" The hour was growing late, and the people who had organized the meeting suggested that the group begin to focus on possible alternatives to the freeway, such as the city expanding its bus system or creating incentives for people to carpool. They also discussed ways to let others know of their concerns. The more cautious members of the group suggested contacting members of the city council and perhaps launching an advertising campaign. Others argued for picketing city council meetings and the site of the new mall, and even laying down in front of bulldozers if necessary.

After the meeting Amy and her friends sat around a table, sipping cokes. Cathy talked about how she would design the advertising signs she had volunteered to make and what she would say when she lobbied city council members. Jill listened politely for a few minutes and then remarked, in a slightly disparaging tone, "But what good will all that do? People can just ignore signs, and the city council won't listen to you anyway. We need to do something that will really grab people's attention and get more people involved."

"Well, I don't think picketing would help," Cathy replied, a bit sharply. "You'll just turn people off. And you sure won't catch me laying down in front of a bulldozer! And what was it with all that shouting in the meeting?"

Then, realizing that Amy seemed lost in her own thoughts and hadn't said anything, both Jill and Cathy turned to her and asked, "What do you think, Amy? What should we do?"

Amy didn't know what to say. She knew that she didn't want the freeway to be built, but what could she and her friends do? She was worried about being part of a demonstration, for she had read about crowds getting unruly and dangerous and some of the people at the meeting appeared pretty agitated. Yet, many of the speakers seemed to have good ideas and she really cared about Forest Park. Could the Committee for Alternative Transportation really make a difference?

The experiences of Amy and her friends involve what sociologists term collective behavior—actions of groups of people, some of whom might not even know each other. Collective behavior can take many forms. It can involve the applause, cheers, and chants of the people who met to discuss the proposed freeway as well as marches to protest the building of the highway. It can involve people spontaneously joining in prayer in the wake of a tragedy, communicating rumors, following the newest fads and fashions of dress, or even performing heroic rescues during a natural disaster such as a flash flood or sudden fire. Collective behavior can also involve mass actions of crowds, such as celebrations after sports victories, spontaneous dancing at a concert, or politically motivated events such as the Boston Tea Party or, more recently, the destruction of the Berlin Wall. Like conventional, everyday behavior, collective behavior occurs in many different settings and involves many different areas of social life. Yet, in contrast to conventional behavior, collective behavior is usually rather unexpected. It appears to happen almost spontaneously and to involve unusual and atypical behaviors.

A special type of collective behavior involves deliberate and organized efforts to effect social change, behaviors that are referred to as social movements. Amy and her friends' work with the Committee for Alternative Transportation is part of a much broader movement to change the way in which people affect the environment, often called the environmental movement. In turn, the environmental movement is just one of many social movements. The civil rights and the feminist movements have helped change the opportunities and constraints traditionally encountered by racial-ethnic minorities and women. Religious movements have worked to convert individuals to new beliefs. Political movements have attempted to shape governmental policies and decisions.

I began this book by describing the paradox of sociology. Sociologists often study areas about which people have passionate, deeply held beliefs and emotions, yet sociologists try to set aside these feelings and use scientific methods to examine these areas. This tension between deep concerns and careful, systematic scientific analysis is especially apparent in analyses of collective behavior. Such actions often embody strong emotions and feelings as diverse as fear, bravery, deep political loyalties, hatred, or love. Collective behaviors also can seem unique, strange, and incomprehensible. Yet, as we will see, when sociologists have dispassionately looked at these events they have found that collective behavior can, in fact, be understood in the same way that we understand other types of behavior.

It is appropriate that we end the book with a discussion of one specific form of collective behavior, social movements. Sociology isn't just about why bad or good things happen to people, but also about what people themselves do to change their social world. As we will see, despite the powerful institutions and vast forces that shape a society, individuals, too, through their deliberate efforts, can and do remake the social structure over time. Not only can sociology provide an understanding of why the world is the way it is, it also can give encouragement to those who want to make it a better place. A sociological imagination can help us see how concerted efforts can begin to alter even the most serious and difficult social problems; it can show us how and why social change occurs.

Our discussion of social movements also reflects, in many ways, our discussion of the paradox of sociology. That is, many sociologists care deeply about the world they study and have strong feelings about social issues, such as the degradation of the environment, the inequalities that face racial-ethnic minorities and women, the gap between the rich and the poor in industrialized countries, and the problems of poverty and inequality in the developing and least developed nations. Thus sociologists often are passionate about the goals of social movements. At the same time, sociologists try to approach the social world dispassionately. They believe that the best way to help solve social problems is by applying the scientific method to understand why they exist.

In the pages to come we explore both collective behavior and social movements. How and why do people behave collectively? Is collective behavior different from other, more conventional, everyday behaviors? Can people really work together to change the social world? What have sociologists learned about effective social movements?

In trying to understand the answers to these questions, I often return to Simmel's notion of the "web of social relations" and the metaphor of a spider web that we used in earlier chapters. As you will see, even though collective behavior differs from conventional, everyday behavior in many ways, both types of behaviors can be understood only by examining our relationships and interactions with other people. These ties bind us to others, but they also give us the resources to bring about social change.

COLLECTIVE BEHAVIOR: PEOPLE ACTING TOGETHER

Over the years sociologists have examined many different kinds of collective behaviors. Examples include political demonstrations; fads in clothing,

How was the Million Man March a form of collective behavior?

hairstyles, and music; large religious gatherings; parades; celebrations after major sports victories or on holidays such as Mardi Gras or New Year's Eve; panic during times of economic crisis; displays of support for political views or candidates; riots and unruly behavior; and actions during times of disaster such as earthquakes, fires, and flood. From analyses of these and other similar events, sociologists have come to understand more about how collective behavior is both different from and similar to other types of group behavior, and they have developed understandings of why people engage in these activities.

Collective Behavior and Other Group Behavior

Throughout your study of sociology, you have learned specific terms used by sociologists to denote sociological concepts. Many of these definitions have been quite straight-forward, easily understood, and generally pretty much agreed on by all sociologists. Defining collective behavior has proven to be more difficult, undoubtedly because the idea of collective behavior is so broad that it could conceivably include any type of group behavior and almost all aspects of social life. Sociologists have spent many years discussing, and sometimes arguing, about just what collective behavior means and have come to some agreements about how collective behavior is different from conventional, or ordinary, behavior and how it is similar.

First, people who are involved in collective behavior seem to be much more engaged and involved than are people performing conventional behavior, even though the behaviors themselves aren't that unusual.[1] For instance, during the Great Depression many people flocked to their local banks to withdraw money; during the Alaska and California gold rushes, thousands of men left their homes to search for riches; in Holland's 1634 tulip mania, many Dutch families invested all their savings in tulip bulbs. There is nothing inherently unusual in withdrawing money from a bank, moving to take a new job, or buying tulip bulbs. What is unusual in each of these cases is the excitement and emotion that accompanies the event. The people involved were very committed and involved: The miners were said to have gold fever, the bank customers were convinced that their banks would fail, and the tulip investors firmly believed that they could make their fortune through their investments.

At the same time collective behavior events are often unexpected; they might move rapidly from one person to another or from one setting to another, or even change form and rapidly disappear.[2] The rise and fall of the phenomenon of streaking—running naked through public places—illustrates this. Streaking first appeared in early 1974. By the end of February, only ten incidents had been reported in news media, but within the first three days of March there were 55 such incidents! Two days later all three national television networks had stories about streaking and in the next few days an average of about 75 incidents a day were reported from all over the country as the practice diffused from place to place. Within a few weeks, however, the phenomenon began to disappear. After this point, there were almost no incidents in the United States, although there were a few reports of streaking in Paris, Rome, and Tokyo.[3]

Both the emotional intensity and the unexpected characteristics of collective behavior reflect the fact that it tends to be governed by emergent norms, norms that develop or emerge from a setting, rather than already being part of the well-known, prespecified expectations established within a culture. Cultural norms are relatively vague or even nonexistent in several areas of social life. For instance, our culture specifies that we should wear clothes, but there is a fair amount of leeway regarding what kinds of clothes are appropriate. Similarly, we don't have strong cultural norms regarding how people should spend their leisure time. Consequently, fashion and leisure fads often sweep the society. Skirt lengths go up and down, body piercing becomes popular, popular hairstyles change, new leisure activities such as hula hoops and in-line skates become widely used and then discarded.[4]

In other areas, we might be relatively unaware of norms because the events are so rare that most people involved have never experienced them. For instance, many people who experience natural disasters, such as a large fire, a flood, or a tornado, are experiencing such an event for the first time. In addition, traditional sources of social control and authority might be unavailable or unable to meet all the needs of the situation. As a result, new patterns of behavior develop. The norms that emerge in such situations can involve helping behavior and even heroic actions, as people brave flames to rescue others or search through piles of rubble for buried survivors. The emerging norms can also involve less altruistic behavior, as people use the opportunity of loosened social controls to engage in looting or other types of criminal activity.[5]

Even though collective behavior tends to be more likely than conventional behavior to be emotionally engaging, unexpected, and to involve emergent, rather than preestablished norms, it is, in most ways, like all other types of social behavior. First, even though the norms in collective behavior settings are emergent and new, they do provide guidance for behaviors. People who violate these emergent norms are likely to be sanctioned much as they would be in other settings. For instance, even streakers conformed to norms: They ran through public settings rather than leisurely strolling along. They bared their bodies in public places such as ball games, but not in public places such as churches. They also tended to run in groups rather than alone.[6] Similarly, the chanting of "Stop the freeway!" at the meeting of the Committee for Alternative Transportation developed spontaneously, but was directly related to the topic of the meeting. Even though many of the participants had come to the meeting with strong emotions, their actions were governed by norms. The words of the chant focused on the issue at hand, and the chanting soon ceased so that actual planning of actions could begin.

Finally, and perhaps most important, most sociologists now agree that individuals' behaviors in crowds or in response to fads and trends result from the same type of processes that govern our behavior in other settings. The theories that we have used to analyze other kinds of behavior, such as symbolic interaction theory and rational choice theory, appear to apply equally well to collective behavior as to conventional behavior.[7] We turn now to these explanations of collective behavior.

Explaining Collective Behavior

Some of the early psychological theorists and writers suggested that people can develop a kind of "group mind" that overrides their usual behavior patterns, leading them to engage in behaviors that they would normally never do. In other words, in a crowd or mass setting, people tend to "lose their minds." The early French psychologist, Gustav LeBon, popularized these ideas when he wrote a book called, *The Psychology of the Crowd,* in the late 1800s. LeBon suggested that people in crowds become somehow transformed. They feel anonymous and as though they are not accountable for their actions. Their usual abilities to reason critically and think independently disappear, and a type of "collective mind" takes over. This collective mind is much more susceptible to the ideas of others, and through a contagious process people come to feel, think, and act in ways that are very different from the way that they would normally behave.[8]

LeBon's book is still in print, more than one hundred years since it was first written. For many years, it had a great influence on other scholars who wrote about collective behavior, and it is still widely believed by the general public.[9] Yet LeBon's ideas have not been at all supported by empirical data. We might want to excuse behaviors away by saying something like, "They were just caught up in the crowd," or "They just lost all judgment and blindly followed the people next to them." But when sociologists have actually looked at what happens in collective behavior settings, they have found that these excuses simply don't hold up.

People who have studied various kinds of collective settings, including political groups, religious gatherings, sports events, demonstrations, and riots, have found that people are not alone or anonymous. Instead, they usually come with or soon find friends, family members, or acquaintances. We are part of a web of social relationships even when we are in such mass gatherings.[10] Nor do people seem to lose their usual reasoning powers or self control when they are in such settings. In fact, studies indicate that when we are in potentially dangerous or problematic settings such as floods, fires, tornadoes, or explosions, we might be quite frightened, but are actually better able to think critically and purposively.[11]

Collective behavior, just as conventional behavior, reacts to mechanisms of social control. As you recall, most social control is actually internal social control, involving our self restraint. As originally described by George Herbert Mead, the scholar who is seen as most central to the development of symbolic interaction, we continually think about, or reflect on, our own actions in the process of reflexive behavior. Mead suggested that, even when we are faced with situations that we have never encountered before—such as the suggestion that we streak through a crowded sports arena, get a pierced earring in an obscure body part, or loot an abandoned store—we interpret this situation, reflect on it using the information that we have, and then reach a decision regarding how we will act.[12] Even though people in situations such as this frequently follow the suggestions of others, they are still in control of their behavior, acting on the information that is available to them.

The nature of their decisions can be understood by using the ideas of exchange and rational choice theory. No matter what kind of situation we are in, we seem to think about the possible rewards and costs that particular actions might bring.[13] For instance, people who contemplate streaking through public areas balance the possible costs of embarrassment and legal sanctions against the perceived benefits of the excitement and novelty of the event. Even people in riot situations think about the possible costs and benefits of actions such as damaging and looting property. Crowd members think about their power relative to that of legal authorities. Are there enough police to stop their actions or to arrest them, or are there so many potential rioters or looters that they will be relatively invincible?

Yet rioting, looting, streaking, and even chanting slogans are only part of the spectrum of collective behaviors. Phenomena such as these tend to be fairly short-lived and often have relatively little impact on the rest of society. In contrast, social movements can be highly organized, last for a number of years, and have strong impacts on the nature of society.

SOCIAL MOVEMENTS: CREATING SOCIAL CHANGE

Social movements have sprung up throughout history as new ideologies have developed and individuals have joined together to promulgate these new worldviews and to change individuals or entire societies. For instance, early Christianity was a social movement: Followers of Jesus spread the gospel throughout the world, establishing churches, and converting others to their faith. The American Revolution dramatically altered the relationship of the colonies to the English government and launched a new country. Likewise, the 1917 Russian Revolution permanently changed the face of Russia, the surrounding countries, and even the world.

Of course, some social movements have also tried to stop change. For instance, the Committee for Alternative Transportation, with which Amy and her friends are involved, wants to halt construction of a freeway through Forest Park. Whether members of a social movement want to implement change or to counter change proposed by others, the important point in sociological terms is that the movement involves planned and concerted actions by groups of individuals that are directed toward specific goals.

Today we can identify many different social movements, which can take a variety of forms. Some aim at changing individuals and their behaviors and beliefs, such as Mothers Against Drunk Driving (MADD) or evangelical religious movements. Others, such as the environmental movement, focus on society as a whole. In addition, some social movements work toward only partial change of either individuals or societies. For instance, MADD seeks only to alter drinking patterns. Many environmental groups, such as Friends of the Earth or the Committee for Alternative Transportation, work within the current political structure to change environmental policy. Other social movements seek a complete and total transformation of individuals or societies. For instance, religious movements like early Christianity or today's Unification church try to convert individuals to a totally new way of life. Political revolutions like the American Revolution or the Russian Revolution seek to change an entire society.[14]

Sociologists sometimes use the term social movement sector to refer to the entire range of activities aimed at changing the social structure. All the various movements noted previously would be seen as part of the social movement sector. Within this sector are specific social movement organizations, formal organizations that work toward the goals of a social movement. The individual groups mentioned earlier such as MADD or the Committee for Alternative Transportation, are examples. A set of such organizations that are all working within the same social movement is referred to as a social movement industry. For instance, the environmental movement industry includes a large number of social movement organizations, from Greenpeace to the Sierra Club to the Committee for Alternative Transportation.[15]

The environmental movement is only one of several recent social movements that sociologists have termed new social movements. Beginning in the United States and in Europe in the 1960s and 1970s, citizen protests of various types became more common. For example, in those decades thousands of citizens in the United States and throughout the world demonstrated against the war in Vietnam. Students in many countries openly challenged university policies, feminist groups protested the treatment and status of women, and environmental groups formed to preserve the planet. More recently, gay and lesbian activists have mobilized in many countries. These movements have been termed "new" for several reasons. First, they tended to use tactics that had rarely been used before, especially the direct involvement of large numbers of citizens in political protest. Second, these movements addressed issues that had not traditionally been the focus of political debate, such as the quality of the environment, gender equality, alternative lifestyles, and human rights in developing countries.[16]

Of course, social movements achieve varying degrees of success. For example, for many years dedicated individuals and organizations on the radical left have attacked capitalism in the United States and tried to replace it with some form of socialism or communism. Despite some successes, especially during the Great Depression of the 1930s, this movement has never attracted enough adherents to fundamentally change the society. In contrast, the environmental, civil rights, and feminist movements all have been more successful in changing individual ideas and behavior, government policies, and the practices of large organizations. The passage of the Clean Air Act, the abolition of legal segregation, and the attention given to previously neglected problems like sexual harassment and unequal pay in the workplace are just a few examples of the ways in which these movements have, over time, effected far-reaching social changes.

Why are some social movements successful whereas others barely influence the larger society? Sociologists have approached this question from three general directions. In recent years several sociologists have tried to explain the success or failure of social movements through resource mobilization theory. First formulated by John McCarthy and Mayer Zald, this theory suggests that social movements are successful to the extent that they can effectively mobilize or activate the resources needed to accomplish their goals.[17] These resources range from the skills and commitment of the individuals involved in the movement to tangible necessities such as work space, access to communication networks, and effective ways of publicizing the movement. Perhaps the most important tangible resource is money because it can be used to purchase many of the other resources. Resource mobilization theorists try to understand why some social movement organizations are more successful than others are at raising the money and mobilizing the other resources needed to meet their aims.

Another group of sociologists has examined social movements from a macrolevel perspective by looking at the broader cultural and political context in which they occur. Still other sociologists have worked at a microlevel of analysis to examine the participation of individuals in social movements. Both of these perspectives focus more on the ideological aspects of social change, such as how the worldviews of societies and individuals change as a result of social movements. These two groups of sociologists also look at how a society's political, cultural, and historical traditions affect the direction that social movements take.

As often happens, these three theoretical orientations are not mutually exclusive; rather, each contributes to our understanding of why social movements succeed and fail. In the next section we focus on the environmental movement. We begin by introducing the study of the environmental movement and the countries we will focus on. We then use a mesolevel analysis to illustrate many of the ideas of resource mobilization theory, a macrolevel analysis to identify the cultural and political context of the environmental movement, and finally a microlevel analysis to explain why individuals choose to participate in social movements and the effect that the environmental movement has had on people around the world.

THE ENVIRONMENTAL MOVEMENT
AS A SOCIAL MOVEMENT

In trying to understand social movements, it is important to look at many kinds of movements in a variety of social contexts. If the theories and concepts sociologists have developed to explain social movements apply across all of these settings, we can have more faith that these ideas are valid. In this section we look at research on the environmental movement that has been conducted in several countries. We will focus especially on the work of Andrew Jamison, Ron Eyerman, and Jacqueline Cramer, sociologists who live and work in Europe, who compared the activities of the environmental movement in Sweden, Denmark, and the Netherlands. By examining the research of Jamison and his associates, as well as that of researchers in the United States, we can see the ways in which resource mobilization theory and other perspectives on social movements apply in different settings.

Unlike other forms of collective behavior, social movements generally occur over a long period of time. Thus sociologists usually investigate the development of social movements with historical research methods, such as the study of artifacts and written records. In addition, sociologists often use case studies to conduct in-depth investigations of a particular organization, movement, or country. In comparing historical accounts across three different societies, Jamison and his associates conducted historical--comparative research, research that uses historical data to compare two or more societies, much as Weber examined the development of economic systems in China, India, and Europe. Only through such comparative research can we see whether findings hold, or are *replicated*, across different settings.

As shown in Table 20-1, the industrialized societies of Sweden, the Netherlands, Denmark, and the United States are similar in many respects. Inhabitants of all four countries enjoy a high standard of living and have long life expectancies. Compared with less developed societies, these nations all have low birth rates and low death rates. In other ways, of course, they are quite different. The Netherlands and Denmark have almost the same land area, but the Netherlands has a much larger population, resulting in a much more densely populated country. Both Sweden and the United States are far less densely populated; forests cover half of Sweden, and much of the United States consists of sparsely populated rural areas. The United States is somewhat less urbanized than the three European countries, but at least three-fourths of the population of all the countries live in urban areas.[18]

The countries we are considering also differ in political and cultural characteristics, as well as in their historical relationship to the environment. For instance, the three European countries have a tradition of democratic socialism, providing proportionately more government funding for services such as medical care than has been common in the United States. The U.S. polity is strongly influenced by business interests. Denmark has a long tradition of citizen protest and social movements that began in the nineteenth century, became active again with the resistance to German occupation in World War II, and is still present. The Netherlands is more divided politically. It also has had a unique experience with the environment, having literally reshaped its natural habitat through dikes and polders (low-lying land areas that have been reclaimed from water). Although they have not sculpted their landscapes, both Sweden and the United States have a long history of using natural resources as the basis of industrialization, through industries such as logging and mining.[19]

By comparing the activities and growth of the environmental movement in these four countries, we can see how social movements are similar from one setting to another and how particular cultural and political characteristics of a society affect the course of a social movement. We begin this analysis by looking at the way social movement organizations mobilize resources to accomplish their aims.

Resource Mobilization: A Mesolevel Analysis

Sociologists who have studied social movement organizations have concluded that the variables that influence their success are similar to those that influence the success of any formal organization, much as we discussed in Chapter 11.[20] We can understand these processes by applying the same ideas as in our discussion of formal organizations: organizational ecology, organizational structure, and organizational leadership.

The Ecology of Social Movement Industries

In Chapter 11 we reviewed sociologists' analyses of how organizations develop, die, and are replaced. A principal finding of the organizational ecologists is the apparent common pattern in the density, or number, of organizations within a population of organizations at any one time. This idea can help us understand changes in social movement industries as new social movement organizations are formed but

TABLE 20-1   *Characteristics of Denmark, the Netherlands, Sweden, and the United States*

COUNTRY	POPULATION (MILLIONS)	BIRTH RATE	DEATH RATE	LIFE EXPECTANCY	DEGREE OF URBANIZATION	PER CAPITA GNP (U.S. $)	POPULATION DENSITY (PER SQ. MILE)
Denmark	5.3	13	11	75 yrs	85%	$32,100	324
Netherlands	15.7	12	9	78	61	25,940	1,197
Sweden	8.9	10	11	79	83	25,710	56
United States	270.2	15	9	76	75	28,020	76

Source: Population Reference Bureau, 1998.

only some succeed while others fail. In looking at the environmental movement in the Netherlands, Sweden, and Denmark, Jamison and his associates distinguished four separate phases in the movement's development. Their description of the life course of environmental organizations is similar for other types of formal organizations.

In the first phase, which lasted from the early-to-mid-1960s, ideas related to environmentalism gradually became more widespread. Much as the organizational ecologists would predict, in this early stage there were relatively few social movement organizations. Many of these organizations were "conservation" or "nature" organizations. For example, Sweden had groups such as the Swedish Conservation Society, the Swedish Ornithological Association, and the Field Biologists.[21] In the Netherlands were groups such as the Society for the Preservation of Natural Monuments, the Contact Committee for the Protection of Nature and Landscape, and the Dutch Youth Organization for the Study of Nature.[22] In the United States early groups included the Sierra Club, which was founded in 1892; the National Audubon Society, founded in 1905; and the Isaak Walton League, founded in 1922.[23]

As environmental issues received increasing media attention in the late 1960s and early 1970s, more organizations began to promote environmental issues. During this second phase, the density of social movement organizations in the environmental movement industry increased. Resource mobilization theorists such as McCarthy and Zald suggest that as more organizations enter a social movement industry, each organization will tend to specialize, focusing on relatively narrow goals and strategies.[24] In the terminology of organizational ecologists, these social movement organizations need to find a *niche,* or a place in the organizational environment, in which they can survive. For instance, since the late 1960s numerous environmental organizations focusing on different issues and specializing in different tactics

have appeared in the United States, including Greenpeace USA, Earth First!, the Rainforest Action Network, Conservation International, Citizen's Clearinghouse for Hazardous Waste, the League of Conservation Voters, and the Sierra Club Legal Defense Fund.[25] A similar proliferation of groups appeared in Sweden, Denmark, and the Netherlands.

Jamison and his associates suggest that the environmental movement in the three European countries reached its high point during the third phase of the movement's history, in the mid-to-late 1970s. For instance, during the 1970s, all the political parties in the Netherlands included environmental issues in their political policy statements.[26] In Denmark, a comprehensive environmental protection law was passed in 1973, and the government increased its activities in this area throughout the rest of the decade.[27] In Sweden, the Centre Party, which was most strongly identified with the environmental movement, doubled its share of the vote between 1958 and 1976.[28] Similar gains were achieved in the United States with the establishment of government agencies such as the Environmental Protection Agency and the passage of legislation directed at curbing water and air pollution.[29]

The fourth phase identified by Jamison and his associates dates from the early 1980s. Since then, there have been a number of changes and regroupings in the environmental movement in Sweden, Denmark, and the Netherlands. The achievements of the earlier period remained, but tensions between environmental organizations became more apparent. As the researchers described it,

> Explosive growth into a fully fledged social movement with organisations, leaders, and more or less coherent convictions brought many internal conflicts out in the open. Forced now to take stands and present programmes on a wide range of complex issues, the movement fragmented into an assortment of more or less disparate groups and organisations, often with their own set of tactics and specific type of activist.[30]

In the United States, the number of environmental groups also proliferated. By the end of the 1980s, the big, established environmental organizations had chosen to specialize in specific areas, such as water pollution or wildlife and energy conservation, as well as in basic tactics such as lobbying, conducting scientific research, or filing lawsuits.[31] In addition to these established organizations, there were many grassroots organizations, social movement groups that develop "from the ground up," away from major political centers, much like the Committee for Alternative Transportation. Although larger, established organizations tend to attract primarily upper-middle-class Euro-Americans, many of these grassroots organizations attract working-class people and members of racial-ethnic minorities, thus adding to the diversity of people involved in the environmental movement.[32] In addition, some groups began to use sit-ins and other direct protest activities, as well as more radical tactics such as "monkey-wrenching," the destruction of equipment deemed damaging to the environment.[33]

What does this proliferation and fragmentation of environmental organizations mean for the future of the movement? Based on their study of previous social movements, some sociologists believe that such proliferation is a sign that the movement is in a declining phase.[34] Others suggest that the proliferation of environmental organizations might instead be a sign of strength, signaling an acceptance of environmental concerns by an ever-increasing segment of the population.[35] In fact, as we will see when we take a microlevel perspective, the research evidence tends to support the latter position, because the commitment to environmental issues in society at large has not only endured but grown over the years.

Based on their examination of numerous social movement organizations, McCarthy and Zald suggest that as new organizations proliferate, more established organizations are more likely than newer organizations to survive and prosper. Just as new businesses have a high likelihood of failure, so, too, do new social movement organizations. For instance, in the Netherlands during the 1970s, membership in the long-established conservation organizations grew tremendously, much more so than did membership in the newly created environmental groups. Membership in one organization alone, the Society for the Preservation of Natural Monuments, swelled from 30,000 in 1960 to 250,000 in 1980.[36] Similarly, the Sierra Club in the United States grew from 15,000 members in 1960 to 560,000 in 1990. The three largest environmental organizations in the United States today—the Sierra Club, the National Audubon Society, and the National Wildlife Federation—all were founded before World War II.[37] As with other types of organizations, some of the suc-

cess of these older environmental groups seems to be related to their structure and culture.

Organizational Structure and Culture According to resource mobilization theorists, the major task facing social movement organizations is obtaining the resources needed to keep the organization going and meet the movement's goals. Resource mobilization theorists distinguish between social movement constituents (organizations and individuals who provide resources for a social movement) and social movement adherents (those who believe in the movement's goals). The task of social movement organizations is threefold: (1) maintaining support from constituents, (2) turning adherents into constituents, and (3) turning nonadherents into adherents.[38]

McCarthy and Zald suggest that a social movement organization's structure and culture affect its success in accomplishing these tasks in three ways. The first two help explain why older organizations are more successful. Older organizations are more likely to have an efficient, tested structure; they have found the structure that works best for their particular niche in the industry. In addition, these older organizations are more likely to be seen as legitimate by people who might provide needed resources. Potential supporters identify with the culture of organizations that they are familiar with, and they are more likely to support these groups. Finally, just as larger businesses are more likely to survive, larger social movement organizations are more likely to succeed, primarily because they obtain more resources, have more staff, and have a greater division of labor among the staff. In other words, larger social movement organizations develop more complex structures, and these more complex structures help ensure their survival.[39]

Leadership In addition to a favorable environment and an effective culture and structure, successful social movement organizations need good leaders. Often the founders of social movements are what Weber termed charismatic leaders, whose authority is based on their personal qualities. An example is Martin Luther King, Jr., who inspired people of all classes, colors, and walks of life to support the civil rights movement. At the same time, however, King proved to be a skilled administrator as the head of the Southern Christian Leadership Conference, a social movement organization that has endured long after his assassination in 1968. According to resource mobilization theorists, skillful administration is just as central (and often more central) to a movement's success as charismatic leadership, because administrative skills are needed to most effectively transmit messages to potential constituents and adherents.

In their analysis of environmental movements in Sweden, Denmark, and the Netherlands, Jamison

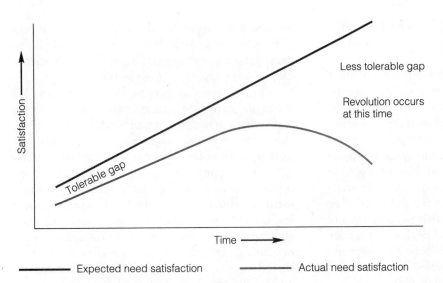

FIGURE 20-1 *The J-Curve of Social Movements*

Less tolerable gap

Revolution occurs
at this time

Tolerable gap

Satisfaction ——→

Time ——→

——— Expected need satisfaction   ——— Actual need satisfaction

and his colleagues found several examples of strong leadership. For instance, they attribute much of the success of the National Organization of Environmental Groups, a Swedish organization, to its leader, a microbiologist named Björn Gillberg. Gillberg appears to have been a charismatic leader, a skilled administrator, and a highly imaginative publicist for the National Organization's aims. As Jamison and his associates described it,

> Gillberg had come onto the national stage in 1969 with a book on the genetic hazards involved in food additives and other industrial chemicals. He had followed the book up by writing a number of debate articles in the leading national newspaper. . . . In the space of a year, Gillberg became a national celebrity; he stood outside supermarkets with a sandwich board, protesting against food-colourings and other additives—and then on a national television programme about his activities in 1971, he demonstratively washed his shirt in a synthetic coffee creamer.... For many Swedes, the environmental movement in Sweden had come to be symbolised by Björn Gillberg.[40]

One of Gillberg's most important achievements was a change in the worldviews of Swedes—a recognition that food additives and industrial chemicals could harm their environment and their health. Changes in people's worldviews are a crucial ingredient in the success of social change efforts, as we can see by looking at social movements from a macrolevel perspective.

Worldviews and Political Environments: A Macrolevel Analysis

Successful social movements involve more than money, organization, and an available ecological niche. They also involve perceptions and ideas. Somehow an issue

has to grab people's attention and win their commitment before a movement can get off the ground. Some sociologists who have looked at social movements from a macrolevel perspective have asked why such movements appear in the first place. Others have explored how the social context of a movement, such as a society's polity and culture, affects its development.

Why Do Social Movements Appear? Where do social movements come from? Why do they appear? Why did the American Revolution occur in the 1770s and not earlier? Why did the civil rights movement become so active in the mid 1950s? Why was environmental degradation a concern to people in 1972 but not in 1952?

Sociologists have formulated three answers to these questions. The first involves the idea of relative deprivation, the gap between people's expectations and the actual state of their lives. The most famous explanation of this idea was developed by James Davies and involves a pattern called a "J-curve," as shown in Figure 20-1. (The term J-curve reflects the fact that one of the lines in the graph is shaped like an upside-down J.) Based on his analysis of a number of historical events, such as the Pullman strike of 1894, the Russian Revolution of 1917, and the Egyptian Revolution of 1953, Davies suggested that revolutionary movements (social movements that seek to overthrow a political regime) are most likely to occur when people's expectations begin to depart drastically from the reality of their lives. As shown in Figure 20-1, the actual state of affairs usually is slightly less pleasant than the way we would like or expect it to be. As long as this difference remains fairly constant, it is tolerable. But if the conditions under which people live suddenly take a turn for the worse and their expectations for life don't change, then this difference may no longer be tolerable. Davies suggests that the sudden widening in the gap between what people expect and what they get is what triggers social revolutions.[41]

Although the idea of relative deprivation might explain the development of many social movements, especially political revolutions, it doesn't do as well in accounting for the new social movements that appeared in the 1960s. These movements occurred during times of economic prosperity and were spearheaded by middle- and upper-middle-class people who were economically quite comfortable. In addition, many of these movements didn't involve protests over injuries or harm that the participants personally suffered, but instead dealt with broader social issues, such as the environment, faraway wars, or human rights in other countries. Why, then, did these movements appear when they did?

The second explanation for the development of social movements applies particularly to these new social movements and comes from the perspective of resource mobilization theory. Theorists suggest that the new social movements arose at least partly because of the ways in which the United States and the Western European countries changed after becoming advanced industrial societies. As more basic human needs were fulfilled and no longer worrisome, people became aware of new kinds of problems, such as environmental degradation and gender inequality. In addition, the ever-increasing education of the population provided more citizens with the ability and resources to interact with government officials.[42] The affluence of the advanced industrial societies also provided the important resource of money. Social movement organizations proliferated in the 1960s in the United States as more money became available. As discretionary income rose—that is, as people had more money left over after meeting their basic needs for food, clothing, and shelter—citizens were able to contribute more to social movement organizations.[43]

But why did the money go to social movement organizations? Why were people concerned with problems they hadn't worried about before? Theorists such as Eyerman and Jamison provide a third perspective, pointing to the importance of opinion leaders—influential authors and speakers who bring issues to the forefront. For instance, in all three countries that Jamison and his colleagues studied, interest in environmental issues was spurred by books like Rachel Carson's *Silent Spring*, which examined the dangers of pesticides, and the Club of Rome's *Limits of Growth*, which examined the dangers of unlimited technological development and population growth. In a sense, the ideas presented in these books are just as much cultural innovations as

are new technologies like the wheel or the steam engine. And like material innovations, new ideas can spread through cultural diffusion, ultimately changing the way people view their world.[44]

The American sociologist, David Snow, and his colleagues have used the term framing to refer to the way social movements present and shape the ideas that underlie social movements. This concept was developed from the dramaturgical theorists' notion of "frames of interaction," the settings in which interaction occurs and the norms that apply to those settings, which was described in Chapter 5. Snow and his colleagues suggest that people involved in social movements, such as the influential opinion leaders, use their writings and speeches to develop shared ways to define social problems and to suggest actions and solutions. Though this framing process can be relatively flexible and emergent in the early stages of a social movement (much like norms emerge in collective behavior), once social movement organizations become well established this framing process can become quite strategic. Movement spokespeople might try to find the ways to present ideas and views that are most acceptable to potential movement constituents and adherents.[45]

If these frames become accepted and a social movement is successful, Eyerman and Jamison suggest, it will eventually lead to a new paradigm, or way of seeing society. For instance, because of the civil rights movement, more and more people who had taken the practice of racial segregation in the United States for granted now saw it as unfair and even evil. Similarly, the environmental movement has changed the way people view their relationship with the environment. When such a new paradigm takes root, it can be a liberating experience in and of itself, for it allows individuals not only to see their society in new ways but also to realize that they can begin to change the world through social actions. Because this perspective tends to focus on ideas and cognitive understandings, it is sometimes called a cognitive theory of social movements.[46]

Of course, new ideas don't arise in a vacuum. Even though people are reading the same books and sharing similar concerns, how they respond to social issues will depend on their particular culture and society. For example, millions of people in the United States and Europe may have been inspired by books like *Silent Spring* to do something to help the environment, but their actions were strongly influenced by the society in which they lived and especially the structure of the political world, or the social institution of the polity. For this reason, we need to consider the broader historical and political context in exploring why social movements appear and the ways they develop.

How Do Political Culture and History Affect Social Movements? In their comparative study of the environmental movement, Jamison and his colleagues found significant differences in the way the movement developed in different countries. For instance, in Sweden political parties incorporated environmental issues into their political platforms at a far earlier stage than did political parties in other areas of Europe. Thus environmental issues became part of government policy debates much sooner than in other countries, and the responses to environmental problems were translated into bureaucratic and technical issues within the Swedish government. Even though nongovernmental environmental organizations exist in Sweden, most environmental initiatives actually have emerged from government agencies, and environmental issues often have been phrased as technical problems that state agencies can handle.[47]

In contrast to Sweden, the Danish political system has been characterized by far more polarization and by a strong tradition of citizen protest groups, so that the environmental movement in Denmark has been much more multifaceted. For example, in the 1970s, numerous grassroots organizations appeared that addressed many different types of environmental issues.[48] Although the number and variety of organizations apparently diminished in the 1980s, social movement organizations, rather than the government, provided the impetus for environmental change in Denmark.[49]

The history of environmental movements in the Netherlands provides yet another example of how political cultures and history affect the development of a social movement. Historically, the Netherlands was strongly divided into three distinct religious groups (orthodox Calvinists, represented by the official Dutch Reformed Church; moderate Protestants; and Roman Catholics), each with its own political parties. Further, the Netherlands has a long tradition of accepting wide differences of opinion.[50] Jamison and his associates suggest that this tradition influenced the development of disparate environmental groups, each focusing on its own issues and employing its own tactics. For instance, some groups attached to the "hippie" counterculture that flourished in Amsterdam for a while, whereas others linked up with the major political parties.[51] In addition, Jamison and his associates suggest, the attitude toward environmentalism in the Netherlands was much more pragmatic than in the other countries. As they put it,

> The general attitude of [the participants in the Dutch movement] was rather down to earth; "Let us do something about the pollution problem." Their environmental view was not intellectualized as much as in Sweden or Denmark. . . . This prag-

matic approach is characteristic not only of the nascent environmental groups but can, in fact, be traced far back into Dutch history. . . . It almost seems an inherent aspect of Dutch society to stress the importance of "doing things" instead of "reflecting upon things."[52]

Jamison and his associates suggest that this cultural emphasis on pragmatism helped spur the development in the Netherlands of several different environmental organizations, each with its own niche and most with professional staff.

The United States differs from each of the European countries in that it lacks a strong tradition of democratic socialism and politically has been favorable to business, with close ties between business interests and government bodies. These ties were especially strong under the Reagan administration in the 1980s, a time in which government agencies were openly hostile toward the environmental movement. This hostility appears to have prompted greater unity among U.S. environmental groups than appeared in the countries studied by Jamison and his associates. Michael McCloskey, the longtime president of the Sierra Club, described the era:

> As the movement became embattled with the Reagan administration, the mainstream groups began to collaborate more. The "Group of Ten" was formed so that the CEOs of the more active groups could meet quarterly to coordinate their strategies, particularly in dealing with a hostile administration. Various ad hoc coalitions and task forces met as needed, with some, such as the Clean Air Coalition, carrying on for years. Quite often, groups outside the movement, such as the American Lung Association, the Steelworkers Union, the American Civil Liberties Union, and the National Taxpayers Union, were brought in to broaden the base. . . . By the end of the 1980s, the movement also found to its surprise that other movements were picking up the environmental theme and institutionalizing it within their organizational structures.[53]

Thus, whereas in Sweden the government itself adopted some of the goals of the environmental movement, in the United States the open hostility of the government actually prompted greater unity among disparate environmental groups. In general, the establishment of social movements and the direction that they take is heavily influenced by the political culture and history of the society in which they develop.[54]

Though a macrolevel analysis allows us to understand the broad context that promotes the development of social movements, the success of a movement depends on the extent to which it can attract adherents and constituents—people who believe in the goals of the movement and people who provide support for it. To understand why individuals like Amy and her friends would choose to participate in a social movement, sociologists have used a microlevel analysis.

Individual Involvement in Social Movements: A Microlevel Analysis

As with religious groups or political activity, within social movements there is always the possibility that individuals can be *free riders*, benefiting from the public goods that an organization or movement brings about without contributing any time or energy themselves. For instance, everyone benefits from cleaner air and water whether or not they participated in the movement that helped to create these changes. Why, then, would people choose to spend time, money, and other resources on social movements?

The theories that we explored in explaining why individuals join religious movements or participate in political activities also can be used to explain why people become committed to social movements. As you will recall from Chapters 12 and 13, an important influence on our joining a particular group is the nature of our social networks, our web of social relationships. Thus, if we have friends and acquaintances who are involved in a social movement, we are much more likely to become involved ourselves.[55] In addition, rational choice theory suggests that in deciding whether to participate in a social movement, we choose the actions that have the lowest costs and provide the greatest benefits.[56] For instance, Amy's decision to work with the Committee for Alternative Transportation was influenced both by the fact that she knew a number of the people involved in the group and by her belief that the costs involved in participating would be more than offset by the possibility of halting the freeway construction.

Social movement organizations can influence people's decision making by helping to define the costs and benefits of getting involved through the messages they communicate to potential adherents and constituents, the way that they frame the issues. For instance, even though members of racial-ethnic minority groups would be most likely to directly benefit from the achievements of the civil rights movement, other citizens could be induced to provide resources by the argument that the movement is working toward a more equitable society, one that would more closely match widely held values and beliefs. Similarly, proponents of the environmental movement have mounted extensive advertising campaigns to promote the idea that a cleaner environ-

ment benefits all people. New social movements have been especially likely to stress how the benefits of participation can go beyond mere self-interest to include ideas of fairness, empathy for less fortunate individuals, and concern for future generations.

One outcome of movement participation is, especially for social movement constituents, the addition of movement membership to one's *self-identity,* the set of categories that we use to define ourselves. Just as when you join religious movements, when you become active in a social movement you become part of the subculture, other members of the group become a reference group, and you identify yourself as a member of the movement—for instance, as an environmentalist or a feminist. In other words, there is a process of socialization through which you come to see yourself as part of the social movement.[57]

And, of course, just as with religious group memberships, people vary in the extent to which movement membership is part of their self identity. Some people are very committed constituents who devote great amounts of time and money, whereas others are only adherents with positive feelings toward a group but who provide few, if any, material resources.

Identity is not just something that belongs to individuals. Instead, people within social movement organizations often develop a collective identity, joint agreements about the nature of the group, who belongs and who doesn't belong, and what the group should and shouldn't do. For instance, in the meeting of the Committee for Alternative Transportation, Amy and others at the meeting were working to develop a collective identity for that group. Through their interactions, their web of social relationships, the meeting members were helping the organization develop its direction and its strength. Participants' development or acceptance of a collective identity can thus become a resource for the group.[58]

The more constituents a social movement has and the more discretionary income these constituents have, the greater the likelihood that a social movement will succeed. Many groups of people around the world and within advanced industrial societies experience severe hardship but are not represented by strong social movements. For instance, even though homeless people in the United States face severe deprivation, few social movements directly address their concerns. This no doubt reflects the fact that the homeless have few discretionary resources themselves and that other people often don't see a benefit in getting involved in promoting their interests.[59]

The true test of a social movement's success is the extent to which it has produced changes in social policies and individual worldviews. In the case of the

BUILDING CONCEPTS

Macrolevel, Mesolevel, and Microlevel Analyses of Social Movements

Below is a summary of some of the different ways sociologists have studied social movements. Notice that many of the theories and concepts introduced in earlier chapters for other topics can be applied to the study of social movements as well. As you have seen throughout this text, different sociological concepts and theories can work together to formulate fuller explanations of the questions being studied.

Level of Analysis	Example of Questions Studied	Sociological Concepts and Theories Used
Macrolevel	Why do social movements develop? How have political and cultural characteristics of a society influenced social movements?	Relative deprivation Cognitive theory of social movements
Mesolevel	How successful have social movement organizations been at resource mobilization?	Resource mobilization theory Organizational ecology Organizational structure and culture Organizational leadership
Microlevel	What motivates individuals to participate in social movements? How have social movements changed individuals' views and behaviors?	Social networks Rational choice theory

environmental movement, Jamison and his associates found widespread policy changes in line with the movement's aims in each of the countries they studied. In Sweden, environmental goals have been adopted and promoted by the major political parties, "the one party trying to outdo the other in its environmentalist rhetoric." In Denmark, the government repudiated nuclear power and passed stringent legislation regulating the release of genetically manipulated organisms. And in the Netherlands, the government adopted a broad-based environmental policy designed to combat soil and air pollution and to promote environmentally friendly behavior among citizens.[60] Despite opposition, especially by

the Reagan administration, similar policy changes appeared in the United States, including legislation such as the Federal Water Pollution Control Act, the Clean Air Act, the Wilderness Act, the Endangered Species Act, and the Resource Conservation and Recovery Act.

Even more striking, perhaps, is the change in individual worldviews and attitudes. Sociologist Riley Dunlap has examined attitudes of citizens toward environmental issues for many years, much as other sociologists have examined trends in attitudes toward racial-ethnic minorities (as we discussed in Chapter 8). His analyses of data from the United States indicate that there has been a steady increase

These New Jersey surfers are part of the environmental movement and illustrate the widespread support for its goals. Based on what you have learned in this chapter, what do you think might have influenced their decision to display these signs?

in the public's support for environmental causes. As he summarized his results,

> [The] evidence on trends in public opinion toward environmental issues strongly indicates that the environmental movement has been extremely successful in attracting and maintaining—for two decades—the public's attention to and endorsement of its cause. . . . Large majorities of the public accept environmentalists' definition of ecological problems as serious (increasingly so) and express support for efforts to ameliorate the problematic conditions. In addition, when asked to do so, growing proportions of the public express positive views of the environmental movement and identify with it.[61]

It is important to realize that this widespread public support for the goals of the environmental movement has changed the terms of the debate over environmental issues. No longer is it fashionable to oppose environmental concerns. Moreover, people who were once regarded as having "extremist" environmental views now are in the mainstream, and those who are willing to sacrifice the environment have been put on the defensive. In the United States, just as in Sweden, both major political parties now must profess support for environmental issues if they want their candidates to win elections.

Dunlap also has looked at the impact of the environmental movement internationally. In collaboration with the Gallup International Institute, he supervised the administration of a major survey on attitudes toward environmental issues. The Gallup firm began in the United States, but, like many other large corporations, it has become multina-

tional in scope, with affiliates throughout the world. These extensive resources allowed the firm to conduct surveys in twenty-four countries around the globe in the early 1990s. About half of the countries were developed countries with high gross national products, such as the United States, Canada, Germany, Japan, and Switzerland. The others were less developed nations with lower gross national products, such as Nigeria, Brazil, Russia, Turkey, Chile, India, and Mexico. Representative samples of residents of each country were carefully selected, and great care was taken in developing and translating the survey questionnaires so that all respondents were asked the same questions, no matter where they lived.

Dunlap and the Gallup organization found a great deal of agreement among people in both developed and less developed countries in their perception of the quality of the environment and the need for change. Citizens in countries throughout the world see environmental quality as a serious national issue, and a majority of people in twenty-one of the twenty-four countries surveyed report that they personally have a "fair amount of concern" about environmental problems.[62] The majority of the respondents in all twenty-four nations believe that environmental problems will affect the health of their children and grandchildren during the next quarter century. In other words, people around the world see environmental degradation as a threat not just to their own quality of life but to the health and well-being of future generations.[63]

Although many people who work with less developed countries stress that economic development can occur without environmental deterioration (the process called *sustainable development*), politicians sometimes try to pit environmental protection against economic development. To see how people around the world prioritize environmental well-being and economic growth, Dunlap and the Gallup organization asked the respondents whether "protecting the environment should be given priority, even at the risk of slowing down economic growth," or whether "economic growth should be given priority, even if the environment suffers to some extent." Surprisingly they found that in every country except Nigeria more people chose environmental protection than chose economic growth. In addition, a majority of citizens in all but three countries (Japan, Nigeria, and the Philippines) say that they would be willing to pay higher prices to protect the environment.[64]

These results are noteworthy because some scholars and political commentators have suggested that concern with environmental degradation and participation in the "new social movements" are largely limited to the richer industrialized countries. These scholars and commentators claim that people in poorer countries, who must be more concerned with problems of day-to-day survival, cannot afford the luxury of thinking about such issues.[65] If this were true, it would be a matter of concern for environmentalists, because international agencies such as the United Nations have documented the fact that environmental degradation is a global issue.[66] Dunlap's work suggests that, in fact, people around the world recognize problems with the environment and are willing to make personal sacrifices to deal with these problems. His analyses, as well as the historical-comparative work of Jamison and his associates, show how individuals worldwide not only are concerned with social issues but can work with one another and with government agencies to try to solve them, much as Amy and her friends are trying to preserve the beauty and tranquillity of Forest Park. To learn more about Riley Dunlap and his work, see Box 20-1.

Although we have focused on studies of the environmental movement in this chapter, similar analyses and conclusions have been reached about other social movements—such as the civil rights movement, the feminist movement, the anti-abortion movement, and various religious movements.[67] In general, whatever the social movement, its success or failure appears to be influenced by the extent to which it can mobilize resources, by its relationship with the larger political and cultural environment, and by the extent to which it can alter worldviews of potential adherents and constituents.

Sociological insights can be used to help develop movements that address severe social problems in developing countries. Just as Dunlap's work suggests that people are willing to work to solve environmental problems, people can join together to solve other types of problems, even those that span entire cultures, as described in Box 20-2. By building webs of social relationships, people can act collectively to promote the welfare of many others and address some of the most daunting problems facing nations today. In this sense, showing how empirical data support the possibility of social change is one way in which sociology not only can provide an understanding of why the world is the way it is but also can give encouragement to those who want to make it a better place.

CRITICAL-THINKING QUESTIONS

1. Suppose that you were a sociologist who wanted to study the actions of the Committee for Alternative Transportation, described at the beginning of this chapter. What kinds of research methods might you use in such a study?
2. What concepts and theories about social movements would help you in this work? What specific hypotheses could you investigate?

Common Sense versus Research Sense

COMMON SENSE: Residents of wealthier nations are more environmentally aware than residents of poorer nations.

RESEARCH SENSE: Dunlap's research found similarly high levels of environmental concern in both rich and poor nations.

A FINAL WORD

Only a few students who take introductory classes become professional sociologists and do research like that of Jamison or Dunlap and their colleagues. But you can apply the ideas you have learned in this book to virtually every area of your life and in every type of job you hold. In the years to come, I hope you will use the knowledge you

BOX 20-1

SOCIOLOGISTS AT WORK

Riley Dunlap

I believe my most valuable contribution is providing useful and scientifically credible information.

Riley Dunlap is professor of sociology at Washington State University and has studied public attitudes toward the environment for many years.

Where did you grow up and go to school? Growing up in Crescent City, a small lumber and fishing community on the Northern California coast, provided an interesting background for my eventual focus on environmental research. I was thoroughly socialized—by family, peers, and work experiences (summer jobs in sawmills)—into seeing redwoods, salmon, and so forth as natural resources, and I never thought much about the need to conserve them. Nonetheless, I also developed a love for the redwood forests that gradually helped change my view of nature.

How did you get interested in studying public attitudes toward the environment? I had just finished a small study of student activists at the University of Oregon for my master's thesis when the campus was mobilizing for the first "Earth Day" in 1970, and I decided to compare "eco-activists" with the political activists. Dick Gale, a new professor, and I surveyed students involved in Earth Day activities, and I became "hooked" on understanding environmentalism and societal reaction to it. I've continued to study these issues as they've evolved over the past quarter century.

Originally I was at best ambivalent about the environmental movement and assumed it would be a passing fad and quickly fade from public atten-

tion. I was wrong, as my own efforts to track the success of the movement in terms of maintaining public support have clearly shown. More personally, as I began to read more by environmental scientists as well as environmentalists, I found my own outlook being challenged. Looking back, I see that during the seventies my view of nature evolved from the "industrial worldview" that I had grown up with into the "ecological worldview" I now embrace.

While it may appear that I just adopted the ideology of the movement I was studying, I think worldviews are more fundamental than ideologies and share much in common with scientific paradigms. In fact, the methodological training I've received in sociology has proven very useful in examining debates over environmental issues. While environmentalists do tend to exaggerate (as do their opponents) and are sometimes proven wrong, over the past quarter century I've been impressed by the consistent accumulation of scientific evidence indicating that industrial societies are having unprecedented impacts on the environment (from local problems such as toxic contamination to global ones such as ozone depletion).

The methodological tools provided by sociology also enable me to do research in a rigorous manner. I know that my personal values affect my decision to study environmental issues, but I try to be as objective as possible in my research. In fact, although I belong to environmental organizations and

have gained to better understand interactions and events in your daily life—with your family and friends, at work, and in your community—as well as in the larger political, economic, and social world around you.

Remember, too, that sociology is a growing and developing social science. As I have noted several times, sociologists do not pretend to have the answers to every question about the social world, and

often the answers they have developed change when more evidence becomes available. Over time sociologists' research techniques have become more sophisticated and more accurate, and research evidence and more developed theories continue to appear at an ever-increasing pace. Even if you do not become a professional social scientist, you can still read about these new findings. Some are reported in the mass media. In addition, soci-

BOX
20-1

SOCIOLOGISTS
AT WORK

Continued

contribute to environmental causes, I believe my most valuable contribution is providing useful and scientifically credible information to those involved in environmental debates and policy-making.

What surprised you most about the results you obtained in your research? While I've been following public attitudes toward environmental issues for a quarter century, clearly my most important study was the twenty-four-nation survey I directed for the Gallup International Institute in 1992. When we began work on the "Health of the Planet Survey," I frankly expected that we would find residents of the wealthier nations to be much more environmentally aware and concerned than those of the poorer nations (as was widely assumed). The fact that we found similarly high levels of environmental concern in both rich and poor nations really surprised me—and most other people as well. This is a good illustration of how social science research sometimes shows "conventional wisdom" to be wrong.

Have you or others replicated the study? What were the results? While there has not yet been another environmental survey covering citizens in such a wide range of nations, recent public opinion surveys in the industrialized nations have also shown continuing strong levels of environmental awareness and concern. Also, several new studies that document growing levels of environmental activism within the poorer nations lend support to our conclusion that environmental concern has become a worldwide phenomenon.

One focus of this book is showing how sociology can help us deal with social issues. What do you think the

implications of your work are for social policy? It has become increasingly clear that environmental problems are "people problems" — they're caused by human actions, they pose threats to human (and other) beings, and solving them requires social change. Sociology can contribute to a better understanding of all of these issues, and both natural scientists and policymakers are gradually recognizing this. The field of environmental sociology deals with such issues and offers a great way for students to apply their sociological training and imaginations to real-world problems.

Another focus of the book is showing students how sociology can help them understand events in their own lives. How does your research apply to students' lives? Most students nowadays are concerned about environmental problems but often don't do much to solve them. Social scientists have shown that things like information, convenience, and peer pressure tend to encourage people to take "ecologically responsible" actions. Think about what you do, and don't do, to help promote environmental quality, and then ask yourself what kinds of things would get you to do more. For example, is it more important to see a TV ad urging you to recycle or to find a recycling bin near your trash? More generally, to reduce gasoline consumption, do you think it's more effective to encourage people to drive less, to develop better mass transportation facilities, or to have regulations requiring new cars to get better gas mileage? Sociologists and other social scientists do research that provides answers to these kinds of questions.

ologists often write books that address social issues in terms that are accessible to the general public. Examples of these books have been cited in the list of recommended sources at the end of each chapter. Such books often are available in local bookstores and libraries, and a few even become popular paperback titles.

As you can see, my hope is that your exploration and use of sociology doesn't stop with this book and

course. I hope that you will be more inquisitive about the world around you and that the ideas and research you have studied will not only help suggest explanations for what you observe but lead you to ask new questions. It has been my pleasure to help introduce you to the discipline I have made my life's work, and I hope you will continue to find sociology both fascinating and useful.

BOX
20-2

APPLYING SOCIOLOGY
TO SOCIAL ISSUES

*Can the Practice of Female
Genital Mutilation Be
Changed?*

In recent years, people throughout the world have become more aware of the practice of female genital mutilation, a custom that is common across about two dozen African countries. The exact nature of the procedure varies from one culture to another, from a pricking of the clitoris with a sharp instrument such as a pin, to the removal and restructuring of genital tissue through suturing of the remaining parts of the organs, a practice termed infibulation. Needless to say, especially in these harsher forms, the practice is extremely painful and results in permanent scarring and lifelong health problems. Yet the practice is very widespread and, when it is present within an area, it is almost universally practiced. It is estimated that as many as 100 million women across Africa have been affected by the practice.[68]

Although the exact origin of female mutilation is not known, reports indicate that it has been part of African cultures for many years. The most common justification involves issues related to marriage and honor. The practice is generally justified as being necessary to ensure a good marriage and to attract better husbands by providing proof of a young woman's virginity. It also reduces the chances of the woman's infidelity by diminishing her sexual desire.[69]

We might expect that this practice would diminish as the African countries become more modernized. In fact, however, the practice is spreading, and observers suggest that it will take many years—even as long as three centuries—to end the practice. Yet, as the social scientist Gerry Mackie has pointed out, the very painful and dangerous practice of binding women's feet was common in China for a thousand years, but it virtually disappeared in one generation. Mackie has looked at the methods that were used to end footbinding to see if they might also be used to end female genital mutilation in Africa.

Footbinding involved gradually bending and reshaping girls' feet, binding them with tight bandages both day and night, to ultimately create a foot that was only about four inches long and was bowed and pointed. The process was extremely painful for the first 6 to 10 years and involved many physical complications, including paralysis and gangrene. About 10 percent of girls did not survive the treatment. The bound feet smelled bad, and the women were crippled and generally housebound. Nevertheless, the practice endured for centuries, beginning with the royalty and court circles, then spreading to the upper classes and gradually throughout other areas of the society.[70]

Mackie notes that the practices of female genital mutilation and footbinding have many similarities:

Both customs are nearly universal where practiced; they are persistent and are practiced even by those who oppose them. Both control sexual access to females and ensure female chastity and fidelity. Both are necessary for proper marriage and family honor. Both are believed to be sanctioned by tradition. . . . Both are exaggerated over time and both increase with status. Both are supported and transmitted by women, are performed on girls about six to eight years old, and are generally not initiation rites. Both are believed to promote health and fertility. Both are defined as aesthetically pleasing compared with the natural alternative.[71]

Like rational choice and exchange theorists, Mackie assumes that people are rational—they think about the possible costs and benefits of their actions and make choices that provide the greater benefits. He also assumes that people want to raise their children successfully, which includes having a good marriage.[72] Given these assumptions, it would be perfectly rational for people in China to bind their daughters' feet and for people in Africa to have their daughters' genitals mutilated. The costs of not doing so—the apparent impossibility of a good marriage—are just too much. People are essentially trapped by the social norms, which are self-enforcing and continuing. A survey of Somali university students reveals the dimensions of this problem. Although a majority of the men and women surveyed

BOX
20-2

APPLYING SOCIOLOGY
TO SOCIAL ISSUES

Continued

believed that female genital mutilation should be abolished, a majority also planned to mutilate their daughters! As Raqiya Abdalla, the scholar who conducted the survey, put it, "No one dares to be the first to abandon it."[73]

Yet, Chinese footbinding, which was just as mutilating and which had existed for just as many, if not more, years disappeared in a single generation at the beginning of the twentieth century. Although the exact year in which the practice disappeared varied from one area of the country to another, the available data indicate that once it started to end it ended quickly. For instance, one sociologist found that in Tinghsien, a conservative rural area about 125 miles south of Beijing, 99 percent of the young women had bound feet in 1899 but none did twenty years later![74] This drastic change occurred long before the modernization of marriage and throughout all areas of the country.[75]

In fact, the end of footbinding in China was the result of a very effective social movement. This movement had three aspects: First, the reformers conducted a modern education campaign and explained that the rest of the world did not bind women's feet and that China was losing face in the world by continuing the practice. Second, the education campaign stressed the health advantages of natural feet and the disadvantages of bound feet. Third, and absolutely critical to the success of the movement, they created "natural-foot soci-

eties," groups whose members promised not to bind their daughters' feet *and* not to allow their sons to marry women with bound feet.[76] The educational campaign was necessary for changing the paradigm or world view of the Chinese people. It helped to frame the issue in a way that they could understand—the health and well-being of their children and national honor. Yet, the natural-foot societies were also necessary for change. These societies allowed parents to know that if they did not bind their daughters' feet they would still be able to have an honorable marriage. These societies also provided social networks that would promote the arrangement of these marriages.

Mackie suggests that a similar, large–scale social movement could also bring about the end of female mutilation in Africa. He suggests that an education campaign needs to let people know about the physical dangers of mutilation and the advantages of natural genitalia and that the people who engage in this practice also need to be tactfully apprised that the international health community deplores the practice. Given the severe health consequences of female mutilation, Mackie rejects the notion that westerners are culturally insensitive by condemning the practice, but stresses the impor-

tance of remembering that practitioners are loving parents and rational actors. As he put it,

The followers of mutilation are good people who love their children; any campaign that insinuates otherwise is doomed to provoke defensive reaction. Because these parents love their children, they will be motivated to change, once they learn of the bad health consequences . . . and once a way to change is found. [Female genital mutilation] will end sooner or later; better that it ends sooner than later.[77]

The most crucial step, suggests Mackie, might be the development of associations of parents, like those in China, that pledge not to mutilate their daughters and not to allow their sons to marry mutilated daughters. These promises could create the social networks and the support that would spring the trap and allow future generations of African women to escape this practice.

CRITICAL-THINKING QUESTIONS

1. What do you think? Could establishing such a social movement end the practice of female mutilation?

2. Should the western, industrialized world be involved in these efforts? Why or why not? If yes, how should they be involved?

SUMMARY

Collective behavior is group behavior that is generally unexpected and spontaneous. Social movements are a specific form of collective behavior and represent concerted efforts by groups of individuals to promote social change. By studying social change and social movements, sociologists hope to make the world a better place. Key chapter topics include the following:

■ Collective behavior differs from conventional group behavior in that participants tend to be more emotionally involved, the behavior is often unexpected and can change or disappear rapidly, and it is governed by emergent, rather than established norms. Yet, the same types of theories that explain other types of behavior can also explain collective behavior.

■ Social movements can take many forms, focusing on individuals or on the society as a whole, on partial change or on more global change, and on efforts to effect change or to stop change.

■ Resource mobilization theory suggests that social movements are successful to the extent that they can effectively activate resources needed to accomplish their goals. With a mesolevel analysis, resource mobilization theorists often look at social movement organizations using many of the concepts and theories described in our analysis of formal organizations in Chapter 11, such as organizational ecology, organization structure and culture, and organizational leadership.

■ At the macrolevel sociologists examine why social movements develop and how political and cultural characteristics of a society influence social movements.

■ At the microlevel, sociologists have looked at why individuals choose to participate in social movements and how social movements change individuals' views and behaviors.

■ Extensive analyses of trends in attitudes toward environmental issues show broad support for the goals of the environmental movement both in the United States and in many other countries.

KEY TERMS

cognitive theory of social movements 542

collective behavior 533

collective identity 544

emergent norms 535

framing 542

grassroots organization 540

historical-comparative research 538

new social movements 537

relative deprivation 541

resource mobilization theory 537

social movement adherents 540

social movement constituents 540

social movement industry 537

social movement organizations 536

social movement sector 536

social movements 533

INTERNET ASSIGNMENTS

1. Think about a current fad or fashion—an example of collective behavior. (Recent possibilities include such diverse behaviors as collecting "beanie babies," trading Internet stocks on-line, and body piercing.) Use a search engine to locate Web sites concerning one contemporary form of such collective behavior. From reading the information on these sites, what can you learn about how these behaviors appeared and how they have spread around the country? What kinds of norms guide the behaviors? What mechanisms of social control regulate the behaviors?

2. Locate Web sites related to contemporary social movement industries, such as environmentalism, gay rights, women's rights, men's rights, racial-ethnic equality, or multiculturalism and cultural diversity, and social movement organizations such as the NAACP, the Sierra Club, or the National Organization for Women. Focus on the Web sites for two specific social movement organizations that are in the same social movement industry. Compare and contrast how these organizations frame the issues that underlie the movement. Develop hypotheses about how the different ways in which

they frame the issues could influence the characteristics of adherents and constituents that the two organizations attract? How could you test these hypotheses?

3. Using the two Web sites located for assignment two, examine how the groups mobilize resources. Can you find indications of the relative success of the two groups? How are their tactics similar and how are they different? Can you propose hypotheses that might account for these strategies and their levels of success?

TIPS FOR SEARCHING
To find social movement industries and organizations, try using both the general name of a movement industry and the names of specific organizations.

INFOTRAC COLLEGE EDITION:
READINGS

Article A20383032 / Chaves, Mark, and James Cavendish. 1997. "Recent changes in women's ordination conflicts: The effect of a social movement on intraorganizational controversy." *Journal for the Scientific Study of Religion, 36,* 574–585. Chaves and Cavendish examine how social movements have contributed to debates regarding the ordination of women in religious organizations.

Article A53436508 / Coombs, W. Timothy. "The Internet as potential equalizer: New leverage for confronting social irresponsibility." *Public Relations Review, 24,* 289–290. Coombs describes how the Internet can give social activists greater leverage and political power than they previously had because it can enhance network connections and the possibilities of effective communication.

Article A21232180 / Stein, Arlene. 1998. "Whose memories? Whose victimhood? Contests for the Holocaust frame in recent social movement discourse." *Sociological Perspectives, 41,* 519–541. Stein shows how several different social movements use the Holocaust in their attempts to frame social issues for potential adherents and constituents.

Article A20836942 / Stott, Clifford, and Steve Reicher. 1998. "How conflict escalates: the inter-group dynamics of collective football crowd 'violence.'" *Sociology, 32,* 353–378. Stott and Reicher analyze incidents of violence by English fans at the 1990 World Cup soccer competition, showing how these incidents fit the criteria of collective behavior as discussed in this chapter.

RECOMMENDED SOURCES

Dunlap, Riley E., and Angela G. Mertig (eds.). *American Environmentalism: The U.S. Environmental Movement, 1970–1990.* Philadelphia: Taylor & Francis. This nice collection of short articles describes different aspects of the environmental movement in the United States and worldwide.

Gamson, William A. 1992. *Talking Politics.* Cambridge: Cambridge University Press. Gamson examines how individuals frame social issues in their own discussions and conversations.

Klandermans, Bert. 1997. *The Social Psychology of Protest.* Cambridge, Mass.: Blackwell. Klandermans focuses on individuals' motivations for involvement in social movements.

Marx, Gary T. and Douglas McAdam. 1994. *Collective Behavior and Social Movements: Process and Structure.* Englewood Cliffs, N.J.: Prentice Hall. This short, readable text covers both collective behavior and social movements.

Miller, G. Tyler, Jr. 1996. *Living in the Environment: Principles, Connections, and Solutions,* 9th ed. Belmont, Calif.: Wadsworth. Tyler's comprehensive environmental science textbook takes an interdisciplinary approach.

Oberschall, Anthony. 1993. *Social Movements: Ideologies, Interests, and Identities.* New Brunswick: Transaction. This is a very readable collection of essays that describe a number of different social movements and incidents of collective action.

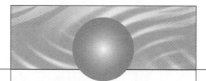

Pulling It Together

Listed below are other chapters where you can find additional discussion or examples
of some topics covered in this chapter. You can also look up topics in the index.

FURTHER READING

Suggestions for additional reading can be found in
Classic Readings in Sociology, bundled with this
book. If you purchased a used copy of this book
that does not include this custom-published reader,
go to www.sociology.wadsworth.com for ordering
information.

INTERNET SOURCE

The American Sociological Association on Social
Movements and Collective Action. This Web site lists
activities and research related to both collective
behavior and social movements..

Notes

Chapter 1

1. Mills, 1959.
2. U.S. Bureau of the Census, 1997, p. 398.
3. U.S. Bureau of the Census, 1997, p. 419.
4. Compare Henslin, 1993, p. 681.
5. Cancian, 1995; Lockwood, 1992, p. 317; Turner, 1998; Wallerstein, 1997.
6. Ritzer, 1988; Turner, Beeghley, and Power, 1997; Coser, 1977.
7. Ritzer, 1988; Turner, Beeghley, and Power, 1997; Coser, 1977.
8. Turner, 1989, p. 421.
9. Peter Bearman, personal communication, August 1992.
10. Coser, 1977, pp. 58–68.
11. For examples of Marx's writings, see Marx, 1964, 1967; also see Turner, Beeghley, and Power, 1997; Ritzer, 1988; Collins, 1985.
12. Buroway, 1990.
13. Coser, 1977, pp. 143–149.
14. See Durkheim 1964a, 1964b, 1965; also see Turner, Beeghley, and Power, 1997; Ritzer, 1988; Collins, 1985; Bearman, 1991.
15. Durkheim, 1951, pp. 241–246; also see Gibbs and Martin, 1964; Bearman, 1991.
16. See especially Gibbs and Martin, 1964.
17. Coser, 1977.
18. Coser, 1977; Gerth and Mills, 1946; Bendix, 1960.
19. See Weber, 1951, 1958a, 1958b; Bendix, 1960; Gerth and Mills, 1946; also see Turner, Beeghley, and Power, 1997; Ritzer, 1988; Collins, 1985, 1997; Landes, 1998.
20. Coser, 1977.
21. Simmel, 1950b, 1955; also see Coser, 1977; Turner, Beeghley, and Power, 1997.
22. Coser, 1977.
23. Alexander, 1988; Barnes, 1995; Coleman, 1990; Collins, 1985; Giddens, 1979; Sewell, 1992; Levine, 1991; Bourdieu, 1990; Hilbert, 1990; also see Chafetz, 1997; Levine, 1995, pp. 328–9; Lockwood, 1992; Ritzer, 1981; Turner, 1982.

Chapter 2

1. Monje, 1992.
2. Cole and Smith, 1998, p. 161.
3. Gelles and Straus, 1989; Kantor and Straus, 1990; Cole, and Smith, 1998, p. 161.
4. Sherman and Berk, 1984.
5. See Worden and Pollitz, 1984, for an example of the use of such field research.
6. See Berk and Loseke, 1980, for an example of this type of work.
7. Kantor and Straus, 1990.
8. Straus and Gelles, 1990.
9. Berk and Loseke, 1980; Worden and Pollitz, 1984.
10. Sherman and Cohn, 1989.
11. Sherman et al., 1991, 1992.
12. Sherman, 1992; Sherman et al., 1992; Berk et al., 1992.

Chapter 3

1. Gore, 1992, provided a number of details used in the following descriptions.
2. See especially Bohannan, 1995; DiMaggio, 1997
3. See White, 1940, 1949, for the classic statements on the centrality of language to culture.
4. Goldschmidt, 1960, pp. 220–221; Linton, 1936; Friedl, 1975.
5. Kroeber, 1952.
6. See Murdock, 1968; Murdock and White, 1969.
7. Friedl, 1975; Stockard and Johnson, 1992.
8. Goodman et al., 1985, pp. 1203–1204.
9. D'Andrade, 1966; Stockard and Johnson, 1992.
10. Stockard and Johnson, 1992, pp. 90–91.
11. Goldschmidt, 1960, pp. 220–221; Linton, 1936.
12. Sanders et al., 1956, pp. 196–198; Reichard, 1928.
13. Sanders et al., 1956, p. 453; Ch'Eng-K'un, 1944; Hsu, 1956.
14. Rokeach, 1973.
15. Rokeach, 1973; Lipset, 1989.
16. Ch'eng-K'un, 1944.
17. Kroeber, 1952; Sanders et al., 1956; Goldschmidt, 1966, pp. 474–482.

18. Wilkie, 1956.

19. Hsu, 1956.

20. See Parsons, 1937, 1951; Parsons, Bales, and Shils, 1953; also see Johnson, 1975; Turner, 1982; Bierstedt, 1981; Alexander, 1987; Camic, 1989.

21. Merton, 1968; Ritzer, 1988.

22. Merton, 1968; Ritzer, 1988.

23. Rokeach, 1973; Lipset, 1989.

24. See especially Harris, 1979.

25. Ekblaw, 1956.

26. See Ogburn, 1964.

27. See Hirsch, 1971; Jordan and Weedon, 1995; Calhoun, Light, and Keller, 1994, p. 67; Page, 1996; Peterson, 1979.

28. Thernstrom, 1980, and Petersen, 1975, p. 98, both cited in Jiobu, 1988, pp. 4–5; also see Royce, 1982; Yetman, 1985, pp. 6–9; Yinger, 1994.

29. Weber, 1968, p. 389, cited in Alba, 1985, p. 17; also see Alba, 1990.

30. Luomala, 1956; Hill, 1956; Kluckhorn, 1956; Wilkie, 1956.

31. Nash, 1989.

32. See Royce, 1982; Alba, 1990.

33. Yetman, 1985, p. 7.

34. Harvey, 1987; Coser, 1977; Ritzer, 1988; Carey, 1975; Ross, 1991; Boskoff, 1969.

35. See Park, 1974b.

36. Alba, 1985, pp. 48–49.

37. Alba, 1985, pp. 53–54.

38. Alba, 1985, pp. 48–49.

39. Alba, 1985, pp.53–54; Marilyn and Jack Whalen, personal communication, March 1993; Park, Burgess, and McKenzie, 1967, pp. 9–12.

40. Park, 1974c.

41. Cooley, 1937.

42. Park, 1967, p. 120; Park, 1974d.

43. Alba, 1985, p. 6; Ravitch, 1985, p. 219.

44. Park, 1974d.

45. See Alba, 1985, pp. 8–9; Jiobu, 1988, pp. 8–10.

46. Lieberson and Waters, 1988, pp. 7–8.

47. Lieberson and Waters, 1988, p. 18; also see Snipp, 1992, pp. 356–357; Jiobu, 1988, pp. 14–26; Nagel, 1996.

48. Waters, 1990, p. 78–81.

49. Waters, 1990, p. 35; see also Hout and Goldstein, 1994.

50. Waters, 1990, p. 127.

51. Waters, 1990, p. 134.

52. Gans, 1985; Lieberson and Waters, 1988; Waters, 1990; Alba, 1990.

53. Tuan, 1998; see also Rumbaut, 1996; Cerulo, 1997 for other analyses and reviews of ethnic identity

54. Kalish, 1995.

55. Kalish, 1995.

56. see Glazer, 1997; Hollinger, 1995; Root, 1996; and Zack, 1995 for discussions of issues related to multiracial heritage as well as multiculturalism.

Chapter 4

1. Vander Zanden, 1987, pp. 106–107; also see Stephan and Stephan, 1985, p. 94.

2. Vander Zanden, 1987, p. 143.

3. Elkin and Handel, 1989, pp. 10–11; see also Goode, 1994.

4. Michener, DeLamater, and Schwartz, 1990, pp. 53–55; Elkin and Handel, 1989, pp. 11–12.

5. Elkin and Handel, 1989, p. 13.

6. Kagan and Moss, 1962.

7. See Stockard and Johnson, 1992, pp. 121–126, for a summary of this research; also Stockard, 1998.

8. See Stockard and Johnson, 1992, pp. 136–151, for an extensive summary of this work.

9. See Stockard and Johnson, 1992, pp. 136–151, for an extensive summary; also see Whiting and Edwards, 1988; Maccoby and Jacklin, 1974; Eaton and Enns, 1986.

10. Rossi, 1977; also see Stockard and Johnson, 1992, pp. 131–132.

11. Eagly and Steffen, 1986; Eagly, 1987; Hyde, 1984, 1986; also see Stockard and Johnson, 1992.

12. Stockard and Johnson, 1992, p. 142.

13. Michener, DeLamater, and Schwartz, 1990, p. 57; Elkin and Handel, 1989, pp. 134–207.

14. Elkin and Handel, 1989, p. 34, citing Yarrow, 1964.

15. Elkin and Handel, 1989, pp. 34–35.

16. Maccoby, 1986, 1990; Maccoby and Jacklin, 1987.

17. Berman, 1997, p. 16; Stockard, 1998.

18. Howard and Hollander, 1997, pp. 45–47; Stephan and Stephan, 1985, pp. 100–101; Vander Zanden, 1987, pp. 120–122.

19. See Stockard and Johnson, 1992, pp. 163–167, and Stockard, 1998, for summaries.

20. Main, Tomasini, and Tolan, 1979; Lynn, 1974, cited in Stephan and Stephan, 1985, p. 101.

21. Michener, Delamater, and Schwartz, 1990, p. 65.

22. Stockard and Johnson, 1992, pp. 163–167; Stockard, 1998

23. Howard and Hollander, 1997, pp. 68–89

24. Piaget, 1954, 1965; Vander Zanden, 1987, pp. 122–127.

25. Vander Zanden, 1987, pp. 122–127, citing Cross, 1976; Elkind, 1968; Flavell, 1977.

26. See, for example, Kohlberg, 1969, 1981.

27. Kohlberg, 1966.

28. Kohlberg, 1966; Ullian, 1976; Stockard and Johnson, 1992, pp. 168–172.

29. Mead, 1934.

30. See Blumer, 1969; also see Howard and Hollander, 1997, pp. 92–98; Stryker, 1980; Michener, DeLamater, and Schwartz, 1990; Stephan and Stephan, 1985.

31. Mead, 1934.

32. Michener, DeLamater, and Schwartz, 1990, pp. 85–87.

33. Elkin and Handel, 1989, p. 67.

34. Michener, DeLamater, and Schwartz, 1990, pp. 85–87.

35. Michener, DeLamater, and Schwartz, 1990, p. 66; Vander Zanden, 1987, pp. 127–128.

36. Michener, DeLamater, and Schwartz, 1990, p. 91; see also Gecas and Burke, 1995.

37. Stryker, 1980, p. 59.

38. Gordon, 1968; also see Michener, DeLamater, and Schwartz, 1990, p. 86.

39. Gordon, 1968; Kuhn and McPartland, 1954; both cited in Michener, DeLamater, and Schwartz, 1990, pp. 92–93.

40. Stryker, 1980; McCall and Simmons, 1978; Michener, DeLamater, and Schwartz, 1990, pp. 96–97.

41. John Bergez, personal communication, 1994.

42. Michener, DeLamater, and Schwartz, 1990, p. 97.

43. Hewitt, 1988; Michener, DeLamater, and Schwartz, 1990, p. 93.

44. Fine, 1987.

45. Broverman et al., 1972; Stockard and Johnson, 1992, p. 135.

46. Fine, 1987, p. 105.

47. See Stockard, 1998 for an extensive review; also see Messner, 1992.

48. Thorne, 1993, p. 100.

49. Michener, DeLamater, and Schwartz, 1990, p. 455; Heise, 1990.

50. Michener, DeLamater, and Schwartz, 1990, pp. 455–456; also see Elder, 1974; Ebaugh, 1988; Elder and O'Rand, 1995.

51. Michener, DeLamater, and Schwartz, 1990, pp. 457–460; Settersten and Mayer, 1997.

52. Michener, DeLamater, and Schwartz, 1990, pp. 458, 462–463; also see Lowenthal et al., 1975.

53. Elkin and Handel, 1989, pp. 56–57; Michener, DeLamater, and Schwartz, 1990, pp. 77–78.

54. See Bengtson, Schaie, and Burton, 1995 and Bengtson and Achenbaum, 1993.

55. Michener, DeLamater, and Schwartz, 1990, pp. 484–487.

56. See Light, 1988, for an extensive discussion of one birth cohort; see also Elder, 1974; Holahan and Sears, 1995.

57. Wald, 1992; Zuckerman, 1991; Kronstadt, 1991.

58. Wald, 1992.

59. Wald, 1992; Kumpfer, 1991.

60. Wald, 1992; Kumpfer, 1991; Center for the Future of Children, 1991.

Chapter 5

1. Marsden, 1990.

2. Wellman, 1983.

3. Vander Zanden, 1987, p. 266.

4. Wellman, 1983.

5. Vander Zanden, 1987, pp. 267–269.

6. Vander Zanden, 1987, pp. 267–269; Granovetter, 1973; Wellman and Wortley, 1990; Freeman, 1992.

7. Coleman, 1990.

8. Homans, 1950, 1961; Howard and Hollander, 1997, pp. 47–64; Cook and Whitmeyer, 1992; Cook, 1991; Frank, 1992; Hechter and Kanazawa, 1997; Michener, DeLamater, and Schwartz, 1990; Molm and Cook, 1995; Stephan and Stephan, 1985; Turner, 1982; Vander Zanden, 1987, pp. 289–293.

9. Stephan and Stephan, 1985, pp. 320–322; Hegtvedt and Markovsky, 1995.

10. Vander Zanden, 1987, p. 435, citing Franklin, 1952, and Genovese, 1974.

11. See Vander Zanden, 1987, pp. 433–458, for an extensive discussion of power in interpersonal interactions.

12. See Cole, 1995, pp. 347–352.

13. Turner, 1982.

14. Vander Zanden, 1987, p. 238; also see Stryker, 1980; Michener, DeLamater, and Schwartz, 1990; Stephan and Stephan, 1985.

15. Vander Zanden, 1987, pp. 239–241.

16. Vander Zanden, 1987, p. 241.

17. Vander Zanden, 1987, pp. 242–243.

18. Vander Zanden, 1987, p. 244.

19. Branaman, 1997; Goffman, 1959; Vander Zanden, 1987, pp. 249–252; Michener, DeLamater, and Schwartz, 1990; Stephan and Stephan, 1985.

20. Branaman, 1997; Goffman, 1974.

21. Goffman, 1963, 1967; Howard and Hollander, 1997, pp. 98–104; Wishman, 1981, cited in Vander Zanden, 1987, pp. 250–251.

22. Goffman, 1959.

23. Vander Zanden, 1987, p. 255.

24. See Mead, 1934; also see Blumer, 1969; Stryker, 1980; Michener, DeLamater, and Schwartz, 1990; Stephan and Stephan, 1985.

25. Samovar and Porter, 1991, p. 175.

26. Stephan and Stephan, 1985, p. 23.

27. Vander Zanden, 1987, pp. 78–89; Michener, DeLamater, and Schwartz, 1990, pp. 194–197.

28. See Ekman and Friesen, 1981; Ekman, Friesen, et al., 1987; Ekman, 1972, 1980; also see Michener, DeLamater, and Schwartz, 1990, pp. 197–199; Vander Zanden, 1987, pp. 89–91.

29. Eibl-Eibesfeldt, 1979.

30. Barna, 1991, pp. 346–347.

31. Maynard and Whalen, 1995; Wooffitt, 1990.

32. See Schegloff, 1968; Michener, DeLamater, and Schwartz, 1990, pp. 208–209.

33. Maynard, 1978, cited in Michener, DeLamater, and Schwartz, 1990.

34. Michener, DeLamater, and Schwartz, 1990, p. 209; Schegloff, 1996b; Ford and Thompson, 1996.

35. Michener, DeLamater, and Schwartz, 1990, pp. 210–211; Schegloff, 1996a.

36. Schegloff, 1992; Fox, Hayashi, and Jasperson, 1996.

37. Whalen and Zimmerman, 1987.

38. Whalen, Whalen, and Henderson, 1995; Marilyn Whalen, personal communication, 1995, 1998; see also Goodwin, 1996.

39. Whalen, 1992; Wardhaugh, 1985.

40. Stephan and Stephan, 1985, pp. 319–320.

41. Stephan and Stephan, 1985, pp. 320–322.

42. See Berger, Cohen, and Zelditch, 1972; Berger et al., 1977; Michener, DeLamater, and Schwartz, 1990, pp. 375–380; Stephan and Stephan, 1985, pp. 320–322.

43. See Berger, Cohen, and Zelditch, 1972; Berger et al., 1977; Berger and Zelditch, 1993; also see Howard and Hollander, 1997, pp. 104–113; Michener, DeLamater, and Schwartz, 1990, pp. 375–380; Ridgeway and Walker, 1995; Stephan and Stephan, 1985, pp. 320–322.

44. Fishman, 1983.

45. Kollock, Blumstein, and Schwartz, 1985; Smith-Lovin and Brody, 1989.

46. West and Zimmerman, 1983, p. 109; West, 1982.

47. Porter and Geis, 1981.

48. Cohen and Roper, 1972.

49. Cohen and Lotan, 1995, 1997, Cohen 1994a, b.

50. See Blau and Duncan, 1967; Duncan, Featherman, and Duncan, 1972; Kerckhoff, 1995; Sewell and Hauser, 1975, 1980; Sewell, Hauser, and Featherman, 1976.

51. Campbell, 1983.

52. Compare Clausen, 1991, and Buchmann, 1989.

Chapter 6

1. See Gibbs, 1981, for an extensive discussion of the definition of these terms; also see Aday, 1990, pp. 49–50; Gibbs, 1989, 1994.
2. Messner, Krohn, and Liska, 1989, p. 140.
3. Adler, Mueller, and Laufer, 1991, pp. 30–32; O'Brien, 1985.
4. Adler, Mueller, and Laufer, 1991, p. 31; O'Brien, 1985.
5. O'Brien, 1990, 1991; Blumstein, Cohen, and Rosenfeld, 1991.
6. Adler, Mueller, and Laufer, 1991, p. 42; Cole and Smith, 1998, p. 61; see also Thornberry, 1997.
7. See Cohen and Land, 1987.
8. O'Brien, 1989; Steffensmeier, Streifel, and Shidaheh, 1992.
9. O'Brien, Stockard, and Isaacson, 1999; Stockard and O'Brien, 1999.
10. Adler, Mueller, and Laufer, 1991, pp. 38, 48.
11. U.S. Bureau of the Census, 1997, p. 103.
12. Compare Messner, Krohn, and Liska, 1989; Farrell and Swigert, 1988, pp. 1–5; Aday, 1990.
13. Durkheim, 1938; Aday, 1990; Farrell and Swigert, 1988, p. 2; Erikson, 1966.
14. Durkheim, 1938, p. 71.
15. Compare Aday, 1990, pp. 27–33; Chambliss and Seidman, 1971; Spitzer, 1975.
16. See Bendix, 1962, pp. 291–294; Weber 1967b, pp. 295–301; also see Aday, 1990, pp. 35–36; Tyler, 1990.
17. Bendix, 1962; Boskoff, 1969, p. 50; Gerth and Mills, 1967, pp. 59–60; Weber, 1967a, pp. 323–324.
18. See Bendix, 1962, pp. 296–297.
19. Emler and Reicher, 1995; Hagan and McCarthy, 1997a,b; Heimer and Matsueda, 1994; Henry and Einstadter, 1998, p. 7; Matsueda, 1982; Osgood et al., 1996.
20. Matsueda, 1992; Lemert, 1951; Becker, 1963; Farrell and Swigert, 1988, pp. 218–221; Hagan and Palloni, 1990.
21. Merton, 1938, 1967.
22. Adler, Mueller, and Laufer, 1991, pp. 45–47.
23. Adler, Mueller, and Laufer, 1991, pp. 283–303; Braithwaite, 1979, 1984; Sutherland, 1983; see also Benson, 1998; Szasz, 1998; Zey, 1993.
24. Agnew, 1998, p. 177.
25. Agnew, 1998, p. 182.
26. Cohen, 1997; Passas, 1997; Hagan and McCarthy 1997a; see Passas and Agnew, 1997; Agnew, 1998, and Agnew, 1997, for extensive discussions of general strain theory.
27. Compare Adler, Mueller, and Laufer, 1991.
28. Cloward and Ohlin, 1960; Becker, 1953.
29. Sutherland, 1939; also see Akers, 1998; Elliott and Menard, 1996; Flannery, Huff, and Manos, 1998; Marcos, Bahr, and Johnson, 1986; Seydlitz and Jenkins, 1998.
30. See, for example, Thrasher, 1927.
31. See Cummings and Monti, 1993; Decker and Van Winkle, 1996; Glaser, 1956; Hagedorn and Macon, 1988; Huff, 1990; Jankowski, 1991; Klein, 1995; Rosecrance, 1998; Shaw and McKay, 1942; Wright and Decker, 1997.
32. Adler, Mueller, and Laufer, 1991, pp. 136–151.
33. Compare Gibbs, 1989; Black, 1984; Horwitz, 1990.
34. Durkheim, 1961, p. 64, quoted in Hirschi, 1988, p. 343.
35. O'Brien, Stockard, and Isaacson, 1999; Stockard and O'Brien, 1999.
36. Kornblum and Julian, 1992; also see Messner and Rosenfeld, 1994.
37. Adler, 1983.
38. Hirschi, 1969; Aday, 1990; Coleman, 1993; Gottfredson and Hirschi, 1990.
39. Loeber and Stouthamer-Loeber, 1986; Loeber, 1996; Gottfredson and Hirschi, 1990, pp. 97–105; McCord, 1979; Seydlitz and Jenkins, 1998; Wells and Rankin, 1998.
40. See Sampson and Laub, 1990, p. 610, citing Olweus, 1979, pp. 854–855; Loeber, 1982; Heusmann, Eron, and Lefkowitz, 1984; also see McCord, 1979; Sampson and Laub, 1992, 1997; Thornberry, 1996; Tracy and Kempf-Leonard, 1996; Wolfgang, Thornberry, and Figlio, 1987, pp. 33–36.
41. Adler, Mueller, and Laufer, 1991, pp. 41–43; Gottfredson and Hirschi, 1990, pp. 124–144.
42. See Gove, 1985, p. 123, cited in Sampson and Laub, 1990; also see Wolfgang, Thornberry, and Figlio, 1987, pp. 37–44.
43. See Laub and Sampson, 1991, for an extensive discussion of the Gluecks' academic careers and conflicts with sociologists of their day.
44. Sampson and Laub, 1990, 1993.
45. Horney, Osgood, and Marshall, 1995.
46. Hagan and Peterson, 1995; Short, 1997.
47. Hagan and McCarthy, 1997b, pp. 8–10
48. Hagan and McCarthy, 1997b, pp. 231–234
49. Hagan and McCarthy, 1997b, p. 232
50. Hagan and McCarthy, 1997b, p. 234

Chapter 7

1. Stark, 1994, p. 235.
2. Hochschild, 1995, pp. 15–38; Schwarz, 1997.
3. Warner and Lunt, 1942, pp. 38–43.
4. Gilbert and Kahl, 1993, p. 22; also see Warner and Lunt, 1942.
5. Warner and Lunt, 1942, pp. 53–76.
6. Warner and Lunt, 1942, p. 81.
7. Warner and Lunt, 1942, pp. 81–84.
8. See, for example, Warner, 1949a; Davis, Gardner, and Gardner, 1941.
9. Cromwell, 1994; Davis, Gardner, and Gardner, 1941; Gatewood, 1990.
10. Gilbert and Kahl, 1993, p. 38.
11. Hodge, Siegal, and Rossi, 1964; Hodge, Treiman, and Rossi, 1966; Gilbert and Kahl, 1993, pp. 38–42; also see Miller, 1991, pp. 323–365, for an extensive discussion of different ways sociologists measure social class.
12. Duncan, 1961, cited in Gilbert and Kahl, 1993.
13. Gilbert and Kahl, 1993, pp. 306–317.
14. Marx and Engels, 1939, pp. 48–49, cited in Bendix and Lipset, 1966, p. 9.
15. See Bendix and Lipset, 1966, for a good summary of Marx's views.
16. See Wright, 1985, 1989a, 1989b, 1997; Grimes, 1991; also see Clement and Myles, 1994.
17. Weber, 1946c, 1968; Collins, 1985; Blau, 1977, pp. 221–224; Scott, 1996.

18. Davis and Moore, 1945; also see Parsons, 1940; Davis, 1942; and, for a historical overview of this tradition, see Grimes, 1991.
19. Stark, 1994, pp. 248–251; also see Wesolowski, 1966.
20. Stark, 1989, pp. 248–251.
21. Benton Johnson, personal communication, 1994; also see Stinchcombe, 1968.
22. See Davis, 1966, p. 59.
23. See especially Tumin, 1953; Wrong, 1945.
24. Lenski, 1966; also see Turner, 1984.
25. Lenski, 1966, pp. 94–95.
26. See Lenski, 1966, pp. 96–101.
27. Lenski, 1966, pp. 119–122, 143–145.
28. Lenski, 1966, pp. 117–188.
29. Lenski, 1966, pp. 62–63.
30. Lenski, 1966, pp. 189–190.
31. Lenski, 1966, pp. 192–210.
32. Lenski, 1966, pp. 297–308.
33. Lenski, 1966, p. 308.
34. Lenski, 1966, pp. 313–318.
35. Gilbert and Kahl, 1993, pp. 145–147.
36. Gilbert and Kahl, 1993, p. 147.
37. Gilbert and Kahl, 1993, pp. 70–74.
38. Gilbert and Kahl, 1993, pp. 145–156; also see Erikson and Goldthorpe, 1992; Hauser and Featherman, 1977; and see Hout, 1989, for an analysis of this process in Ireland.
39. Gilbert and Kahl, 1993, pp. 78–82, 278.
40. Gilbert and Kahl, 1993, p. 155; Gilbert, 1998, p. 153; U.S. Bureau of the Census, 1998a.
41. Gilbert and Kahl, 1993, pp. 274–275.
42. Gilbert, 1998, pp. 256–257.
43. Dalaker and Naifeh, 1998, p. xi.
44. Gilbert, 1998, pp. 257–258, citing Ornati, 1966.
45. Gilbert and Kahl, 1993, p. 276, citing Rainwater, 1974.
46. Gilbert, 1998, p. 259; see also Edin and Lein, 1997, pp. 20–59.
47. Gilbert and Kahl, 1993, pp. 276–278; also see Ruggles, 1990; National Research Council, 1996.
48. Calculated from U.S. Bureau of the Census, 1997, p. 465.
49. Gilbert and Kahl, 1993, p. 272.
50. Dalaker and Naifeh, 1998, pp. viii–xi.
51. Rosenblatt, 1994.
52. Gilbert and Kahl, 1993, pp. 286–289.
53. U.S. Bureau of the Census, 1997, p. 472.
54. Danziger and Gottschalk, 1995; Gilbert, 1998, pp. 101–113; Hacker, 1997; Levy, 1995; Nielsen and Alderson, 1997; Passell, 1998; Pfaff, 1995; Farley, 1996, pp. 64–107; U.S. Bureau of the Census, 1998b; Weinberg, 1996.
55. Gilbert and Kahl, 1993, p. 12.
56. Gilbert, 1998, p. 103.
57. Gilbert, 1998, p. 104.
58. Allen, 1987, p. 23, quoting Nathaniel Burt.
59. See Allen, 1987, p. 23.
60. Allen, 1987, p. 2.
61. See, for example, Bourdieu, 1977.
62. Allen, 1987, p. 23; see also Erickson, 1996, for a discussion of cultural and social capital.
63. Allen, 1987, pp. 2–3.
64. Allen, 1987, pp. 26–31, 307–309, 420–421.
65. Allen, 1987, p. 168.
66. Allen, 1987, pp. 66, 250; also see Kadushin, 1995 for an analysis of the role of friendship patterns, or social capital, among the financial elite in France.
67. Allen, 1987, p. 299; also see Allen, 1991; Allen and Broyles, 1989, pp. 297–302.
68. Allen, 1987, p. 34.
69. Allen, 1987, pp. 38–41.
70. Allen, 1987, p. 284.
71. Allen, 1987, p. 286.
72. Allen, 1987, p. 290.
73. Allen, 1987, p. 302.
74. Allen, 1987, pp. 290–296.
75. Allen, 1987, p. 303.
76. See Hogan and Lichter, 1995; Hernandez, 1997; Blank, 1997.
77. Dalaker and Naifeh, 1998, pp. C-6 to C-8.
78. Caspi et al., 1998; Duncan and Brooks-Gunn, 1997; Duncan, Yeung, Brooks-Gunn, and Smith, 1998; Mayer, 1997.
79. See Bergmann, 1996, pp. 5–6.
80. Chase-Lansdale and Brooks Gunn, 1995, p. 1.
81. Bergmann, 1996; McFate, Lawson, and Wilson, 1995.
82. Bergmann, 1996, pp. 6–7.

Chapter 8

1. Compare Marger, 1994, p. 7; Marx, 1998.
2. See Burkey, 1978; Pettigrew, 1964; Marger, 1994, pp. 17–26.
3. Marger, 1994, pp. 87–88.
4. Marger, 1997, p. 71; Duckitt, 1992.
5. Myrdal, 1962, p. li.
6. Myrdal, 1962, p. xxv.
7. See Hacker, 1992.
8. Gittler, 1956; Yetman, 1985, p. 1.
9. U.S. Bureau of the Census, 1997, p. 31.
10. del Pinal and Singer, 1997.
11. Marger, 1994, p. 338; Lee, 1998.
12. U.S. Bureau of the Census, 1995, p. 31.
13. U.S. Bureau of the Census, 1997, p. 51; Marger, 1994, pp. 165–166.
14. Marger, 1994, p. 340; Lee, 1998.
15. Marger, 1994, pp. 299–300.
16. Marger, 1994, p. 166.
17. O'Neill, 1992.
18. Horne, 1997.
19. Marger, 1994, pp. 344–345; Marger, 1997, p. 337.
20. U.S. Department of Education, 1997, p. 17.
21. U.S. Department of Education, 1997, p. 111.
22. Marger, 1994, pp. 341–344.
23. Eller and Fraser, 1995; see Oliver and Shapiro, 1995, for an extensive discussion of these differences.
24. Wilson, 1980.
25. Bonacich, 1972, 1973, 1976.
26. Wilson, 1980, pp. 78–82.
27. Marger, 1994, pp. 240–241.
28. Bonacich, 1972, 1973, 1976; see also Min, 1996.

29. See Hill and Jones, 1993; Clayton, 1996, for discussions of these effects and the changes that occurred.

30. Farley, 1996, pp. 208–271, Thernstrom and Thernstrom, 1997.

31. O'Neill, 1992, p. 142.

32. U.S. Bureau of the Census, 1997, p. 285.

33. Marger, 1994, pp. 262–264.

34. See Wilson, 1980; Landry, 1987.

35. Dalaker and Naifeh, 1998, p. vii, and DeBarros and Bennett, 1998, Table 10.

36. See Wilson, 1980; also see Brooks, 1990; Hochschild, 1995.

37. Wilson, 1987, p. 39; also see Wilson, 1997.

38. Compare Willie, 1979, 1983; Omi and Winant, 1986; for more sympathetic critiques, see Jencks, 1992; Jencks and Peterson, 1991; Massey, 1990; Shulman and Darity, 1989.

39. U.S. Bureau of the Census, 1994, p. 418; 1997, p. 420; also see Oliver and Shapiro, 1995; Zwerling and Silver, 1992.

40. Collins, 1997; O'Neill, 1992.

41. Marger, 1994, pp. 169–170.

42. Marger, 1994, p. 334.

43. Marger, 1994, p. 315, citing Westerman, 1989.

44. Marger, 1994, pp. 237–243.

45. Schuman, Steeh, and Bobo, 1985; Schuman and Bobo, 1988; Steeh and Schuman, 1992.

46. Schuman, Steeh, and Bobo, 1985, pp. 74–76; Marger, 1994, p. 276, for 1989 data.

47. Firebaugh and Davis, 1988; Steeh and Schuman, 1992.

48. Schuman, Steeh, and Bobo, 1985, p. 137; Kinder and Sanders, 1996.

49. Schuman, Steeh, and Bobo, 1985, pp. 109–110.

50. Firebaugh and Davis, 1988.

51. Pearlin, 1954, p. 48.

52. Pearlin, 1954.

53. Merton, 1949; Marger, 1994, pp. 100–101.

54. Feagin, 1991; also see Feagin and Sikes, 1994; Cose, 1993.

55. Feagin, 1991, p. 105.

56. Feagin, 1991, p. 106.

57. Feagin, 1991, p. 113.

58. Feagin, 1991, p. 114.

59. Feagin, 1991, p. 105.

60. Feagin, 1991, p. 115.

61. Feagin, 1991, p. 115.

62. Quillian, 1995; also see Blalock, 1956; Blumer, 1958; and Bobo, 1983; 1988, all cited in Quillian, 1995.

63. Quillian, 1995, 1996.

64. Quillian, 1995, p. 602.

65. Quillian, 1996.

66. See Miller and Brewer, 1984, pp. 2–3, for a discussion of this heritage; also see Blalock, 1967, and Blalock and Wilken, 1979, for a more sociological approach and Forbes, 1997 for a critique.

67. Miller and Brewer, 1984, p. 2.

68. Moskos and Butler, 1996, pp. 2–6.

69. Moskos and Butler, 1996, pp. 16–35.

70. Moskos and Butler, 1996, pp. 70–71.

Chapter 9

1. U.S. Bureau of the Census, 1998b, calculated from pp. 31, 33.

2. See Hewlett, 1991; Hochschild, 1989.

3. Stockard and Johnson, 1992, pp. 3–4.

4. See, for example, Kulis, 1988; Kulis and Miller, 1989; Kulis et al., 1986.

5. See Stockard and Johnson, 1992, p. 10.

6. D'Andrade, 1966.

7. Mukhopdhyay and Higgins, 1988, pp. 468–481.

8. Weiner, 1976; Stockard and Johnson, 1992, pp. 90–91.

9. See Whyte, 1978, p. 89.

10. Compare Quinn, 1977; Whyte, 1978.

11. Whyte, 1978.

12. Snyder and Hoffman, 1993, p. 186.

13. Lawton, 1994.

14. Blau and Ferber, 1992, p. 144; U.S. Department of Education, National Center for Education Statistics, 1998, pp. 12–14.

15. Bielby and Baron, 1984, 1986; see also Petersen and Morgan, 1995.

16. See Tang and Smith, 1996, for discussion of these trends.

17. Reskin and Roos, 1990; Blau and Ferber, 1992, p. 129; U.S. Bureau of the Census, 1995, p. 412; see also Hagan and Kay, 1995.

18. Stockard, 1985; U.S. Bureau of the Census, 1998b.

19. Bernhardt, Morris, and Handcock, 1995, 1997; Bianchi and Spain, 1996; Bianchi, 1995; Morgan, 1998; Spain and Bianchi, 1996.

20. Sørensen and Trappe, 1995; Stockard and Johnson, 1992, pp. 79–83; Wright, Baxter, and Birkelund, 1995.

21. Beeghley, 1996; Casper, McLanahan, and Garfinkel, 1994; Spain and Bianchi, 1996, pp. 136–138.

22. See Ridgeway, 1997.

23. U.S. Bureau of the Census, 1997, p. 288; also see Manza and Brooks, 1998, p. 1236.

24. Center for the American Woman and Politics, 1998a, b; U.S. Bureau of the Census, 1997, p. 285.

25. Stockard and Johnson, 1992, pp. 85–87; United Nations, Department for Economic and Social Information and Policy Analysis, 1997, p. 157.

26. Cook, Thomas and Wilcox, 1994, p. 18.

27. Stockard and Johnson, 1992, pp. 22–24, Cook, Thomas and Wilcox, 1994; Seltzer, Newman, and Leighton, 1997; Thomas and Wilcox, 1998.

28. Acock, 1994, pp. 74–88,102–103; Stockard and Johnson, 1992, p. 56; Berk, 1985, pp. 70–77; Hochschild, 1989; Spain and Bianchi, 1996.

29. Stockard and Johnson, 1992, p. 56; England and Farkas, 1986, pp. 53–55; see also Potuchek, 1997.

30. U.S. Bureau of the Census, 1997, p. 404.

31. Bianchi and Spain, 1996, p. 32; Spain and Bianchi, 1996, pp. 169–171.

32. Coverman, 1985; Barnett and Baruch, 1987; Gerson, 1993; England and Farkas, 1986, p. 99; Pleck, 1985, p. 50; Risman, 1998; Ybarra, 1982; Glass, 1998; Coltrane, 1996.

33. Spain and Bianchi, 1996, pp. 169–71; Bianchi and Spain, 1996, p. 32; Stockard and Johnson, 1992, p. 57; Pleck, 1985; Huber and Spitze, 1983; Barnett and Baruch, 1987; Baruch and Barnett, 1986.

34. Stockard and Johnson, 1992, p. 56; also see Bergmann, 1986, p. 269; Blumstein and Schwartz, 1983, p. 139; McDonald, 1980; Risman, 1998.

35. Bergmann, 1996; Blau and Ehrenberg, 1997.

36. Described in Reskin and Padavic, 1994, p. xiii.

37. Doeringer and Piore, 1971; Piore, 1971; Blau and Ferber, 1992, p. 216; Stockard and Johnson, 1992, p. 39.

38. Stockard and Johnson, 1992, pp. 39–40; Blau and Ferber, 1992, p. 215; see also Marini and Fan, 1997.

39. Jencks, Perman, and Rainwater, 1988; see also Jacobs and Steinberg, 1990, 1995.

40. Johnson, 1988; Stockard and Johnson, 1979; Chodorow, 1978, 1989; also see Smelser, 1998, for a more general discussion of the role of rationality, especially as embodied in rational choice theory, in sociology.

41. Chehrazi, 1986; Fliegel, 1986, pp. 19–22; Alpert and Spencer, 1986; Fast, 1984; Mendell, 1982.

42. See especially Jones, 1933, 1935; Klein, 1960; Horney, 1967; Deutsch, 1944–45; Fairbairn, 1952; Mead, 1974.

43. See Stockard and Johnson, 1979, for a summary of this tradition.

44. Mead, 1949, pp. 168–169.

45. Williams, 1989, p. 16.

46. Williams, 1989, p. 14.

47. Williams, 1989, pp. 91, 114.

48. Williams, 1989, p. 116.

49. Williams, 1989, p. 66.

50. Williams, 1989, p. 66.

51. Williams, 1989, p. 63.

52. Schulte, 1994.

53. Williams, 1989, pp. 140–141.

54. Wood, 1994, p. 129.

55. Stockard and Johnson, 1992.

Chapter 10

1. Content inspired to some extent by Cherlin, 1992, p. 1, and Ahrons and Rodgers, 1987.

2. See Reiss, 1976, p. 21, citing Davis, 1950.

3. Leslie, 1967, pp. 5–6, citing Hertzler, 1961, and Merton, 1967.

4. See Parsons, 1966; also see Johnson, 1975; Glass and Klein, 1990, p. 9.

5. Leslie, 1967, pp. 106, 169.

6. Form, 1990.

7. Leslie, 1967, p. 12.

8. Hernandez, 1993, pp. 64–66.

9. See Blumstein and Schwartz, 1983, and Tasker and Golombok, 1997, for exceptions.

10. The distinction between the two ways nuclear families can be combined was originally developed by Murdock, 1949, pp. 1–2; also see Saxton, 1993, p. 224. See Stephens, 1963, for an extensive discussion of cross-cultural forms of the family with many examples.

11. Leslie, 1967, pp. 26–34; see also Jones, 1994, pp. 268–287.

12. Woolf, 1975, p. 240.

13. Goode, 1963.

14. Leslie, 1967, pp. 34–36.

15. Lamanna and Riedmann, 1997, pp. 30–35.

16. Leslie, 1967; Goode, 1963.

17. Leslie, 1967, p. 34.

18. Leslie, 1967, pp. 35–37.

19. Compare Rodgers and White, 1993, p. 236.

20. Kalmijn, 1991a, 1991b, 1994a; Pagnini and Morgan, 1990.

21. See, for example, Reiss, 1976, pp. 18–25, citing Yarrow et al., 1964, and Bowlby, 1951; Glenn, 1993; Popenoe, 1993a, 1993b.

22. See Saxton, 1993, p. 227; Reiss, 1976, pp. 12–14.

23. Mintz and Kellogg, 1988; Skolnick, 1991; Uhlenberg, 1992.

24. Cherlin, 1992, pp. 8–11.

25. Aulette, 1994, p. 264.

26. Cherlin, 1992, pp. 94–95; Schoen, 1995; Staples and Johnson, 1993, pp. 94–120; Tucker and Mitchell-Kernan, 1995.

27. See Wilson, 1980, 1987; Cherlin, 1992; Testa and Krogh, 1995; also see Lichter, LeClere, and McLaughlin, 1991.

28. Cherlin, 1992, pp. 11–13.

29. Cherlin, 1992, p. 14; also see Sweet and Bumpass, 1992; Rodgers and White, 1993.

30. See Lamanna and Riedmann, 1997, also Jacques, 1998.

31. See Aulette, 1994, pp. 50–51; Cherlin, 1992, pp. 18–20; also see Blumin, 1977; Hareven and Vinovskis, 1978.

32. Calculated from Ventura, Martin, Mathews, and Clarke, 1996, p. 29.

33. U.S. Bureau of the Census, 1997, p. 79.

34. Cherlin, 1992, pp. 96–98; U.S. Bureau of the Census, 1995, p. 73; also see Luker, 1992.

35. See McLanahan and Sandefur, 1994, pp. 139–143.

36. Calculated from U.S. Bureau of the Census, 1997, p. 59; also see Goldscheider and Waite, 1991; Cherlin, 1992, pp. 68–70.

37. Cherlin, 1992, pp. 20–25; Clarke, 1995, p. 9, U.S. Bureau of the Census, 1997, p. 105; Jacobson, 1959, 90–95; also see Riley, 1991, for an extensive narrative history of divorce in the United States; Phillips, 1988, for a historical analysis over an even longer period of time; and Goode, 1992, 1993, for cross-cultural comparisons.

38. Cherlin, 1992, pp. 94–95.

39. Furstenberg, 1990; Goode, 1992, 1993.

40. Cherlin, 1992, pp. 43–61; Ruggles, 1997; but see Oppenheimer, 1997 for a contrary argument.

41. Cherlin, 1992, p. 25; also see Aulette, 1994, p. 287.

42. Cherlin, 1992, p. 28; U.S. Bureau of the Census, 1997, pp. 105–107.

43. See, for example, Popenoe, 1993a, 1993b; Glenn, 1993; Stacey, 1993; Cowan, 1993.

44. Uhlenberg, 1992.

45. Cherlin, 1992, p. 28.

46. Cherlin, 1992, p. 26; see also Ruggles, 1994.

47. Beeghley, 1996, p. 27; Donovan, 1998; McLanahan and Sandefur, 1994, pp. 138–139.

48. See Goode, 1992, pp. 30–47; Goode, 1993; Jones, 1994.

49. See, for example, Furstenberg, 1990, 1992; Furstenberg and Harris, 1992; Teachman, 1992.

50. Also see Toman, 1993, for a similar perspective, but from a psychologist rather than a sociologist.

51. Cherlin, 1992, p. 69.

52. See, for example, Skolnick, 1991.

53. Compare Wilkie, 1993.

54. Ahrons and Rodgers, 1987, p. 23.

55. Also see Johnson, 1988.

56. See Ahrons and Rodgers, 1987; also see Cherlin, 1992, pp. 80–86; Pasley and Ihinger-Tallman, 1982, cited in Ahrons and Rodgers, 1987.

57. See White, 1991, pp. 50–52; Rodgers and White, 1993, p. 238; Aldous, 1978; also see Hareven, 1978.

58. See especially Rodgers, 1973; White, 1991; Rodgers and White, 1993; Aldous, 1978, 1996; also see Klein and White, 1996, pp. 119–148.

59. See Rodgers and White, 1993; White, 1991; Rodgers, 1973; for examples of research that utilizes a developmental perspective, see Riley, 1992; Rossi and Rossi, 1990; Chiriboga, Catron, and associates, 1991; Kitson and Holmes, 1992.

60. Rodgers, 1973; White, 1991; Rodgers and White, 1993, p. 234.

61. Rodgers and White, 1993, p. 240.

62. See Whyte, 1990, pp. 55–63.

63. See Ahrons and Rodgers, 1987, especially pp. 103–137.

64. Cherlin, 1992.

65. Compare Hareven, 1982.

66. See Rodgers and White, 1993; White, 1991; also see Waite and Lillard, 1991, and Morgan, Lye, and Condran, 1988, for empirical examples.

67. This discussion is largely based on McLanahan and Sandefur, 1994, pp. 7–17.

68. Moynihan, 1965.

69. See McLanahan and Sandefur, 1994, pp. 7–8.

70. See especially Herzog and Sudia, 1973.

71. See, for example, Stack, 1974.

72. See, for example, Wallerstein and Kelly, 1980; Hetherington, Cox, and Cox, 1978; also see Wallerstein and Blakeslee, 1989.

73. See, for example, Stacey, 1993; Cowan, 1993.

74. McLanahan and Sandefur, 1994, pp. 2–3; also see McLanahan, 1991.

75. Compare Popenoe, 1993a, 1993b; Glenn, 1993; also see Thornton, 1991, 1992; Krantz, 1992; Biblarz and Raftery, 1993; McLanahan and Bumpass, 1988.

76. See also Morgan, 1991; Weitzman and Maclean, 1992; Peterson, 1989.

77. McLanahan and Sandefur, 1994, pp. 157–161.

78. McLanahan and Sandefur, 1994, pp. 40–48.

79. McLanahan and Sandefur, 1994, pp. 48–51.

80. See also Cherlin, Kiernan, and Chase-Lansdale, 1995; Wu, 1996.

81. McLanahan and Sandefur, 1994, pp. 39–78.

82. McLanahan and Sandefur, 1994, pp. 39–43, 60–61.

83. McLanahan and Sandefur, 1994, p. 82.

84. McLanahan and Sandefur, 1994, p. 134.

85. McLanahan and Sandefur, 1994, p. 127; for further research in this area see Acock and Demo, 1994; Buchanan, Maccoby, and Dornbusch, 1996; Amato and Booth, 1997.

Chapter 11

1. McNall and McNall, 1992, p. 164; also see Morgan, 1990, pp. 4–6; Hage and Aiken, 1970, pp. 6–11.

2. Compare Morgan, 1990.

3. John Clark, personal communication, September 1994.

4. Compare Scott, 1987.

5. See Blau and Meyer, 1987, pp. 3–4.

6. Blau and Meyer, 1987, pp. 3–4.

7. Blau and Meyer, 1987, pp. 7–9; Bendix, 1962, p. 424; Weber, 1946a, pp. 196–204.

8. Weber, 1946a, pp. 221–222.

9. Weber, 1946a, pp. 212–213, 215.

10. Weber, 1946a, p. 214; also see Meyer and Rowan, 1991, p. 23.

11. Weber, 1946a, pp. 224–235; also see Blau and Meyer, 1987, pp. 24–25.

12. Compare Blau and Meyer, 1987, pp. 4–5.

13. See Blau, 1974; Blau and Meyer, 1987; and Ranger-Moore, 1997, as examples of discussions of the role of size in organizational structure and success.

14. Compare Blau and Meyer, 1987; Blau, 1974; Hall, 1991, pp. 48–83; also see Pugh et al., 1968; Scott, 1987.

15. Hall, 1991, pp. 52–55.

16. Hall, 1991, p. 54.

17. Hall, 1991, p. 196; Hage and Aiken, 1970, pp. 33–38; Hage, 1980, p. 44.

18. Hall, 1991, pp. 62–64.

19. Hage and Aiken, 1970, pp. 43–45; also see Hall, 1991, p. 196; Hage, 1980, p. 44.

20. Compare Peters, 1993.

21. Blau and Meyer, 1987, p. 100; Hage, 1980, p. 44.

22. See Blau and Meyer, 1987; Blau, 1974; Scott, 1987, pp. 53–56.

23. Scott, 1987, p. 274, citing Hickson et al., 1971, p. 221.

24. Reeves, 1993, p. 416.

25. Scott, 1987, pp. 212–215; also see Blau and Meyer, 1987, pp. 108–109.

26. Scott, 1987, pp. 210–223.

27. Scott, 1987, p. 214.

28. See especially Trice and Beyer, 1993; Ott, 1989b; also see Martin, 1992; Morrill, 1991; Moos, 1987; Schein, 1990.

29. Compare Blau and Meyer, 1987, pp. 54–58.

30. Janis, 1983.

31. March and Simon, 1958; Scott, 1987, pp. 47–48; Perrow, 1986, pp. 122–123; Hall, 1991.

32. Hage and Aiken, 1970, p. 74, citing Sills, 1958.

33. See Jacobs, 1974; Hall, 1991, pp. 216–243; Lawrence and Lorsch, 1992; Hall et al., 1977; Perrucci and Potter, 1989; Perrow, 1986, pp. 192–208.

34. Mayberry et al., 1995.

35. See Meyer and Scott, 1983, and Powell and DiMaggio, 1991, for classic collections of articles related to this perspective; also see Scott, 1995; Scott and Christensen, 1995; and Scott and Meyer et al., 1994.

36. Blau and Meyer, 1987, pp. 129–132.

37. Blau and Meyer, 1987, pp. 123–124.

38. See, for example, Hannan and Freeman, 1977, 1984, 1989; Freeman and Hannan, 1989; Singh and Lumsden, 1990; Aldrich and Marsden, 1988.

39. Hannan and Carroll, 1992.

40. Hannan and Carroll, 1992, p. 219.

41. Hannan and Carroll, 1992, p. 226.

42. See, however, Delacroix, Swaminathan, and Solt, 1989, for an example of results contrary to the organizational ecologists' model.

43. See, for example, Hannan et al., 1995.

44. Scott, 1987, pp. 286–289; also see Gortner, Mahler, and Bell Nicholson, 1989, pp. 294–299; Ott, 1989a, pp. 243–355; Kanter and Stein, 1979, pp. 3–20.

45. Bales and Slater, 1955; Trice and Beyer, 1993, pp. 255–256; Hall, 1991, pp. 136–137.

46. Hall, 1991, p. 151.

47. See Hall, 1991, pp. 146–148; Jacobs and Singell, 1993, p. 168.

48. Trice and Beyer, 1993, pp. 271–272.

49. See Stockard and Mayberry, 1992, for a summary; also see Eberts and Stone, 1988.

50. Jacobs and Singell, 1993.

51. Perrow, 1986, pp. 86–96.

52. Trice and Beyer, 1993, pp. 256–257; Scott, 1987, pp. 59–61; Smith and Peterson, 1988.

53. Compare Hage, 1980, p. 9.

54. Brüderl, Preisendörfer, and Ziegler, 1992.

55. U.S. Bureau of the Census, 1994, p. 547.

56. Variables regrouped slightly to reflect the theories somewhat more accurately than I think Brüderl, Preisendörfer, and Ziegler did. This mainly involved moving some of the organizational characteristics, such as specialist, innovative, follower business, and scope of market to the environmental section. I think these reflect issues related to legitimacy, density, and competition more than issues of organizational structure.

57. Brüderl, Preisendörfer, and Ziegler, 1992, p. 237; see also Ehlers and Main, 1998.

58. Brüderl, Preisendörfer, and Ziegler, 1992, p. 237.

59. Brüderl, Preisendörfer, and Ziegler, 1992, p. 238.

60. Brüderl, Preisendörfer, and Ziegler, 1992, p. 239.

61. Brüderl, Preisendörfer, and Ziegler, 1992, p. 239.

62. Lipset, Trow, and Coleman, 1956.

63. Blau and Meyer, 1987, p. 191, n. 6.

64. See Blau and Meyer, 1987, pp. 193–194.

65. See Blau, 1974, p. 20.

Chapter 12

1. Kluckhorn, 1956; Hill, 1956.

2. Compare Roberts, 1995, pp. 20–21.

3. Durkheim, 1965.

4. See Becker and Barnes, 1961, pp. 579–580.

5. See especially Weber, 1946, for a fascinating discussion of his firsthand observations of American religion that also demonstrates his views toward religion. Also see Roberts, 1995, p. 338; Finke and Stark, 1992; Warner, 1993; Wilson, 1985.

6. See, for example, Wilson, 1976; Bruce, 1992, 1996; Barker, Beckford, and Dobbelaere, 1993; also see Greeley, 1995, for an overview of these views.

7. Babbie, 1992, p. 56.

8. Kuhn, 1970.

9. See Warner, 1993, 1997; also see Roof, 1985, pp. 77–80; Mol, 1985; Richardson, 1985; Hunter, 1985; Wuthnow, 1988, p. 475.

10. For examples of competing views see Lechner, 1997; Olson, 1998.

11. Compare Roberts, 1995, pp. 27–28, 188.

12. Famighetti, 1994, p. 731.

13. Famighetti, 1996, p. 646; U.S. Bureau of the Census, 1997, p. 69.

14. Stark and Iannaccone, 1994; also see Tiryakian, 1993.

15. e.g. Stark and Bainbridge, 1987.

16. Stark and Iannaccone, 1994, pp. 236–237.

17. Hamberg and Pettersson, 1994; Stark and Iannaccone, 1994, pp. 237–238.

18. See Stark and Iannaccone, 1994; also see Hamberg and Pettersson, 1994.

19. See Stoll, 1990; Martin, 1990; Bruce, 1996, pp. 113–114.

20. Greeley, 1994, p. 253; also see Ballis, McLane, and Shabad, 1962, p. 492.

21. Greeley, 1994, p. 257.

22. Greeley, 1994, p. 271.

23. Finke and Stark, 1992, pp. 23, 40, 59–60; Warner, 1993, pp. 1050–1051.

24. Calculated from Finke and Stark, 1992, p. 25.

25. Melton, 1989.

26. See Finke, Guest, and Stark, 1996, for further tests of the religious economy hypothesis with data from the United States and Finke and Stark, 1998, for a lengthy list of supporting studies; also see Greeley, 1989, for an analysis of 40 years of survey data on religious beliefs that indicate no support for the secularization thesis.

27. Finke and Stark, 1992.

28. Warner, 1993, p. 1059.

29. See Marsden, 1991, pp. 44–46; Warner, 1993, p. 1059.

30. See, for example, Greeley, 1972, pp. 108–126; Greeley, 1974.

31. Roof, 1993; Warner, 1993, p. 1061; and see Kosmin and Lachman, 1993, for an extended discussion of all religious groups in the United States.

32. Wilmore, 1972, cited in Roberts, 1995, p. 277; Kosmin and Lachman, 1993, pp. 31–33.

33. U.S. Bureau of the Census, 1994, p. 71; figures for Mormons combine The Church of Jesus Christ of Latter-day Saints and Reorganized Church of Jesus Christ of Latter-day Saints; Finke and Stark, 1992, pp. 163–166; Marsden, 1991, pp. 39–44; Marty, 1986, pp. 238–247.

34. Roberts, 1995, p. 68; also see Bellah and Greenspahn, 1987.

35. Lincoln and Mamiya, 1990, and Roof and McKinney, 1987, both cited in Roberts, 1995, p. 276.

36. Finke and Stark, 1992.

37. See Weber, 1946b, pp. 305–308.

38. See Roberts, 1995, pp. 203–214, for an extensive summary of these attempts.

39. Johnson, 1963, p. 542; Johnson, 1971; also see Bainbridge, 1997, pp. 31–59.

40. Iannaccone, 1994, pp. 1190–1192.

41. See Roberts, 1995, p. 208.

42. See Gaer, 1963 pp. 39–44.

43. Finke and Stark, 19922, pp. 72–108.

44. Finke and Stark, 19932, pp. 145–163.

45. Finke and Stark, 1992, pp. 163–166.

46. Melton, 1993, p. 105.

47. Finke and Stark, 1992; Stark and Bainbridge, 1987.

48. See Roberts, 1995, p. 217.

49. See Melton, 1989, 1993.

50. U.S. Bureau of the Census, 1994, pp. 70–71; Finke and Stark, 1992, p. 241.

51. U.S. Bureau of the Census, 1997, p. 70; see also Kosmin and Lachman, 1993; Newport and Saad, 1997.

52. Stark, 1985, p. 146; Bainbridge, 1989.

53. Melton, 1993, p. 102; Bainbridge, 1997, pp. 179–207.

54. See Finke and Stark, 1992, pp. 239–249; Melton, 1993; also see Finke and Iannaccone, 1993, pp. 36–37.

55. See Melton, 1993, pp. 104–105; also see Bainbridge, 1985, cited in Melton, 1993; Albanese, 1993; Finke and Stark, 1992, pp. 244–245; Barker, 1985.

56. See, for example, Bibby and Brinkerhoff, 1994.

57. Hoge, Johnson, and Luidens, 1994, p. 41.

58. Hoge, Johnson, and Luidens, 1994, pp. 42, 221–223.

59. See Kelley and DeGraaf, 1997, and Myers, 1996, for analyses of the influence of parental religiousity on the religious of children.

60. Roof and McKinney, 1987, p. 165; also Greeley, 1995.

61. Roof and McKinney, 1987.

62. Hoge, Johnson, and Luidens, 1994, p. 102.

63. Hoge, Johnson, and Luidens, 1994, p. 126; also see Greeley, 1995.

64. Hoge, Johnson, and Luidens, 1994, p. 130.

65. Lofland, 1977; see also Zellner, 1995, pp. 136–140.

66. See, for example, Snow and Phillips, 1980; Roof and McKinney, 1987; Stark and Bainbridge, 1985; all cited in Roberts, 1995, pp. 126–128; also see Van Zandt, 1991.

67. Hoge, Johnson, and Luidens, 1994, pp. 176–178.

68. Quoted in Lofland, 1977, p. 311, and cited in Roberts, 1995, p. 129.

69. See Richardson, 1985, p. 107; Young, 1997.

70. See Iannaccone, 1994, p. 1182.

71. See Kelley, 1972.

72. Johnson and Stanley, personal communication, 1995.

73. Hoge, Johnson, and Luidens, 1994, pp. 75–87.

74. Iannaccone, 1994, pp. 1200–1201, 1192–1197.

75. Iannaccone, 1994; see also Blau et al., 1997, pp. 573–574.

76. See, for example, Poloma, 1989; also see Ammerman, 1987; Neitz, 1987; Warner, 1988, 1993; and McGuire, 1982; all cited in Roberts, 1995, p. 250; and Miller, Donald E., 1997, Shibley, 1996, pp. 114–115.

77. See Roberts, 1995, p. 250.

78. See Farley, 1988, and Roberts, 1995, especially pp. 308–309.

79. See especially Roberts, 1995, p. 309.

Chapter 13

1. Lasswell, 1936.

2. Compare Clawson, Neustadtl, and Scott, 1992, pp. 20–21; also see Wartenberg, 1990.

3. Lenski, 1966, p. 151.

4. Lenski, 1966, pp. 304–318.

5. Compare Knoke, 1990, pp. 3–7.

6. Gilbert and Kahl, 1993, p. 106.

7. See Hechter, 1983, ch. 2; also see Coleman, 1990.

8. See Niemi and Weisberg, 1992; Downs, 1957; Grofman, 1993; Mueller, 1979; Tallman and Gray, 1990; Coleman, 1990.

9. U.S. Bureau of the Census, 1994, p. 288.

10. Wasburn, 1982, pp. 147–148; Orum, 1988, p. 413, citing Verba and Nie, 1972.

11. Dalton, 1988, pp. 38–39.

12. Maslow, 1943, 1962; Davies, 1963; also see Wasburn, 1982; Peterson, 1990.

13. See Davidson and Grofman, 1994.

14. See Rosenstone and Wolfinger, 1984.

15. Federal Election Commission, 1997.

16. Wolfinger and Rosenstone, 1980, pp. 20–26.

17. Nie, Junn, and Stehlik-Barry, 1996.

18. Firebaugh and Chen, 1995.

19. Luttbeg and Gant, 1995; cited by DeLuca, 1995, p. 5.

20. See Wolfinger and Rosenstone, 1980, pp. 37–60.

21. Wolfinger and Rosenstone, 1980, pp. 89–93; Orum, 1988, pp. 413–414; Bobo and Gilliam, 1990, cited in DeLuca, 1995; see also Garcia, F.C. 1997; Garcia, J. 1997.

22. Compare Kingdon, 1993.

23. Compare Walker, 1990.

24. Zipp and Smith, 1979, cited in Wasburn, 1982, p. 148; also see Knoke, 1990, 1993; Verba, Schlozman, and Brady, 1995.

25. Verba, Schlozman, and Brady, 1995.

26. Gilbert and Kahl, 1993, p. 216; Gilbert, 1998, p. 206.

27. Gilbert and Kahl, 1993, p. 214, citing Mintz, 1975; also Gilbert, 1998, pp. 205–206.

28. Quoted in Gilbert and Kahl, 1993, p. 215, citing *New York Times,* 25 November 1984.

29. Dahl, 1961; Vogel, 1996.

30. Compare Greider, 1992, p. 19; Rothenberg, 1992.

31. Wasburn, 1982, pp. 300–301, citing Dahl, 1961; also see Orum, 1988, pp. 405–406.

32. See Riesman, 1961, cited in Orum, 1988, p. 403.

33. See Lukes, 1974, cited in Clawson, Neustadtl, and Scott, 1992, p. 20.

34. Rosenstiel and Thomas, 1995; Dahl, 1985, 1989, 1990; and DeLuca, 1995, pp. 217–233.

35. See Rose, 1967, cited in Wasburn, 1982, p. 301.

36. Compare Roy, 1992, p. 762.

37. See Mills, 1956; also see especially Domhoff, 1970, 1978, 1979, 1990; Wasburn, 1982, p. 297.

38. Tiger, 1995.

39. Carroll, 1995, p. 30.

40. Tiger, 1995; Harding, 1998.

41. See, for example, Mizruchi, 1992; Allen, 1974; Jacobs, 1974.

42. See Mills, 1956.

43. See Wartenberg, 1990, cited in Clawson, Neustadtl, and Scott, 1992, pp. 20–21.

44. Gramsci, 1972.

45. Compare Clawson, Neustadtl, and Scott, 1992, p. 23.

46. See Bergmann, 1993; Hewlett, 1991.

47. Clawson, Neustadtl, and Scott, 1992, p. 21.

48. Simpson, 1964, p. 8.

49. See Clawson, Neustadtl, and Scott, 1992, pp. 24–26, for a succinct statement regarding this issue.

50. Coleman, 1990.

51. See, for example, Clawson, Neustadtl, and Scott, 1992; Etzioni, 1990.

52. Clawson, Neustadtl, and Scott, 1992, p. 28.

53. U.S. Bureau of the Census, 1994, p. 291.

54. U.S. Bureau of the Census, 1998c, p. 302.

55. Compare Clawson, Neustadtl, and Scott, 1992, pp. 34–43.
56. U.S. Bureau of the Census, 1998c, p. 302.
57. Clawson, Neustadtl, and Scott, 1992, pp. 12–13; Clawson, Neustadtl, and Weller, 1998, p. 39; also see Clawson and Neustadtl, 1989; Clawson, Neustadtl, and Bearden, 1986; Clawson and Su, 1990.
58. Clawson, Neustadtl, and Scott, 1992, p. 15.
59. Clawson, Neustadtl, and Scott, 1992, p. 17.
60. Clawson, Neustadtl, and Scott, 1992, p. 53.
61. Clawson, Neustadtl, and Scott, 1992, pp. 56–57, 75–79, 85–86.
62. Clawson, Neustadtl, and Scott, 1992, pp. 59–61.
63. Clawson, Neustadtl, and Scott, 1992, p. 91.
64. Clawson, Neustadtl, and Scott, 1992, p. 95.
65. Clawson, Neustadtl, and Scott, 1992, p. 4.
66. Clawson, Neustadtl, and Weller, 1998, pp. 69–70.
67. Clawson, Neustadtl, and Scott, 1992, pp. 119–125.
68. Clawson, Neustadtl, and Scott, 1992, p. 145.
69. U.S. Bureau of the Census, 1998c, p. 302; Clawson, Neustadtl, and Weller, 1998, p. 39.
70. Clawson, Neustadtl, and Scott, 1992, pp. 147, 151.
71. Quoted in Clawson, Neustadtl, and Scott, 1992, p. 150.
72. Clawson, Neustadtl, and Scott, 1992, p. 151; also see Clawson, Neustadtl, and Weller, 1998, pp. 139–166.
73. Clawson, Neustadtl, and Scott, 1992, p. 157.
74. Clawson, Neustadtl, and Weller, 1998, pp. 108–10.
75. Clawson, Neustadtl, and Weller, 1998, calculated from p. 109.
76. Clawson, Neustadtl, and Weller, 1998, pp. 116–123.
77. Clawson, Neustadtl, and Weller, 1998, pp. 128–130.
78. Clawson, Neustadtl, and Weller, 1998, pp. 134–137.
79. Also see Akard, 1992; Dreiling, 1998.
80. Clawson, Neustadtl, and Scott, 1992, p. 160; also see Mizruchi, 1992.
81. Clawson, Neustadtl, and Scott, 1992, p. 173.
82. Clawson, Neustadtl, and Weller, 1998, pp. 167–196; also see Edsall, 1984.
83. Clawson, Neustadtl, and Scott, 1992, p. 218; see also Mitchell, 1997, and Vogel 1996.
84. Clawson, Neustadtl, and Scott, 1992, p. 26.
85. Dalton, 1988, p. 40.
86. Crewe, 1981, pp. 234–237.
87. Dalton, 1988; Rose, 1974; Crewe, 1981.
88. Federal Election Commission, 1997.
89. Dalton, 1988, p. 39.
90. See DeLuca, 1995, and Phillips, 1994 for discussions of ways to reduce voter apathy in the United States.
91. Vanneman and Cannon, 1987.

Chapter 14
1. Swedberg, 1990, p. 9.
2. See Swedberg, 1998, for an explanation of Weber's views in this area.
3. Compare Smelser, 1963, pp. 22–35.
4. Duesenberry, 1960, cited in Granovetter, 1985, p. 485.
5. See especially Coleman, 1990, for the fullest development of rational choice theory in sociology.
6. See especially Swedberg, 1990, 1991, 1993; Granovetter and Swedberg, 1992; Granovetter, 1985; Etzioni and Lawrence, 1991; Mizruchi and Schwartz, 1987; Brinton and Nee, 1998.
7. Misztal, 1996.
8. See Friedl, 1975.
9. See Lenski, 1966; Lenski and Lenski, 1982.
10. Henslin, 1993, pp. 371–372.
11. See especially Granovetter, 1985; also see Swedberg, 1993; Burt, 1993; Sabel, 1993; Granovetter, 1985.
12. World Bank, 1998b; Haub and Yanagishita, 1995.
13. U.S. Bureau of the Census, 1998c, p. 460.
14. See Stinchcombe, 1983; also see Lenski, 1966.
15. Lenski, 1966, pp. 150, 160–162.
16. See Stinchcombe, 1983, pp. 66–67; he also cites Duncan and Lieberson, 1970.
17. Stinchcombe, 1983, p. 81.
18. U.S. Bureau of the Census, 1998c, p. 460.
19. Famighetti, 1994, p. 800; Wright, 1998, pp. 613–615.
20. Haub and Yanagishita, 1995; Grant, 1994, p. 85.
21. World Bank, 1998b.
22. Henslin, 1993, p. 378; Calhoun, Light, and Keller, 1994, p. 398.
23. Henslin, 1993, p. 380.
24. Henslin, 1993, pp. 378–379.
25. Henslin, 1993, p. 380.
26. Kerr, 1983; Form, 1979, cited in Henslin, 1993, p. 383; also see Langlois et al., 1994.
27. See Lenski, 1966, pp. 195–196.
28. Hodson and Sullivan, 1990, pp. 384–389.
29. Hodson and Sullivan, 1990, pp. 389–392.
30. U.S. Bureau of the Census, 1998c, p. 567; data are for 1995.
31. Hoopes, 1994.
32. Wallerstein, 1974, 1979, 1984; Henslin, 1993, pp. 240–241; Calhoun, Light, and Keller, 1994, p. 47.
33. Hodson and Sullivan, 1990, pp. 384–389.
34. Calhoun, Light, and Keller, 1994, p. 477; Frank, 1967, 1980; also see Kowalewski, 1991.
35. See Wallerstein, 1984, pp. 6–7; also see Chirot, 1986.
36. Hodson and Sullivan, 1990, p. 389; also see Vogel, 1991.
37. So and Chiu, 1995.
38. Famighetti, 1994, p. 165.
39. Famighetti, 1994, pp. 844–845.
40. Encarta, 1996a, b.
41. Hodson and Sullivan, 1990, pp. 146–156; Sutton et al., 1994.
42. Hodson and Sullivan, 1990, p. 162, citing Lipset, 1986.
43. Hodson and Sullivan, 1990, pp. 162–163; also see Dudley, 1994; Goldfield, 1987.
44. U.S. Bureau of the Census, 1975, p. 139; U.S. Bureau of the Census, 1994, p. 412.
45. U.S. Bureau of the Census, 1975, p. 139.
46. U.S. Bureau of the Census, 1975, p. 139; U.S. Bureau of the Census, 1994, p. 412.
47. Danziger and Gottschalk, 1995; Ferguson, 1995; Morales and Bonilla, 1993; Tilly, 1996.
48. Hodson and Sullivan, 1995, p. 206.
49. U.S. Bureau of the Census, 1997, p. 414; 1998, p. 420.
50. U.S. Bureau of the Census, 1998c, p. 547.
51. Hodson and Sullivan, 1990, p. 379, citing Granovetter, 1984.
52. See Roy, 1997, and Stearns and Allan, 1996 for analyses of this process in particular historical eras.

53. Hodson and Sullivan, 1990, p. 365.
54. See Allen, 1987, p. 34; also see Henslin, 1993, p. 379, citing Josephson, 1949.
55. Hodson and Sullivan, 1990, p. 365.
56. Hodson and Sullivan, 1990, p. 365.
57. See Allen, 1987, pp. 43–45.
58. Hodson and Sullivan, 1990, p. 367.
59. Hodson and Sullivan, 1995, pp. 396–401; U.S. Bureau of the Census, 1997, p. 550; Davis, Diekmann, and Tinsley, 1994.
60. Disney, 1998.
61. http://www.citigroupinfo.com/
62. Bachmann, 1995.
63. See Hodson and Sullivan, 1990, pp. 365–372; 1995, p. 397; also see Chandler, 1990; Fligstein, 1990; Hirsch, 1986.
64. Hodson and Sullivan, 1990, pp. 390–391.
65. Stinchcombe, 1983, pp. 85–86; Marx, 1959.
66. Hodson and Sullivan, 1990, pp. 372–375; also see Rosen, 1987; Fuechtmann, 1989.
67. Hodson and Sullivan, 1990, p. 395; also see O'Hearn, 1989; Bradshaw, 1988.
68. Famighetti, 1995, p. 718.
69. Granovetter, 1974, 1985.
70. Faulkner, 1983.
71. Faulkner, 1983, pp. 24–25.
72. Faulkner, 1983, p. 49.
73. Faulkner, 1983, p. 25.
74. Faulkner, 1983; also see Faulkner and Anderson, 1987.
75. Faulkner, 1983, pp. 27–31.
76. Faulkner, 1983, pp. 50–51.
77. For example, see Henson, 1996.
78. Compare Faulkner, 1983, pp. 92–93.
79. Faulkner, 1983, pp. 98, 121–145.
80. Faulkner, 1983, p. 70, emphasis in original.
81. Faulkner, 1983, pp. 69–88.
82. Faulkner, 1983, p. 117.
83. Monaco, 1991.
84. Grannovetter, 1995; Bian, 1997; Beggs and Hurlbert, 1997; Fernandez and Weinberg, 1997.
85. Portes, 1981; Portes and Rumbaut, 1990; Portes, 1995; Sanders and Nee, 1996.
86. See Nee, Sanders, and Sernau, 1994.

Chapter 15

1. Boli, 1989, pp. 27, 58.
2. Boli, 1989; Ramirez and Boli, 1987.
3. Ramirez and Boli, 1989; Meyer et al., 1977; Meyer and Hannan, 1979.
4. UNESCO, 1998.
5. Hurn, 1993, pp. 15–16.
6. See Brint and Karabel, 1989; Brown, 1995.
7. Hurn, 1993, p. 26; Ballantine, 1993, pp. 274–275; U.S. Bureau of the Census, 1994, p. 155; U.S. Bureau of the Census, 1998c, p. 157.
8. See Hurn, 1993, pp. 15–17.
9. See, for example, Clark, 1961, and Trow, 1961, both cited in Hurn, 1993, pp. 76–77.
10. See Durkheim, 1961; Parsons, 1959; Hurn, 1993, pp. 78–79; and Bidwell and Friedkin, 1988; for descriptions of this view.
11. See especially Bowles and Gintis, 1976.
12. See, for example, Bowles and Gintis, 1976.
13. Boli, 1989, pp. 27–28; Meyer et al., 1979; Richardson, 1980.
14. See Hurn, 1993, p. 82.
15. Compare Boli, 1989.
16. See Cremin, 1976, pp. 37–38.
17. See Ramirez and Boli, 1987; Hurn, 1993, p. 89; Meyer and Hannan, 1979; Meyer, Kamens, and Benavot, 1992.
18. See Ramirez and Boli, 1987; Boli, 1989, pp. 34–44; Meyer et al., 1977.
19. Boli, 1989, p. 50.
20. Boli, 1989, p. 34; also see Cremin, 1990; Fass, 1989.
21. Brint and Karabel, 1989; Hurn, 1993; Shavit and Blosfeld, 1993; Tyack and Hansot, 1990; Gerber and Hout, 1995.
22. U.S. Bureau of the Census, 1998c, p. 178.
23. See Ballantine, 1993, p. 90; also see Schneider and Coleman, 1993, Kalmijn, 1994b.
24. Clark, 1983, p. 18.
25. Clark, 1983, p. 111.
26. Hrabowski, Maton, and Greif, 1998; Lam, 1997.
27. Steinberg, Brown, and Dornbusch, 1996, pp. 138–141.
28. Steinberg, Brown, and Dornbusch, 1996, pp. 143–146.
29. Steinberg, Brown, and Dornbusch, 1996, pp. 148–149.
30. Steinberg, Brown, and Dornbusch, 1996, pp. 150–155.
31. Pedersen, Faucher, and Eaton, 1978, pp. 2–3.
32. Pedersen, Faucher, and Eaton, 1978.
33. Brophy and Good, 1986; Brophy, 1986; Stockard and Mayberry, 1992, pp. 54–57.
34. Pedersen, Faucher, and Eaton, 1978, pp. 19–20.
35. Ballantine, 1993, pp. 227–235; Halsey, 1989; Stockard and Mayberry, 1992.
36. See series of articles in these special issues: *Harvard Educational Review* 51(4) (1981); *Sociology of Education* 53(2/3) (1982) and 58(2) (1985).
37. See Coleman, Hoffer, and Kilgore, 1982; Greeley, 1982; Jencks, 1985.
38. Bryk, Lee, and Holland, 1993, p. 73.
39. See Bryk, Lee, and Holland, 1993, pp. ix–xi.
40. Bryk, Lee, and Holland, 1993, p. 63.
41. Bryk, Lee, and Holland, 1993, p. 60.
42. Bryk, Lee, and Holland, 1993, p. xi.
43. Bryk, Lee, and Holland, 1993, p. 125.
44. Bryk, Lee, and Holland, 1993, p. 141.
45. Bryk, Lee, and Holland, 1993, p. 77.
46. Bryk, Lee, and Holland, 1993, p. 119.
47. Barker and Gump, 1964; Gregory and Smith, 1987; Stockard and Mayberry, 1992.
48. Bryk, Lee, and Holland, 1993, p. 275.
49. Bryk, Lee, and Holland, 1993, p. 275.
50. Bryk, Lee, and Holland, 1993, p. 275.
51. Bryk, Lee, and Holland, 1993, pp. 277–278.
52. Stockard and Mayberry, 1992.
53. Bryk, Lee, and Holland, 1993, p. 321.

Chapter 16

1. Cockerham, 1998, p. 2.
2. Rockett, 1994, p. 2; Cockerham, 1998, pp. 12–16.
3. Cockerham, 1998.

4. See Idler and Benyamini, 1997.

5. See especially Braithwaite and Taylor, 1992, also Ferraro, Farmer, and Wybraniec, 1997.

6. Floyd, 1992, p. 167; also see Gortmaker and Wise, 1997.

7. National Center for Health Statistics, 1998, p. 181.

8. Floyd, 1992, p. 167.

9. Cockerham, 1998, p. 29.

10. Rockett, 1994, p. 9, Cockerham, 1998, p. 27; National Center for Health Statistics, 1998, p. 212; Ventura et al., 1998, p. 7.

11. U.S. Bureau of the Census, 1997, p. 100.

12. See Cockerham, 1998, pp. 45–53; also Cramer, 1995; McDonough et al., 1997; Gregorio, Walsh, and Paturzo, 1997.

13. Adler et al., 1994, cited by Cockerham, 1997, p. 123; Cockerham, 1998, pp. 53–61; Mechanic, 1994, pp. 137–150; Wilkinson, 1986; Gregorio, Walsh, and Paturzo, 1997; McDonough et al., 1997; Hemingway et al., 1997; Marks and Shinberg, 1997; Power et al., 1997; Marmot, Bobak, and Smith, 1995.

14. Bartley and Plewis, 1997.

15. Adler, et al., 1994, cited by Cockerham, 1997, p. 123; Evans, 1994; Marmot, Bobak, and Smith, 1995; Cockerham, 1998, p. 55.

16. Rice, 1992, p. 3.

17. Kanner et al., 1981, p. 3; Nakano, 1989; both cited in Rice, 1992, p. 10; see also Lovallo, 1997, p. 28; Reynolds, 1997; Kessler et al., 1995, p. 551.

18. Cockerham, 1998, pp. 69–70, Rice, 1992; Lovallo, 1997, pp. 35–73.

19. Lovallo, 1997, p. 117; Rice, 1992; Cockerham, 1998, pp. 70–71.

20. Cockerham, 1998, p. 65; Rice, 1992, p. 65; see also Farmer and Ferraro, 1997.

21. Rice, 1992, pp. 63–68.

22. Cockerham, 1998, p. 68; Kessler et al., 1995, p. 553–555.

23. Cockerham, 1998, p. 73, citing Moss, 1973; see also Wickrama et al., 1997; Ross and Willigen, 1997; Wickrama, Lorenz, and Conger, 1997.

24. e.g. Waite, 1995.

25. Rockett, 1994, pp. 7–9; Cockerham, 1998, p. 85.

26. McGinnis and Foege, 1993, cited in Weitz, 1996, p. 31.

27. Cockerham, 1998, pp. 89–90 citing Green, 1986.

28. Cockerham, 1998, p. 93, citing Harris and Guten, 1979, and Kronenfeld et al., 1988.

29. Cockerham, 1998, p. 97.

30. National Center for Health Statistics, 1998, p. 176.

31. Cockerham, 1998, pp. 86–89; and Bourdieu and Wacquant, 1992, and Cockerham, Rütten, and Abel, 1997, both cited by Cockerham, 1997, p. 125.

32. Bartley and Plewis, 1997.

33. Cockerham, 1998, pp. 120–121; 170–173.

34. Cockerham, 1998, pp. 173–74.

35. Cockerham, 1998, pp. 114–116, citing especially Dutton, 1978, 1979, 1986.

36. National Center for Health Statistics, 1998, p. 288.

37. Cockerham, 1998, p. 115; see also Weiss, 1997.

38. National Center for Health Statistics, 1998, p. 341.

39. National Center for Health Statistics, 1998, p. 342.

40. Calculated from pp. 122 and 424 of U.S. Bureau of the Census, 1997.

41. Cockerham, 1998, p. 283.

42. Cockerham, 1998, p. 264.

43. See Stevens, 1971, for an extensive history of medicine in the United States.

44. Brown, 1979, pp. 62–63, quoted in Cockerham, 1998, p. 184.

45. Weitz, 1996, pp. 225–227.

46. Quah 1989, cited in Cockerham, 1998, p. 182.

47. Cockerham, 1998, pp. 184–185.

48. Cockerham, 1998, 185–186; see also Starr, 1982 for an extended analysis of the history of the medicine in the United States.

49. Cockerham, 1998, pp. 188–189.

50. Cockerham, 1998, p. 189.

51. Weitz, 1996, pp. 328–330; see also Stevens, 1989, for an extensive historical overview of this period.

52. Weitz, 1996, pp. 330–331.

53. Calculated from pp. 18 and 121 of U.S.Bureau of the Census, 1997.

54. Weitz, 1996, p. 335; Cockerham, 1998, p. 268.

55. Cockerham, 1998, pp. 265–266.

56. Cockerham, 1998, pp. 268–269.

57. White House Fact Sheet, 1998.

58. Cockerham, 1998, p. 179.

59. Starr, 1994, cited by Weitz, pp. 340–341.

60. Cockerham, 1998, pp. 267–268.

61. http://cnn.com/health/specials/HMO/ and U.S. Bureau of the Census, 1997, p. 121.

62. http://www.mahmo.org/hmoterms.html.

63. U.S. Bureau of the Census, 1998c, p. 120; National Center for Health Statistics, 1998, p. 362.

64. Cockerham, 1998, p. 268.

65. Cockerham, 1998, pp. 283–284.

66. Cockerham, 1998, pp. 285–287.

67. Leeman, 1998.

68. Cockerham, 1998, p. 283.

69. Cockerham, 1998, p. 311.

70. Calculated from U.S. Bureau of the Census, 1998c, pp. 8 and 122.

71. U. S. Bureau of the Census, 1975, p. 55 and U.S. Bureau of the Census, 1997, p. 88.

72. U.S. Bureau of the Census, 1975, p. 57; National Center for Health Statistics, 1998, p. 245.

73. Cockerham, 1998, pp. 19–20.

74. Rockett, 1994, pp. 8–9; Olshansky et al., 1997; McKeown, 1979; Rosner, 1995.

75. Cockerham, 1998, pp. 59–60.

76. Population Reference Bureau, 1998.

77. Population Reference Bureau, 1998.

78. Olshansky et al., 1997, pp. 6–7.

79. Goliber, 1997, pp. 28–31.

80. Olshansky et al., 1997, p. 20.

81. Altman, 1998; Haney, 1998; Ventura et al., 1998.

82. Bellamy, 1997, p. 98.

83. Wilkinson, 1996, pp. 29–42.

84. Calculated from Population Reference Bureau, 1998 and U.S. Bureau of the Census, 1997, pp. 832–833, 835.

85. Wilkinson, 1996, pp. 72–81; Wilkinson, 1989, 1992.

86. Wilkinson, 1996, 1997; Kawachi et al., 1997; see also Hayward, Pienta, and McLaughlin, 1997.
87. Wilkinson, 1996, pp. 18, 131–134.
88. Bruhn and Wolf, 1979, pp. 81–82, 136, cited by Wilkinson, 1996, p. 117.
89. Wilkinson, 1996, pp. 116–118, citing especially Bruhn and Wolf, 1979.
90. Cockerham, 1998, p. 273.
91. Cockerham, 1998, pp. 269–270.
92. Cockerham, 1998, pp. 271–272; also Skocpol, 1996.

Chapter 17
1. Ehrlich, 1986, p. 19.
2. Weeks, 1994, pp. 7–11.
3. Weeks, 1994, pp. 18–21.
4. Weeks, 1994, p. 22.
5. Weeks, 1994, pp. 11–17.
6. Weeks, 1994, pp. 241–244.
7. U.S. Bureau of the Census, 1998c, p. 9.
8. Lutz, 1994, p. 2; United Nations Population Division, 1999.
9. United Nations Population Division, 1999.
10. See Lutz, 1994, pp. 2–7; Ashford, 1995, pp. 4–5; Weeks, 1994, pp. 28–54; Population Reference Bureau, 1998.
11. McFalls, 1998, p. 34.
12. Grant, 1994; Weeks, 1994; Ashford, 1995; Lutz, 1994, p. 23.
13. Lenski, 1966, p. 98.
14. Weeks, 1994, p. 32; also see Freidl, 1975.
15. Weeks, 1994, p. 32.
16. Weeks, 1994, p. 32.
17. Weeks, 1994.
18. See Poston and Yaukey, 1992, for an extensive discussion of population issues in China.
19. Lutz, 1994, p. 7; Visaria and Visaria, 1995.
20. Davis, 1945; Weeks, 1994, pp. 75–77.
21. See Lutz, 1994, pp. 9–10; also see Weeks, 1994, pp. 76–82.
22. See Grant, 1995, pp. 74–75.
23. See, for example, Aly and Grabowski, 1990; Victora et al., 1992; Riley, 1997; London, 1992; Caldwell, 1982.
24. See Weeks, 1994; Lutz, 1994; Ashford, 1995; Mason, 1997; Gallagher, Stokes, and Anderson, 1996.
25. Haub and Yanagishita, 1995; Grant, 1995, pp. 78–79; Goliber, 1997.
26. Ashford, 1995, p. 29.
27. Ashford, 1995, p. 13; Lutz, 1994, p. 7; Haub and Yanagishita, 1995.
28. Crenshaw, Ameen, and Christenson, 1997; see also Blau, 1994, for a theoretical overview of this influence.
29. Bellamy, 1997, pp. 84–86.
30. Bellamy, 1997, pp. 82, 90.
31. See especially Grant, 1994, p. 25.
32. Theroux, 1989, p. 386.
33. Weeks, 1994, p. 425.
34. Grant, 1994, p. 2.
35. Sherbinin and Kalish, 1994, p. 2.
36. Crews and Stauffer, 1997; Livernash and Rodenburg, 1998.
37. See Grant, 1994.

38. Weeks, 1994, p. 29.
39. Compare Serow et al., 1990.
40. Martin and Widgren, 1996, p. 2; also see Kalish, 1994, p. 1.
41. Theroux, 1989.
42. United Nations High Commissioner for Refugees, Statistical Unit, 1998, pp. 5–6.
43. Martin and Widgren, 1996, p. 16.
44. See, for example, Simon, 1989.
45. Martin and Widgren, 1996, pp. 2, 16–17; Weeks, 1996, pp. 234–235.
46. See Ferrante, 1995, pp. 346–349.
47. Lutz, 1994, p. 24; McFalls, 1998, p. 19.
48. Lutz, 1994, p. 25; Ferrante, 1995, p. 349.
49. Portes and Rumbaut, 1990, p. 333; Chiswick and Sullivan, 1995; U. S. Bureau of the Census, 1998c, pp. 10–11, 13, 55.
50. U.S. Bureau of the Census, 1998c, pp. 10,12.
51. Lutz, 1994, p. 25; Ferrante, 1995, p. 349.
52. Massey and Espinosa, 1997; see also Gold, 1997.
53. See Weeks, 1994, pp. 214–217; Portes and Rumbaut, 1990; Marcelli and Heer, 1998; Hamermesh and Bean, 1998; Waldinger, 1997; Smith and Edmonston, 1997.
54. See Weeks, 1994, pp. 197–198; DeJong and Gardner, 1981.
55. See, for example, Aly and Grabowski, 1990; Victora et al., 1992; Riley, 1997; London, 1992; Caldwell, 1982.
56. LeVine et al., 1991.
57. LeVine et al., 1991, pp. 464–466.
58. LeVine et al., 1991, pp. 474–482.
59. LeVine et al., 1991, p. 488.
60. LeVine et al., 1991, p. 492.
61. Calculated from U.S. Bureau of the Census, 1998c, p. 17.
62. Treas, 1995.

Chapter 18
1. Calhoun, Light, and Keller, 1994, p. 542.
2. U.S. Bureau of the Census, 1994; Garkovich, 1989.
3. U.S. Bureau of the Census, 1994, p. 926.
4. U.S. Bureau of the Census, 1994, p. 37.
5. Calhoun, Light, and Keller, 1994, p. 528.
6. Palen, 1995; also see Long, 1988, p. 199.
7. Johnson and Beale, 1995.
8. See Calhoun, Light, and Keller, 1994, pp. 535–536; also see Fuguitt, Brown, and Beale, 1989; Wilkinson, 1991, p. 9.
9. Michael Hibbard, personal communication, 1995.
10. Bellamy, 1997, p. 99.
11. Bellamy, 1997, p. 99.
12. Bellamy, 1997, p. 99; see also Gugler, 1996, 1997.
13. Calculated from Famighetti, 1994, p. 840; and U.S. Bureau of the Census, 1994, pp. xii, 45.
14. See Weeks, 1994, p. 358.
15. World Bank, 1998a; also see Smith, 1996.
16. See Grant, 1994, 1995; also see Edel and Hellman, 1989.
17. Tönnies, 1963.
18. Simmel, 1950a.
19. Wirth, 1938, p. 22.
20. Wirth, 1938, p. 21.

21. Wirth, 1938, p. 24.
22. See Henslin, 1993, pp. 582–583.
23. Gans, 1962.
24. Gans, 1970, pp. 158–160; Gans, 1991, pp. 54–55.
25. Gans, 1970, p. 160.
26. Gans, 1970; also see Fischer et al., 1977.
27. Fischer, 1982, p. 17.
28. Fischer, 1982, p. 36.
29. Fischer, 1982, p. 46.
30. Fischer, 1982, p. 49.
31. Fischer, 1982, p. 80.
32. Fischer, 1982, pp. 193–198; also see Fischer, 1975, 1995.
33. See Fischer, 1975, 1982, 1995.
34. Fischer, 1982, p. 234.
35. Fischer, 1982, p. 235.
36. Sampson, 1988.
37. Fischer, 1982, p. 63.
38. Fischer, 1982, p. 65.
39. Fischer, 1982, p. 75.
40. See Wilson, Thomas C., 1985, 1991; Tuch, 1987.
41. White, 1987, pp. 118–120; Burgess, 1967; Massey and Denton, 1993, p. 27.
42. Orum, 1995; Sugrue, 1996.
43. White, 1987, p. 120; Hoyt, 1939.
44. White, 1987, p. 121; Harris and Ullman, 1945; Palen, 1995; Waldinger and Bozorgmehr, 1996; Scott and Soja, 1996.
45. Massey and Denton, 1993, p. 32.
46. Massey and Denton, 1993, pp. 32–33; Lieberson, 1980.
47. White, 1987, p. 84.
48. Massey and Denton, 1993, p. 20.
49. Massey and Denton, 1993, p. 222.
50. Massey and Denton, 1993, pp. 19–31.
51. Massey and Denton, 1993, pp. 46–47; also see Palen, 1995; Farley and Frey, 1994.
52. Massey and Denton, 1993, pp. 85–87.
53. Massey and Denton, 1993, p. 152.
54. Massey and Denton, 1993, p. 147; Massey and Denton, 1985.
55. Massey and Denton, 1993, p. 77; also see Massey, 1996.
56. Also see Crane, 1991; Bursik and Grasmick, 1993; Kasarda, 1995; Brooks-Gunn, Duncan, and Aber, 1997a, 1997b; Brewster, 1994; Wilson, 1997.
57. See Palen, 1995; Jargowsky 1996, 1997; but also see Massey, Gross, and Shibuya, 1994, for a slightly different view.
58. Massey and Denton, 1993, pp. 118–125.
59. Massey and Denton, 1993, p. 126; also see Massey, 1990.
60. Farley et al., 1994.
61. Farley et al., 1994, p. 757.
62. Farley et al., 1994, p. 776; see also Yinger, 1995.
63. Farley et al., 1994, p. 777.

Chapter 19
1. Mills, 1959, p. 8.
2. Lenski, Nolan, and Lenski, 1995; Couch, 1996.
3. Lenski et al., 1995, pp. 67, 130.
4. Diamond, 1997; Landes, 1998.
5. Lenski and Lenski, 1982, pp. 109–110.
6. Lenski and Lenski, 1982.
7. Lenski and Lenski, 1982, pp. 106–107.
8. Lenski and Lenski, 1982, pp. 135–136; Lenski, Nolan, and Lenski, 1995, p. 137.
9. Lenski and Lenski, 1982, pp. 88–89, 136–137.
10. Lenski and Lenski, 1982, p. 137.
11. Also see Chirot, 1994.
12. Lenski and Lenski, 1982; Chirot, 1994.
13. Lenski and Lenski, 1982, pp. 142–144.
14. Lenski and Lenski, 1982, pp. 88–90.
15. Lenski and Lenski, 1982, p. 217; Henslin, 1993, pp. 144–145; Lenski, 1966; Maryanski and Turner, 1992.
16. Lenski and Lenski, 1982, pp. 178–189.
17. Lenski and Lenski, 1982, pp. 236–240.
18. See Chirot, 1985, 1986, 1994; Landes, 1998.
19. Lenski and Lenski, 1982, pp. 248–252; Lubrano, 1997; Maryanski and Turner, 1992.
20. e.g. Naisbitt, 1982; Toffler, 1990.
21. Rosenblum, 1998; Wysocki, 1998.
22. Rheingold, 1993, p. 3; cited by Wysocki, 1996.
23. Rosenblum, 1998; Wysocki, 1996.
24. Henslin, 1993, p. 241.
25. Albrow, 1997; Castells, 1997.
26. Holmes, 1997, p. 4; Schaeffer, 1997.
27. Hallinan, 1997.
28. Holmes, 1997; see also Skocpol, 1994, 1979, for an extensive discussion of the reasons underlying the collapse of state organizations that, in many ways, parallels Holmes' analysis.
29. Stokes, 1993; Skidelsky, 1995; Schaeffer, 1997; Lipset, 1994; Crenshaw, 1995.
30. Boli and Thomas, 1997; Meyer et al., 1997; MacIntosh, 1998.
31. Gerber and Hout, 1998.
32. Fischer, 1992, pp. 20–21.
33. Fischer, 1992, p. 8.
34. Fischer, 1992, pp. 260–263.
35. Fischer, 1992, pp. 28–31.
36. Fischer, 1992, pp. 35–42.
37. U.S. Bureau of the Census, 1975, p. 783.
38. Fischer, 1992, pp. 43, 94.
39. Fischer, 1992, p. 1.
40. Fischer, 1992, p. 260.
41. Fischer, 1992, p. 263.
42. Fischer, 1992, pp. 265–266.
43. U.S. Bureau of the Census, 1998c, p. 840.
44. Wresch, 1996.
45. Grant, 1995, p. 39.
46. Calhoun, Light, and Keller, 1994, p. 475.
47. Calhoun, Light, and Keller, 1994, pp. 475–476, citing Rostow, 1952, 1990.
48. Vogel, 1991; also see Calhoun, Light, and Keller, 1994; p. 475; Haub and Yanagishita, 1995.
49. Calhoun, Light, and Keller, 1994, p. 476.
50. Grant, 1995.
51. Calhoun, Light, and Keller, 1994, p. 466.
52. Grant, 1995, p. 43.
53. See Grant, 1995, pp. 44–45.
54. See Grant, 1995.
55. See Caldwell, 1996.

56. See Catton and Dunlap, 1980; Olsen, Lodwick, and Dunlap, 1991, p. 5; Burch, 1998; also Miller, G. Tyler, 1996.

57. Miller, G. Tyler, 1997, p. 179.

58. Grant, 1995, p. 14; Bellamy, 1998, p. 54.

59. Grant, 1995, p. 12.

60. Grant, 1995, p. 13.

61. Grant, 1995, pp. 36–37.

62. Grant, 1995, pp. 38–39.

63. Grant, 1995, pp. 67–71.

64. U.S. Bureau of the Census, 1997, p. 91; Grant, 1995, p. 67.

65. U.S. Bureau of the Census, 1998c, p. 125.

Chapter 20

1. Marx and McAdam, 1994, p. 11.

2. Marx and McAdam, 1994, p. 11.

3. Marx and McAdam, 1994, p. 50, citing Aguirre, Quarantelli, and Mendoza, 1988.

4. Marx and McAdam, 1994, p. 18; Turner and Killian, 1972.

5. Marx and McAdam, 1994, p. 18.

6. Marx and McAdam, 1994, p. 49.

7. See especially Oberschall, 1993.

8. LeBon, 1960, see especially discussion by McPhail, 1991, pp. 2–5, also Agguire, 1994, p. 262, Killian, 1994, p. 276; Snow and Oliver, 1995, pp. 571–572.

9. McPhail, 1991, suggests that Robert Park and Herbert Blumer are most responsible for this trend in sociology.

10. McPhail, 1991, pp. 14–15; Johnston and Snow, 1998; Snow and Oliver, 1995, pp. 574–575.

11. McPhail, 1991, pp. 14–15.

12. McPhail, 1991, pp. 192–198.

13. Oberschall, 1993, pp. 11–16; McPhail, 1991, pp. 14, 121–126, Berk, 1974; Snow and Oliver, 1995, pp. 583–586.

14. See Aberle, 1966, cited in Henslin, 1993, p. 612.

15. McCarthy and Zald, 1977, pp. 1219–1220.

16. Dalton, 1993, pp. 8–9; also see Berry, 1993; Kriesi et al., 1995; Pichardo, 1997.

17. See McCarthy and Zald, 1973, 1977; also see Gamson, 1975; Frey, Dietz, and Kalof, 1992.

18. Haub and Yanagishita, 1995; Famighetti, 1994.

19. Famighetti, 1994; Jamison et al., 1990.

20. For example see Minkoff, 1997.

21. Jamison et al., 1990, pp. 17–18.

22. Jamison et al., 1990, p. 128.

23. Mitchell, Mertig, and Dunlap, 1992, p. 13.

24. McCarthy and Zald, 1977, p. 1234.

25. Mitchell, Mertig, and Dunlap, 1992, pp. 18–19.

26. Jamison et al., 1990, p. 138.

27. Jamison et al., 1990, p. 94.

28. Jamison et al., 1990, p. 41.

29. Dunlap and Mertig, 1992.

30. Jamison et al., 1990, p. 11.

31. Dunlap and Mertig, 1992, p. 6.

32. Bullard and Wright, 1992.

33. Dunlap and Mertig, 1992, p. 6; also see McCloskey, 1992, and Gale, 1986, for a discussion of stages in the environmental movement.

34. See, for example, Mauss, 1975.

35. Dunlap and Mertig, 1992.

36. Jamison et al., 1990, p. 141.

37. Mitchell, Mertig, and Dunlap, 1992, p. 13.

38. McCarthy and Zald, 1977, p. 1221.

39. McCarthy and Zald, 1977, pp. 1233–1234.

40. Jamison et al., 1990, pp. 27–28.

41. Davies, 1962, 1974; Calhoun, Light, and Keller, 1994, p. 555.

42. Dalton, 1993, pp. 8–9; Jamison et al., 1990, p. 192.

43. McCarthy and Zald, 1973, 1977, p. 1225.

44. Jamison et al., 1990, pp. 9, 136; Eyerman and Jamison, 1991, p. 30.

45. Snow et al., 1986; Snow and Benford, 1992; McAdam, McCarthy, and Zald, 1996b; Zald, 1996; Babb, 1996; Melucci, 1996; Kebede and Knottnerus, 1998; Stein, 1998; Gamson, 1992.

46. Eyerman and Jamison, 1991; also see McAdam, 1982, pp. 48–54; Shin, 1994; Morris and Mueller, 1992.

47. Jamison et al., 1990, pp. 41–44.

48. Jamison et al., 1990, p. 109.

49. Jamison et al., 1990, pp. 116–119.

50. Jamison et al., 1990, pp. 122–124.

51. See Jamison et al., 1990, pp. 138–141.

52. Jamison et al., 1990, p. 142.

53. McCloskey, 1992, p. 84.

54. McAdam, McCarthy, and Zald, 1996a; Jenkins and Klandermans, 1995; Tarrow, 1996.

55. See Morris and Mueller, 1992; Gould, 1993; Zhao, 1998; Snow and Oliver, 1995.

56. See Chong, 1991; also see Morris and Mueller, 1992; Snow and Oliver, 1995.

57. Snow and Oliver, 1995, p. 579.

58. See Johnston, Laraña, and Gusfield, 1994; Melluci, 1996; Calhoun, 1994; Klandermans, 1997.

59. See McCarthy and Zald, 1977, pp. 1225–1226.

60. Jamison et al., 1990, pp. 59, 92, 166–167.

61. Dunlap, 1992, p. 112; also see Jones and Dunlap, 1992; Dunlap and Scarce, 1991; Rosa and Dunlap, 1994; Dunlap, 1991; Rosa and Freudenburg, 1993.

62. Dunlap, Gallup, and Gallup, 1993, p. 11.

63. Dunlap, Gallup, and Gallup, 1993, p. 15.

64. Dunlap, Gallup, and Gallup, 1993, pp. 33–35.

65. Ingelhart, 1990, cited in Dunlap, Gallup, and Gallup, 1993.

66. Grant, 1994.

67. For discussion of specific movements see, for example, Calhoun, 1997, on nationalism; Haines, 1996, on the anti-death penalty movement; McCarthy and Wolfson, 1996, on anti-drunk driving; Giele, 1995; Whittier, 1995, 1997, Rupp and Taylor, 1987, Ferree and Martin, 1995, on the feminist movement; Andrews, 1997, on the civil rights movement; and Diani, 1995, on the Italian environmental movement.

68. Mackie, 1996, pp. 999, 1002–1004.

69. Mackie, 1996, pp. 1003–1005.

70. Mackie,1996, pp.1000–1001.

71. Mackie, 1996, pp. 999–1000.

72. Mackie, 1996, p. 1010.

73. Abdalla, 1982, pp. 94–95, quoted in Mackie, 1996, p. 1014.

74. Mackie, 1996, p. 1001, citing Gamble, 1943.

75. Mackie, 1996, p. 1013.

76. Mackie, 1996, p. 1011.

77. Mackie, 1996, p. 1015.

Glossary

accommodation the norms that develop between ethnic groups regarding how they will interact with one another; the second stage of Park's race relations cycle

achieved status position that is attained through individual efforts, such as one's education or occupation

advanced industrial societies term used to refer to societies that have very productive extractive and manufacturing sectors and a growing service sector and a complex and productive communications industry

age-based norms expectations and rules about behaviors that should occur at various ages

age effect phenomenon whereby age-based norms influence our behaviors and actions

age/sex structure see *population structure*

agent of socialization the person or group that provides information about social roles

aggregates large groups of people who actually have no relationship to one another except that they might happen to be in the same place at the same time

agrarian societies societies with intense agricultural production made possible by the plow

alienation term, first used by Marx, that refers to the separation of workers from the product or result of their work, which can result in feelings of powerlessness

anomie Durkheim's term describing a situation of uncertainty over norms, or normlessness; the opposite of *synnomie*

anticipatory socialization the process by which we prepare ourselves for future roles through thinking about and rehearsing the actions, emotions, and skills that may be involved in these new roles

ascribed status position that is attained through circumstances of birth and that cannot be changed, such as one's race-ethnicity or gender

assimilation the merger of two ethnic groups; the final stage of Park's race relations cycle

attributes individual characteristics of persons or things; the values of variables

back-channel feedback verbal and nonverbal conversational techniques used to let others know whether their spoken messages are being understood

backstage Goffman's term for the setting, or frame, in which impression management is not needed; contrast with *frontstage*

base term used by Marx and Marxian scholars to refer to the economy and the dynamics surrounding the means of production, which he saw as the basis of social structure

birth cohort people who are born in the same year; see also *cohort*

birth rates measures of the number of births in a society relative to the total number of people in the society

body language physical movements, postures, and gestures

bounded rationality term used to refer to how decisions made within organizations are constrained, or bounded, by the structure, technology, and culture of the organization

bourgeoisie term used by Marx and Marxian scholars to describe capitalists, people who own the factories and mills; contrast with *proletariat*

bureaucracies highly structured and formalized organizations that are governed by laws and rules

capitalism the free-enterprise economic system in which private individuals or corporations develop, own, and control business enterprises; contrast with *socialism*

capitalists people who own large industries, who continually reinvest the profits from these industries to increase their wealth, and whose power and social standing come from their control of capital (money and other tangible resources)

career the sequence of roles that we enact during our lifetimes

cartel organization of nations that produce similar products and cooperate to limit competition among one another and jointly set prices for their products

case study a research technique that involves an in-depth look at one case, such as one person, one group, or one organization

caste system a system of stratification like that traditionally found in India, whereby the family into which one is born determines one's social status and the types of occupations one can hold

causal model graphic device that illustrates sociological theories involving causal relationships between variables

causal relationship an association between variables in which one influences or causes the other

census a complete count of a group of people

census tracts small, relatively permanent areas within cities and towns that are designed by the Census Bureau to be fairly homogeneous in population, economic situation, and living conditions and that represent, as much as possible, a city's neighborhood boundaries

charismatic authority Weber's term for authority that is based on the personal qualities of a particular individual

church a religious group that accepts the social environment in which it exists; compare with *sect*

class consciousness Marx's term for people in a social class who are aware of their common interests and concerns and that these interests conflict with those of another class group

closed-ended questions survey questions that give respondents only certain possible options from which to choose their answers

cognitive development emerging ability to understand and interpret the world

cognitive theory of social movements a theory of social movements that focuses on individuals' ideas and cognitive understandings and how these both prompt involvement in social movements and are shaped and developed by the actions of social movements

cohabitation the practice of men and women living together without marriage

cohort group of individuals who have some type of characteristic in common, such as the year in which they were born

cohort effect the common influence on attitudes and behaviors stemming from similar experiences based on cohort membership

collective behavior actions of groups of people, some of whom may not even know each other, that are often unexpected and spontaneous and can involve unusual or atypical behaviors

collective consciousness Durkheim's term describing the common beliefs, values, and norms of people within a society

collective identity joint agreements about the nature of the group, who belongs and who doesn't belong, and what the group should and should not do

command economy an economic system in which the political elite determine how goods and services will be produced, distributed, and consumed and in which workers' labor is coerced rather than freely hired

common schools schools held "in common," that is, for all children to attend

community a group of people who have frequent face-to-face interactions and common values and interests, relatively enduring ties, and a sense of personal closeness to one another

comparative research research method that involves the comparison of data from a variety of groups or settings, such as nations or historical eras

compensating differentials theory that employers compensate their employees for working in undesirable working conditions by paying them more

competition the inevitable conflicts between new and entrenched ethnic groups; the first stage in Park's race relations cycle

computer simulation the use of computer programs to simulate, or model, selected characteristics of real life; can be used by sociologists to conduct hypothetical experiments

concentric-zone model model of urban growth suggesting that cities expand outward from the central business district in a series of concentric circles or zones

concrete operational stage Piaget's term for the developmental stage, from about ages seven to eleven, in which children become able to handle logical operations

conglomerate corporate structure that includes many different firms specializing in entirely different types of products

conjugal family system a kinship system that emphasizes marriage ties over blood ties

consanguine family system a kinship system that emphasizes blood ties over marriage ties

conscience an internal set of ethical and moral principles that guide actions

construction industry industry that uses raw materials to build homes, offices, roads, and other structures

contact hypothesis hypothesis suggesting that individual discrimination and prejudice toward members of a minority group will diminish once members of the two groups have direct interpersonal contact

content analysis research method that involves the study of written documents or other types of recorded materials such as pictures, movies, or television shows

contingency approach theory of organizational leadership that suggests effective leadership traits and behaviors vary depending on a particular situation

control theory theory emphasizing that one's connections to others within a society are the major influence on one's desire to conform to society's norms

control variable variable that is added to an analysis to see if it affects the relationship between an independent and a dependent variable

convergence theory theory that most countries around the world are moving toward the establishment of a mixed economic system

conversation analysis research technique involving close examination of patterns in verbal interactions

core nations term used by world systems theorists to refer to the most powerful, dominant countries in the global economy

corporate actors a unified group, such as an organization or nation, that engages in social actions

correlation the ways in which two variables may be related to each other in a predictable pattern

crime actions that a society explicitly prohibits and that are sanctioned through official means

crime rate a calculation of how frequently crime occurs for every 100,000 people within a population

cult a religion that is in its beginning stages and is new to a society; the same as a *new religious movement*

cultural capital Bourdieu's term describing how people behave, dress, and talk and how these manners and styles differentiate those in one class group from those in another

cultural diffusion process whereby elements of one culture or subculture spread from one society or culture to another

cultural elements basic aspects or characteristics of a culture, both material and nonmaterial

cultural gatekeepers powerful organizations and individuals that control the entrance of new cultural elements into a society

cultural identity how people come to see themselves as belonging to or being part of a culture or cultural group

cultural innovation process whereby new cultural elements are created

cultural lag period of delay when one part of a society changes but other parts of a society have not yet readjusted

cultural materialism theoretical perspective suggesting that societies develop differently primarily because they exist in different environments and thus tap different resources or materials

cultural universals characteristics of a society that appear in all or virtually all cultures

culture a common way of life; the complex pattern of living that humans develop and pass on from generation to generation

curvilinear a relationship between two variables that is not well represented by a straight line and is better represented by some type of curved line

data factual information that is used as the basis for making decisions and drawing conclusions; the plural form of the word datum

deductive reasoning logical process of reasoning that moves from general theories or ideas to specific hypotheses or expectations; the opposite of *inductive reasoning*

deforestation the disappearance of forested land, usually caused by overcutting of trees

democracy government by the people, especially majority rule

democratic ideology the belief that the state belongs to the people and should serve the interests of all citizens

democratic organization an organization that is responsive to the desires of its members and does not degenerate into an oligarchy

democratic socialism a mixture of capitalism and socialism whereby the government owns and operates some aspects of the economy and private industry owns and operates others; also called welfare socialism

demographic momentum the growth of a society that occurs simply because of the people, or demographic characteristics, that are already there

demographic (population) processes ways in which populations change in composition and size

demographic transition theory theory that summarizes the changes, or transitions, in population that societies experience as they move from the high fertility and mortality typical of pre-industrial societies to the low fertility and mortality rates seen in highly industrialized societies

demography the scientific study of populations and their effects

dependency ratio the ratio of the dependent-age population, considered to be those younger than age 15 or older than age 64, to those of the traditional working ages (ages 15 to 64), used as a measure of how many people the average working person has to support

dependent variable variable that is said to be influenced or caused by another variable (the independent variable)

desertification an increase in desert land, usually caused by overgrazing

developing countries countries that are in the process of industrializing and that are in the middle of the global stratification system

deviance behaviors that violate social norms

differential association theory theory that focuses on how people learn deviant roles from their association or relationships with others engaged in deviant activities

diffuse status characteristics status characteristics that go beyond specific task-related skills and reflect more general statuses, such as occupation, education, gender, age, or race-ethnicity

discrimination differential treatment accorded to a group of people based solely on ascribed characteristics such as race-ethnicity; contrast with *specific status characteristics*

distributive justice see *norm of equity*

diversification form of corporate merger that results in a corporation specializing in many different aspects of one industry

divorce rate the number of divorces in a year relative to every 1,000 existing marriages

domination form of power in which one party controls the behavior of others through sanctions; compare with *influence*

dramaturgical theory theory derived from the work of Goffman that uses the metaphor of a drama to explain how individuals play social roles and thus produce social structure

dysfunctions negative consequences of a structure of society for the whole of society

ecological worldview set of beliefs suggesting that new technology may create as many problems as it solves, that the earth's resources are limited, that the world's population has reached the earth's carrying capacity, and that population growth must be curbed

economic capital income-producing wealth

economic exchanges economic transactions; the transfer of goods and services in return for some type of valued item

economy social institution that includes all the norms, organizations, roles, and activities involved in the production, distribution, and consumption of goods and services

education social institution that includes the groups, organizations, norms, roles, and statuses associated with a society's transmission of knowledge and skills to its members

educational attainment the number of years of education a person has

elites powerful people who are able to influence the political process

elitism view that political power and influence are dominated by a small handful of people who are relatively unified and form a comparatively small, tight-knit social network

emergent norms norms that develop or emerge from a setting, rather than already being part of the well-known, and specified expectations established within a culture.

empirical based on experience and observation rather than preexisting ideas

endogamy marriage rule requiring people to select partners from within their own tribe, community, social class, or racial-ethnic or other such group

epidemiology literally the study of epidemics, widespread outbreaks of communicable disease; today it also includes the study of other conditions that threaten health

ethnic enclave area of a community in which there is a great deal of entrepreneurial business activity by members of a particular racial-ethnic group

ethnic group group of people who have a common geographical origin and biological heritage and who share cultural elements, such as language; traditions, values, and symbols; religious beliefs; and aspects of everyday life, such as food preferences

ethnic identity feeling of belonging specifically to an ethnic group

ethnocentrism the belief that one's own culture or way of life is superior to that of others

evaluation research research designed to examine the effectiveness of social change or intervention projects

exchange theory Homans' theory that social action is an ongoing interchange of activity between rational individuals who decide whether they will perform a given action based on its relative rewards or costs

exclusion term describing attempts to entirely remove lower-paid groups from a labor market, as, for example, through restrictive immigration laws

exogamy marriage rule requiring people to select partners from outside their own tribe, community, social class, or racial-ethnic or other such group

expectation states beliefs about what others within a group can do

expectation states theory theory that group members' beliefs about what other members can do influence group interactions and the power structure within groups

experiment research method that uses control groups and experimental groups to assess whether a causal relationship exists between an independent and a dependent variable

experimental variable see *independent variable*

expressive leadership leadership that deals with interpersonal relationships within a group or organization

extended family family that includes relatives besides parents and children; contrast with *nuclear family*

external area term used by world systems theorists to refer to countries that are largely excluded from the global economy and that are much poorer than other countries

external social control attempts by others to control one's behavior

extractive industries industries involved in the removal of raw materials from the environment, such as agriculture, fishing, forestry, and mining

falsification the logic that underlies the testing of hypotheses; we can never prove that a theory is true, but can only say that it has either been falsified (shown to be untrue) or not yet been falsified

family development theory theoretical perspective that looks at changes in family structure to understand how family life changes and how these changes affect individuals

family of orientation the family in which one grows up

family of procreation the family in which one lives as an adult and in which one has children (procreates)

family structure the combination of statuses within a family, particularly the number and type of statuses that are included

feminism an ideology that directly challenges gender stratification and male dominance and promotes the development of a society in which men and women have equality in all areas of life

feminist movement   a social movement designed to promote the interests of women, much as the civil rights movement promoted racial-ethnic equality

fertility rates   measures of the number of births in a society relative to the total number of women of child-bearing age

field experiment   experiment that takes place in a real-life setting rather than in a laboratory

field research   research method in which a researcher directly observes behaviors or other phenomena in their natural setting

firm   the parent company or organization of work establishments

folklore   myths and stories that are passed from one generation to the next within a culture

folkways   norms that govern the customary way of doing everyday things

formal operations stage   Piaget's term for the final developmental stage in which children become capable of more complex and abstract thought

formal organizations   groups that have been deliberately created to accomplish certain goals

formal task groups   groups with specific goals, established norms, and recognized membership

frame   term used by dramaturgical theorists to refer to the setting in which interaction occurs and the norms that apply to a given situation

framing   the way social movements present and shape the ideas that underlie social movements

free riders   individuals who benefit from the public goods of a group or society even if they don't take the time or energy to participate

front-stage   Goffman's term for the setting, or frame, in which behaviors are designed to impress or influence others and in which impression management is important; contrast with *backstage*

function   the part a structure plays in maintaining or altering the society

game stage   Mead's term for the stage in which children begin to understand how different roles go together and become able to take the role of the other

Gemeinschaft   Tönnies's term describing relationships that might appear in small, close-knit communities in which people are involved in social networks with relatives and long-time friends and neighbors, much like those that appear in primary groups

gender-based division of labor   rules about what tasks members of each sex should perform

gender discrimination   differential treatment of women because of their sex

gender identity   one's gut-level belief that one is a male or a female

gender roles   the norms and expectations associated with being male or female

gender schema   cognitive framework that is used to organize information as relevant to one sex or the other

gender segregation   the restriction of members of each sex group to different statuses and roles

gender segregation of the labor force   the gender-based division of labor in the occupational world, or the phenomenon of men and women holding very different jobs; also called *occupational gender segregation*

gender socialization   learning to see oneself as a male or female and learning the roles and expectations associated with that sex group

gender stratification   the organization of society such that members of one sex group have more access to wealth, prestige, and power than members of the other sex group do

general fertility rate   the number of births per every 1,000 women of childbearing age in a given year

general strain theory   theory of deviance that expands on Merton's original formulation and suggests that the sources of structural strain can be much broader than Merton suggested and that the response that individuals choose is influenced by their perceptions of the fairness of the situation and the nature of their social networks.

generalized other   Mead's term for the conception people have of the expectations and norms that others generally hold; the basis of the "me"

Gesellschaft   Tönnies's term describing relationships that come about through formal organizations and economic relationships rather than through kinship or friendship, similar to those that appear in secondary groups

ghetto   highly segregated neighborhood populated primarily by people of one racial-ethnic heritage

Gini index   a measure of income inequality within societies that can vary from zero, indicating that everyone receives an equal share of the country's income, to 100, indicating that all the income is received by just one person

global economy   economic exchanges and markets that include the entire world

globalization   the process by which all areas of the world are becoming interdependent and linked with one another

global stratification system   differences between nations that result in some societies having more power, prestige, and property than others

goods-producing industries   industries involved in either the extraction, manufacture, or construction of products

grassroots organizations   social movement groups that develop "from the ground up," away from major political centers

gross national product   the value of all goods and services that a nation produces; GNP

group   two or more people who regularly and consciously interact with each other through engaging in some common activity and having some relatively stable social relationship

group boundaries   patterns of behavior that define who does and who does not belong to a given group

groupthink process by which members of a group become so oriented to maintaining the cohesiveness of the group that they ignore or suppress information critical of their decisions

group threat the extent to which members of a dominant group perceive that minority groups threaten their well-being; used as an explanation of variations in prejudice from one area or time period to another

growth rate a measure of how many people are added to a population each year divided by the number of people who are already in the population

hegemony a hidden but pervasive power involving such extreme domination of social life that we seldom recognize it or question its legitimacy

hierarchy of identities ordering of identities such that some identities are more important than others

historical-comparative research research that uses historical data to compare two or more societies

historical research research method that involves the examination of data from the past, often written artifacts and records

horizontal differentiation the extent to which the work of an organization is divided up among different units or subgroups; part of *organizational complexity*

horticultural societies societies that are based on a gardening economy

human capital resources that we have as individuals and workers, such as our education, skills, and work experience

human capital theory theory that suggests that different occupations and incomes reflect different amounts of human capital

hunting and gathering societies societies that obtain their food through hunting animals and gathering plants and that, compared with other types of societies, use the simplest, most primitive tools and work techniques

hypersegregation extreme isolation from other areas of the community

hypotheses statements about the expected relationship between two or more variables; often derived from theories

ideal types concepts or descriptions of phenomena that might not exist in a pure form in the real world but that define basic aspects of a given situation

ideologies complex and involved cultural belief systems

imperialism practice whereby one powerful country forcibly acquires territories in other areas of the world, creating vast colonial empires

impression management Goffman's term describing how individuals can manipulate the impression or view that others have of them and give out cues to guide interactions in a particular direction

income how much money a person receives in a given time, such as $20,000 a year

independent variable variable that is said to cause or influence another variable (the dependent variable); also called the experimental variable in experiments

Index crimes set of eight serious crimes used in the Uniform Crime Reports as the basis to calculate crime rates; see also *property crimes* and *violent crimes*

index of dissimilarity a measure of segregation that is calculated from data on the characteristics of people living in different census tracts; also sometimes called a *segregation index*

individual discrimination actions of individuals or small groups involving both discrimination and prejudice

inductive reasoning logical process of reasoning that moves from specific ideas and observations to more general hypotheses and theories; the opposite of *deductive reasoning*

industrialized countries countries that have a highly industrialized economy and that are at the top of the global stratification system

industrial worldview the set of beliefs involving the notion that the earth's resources are valuable because they provide for humans' needs

industry branches or areas of economic activity

infant mortality rate the number of babies in 1,000 who are expected to die before the age of one in a given year

influence form of power whereby providing information and knowledge leads others to take different actions; compare with *domination*

informal groups groups with no specified goals or formalized norms and no set membership

informal structure social networks within an organization that exist alongside the organization's formal structure

institutional norms norms that prescribe appropriate structures and behaviors of organizations and other aspects of social institutions

instrumental leadership leadership that deals with the tasks or work of an organization

interaction effect a pattern of influence in which the influence of one variable changes depending upon the status of another variable

interest groups political organizations that concentrate their activities on specific policy issues or concerns

intergenerational mobility holding a different occupational or social class position than one's parents

intergenerational succession holding the same occupational or social class position as one's parents

internal migration migration within national borders

internalization acceptance of the norms and behaviors of the group; the development of a conscience

internal labor market the series of jobs that people may hold within an organization throughout their work careers

internal social control control over one's behavior that is based on internalized standards; self-control

international migration migration across national borders

intervening variable variable that comes between a dependent variable and an independent variable in a causal relationship

kin   group of relatives beyond the nuclear family

labeling theory   theory suggesting that definitions of deviant behavior develop from social interactions and that the key element in becoming deviant is how others respond to people's behavior, rather than how they actually behave

labor force   people employed within a society and, sometimes, those actively seeking work; work force

labor market   set of jobs within a society

labor market segmentation theory   theory that the labor market is divided into primary and secondary sectors and that women and members of racial-ethnic minorities earn less than others because they more often hold jobs in the secondary sector

labor unions   organizations of workers who have joined together to collectively promote their interests as workers

laissez-faire capitalism   capitalistic system in which the government has no control or influence over the operation of business

land reform   a policy that calls for redistributing very large parcels of land, many of which are owned by extremely wealthy landowners or multinational corporations, to landless peasants

latent functions   functions that are less obvious and often unintended and that generally are unnoted by the people involved; contrast with *manifest functions*

laws   norms that are codified or written down; can be either folkways or mores

least developed countries   countries that are the least industrialized and that are at the bottom of the global stratification system

legal authority   Weber's term for authority based on written rules

life expectancy   the number of years a person is expected to live

literacy   the ability to read and write

longitudinal study   study involving data that have been collected at different times

macrolevel   theories and analyses that deal with relatively broad areas of society rather than with individuals

mainstreaming   the practice of integrating special education students into the general curriculum and school day

male dominance   cultural beliefs that give greater value and prestige to men and to their roles and activities

manifest functions   functions that are easily seen and obvious; contrast with *latent functions*

manufacturing industries   industries that process raw materials into more usable forms

market   process of buying and selling; how values are established for goods and services

mass schooling   elementary-level schooling for all children; common schools

material culture   the physical objects that are distinctive to a group of people, such as their food, clothes, houses, and hairstyles

means of production   term used by Marx and Marxian scholars to refer to the way in which people produce their living, such as by farming or manufacturing or hunting and gathering

measure   how concepts involved in a theory are translated into actual data

mechanical solidarity   Durkheim's term describing a society in which there is little differentiation, with people performing similar tasks, sharing similar responsibilities, and having similar behaviors; contrast with *organic solidarity*

median   the midway point in a distribution; the point at which 50 percent of the cases are larger and 50 percent are smaller

megalopolis   literally, "great city"; a term used to designate a vast area of land with a dense population

megamergers   form of corporate merger common since the 1980s that involves the union of some of the largest companies in the economy

mesolevel   theories and analyses that deal with social groups and organizations, such as individual families, classrooms, and work groups, rather than with very broad areas of society or with individuals

meta-analyses   the review and summary of a large number of studies of the same phenomenon to see what patterns emerge

methodology   the rules and procedures that guide research and help make it valid

metropolitan area (MA)   a region designated by the Census Bureau for statistical purposes that usually includes both the core area of a city with a large population and nearby communities

microlevel   theories and analyses that deal with relatively narrow aspects of social life, such as individuals' day-to-day activities and relations with other people

middle-range theories   theories that focus on relatively limited areas of the social world, as opposed to grand theories; often incorporate aspects of grand theories but are much more directed and applied toward specific research problems and can thus be more easily tested

migration rates   measures of the number of people who have moved into or out of an area relative to its total population

minority group   subculture that is subordinate to another group or groups within the society and that has less power, privilege, wealth, or prestige

mixed economic system   an economic system that blends elements of socialism and capitalism

mobility table   table of data used to illustrate patterns of intergenerational mobility and succession, typically involving the relationship between the occupations of parents and their offspring

modal   the most common or frequently occurring category or case

modernization theory perspective suggesting that the majority of less developed countries will eventually industrialize in the manner of countries such as England and the United States

monogamy the marriage of one man and one woman

monopoly situation in which one company controls an entire market or industry

mores norms that are vital to society, and violation of which is seen as morally offensive

mortality rates measures of the number of deaths in a society relative to certain characteristics of the population

multinational corporations business corporations that have outlets around the globe

multiple-nuclei model model of urban growth suggesting that cities may have a number of different core areas (or nuclei), often influenced by the geographic and physical characteristics of an area

nation-state political entity that unites large groups of people under agreed-upon laws and regulations

natural increase the excess of births over deaths; the change in population size that comes from the difference in the number of births and deaths

neo-imperialism system whereby one country dominates another through economic means rather than military or political control; said to characterize the post–World War II relationships of the United States with other countries

neo-Marxist term used to describe recently developed theories that are in the Marxian tradition although they may depart from Marx's thought in certain ways

new religious movement a religion that is in its beginning stages and is new to a society; the same as a *cult*

new social movements recently developed social movements that tend to use tactics that have rarely been used before, especially the direct involvement of large numbers of citizens in political protest, and to address issues that have not been the focus of recent political debate, such as the quality of the environment, gender equality, alternative lifestyles, and human rights in developing countries

nodes see *social units*

nonmaterial culture the way of thinking of a group, including norms, values, ideology, folklore, and language

nonparticipant observation type of field research in which a researcher studies a group through observations without actually participating

nonverbal communication all the ways in which we send messages to others without words, including posture and movements, facial expressions, clothes and hairstyles, and manner of speaking

normative life stages periods during the life course when people are expected to perform certain activities

norm of equity belief that things should be "fair," that we and others should receive rewards that equal what we contribute to a relationship or interaction; also called *distributive justice*

norms cultural rules defining behavior that is expected or required within a group or situation; includes folkways, mores, and laws

nuclear family family group consisting of a mother, a father, and their children; contrast with *extended family*

occupational gender segregation see *gender segregation of the labor force*

occupational prestige scores scores given to occupations that indicate their relative prestige ranking

occupational socialization the process of learning and identifying with the norms and roles associated with a particular occupation

occupational status the occupation that one holds; often used in reference to the relative importance of one occupation over another

occupational structure occupations available within a society in a given time period

open-ended questions survey questions that allow respondents to give whatever responses they desire

organic solidarity Durkheim's term describing a society in which tasks, responsibilities, and behaviors are more highly differentiated, so that people are more dependent on one another; contrast with *mechanical solidarity*

organizational centralization the extent to which organizational power is centralized and decisions are made hierarchically

organizational complexity the extent to which the work of an organization is broken up and differentiated among various units; can be horizontal, vertical, or spatial

organizational culture the way of life, or culture, of an organization

organizational ecology theoretical perspective that looks at the relationship between organizations and their environments

organizational formalization the extent to which an organization uses written rules and procedures to control individuals within it

organizational leaders people within organizations who are able to influence the things others do and believe

organizational structure how the various parts of an organization are arranged; typically includes the elements of complexity, formalization, and centralization

organizational technology the work of organizations, including the skills and knowledge of workers in an organization, as well as the characteristics of the materials they work with and the machines they use

panel study a longitudinal study that includes information on the same people over a long period of time

paradigm a fundamental model or scheme that guides people's thinking about a particular subject

paradigm crisis the breakdown of an explanatory scheme, requiring a fundamental reorientation or rethinking of basic assumptions

paralanguage nonverbal aspects of speech, such as tone of voice and emphasis on words

participant observation type of field research in which a researcher studies a group or event while actually participating in it

peer review process by which research findings are evaluated by other experts in the field before publication to see if they meet the standard rules of research methodology

people of color people whose ancestors or who themselves came from non-European areas of the world and who can be identified through the color of their skin

per capita GNP the gross national product adjusted for the number of people who live in the country

period effect the influence of the norms and events of a particular historical time on individuals' lives

peripheral nations term used by world systems theorists to refer to countries that sell raw materials to core nations and that are at the edge of the global economy

play stage Mead's term for the developmental stage in which children "play at" or assume various roles one at a time

pluralism view that the political power structure involves a number of powerful groups and individuals, all of which can potentially influence the decision-making process

plutocracy government by the wealthy

political economy the interrelationship of economic and political processes

polity the social institution that includes all the various ways that societies have developed to maintain social order and control; the political world

polyandry a system of marriage whereby the wife can have more than one husband; the opposite of *polygyny*

polygyny a system of plural marriage whereby the husband has more than one wife

population the entire group or set of cases that a researcher is interested in generalizing to; see also sample

population projections estimates of the size and composition of a population in future years

population pyramid diagram used by demographers to illustrate the population or age/sex structure of a society

population structure the age and sex composition of a society; the age/sex structure

poverty line the income level under which families are officially defined as being poor

poverty rate the percentage of the total population that lives in families with incomes under the official poverty level

power the ability of one social element, either a group or a person, to compel another social element to do what it wants

power structure the power differences within a group; the way that power is distributed in a society

PPE spiral term used to refer to the intertwining influences of poverty, population growth, and environmental degradation

prejudice preconceived hostile attitudes toward a group of people simply on the basis of their group membership

preoperational stage Piaget's term for the developmental stage, from about eighteen months to seven years of age, in which children learn to use symbols and communicate with language but don't yet have the mental flexibility to perform mental operations that involve complex relationships

prestige classes aggregates or clusters of people who possess similar characteristics, who are perceived by their fellow citizens as being similar, and who are accorded similar levels of respect and esteem

primary groups groups that include only a few people and that are characterized by intimate, face-to-face interaction

primary labor market "good" jobs with good wages and benefits, comfortable working conditions, job security, and chances for advancement; contrast with *secondary labor market*

probability sample sample that can be generalized to a larger group, typically chosen through some type of random selection process

profession work group with a highly specialized body of knowledge and a service orientation

proletariat term used by Marx and Marxian scholars to describe the workers; contrast with *bourgeoisie*

property crimes crimes that don't involve the use of violence or force against individual people, such as burglary, larceny-theft, motor vehicle theft, and arson

public goods necessities of group life that individuals cannot provide by themselves but must obtain through cooperation with others; common goods

qualitative data measures of data that cannot be assigned real numbers; contrast with *quantitative data*

quantitative data measures that can be assigned real, or meaningful, numbers—for example, income or age

racial-ethnic group subculture that can be distinguished on the basis of skin color and ethnic heritage

racial-ethnic stratification the organization of society such that people in some racial-ethnic groups have more property, power, or prestige than do people in other groups

random selection process that gives each member of a population an equal chance of being included in a sample

rational choice theory theory suggesting that people make decisions by balancing the costs and benefits involved and choosing the actions that provide the lowest costs and the greatest benefits

reference group group of people that we look to, or use for reference, in evaluating ourselves and our position in life

reflexive behavior   the process of being able to think about one's own actions and mentally assume the role of other people

regional stratification   differences within a country that result in some regions having more power, prestige, and property than others

relative deprivation   a gap between people's expectations and the actual state of their lives

reliability   the extent to which a measure yields the same results when used by different researchers on the same subject at different times

religion   the social institution that deals with the area of life people regard as holy or sacred; it involves the statuses, roles, organizations, norms, and beliefs that are related to humans' relationship with the supernatural, including shared beliefs, ethical rules, rituals and ceremonies, and communities of people with common beliefs and standards

religious economy   the various religious options or choices that are available within a society

replacement rate   the fertility rate that is needed for a population to replace itself or to remain stable in size (net of migration)

replication   repetition of an earlier study to see if the same results occur and if they hold in other settings

representative democracy   a political system in which elected officials represent the citizens and make important governmental decisions

repression   psychoanalytic term for the notion that people repress, or push out of conscious awareness, ideas that are uncomfortable or painful to think about

resource mobilization theory   theory suggesting that social movements are successful to the extent that they can effectively mobilize or activate the resources needed to accomplish their goals

rite of passage   ceremony or ritual that helps individuals deal with role transitions through the life course

rituals   cultural ceremonies that often mark important life events

role conflict   situation in which a person holds roles with incompatible norms or obligations

role theory   perspective that social structure is created and maintained because people generally act in ways that conform to social roles

role transition   moves, or transitions, from one role to another during the life course

rural development   industrialization in rural areas

sample   subset of a larger group or population; see also *probability sample*

sanctions   social reactions to an individual's behavior, generally reflecting attempts to control the behavior; rewards and punishments

scheme   a mental framework or set of rules to understand how the world works

schooling   formal instruction by trained teachers, usually within schools

secondary analysis   analysis of data that have already been gathered by other researchers

secondary deviance   deviance that results from the process of being labeled as a deviant

secondary groups   groups that involve more than just a few people and relatively distant social relations

secondary labor market   "bad" jobs with low pay, few if any benefits, little job security, and few chances for advancement; contrast with *primary labor market*

sect   a religious group that rejects the social environment in which it exists; compare with *church*

sector model   model of urban growth suggesting that cities are often composed of sectors or segments that extend out from the central business district

secularization   process of transformation to an outlook on life based on science and reason rather than on faith and supernatural explanations

segmented occupational structure   division of occupations into "good jobs," characterized by stable employment, benefits, and higher salaries, and "not-so-good jobs," which lack these characteristics

segregation   the division or separation of neighborhoods in ways that lead to the inclusion of some groups and the exclusion of others

segregation index   see *index of dissimilarity*

self   one's view of oneself as a distinct person with a clear identity

self-concept   the thoughts and feelings we have about ourselves

self-identity   a set of categories used to define the self; the way we think about ourselves

semiperiphery   term used by world systems theorists to refer to nations that fall between the core and periphery

sensorimotor stage   Piaget's term for the developmental stage, from birth to about eighteen months, in which infants learn through their senses and their movements

serial monogamy   a pattern of marriage in which a person has several spouses over a lifetime, but only one at a time

service industries   global term used to describe the entire set of non-goods-producing industries; also used to refer to a specific set of industries that provide various nonmaterial things people want or need, such as medical care, education, social welfare, and entertainment

significant others   people with whom you interact and who are emotionally important to you

situated self   the subset of self-concepts or self-identities that apply to one's self-view and behaviors in a given situation

social action   day-to-day decisions and actions of individuals within the social world; social actions both influence and are patterned and influenced by social structure

social capital   resources or benefits people gain from their social networks

social change   the way in which societies and cultures alter over time

social class groups of people who occupy a similar level in the stratification system

social control efforts to help ensure conformity to norms

social distance the types of social ties or interactions people are willing to establish with others

social ecology the study of how people interact with their physical and geographic environment

social institutions the complex sets of statuses, roles, organizations, norms, and beliefs that meet people's basic needs within a society

socialism an economic system that involves public rather than private ownership of the means of production; contrast with *capitalism*

socialization how we develop, through interactions with others, the ability to relate to other people and to play a part in society

social learning perspective theory of socialization suggesting that we learn behaviors by being rewarded for those that conform to norms and by imitating those that we think conform to roles we want to fill

social mobility movement between social class groups

social movement adherents organizations and individuals that believe in the goals of a social movement

social movement constituents organizations and individuals that provide resources for a social movement

social movement industry the set of social movement organizations that are all working within the same social movement

social movement organizations formal organizations that work toward the goals of a social movement

social movements organized and concerted efforts to promote social change by groups of individuals

social movement sector the entire range of activities that are aimed at changing the social structure.

social network patterns or webs of social relationships; the linkages between individuals or organizations formed by social interactions

social role expectations, obligations, and norms that are associated with a particular position in a social network

social status positions that individuals occupy within the social structure

social stratification the organization of society in a way that results in some people having more and some people having less; divisions in a society based on social class

social structure relatively stable patterns that underlie social life; the ways in which people and groups are related to each other, and the characteristics of groups that influence our behavior

social ties relationships between units of a social network

social units elements of a social network; also called *nodes*

society a group of people who live within a bounded territory and who share a common way of life

sociological imagination Mills's term describing the ability to discern patterns in social events and view personal experiences in the light of these patterns

sociology the science of society; the scientific study of the social world and social interactions

spatial dispersion the extent to which the various units of an organization are spread out in different locations; an element of organizational complexity

specific status characteristics status characteristics that are closely related to the given tasks of a group; contrast with *diffuse status characteristics*

split labor market the division of the labor market into three class groups: business owners and employers, higher-paid workers, and lower-paid workers

spurious correlation correlation between two variables that only occurs because of the influence of a third variable

state the organized monopoly or control of the use of force in a society; synonymous with government or the polity

statistical discrimination theory that employers have beliefs, or stereotypes, about the relative stability and productivity of men and women workers, which they rely on in hiring employees; when these average characteristics don't apply to individual women, statistical discrimination is said to have occurred

status term used in the Weberian tradition to designate one dimension of stratification, that involving communities or social networks of people with similar lifestyles and viewpoints; synonymous with prestige

status attainment model theoretical model that describes the way variables, such as family background, individual motivations and ability, and interactions with significant others, influence the educational levels and occupations and incomes we have in adulthood

status characteristics the social statuses that people hold and the evaluations and beliefs (characteristics) that are attached to these statuses

status generalization the process by which diffuse status characteristics influence group interactions

strain theory Merton's theory that deviant behavior results when individuals accept culturally defined goals but do not have the institutionalized means of attaining these goals

stratification system see *social stratification*

structural discrimination discrimination that results from the normal and usual functioning of the society (the social structure) rather than from prejudice or from laws and norms that promote segregation or exclusion

structural functionalism sociological theory that tries to account for the nature of social order and the relationship between different parts of society by noting the ways in which these parts or structures function to maintain the entire society

structural mobility intergenerational mobility resulting from changes in the occupational structure

subcultural theory of deviance   theory suggesting that deviant behavior can reflect conformity to the norms of subcultures that support deviant lifestyles

subculture   group or culture that exists within a broader culture

suburb   a city or town outside a city's boundaries but either adjacent to the city or within commuting distance

superstructure   term used by Marx and Marxian scholars to refer to the areas of social life that they believe are influenced by the economy, such as religious beliefs, family relations, and political life

survey research   method of data gathering that involves asking people questions, through either interviews or written questionnaires

sustainable development   economic development that does not deplete or degrade natural resources.

symbolic interactionism   theory that social interaction involves a constant process of presenting and interpreting symbols through thinking about what another person is trying to communicate through the use of symbols

symbols   anything that people use to represent something else; for example, language uses the symbols of words to represent objects and ideas

synnomie   the state of a society that has a strong collective consciousness and a high degree of cohesion; the opposite of *anomie*

target of socialization   the person who is learning a social role, who is being socialized

technological innovations   the development of new material elements of culture that help people deal with the problems of day-to-day life

technology   the application of knowledge to solve problems

theories   broad systems of ideas that help explain patterns in the social world

traditional authority   Weber's term for authority that is accepted because it is an integral part of the social structure and it is impossible to conceive of any other way of being

trend analysis   study that examines data collected at different time points but that uses a different sample each time

turn taking   term used by conversation analysts to describe the conversation process whereby one person talks and then the other talks; results from both verbal and nonverbal cues

typology   a classification of a group or phenomenon into discrete categories

unconscious   psychoanalytic term for thoughts and impulses individuals might have that they aren't consciously aware of

unobtrusive research   research method in which a researcher obtains data without directly talking to or watching people

upward mobility   social mobility that involves movement to a social class position that is higher than one's parents occupied

urbanization   the process of societal change that involves the movement of people from rural areas or small towns to metropolitan areas

validity   the extent to which a measure actually represents the concept it is said to be measuring; when applied to a research design, indicates that we can trust the conclusions

values   general standards about what is important to a group

variables   logical groupings of attributes; literally, things that vary or have more than one value

vertical differentiation   the number of supervisory levels in an organization; part of organizational complexity

vertical integration   form of corporate merger producing companies that control the entire extraction, manufacturing, and distribution process

violent crimes   crimes that involve the actual use or threat of force against people, including murder, forcible rape, robbery, and aggravated assault

vital statistics   data that record the "vital events" of life, such as births, deaths, marriages, divorces, and migration

waves of data collection   times of data collection in a panel study

wealth   assets resulting from the accumulation of income, such as houses, cars, real estate, and stocks and bonds

welfare capitalism   form of capitalism in which a broad system of laws protects the welfare of workers and consumers in their economic transactions

welfare socialism   see *democratic socialism*

white-collar crime   nonviolent crimes that generally involve fraud and deception and are committed in the workplace

white ethnics   people who identify themselves as belonging to a group with European origins, usually other than Great Britain

wholesale and retail trade industry   industry involving the sale of goods to stores and directly to individual consumers

work establishment   the actual place where someone works; compare with *firm*

work force   see *labor force*

world systems theory   perspective developed by Wallerstein to explain global stratification; suggests that a set of core, highly developed nations maintain their wealth by dominating and exploiting other nations, said to be on the periphery of the world economy

zero population growth   the idea that the world population will stop growing when birth rates equal death rates and when, on average, each woman has only two children

References

Abdalla, Raqiya Havi Dualeh. 1982. *Sisters in Affliction: Circumcision and Infibulation of Women in Africa.* London: Zed.

Aberle, David. 1966. *The Peyote Religion Among the Navajo.* Chicago: Aldine.

Acock, Alan C., and David H. Demo. 1994. *Family Diversity and Well-Being.* Thousand Oaks, Calif.: Sage.

Aday, David P., Jr. 1990. *Social Control at the Margins: Toward a General Understanding of Deviance.* Belmont, Calif.: Wadsworth.

Adler, Freda. 1983. *Nations Not Obsessed with Crime.* Littleton, Colo.: Rothman.

Adler, Freda, Gerhard O. W. Mueller, and William S. Laufer. 1991. *Criminology.* New York: McGraw-Hill.

Adler, Nancy E., Thomas Boyce, Margaret A. Chesney, Sheldon Cohen, Susan Folkman, Robert L. Kahn, and S. Leonard Syme. 1994. "Socioeconomic Status and Health: The Challenge of the Gradient." *American Psychologist* 49: 15–24.

Agnew, Robert. 1997."Stability and Change in Crime over the Life Course: A Strain Theory Explanation." Pp. 101–132 in Terence P. Thornberry (ed.), *Developmental Theories of Crime and Delinquency, Advances in Criminological Theory,* Vol. 7. New Brunswick, N.J.: Transaction.

Agnew, Robert. 1998. "Foundation for a General Strain Theory of Crime and Delinquency." Pp. 177–194 in Stuart Henry and Werner Einstadter (eds.), *The Criminology Theory Reader.* New York: New York University Press.

Aguirre, Benigno E. 1994. "Collective Behavior and Social Movement Theory." Pp. 257–272 in Russell R. Dynes and Kathleen J. Tierney. (eds.). *Disasters, Collective Behavior, and Social Organization.* Newark, N.J.: University of Delaware Press.

Aguirre, Benigno E., E. L. Quarantelli, and J.L. Mendora. 1988. "The Collective Behavior of Fads: The Characteristics, Effects, and Career of Streaking." *American Sociological Review* 53: 569–584.

Ahrons, Constance R., and Roy H. Rodgers. 1987. *Divorced Families: A Multidisciplinary Developmental View.* New York: Norton.

Akard, Patrick J. 1992. "Corporate Mobilization and Political Power: The Transformation of U.S. Economic Policy in the 1970s." *American Sociological Review* 57: 597–615.

Akers, Ronald L. 1998. *Social Learning and Social Structure: A General Theory of Crime and Deviance.* Boston: Northeastern University Press.

Alba, Richard D. 1985. *Italian Americans: Into the Twilight of Ethnicity.* Englewood Cliffs, N.J.: Prentice-Hall.

Alba, Richard D. 1990. *Ethnic Identity: The Transformation of White America.* New Haven, Conn.: Yale University Press.

Albanese, Catherine L. 1993. "Fisher Kings and Public Places: The Old New Age in the 1990s." *Annals, AAPSS* 527: 131–143.

Albrow, Martin. 1997. *The Global Age: State and Society Beyond Modernity.* Stanford, Calif.: Stanford University Press.

Aldous, Joan. 1978. *Family Careers.* New York: Wiley.

Aldous, Joan. 1996. *Family Careers: Rethinking the Developmental Perspective.* Thousand Oaks, Calif.: Sage.

Aldrich, Howard E., and Peter V. Marsden. 1988. "Environments and Organizations." Pp. 361–392 in Neil Smelser (ed.), *Handbook of Sociology.* Newbury Park, Calif.: Sage.

Alexander, Jeffrey C. 1987. *Twenty Lectures: Sociological Theory Since World War II.* New York: Columbia University Press.

Alexander, Jeffrey C. 1988. *Action and Its Environments: Toward a New Synthesis.* New York: Columbia University Press.

Alexander, Jeffrey C. 1998. *Neofunctionalism and After.* Malden, Massachusetts: Blackwell.

Allen, Michael Patrick. 1974. "The Structure of Interorganizational Elite Cooptation: Interlocking Corporate Directorates." *American Sociological Review* 39: 393–406.

Allen, Michael Patrick. 1987. *The Founding Fortunes: A New Anatomy of the Super-Rich Families in America.* New York: Truman Talley.

Allen, Michael Patrick. 1991. "Capitalist Response to State Intervention: Theories of the State and Political Finance in the New Deal." *American Sociological Review* 56: 679–689.

Allen, Michael Patrick, and Philip Broyles. 1989. "Class Hegemony and Political Finance: Presidential Campaign Contributions of Wealthy Capitalist Families." *American Sociological Review* 54: 275–287.

Almanac of Federal PACs. 1990. Washington, D.C.: Amward.

Alpert, Judith L., and Jody Boghossian Spencer. 1986. "Morality, Gender, and Analysis." Pp. 83–111 in Judith L. Alpert (ed.), *Psychoanalysis and Women: Contemporary Reappraisals.* New York: Analytic.

Altman, Lawrence K. 1998. "Prevention Still Paramount in Fighting Spread of AIDS." *Sunday Oregonian*, July 5, 1998, p. A6.

Aly, Hassan Y., and Richard Grabowski. 1990. "Education and Child Mortality in Egypt." *World Development* 18: 733–742.

Amato, Paul R,. and Alan Booth. 1997. *A Generation at Risk: Growing Up in an Era of Family Upheaval.* Cambridge, Mass.: Harvard University Press.

Ammerman, Nancy. 1987. *Bible Believers: Fundamentalists in the Modern World.* New Brunswick, N.J.: Rutgers University Press.

Andrews, Kenneth T. 1997. "The Impacts of Social Movements on the Political Process: The Civil Rights Movement and Black Electoral Politics in Mississippi." *American Sociological Review* 62: 800–819.

Archer, Dane, and Rosemary Gartner. 1984. *Violence and Crime in Cross-National Perspective.* New Haven, Conn.: Yale University Press.

Ashford, Lori S. 1995. "New Perspectives on Population: Lessons from Cairo." *Population Bulletin* 50/1 (March).

Aulette, Judy Root. 1994. *Changing Families.* Belmont, Calif.: Wadsworth.

Babb, Sarah. 1996. " 'A True American System of Finance' Frame Resonance in the U.S. Labor Movement, 1866–1886." *American Sociological Review* 61: 1033–1052.

Babbie, Earl. 1992. *The Practice of Social Research,* 6th ed. Belmont, Calif.: Wadsworth.

Bachmann, Thomas (ed.). 1995. *Directory of Corporate Affiliations,* Vol. 3: *U.S. Public Companies: "Who Owns Whom."* New Providence, N.J.: National Register.

Bainbridge, William Sims. 1985. "Cultural Genetics." In Rodney Stark (ed.), *Religious Movements: Genesis, Exodus, and Numbers.* New York: Paragon.

Bainbridge, William Sims. 1989. "The Religious Ecology of Deviance." *American Sociological Review* 54: 288–295.

Bainbridge, William Sims. 1997. *The Sociology of Religious Movements.* New York and London: Routledge.

Bales, R. F., and P. Slater. 1955. "Role Differentiation in Small Social Groups." Pp. 259–306 in Talcott Parsons, Robert F. Bales and E. A. Shils (eds.), *Family, Socialization, and Interaction Process.* Glencoe, Ill.: Free Press.

Ballantine, Jeanne H. 1993. *The Sociology of Education: A Systematic Analysis,* 3rd ed. Englewood Cliffs, N.J.: Prentice-Hall.

Ballis, William B., Charles R. McLane, and Theodore Shabad. 1962. "Russia." Pp. 486–507 in *The World Book Encyclopedia*, Vol. 15. Chicago: Field Enterprises.

Barker, Eileen. 1985. "New Religious Movements: Yet Another Great Awakening?" Pp. 36–57 in Phillip E. Hammond (ed.), *The Sacred in a Secular Age: Toward Revision in the Scientific Study of Religion.* Berkeley: University of California Press.

Barker, Eileen, James A. Beckford, and Karel Dobbelaere (eds.). 1993. *Secularization, Rationalism, and Sectarianism: Essays in Honour of Bryan R. Wilson.* Oxford: Clarendon.

Barker, R. G., and P. V. Gump. 1964. *Big School, Small School.* Stanford, Calif.: Stanford University Press.

Barna, Laray M. 1991. "Stumbling Blocks in Intercultural Communication." Pp. 345–352 in Larry A. Samovar and Richard E. Porter (eds.), *Intercultural Communication: A Reader.* Belmont, Calif.: Wadsworth.

Barnes, Barry. 1995. *The Elements of Social Theory.* Princeton, N.J.: Princeton University Press.

Barnett, Rosalind C., and Grace K. Baruch. 1987. "Determinants of Fathers' Participation in Family Work." *Journal of Marriage and the Family* 49: 29–40.

Bartley, Mel, and Ian Plewis. 1997. "Does Health-Selective Mobility Account for Socioeconomic Differences in Health? Evidence from England and Wales, 1971 to 1991." *Journal of Health and Social Behavior* 38: 376—386.

Baruch, Grace K., and Rosalind C. Barnett. 1986. "Consequences of Fathers' Participation in Family Work: Parents' Role Strain and Well-Being." *Journal of Personality and Social Psychology* 51: 981–992.

Bearman, Peter S. 1991. "The Social Structure of Suicide." *Sociological Forum* 6: 501–524.

Becker, Howard. 1953. "Becoming a Marihuana User." *American Journal of Sociology* 59: 235–242.

Becker, Howard S. 1963. *Outsiders.* New York: Free Press.

Becker, Howard, and Harry Elmer Barnes. 1961. *Social Thought from Lore to Science,* Vol. 2, 3rd ed. New York: Dover.

Beeghley, Leonard. 1996. *What Does Your Wife Do? Gender and the Transformation of Family Life.* Boulder, Colo.: Westview.

Beggs, John J., and Jeanne S. Hurlbert. 1997. "The Social Context of Men's and Women's Job Search Ties: Membership in Voluntary Organizations, Social Resources, and Job Search Outcomes." *Sociological Perspectives* 40: 601–622.

Bellah, Robert N., and Frederick E. Greenspahn. 1987. *Uncivil Religion: Interreligious Hostility in America.* New York: Crossroad.

Bellamy, Carol. 1997. *The State of the World's Children, 1997.* Oxford: Oxford University Press.

Bellamy, Carol. 1998. *The State of theWorld's Children, 1998.* Oxford: Oxford University Press.

Bem, Sandra Lipsitz. 1981. "Gender Schema Theory: A Cognitive Account of Sex Typing." *Psychological Review* 88: 354–364.

Bendix, Reinhard. 1960. *Max Weber: An Intellectual Portrait*. Garden City, N.Y.: Doubleday.

Bendix, Reinhard. 1962. *Max Weber: An Intellectual Portrait*. New York: Doubleday Anchor.

Bendix, Reinhard, and Seymour Martin Lipset. 1966. "Karl Marx's Theory of Social Classes." Pp. 6–11 in *Class, Status, and Power: Social Stratification in Comparative Perspective,* 2nd ed. New York: Free Press.

Bengtson, Vern L., and W. Andrew Achenbaum (eds.). 1993. *The Changing Contract Across Generations*. New York: Aldine de Gruyter.

Bengtson, Vern L., K. Warner Schaie, and Linda M. Burton (eds.). 1995. *Adult Intergenerational Relations: Effects of Societal Change*. New York: Springer.

Benson, Michael L. 1998. "Denying the Guilty Mind: Accounting for Involvement in a White-Collar Crime." Pp. 247–264 in Stuart Henry and Werner Einstadter (eds.), *The Criminology Theory Reader*. New York: New York University Press.

Berger, Joseph, Bernard P. Cohen, and Morris Zelditch, Jr. 1972. "Status Conceptions and Social Interaction." *American Sociological Review* 37: 241–255.

Berger, Joseph, M. Hamit Fisek, Robert Z. Norman, and Morris Zelditch, Jr. 1977. *Status Characteristics and Social Interaction: An Expectation-States Approach*. New York: Elsevier.

Berger, Joseph, and Morris Zelditch, Jr. (eds.). 1993. *Theoretical Research Programs: Studies in the Growth of Theory*. Stanford, Calif.: Stanford University Press.

Bergin, Allen E. 1983. "Religiosity and Mental Health: A Critical Reevaluation and Meta-Analysis." *Professional Psychology: Research and Practice* 14: 170–184.

Bergmann, Barbara R. 1986. *The Economic Emergence of Women*. New York: Basic Books.

Bergmann, Barbara R. 1993. "The French Child Welfare System: An Excellent System We Could Adapt and Afford." Pp. 341–350 in William Julius Wilson (ed.), *Sociology and the Public Agenda*. Newbury Park, Calif.: Sage.

Bergmann, Barbara R. 1996. *Saving Our Children From Poverty: What the United States Can Learn from France*. New York: Russell Sage Foundation.

Berk, Richard. 1974. *Collective Behavior*. Dubuque, Iowa: William C. Brown.

Berk, Richard A., Alex Campbell, Ruth Klap, and Bruce Western. 1992. "The Deterrent Effect of Arrest: A Bayesian Analysis of Four Field Experiments." *American Sociological Review* 57: 698–708.

Berk, Sarah Fenstermaker. 1985. *The Gender Factory: The Apportionment of Work in American Households*. New York: Plenum.

Berk, Sarah Fenstermaker, and Donileen R. Loseke. 1980. "'Handling' Family Violence: Situational Determinants of Police Arrest in Domestic Disturbances." *Law and Society Review* 15: 317–346.

Berman, Sheldon. 1997. *Children's Social Consciousness and the Development of Social Responsibility*. Albany, N.Y.: State University of New York Press.

Bernhardt, Annette, Martina Morris, and Mark S. Handcock. 1995. "Women's Gains or Men's Losses? A Closer Look at the Shrinking Gender Gap in Earnings." *American Journal of Sociology*. 101: 302–328.

Bernhardt, Annette, Martina Morris, and Mark S. Handcock. 1997. "Percentages, Odds, and the Meaning of Inequality: Reply to Cotter, et al." *American Journal of Sociology* 102: 1154–1162.

Berry, Jeffrey M. 1993. "Citizen Groups and the Changing Nature of Interest Group Politics in America." Pp. 30–41 in Russell J. Dalton (ed.), *Citizens, Protest, and Democracy. The Annals of the American Academy of Political and Social Science* 528 (July).

Bian, Yanjie. 1997. "Bringing Strong Ties Back In: Indirect Ties, Network Bridges, and Job Searches in China." *American Sociological Review* 62: 366–385.

Bianchi, Suzanne. 1990. "America's Children: Mixed Prospects." *Population Bulletin* 45/1 (June).

Bianchi, Suzanne M. 1995. "Changing Economic Roles of Women and Men." Pp. 107–154 in Reynolds Farley (ed.), *State of the Union: America in the 1990s,* Vol. 1: *Economic Trends*. New York: Russell Sage.

Bianchi, Suzanne M., and Daphne Spain. 1996. "Women, Work, and Family in America." *Population Bulletin* 51(3). Washington, D.C.: Population Preference Bureau.

Bibby, Reginald W., and Merlin B. Brinkerhoff. 1994. "Circulation of the Saints 1966–1990: New Data, New Reflections." *Journal for the Scientific Study of Religion* 33: 273–280.

Biblarz, Timothy J., and Adrian E. Raftery. 1993. "The Effects of Family Disruption on Social Mobility." *American Sociological Review* 58: 97–109.

Bidwell, Charles E., and Noah E. Friedkin. 1988. "The Sociology of Education." Pp. 449–471 in Neil Smelser (ed.), *Handbook of Sociology*. Newbury Park, Calif.: Sage.

Bielby, William T., and James N. Baron. 1984. "A Woman's Place is with Other Women: Sex Segregation within Organizations." Pp. 27–55 in Barbara F. Reskin (ed.), *Sex Segregation in the Workplace: Trends, Explanations, Remedies*. Washington, D.C.: National Academy.

Bielby, William T., and James N. Baron. 1986. "Men and Women at Work: Sex Segregation and Statistical Discrimination." *American Journal of Sociology* 91: 759–799.

Bierstedt, Robert. 1981. *American Sociological Theory: A Critical History*. New York: Academic.

Black, Donald (ed.). 1984. *Toward a General Theory of Social Control,* Vol. 1: *Fundamentals*. Orlando, Fla.: Academic.

Blalock, Hubert M. 1956. "Economic Discrimination and Negro Increase." *American Sociological Review* 21: 548–588.

Blalock, Hubert M., Jr. 1967. *Toward a Theory of Minority Group Relations*. New York: Wiley.

Blalock, Hubert M., Jr., and Paul H. Wilken. 1979. *Intergroup Processes: A Micro-Macro Perspective*. New York: Free Press.

Blank, Rebecca M. 1997. *It Takes a Nation: A New Agenda for Fighting Poverty*. New York: Russell Sage.

Blau, Francine D. and Ronald G. Ehrenberg (eds.). 1997. *Gender and Family Issues in the Workplace*. New York: Russell Sage.

Blau, Francine D., and Marianne A. Ferber. 1992. *The Economics of Women, Men, and Work,* 2nd ed. Englewood Cliffs, N.J.: Prentice-Hall.

Blau, Judith R., Kent Reading, Walter R. Davis, and Kenneth C. Land. 1997. "Spatial Processes and the Duality of Church and Faith: A Simmelian Perspective on U.S. Denominational Growth, 1900–1930." *Sociological Perspectives* 40: 557–580.

Blau, Peter M. 1974. *On the Nature of Organizations.* New York: Wiley.

Blau, Peter M. 1977. *Inequality and Heterogeneity: A Primitive Theory of Social Structure.* New York: Free Press.

Blau, Peter M. 1994. *Structural Context of Opportunities.* Chicago: University of Chicago Press.

Blau, Peter M., and Otis Dudley Duncan. 1967. *The American Occupational Structure.* New York: Wiley.

Blau, Peter M., and Marshall W. Meyer. 1987. *Bureaucracy in Modern Society,* 3rd ed. New York: Random House.

Blumer, Herbert. 1958. "Race Prejudice as a Sense of Group Position." *Pacific Sociological Review* 1: 3–7.

Blumer, Herbert. 1969. *Symbolic Interactionism.* Englewood Cliffs, N.J.: Prentice-Hall.

Blumin, Stuart M. 1977. "Rip Van Winkle's Grandchildren: Family and Household in the Hudson Valley, 1800–1860." Pp. 100–121 in Tamara K. Hareven (ed.), *Family and Kin in Urban Communities, 1700–1930.* New York: New Viewpoints/Franklin Watts.

Blumstein, Alfred, Jacqueline Cohen, and Richard Rosenfeld. 1991. "Trend and Deviation in Crime Rates: A Comparison of UCR and NCS Data for Burglary and Robbery." *Criminology* 29: 237–263.

Blumstein, Philip W., and Pepper Schwartz. 1983. *American Couples: Money, Work, Sex.* New York: Morrow.

Bobo, Lawrence. 1983. "'Whites' Opposition to Busing: Symbolic Racism or Realistic Group Conflict?" *Journal of Personality and Social Psychology* 45: 1196–1210.

Bobo, Lawrence. 1988. "Group Conflict, Prejudice and the Paradox of Contemporary Racial Attitudes." Pp. 85–109 in P. Katz and D. Taylor (eds.), *Eliminating Racism: Profiles in Controversy.* New York: Plenum.

Bobo, Lawrence, and Franklin D. Gilliam Jr. 1990. "Race, Sociopolitical Participation, and Black Empowerment." *American Political Science Review* 84: 377–394.

Bohannan, Paul. 1995. *How Culture Works.* New York: Free Press.

Boli, John. 1989. *New Citizens for a New Society: The Institutional Origins of Mass Schooling in Sweden.* Oxford: Pergamon.

Boli, John, and George M. Thomas. 1997. "World Culture in the World Polity: A Century of International Non-Governmental Organization." *American Sociological Review* 62: 171–190.

Bonacich, Edna. 1972. "A Theory of Ethnic Antagonism: The Split Labor Market." *American Sociological Review* 37: 547–559.

Bonacich, Edna. 1973. "A Theory of Middleman Minorities." *American Sociological Review* 38: 583–594.

Bonacich, Edna. 1976. "Advanced Capitalism and Black/White Relations in the United States: A Split Labor Market Interpretation." *American Sociological Review* 41: 34–51.

Boskoff, Alvin. 1969. *Theory in American Sociology: Major Sources and Applications.* New York: Crowell.

Bourdieu, Pierre. 1977. *Outline of a Theory of Practice.* Trans. Richard Nice. New York: Cambridge University Press.

Bourdieu, Pierre. 1990. *In Other Words: Essays Toward a Reflexive Sociology.* Trans. Matthew Adamson. Stanford, Calif.: Stanford University Press.

Bourdieu, Pierre, and Löic J. D. Wacquant. 1992. *An Invitation to Reflexive Sociology.* Chicago: University of Chicago Press.

Bowlby, John. 1951. *Maternal Care and Mental Health.* Geneva: World Health Organization.

Bowles, Samuel, and Herbert Gintis. 1976. *Schooling in Capitalist America.* New York: Basic Books.

Bradshaw, York W. 1988. "Reassessing Economic Dependency and Uneven Development: The Kenyan Experience." *American Sociological Review* 53: 693–708.

Braithwaite, John. 1979. *Inequality, Crime, and Public Policy.* London: Routledge & Kegan Paul.

Braithwaite, John. 1984. *Corporate Crime in the Pharmaceutical Industry.* London: Routledge & Kegan Paul.

Braithwaite, Ronald L., and Sandra E. Taylor (eds.), 1992. *Health Issues in the Black Community.* San Francisco: Jossey-Bass.

Branaman, Ann. 1997. "Goffman's Social Theory." Pp. xlv–lxxxii in Charles Lemert and Ann Branaman (eds.), *The Goffman Reader.* Malden, Mass.: Blackwell.

Brewster, Karin L. 1994. "Race Differences in Sexual Activity Among Adolescent Women: The Role of Neighborhood Characteristics." *American Sociological Review* 59: 408–424.

Brint, Steven, and Jerome Karabel. 1989. *The Diverted Dream: Community Colleges and the Promise of Educational Opportunity in America, 1900–1985.* New York: Oxford University Press.

Brinton, Mary C., and Victor Nee. (eds.). 1998. *The New Institutionalism in Sociology.* New York: Russell Sage.

Brody, Jane. 1992. "Spouse Abuse Triggers Cries for Reform." *The Sunday Oregonian,* 5 April, p. L5.

Brooks-Gunn, Jeanne, Greg J. Duncan, and J. Lawrence Aber (eds.). 1997. *Neighborhood Poverty,* Vol. I: *Context and Consequences for Children.* New York: Russell Sage.

Brooks-Gunn, Jeanne, Greg J. Duncan, and J. Lawrence Aber (eds.). 1997. *Neighborhood Poverty,* Vol. II: *Policy Implications in Studying Neighborhoods.* New York: Russell Sage.

Brooks, Roy L. 1990. *Rethinking the American Race Problem.* Berkeley: University of California Press.

Brophy, J. E. 1986. "Teacher Influences on Student Achievement." *American Psychologist* 41: 1069–1077.

Brophy, J. E., and T. L. Good. 1986. "Teacher Behavior and Student Achievement." Pp. 328–375 in M. C. Wittrock (ed.), *Handbook of Research on Teaching.* New York: Macmillan.

Broverman, Inge K., Susan R. Vogel, Donald M. Broverman, Frank E. Clarkson, and Paul S. Rosenkrantz. 1972. "Sex-Role Stereotypes: A Current Appraisal." *Journal of Social Issues* 28: 59–78.

Brown, David K. 1995. *Degrees of Control: A Sociology of Educational Expansion and Occupational Credentialism.* New York: Teachers' College Press.

Brown, E. Richard. 1979. *Rockefeller Medicine Men: Medicine and Capitalism in America.* Berkeley: University of California Press.

Bruce, Steve. 1992. *Religion and Modernization: Sociologists and Historians Debate the Secularization Thesis.* Oxford: Clarendon.

Bruce, Steve. 1996. *Religion in the Modern World: From Cathedrals to Cults.* Oxford and New York: Oxford University Press.

Brüderl, Josef, Peter Preisendörfer, and Rolf Ziegler. 1992. "Survival Chances of Newly Founded Business Organizations." *American Sociological Review* 57: 227–242.

Bruhn, J. G., and S. Wolf. 1979. *The Roseto Story.* Norman: University of Oklahoma Press.

Bryk, Anthony S., Valerie E. Lee, and Peter B. Holland. 1993. *Catholic Schools and the Common Good.* Cambridge, Mass.: Harvard University Press.

Buchanan, Christy M., Eleanor E. Maccoby, and Sanford M. Dornbusch. 1996. *Adolescents after Divorce.* Cambridge, Mass.: Harvard University Press.

Buchmann, Marlis. 1989. *The Script of Life in Modern Society: Entry into Adulthood in a Changing World.* Chicago: University of Chicago Press.

Bullard, Robert D., and Beverly H. Wright. 1992. "The Quest for Environmental Equity: Mobilizing the African-American Community for Social Change." Pp. 39–50 in Riley E. Dunlap and Angela G. Mertig (eds.), *American Environmentalism: The U.S. Environmental Movement, 1970–1990.* Philadelphia: Taylor & Francis.

Burch, David. 1998. "Science, Technology, and the Less-Developed Countries," Pp. 206–231 in Martin Bridgestock, David Burch, John Forge, John Laurent, and Ian Lowe. *Science, Technology, and Society: An Introduction.* Cambridge: Cambridge University Press.

Bureau of Labor Statistics. 1998. "Labor Force Statistics from the Current Population Survey," Annual Average Tables from the January 1998 Issue of Employment and Earnings. (ftp://ftp.bls.gov/pub/special.requests/lf/aat10.txt and ftp://ftp.bls.gov/pub/special.requests/lf/aat11.txt)

Burgess, Ernest W. 1967. "The Growth of the City." Pp. 47–62 in R. E. Park and E. W. Burgess (eds.), *The City.* Chicago: University of Chicago Press.

Burkey, Richard M. 1978. *Ethnic and Racial Groups: The Dynamics of Dominance.* Menlo Park, Calif.: Cummings.

Buroway, Michael. 1990. "Marxism as Science: Historical Challenges and Theoretical Growth." *American Sociological Review* 55: 775–793.

Bursik, Robert J., Jr. and Harold G. Grasmick. 1993. *Neighborhoods and Crime: The Dimensions of Effective Community Control.* New York: Lexington.

Burt, Ronald S. 1993. "The Social Structure of Competition." Pp. 65–103 in Richard Swedberg (ed.), *Explorations in Economic Sociology.* New York: Russell Sage Foundation.

Caldwell, John C. 1982. *Theory of Fertility Decline.* New York: Academic.

Caldwell, Lynton Keith. 1996. *International Environmental Policy: From the Twentieth to the Twenty-First Century.* Durham, N.C.: Duke University Press.

Calhoun, Craig (ed.), 1994. *Social Theory and the Politics of Identity.* Cambridge, Mass.: Blackwell.

Calhoun, Craig. 1997. *Nationalism.* Minneapolis: University of Minnesota Press.

Calhoun, Craig, Donald Light, and Suzanne Keller. 1994. *Sociology,* 6th ed. New York: McGraw-Hill.

Calhoun, Craig, and W. Richard Scott. 1990. "Introduction: Peter Blau's Sociological Structuralism." Pp. 1–36 in Craig Calhoun, Marshall W. Meyer, and W. Richard Scott (eds.), *Structures of Power and Constraint: Papers in Honor of Peter M. Blau.* Cambridge: Cambridge University Press.

Camic, Charles. 1989. "Structure After 50 Years: The Anatomy of a Charter." *American Journal of Sociology* 95: 38–107.

Campbell, R. T. 1983. "Status Attainment Research: End of the Beginning or Beginning of the End." *Sociology of Education* 56: 47–62.

Cancian, Francesca M. 1995. "Truth and Goodness: Does the Sociology of Inequality Promote Social Betterment?" *Sociological Perspectives* 38: 339–356.

Capps, Donald. 1985. "Religion and Psychological Well-Being." Pp. 237–256 in Phillip E. Hammond (ed.), *The Sacred in a Secular Age: Toward Revision in the Scientific Study of Religion.* Berkeley: University of California Press.

Carbaugh, Donal. 1996. *Situating Selves: The Communication of Social Identities in American Scenes.* Albany: State University of New York Press.

Carey, James T. 1975. *Sociology and Public Affairs: The Chicago School.* Beverly Hills, Calif.: Sage.

Carroll, Ginny. 1995. "Get Ready for Mr. Relentless." *Newsweek,* 20 February, pp. 28–30.

Carter, D. B. 1987. "The Roles of Peers in Sex Role Socialization." Pp. 101–121 in D. Bruce Carter (ed.), *Current Conceptions of Sex Roles and Sex Typing: Theory and Research.* New York: Praeger.

Casper, Lynne M., Sara S. McLanahan, and Irwin Garfinkel. 1994. "The Gender-Poverty Gap: What We Can Learn From Other Countries." *American Sociological Review* 59: 594–605.

Caspi, Avshalom, Bradley R. Entner Wright, Terrie E. Moffitt, and Phil A. Silva. 1998. "Early Failure in the Labor Market: Childhood and Adolescent Predictors of Unemployment in the Transition to Adulthood." *American Sociological Review* 63: 424–451.

Castells, Manuel. 1997. *The Power of Identity.* Malden, Mass.: Blackwell.

Castells, Manuel. 1998. *End of Millennium.* Malden, Mass.: Blackwell.

Catton, William R., Jr., and Riley Dunlap. 1980. "A New Ecological Paradigm for Post-Exuberant Sociology." *American Behavioral Scientist* 24: 15–47.

Center for the American Woman in Politics. 1998a. "Women's Electoral Success: A Familiar Formula." Press Release. November 25, 1998. http://www.rci.rutgers.edu/~cawp/98electpress.html

Center for the American Woman in Politics. 1998b. "Women in State Legislatures: Modest Gains Set New Record in 1998 Elections, Exciting Opportunities Lie Ahead." Press Release. November 24, 1998. http://www.rci.rutgers.edu/~cawp/legpress98.html

Center for the Future of Children. 1991. "Analysis." *The Future of Children* 1(1): 9–16.

Cerulo, Karen A. 1997. "Identity Construction: New Issues, New Directions." *Annual Review of Sociology* 23: 385–409.

Chafetz, Janet Saltzman. 1997. "Feminist Theory and Sociology. Underutilized Contributions for Mainstream Theory." *Annual Review of Sociology* 23: 97–120.

Chambliss, William J., and Robert B. Seidman. 1971. *Law, Order, and Power*. Reading, Mass.: Addison-Wesley.

Chandler, Alfred D., Jr. 1990. *Scale and Scope: The Dynamics of Industrial Capitalism*. Cambridge, Mass.: Belknap.

Chase-Lansdale, P. Lindsay, and Jeanne Brooks-Gunn. 1995. "Introduction." Pp. 1–10 in P. Lindsay Chase-Lansdale and Jeanne Brooks-Gunn (eds.), *Escape from Poverty: What Makes a Difference for Children?* New York: Cambridge University Press.

Chehrazi, Shahla. 1986. "Female Psychology: A Review." *Journal of the American Psychoanalytic Association* 34: 111–162.

Ch'Eng-K'un, Cheng. 1944. "Familism the Foundation of Chinese Social Organization." *Social Forces* 23: 50–59.

Cherlin, Andrew J. 1992. *Marriage, Divorce, Remarriage* (rev. & enl. ed.), Cambridge, Mass.: Harvard University Press.

Cherlin, Andrew J., Kathleen E. Kiernan, and P. Lindsay Chase-Lansdale. 1995. "Parental Divorce in Childhood and Demographic Outcomes in Young Adulthood." *Demography* 32: 299–318.

Chiriboga, David A., Linda S. Catron, and associates. 1991. *Divorce: Crisis, Challenge or Relief?* New York: New York University Press.

Chirot, Daniel. 1985. "The Rise of the West." *American Sociological Review* 50: 181–195.

Chirot, Daniel. 1986. *Social Change in the Modern Era*. San Diego: Harcourt Brace Jovanovich.

Chirot, Daniel. 1994. *How Societies Change*. Thousand Oaks, Calif.: Pine Forge.

Chiswick, Barry R., and Teresa A. Sullivan. 1995. "The New Immigrants." Pp. 211–270 in Reynolds Farley (ed.), *State of the Union: America in the 1990s*, Vol. 2: *Social Trends*. New York: Russell Sage.

Chodorow, Nancy J. 1978. *The Reproduction of Mothering*. Berkeley: University of California Press.

Chodorow, Nancy J. 1989. *Feminism and Psychoanalytic Theory*. New Haven, Conn.: Yale University Press.

Chong, Dennis. 1991. *Collective Action and the Civil Rights Movement*. Chicago: University of Chicago Press.

Clark, Burton. 1961. *Educating the Expert Society*. San Francisco: Chandler.

Clark, Reginald. 1983. *Family Life and School Achievement: Why Poor Black Children Succeed or Fail*. Chicago: University of Chicago Press.

Clarke, Sally C. 1995. "Advance Report of Final Divorce Statistics, 1989 and 1990." *Monthly Vital Statistics Report*, 43(9), Supplement. Hyattsville, Md.: National Center for Health Statistics.

Clausen, John S. 1991. "Adolescent Competence and the Shaping of the Life Course." *American Journal of Sociology* 96: 805–842.

Clawson, Dan, and Alan Neustadtl. 1989. "Interlocks, PACs, and Corporate Conservatism." *American Journal of Sociology* 94: 749–773.

Clawson, Dan, Alan Neustadtl, and James Bearden. 1986. "The Logic of Business Unity: Corporate Contributions to the 1980 Congressional Elections." *American Sociological Review* 51: 797–811.

Clawson, Dan, Alan Neustadtl, and Denise Scott. 1992. *Money Talks: Corporate PACs and Political Influence*. New York: Basic.

Clawson, Dan, Alan Neustadtl, and Mark Weller. 1998. *Dollars and Votes: How Business Campaign Contributions Subvert Democracy*. Philadelphia: Temple University Press.

Clawson, Dan, and Tie-Ting Su. 1990. "Was 1980 Special? A Comparison of 1980 and 1986 Corporate PAC Contributions." *Sociological Quarterly* 31: 371–388.

Clayton, Obie, Jr. (ed.). 1996. *An American Dilemma Revisited: Race Relations in a Changing World*. New York: Russell Sage Foundation.

Clement, Wallace, and John Myles. 1994. *Relations of Ruling: Class and Gender in Postindustrial Societies*. Montreal: McGill-Queen's University Press.

Cloward, Richard A., and Lloyd E. Ohlin. 1960. *Delinquency and Opportunity: A Theory of Delinquent Gangs*. New York: Free Press.

Cockerham, William C. 1997. "The Social Determinants of the Decline of Life Expectancy in Russia and Eastern Europe: A Lifestyle Explanation." *Journal of Health and Social Behavior* 38: 117–130.

Cockerham, William C. 1998. *Medical Sociology*. Upper Saddle River, N.J.: Prentice Hall.

Cockerham, William C., Alfred Rütten, and Thomas Abel. 1997. "Conceptualizing Contemporary Health Lifestyles: Moving Beyond Weber." *Sociological Quarterly* 38: 601–622.

Cohen, Albert K. 1997. "An Elaboration of Anomie Theory." Pp. 52–61 in Nikos Passas and Robert Agnew (eds.), *The Future of Anomie Theory*. Boston: Northeastern University Press.

Cohen, Elizabeth G. 1994a. *Designing Group Work: Strategies for the Heterogeneous Classroom*, 2nd ed. New York: Teachers College Press.

Cohen, Elizabeth G. 1994b. "Restructuring the Classroom: Conditions for Productive Small Groups." *Review of Educational Research*. 64: 1–36.

Cohen, Elizabeth G., and Rachel A. Lotan. 1995. "Producing Equal-Status Interaction in the Heterogeneous Classroom." *American Educational Research Journal* 32: 99–120.

Cohen, Elizabeth G., and Rachel A. Lotan (eds.). 1997. *Working for Equity in Heterogeneous Classrooms: Sociological Theory in Practice*. New York: Teachers College Press.

Cohen, Elizabeth G., and Susan S. Roper. 1972. "Modification of Interracial Interaction Disability: An Application of Status Characteristic Theory." *American Sociological Review* 37: 643–657.

Cohen, Lawrence E., and Kenneth C. Land. 1987. "Age Structure and Crime: Symmetry Versus Asymmetry and the Projection of Crime Rates Through the 1990s." *American Sociological Review* 52: 170–183.

Cole, George F. 1995. *The American System of Criminal Justice*. Belmont, Calif.: Wadsworth.

Cole, George F., and Christopher E. Smith. 1998. *The American System of Criminal Justice*, 8th ed. Belmont, Calif.: West/Wadsworth.

Coleman, James S. 1990. *Foundations of Social Theory.* Cambridge, Mass.: Belknap.

Coleman, James S. 1993. "The Rational Reconstruction of Society." *American Sociological Review* 58: 1–15.

Coleman, James, Thomas Hoffer, and Sally Kilgore. 1982. *High School Achievement: Public and Private Schools Compared.* New York: Basic.

Collins, Randall. 1985. *Three Sociological Traditions.* New York: Oxford University Press.

Collins, Randall. 1997. "An Asian Route to Capitalism: Religious Economy and the Origins of Self-Transforming Growth in Japan." *American Sociological Review* 62: 843–865.

Collins, Sharon M. 1997. *Black Corporate Executives: The Making and Breaking of a Black Middle Class.* Philadelphia: Temple University Press.

Coltrane, Scott. 1996. *Family Man: Fatherhood, Housework, and Gender Equity.* New York: Oxford University Press.

Cook, Elizabeth Adell, Sue Thomas, and Clyde Wilcox. 1994. *The Year of the Woman: Myths and Realities.* Boulder, Colo.: Westview.

Cook, Karen S. (ed.). 1987. *Social Exchange Theory.* Newbury Park, Calif.: Sage.

Cook, Karen S. 1991. "The Microfoundations of Social Structure: An Exchange Perspective." Pp. 29–45 in Joan Huber (ed.), *Macro-Micro Linkages in Sociology.* Newbury Park, Calif.: Sage.

Cook, Karen S., and J. M. Whitmeyer. 1992. "Two Approaches to Social Structure: Exchange Theory and Network Analysis." *Annual Review of Sociology* 18: 109–127.

Cooley, Charles Horton. 1937. *Social Organization.* New York: Scribner. (Orig. pub. 1909.)

Coontz, Stephanie, and Peta Henderson. 1986. "Property Forms, Political Power, and Female Labour in the Origins of Class and State Societies." Pp. 108–155 in Stephanie Coontz and Peta Henderson (eds.). *Women's Work, Men's Property: The Origins of Gender and Class.* London: Verso.

Cose, Ellis. 1993. *The Rage of a Privileged Class.* New York: Harper Collins.

Coser, Lewis A. 1977. *Masters of Sociological Thought: Ideas in Historical and Social Context,* 2nd ed. New York: Harcourt Brace Jovanovich.

Cota, A. A., and K. L. Dion. 1986. "Salience of Gender and Sex Composition of Ad Hoc Groups: An Experimental Test of Distinctiveness Theory." *Journal of Personality and Social Psychology* 50: 770–776.

Couch, Carl J. 1996. *Information Technologies and Social Orders.* Ed.& intro. by David R. Maines and Shing-Ling Chen. New York: Aldine.

Coverman, Shelley. 1985. "Explaining Husbands' Participation in Domestic Labor." *Sociological Quarterly* 26: 81–97.

Cowan, Philip A. 1993. "The Sky Is Falling, but Popenoe's Analysis Won't Help Us Do Anything About It." *Journal of Marriage and the Family* 55: 548–553.

Cramer, James C. 1995. "Racial and Ethnic Differences in Birthweight: The Role of Income and Financial Assistance." *Demography* 32: 231–247.

Crane, Jonathan. 1991. "The Epidemic Theory of Ghettos and Neighborhood Effects on Dropping Out and Teenage Childbearing." *American Journal of Sociology* 96: 1226–1259.

Cremin, Lawrence A. 1976. *Traditions of American Education.* New York: Basic.

Cremin, Lawrence A. 1990. *Popular Education and Its Discontents.* New York: Harper & Row.

Crenshaw, Edward M. 1995. "Democracy and Demographic Inheritance: The Influence of Modernity and Proto-Modernity on Political and Civil Rights, 1965 to 1980." *American Sociological Review* 60: 702–718.

Crenshaw, Edward M., Ansari Z. Ameen, and Matthew Christenson. 1997. "Population Dynamics and Economic Development: Age-Specific Population Growth Rates and Economic Growth in Developing Countries, 1965 to 1990." *American Sociological Review* 62: 974–984.

Crewe, Ivor. 1981. "Electoral Participation." Pp. 216–263 in David Butler, Howard R. Penniman, and Austin Ranney (eds.), *Democracy at the Polls: A Comparative Study of Competitive National Elections.* Washington, D.C.: American Enterprise Institute for Public Policy Research.

Crews, Kimberly A., and Cheryl Lynn Stauffer. 1997. *World Population and the Environment.* Washington, D.C.: Population Reference Bureau.

Cromwell, Adelaide. 1994. *The Other Brahmins: Boston's Black Upper Class, 1750–1950.* Fayetteville: University of Arkansas Press.

Cross, K. P. 1976. *Accent on Learning.* San Francisco: Jossey-Bass.

Cummings, Scott, and Daniel J. Monti (eds.). 1993. *Gangs: The Origins and Impact of Contemporary Youth Gangs in the United States.* Albany: State University of New York Press.

Dahl, Robert A. 1985. *A Preface to Economic Democracy.* Berkeley: University of California Press.

Dahl, Robert A. 1989. *Democracy and Its Critics.* New Haven, Conn.: Yale University Press.

Dahl, Robert A. 1990. *After the Revolution?* New Haven, Conn.: Yale University Press.

Dalaker, Joseph, and Mary Naifeh. 1998. *Poverty in the United States: 1997,* U.S. Bureau of the Census, Current Population Reports, Series P60–201. Washington, D.C.: U.S. Government Printing Office.

Dalton, Russell J. 1988. *Citizen Politics in Western Democracies: Public Opinion and Political Parties in the United States, Great Britain, West Germany, and France.* Chatham, N.J.: Chatham House.

Dalton, Russell J. 1993. "Preface." Pp. 8–12 in Russell J. Dalton (ed.), *Citizens, Protest, and Democracy. The Annals of the American Academy of Political and Social Science* 528 (July).

D'Andrade, Roy G. 1966. "Sex Differences and Cultural Institutions." Pp. 173–203 in Eleanor E. Maccoby (ed.), *The Development of Sex Differences.* Stanford, Calif.: Stanford University Press.

Danziger, Sheldon, and Peter Gottschalk. 1995. *America Unequal.* Cambridge, Mass.: Harvard University Press.

Davidson, Chandler, and Bernard Grofman (eds.). 1994. *Quiet Revolution in the South: The Impact of the Voting Rights Act, 1965–1990.* Princeton, N.J.: Princeton University Press.

Davies, James C. 1963. *Human Nature in Politics: The Dynamics of Political Behavior.* New York: Wiley.

Davis, Allison, Burleigh B. Gardner, and Mary R. Gardner. 1941. *Deep South: A Social Anthropological Study of Caste and Class.* Chicago: University of Chicago Press.

Davis, Gerald F., Kristina A. Diekmann, and Catherine H. Tinsley. 1994. "The Decline and Fall of the Conglomerate Firm in the 1980s: The Deinstitutionalization of an Organizational Form." *American Sociological Review* 59: 547–570.

Davis, Kingsley. 1942. "A Conceptual Analysis of Stratification." *The American Sociological Review* 7: 309–321.

Davis, Kingsley. 1945. "The World Demographic Transition." *The Annals of the American Academy of Political and Social Science* 237 (January): 1–11.

Davis, Kingsley. 1950. *Human Society.* New York: Macmillan.

Davis, Kingsley. 1953. "Reply to Tumin." *The American Sociological Review* 18: 394–397. Reprinted in Reinhard Bendix and Seymour Martin Lipset (eds.). 1966. *Class, Status, and Power: Social Stratification in Comparative Perspective,* 2nd ed. New York: Free Press, pp. 59–62.

Davis, Kingsley. 1955. "The Origin and Growth of Urbanization in the World." *American Journal of Sociology* 60: 429–437.

Davis, Kingsley, and Wilbert E. Moore. 1945. "Some Principles of Stratification." *The American Sociological Review* 10: 242–249.

Day, Jennifer, and Andrea Curry. 1998. "Educational Attainment in the United States: March 1997, Detailed Tables." *Current Population Reports,* Series P20-505. Washington, D.C.: U.S. Bureau of the Census, U.S. Government Printing Office.

DeBarros, Kymberly, and Claudette Bennett. 1998. "The Black Population in the United States: March 1997 (Update)." *Current Population Reports, Population Characteristics,* P20–508. Washington, D.C. U.S. Census Bureau, U.S. Department of Commerce.

Decker, Scott H., and Barrik Van Winkle. 1996. *Life in the Gang: Family, Friends, and Violence.* New York: Cambridge University Press.

DeJong, Gordon F., and Robert W. Gardner (eds.). 1981. *Migration Decision Making: Multidisciplinary Approaches to Microlevel Studies in Developed and Developing Countries.* New York: Pergamon.

Delacroix, Jacques, Anand Swaminathan, and Michael E. Solt. 1989. "Density Dependence Versus Population Dynamics: An Ecological Study of Failings in the California Wine Industry." *American Sociological Review* 54: 245–262.

del Pinal, Jorge, and Audrey Singer. 1997. "Generations of Diversity: Latinos in the United States." *Population Bulletin* 52(3). Washington, D.C.: Population Reference Bureau.

DeLuca, Tom. 1995. *The Two Faces of Political Apathy.* Philadelphia: Temple University Press.

DeMott, Benjamin. 1990. *The Imperial Middle: Why Americans Can't Think Straight About Class.* New York: Morrow.

Deutsch, Helene. 1944–45. *The Psychology of Women: A Psychoanalytic Interpretation,* Vol. 1: *Girlhood;* Vol. 2: *Motherhood.* New York: Bantam.

Diamond, Jared. 1997. *Guns, Germs, and Steel: The Fates of Human Societies.* New York: W.W. Norton.

Diani, Mario. 1995. *Green Networks: A Structural Analysis of the Italian Environmental Movement.* Edinburgh: Edinburgh University Press.

DiMaggio, Paul. 1997. "Culture and Cognition." *Annual Review of Sociology* 23: 263–287.

Disney. 1998. Walt Disney Corporation. Annual Report, 1997. http://www.disney.com/investors/annual97

Doeringer, Peter B., and Michael J. Piore. 1971. *Internal Labor Markets and Manpower Analysis.* Lexington, Mass.: Heath.

Domhoff, G. William. 1970. *The Higher Circles: The Governing Class in America.* New York: Random House.

Domhoff, G. William. 1978. *Who Really Rules?* New Brunswick, N.J.: Transaction.

Domhoff, G. William. 1979. *The Powers That Be: Processes of Ruling-Class Domination in America.* New York: Random House.

Domhoff, G. William. 1990. *The Power Elite and the State: How Policy Is Made in America.* New York: Aldine.

Donovan, Patricia. 1998. "Falling Teen Pregnancy, Birthrates: What's Behind the Declines?" *The Guttmacher Report on Public Policy* 1(No. 5, October).

Downs, Anthony. 1957. *An Economic Theory of Democracy.* New York: Harper.

Dreiling, Michael C. 1998. "The Class Embeddedness of Corporate Political Action: Leadership in Defense of the NAFTA." Unpublished Paper, University of Oregon Department of Sociology.

DuBois, W. E. B. 1904. "The Atlanta Conferences." *Voice of the Negro* 1: 85–89.

Duckitt, John. 1992. *The Social Psychology of Prejudice.* New York: Praeger.

Dudley, Kathryn Marie. 1994. *The End of the Line: Lost Jobs, New Lives in Postindustrial America.* Chicago: University of Chicago Press.

Duesenberry, James. 1960. "Comment on 'An Economic Analysis of Fertility.'" In Universities–National Bureau Committee for Economic Research (ed.), *Demographic and Economic Change in Developed Countries.* Princeton, N.J.: Princeton University Press.

Duncan, Beverly, and Stanley Lieberson. 1970. *Metropolis and Region in Transition.* Beverly Hills, Calif.: Sage.

Duncan, Greg J., and Jeanne Brooks-Gunn (eds.), 1997. *Consequences of Growing Up Poor.* New York: Russell Sage.

Duncan, Greg J., W. Jean Yeung, Jeanne Brooks-Gunn, and Judith R. Smith. 1998. "How Much Does Childhood Poverty Affect the Life Chances of Children?" *American Sociological Review* 63: 406–424.

Duncan, Gregg, Richard Coe, et al. 1984. *Years of Poverty, Years of Plenty: The Changing Economic Fortunes of American Workers and Families.* Ann Arbor: Institute for Social Research, University of Michigan.

Duncan, Otis Dudley. 1961. "A Socio-economic Index for All Occupations," and "Properties and Characteristics of the Socioeconomic Index." Pp. 109–161 in Albert Reiss (ed.), *Occupations and Social Status.* Glencoe, Ill.: Free Press.

Duncan, Otis Dudley, David L. Featherman, and Beverly Duncan. 1972. *Socioeconomic Background and Achievement.* New York: Seminar Press.

Dunlap, Riley E. 1991. "Public Opinion in the 1980s: Clear Consensus, Ambiguous Commitment." *Environment* 33 (October): 10–15, 32–37.

Dunlap, Riley E. 1992. "Trends in Public Opinion Toward Environmental Issues: 1965–1990." Pp. 89–116 in Riley E. Dunlap and Angela G. Mertig (eds.), *American Environmentalism: The U.S. Environmental Movement, 1970–1990.* Philadelphia: Taylor & Francis.

Dunlap, Riley E., George H. Gallup, Jr., and Alec M. Gallup. 1993. *Environment* 35(9): 7–15, 33–39.

Dunlap, Riley E., and Angela G. Mertig. 1992. "The Evolution of the U.S. Environmental Movement from 1970 to 1990: An Overview." Pp. 1–10 in Riley E. Dunlap, and Angela G. Mertig (eds.), *American Environmentalism: The U.S. Environmental Movement, 1970–1990.* Philadelphia: Taylor & Francis.

Dunlap, Riley E., and Rik Scarce. 1991. "The Polls—Poll Trends: Environmental Problems and Protection." *Public Opinion Quarterly* 55: 651–672.

Durkheim, Émile. 1965. *The Elementary Forms of Religious Life.* New York: Free Press. (Orig. pub. 1912.)

Durkheim, Émile. 1951. *Suicide.* New York: Free Press. (Orig. pub. 1897.)

Durkheim, Émile. 1961. *Moral Education.* Glencoe, Ill.: Free Press. (Orig. pub. 1925.)

Durkheim, Émile. 1964. *The Division of Labor in Society.* New York: Free Press. (Orig. pub. 1893.)

Durkheim, Émile. 1938. *The Rules of Sociological Method.* New York: Free Press. (Orig. pub. 1895.)

Dutton, Diana B. 1978. "Explaining the Low Use of Health Services by the Poor: Costs, Attitudes, or Delivery Systems?" *American Sociological Review* 43: 348–368.

Dutton, Diana B. 1979. "Patterns of Ambulatory Health Care in Five Different Delivery Systems." *Medical Care* 17: 221–241.

Dutton, Diana B. 1986. "Social Class, Health, and Illness," Pp. 31–62 in L. Aiken and D. Mechanic (eds.), *Applications of Social Science to Clinical Medicine and Health Policy.* New Brunswick, N.J.: Rutgers University Press.

Eagly, Alice H. 1987. *Sex Differences in Social Behavior: A Social-Role Interpretation.* Hillsdale, N.J.: Erlbaum.

Eagly, Alice H., and Valerie J. Steffen. 1986. "Gender and Aggressive Behavior: A Meta-Analytic Review of the Social Psychological Literature." *Psychological Bulletin* 110: 309–330.

Eaton, Warren O., and Lesley Reid Enns. 1986. "Sex Differences in Human Motor Activity Level." *Psychological Bulletin* 100: 19–28.

Ebaugh, Helen Rose Fuchs. 1988. *Becoming an Ex: The Process of Role Exit.* Chicago: University of Chicago Press.

Eberts, Randall W., and Joe A. Stone. 1988. "Student Achievement in Public Schools: Do Principals Make a Difference?" *Economics of Education Review* 7: 291–299.

Edel, Matthew, and Ronald G. Hellman (eds.). 1989. *Cities in Crisis: The Urban Challenge in the Americas.* Graduate School and University Center of the City University of New York: Bildner Center for Western Hemisphere Studies.

Edin, Kathryn, and Laura Lein. 1997. *Making Ends Meet: How Single Mothers Survive Welfare and Low-Wage Work.* New York: Russell Sage.

Edsall, Thomas Byrne. 1984. *The New Politics of Inequality.* New York: Norton.

Ehlers, Tracy Bachrach, and Karen Main. 1998. "Women and the False Promise of Microenterprise." *Gender and Society* 12: 424–440.

Ehrenreich, Barbara. 1989. *Fear of Falling: The Inner Life of the Middle Class.* New York: Pantheon Books.

Ehrlich, Anne H., and Paul R. Ehrlich. 1987. *Earth.* London: Thames Methuen.

Ehrlich, Paul R. 1986. *The Machinery of Nature.* New York: Simon & Schuster.

Eibl-Eibesfeldt, I. 1979. "Universals in Human Expressive Behavior." In A. Wolfgang (ed.), *Nonverbal Behavior: Applications and Cultural Implications.* New York: Academic.

Ekblaw, 1956. "Settlements of the Polar Eskimo." Pp. 41–44 in Irwin T. Sanders, Richard B. Woodbury, Frank J. Essene, Thomas P. Field, Joseph R. Schwendeman, and Charles, *Societies Around the World.* New York: Holt, Rinehart, & Winston.

Ekman, Paul. 1972. "Universals and Cultural Differences in Facial Expression of Emotion." Pp. 207–284 in J. K. Cole (ed.), *Nebraska Symposium on Motivation, 1971.* Lincoln: University of Nebraska Press.

Ekman, Paul. 1980. *The Face of Man: Expressions of Universal Emotions in a New Guinea Village.* New York: Garland STPM.

Ekman, Paul, and Wallace V. Friesen. 1981. "The Repertoire of Nonverbal Behavior: Categories, Origins, Usage, and Coding." Pp. 57–105 in Adam Kendon (ed.), *Nonverbal Communication, Interaction, and Gesture.* The Hague: Mouton.

Ekman, Paul, Wallace V. Friesen, et al. 1987. "Universals and Cultural Differences in the Judgments of Facial Expressions of Emotion." *Journal of Personality and Social Psychology* 53: 712–717.

Elder, Glen H., Jr. 1974. *Children of the Great Depression: Social Change in Life Experience.* Chicago: University of Chicago Press.

Elder, Glen H., Jr., and Angela M. O'Rand. 1995. "Adult Lives in a Changing Society." Pp. 452–475 in Karen S. Cook, Gary Alan Fine, and James S. House (eds.), *Sociological Perspectives on Social Psychology.* Boston: Allyn & Bacon.

Elkin, Frederick, and Gerald Handel. 1989. *The Child and Society: The Process of Socialization,* 5th ed. New York: Random House.

Elkind, D. 1968. "Giant in the Nursery—Jean Piaget." *New York Times Magazine,* 26 May, p. 25ff.

Elkind, David. 1994. *Ties That Stress: The New Family Imbalance.* Cambridge, Mass.: Harvard University Press.

Eller, T. J., and Wallace Fraser. 1995. *Asset Ownership of Households: 1993.* U.S. Bureau of the Census. Current Population Reports, P70–47. Washington, D.C.: U.S. Government Printing Office.

Elliott, Delbert S., and Scott Menard. 1996. "Delinquent Friends and Delinquent Behavior: Temporal and Developmental Patterns." Pp. 28–67 in J. David Hawkins (ed.), *Delinquency and Crime: Current Theories.* New York: Cambridge University Press.

Ellwood, David T. 1988. *Poor Support: Poverty in the American Family.* New York: Basic.

Emler, Nicholas, and Stephen Reicher. 1995. *Adolescence and Delinquency.* Oxford: Blackwell.

Encarta, 1996a. "International Bank for Reconstruction and Development." *Microsoft Encarta 96 Encyclopedia.* 1993–1995 Microsoft Corporation. Funk and Wagnalls Corporation.

Encarta, 1996b. "International Monetary Fund." *Microsoft Encarta 96 Encyclopedia.* 1993–1995 Microsoft Corporation. Funk and Wagnalls Corporation.

England, Paula, and George Farkas. 1986. *Households, Employment, and Gender.* New York: Aldine.

Erickson, Bonnie H. 1996. "Culture, Class and Connections." *American Journal of Sociology* 102: 217–251.

Erikson, Kai T. 1966. *Wayward Puritans: A Study in the Sociology of Deviance.* New York: Macmillan.

Erikson, Robert, and John H. Goldthorpe. 1992. *The Constant Flux: A Study of Class Mobility in Industrial Societies.* Oxford: Clarendon.

Etzioni, Amitai. 1990. "Status-Separation and Status-Fusion: The Role of PACs in Contemporary American Democracy." Pp. 67–87 in *The Political Sociology of the State: Essays on the Origins, Structure, and Impact of the Modern State.* Greenwich, Conn.: JAI.

Etzioni, Amitai, and Paul R. Lawrence (eds.). 1991. *Socio-Economics: Toward a New Synthesis.* Armonk, N.Y.: Sharpe.

Evans, Robert G. 1994. "Introduction." Pp. 3–26 in Robert G. Evans, Morris L. Barer, and Theodore R. Marmor (eds.), *Why Are Some People Healthy and Others Not? The Determinants of the Health of Populations.* New York: Aldine.

Eyerman, Ron, and Andrew Jamison. 1991. *Social Movements: A Cognitive Approach.* University Park: The Pennsylvania State University Press.

Fagot, Beverly. 1981. "Continuity and Changes in Play Styles as a Function of Sex of the Child." *International Journal of Behavioral Development* 4: 37–43.

Fagot, Beverly. 1985. "Beyond the Reinforcement Principle: Another Step Toward Understanding Sex Role Development." *Developmental Psychology* 21: 1097–1104.

Fairbairn, Ronald. 1952. *An Object-Relations Theory of the Personality.* New York: Basic Books.

Famighetti, Robert (ed.). 1994. *The World Almanac and Book of Facts, 1995.* Mahwah, N.J.: Funk & Wagnalls.

Famighetti, Robert (ed.). 1995. *The World Almanac and Book of Facts, 1995.* Mahwah, N.J.: Funk & Wagnalls.

Famighetti, Robert (ed.). 1996. *The World Almanac and Book of Facts, 1997.* Mahwah, New Jersey: World Almanac Books.

Farley, John E. 1988. *Majority-Minority Relatives,* 2nd ed. Englewood Cliffs, NJ: Prentice-Hall.

Farley, Reynolds. 1996. *The New American Reality: Who We Are, How We Got Here, Where We Are Going.* New York: Russell Sage.

Farley, Reynolds, and Walter R. Allen. 1987. *The Color Line and the Quality of Life in America.* New York: Russell Sage.

Farley, Reynolds, and William H. Frey. 1994. "Changes in the Segregation of Whites from Blacks During the 1980s: Small Steps Toward a More Integrated Society." *American Sociological Review* 59: 23–45.

Farley, Reynolds, Charlotte Steeh, Maria Krysan, Tara Jackson, and Keith Reeves. 1994. "Stereotypes and Segregation: Neighborhoods in the Detroit Area." *American Journal of Sociology* 100: 750–800.

Farmer, Melissa M., and Kenneth F. Ferraro. 1997. "Distress and Perceived Health: Mechanisms of Health Decline." *Journal of Health and Social Behavior* 38: 298–311.

Farrell, Ronald A., and Victoria Lynn Swigert (eds.). 1988. *Social Deviance,* 3rd ed. Belmont, Calif.: Wadsworth.

Fass, Paula S. 1989. *Outside In: Minorities and the Transformation of American Education.* New York: Oxford University Press.

Fast, Irene. 1984. *Gender Identity: A Differentiation Model.* Hillsdale, N.J.: Erlbaum.

Faulkner, Robert R. 1983. *Music on Demand: Composers and Careers in the Hollywood Film Industry.* New Brunswick, N.J.: Transaction.

Faulkner, Robert R., and Andy B. Anderson. 1987. "Short-term Projects and Emergent Careers: Evidence from Hollywood." *American Journal of Sociology* 92: 879–909.

Feagin, Joe R. 1991. "The Continuing Significance of Race: Antiblack Discrimination in Public Places." *American Sociological Review* 56: 101–116.

Feagin, Joe R., and Melvin P. Sikes. 1994. *Living With Racism: The Black Middle-Class Experience.* Boston: Beacon.

Feather, Norman T. 1990. *The Psychological Impact of Unemployment.* New York: Springer-Verlag.

Federal Election Commission. 1997. *Executive Summary of the Federal Election Commission's Report to the Congress on the Impact of the National Voter Registration Act of 1993 on the Administration of Federal Elections. June, 1997.* Washington, D.C.: Federal Election Commission. (http://www.fec.gov/voteregis/nvrasum.htm)

Ferguson, Ronald F. 1995. "Shifting Challenges: Fifty Years of Economic Change Toward Black-White Earnings Equality." *Daedalus.* 124(1): 37–76.

Fernandez, Roberto M., and Nancy Weinberg. 1997. "Sifting and Sorting: Personal Contacts and Hiring in a Retail Bank." *American Sociological Review* 62: 883–902.

Ferrante, Joan. 1995. *Sociology: A Global Perspective,* 2nd ed. Belmont, Calif.: Wadsworth.

Ferraro, Kenneth F., Melissa M. Farmer, and John A. Wybraniec. 1997. "Health Trajectories: Long-Term Dynamics Among Black and White Adults." *Journal of Health and Social Behavior* 38: 38–54.

Ferree, Myra Marx, and Patricia Yancey Martin (eds.). 1995. *Feminist Organizations: Harvest of the New Women's Movement.* Philadelphia: Temple University Press.

Fine, Gary Alan. 1987. *With the Boys: Little League Baseball and Preadolescent Culture.* Chicago: University of Chicago Press.

Fineman, Howard. 1995. "Powell on the March." *Newsweek,* 11 September, pp. 26–31.

Finke, Roger, Avery M. Guest, and Rodney Stark. 1996. "Mobilizing Local Religious Markets: Religious Pluralism in the Empire State, 1855–1865." *American Sociological Review* 61: 203–218.

Finke, Roger, and Laurence R. Iannaccone. 1993. "Supply-Side Explanations for Religious Change." *Annals, AAPSS* 527: 27–39.

Finke, Roger, and Rodney Stark. 1992. *The Churching of America, 1776–1990: Winners and Losers in Our Religious Economy.* New Brunswick, N.J.: Rutgers University Press.

Finke, Roger, and Rodney Stark. 1998. "Religious Choice and Competition." *American Sociological Review* 63: 761–766.

Firebaugh, Glenn, and Kevin Chen. 1995. "Vote Turnout of Nineteenth Amendment Women: The Enduring Effect of Disenfranchisement." *American Journal of Sociology* 100: 972–996.

Firebaugh, Glenn, and Kenneth E. Davis. 1988. "Trends in Antiblack Prejudice, 1972–1984: Region and Cohort Effects." *American Journal of Sociology* 94: 251–272.

Fischer, Claude S. 1975. "Toward a Subcultural Theory of Urbanism." *American Journal of Sociology* 80: 1319–1341.

Fischer, Claude S. 1982. *To Dwell Among Friends: Personal Networks in Town and City.* Chicago: University of Chicago Press.

Fischer, Claude S. 1992. *America Calling: A Social History of the Telephone to 1940.* Berkeley: University of California Press.

Fischer, Claude S. 1995. "The Subcultural Theory of Urbanism: A Twentieth-Year Assessment." *American Journal of Sociology* 101: 543–577.

Fischer, Claude S., Robert Max Jackson, C. Ann Stueve, Kathleen Gerson, and Lynne McCallister Jones, with Mark Baldassare. 1977. *Networks and Places: Social Relations in the Urban Setting.* New York: Free Press.

Fishman, Pamela M. 1983. "Interaction: The Work Women Do." Pp. 89–101 in Barrie Thorne, Cheris Kramarae, and Nancy Henley (eds.), *Language, Gender, and Society.* Rowley, Mass.: Newbury House.

Fitzgerald, Francis. 1986. *Cities on a Hill: A Journey Through Contemporary American Cultures.* New York: Simon & Schuster.

Flannery, Daniel J., C. Ronald Huff, and Michael Manos. 1998. "Youth Gangs: A Developmental Perspective." Pp. 175–204 in Thomas P. Gullotta, Gerald R. Adams, and Raymond Montemayor (eds.), *Delinquent Violent Youth: Theory and Interventions.* Thousand Oaks, Calif.: Sage.

Flavell, J. H. 1977. *Cognitive Development.* Englewood Cliffs, N.J.: Prentice-Hall.

Fliegel, Zenia Odes. 1986. "Women's Development in Analytic Theory: Six Decades of Controversy." Pp. 3–31 in Judith L. Alpert (ed.), *Psychoanalysis and Women: Contemporary Reappraisals.* New York: Analytic.

Fligstein, Neil. 1990. *The Transformation of Corporate Control.* Cambridge, Mass.: Harvard University Press.

Floyd, Virginia Davis. 1992. "'Too Soon, Too Small, Too Sick': Black Infant Mortality." Pp. 165–177 in Ronald L. Braithwaite and Sandra E. Taylor (eds.), *Health Issues in the Black Community.* San Francisco: Jossey-Bass.

Forbes, H.D. 1997. *Ethnic Conflict: Commerce, Culture, and the Contact Hypothesis.* New Haven: Yale University Press.

Ford, Cecilia E., and Sandra A. Thompson. 1996. "Interactional Units in Conversation: Syntactic, Intonational, and Pragmatic Resources for the Management of Turns." Pp. 134–184 in Elinor Ochs, Emanuel A. Schegloff, and Sandra A. Thompson (eds.), *Interaction and Grammar.* Cambridge: Cambridge University Press.

Form, William. 1979. "Comparative Industrial Sociology and the Convergence Hypothesis." *Annual Review of Sociology* 5: 1–25.

Form, William. 1990. "Institutional Analysis: An Organizational Approach." Pp. 257–271 in Maureen T. Hallinan, David M. Klein, and Jennifer Glass (eds.), *Change in Societal Institutions.* New York: Plenum.

Fox, Barbara A., Makoto Hayashi, and Robert Jasperson. 1996. "Resources and Repair: A Cross-Linguistic Study of Syntax and Repair." Pp. 185–237 in Elinor Ochs, Emanuel A. Schegloff, and Sandra A. Thompson (eds.), *Interaction and Grammar.* Cambridge: Cambridge University Press.

Frank, André Gunder. 1967. *Capitalism and Underdevelopment in Latin America.* New York: Monthly Review.

Frank, André Gunder. 1980. *Crisis in the Third World.* New York: Holmes & Meier.

Frank, Robert H. 1992. "Melding Sociology and Economics: James Coleman's Foundations of Social Theory." *Journal of Economic Literature* 30: 147–170.

Franklin, J. H. 1952. *From Slavery to Freedom.* New York: Knopf.

Freeman, John, and Michael T. Hannan. 1989. "Setting the Record Straight on Organizational Ecology: Rebuttal to Young." *American Journal of Sociology* 95: 425–439.

Freeman, Linton C. 1992. "The Sociological Concept of 'Group': An Empirical Test of Two Models." *American Journal of Sociology* 98: 152–166.

Freidl, Ernestine. 1975. *Women and Men: An Anthropologist's View.* New York: Holt, Rinehart & Winston.

Frey, R. Scott, Thomas Dietz, and Linda Kalof. 1992. "Characteristics of Successful American Protest Groups: Another Look at Gamson's Strategy of Social Protest." *American Journal of Sociology* 98: 368–387.

Fuechtmann, Thomas G. 1989. *Steeples and Stacks: Religion and Steel Crisis in Youngstown.* Cambridge: Cambridge University Press.

Fuguitt, Glenn V., David L. Brown, and Calvin L. Beale. 1989. *Rural and Small Town America.* New York: Russell Sage.

Furstenberg, Frank F., Jr. 1990. "Divorce and the American Family." *Annual Review of Sociology* 16: 379–403.

Furstenberg, Frank F., Jr. 1992. "Good Dads—Bad Dads: Two Faces of Fatherhood." Pp. 342–362 in Arlene S. Skolnick and Jerome H. Skolnick (eds.), *Family in Transition: Rethinking Marriage, Sexuality, Child Rearing, and Family Organization*. New York: Harper Collins.

Furstenberg, Frank F., Jr., and Andrew J. Cherlin. 1991. *Divided Families: What Happens to Children When Parents Part*. Cambridge, Mass.: Harvard University Press.

Furstenberg, Frank F., Jr., and Kathleen Mullan Harris. 1992. "The Disappearing American Father? Divorce and the Waning Significance of Biological Parenthood." Pp. 197–223 in Scott J. South and Stewart E. Tolnay (eds.), *The Changing American Family: Sociological and Demographic Perspectives*. Boulder, Colo.: Westview.

Gaer, Joseph. 1963. *What the Great Religions Believe*. New York: Signet.

Gale, Richard P. 1986. "Social Movements and the State: The Environmental Movement, Countermovement, and Government Agencies." *Sociological Perspectives* 29: 202–240.

Gallagher, Sally K., Randall G. Stokes, and Andy B. Anderson. 1996. "Economic Disarticulation and Fertility in Less Developed Societies." *The Sociological Quarterly* 37: 227–244.

Gamble, Sidney D. 1943. "The Disapppearance of Foot-Binding in Tinghsien." *American Journal of Sociology* 49: 181–183.

Gamson, William A. 1975. *The Strategy of Social Protest*. Homewood, Ill.: Dorsey.

Gamson, William A. 1992. *Talking Politics*. Cambridge: Cambridge University Press.

Gans, Herbert J. 1962. *The Urban Villagers*. New York: Free Press.

Gans, Herbert J. 1970. "Urbanism and Suburbanism." Pp. 157–164 in Albert N. Cousins and Hans Nagpaul (eds.), *Urban Man and Society: A Reader in Urban Sociology*. New York: Knopf.

Gans, Herbert J. 1985. "Symbolic Ethnicity: The Future of Ethnic Groups and Cultures in America." Pp. 429–442 in Norman R. Yetman (ed.), *Majority and Minority: The Dynamics of Race and Ethnicity in American Life*. Boston: Allyn & Bacon. (Orig. pub. in 1979.)

Gans, Herbert J. 1991. *People, Plans, and Policies: Essays on Poverty, Racism, and Other National Urban Problems*. New York: Columbia University Press.

Garcia, F. Chris. 1997. "Input to the Political System: Participation." Pp. 31–43 in H. Chris Garcia (ed.), *Pursuing Power: Latinos and the Political System*. Notre Dame, Ind.: University of Notre Dame Press.

Garcia, John A. 1997. "Political Participation: Resources and Involvement among Latinos in the American Political System." Pp. 44–71 in H. Chris Garcia (ed.), *Pursuing Power: Latinos and the Political System*. Notre Dame, Ind.: University of Notre Dame Press.

Garkovich, Lorraine. 1989. *Population and Community in Rural America*. New York: Greenwood.

Gatewood, Willard B. 1990. *Aristocrats of Color: The Black Elite, 1880–1920*. Bloomington: Indiana University Press.

Gecas, Victor, and Peter J. Burke. 1995. "Self and Identity." Pp. 41–67 in Karen S. Cook, Gary Alan Fine, and James S. House (eds.), *Sociological Perspectives on Social Psychology*. Boston: Allyn & Bacon.

Gelles, Richard J., and Murray A. Straus. 1989. *Intimate Violence: The Causes and Consequences of Abuse in the American Family*. New York: Touchstone.

Genovese, E. D. 1974. *Roll, Jordan, Roll: The World the Slaves Made*. New York: Pantheon.

Gerber, Theodore P., and Michael Hout. 1995. "Educational Stratification in Russia During the Soviet Period." *American Journal of Sociology* 101: 611–660.

Gerber, Theodore P., and Michael Hout. 1998. "More Shock than Therapy: Market Transition, Employment, and Income in Russia, 1991–1995." *American Journal of Sociology* 104: 1–50.

Gerson, Kathleen. 1993. *No Man's Land: Men's Changing Commitments to Family and Work*. New York: Basic.

Gerth, H. H., and C. Wright Mills (eds. and trans.). 1946. *From Max Weber: Essays in Sociology*. New York: Oxford University Press.

Gerth, H. H., and C. Wright Mills (eds. and trans.). 1967. *From Max Weber: Essays in Sociology*. New York: Oxford University Press

Gibbs, Jack P. 1981. *Norms, Deviance, and Social Control: Conceptual Matters*. New York: Elsevier.

Gibbs, Jack P. 1989. *Control: Sociology's Central Notion*. Urbana and Chicago: University of Illinois Press.

Gibbs, Jack P. 1994. *A Theory About Control*. Boulder, Colo.: Westview.

Gibbs, Jack, and Walter T. Martin. 1964. *Status Integration and Suicide: A Sociological Study*. Eugene: University of Oregon Press.

Giddens, Anthony. 1979. *Central Problems in Social Theory*. Berkeley and Los Angeles: University of California Press.

Giele, Janet Zollinger. 1995. *Two Paths to Women's Equality: Temperance, Suffrage, and the Origins of Modern Feminism*. New York: Twayne.

Gilbert, Dennis. 1998. *The American Class Structure: In an Age of Growing Inequality*, 5th ed. Belmont, Calif.: Wadsworth.

Gilbert, Dennis, and Joseph A. Kahl. 1993. *The American Class Structure: A New Synthesis*, 4th ed. Belmont, Calif.: Wadsworth.

Gittler, Joseph B. 1956. *Understanding Minority Groups*. New York: Wiley.

Glasco, Laurence A. 1977. "The Life Cycles and Household Structure of American Ethnic Groups: Irish, Germans, and Native-born Whites in Buffalo, New York, 1855." Pp. 122–143 in Tamara K. Hareven (ed.), *Family and Kin in Urban Communities, 1700–1930*. New York: New Viewpoints/Franklin Watts.

Glaser, Daniel. 1956. "Criminality Theories and Behavioral Images." *American Journal of Sociology* 61: 433–444.

Glass, Jennifer. 1998. "Gender Liberation, Economic Squeeze, or Fear of Strangers: Why Fathers Provide Infant Care in Dual-Earner Families." *Journal of Marriage and the Family* 60: 821–834.

Glass, Jennifer, and David M. Klein. 1990. "Introduction." Pp. 1–11 in Maureen T. Hallinan, David M. Klein, and Jennifer Glass (eds.), *Change in Societal Institutions.* New York: Plenum.

Glazer, Nathan. 1997. *We Are All Multiculturalists Now.* Cambridge, Mass.: Harvard University Press.

Glazer, Nathan, and Daniel P. Moynihan. 1970. *Beyond the Melting Pot: The Negroes, Puerto Ricans, Jews, Italians, and Irish of New York City,* 2nd ed. Cambridge, Mass.: M.I.T. Press.

Glenn, Norval D. 1993. "A Plea for Objective Assessment of the Notion of Family Decline." *Journal of Marriage and the Family* 55: 542–544.

Goffman, Erving. 1959. *The Presentation of Self in Everyday Life.* Garden City, N.Y.: Doubleday.

Goffman, Erving. 1963. *Behavior in Public Places.* New York: Free Press.

Goffman, Erving. 1967. *Interaction Ritual.* Garden City, N.Y.: Doubleday.

Goffman, Erving. 1974. *Frame Analysis: An Essay on the Organization of Experience.* Cambridge, Mass.: Harvard University Press.

Gold, Steven J. 1997. "Transnationalism and Vocabularies of Motive in International Migration: The Case of Israelis in the United States." *Sociological Perspectives* 40: 409–426.

Goldfield, Michael. 1987. *The Decline of Organized Labor in the United States.* Chicago: University of Chicago Press.

Goldscheider, Frances K., and Linda J. Waite. 1991. *New Families, No Families? The Transformation of the American Home.* Berkeley: University of California Press.

Goldschmidt, Walter. 1960. *Exploring the Ways of Mankind.* New York: Holt, Rinehart & Winston.

Goliber, Thomas J. 1997. "Population and Reproductive Health in Sub-Saharan Africa." *Population Bulletin* 52(4). Washington, D.C.: Population Reference Bureau, Inc., November 1997.

Goode, David. 1994. *A World Without Words: The Social Construction of Children Born Deaf and Blind.* Philadelphia: Temple University Press.

Goode, William J. 1963. *World Revolution and Family Patterns.* New York: Free Press.

Goode, William J. 1992. "World Changes in Divorce Patterns." Pp. 11–49 in Lenore J. Weitzman and Mavis Maclean (eds.), *Economic Consequences of Divorce: The International Perspective.* Oxford: Clarendon.

Goode, William J. 1993. *World Changes in Divorce Patterns.* New Haven, Conn.: Yale University Press.

Goodman, Madeleine J., P. Bion Griffin, Agnes A. Estioko-Griffin, and John S. Grove. 1985. "The Compatibility of Hunting and Mothering Among the Agta Hunter-Gatherers of the Philippines." *Sex Roles* 12: 1199–1209.

Goodwin, Charles. 1996. "Transparent Vision." Pp. 370–404 in Elinor Ochs, Emanuel A. Schegloff, and Sandra A. Thompson (eds.), *Interaction and Grammar.* Cambridge: Cambridge University Press.

Gordon, C. 1968. "Self-Conceptions: Configurations of Content." Pp. 115–136 in C. Gordon and K. J. Gergen (eds.), *The Self in Social Interaction, I: Classic and Contemporary Perspectives.* New York: Wiley.

Gore, Rick. 1992. "Sunset Boulevard: Street to the Stars." *National Geographic,* June, pp. 40–69.

Gortmaker, Steven L., and Paul H. Wise. 1997. "The First Injustice: Socioeconomic Disparities, Health Services Technology, and Infant Mortality." *Annual Review of Sociology* 23: 147–170.

Gortner, Harold F., Julianne Mahler, and Jeanne Bell Nicholson. 1989. *Organization Theory: A Public Perspective.* Pacific Grove, Calif.: Brooks/Cole.

Gottfredson, Michael R., and Travis Hirschi. 1990. *A General Theory of Crime.* Stanford, Calif.: Stanford University Press.

Gould, Roger V. 1993. "Collective Action and Network Structure." *American Sociological Review* 58: 182–196.

Gove, Walter R. 1985. "The Effect of Age and Gender on Deviant Behavior: A Biopsychosocial Perspective." Pp. 115–144 in Alice S. Rossi (ed.), *Gender and the Life Course.* New York: Aldine.

Gramsci, Antonio. 1972. *Selections from "The Prison Notebooks of Antonio Gramsci."* Quintin Hoare and Geoffrey Nowell Smith (trans. and ed.). New York: International.

Granovetter, Mark. 1973. "The Strength of Weak Ties." *American Journal of Sociology* 78: 1360–1380.

Granovetter, Mark. 1974. *Getting a Job: A Study of Contacts and Careers.* Cambridge, Mass.: Harvard University Press.

Granovetter, Mark. 1984. "Small is Beautiful: Labor Markets and Estabishment Size." *American Sociological Review* 49: 323–334.

Granovetter, Mark. 1985. "Economic Action and Social Structure: The Problem of Embeddedness." *American Journal of Sociology* 91: 481–510.

Granovetter, Mark. 1995. *Getting a Job: A Study of Contacts and Careers,* 2nd ed. Chicago: University of Chicago Press.

Granovetter, Mark, and Richard Swedberg (eds.). 1992. *The Sociology of Economic Life.* Boulder, Colo.: Westview.

Grant, James P. 1994. *The State of the World's Children, 1994.* New York: Oxford University Press.

Grant, James P. 1995. *The State of the World's Children, 1995.* New York: Oxford University Press.

Greeley, Andrew. 1972. *The Denominational Society.* Glenview, Ill.: Scott, Foresman.

Greeley, Andrew. 1974. *Ethnicity in the United States.* New York: Wiley.

Greeley, Andrew M. 1982. *Catholic High Schools and Minority Students.* New Brunswick, N.J.: Transaction.

Greeley, Andrew M. 1989. *Religious Change in America.* Cambridge, Mass.: Harvard University Press.

Greeley, Andrew. 1994. "A Religious Revival in Russia?" *Journal for the Scientific Study of Religion* 33: 253–272.

Greeley, Andrew M. 1995. *Religion as Poetry.* New Brunswick, N.J.: Transaction.

Green, Harvey. 1986. *Fit for America: Health, Fitness, Sport, and American Society.* New York: Pantheon.

Gregorio, David I., Stephen J. Walsh, and Deborah Paturzo. 1997. "The Effects of Occupation-Based Social Position on Mortality in a Large American Cohort." *American Journal of Public Health* 87: 1472–1475.

Gregory, T. B., and G. R. Smith. 1987. *High Schools as Communities: The Small School Reconsidered.* Bloomington, Ind.: Phi Delta Kappa Educational Foundation.

Greider, William B. 1992. *Who Will Tell the People: The Betrayal of American Democracy.* New York: Simon & Schuster.

Grimes, Michael D. 1991. *Class in Twentieth-Century American Sociology: An Analysis of Theories and Measurement Strategies.* New York: Praeger.

Grofman, Bernard (ed.). *Information, Participation, and Choice: An Economic Theory of Democracy in Perspective.* Ann Arbor: University of Michigan Press.

Gugler, Josef (ed.). 1996. *The Urban Transformation of the Developing World.* Oxford: Oxford University Press.

Gugler, Josef (ed.). 1997. *Cities in the Developing World: Issues, Theory, and Policy.* Oxford: Oxford University Press.

Hacker, Andrew. 1992. *Two Nations: Black and White, Separate, Hostile, Unequal.* New York: Charles Scribners.

Hacker, Andrew. 1997. *Money: Who Has How Much and Why.* New York: Scribner.

Hagan, John, and Fiona Kay. 1995. *Gender in Practice: A Study of Lawyers' Lives.* New York: Oxford University Press.

Hagan, John, and Bill McCarthy. 1997a. "Anomie, Social Capital, and Street Criminology." Pp. 124–141 in Nikos Passas and Robert Agnew (eds.), *The Future of Anomie Theory.* Boston: Northeastern University Press.

Hagan, John, and Bill McCarthy. 1997b. *Mean Streets: Youth Crime and Homelessness.* New York: Cambridge University Press.

Hagan, John, and Alberto Palloni. 1990. "The Social Reproduction of a Criminal Class in Working-Class London, Circa 1950–1980." *American Journal of Sociology* 96: 265–299.

Hagan, John, and Ruth D. Peterson. 1995. *Crime and Inequality.* Stanford, Calif.: Stanford University Press.

Hage, Jerald. 1980. *Theories of Organizations: Form, Process, and Transformation.* New York: Wiley.

Hage, Jerald, and Michael Aiken. 1970. *Social Change in Complex Organizations.* New York: Random House.

Hagedorn, John, with Perry Macon. 1988. *People and Folks: Gangs, Crime and the Underclass in a Rustbelt City.* Chicago: Lakeview.

Haines, Herbert H. 1996. *Against Capital Punishment: The Anti-Death Penalty Movement in America, 1972–1994.* New York: Oxford University Press.

Hall, John R. 1988. "Social Organization and Pathways of Commitment: Types of Communal Groups, Rational Choice Theory, and the Kanter Thesis." *American Sociological Review* 53: 679–692.

Hall, Richard H. 1991. *Organizations: Structures, Processes, and Outcomes,* 5th ed. Englewood Cliffs, N.J.: Prentice-Hall.

Hall, Richard H., John P. Clark, Peggy C. Giordano, Paul V. Johnson, and Martha Van Roekel. 1977. "Patterns of Interorganizational Relationships." *Administrative Science Quarterly* 22: 457–474.

Hallinan, Maureen T., 1997. The Sociological Study of Social Change: 1996 Presidential Address. *American Sociological Review* 62: 1–11.

Halsey, A. H.. 1989. "Education Can Compensate." Pp. 298–302 in Jeanne H. Ballantine (ed.), *Schools and Society: A Unified Reader,* 2nd ed. Mountain View, Calif.: Mayfield.

Hamberg, Eva M., and Thorleif Pettersson. 1994. "The Religious Market: Denominational Competition and Religious Participation in Contemporary Sweden." *Journal for the Scientific Study of Religion* 33: 205–216.

Hamermesh, Daniel S., and Frank D. Bean (eds.). 1998. *Help or Hindrance? The Economic Implications of Immigration for African Americans.* New York: Russell Sage.

Haney, Daniel. 1998. "Gap in Care Quality Widens as HIV Spreads." *The Eugene Register Guard,* July 4, 1998, p. 9a.

Hannan, Michael T., and Glenn R. Carroll. 1992. *Dynamics of Organizational Populations: Density, Legitimation, and Competition.* New York: Oxford University Press.

Hannan, Michael T., Glenn R. Carroll, Elizabeth A. Dundon, and John Charles Torres. 1995. "Organizational Evolution in a Multinational Context: Entries of Automobile Manufacturers in Belgium, France, Germany, and Italy." *American Sociological Review* 60: 509–528.

Hannan, Michael T., and John Freeman. 1977. "The Population Ecology of Organizations." *American Journal of Sociology* 82: 929–964.

Hannan, Michael T., and John Freeman. 1984. "Structural Inertia and Organizational Change." *American Sociological Review* 49: 149–164.

Hannan, Michael T., and John Freeman. 1989. *Organizational Ecology.* Cambridge, Mass.: Harvard University Press.

Harding, Ted. 1998. "People February 1, 1998: Profile of Clinton confidant Vernon Jordan." *The Sunday Business Post: Ireland's Financial, Political, and Economic News on Line.* February 1, 1998. http://www.sbpost.ie/newspaper/010298/people/jordan.html.

Hareven, Tamara K. (ed.). 1978. *Transitions: The Family and the Life Course in Historical Perspective.* New York: Academic.

Hareven, Tamara K. 1982. *Family Time and Industrial Time: The Relationship Between the Family and Work in a New England Industrial Community.* Cambridge: Cambridge University Press.

Hareven, Tamara K., and Maris A. Vinovskis (eds.). 1978. *Family and Population in Nineteenth-Century America.* Princeton, N.J.: Princeton University Press.

Harris, Chauncy D., and Edward Ullman. 1945. "The Nature of Cities." *Annals of the American Academy of Political and Social Science* 142: 7–17.

Harris, Daniel M., and Sharon Guten. 1979. "Health-Protective Behavior: An Exploratory Study." *Journal of Health and Social Behavior* 20: 17–29.

Harris, Kathleen Mullan. 1993. "Work and Welfare Among Single Mothers in Poverty." *American Journal of Sociology* 99: 317–352.

Harris, Marvin. 1979. *Cultural Materialism: The Struggle for a Science of Culture.* New York: Random House.

Harrison, R. J., and D. H. Weinberg. 1992. "Racial and Ethnic Segregation in 1990." Presented at the annual meeting of the Population Association of America, April 30–May 2, Denver, Colo.

Harvey, Lee. 1987. *Myths of the Chicago School of Sociology.* Aldershot, England: Avebury.

Haub, Carl, and Machiko Yanagishita. 1995. *World Population Data Sheet, 1995.* Washington, D.C.: Population Reference Bureau.

Hauser, Robert M., and David L. Featherman. 1977. *The Process of Stratification: Trends and Analysis.* New York: Academic.

Hayward, Mark D., Amy M. Pienta, and Diane K. McLaughlin. 1997. "Inequality in Men's Mortality: The Socioeconomic Status Gradient and Geographic Context." *Journal of Health and Social Behavior* 38: 313–330.

Hechter, Michael. 1983. "A Theory of Group Solidarity." Pp. 16–57 in Michael Hechter (ed.), *The Microfoundations of Macrosociology.* Philadelphia: Temple University Press.

Hechter, Michael, and Satoshi Kanazawa. 1997. "Sociological Rational Choice Theory." *Annual Review of Sociology* 23: 191–214.

Hegtvedt, Karen A., and Barry Markovsky. 1995. "Justice and Injustice." Pp. 257–280 in Karen S. Cook, Gary Alan Fine, and James S. House (eds.), *Sociological Perspectives on Social Psychology.* Boston: Allyn & Bacon.

Heimer, Karen, and Ross L. Matsueda. 1994. "Role-Taking, Role Commitment, and Delinquency: A Theory of Differential Social Control." *American Sociological Review* 59: 365–390.

Heise, David R. 1990. "Careers, Career Trajectories, and the Self." Pp. 59–84 in Judith Rodin, Carmi Schooler, and K. Warner Schaie (eds.), *Self-Directedness: Cause and Effects Throughout the Life Course.* Hillsdale, N.J.: Erlbaum.

Hemingway, Harry, Amanda Nicholson, Ron Roberts, and Michael Marmot. 1997. "The Impact of Socioeconomic Status on Health Functioning as Assessed by the SF-36 Questionnaire: The Whitehall II Study." *American Journal of Public Health* 87: 1484–1490.

Henry, Stuart, and Werner Einstadter. 1998. "Introduction: Criminology and Criminological Theory." Pp. 1–13 in Stuart Henry and Werner Einstadter (eds.). *The Criminology Theory Reader.* New York: New York University Press.

Henslin, James M. 1993. *Sociology: A Down-to-Earth Approach.* Boston: Allyn & Bacon.

Henson, Kevin D. 1996. *Just a Temp.* Philadelphia: Temple University Press.

Hernandez, Donald J. 1993. *America's Children: Resources from Family, Government, and the Economy.* New York: Russell Sage.

Hernandez, Donald J. 1997. "Poverty Trends." Pp. 18–34 in Greg J. Duncan and Jeanne Brooks-Gunn (eds.), 1997. *Consequences of Growing Up Poor.* New York: Russell Sage.

Hertzler, J. O. 1961. *American Social Institutions: A Sociological Analysis.* Boston: Allyn & Bacon.

Herzog, Elizabeth, and Cecilia E. Sudia. 1973. "Children in Fatherless Families." Pp. 141–232 in B. Caldwell and H. N. Ricciuti (eds.), *Review of Child Development Research,* Vol. 3. Chicago: University of Chicago Press.

Hetherington, E. Mavis, Martha Cox, and Roger Cox. 1978. "The Aftermath of Divorce." Pp. 148–176 in Joseph H. Stevens and Marilyn Mathews (eds.), *Mother-Child, Father-Child Relations.* Washington, D.C.: National Association for the Education of Young Children.

Heusmann, L. Rowell, Leonard D. Eron, and Monroe M. Lefkowitz. 1984. "Stability of Aggression over Time and Generations." *Developmental Psychology* 20: 1120–1134.

Hewitt, John P. 1988. *Self and Society,* 4th ed. Boston: Allyn & Bacon.

Hewlett, Sylvia Ann. 1991. *When the Bough Breaks: The Cost of Neglecting Our Children.* New York: Harper Collins.

Hickson, David J., C. R. Hinings, C. A. Lee, R. E. Schneck, and J. M. Pennings. 1971. "A Strategic Contingencies Theory of Organizational Power." *Administrative Science Quarterly* 16: 216–229.

Hilbert, Richard A. 1990. "Ethnomethodology and the Micro-Macro Order." *American Sociological Review* 55: 794–808.

Hill, Herbert, and James E. Jones. (eds.). 1993. *Race in America: The Struggle for Equality.* Madison: University of Wisconsin Press.

Hill, W. W. 1956. "The Ritualization of Everyday Behavior." Pp. 225–229 in Irwin T. Sanders, Richard B. Woodbury, Frank J. Essene, Thomas P. Field, Joseph R. Schwendeman, and Charles E. Snow (eds.), *Societies Around the World.* New York: Holt, Rinehart & Winston. (Orig. pub. 1938.)

Hirsch, Paul M. 1971. "Processing Fads and Fashions: An Organization-Set Analysis of Cultural Industry Systems." *American Journal of Sociology* 77: 639–659.

Hirsch, Paul M. 1986. "From Ambushes to Golden Parachutes: Corporate Takeovers as an Instance of Cultural Framing and Institutional Integration." *American Journal of Sociology* 4: 800–837.

Hirschi, Travis. 1969. *Causes of Delinquency.* Berkeley: University of California Press.

Hirschi, Travis. 1988. "A Control Theory of Delinquency." Pp. 342–347 in Ronald A. Farrell and Victoria Lynn Swigert (eds.), *Social Deviance,* 3rd ed. Belmont, Calif.: Wadsworth.

Hitchings, Thomas E. (ed.). 1994. *Facts on File: World News Digest with Index.* New York: Facts on File.

Hochschild, Arlie, with Anne Machung. 1989. *The Second Shift.* New York: Avon.

Hochschild, Jennifer L. 1995. *Facing Up to the American dream : Race, Class, and the Soul of the Nation.* Princeton, N.J.: Princeton University Press.

Hodge, Robert W., P. M. Siegal, and Peter Rossi. 1964. "Occupational Prestige in the United States, 1925–1963." *American Journal of Sociology* 70: 286–302.

Hodge, Robert W., Donald J. Treiman, and Peter H. Rossi. 1966. "A Comparative Study of Occupational Prestige." Pp. 309–321 in Reinhard Bendix and Seymour Martin Lipset (eds.), *Class, Status, and Power,* 2nd ed. New York: Free Press.

Hodson, Randy, and Teresa A. Sullivan. 1990. *The Social Organization of Work.* Belmont, Calif.: Wadsworth.

Hodson, Randy, and Teresa A. Sullivan. 1995. *The Social Organization of Work,* 2nd ed. Belmont, Calif.: Wadsworth.

Hogan, Dennis P., and Daniel T. Lichter. 1995. "Children and Youth: Living Arrangements and Welfare." Pp. 93–140 in Reynolds Farley (ed.), *State of the Union: America in the 1990s,* Vol. 2: *Social Trends.* New York: Russell Sage.

Hoge, Dean R., Benton Johnson, and Donald A. Luidens. 1994. *Vanishing Boundaries: The Religion of Mainline Protestant Baby Boomers.* Louisville, Ky.: Westminster/John Knox.

Holahan, Carole K., and Robert R. Sears. 1995. *The Gifted Group in Later Maturity.* Stanford, Calif.: Stanford University Press.

Hollinger, David A. 1995. *Postethnic America: Beyond Multiculturalism.* New York: Basic.

Holmes, Leslie. 1997. *Post-Communism: An Introduction.* Durham, N.C.: Duke University Press.

Homans, George C. 1950. *The Human Group.* New York: Harcourt Brace Jovanovich.

Homans, George C. 1961. *Social Behavior: Its Elementary Forms.* New York: Harcourt Brace Jovanovich.

Hooper, Linda M., and Claudette E. Bennett. 1998. "The Asian and Pacific Islander Population in the United States: March 1997 (Update)." *Current Population Reports, Population Characteristics,* P20–512. Washington, D.C.: Census Bureau, U.S. Department of Commerce.

Hoopes, David S. (ed.). 1994. *Worldwide Branch Locations of Multinational Companies.* Detroit: Gale Research.

Horne, Gerald. 1997. "New Racial Divide Now East-West, Not North-South." *Eugene Register Guard,* Sunday, August 31, 1997, pp. F1, F4.

Horney, Julie, D. Wayne Osgood, and Ineke Haen Marshall. 1995. "Criminal Careers in the Short-Term: Intra-Individual Variability in Crime and Its Relation to Local Life Circumstances." *American Sociological Review* 60: 655–673.

Horney, Karen. 1967. *Feminine Psychology.* Harold Kelman (ed.). New York: Norton.

Horwitz, Allan V. 1990. *The Logic of Social Control.* New York: Plenum.

Hout, Michael. 1989. *Following in Father's Footsteps: Social Mobility in Ireland.* Cambridge, Mass.: Harvard University Press.

Hout, Michael, and Joshua R. Goldstein. 1994. "How 4.5 Million Irish Immigrants Became 40 Million Irish Americans: Demographic and Subjective Aspects of the Ethnic Composition of White Americans." *American Sociological Review* 59: 64–82.

Howard, Judith A., and Jocelyn Hollander. 1997. *Gendered Situations, Gendered Selves.* Thousand Oaks, Calif.: Sage.

Hoyt, Homer. 1939. *The Structure and Growth of Residential Neighborhoods in American Cities.* Washington, D.C.: U.S. Government Printing Office.

Hrabowski, Freeman A., III, Kenneth I. Maton, and Geoffrey L. Greif. 1998. *Beating the Odds: Raising Academically Successful African American Males.* New York: Oxford University Press.

Hsu, Francis L. K. 1956. "Chinese Religion and Ancestor Worship." Pp. 460–463 in Irwin T. Sanders, Richard B. Woodbury, Frank J. Essene, Thomas P. Field, Joseph R. Schwendeman, and Charles E. Snow (eds.), *Societies Around the World.* New York: Holt, Rinehart & Winston.

Huber, Joan, and Glenna Spitze. 1983. *Sex Stratification: Children, Housework, and Jobs.* New York: Academic.

Huff, C. Ronald (ed.). 1990. *Gangs in America.* Newbury Park, Calif.: Sage.

Hunter, James Davison. 1985. "Conservative Protestantism." Pp. 150–166 in Phillip E. Hammond (ed.), *The Sacred in a Secular Age: Toward Revision in the Scientific Study of Religion.* Berkeley: University of California Press.

Hurn, Christopher J. 1993. *The Limits and Possibilities of Schooling: An Introduction to the Sociology of Education.* Boston: Allyn & Bacon.

Hyde, Janet Shibley. 1984. "How Large Are Gender Differences in Aggression? A Meta-Analysis." *Developmental Psychology* 20: 722–736.

Hyde, Janet Shibley. 1986. "Gender Differences in Aggression." In Janet Shibley Hyde and Marcia C. Linn (eds.), *The Psychology of Gender: Advances Through Meta-Analysis.* Baltimore: Johns Hopkins University Press.

Iannaccone, Laurence R. 1990. "Religious Practice: A Human Capital Approach." *Journal for the Scientific Study of Religion* 29: 297–314.

Iannaccone, Laurence R. 1994. "Why Strict Churches Are Strong." *American Journal of Sociology* 99: 1180–1211.

Idler, Ellen L., and Yael Benyamini. 1997. "Self-Rated Health and Mortality: A Review of Twenty-Seven Community Studies." *Journal of Health and Social Behavior* 38: 21–37.

Idler, Ellen L., and Stanislav V. Kasl. 1992. "Religion, Disability, Depression, and the Timing of Death." *American Journal of Sociology* 97: 1052–1079.

Inglehart, R. 1990. *Culture Shift in Advanced Industrial Society.* Princeton, N.J.: Princeton University Press.

Jacobs, David. 1974. "Dependency and Vulnerability: An Exchange Approach to the Control of Organizations." *Administrative Science Quarterly* 19: 45–59.

Jacobs, David, and Larry Singell. 1993. "Leadership and Organizational Performance: Isolating Links Between Managers and Collective Success." *Social Science Research* 22: 165–189.

Jacobs, Jerry A., and Ronnie J. Steinberg. 1990. "Compensating Differentials and the Male-Female Wage Gap: Evidence from the New York State Pay Equity Study." *Social Forces* 69: 439–468.

Jacobs, Jerry A., and Ronnie J. Steinberg. 1995. "Further Evidence on Compensating Differentials and the Gender Gap in Wages." Pp. 93–124 in Jerry A. Jacobs (ed.), *Gender Inequality at Work.* Thousand Oaks, Calif.: Sage.

Jacobson, Paul H. 1959. *American Marriage and Divorce.* New York: Rinehart.

Jacques, Jeffrey M. 1998. "Changing Marital and Family Patterns: A Test of the Post-modern Perspective." *Sociological Perspectives* 41: 381–413.

Jamison, Andrew, Ron Eyerman, and Jacqueline Cramer, with Jeppe Læssøe. 1990. *The Making of the New Environmental Consciousness: A Comparative Study of the Environmental Movements in Sweden, Denmark, and the Netherlands.* Edinburgh: Edinburgh University Press.

Janis, Irving L. 1983. *Group Think: Psychological Studies of Policy Decisions and Fiascoes,* 2nd ed. Boston: Houghton Mifflin.

Jankowski, Martin Sanchez. 1991. *Islands in the Street: Gangs and American Urban Society.* Berkeley: University of California Press.

Jargowsky, Paul A. 1996. "Take the Money and Run: Economic Segregation in U.S. Metropolitan Areas." *American Sociological Review* 61: 984–998.

Jargowsky, Paul A. 1997. *Poverty and Place: Ghettos, Barrios, and the American City.* New York: Russell Sage.

Jencks, Christopher. 1985. "How Much Do High School Students Learn?" *Sociology of Education* 58: 128–135.

Jencks, Christopher. 1992. *Rethinking Social Policy: Race, Poverty, and the Underclass.* Cambridge, Mass.: Harvard University Press.

Jencks, Christopher, Lauri Perman, and Lee Rainwater. 1988. "What Is a Good Job? A New Measure of Labor-Market Success." *American Journal of Sociology* 93: 1322–1357.

Jencks, Christopher, and Paul E. Peterson (eds.). 1991. *The Urban Underclass.* Washington, D.C.: Brookings.

Jenkins, J. Craig, and Bert Klandermans (eds.). 1995. *The Politics of Social Protest: Comparative Perspectives on States and Social Movements, Social Movements, Protest, and Content,* Vol. 3. Minneapolis: University of Minnesota Press.

Jiobu, Robert M. 1988. *Ethnicity and Assimilation: Blacks, Chinese, Filipinos, Japanese, Koreans, Mexicans, Vietnamese, and Whites.* Albany: State University of New York Press.

Johnson, Benton. 1963. "On Church and Sect." *American Sociological Review* 28: 539–549.

Johnson, Benton. 1971. "Church and Sect Revisited." *Journal for the Scientific Study of Religion* 10: 124–137.

Johnson, Benton. 1975. *Functionalism in Modern Sociology: Understanding Talcott Parsons.* Morristown, N.J.: General Learning.

Johnson, Colleen Leahy. 1988. *Ex Familia: Grandparents, Parents, and Children Adjust to Divorce.* New Brunswick, N.J.: Rutgers University Press.

Johnson, Kenneth M., and Calvin L. Beale. 1995. "The Rural Rebound Revisited." *American Demographics.* July, pp. 46–54.

Johnson, Miriam M. 1988. *Strong Mothers, Weak Wives: The Search for Gender Equality.* Berkeley: University of California Press.

Johnson, Otto (ed.). 1995. *Information Please Almanac 1996,* 49th ed. Boston: Houghton Mifflin.

Johnston, Hank, Enrique Laraña, and Joseph R. Gusfield. 1994. Identities, Grievances, and New Social Movements. Pp. 3–35 in Enrique Laraña, Hank Johnston, and Joseph R. Gusfield. 1994. *New Social Movements: From Ideology to Identity.* Philadelphia: Temple University Press.

Johnston, Hank, and David A. Snow. 1998. "Subcultures of Opposition and Social Movements." *Sociological Perspectives* 41: 473–498.

Jones, Ernest. 1933. "The Phallic Phase." *International Journal of Psychoanalysis* 14: 1–33.

Jones, Ernest. 1935. "Early Female Sexuality." *International Journal of Psychoanalysis* 16: 263–273.

Jones, Gavin W. 1994. *Marriage and Divorce in Islamic South-East Asia.* Kuala Lumpur: Oxford University Press.

Jones, Robert Emmet, and Riley E. Dunlap. 1992. "The Social Bases of Environmental Concern: Have They Changed over Time?" *Rural Sociology* 57: 28–47.

Jordan, Glenn, and Chris Weedon. 1995. *Cultural Politics: Class, Gender, Race, and the Postmodern World.* Oxford: Blackwell.

Josephson, Matthew. 1949. "The Robber Barons." Pp. 34–48 in Earl Latham (ed.), *John D. Rockefeller: Robber Baron or Industrial Statesman?* Boston: Heath.

Kadushin, Charles. 1995. "Friendship Among the French Financial Elite." *American Sociological Review* 60: 202–221.

Kagan, Jerome, and Howard A. Moss. 1962. *Birth to Maturity: A Study in Psychological Development.* New York: Wiley.

Kalish, Susan. 1994. "International Migration: New Findings on Magnitude, Importance." *Population Today* 22(3).

Kalish, Susan. 1995. "Multiracial Births Increase as U.S. Ponders Racial Definitions." *Population Today* 23(4): 1–2.

Kalmijn, Matthijs. 1991a. "Shifting Boundaries: Trends in Religious and Educational Homogamy." *American Sociological Review* 56: 786–800.

Kalmijn, Matthijs. 1991b. "Status Homogamy in the United States." *American Journal of Sociology* 97: 496–523.

Kalmijn, Matthijs. 1994a. "Assortative Mating by Cultural and Economic Occupational Status." *American Journal of Sociology* 100: 422–452.

Kalmijn, Matthijs. 1994b. "Mother's Occupational Status and Children's Schooling." *American Sociological Review* 59: 257–275.

Kanner, A. D., J. C. Coyne, C. Schaefer, and R.S. Lazarus. 1981. "Comparison of Two Modes of Stress Measurement: Daily Hassles and Uplifts versus Major Life Events." *Journal of Behavioral Medicine* 4: 1–39.

Kanter, Rosabeth Moss, and Barry A. Stein (eds.). 1979. *Life in Organizations: Workplaces As People Experience Them.* New York: Basic.

Kantor, Glenda Kaufman, and Murray A. Straus. 1990. "Response of Victims and the Police to Assaults on Wives." Pp. 473–487 in Murray A. Straus and Richard J. Gelles (eds.), *Physical Violence in American Families: Risk Factors and Adaptations to Violence in 8,145 Families.* New Brunswick, N.J.: Transaction.

Kasarda, John D. 1995. "Industrial Restructuring and the Changing Location of Jobs." Pp. 215–268 in Reynolds Farley (ed.). *State of the Union: America in the 1990s.* Vol. 1: *Economic Trends.* New York: Russell Sage.

Kawachi, Ichiro, Bruce P. Kennedy, Kimberly Lchner, and Deborah Prothrow-Stith. 1997. "Social Capital, Income Inequality, and Mortality." *American Journal of Public Health.* 87: 1491–1498.

Kebede, Alem Seghed, and David Knottnerus. 1998. "Beyond the Pales of Babylon: The Ideational Components and Social Psychological Foundations of Rastafari." *Sociological Perspectives* 41: 499–518.

Kelley, Dean M. 1927. *Why Conservative Churches Are Growing.* New York: Harper & Row.

Kelley, Jonathan, and Nan Dirk DeGraaf. 1997. "National Context, Parental Socialization, and Religious Belief: Results from 15 Nations." *American Sociological Review* 62: 639–659.

Kerckhoff, Alan C. 1995. "Social Stratification and Mobility Processes: Interaction between Individuals and Social Structures." Pp. 476–498 in Karen S. Cook, Gary Alan Fine, and James S. House (eds.), *Sociological Perspectives on Social Psychology.* Boston: Allyn & Bacon.

Kerr, Clark. 1983. *The Future of Industrialized Societies.* Cambridge, Mass.: Harvard University Press.

Kessler, Ronald C., James S. House, Renee R. Anspach, and David R. Williams. 1995. "Social Psychology and Health." Pp. 548–570 in Karen S. Cook, Gary Alan Fine, and James S. House (eds.), *Sociological Perspectives on Social Psychology.* Boston: Allyn & Bacon.

Killian, Lewis M. 1994. "Are Social Movements Irrational or Are They Collective Behavior?" Pp. 273–280 in Russell R. Dynes and Kathleen J. Tierney. (eds.), *Disasters, Collective Behavior, and Social Organization.* Newark, N.J.: University of Delaware Press.

Kinder, Donald R., and Lynn M. Sanders. 1996. *Divided by Color: Racial Politics and Democratic Ideals.* Chicago: University of Chicago Press.

Kingdon, John W. 1993. "How Do Issues Get on Public Policy Agendas?" Pp. 40–50 in William Julius Wilson (ed.), *Sociology and the Public Agenda.* Newbury Park, Calif.: Sage.

Kitson, Gay C., with William M. Holmes. 1992. *Portrait of Divorce: Adjustment to Marital Breakdown.* New York: Guilford.

Klandermans, Bert. 1997. *The Social Psychology of Protest.* Cambridge, Mass.: Blackwell.

Klein, David M. and James White. 1996. *Family Theories: An Introduction.* Thousand Oaks, Calif.: Sage.

Klein, Malcolm W. 1995. *The American Street Gang: Its Nature, Prevalence, and Control.* New York: Oxford University Press.

Klein, Melanie. 1960. *The Psychoanalysis of Children.* New York: Grove. (Orig. pub. 1932.)

Kluckhorn, Clyde. 1956. "The Great Chants of the Navajo." Pp. 229–232 in Irwin T. Sanders, Richard B. Woodbury, Frank J. Essene, Thomas P. Field, Joseph R. Schwendeman, and Charles E. Snow (eds.), *Societies Around the World.* New York: Holt, Rinehart & Winston. (Orig. pub. 1933.)

Knoke, David. 1990. *Political Networks: The Structural Perspective.* Cambridge: Cambridge University Press.

Knoke, David. 1993. "Networks as Political Glue: Explaining Public Policy-Making." Pp. 164–184 in William Julius Wilson (ed.), *Sociology and the Public Agenda.* Newbury Park, Calif.: Sage.

Kohlberg, L. 1966. "A Cognitive-Developmental Analysis of Children's Sex-Role Concepts and Attitudes." In E. E. Maccoby (ed.), *The Development of Sex Differences.* Stanford, Calif.: Stanford University Press.

Kohlberg, L. 1969. *Stages in the Development of Moral Thought and Action.* New York: Holt, Rinehart & Winston.

Kohlberg, L. 1981. *The Philosophy of Moral Development.* New York: Harper & Row.

Kollock, Peter, Philip Blumstein, and Pepper Schwartz. 1985. "Sex and Power in Interaction: Conversational Privileges and Duties." *American Sociological Review* 50: 34–46.

Kornblum, William, and Joseph Julian. 1992. *Social Problems,* 7th ed. Englewood Cliffs, N.J.: Prentice-Hall.

Kosmin, Barry A., and Seymour P. Lachman. 1993. *One Nation Under God: Religion in Contemporary American Society.* New York: Harmony.

Kowalewski, David. 1991. "Core Intervention and Periphery Revolution, 1821–1985." *American Journal of Sociology* 97: 70–95.

Krantz, Susan E. 1992. "The Impact of Divorce on Children." Pp. 242–265 in Arlene S. Skolnick and Jerome H. Skolnick (eds.), *Family in Transition: Rethinking Marriage, Sexuality, Child Rearing, and Family Organization.* New York: Harper Collins.

Kriesi, Hanspeter, Ruud Koopmans, Jan Willem Dyvendak, and Marco G. Giugni. 1995. *New Social Movements in Western Europe: A Comparative Analysis.* Minneapolis: University of Minnesota Press.

Kroeber, Alfred L. 1952. *The Nature of Culture.* Chicago: University of Chicago Press.

Kronenfeld, Jennie J., Kirby L. Jackson, Keith E. Davis, and Steven N. Blair. 1988. "Changing Health Practices: The Experience from a Worksite Health Promotion Project." *Social Science and Medicine* 26: 515–524.

Kronstadt, Diana. 1991. "Complex Developmental Issues of Prenatal Drug Exposure." *The Future of Children* 1(1): 36–49.

Kuhn, M. H., and T. McPartland. 1954. "An Empirical Investigation of Self-Attitudes." *American Sociological Review* 19: 68–76.

Kuhn, Thomas S. 1970. *The Structure of Scientific Revolutions,* 2nd ed., enl. Chicago: University of Chicago Press.

Kulis, Stephen. 1988. "The Representation of Women in Top Ranked Sociology Departments." *American Sociologist* 19: 203–217.

Kulis, Stephen, and Karen A. Miller. 1989. "Are Minority Women Sociologists in Double Jeopardy?" *American Sociologist* 19: 323–339.

Kulis, Stephen, Karen A. Miller, Morris Alexrod, and Leonard Gordon. 1986. "Minorities and Women in the Pacific Sociological Association Region: A Five-Year Progress Report." *Sociological Perspectives* 29: 147–170.

Kumpfer, Karol. 1991. "Treatment Programs for Drug-Abusing Women." *The Future of Children* 1(1): 50–60.

Laba, Richard D., and John R. Logan. 1993. "Minority Proximity to Whites in Suburbs: An Individual-Level Analysis of Segregation." *American Journal of Sociology* 6: 1388–1427.

Lam, Shui Fong. 1997. *How the Family Influences Children's Academic Achievement.* New York: Garland.

Lamanna, Mary Ann, and Agnes Riedmann. 1997. *Marriages and Families: Making Choices and Facing Change,* 5th ed. Belmont, Calif.: Wadsworth.

Landes, David S. 1998. *The Wealth and Poverty of Nations: Why Some are So Rich and Some So Poor.* New York: W.W. Norton.

Landry, Bart. 1987. *The New Black Middle Class.* Berkeley: University of California Press.

Langan, Patrick A., and Christopher A. Innes. 1985. "The Risk of Violent Crime." *Bureau of Justice Statistics Special Report* (May).

Langlois, Simon, with Theodore Caplow, Henri Mendras, and Wolfgang Glatzer. 1994. *Convergence or Divergence? Comparing Recent Social Trends in Industrial Societies.* Montreal: Campus Verlag.

Lasswell, Harold D. 1936. *Politics: Who Gets What, When, How.* New York: Whittlesey.

Laub, John H., and Robert J. Sampson. 1991. "The Sutherland-Glueck Debate: On the Sociology of Criminological Knowledge." *American Journal of Sociology,* 96: 1402–1440.

Lawrence, Paul R., and Jay W. Lorsch. 1992. "Organization-Environment Interface." Pp. 229–233 in Jay M. Shafritz and J. Steven Ott (eds.), *Classics of Organization Theory,* 3rd ed. Belmont, Calif.: Wadsworth. (Orig. pub. 1969.)

Lawton, Millicent. 1994. "Female Seniors in 1992 More Ambitious Than in '72, Study Finds." *Education Week,* 9 March, citing National Center for Education Statistics data.

LeBon, Gustav. 1960. *The Crowd: A Study of the Popular Mind.* New York: Viking (originally published in 1895).

Lechner, Frank J. 1997. "The 'New Paradigm' in the Sociology of Religion: Comment on Warner." *American Journal of Sociology* 103: 182–192.

Lee, Sharon M. 1998. "Asian Americans: Diverse and Growing." *Population Bulletin* 53(2) Washington, D.C.: Population Reference Bureau.

Leeman, Sue. 1998. "Britons Keen on System of Free, Public Health Care." *Eugene Register Guard,* Sunday, July 26, 1998, p. 13A.

Lemert, Edwin M. 1951. *Social Pathology: A Systematic Approach to the Theory of Sociopathic Behavior.* New York: McGraw-Hill.

Lenski, Gerhard. 1966. *Power and Privilege: A Theory of Social Stratification.* New York: McGraw-Hill.

Lenski, Gerhard, and Jean Lenski. 1982. *Human Societies: An Introduction to Macrosociology.* New York: McGraw-Hill.

Lenski, Gerhard, Patrick Nolan, and Jean Lenski. 1995. *Human Societies: An Introduction to Macrosociology,* 7th ed. New York: McGraw Hill.

Leslie, Gerald R. 1967. *The Family in Social Context.* New York: Oxford University Press.

Levine, Donald N. 1991. "Simmel and Parsons Reconsidered." *American Journal of Sociology* 96: 1097–1116.

Levine, Donald N. 1995. *Visions of the Sociological Tradition.* Chicago: University of Chicago Press.

LeVine, Robert, Sarah E. LeVine, Amy Richman, F. Medardo Tapia Uribe, Clara Sunderland Correa, and Patrice M. Miller. 1991. "Women's Schooling and Child Care in the Demographic Transition: A Mexican Case Study." *Population and Development Review* 17: 459–496.

Levy, Frank. 1995. "Incomes and Income Inequality." Pp. 1–57 in Reynolds Farley (ed.), *State of the Union. America in the 1990s.* Vol. 1: *Economic Trends.* New York: Russell Sage.

Lichter, Daniel T., Felicia B. LeClere, and Diane K. McLaughlin. 1991. "Local Marriage Markets and the Marital Behavior of Black and White Women." *American Journal of Sociology* 96: 843–867.

Lieberson, Stanley. 1980. *A Piece of the Pie.* Berkeley: University of California Press.

Lieberson, Stanley, and Mary C. Waters. 1988. *From Many Strands: Ethnic and Racial Groups in Contemporary America.* New York: Russell Sage.

Light, Paul C. 1988. *Baby Boomers.* New York: Norton.

Lincoln, C. Eric, and Lawrence H. Mamiya. 1990. *The Black Church in the African American Experience.* Durham, N.C.: Duke University Press.

Linton, Ralph. 1936. *The Study of Man.* New York: Appleton-Century.

Lipset, Seymour Martin. 1986. *Unions in Transition.* San Francisco: ICS.

Lipset, Seymour Martin. 1989. *Continental Divide: The Values and Institutions of the United States and Canada.* Toronto: C. D. Howe.

Lipset, Seymour Martin. 1994. "The Social Requisites of Democracy Revisited." *American Sociological Review* 59: 1–22.

Lipset, Seymour Martin, Martin Trow, and James Coleman. 1956. *Union Democracy.* New York: Free Press.

Livernash, Robert and Eric Rodenburg. 1998. "Population Change, Resources, and the Environment." *Population Bulletin* 53(1). Washington, D.C.: Population Reference Bureau.

Lockwood, David. 1992. *Solidarity and Schism: "The Problem of Disorder" in Durkheimian and Marxist Sociology.* Oxford: Clarendon.

Loeber, Rolf. 1982. "The Stability of Antisocial Child Behavior: A Review." *Child Development* 53: 431–446.

Loeber, Rolf. 1996. "Developmental Continuity, Change, and Pathways in Male Juvenile Problem Behaviors and Delinquency." Pp. 1–27 in J. David Hawkins (ed.). *Delinquency and Crime: Current Theories.* New York: Cambridge University Press.

Loeber, Rolf, and Magda Stouthamer-Loeber. 1986. "Family Factors as Correlates and Predictors of Juvenile Conduct Problems and Delinquency." Pp. 29–149 in Michael Tonry and Norval Morris (eds.), *Crime and Justice: An Annual Review of Research,* Vol. 7. Chicago: University of Chicago Press.

Lofland, John. 1977. *Doomsday Cult* (enl. ed.). New York: Irvington.

London, Bruce. 1992. "School-Enrollment Rates and Trends, Gender, and Fertility: A Cross-National Analysis." *Sociology of Education* 65: 306–316.

Long, Larry. 1988. *Migration and Residential Mobility in the United States.* New York: Russell Sage.

Lovallo, William R. 1997. *Stress and Health: Biological and Psychological Interactions.* Thousand Oaks, Calif.: Sage.

Lowenthal, Marjorie Fiske, Majda Thurnher, David Chiriboga, and associates. 1975. *Four Stages of Life.* San Francisco: Jossey-Bass.

Lubrano, Annteresa. 1997. *The Telegraph: How Technology Innovation Caused Social Change.* New York: Garland.

Luker, Kristin. 1992. "Dubious Conceptions: The Controversy over Teen Pregnancy." Pp. 160–172 in Arlene S. Skolnick and Jerome H. Skolnick (eds.), *Family in Transition: Rethinking Marriage, Sexuality, Child Rearing, and Family Organization.* New York: Harper Collins.

Lukes, Steven. 1974. *Power: A Radical View.* New York: Macmillan.

Luomala, Katherine. 1956. "Clan Origins." Pp. 199–200 in Irwin T. Sanders, Richard B. Woodbury, Frank J. Essene, Thomas P. Field, Joseph R. Schwendeman, and Charles E. Snow (eds.), *Societies Around the World.* New York: Holt, Rinehart & Winston. (Orig. pub. 1938.)

Luttbeg, Norman R., and Michael M. Gant. 1995. *American Electoral Behavior 1952–1992,* 2nd ed. Itasca, Ill.: F.E. Peacock.

Lutz, Wolfgang. 1994. "The Future of World Population." *Population Bulletin* 49(1). Washington, D.C.: Population Reference Bureau.

Lynn, David B. 1969. *Parental and Sex-Role Identification.* Berkeley, Calif.: McCutchan.

Lynn, David B. 1974. *The Father: His Role in Child Development.* Belmont, Calif.: Wadsworth.

Maccoby, Eleanor E. 1980. *Social Development.* New York: Harcourt Brace Jovanovich.

Maccoby, Eleanor E. 1986. "Social Groupings in Childhood: Their Relationship to Prosocial and Antisocial Behavior in Boys and Girls." In Dan Olweus, Jack Block, and Marian Radke-Yarrow (eds.), *Development of Antisocial and Prosocial Behavior: Research Theories and Issues.* Orlando, Fla.: Academic.

Maccoby, Eleanor E. 1988. "Gender as a Social Category." *Developmental Psychology* 24: 755–765.

Maccoby, Eleanor E. 1990. "Gender and Relationships: A Developmental Account." *American Psychologist* 45: 513–520.

Maccoby, Eleanor E., and Carol N. Jacklin. 1974. *The Psychology of Sex Differences.* Stanford, Calif.: Stanford University Press.

Maccoby, Eleanor E., and Carol N. Jacklin. 1987. "Gender Segregation in Childhood." *Advances in Child Development and Behavior* 20: 239–287.

MacIntosh, Randall. 1998. "Global Attitude Measurement: An Assessment of the World Values Survey Postmaterialism Scale." *American Sociological Review* 63: 452–464.

Mackie, Gerry. 1996. Ending Footbinding and Infibulation: A Convention Account. *American Sociological Review* 61: 999–1017.

Main, M., L. Tomasini, and W. Tolan. 1979. "Differences Among Mothers of Infants Judged to Differ in Security." *Developmental Psychology* 15: 472–473.

Manza, Jeff, and Clem Brooks. 1998. "The Gender Gap in U.S. Presidential Elections: When? Why? Implications?" *American Journal of Sociology* 103: 1235–1266.

Marcelli, Enrico A., and David M. Heer. 1998. "The Unauthorized Mexican Immigrant Population and Welfare in Los Angeles: A Comparative Statistical Analysis." *Sociological Perspectives* 41: 279–302.

March, James G., and Herbert A. Simon. 1958. *Organizations.* New York: Wiley.

Marcos, Anastasios C., Stephen J. Bahr, and Richard E. Johnson. 1986. "Test of a Bonding/Association Theory of Adolescent Drug Use." *Social Forces* 65: 135–161.

Marger, Martin N. 1994. *Race and Ethnic Relations: American and Global Perspectives,* 3rd ed. Belmont, Calif.: Wadsworth.

Marger, Martin N. 1997. *Race and Ethnic Relations: American and Global Perspectives,* 4th ed. Belmont, Calif.: Wadsworth.

Marini, Margaret Mooney, and Pi-Ling Fan. 1997. "The Gender Gap in Earnings at Career Entry." *American Sociological Review* 62: 588–604.

Marks, Nadine F., and Diane S. Shingerg. 1997. "Socioeconomic Differences in Hysterectomy: The Wisconsin Longitudinal Study." *American Journal of Public Health* 87: 1507–1514.

Markus, Helen. 1977. "Self-Schemata and Processing Information About the Self." *Journal of Personality and Social Psychology* 35: 63–78.

Markus, Helen, M. Crane, S. Bernstein, and M. Saladi. 1982. "Self Schemas and Gender." *Journal of Personality and Social Psychology* 42: 38–50.

Marini, Margaret Mooney, and Pi-Ling Fan. 1997. "The Gender Gap in Earnings at Career Entry." *American Sociological Review* 62: 588–604.

Marks, Nadine F., and Diane S. Shingerg. 1997. "Socioeconomic Differences in Hysterectomy: The Wisconsin Longitudinal Study." *American Journal of Public Health* 87: 15071514.

Marmot, Michael, Martin Bobak, and George Davey Smith. 1995. "Explanations for Social Inequalities in Health." Pp. 172–210 in Benjamin C. Amick III, Sol Levine, Alvin R. Tarlov, and Diana Chapman Walsh (eds.), *Society and Health.* New York: Oxford University Press.

Marsden, George M. 1991. *Understanding Fundamentalism and Evangelicalism.* Grand Rapids, Mich.: Eerdmans.

Marsden, Peter V. 1990. "Network Data and Measurement." *Annual Review of Sociology* 16: 435–463.

Martin, Carol Lynn, and Charles S. Halvorson. 1981. "A Schematic Processing Model of Sex Typing and Stereotyping in Children." *Child Development* 52: 1119–1134.

Martin, David. 1990. *Tongues of Fire: The Explosion of Protestantism in Latin America.* Cambridge, Mass.: Blackwell.

Martin, JoAnne. 1992. *Cultures in Organizations: Three Perspectives.* New York: Oxford University Press.

Martin, Philip, and Jonas Widgren. 1996. "International Migration: A Global Challenge." *Population Bulletin* 51(1), Washington, D.C.: Population Reference Bureau.

Marty, Martin E. 1986. *Modern American Religion,* Vol. 1: *The Irony of It All, 1893–1919.* Chicago: University of Chicago Press.

Marx, Anthony W. 1998. *Making Race and Nation: A Comparison of South Africa, the United States, and Brazil.* Cambridge: Cambridge University Press.

Marx, Gary T., and Douglas McAdam. 1994. *Collective Behavior and Social Movements: Process and Structure.* Englewood Cliffs, N.J.: Prentice Hall.

Marx, Karl. 1959. *The Economic and Philosophic Manuscripts of 1844.* Trans. Martin Milligan. Moscow: International. (Orig. pub. 1844.)

Marx, Karl. 1964. *Karl Marx: Early Writing.* Trans. and ed. T. B. Bottomore. New York: McGraw-Hill.

Marx, Karl. 1967. *Capital: A Critique of Political Economy,* Vol. 1. New York: International. (Orig. pub. 1867.)

Marx, Karl, and Friedrich Engels. 1939. *The German Ideology.* New York: International.

Maryanski, Alexandra, and Jonathan H. Turner. 1992. *The Social Cage: Human Nature and the Evolution of Society.* Stanford, Calif.: Stanford University Press.

Maslow, Abraham H. 1943. "A Theory of Motivation." *Psychological Review* 50: 370–396.

Maslow, Abraham H. 1962. *Toward a Psychology of Being.* Princeton, N.J.: Van Nostrand.

Mason, Karen Oppenheim. 1997. "Explaining Fertility Transitions." *Demography* 34: 443–454.

Massey, Douglas S. 1990. "American Apartheid: Segregation and the Making of the Underclass." *American Journal of Sociology* 96: 329–357.

Massey, Douglas S. 1996. "The Age of Extremes: Concentrated Affluence and Poverty in the Twenty-First Century." *Demography* 33: 395–412.

Massey, Douglas S., Gretchen A. Condran, and Nancy A. Denton. 1987. "The Effect of Residential Segregation on Black Social and Economic Well-Being." *Social Forces* 66: 29–57.

Massey, Douglas S., and Nancy A. Denton. 1985. "Spatial Assimilation as a Sociological Outcome." *American Sociological Review* 50: 94–106.

Massey, Douglas S., and Nancy A. Denton. 1993. *American Apartheid: Segregation and the Making of the Underclass.* Cambridge, Mass.: Harvard University Press.

Massey, Douglas, and Kristen Espinosa. 1997. "What's Driving Mexico-U.S. Migration? A Theoretical, Empirical, and Policy Analysis," *American Journal of Sociology* 102: 939–999.

Massey, Douglas S., Andrew B. Gross, and Kukiko Shibuya. 1994. "Migration, Segregation, and the Geographic Concentration of Poverty." *American Sociological Review* 59: 425–445.

Matsueda, Ross L. 1982. "Testing Control Theory and Differential Association." *American Sociological Review* 47: 489–504.

Matsueda, Ross L. 1992. "Reflected Appraisals, Parental Labeling, and Delinquency: Specifying a Symbolic Interactionist Theory." *American Journal of Sociology* 97: 1577–1611.

Mauss, A. L. 1975. *Social Problems as Social Movements.* Philadelphia: Lippincott.

Mayberry, Maralee, J. Gary Knowles, Brian Ray, and Stacey Marlow. 1995. *Home Schooling: Parents as Educators.* Thousand Oaks, Calif.: Corwin/Sage.

Mayer, Susan E. 1997. *What Money Can't Buy: Family Income and Children's Life Chances.* Cambridge, Mass.: Harvard University Press.

Maynard, Douglas. 1978. "Placement of Topic Changes in Conversation." *Semiotica* 30: 263–290.

Maynard, Douglas W., and Marilyn R. Whalen. 1995. "Language, Action, and Social Interaction." Pp. 149–175 in Karen S. Cook, Gary Alan Fine, and James S. House (eds.), *Sociological Perspectives on Social Psychology.* Boston: Allyn & Bacon.

McAdam, Doug. 1982. *Political Process and the Development of Black Insurgency, 1930–1970.* Chicago: University of Chicago Press.

McAdam, Doug, John D. McCarthy, and Mayer N. Zald. (eds.). 1996a. *Comparative Perspectives on Social Movements: Political Opportunities, Mobilizing Structures, and Cultural Framings.* Cambridge: Cambridge University Press.

McAdam, Doug, John D. McCarthy, and Mayer N. Zald. 1996b. "Introduction: Opportunities, Mobilizing Structures, and Framing Processes—Toward a Synthetic, Comparative Perspective on Social Movements." Pp. 1–20 in Doug McAdam, John D. McCarthy, and Mayer N. Zald. (eds.), *Comparative Perspectives on Social Movements: Political Opportunities, Mobilizing Structures, and Cultural Framings.* Cambridge: Cambridge University Press.

McCall, G. J., and J. L. Simmons. 1978. *Identities and Interactions.* New York: Free Press.

McCarthy, John D., and Mark Wolfson. 1996. "Resource Mobilization by Local Social Movement Organizations: Agency, Strategy, and Organiztion in the Movement Against Drinking and Driving." *American Sociological Review* 61: 1070–1088

McCarthy, John D., and Mayer N. Zald. 1973. *The Trend of Social Movements in America: Professionalizaton and Resource Mobilization.* Morristown, N.J.: General Learning Press.

McCarthy, John D., and Mayer N. Zald. 1977. "Resource Mobilization and Social Movements: A Partial Theory." *American Journal of Sociology* 82: 1212–1241.

McCloskey, Michael. 1992. "Twenty Years of Change in the Environmental Movement: An Insider's View." Pp. 77–88 in Riley E. Dunlap and Angela G. Mertig (eds.), *American Environmentalism: The U.S. Environmental Movement, 1970–1990.* Philadelphia: Taylor & Francis.

McCord, Joan. 1979. "Some Child-Rearing Antecedents of Criminal Behavior in Adult Men." *Journal of Personality and Social Psychology* 37: 1477–1486.

McDonald, Gerald W. 1980. "Family Power: The Assessment of a Decade of Theory and Research, 1970–1979." *Journal of Marriage and the Family* 42: 841–852.

McDonough, Peggy, Greg J. Duncan, David Williams, and James House. 1997. "Income Dynamics and Adult Mortality in the United States, 1972 through 1989." *American Journal of Public Health* 87: 1476–1483.

McFalls, Joseph A., Jr. 1998. "Population: A Lively Introduction." *Population Bulletin* 53(3). Washington, D.C.: Population Reference Bureau.

McFate, Katherine, Roger Lawson, and William Julius Wilson. (eds.). 1995. *Poverty, Inequality, and the Future of Social Policy: Western States in the New World Order.* New York: Russell Sage.

McGinnis, J. Michael, and William H. Foege. 1993. "Actual Causes of Death in the United States." *Journal of the American Medical Association* 270: 2207–2212.

McGuire, Kathleen, and Ann L. Pastore. 1997. *Bureau of Justice Statistics Sourcebook of Criminal Justice Statistics, 1996.* Washington, D.C.: U.S. Department of Justice.

McGuire, Meredith B. 1982. *Pentecostal Catholics.* Philadelphia: Temple University Press.

McGuire, W. J., and C. McGuire. 1982. "Significant Others in Self-Space: Sex Differences and Developmental Trends in the Social Self." Pp. 71–96 in J. Suls (ed.), *Psychological Perspectives on the Self,* Vol. 1. Hillsdale, N.J.: Erlbaum.

McKeown, Thomas. 1979. *The Role of Medicine: Dream, Mirage or Nemesis?* Princeton, N.J.: Princeton University Press.

McLanahan, Sara S. 1991. "The Two Faces of Divorce: Women's and Children's Interests." Pp. 193–207 in Joan Huber (ed.), *Macro-Micro Linkages in Sociology.* Newbury Park, Calif.: Sage.

McLanahan, Sara, and Larry Bumpass. 1988. "Intergenerational Consequences of Family Disruption." *American Journal of Sociology* 94: 130–152.

McLanahan, Sara, and Gary Sandefur. 1994. *Growing Up with a Single Parent: What Hurts, What Helps.* Cambridge, Mass.: Harvard University Press.

McNall, Scott G., and Sally A. McNall. 1992. *Sociology.* Englewood Cliffs, N.J.: Prentice-Hall.

McPhail, Clark. 1991. *The Myth of the Madding Crowd.* New York: Aldine.

Mead, George Herbert. 1934. *Mind, Self and Society.* Chicago: University of Chicago Press.

Mead, Margaret. 1949. *Male and Female: A Study of the Sexes in a Changing World.* New York: Dell.

Mead, Margaret. 1974. "On Freud's View of Female Psychology." Pp. 95–106 in Jean Strouse (ed.), *Women and Analysis.* New York: Grossman.

Mechanic, David. 1994. *Inescapable Decisions: The Imperatives of Health Reform.* New Brunswick, N.J.: Transaction.

Melucci, Alberto 1989. *Nomads of the Present: Social Movements and Individual Needs in Contemporary Society.* John Keane and Paul Mier (eds.). Philadelphia: Temple University Press.

Melucci, Alberto. 1996. *Challenging Codes: Collective Action in the Information Age.* Cambridge: Cambridge University Press.

Melton, J. Gordon. 1989. *Encyclopedia of American Religions,* 3rd ed. Detroit: Gale.

Melton, J. Gordon. 1993. "Another Look at New Religions." *Annals of the American Academy of Political and Social Science* 527: 97–112.

Mendell, Dale (ed.). 1982. *Early Female Development: Current Psychoanalytic Views.* New York: Spectrum.

Merton, Robert. 1938. "Social Structure and Anomie." *American Sociological Review* 3: 672–682.

Merton, Robert K. 1949. "Discrimination and the American Creed." Pp. 99–126 in R. H. MacIver (ed.), *Discrimination and National Welfare.* New York: Harper & Row.

Merton, Robert K. 1967. *Social Theory and Social Structure.* New York: Free Press.

Merton, Robert K. 1968. "Manifest and Latent Functions." Pp. 73–138 in *Social Theory and Social Structure.* New York: Free Press. (Orig. pub. 1949.)

Messner, Michael A. 1992. *Power at Play: Sports and the Problem of Masculinity.* Boston: Beacon.

Messner, Steven F., and Richard Rosenfeld. 1994. *Crime and the American Dream.* Belmont, Calif.: Wadsworth.

Messner, Steven F., Marvin D. Krohn, and Allen E. Liska. 1989. *Theoretical Integration in the Study of Deviance and Crime: Problems and Prospects.* Albany: State University of New York Press.

Meyer, John W., John Boli, George M. Thomas, and Francisco O. Ramirez. 1997. "World Society and the Nation-State." *American Journal of Sociology* 103: 144–181.

Meyer, John W., and Michael T. Hannan (eds.). 1979. *National Development and the World System: Educational, Economic, and Political Change, 1950–1970.* Chicago: University of Chicago Press.

Meyer, John W., David H. Kamens, and Aaron Benavot. 1992. *School Knowledge for the Masses: World Models and National Primary Curricular Categories in the Twentieth Century.* Washington, D.C.: Falmer.

Meyer, John W., and Brian Rowan. 1991. "Institutionalized Organizations: Formal Structure as Myth and Ceremony." Pp. 41–62 in Walter W. Powell and Paul J. DiMaggio (eds.), *The New Institutionalism in Organizational Analysis.* Chicago: University of Chicago Press.

Meyer, John W., Richard Rubinson, Francisco O. Ramirez, and John Boli-Bennett. 1977. "The World Educational Revolution, 1950–1970." *Sociology of Education* 50: 242–258.

Meyer, John W., and W. Richard Scott. 1983. *Organizational Environments: Ritual and Rationality.* Beverly Hills, Calif.: Sage.

Meyer, John W., David Tyack, Joane Nagel, and Audri Gordon. 1979. "Public Education as Nation-Building in America: Enrollments and Bureaucratization in the American States, 1870–1930." *American Journal of Sociology* 85: 591–613.

Michels, Robert. 1949. *Political Parties.* New York: Free Press. (Orig. pub. 1915.)

Michener, H. Andrew, John D. DeLamater, and Shalom H. Schwartz. 1990. *Social Psychology,* 2nd ed. San Diego: Harcourt Brace Jovanovich.

Miller, Delbert C. 1991. *Handbook of Research Design and Social Measurement,* 5th ed. Newbury Park, Calif.: Sage.

Miller, Donald E. 1997. *Reinventing American Protestantism: Christianity in the New Millennium.* Berkeley: University of California Press.

Miller, G. Tyler, Jr. 1996. *Living in the Environment: Principles, Connection, and Solutions.* Belmont, Calif.: Wadsworth.

Miller, G. Tyler, Jr. 1997. *Environmental Science: Working with the Earth,* 6th ed. Belmont, Calif.: Wadsworth.

Miller, Norman, and Marilynn B. Brewer. 1984. "The Social Psychology of Desegregation: An Introduction." Pp. 1–8 in Norman Miller and Marilynn B. Brewer (eds.), *Groups in Contact: The Psychology of Desegregation*. Orlando, Fla.: Academic.

Mills, C. Wright. 1956. *The Power Elite*. New York: Oxford University Press.

Mills, C. Wright. 1959. *The Sociological Imagination*. London: Oxford University Press.

Min, Pyong Gap. 1996. *Caught in the Middle: Korean Merchants in America's Multiethnic Cities*. Berkeley: University of California Press.

Minkoff, Debra C. 1997. "The Sequencing of Social Movements." *American Sociological Review* 62: 779–799.

Mintz, Beth. 1975. "The President's Cabinet, 1897–1972." *Insurgent Sociologist* 5: 131–149.

Mintz, Steven, and Susan Kellogg. 1988. *Domestic Revolutions: A Social History of American Family Life*. New York: Free Press.

Misztal, Barbara A. 1996. *Trust in Modern Societies: The Search for the Bases of Social Order*. Cambridge, U.K.: Polity.

Mitchell, Neil J. 1997. *The Conspicuous Corporation: Business, Public Policy, and Representative Democracy*. Ann Arbor: University of Michigan Press.

Mitchell, Robert Cameron, Angela G. Mertig, and Riley E. Dunlap. 1992. "Twenty Years of Environmental Mobilization: Trends Among National Environmental Organizations." Pp. 11–26 in Riley E. Dunlap and Angela G. Mertig (eds.), *American Environmentalism: The U.S. Environmental Movement, 1970–1990*. Philadelphia: Taylor & Francis.

Mizruchi, Mark S. 1992. *The Structure of Corporate Political Action: Interfirm Relations and Their Consequences*. Cambridge, Mass.: Harvard University Press.

Mizruchi, Mark S., and Michael Schwartz. 1987. *Intercorporate Relations: The Structural Analysis of Business*. Cambridge: Cambridge University Press.

Mol, Hans. 1985. "New Perspectives from Cross-Cultural Studies." Pp. 90–103 in Phillip E. Hammond (ed.), *The Sacred in a Secular Age: Toward Revision in the Scientific Study of Religion*. Berkeley: University of California Press.

Molm, Linda D., and Karen S. Cook 1995. "Social Exchange and Exchange Networks." Pp. 209–235 in Karen S. Cook, Gary Alan Fine, and James S. House (eds.), *Sociological Perspectives on Social Psychology*. Boston: Allyn & Bacon.

Monaco, James. 1991. *The Encyclopedia of Film*. New York: Baseline/Perigree.

Monje, Kathleen. 1992. "Rise in Domestic Violence Linked to Sinking Economy." *The Oregonian*, 4 April, p. D1.

Moos, Rudolf H. 1987. "Person-Environment Congruence in Work, School, and Health Care Settings." *Journal of Vocational Behavior* 31: 231–247.

Morales, Rebecca, and Frank Bonilla (eds.). *Latinos in a Changing U.S. Economy: Comparative Perspectives on Growing Inequality*. Newbury Park, Calif.: Sage.

Morgan, Glenn. 1990. *Organizations in Society*. New York: St. Martin's.

Morgan, Laurie A. 1998. "Glass-Ceiling Effect or Cohort Effect? A Longitudinal Study of the Gender Earnings Gap for Engineers, 1982 to 1989." *American Sociological Review* 63: 479–483.

Morgan, Leslie A. 1991. *After Marriage Ends: Economic Consequences for Midlife Women*. Newbury Park, Calif.: Sage.

Morgan, S. Philip, Diane N. Lye, and Gretchen A. Condran. 1988. "Sons, Daughters, and the Risk of Marital Disruption." *American Journal of Sociology* 94: 110–129.

Morganthau, Tom. 1995. "What Color Is Black?" *Newsweek*, 13 February, pp. 63–65.

Morin, Richard. 1996. "Wanna Get Down, Closer to the Ground? Americans Fall to Their Knees More Than Ever." *The Portland Oregonian*, 6 January, p. D1.

Morrill, Calvin. 1991. "Conflict Management, Honor, and Organizational Change." *American Journal of Sociology* 97: 585–621.

Morris, Aldon D., and Carol McClurg Mueller. 1992. *Frontiers in Social Movement Theory*. New Haven, Conn.: Yale University Press.

Moskos, Charles C., and John Sibley Butler. 1996. *All That We Can Be: Black Leadership and Racial Integration the Army Way*. New York: Basic Books

Moss, Gordon E. 1973. *Illness, Immunity, and Social Interaction*. New York: John Wiley.

Moynihan, Daniel P. 1965. *The Negro Family: The Case for National Action*. Washington, D.C.: Office of Planning and Research, U.S. Department of Labor.

Mueller, Dennis C. 1979. *Public Choice*. Cambridge: Cambridge University Press.

Mukhopadhyay, Carol C., and Patricia J. Higgins. 1988. "Anthropological Studies of Women's Status Revisited: 1977–1987." *Annual Review of Anthropology* 17: 461–495.

Murdock, George P. 1949. *Social Structure*. New York: Macmillan.

Murdock, George P. 1968. "World Sampling Provinces." *Ethnology* 7: 304–326.

Murdock, George P., and D. R. White. 1969. "Standard Cross-Cultural Sample." *Ethnology* 8: 329–369.

Myers, Scott M. 1996. "An Interactive Model of Religiosity Inheritance: The Importance of Family Context." *American Sociological Review* 61: 858–866.

Myrdal, Gunnar. 1962. *An American Dilemma: The Negro Problem and Modern Democracy*, 20th anniversary ed. New York: Harper & Row.

Nagel, Joane. 1996. *American Indian Ethnic Renewal: Red Power and the Resurgence of Identity and Culture*. New York: Oxford University Press.

Naisbitt, John. 1982. *Megatrends: Ten New Directions Transforming Our Lives*. New York: Warner.

Nakano, K. 1989. "Intervening Variables of Stress, Hassles, and Health." *Japanese Psychological Research* 31: 143–148.

Nash, Manning. 1989. *The Cauldron of Ethnicity in the Modern World*. Chicago: University of Chicago Press.

National Center for Health Statistics. 1998. *Health, United States, 1998 With Socioeconomic Status and Health Chartbook*. Hyattsville, Md.: National Center for Health Statistics.

National Research Council. 1996. *Measuring Poverty: A New Approach*. Washington, D.C.: National Academy.

Nee, Victor, Jimy M. Sanders, and Scott Sernau. 1994. "Job Transitions in an Immigrant Metropolis: Ethnic Boundaries and the Mixed Economy." *American Sociological Review* 59: 849–871.

Neilsen, Francois, and Arthur S. Alderson. 1997. "The Kuznets Curve and the Great U-Turn: Income Inequality in U.S. Counties, 1970 to 1990." *American Sociological Review* 62: 12–33.

Neitz, Mary J. 1987. *Charisma and Community: A Study of Religious Commitment Within the Charismatic Renewal*. New Brunswick, N.J.: Transaction.

Newman, Katherine S. 1988. *Falling from Grace: The Experience of Downward Mobility in the American Middle Class*. New York: Free Press.

Newman, Katherine S. 1993. *Declining Fortunes: The Withering of the American Dream*. New York: Basic Books.

Newport, Frank, and Lydia Saad. 1997. "Religious Faith is Widespread, But Many Skip Church: Little Change in Recent Years." *Gallup Poll Archives*. Princeton, N.J.: Gallup. (http://198.175.140.8/poll%5Farchives/1997/970329.htm).

Nie, Norman H., Jane Junn, and Kenneth Stehlik-Barry. 1996. *Education and Democratic Citizenship in America*. Chicago: University of Chicago Press.

Niemi, Richard G., and Herbert F. Weisberg. 1992. *Classics in Voting Behavior*. Washington, D.C.: Congressional Quarterly Press.

Oberschall, Anthony. 1993. *Social Movements: Ideologies, Interests, and Identities*. New Brunswick, N.J.: Transaction.

O'Brien, Robert M. 1985. *Crime and Victimization Data*. Beverly Hills, Calif.: Sage.

O'Brien, Robert M. 1989. "Relative Cohort Size and Age-Specific Crime Rates: An Age-Period-Relative-Cohort-Size Model." *Criminology* 27: 57–78.

O'Brien, Robert M. 1990. "Comparing Detrended UCR and NCS Crime Rates over Time: 1973–1986." *Journal of Criminal Justice* 18: 229–238.

O'Brien, Robert M. 1991. "Detrended UCR and NCS Crime Rates: Their Utility and Meaning." *Journal of Criminal Justice* 19: 569–574.

O'Brien, Robert M., Jean Stockard, and Lynne Isaacson. 1999. "The Enduring Effects of Cohort Size and Percentage of Nonmarital Births on Age-Specific Homicide Rates, 1960–1995." *American Journal of Sociology,* 104: 1061–1095.

Ogburn, William F. 1964. *On Culture and Social Change*. Chicago: University of Chicago Press.

O'Hearn, Denis. 1989. "The Irish Case of Dependency: An Exception to the Exception?" *American Sociological Review* 54: 578–596.

Olsen, Marvin E., Dora G. Lodwick, and Riley E. Dunlap. 1991. *Viewing the World Ecologically*. Boulder, Colo.: Westview.

Oliver, Melvin L., and Thomas M. Shapiro. 1995. *Black Wealth/White Wealth: A New Perspective on Racial Inequality*. New York: Routledge.

Olshansky, S. Jay, Bruce Carnes, Richard G. Rogers, and Len Smith. 1997. "Infectious Diseases—New and Ancient Threats to World Health." *Population Bulletin* 52(2). Washington, D.C.: Population Reference Bureau.

Olson, Daniel V. A. 1998. "Religious Pluralism in Contemporary U.S. Counties." *American Sociological Review* 63: 759–761.

Olweus, Daniel. 1979. "Stability of Aggressive Reaction Patterns in Males: A Review." *Psychological Bulletin* 86: 852–875.

Omi, Michael, and Howard Winant. 1986. *Racial Formation in the United States: From the 1960s to the 1980s*. New York: Routledge.

O'Neill, June. 1992. "The Changing Economic Status of Black Americans." *The American Enterprise,* September/October, pp. 71–79.

Oppenheimer, Valerie Kincade. 1997. "Women's Employment and the Gain to Marriage: The Specialization and Trading Model." *Annual Review of Sociology* 23: 431–453.

Ornati, Oscar. 1966. *Poverty Amidst Affluence*. New York: Twentieth Century.

Orum, Anthony M. 1988. "Political Sociology." Pp. 393–423 in Neil J. Smelser (ed.), *Handbook of Sociology*. Newbury Park, Calif.: Sage.

Orum, Anthony M. 1995. *City-Building in America*. Boulder, Colo.: Westview.

Osgood, D. Wayne, Janet K. Wilson, Patrick M. O'Malley, Jerald G. Bachman, and Lloyd D. Johnston. 1996. "Routine Activities and Individual Deviant Behavior." *American Sociological Review* 61: 635–655.

Ott, J. Steven. 1989a. *Classic Readings in Organizational Behavior*. Pacific Grove, Calif.: Brooks/Cole.

Ott, J. Steven. 1989b. *The Organizational Culture Perspective*. Pacific Grove, Calif.: Brooks/Cole.

Page, Benjamin I. 1996. *Who Deliberates? Mass Media in Modern Democracy*. Chicago: University of Chicago Press.

Pagnini, Deanna L., and S. Phillip Morgan. 1990. "Intermarriage and Social Distance Among U.S. Immigrants at the Turn of the Century." *American Journal of Sociology* 96: 405–432.

Palen, J. John. 1995. *The Suburbs*. New York: McGraw-Hill.

Park, Robert E. 1974a. "An Autobiographical Note." Pp. v–ix in Everett Cherrington Hughes, Charles S. Johnson, Jitsuichi Masuoka, Robert Redfield, and Louis Wirth (eds.), *The Collected Papers of Robert Ezra Park,* Vol. 1. New York: Arno.

Park, Robert E. 1974b. "The Etiquette of Race Relations in the South." Pp. 177–188 in Everett Cherrington Hughes, Charles S. Johnson, Jitsuichi Masuoka, Robert Redfield, and Louis Wirth (eds.), *The Collected Papers of Robert Ezra Park,* Vol. 1. New York: Arno.

Park, Robert E. 1974c. "Our Racial Frontier on the Pacific." Pp. 138–151 in Everett Cherrington Hughes, Charles S. Johnson, Jitsuichi Masuoka, Robert Redfield, and Louis Wirth (eds.), *The Collected Papers of Robert Ezra Park,* Vol. 1. New York: Arno.

Park, Robert E. 1974d. "Racial Assimilation in Secondary Groups: With Particular Reference to the Negro." Pp. 204–220 in Everett Cherrington Hughes, Charles S. Johnson, Jitsuichi Masuoka, Robert Redfield, and Louis Wirth (eds.), *The Collected Papers of Robert Ezra Park,* Vol. 1. New York: Arno.

Park, Robert E., Ernest W. Burgess, and Roderick D. McKenzie. 1967. *The City.* Chicago: University of Chicago Press. (Orig. pub. 1925.)

Parsons, Talcott. 1937. *The Structure of Social Action.* New York: McGraw-Hill.

Parsons, Talcott. 1940. "An Analytic Approach to the Theory of Social Stratification." *American Journal of Sociology* 45: 841–862.

Parsons, Talcott. 1951. *The Social System.* Glencoe, Ill.: Free Press.

Parsons, Talcott. 1954. "Certain Primary Sources and Patterns of Aggression in the Social Structure of the Western World." Pp. 298–322 in *Essays in Sociological Theory,* rev. ed. Glencoe, Ill.: Free Press.

Parsons, Talcott. 1959. "The School Class as a Social System." *Harvard Educational Review* 29: 297–308.

Parsons, Talcott. 1966. *Societies: Evolutionary and Comparative Perspectives.* Englewood Cliffs, N.J.: Prentice-Hall.

Parsons, Talcott, Robert F. Bales, and Edward A. Shils. 1953. *Working Papers in the Theory of Action.* Glencoe, Ill.: Free Press.

Pasley, K., and Marilyn Ihinger-Tallman. 1982. "Remarried Family Life: Supports and Constraints." Pp. 367–384 in G. Rose (ed.), *Building Family Strengths,* Vol. 4. Lincoln: University of Nebraska Press.

Passas, Nikos. 1997. "Anomie, Reference Groups, and Relative Deprivation." Pp. 62–94 in Nikos Passas and Robert Agnew (eds.), *The Future of Anomie Theory.* Boston: Northeastern University Press.

Passas, Nikos, and Robert Agnew (eds.). 1997. *The Future of Anomie Theory.* Boston: Northeastern University Press.

Passell, Peter. 1998. "Unskilled Workers Miss Out on More than High Wages." *The Sunday Oregonian,* June 14, 1998, p. A16.

Pate, Antony M., and Edwin E. Hamilton. 1992. "Formal and Informal Deterrents to Domestic Violence: The Dade County Spouse Assault Experiment." *American Sociological Review* 57: 691–697.

Patten, Simon N. 1895. "Discussion of Albion Small, 'The Relation of Sociology to Economics.'" *Publications of the American Economic Association* 10, Supplement (March): 107–110.

Pearlin, Leonard I. 1954. "Shifting Group Attachments and Attitudes Toward Negroes." *Social Forces* 33: 47–50.

Pedersen, Eigel, and Thérèse Annette Faucher, with William W. Eaton. 1978. "A New Perspective on the Effects of First-Grade Teachers on Children's Subsequent Adult Status." *Harvard Educational Review* 48: 1–31.

Perrow, Charles. 1986. *Complete Organizations: A Critical Essay.* New York: McGraw-Hill.

Perrucci, Robert, and Harry R. Potter (eds.). 1989. *Networks of Power: Organizational Actors at the National, Corporate and Community Levels.* New York: Aldine.

Peters, Charles. 1993. "From Quogadougou to Cape Canaveral: Why the Bad News Doesn't Travel Up." Pp. 173–177 in Kurt Finsterbusch (ed.), *Annual Editions: Sociology 93/94.* Guilford, Conn.: Dushkin.

Petersen, William. 1975. *Population,* 3rd ed. New York: Macmillan.

Petersen, Trond, and Laurie A. Morgan. 1995. "Separate and Unequal: Occupation-Establishment Sex Segregation and the Gender Wage Gap." *American Journal of Sociology* 101: 329–365.

Peterson, Richard A. 1979. "Revitalizing the Culture Concept." *Annual Review of Sociology* 5: 137–166.

Peterson, Richard R. 1989. *Women, Work, and Divorce.* Albany: State University of New York Press.

Peterson, Steven A. 1990. *Political Behavior: Patterns in Everyday Life.* Newbury Park, Calif.: Sage.

Pettigrew, Thomas F. 1964. *A Profile of the Negro American.* Princeton, N.J.: Van Nostrand.

Pfaff, William. 1995. "French Distrust U.S. Capitalism." *Eugene Register Guard,* 15 December, p. A17.

Phillips, Kevin. 1994. *Arrogant Capital: Washington, Wall Street, and the Frustration of American Politics.* Boston: Little, Brown.

Phillips, Roderick. 1988. *Putting Asunder: A History of Divorce in Western Society.* Cambridge: Cambridge University Press.

Piaget, Jean. 1954. *The Construction of Reality in the Child.* New York: Basic.

Piaget, Jean. 1965. *The Moral Judgment of the Child.* New York: Free Press.

Pichardo, Nelson A. 1997. "New Social Movements: A Critical Review." *Annual Review of Sociology* 23: 411–430.

Piore, Michael J. 1971. "The Dual Labor Market: Theory and Implications." Pp. 90–94 in D. M. Gordon (ed.), *Problems in Political Economy: An Urban Perspective.* Lexington, Mass.: Heath.

Pleck, Joseph H. 1985. *Working Wives/Working Husbands.* Beverly Hills, Calif.: Sage.

Poloma, Margaret M. 1989. *The Assemblies of God at the Crossroads: Charisma and Institutional Dilemmas.* Knoxville: University of Tennessee Press.

Popenoe, David. 1993a. "American Family Decline, 1960–1990: A Review and Appraisal." *Journal of Marriage and the Family* 55: 527–542.

Popenoe, David. 1993b. "The National Family Wars." *Journal of Marriage and the Family* 55: 553–555.

Population Reference Bureau. 1994. "OMB Reviews Racial Categories." *Population Today: News, Numbers, and Analysis* 22(7/8): 8.

Population Reference Bureau. 1995. *1995 World Population Data Sheet.* Washington, D.C.: World Population Bureau.

Population Reference Bureau. 1998. *1998 World Population Data Sheet.* Washington, D.C.: Population Reference Bureau.

Porter, Natalie, and Florence Geis. 1981. "Women and Nonverbal Leadership Cues: When Seeing Is Not Believing." Pp. 39–61 in Clara Mayo and Nancy M. Henley (eds.), *Gender and Nonverbal Behavior.* New York: Springer-Verlag.

Portes, Alejandro. 1981. "Modes of Structural Incorporation and Present Theories of Immigration." Pp. 279–297 in Mary M. Kritz, Charles B. Keely, and Sylvano M. Tomasi (eds.), *Global Trends in Migration.* Staten Island, N.Y.: CMS.

Portes, Alejandro (ed.). 1995. *The Economic Sociology of Immigration. Essays on Networks, Ethnicity, and Entrepreneurship.* New York: Russell Sage.

Portes, Alejandro, and Rubèn G. Rumbaut. 1990. *Immigrant America: A Portrait.* Berkeley: University of California Press.

Poston, Dudley L., Jr., and David Yaukey. 1992. *The Population of Modern China.* New York: Plenum.

Potuchek, Jean L. 1997. *Who Supports the Family? Gender and Breadwinning in Dual-Earner Marriages.* Stanford, Calif.: Stanford University Press.

Powell, Walter W., and Paul J. DiMaggio. 1991. *The New Institutionalism in Organizational Analysis.* Chicago: University of Chicago Press.

Power, Chris, Clyde Hertzman, Sharon Matthews, and Orly Manor. 1997. "Social Differences in Health: Life-Cycle Effects Between Ages 23 and 33 in the 1958 British Birth Cohort." *American Journal of Public Health* 87: 1499–1503.

Powers, Daniel A. 1994. "Transitions into Idleness Among White, Black, and Hispanic Youth: Some Determinants and Policy Implications of Weak Labor Force Attachment." *Sociological Perspectives* 37: 183–201.

Pugh, D. S., D. J. Hickson, C. R. Hinings, and C. Turner. 1968. "Dimensions of Organization Structure." *Administrative Science Quarterly* 13: 65–104.

Quah, Stella R. 1989. "The Social Position and Internal Organization of the Medical Profession in the Third World: The Case of Singapore." *Journal of Health and Social Behavior* 30: 450–466.

Quillian, Lincoln. 1995. "Prejudice as a Response to Perceived Group Threat: Population Composition and Anti-Immigrant and Racial Prejudice in Europe." *American Sociological Review* 60: 586–611.

Quillian, Lincoln. 1996. "Group Threat and Regional Change in Attitudes toward African-Americans." *American Journal of Sociology* 102: 816–860.

Quinn, Naomi. 1977. "Anthropological Studies on Women's Status." *Annual Review of Anthropology* 6: 181–225.

Rainwater, Lee. 1974. *What Money Buys.* New York: Basic.

Ramirez, Francisco O., and John Boli. 1987. "The Political Construction of Mass Schooling: European Origins and Worldwide Institutionalization." *Sociology of Education* 60: 2–17.

Ramirez, Francisco O., and John Boli-Bennett. 1989. "Global Patterns of Educational Institutions." Pp. 479–486 in Jeanne H. Ballantine (ed.), *Schools and Society: A Unified Reader,* 2nd ed. Mountain View, Calif.: Mayfield.

Ranger-Moore, James. 1997. "Bigger May be Better, but is Older Wiser? Organizational Age and Size in the New York Life Insurance Industry." *American Sociological Review* 62: 903–920.

Ravitch, Diane. 1985. *The Schools We Deserve: Reflections on the Educational Crises of Our Times.* New York: Basic.

Reed, John, and Roberto R. Ramirez. 1998. "The Hispanic Population in the United States: March 1997 (Update)." *Current Population Reports, Population Characteristics,* P20-511. Washington, D.C.: Census Bureau, U.S. Department of Commerce.

Reeves, Richard. 1993. *President Kennedy: Profile of Power.* New York: Touchstone.

Reichard, Gladys A. 1928. "Social Life of the Navajo Indians with Some Attention to Minor Ceremonies." *Columbia University Contributions to Anthropology* 7.

Reiss, Ira L. 1976. *Family Systems in America,* 2nd ed. Hinsdale, Ill.: Dryden.

Reskin, Barbara, and Irene Padavic. 1994. *Women and Men at Work.* Thousand Oaks, Calif.: Pine Forge.

Reskin, Barbara F., and Patricia A. Roos. 1990. *Job Queues, Gender Queues: Explaining Women's Inroads into Male Occupations.* Philadelphia: Temple University Press.

Reynolds, John R. 1997. "The Effects of Industrial Employment Conditions on Job-Related Distress." *Journal of Health and Social Behavior* 38: 105–116.

Rheingold, Howard. 1993. *The Virtual Community: Homesteading on the Electronic Frontier.* Reading, Mass.: Addison-Wesley .

Rice, Phillip L. 1992. *Stress and Health,* 2nd ed. Pacific Grove, Calif.: Brooks/Cole.

Richardson, James T. 1985. "Studies of Conversion: Secularization or Re-enchantment?" Pp. 104–121 in Phillip E. Hammond (ed.), *The Sacred in a Secular Age: Toward Revision in the Scientific Study of Religion.* Berkeley: University of California Press.

Richardson, John G. 1980. "Variation in Date of Enactment of Compulsory School Attendance Laws: An Empirical Inquiry." *Sociology of Education* 53: 153–163.

Ridgeway, Cecilia L. 1997. "Interaction and the Conservation of Gender Inequality: Considering Employment." *American Sociological Review* 62: 218–235.

Ridgeway, Cecelia L., and Henry A. Walker. 1995. "Status Structures." Pp. 281–310 in Karen S. Cook, Gary Alan Fine, and James S. House (eds.), *Sociological Perspectives on Social Psychology.* Boston: Allyn & Bacon.

Riesman, David. 1961. *The Lonely Crowd.* New Haven, Conn.: Yale University Press.

Riley, Glenda. 1991. *Divorce: An American Tradition.* New York: Oxford University Press.

Riley, Matilda White. 1992. "The Family in an Aging Society: A Matrix of Latent Relationships." Pp. 524–535 in Arlene S. Skolnick and Jerome H. Skolnick (eds.), *Family in Transition: Rethinking Marriage, Sexuality, Child Rearing, and Family Organization.* New York: Harper Collins.

Riley, Nancy E. 1997. "Gender, Power, and Population Change." *Population Bulletin* 52(1). Washington, D.C.: Population Reference Bureau.

Risman, Barbara J. 1998. *Gender Vertigo: American Families in Transition.* New Haven: Yale University Press.

Ritzer, George. 1981. *Toward an Integrated Sociological Paradigm: The Search for an Exemplar and an Image of the Subject Matter.* Boston: Allyn & Bacon.

Ritzer, George. 1988. *Contemporary Sociological Theory,* 2nd ed. New York: Knopf.

Roberts, Keith A. 1995. *Religion in Sociological Perspective,* 3rd ed. Belmont, Calif.: Wadsworth.

Rockett, Ian R. H. 1994. "Population and Health: An Introduction to Epidemiology." *Population Bulletin* 49(3). Washington, D.C.: Population Reference Bureau.

Rodgers, Roy H. 1973. *Family Interaction and Transaction: The Developmental Approach.* Englewood Cliffs, N.J.: Prentice-Hall.

Rodgers, Roy H., and James M. White. 1993. "Family Development Theory." Pp. 225–254 in Pauline G. Boss, William J. Doherty, Ralph LaRossa, Walter R. Schumm, and Suzanne K. Steinmetz (eds.), *Sourcebook of Family Theories and Methods: A Contextual Approach.* New York: Plenum.

Rokeach, Milton. 1973. *The Nature of Human Values.* New York: Free Press.

Roof, Wade Clark. 1985. "The Study of Social Change in Religion." Pp. 75–89 in Phillip E. Hammond (ed.), *The Sacred in a Secular Age: Toward Revision in the Scientific Study of Religion.* Berkeley: University of California Press.

Roof, Wade Clark. 1993. "Toward the Year 2000: Reconstructions of Religious Space." *Annals, AAPSS* 527: 155–170.

Roof, Wade Clark, and William McKinney. 1987. *American Mainline Religion: Its Changing Shape and Future.* New Brunswick, N.J.: Rutgers University Press.

Root, Maria P. P. (ed.). 1996. *The Multiracial Experience: Racial Borders as the New Frontier.* Thousand Oaks, Calif.: Sage.

Rosa, Eugene A., and Riley E. Dunlap. 1994. "The Polls—Poll Trends: Nuclear Power: Three Decades of Public Opinion." *Public Opinion Quarterly* 58: 295–325.

Rosa, Eugene A., and William R. Freudenburg. 1993. "The Historical Development of Public Reactions to Nuclear Power: Implications for Nuclear Waste Policy." Pp. 32–63 in Riley E. Dunlap, Michael E. Kraft, and Eugene A. Rosa (eds.), *Public Reactions to Nuclear Waste: Citizens' Views of Repository Siting.* Durham, N.C.: Duke University Press.

Rose, Arnold M. 1967. *The Power Structure.* New York: Oxford University Press.

Rose, Richard (ed.). 1974. *Electoral Behavior: A Comparative Handbook.* New York: Free Press.

Rosecrance, John. 1998. "The Stooper: A Professional Thief in the Sutherland Manner." Pp. 217–227 in Stuart Henry and Werner Einstadter (eds.), *The Criminology Theory Reader.* New York: New York University Press.

Rosen, Ellen Israel. 1987. *Bitter Choices: Blue-Collar Women in and out of Work.* Chicago: University of Chicago Press.

Rosenblatt, Robert A. 1994. "Working Poor on the Rise." *The Oregonian,* 31 March, pp. D1, D3.

Rosenblum, Ava. 1998. *Connecting On Line: An Ethnomethodologically Informed Study of Communication Between Couples on the Internet.* Unpublished Ph.D. Dissertation, University of Oregon, Summer, 1998.

Rosenstiel, Thomas, with Rich Thomas. "Senior Power Rides Again." *Newsweek,* 20 February, p. 31.

Rosenstone, Steven J., and Raymond E. Wolfinger. 1984. "The Effect of Registration Laws on Voter Turnout." Pp. 54–86 in Richard G. Niemi and Herbert F. Weisberg (eds.), *Controversies in Voting Behavior,* 2nd ed. Washington, D.C.: Congressional Quarterly.

Rosner, David. (ed.). 1995. *Hives of Sickness: Public Health and Epidemics in New York City.* New Brunswick, N.J.: Rutgers University Press.

Ross, Catherine E., and Marieke Van Willigen. 1997. "Education and the Subjective Quality of Life." *Journal of Health and Social Behavior* 38: 275–297.

Ross, Dorothy. 1991. *The Origins of American Social Science.* Cambridge: Cambridge University Press.

Rossi, Alice S. 1977. "A Biosocial Perspective on Parenting." *Daedalus* 106: 1–31.

Rossi, Alice S., and Peter H. Rossi. 1990. *Of Human Bonding: Parent-Child Relations Across the Life Course.* New York: Aldine.

Rostow, Walt Whitman. 1952. *The Process of Economic Growth.* New York: Norton.

Rostow, Walt Whitman. 1980. *The Stages of Economic Growth: A Noncommunist Manifesto,* 3rd ed. New York: Cambridge University Press.

Rothenberg, Lawrence S. 1992. *Linking Citizens to Government: Interest Group Politics at Common Cause.* Cambridge: Cambridge University Press.

Roy, William G. 1992. "Money Talks . . . to Whom?" *Contemporary Sociology* 21: 762–763.

Roy, William G. 1997. *Socializing Capital: The Rise of the Large Industrial Corporation in America.* Princeton, N.J.: Princeton University Press.

Royce, Anya Peterson. 1982. *Ethnic Identity: Strategies of Diversity.* Bloomington: Indiana University Press.

Ruggles, Patricia. 1990. *Drawing the Line: Alternative Poverty Measures and Their Implications for Public Policy.* Washington, D.C.: Urban Institute.

Ruggles, Steven. 1994. "The Origins of African-American Family Structure." *American Sociological Review* 59: 136–151.

Ruggles, Steven. 1997. "The Rise of Divorce and Separation in the United States: 1880–1990." *Demography* 34: 455–466.

Rumbaut, Ruben G. 1996. "The Crucible Within: Ethnic Identity, Self Esteem, and Segmented Assimilation Among Children of Immigrants." Pp. 119–170 in Alejandro Portes (ed.), *The New Second Generation.* New York: Russell Sage.

Rupp, Leila J., and Verta Taylor. 1987. *Survival in the Doldrums: The American Woman's Rights Movement, 1945 to the 1960s.* New York: Oxford University Press.

Sabel, Charles F. 1993. "Studied Trust: Building New Forms of Cooperation in a Volatile Economy." Pp. 104–144 in Richard Swedberg (ed.), *Explorations in Economic Sociology.* New York: Russell Sage.

Samovar, Larry A., and Richard E. Porter. 1991. *Communication Between Cultures.* Belmont, Calif.: Wadsworth.

Sampson, Robert J. 1988. "Local Friendship Ties and Community Attachment in Mass Society: A Multilevel Systemic Model." *American Sociological Review* 53: 766–779.

Sampson, Robert J., and John H. Laub. 1990. "Crime and Deviance over the Life Course: The Salience of Adult Social Bonds." *American Sociological Review* 55: 609–627.

Sampson, Robert J., and John H. Laub. 1992. "Crime and Deviance in the Life Course." *Annual Review of Sociology* 18: 63–84.

Sampson, Robert J., and John H. Laub. 1993. *Crime in the Making: Pathways and Turning Points Through Life.* Cambridge, Mass.: Harvard University Press.

Sampson, Robert J., and John H. Laub. 1997. "A Life-Course Theory of Cumulative Disadvantage and the Stability of Delinquency." Pp. 133–162 in Terence P. Thornberry (ed.), *Developmental Theories of Crime and Delinquency, Advances in Criminological Theory,* Vol. 7. New Brunswick, N.J.: Transaction.

Sanders, Irwin T., Richard B. Woodbury, Frank J. Essene, Thomas P. Field, Joseph R. Schwendeman, and Charles E. Snow (eds.). 1956. *Societies Around the World.* New York: Holt, Rinehart & Winston.

Sanders, Jimy M. and Victor Nee. 1996. "Immigrant Self-Employment: The Family as Social Capital and the Value of Human Capital." *American Sociological Review* 61: 231–249.

Saxton, Lloyd. 1993. *The Individual, Marriage, and the Family,* 8th ed. Belmont, Calif.: Wadsworth.

Schaeffer, Robert K. 1997. *Power to the People: Democratization Around the World.* Boulder, Colo.: Westview.

Schegloff, Emanuel A. 1968. "Sequencing in Conversational Openings." *American Anthropologist* 70: 1075–1095.

Schegloff, Emanuel A. 1992. "Repair After Next Turn: The Last Structurally Provided Defense of Intersubjectivity in Conversation." *American Journal of Sociology* 97: 1295–1345.

Schegloff, Emanuel A. 1996a. "Confirming Allusions: Toward an Empirical Account of Action." *American Journal of Sociology* 102: 161–216.

Schegloff, Emanuel A. 1996b. "Turn Organization: One Intersection of Grammar and Interaction." Pp. 52–133 in Elinor Ochs, Emanuel A. Schegloff, and Sandra A. Thompson (eds.), *Interaction and Grammar.* Cambridge: Cambridge University Press.

Schein, Edgar H. 1990. "Organizational Culture." *American Psychologist* 45: 109–119.

Schneider, Barbara, and James S. Coleman (eds.). 1993. *Parents, Their Children, and Schools.* Boulder, Colo.: Westview.

Schoen, Robert. 1995. "The Widening Gap Between Black and White Marriage Rates: Context and Implications." Pp. 103–116 in M. Belinda Tucker and Claudia Mitchell-Kernan (eds.), *The Decline in Marriage Among African Americans: Causes, Consequences, and Policy Implications.* New York: Russell Sage.

Schulte, Brigid. 1994. "First Woman Marine Combat Pilot Bucks Odds, Critics to Achieve Dream." *The Sunday Oregonian,* 28 August, p. A24.

Schuman, Howard, and Lawrence Bobo. 1988. "Survey-Based Experiments on White Racial Attitudes Toward Residential Integration." *American Journal of Sociology* 94: 273–299.

Schuman, Howard, Charlotte Steeh, and Lawrence Bobo. 1985. *Racial Attitudes in America: Trends and Interpretations.* Cambridge, Mass.: Harvard University Press.

Schwarz, John E. 1997. *Illusions of Opportunity: The American Dream in Question.* New York: W.W. Norton.

Scott, Alan J., and Edward W. Soja (eds.). 1996. *The City: Los Angeles and Urban Theory at the End of the Twentieth Century.* Berkeley: University of California Press.

Scott, John. 1996. *Stratification and Power: Structures of Class, Status and Command.* Cambridge, U.K.: Polity.

Scott, W. Richard. 1987. *Organizations: Rational, Natural, and Open Systems,* 2nd ed. Englewood Cliffs, N.J.: Prentice-Hall.

Scott, W. Richard. 1995. *Institutions and Organizations.* Thousand Oaks, Calif.: Sage.

Scott, W. Richard, and Søren Christensen (eds.). 1995. *The Institutional Construction of Organizations: International and Longitudinal Studies.* Thousand Oaks, Calif.: Sage.

Scott, W. Richard, John W. Meyer, and Associates. 1994. *Institutional Environments and Organizations: Structural Complexity and Individualism.* Thousand Oaks, Calif.: Sage.

Seltzer, Judith A. 1991. "Legal Custody Arrangements and Children's Economic Welfare." *American Journal of Sociology* 96: 895–929.

Seltzer, Richard A., Jody Newman, and Melissa Voorhees Leighton. 1997. *Sex as a Political Variable: Women as Candidates and Voters in U.S. Elections.* Boulder, Colo.: Lynne Rienner.

Serow, William J., Charles B. Nam, David F. Sly, and Robert H. Weller. 1990. *Handbook on International Migration.* New York: Greenwood.

Settersten, Richard A., Jr., and Karl Ulrich Mayer. 1997. "The Measurement of Age, Age Structuring, and the Life Course." *Annual Review of Sociology* 23: 233–261.

Sewell, William H., Jr. 1992. "A Theory of Structure: Duality, Agency and Transformation." *American Journal of Sociology* 98: 1–29.

Sewell, W. H., and R. M. Hauser. 1975. *Education, Occupation and Earnings: Achievement in the Early Career.* New York: Academic.

Sewell, W. H., and R. M. Hauser. 1980. "The Wisconsin Longitudinal Study of Social and Psychological Factors in Aspirations and Achievement." Pp. 59–101 in Alan Kerckhoff (ed.), *Research in Sociology of Education and Socialization.* Greenwich, Conn.: JAI.

Sewell, W. H., R. M. Hauser, and D. L. Featherman (eds.). 1976. *Schooling and Achievement.* New York: Academic.

Seyditlz, Ruth, and Pamela Jenkins. 1998. "The Influence of Families, Friends, Schools, and Community on Delinquent Behavior." Pp. 53–97 in Thomas P. Gullotta, Gerald R. Adams, and Raymond Montemayor (eds.), *Delinquent Violent Youth: Theory and Interventions.* Thousand Oaks, Calif.: Sage.

Shavit, Yossi, and Hans-Peter Blossfeld (eds.). 1993. *Persistent Inequality: Changing Educational Attainment in Thirteen Countries.* Boulder, Colo.: Westview.

Shaw, Clifford R., and Henry D. McKay. 1942. *Juvenile Delinquency in Urban Areas.* Chicago: University of Chicago Press.

Sherbinin, Alex de, and Susan Kalish. 1994. "Population-Environment Links: Crucial, but Unwieldy." *Population Today* 22 (January): 1–2.

Sherman, Lawrence W. 1992. *Policing Domestic Violence: Experiments and Dilemmas.* New York: Free Press.

Sherman, Lawrence W., and Richard A. Berk. 1984. "The Specific Deterrent Effects of Arrest for Domestic Assault." *American Sociological Review* 49: 261–272.

Sherman, Lawrence W., and Ellen G. Cohn. 1989. "The Impact of Research on Legal Policy: The Minneapolis Domestic Violence Experiment." *Law and Society Review* 23: 117–144.

Sherman, Lawrence W., Janell D. Schmidt, Dennis P. Rogan, Patrick R. Gartin, Ellen G. Cohn, Dean J. Collins, and Anthony R. Bacich. 1991. "From Initial Deterrence to Long-Term Escalation: Short-Custody Arrest for Poverty Ghetto Domestic Violence." *Criminology* 29: 821–849.

Sherman, Lawrence W., and Douglas A. Smith, with Janell D. Schmidt, and Dennis P. Rogan. 1992. "Crime, Punishment, and Stake in Conformity: Legal and Informal Control of Domestic Violence." *American Sociological Review* 57: 680–690.

Shibley, Mark A. 1996. *Resurgent Evangelicalism in the United States: Mapping Cultural Change since 1970.* Columbia: University of South Carolina Press.

Shin, Gi-Wook. 1994. "The Historical Making of Collective Action: The Korean Peasant Uprisings of 1946." *American Journal of Sociology* 99: 1596–1624.

Short, James F., Jr. 1997. *Poverty, Ethnicity, and Violent Crime.* Boulder, Colo.: Westview.

Shulman, Steven, and William Darity, Jr. (eds.). 1989. *The Question of Discrimination: Racial Inequality in the U.S. Labor Market.* Middletown, Conn.: Wesleyan University Press.

Sills, David. 1958. *The Volunteers.* Glencoe, Ill.: Free Press.

Simmel, Georg. 1950a. "The Metropolis and Mental Life." Pp. 409–424 in Kurt Wolff (trans. and ed.), *The Sociology of Georg Simmel.* New York: Free Press.

Simmel, Georg. 1950b. *The Sociology of Georg Simmel.* Trans. and ed. Kurt H. Wolff. Glencoe, Ill.: Free Press.

Simmel, Georg. 1955. *Conflict and the Web of Group-Affiliations.* Trans. and ed. Kurt H. Wolff. Glencoe, Ill.: Free Press.

Simon, Julian L. 1989. *The Economic Consequences of Migration.* Oxford: Basil Blackwell.

Simpson, James B. (compiler). 1964. *Contemporary Quotations.* New York: Crowell.

Singh, Jitendra V., and Charles J. Lumsden. 1990. "Theory and Research in Organizational Ecology." *Annual Review of Sociology* 16: 161–195.

Skidelsky, Robert. 1995. *The Road from Serfdom: The Economic and Political Consequences of the End of Communisim.* New York: Allen Lane, Penguin.

Skocpol, Theda. 1979. *States and Social Revolutions: A Comparative Analysis of France, Russia, and China.* Cambridge: Cambridge University Press.

Skocpol, Theda. 1994. *Social Revolutions in the Modern World.* Cambridge: Cambridge University Press.

Skocpol, Theda. 1996. *Boomerang: Clinton's Health Security Effort and the Turn against Government in U.S. Politics.* New York: W.W. Norton.

Skolnick, Arlene. 1991. *Embattled Paradise: The American Family in an Age of Uncertainty.* New York: Basic.

Smelser, Neil J. 1963. *The Sociology of Economic Life.* Englewood Cliffs, N.J.: Prentice-Hall.

Smelser, Neil J. 1998. "The Rational and the Ambivalent in the Social Sciences." *American Sociological Review* 63: 1–16.

Smith, David A. 1996. *Third World Cities in Global Perspective: The Political Economy of Uneven Urbanization.* Boulder, Colo.: Westview.

Smith, James P., and Barry Edmonston (eds.). 1997. *The New Americans: Economic, Demographic, and Fiscal Effects of Immigration.* Washington, D.C.: National Academy.

Smith, Peter B., and Mark F. Peterson. 1988. *Leadership, Organizations and Culture.* Newbury Park, Calif.: Sage.

Smith-Lovin, Lynn, and Charles Brody. 1989. "Interruptions in Group Discussions: The Effects of Gender and Group Composition." *American Sociological Review* 54: 424–435.

Snipp, C. Matthew. 1992. "Sociological Perspectives on American Indians." *Annual Review of Sociology* 18: 351–371.

Snow, David A., and Robert D. Benford. 1992. "Master Frames and Cycles of Protest." Pp. 133–155 in Aldon Morris and Carol M. Mueller (eds.), *Frontiers in Social Movement Theory.* New Haven, Conn: Yale University Press.

Snow, David A., and Pamela E. Oliver. 1995. "Social Movements and Collective Behavior: Social Psychological Dimensions and Considerations." Pp. 571–599 in Karen S. Cook, Gary Alan Fine, and James S. House (eds.), *Sociological Perspectives on Social Psychology.* Boston: Allyn & Bacon.

Snow, David A., and Cynthia L. Phillips. 1980. "The Lofland-Stark Conversion Model: A Critical Reassessment." *Social Problems* 27: 430–447.

Snow, David A., E. Burke Rochford, Jr.; Steven K. Worden, and Robert D. Benford. 1986. "Frame Alignment Processes, Micromobilization, and Movement Participation." *American Sociological Review* 51: 464–481.

Snow, E. (ed.), *Societies Around the World.* New York: Holt, Rinehart & Winston.

Snyder, Thomas D., and Charlene M. Hoffman. 1993. *Digest of Education Statistics, 1993.* Washington, D.C.: National Center for Education Statistics, U.S. Department of Education.

Snyder, Thomas D., and Charlene M. Hoffman. 1995. *Digest of Education Statistics, 1995.* Washington, D.C.: National Center for Education Statistics, U.S. Department of Education.

Snyder, Thomas D., Charlene M. Hoffman, and Claire M. Geddes. 1997. *Digest of Education Statistics, 1997,* NCES 98-015. Washington, D.C.: National Center for Education Statistics. U.S. Department of Education.

So, Alvin Y. and Stephen W. K. Chiu. 1995. *East Asia and the World Economy.* Thousand Oaks, Calif.: Sage.

Sørenson, Annemette, and Heike Trappe. 1995. "The Persistence of Gender Inequality in Earnings in the German Democratic Republic." *American Sociological Review* 60: 398–406.

Spain, Daphne, and Suzanne M. Bianchi. 1996. *Balancing Act: Motherhood, Marriage, and Employment Among American Women.* New York: Russell Sage.

Spergel, Irving A. 1995. *The Youth Gang Problem: A Community Approach.* New York: Oxford University Press.

Spitzer, Steven. 1975. "Toward a Marxian Theory of Deviance." *Social Problems* 22: 641–651.

Stacey, Judith. 1993. "Good Riddance to 'The Family': A Response to David Popenoe." *Journal of Marriage and the Family* 55: 544–547.

Stack, Carol. 1974. *All Our Kin: Strategies for Survival in a Black Community.* New York: Harper & Row.

Staples, Robert, and Leanor Boulin Johnson. 1993. *Black Families at the Crossroads: Challenges and Prospects.* San Francisco: Jossey-Bass.

Stark, Rodney. 1985. "Church and Sect." Pp. 139–149 in Phillip E. Hammond (ed.), *The Sacred in a Secular Age: Toward Revision in the Scientific Study of Religion.* Berkeley: University of California Press.

Stark, Rodney. 1989. *Sociology,* 3rd ed. Belmont, Calif.: Wadsworth.

Stark, Rodney. 1994. *Sociology,* 5th ed. Belmont, Calif.: Wadsworth.

Stark, Rodney, and William Sims Bainbridge. 1979. "Of Churches, Sects, and Cults: Preliminary Concepts for a Theory of Religious Movements." *Journal for the Scientific Study of Religion* 18: 117–133.

Stark, Rodney, and William Sims Bainbridge. 1985. *The Future of Religion: Secularization, Renewal, and Cult Formation.* Berkeley: University of California Press.

Stark, Rodney, and William Sims Bainbridge. 1987. *A Theory of Religion.* New York: Peter Lang.

Stark, Rodney, and Laurence R. Iannaccone. 1994. "A Supply-Side Reinterpretation of the 'Secularization' of Europe." *Journal for the Scientific Study of Religion* 33: 230–252.

Starr, Paul. 1982. *The Social Transformation of American Medicine.* New York: Basic

Starr, Paul. 1994. *The Logic of Health Care Reform: Why and How the President's Plan Will Work.* New York: Penguin.

Stearns, Linda Brewster, and Kenneth D. Allan. 1996. "Economic Behavior in Institutional Environments: The Corporate Merger Wave of the 1980s." *American Sociological Review* 61: 699–718.

Steeh, Charlotte, and Howard Schuman. 1992. "Young White Adults: Did Racial Attitudes Change in the 1980s?" *American Journal of Sociology* 98: 340–367.

Steffensmeier, Darrell, Cathy Streifel, and Edward S. Shihadeh. 1992. "Cohort Size and Arrest Rates over the Life Course: The Easterlin Hypothesis Reconsidered." *American Sociological Review* 57: 306–314.

Stein, Arlene. 1998. "Whose Memories? Whose Victimhood? Contests for the Holocaust Frame in Recent Social Movements Discourse." *Sociological Perspectives* 41: 519–540.

Steinberg, Laurence, with B. Bradford Brown, and Sanford M. Dornbusch. 1996. *Beyond the Classroom: Why School Reform has Failed and What Parents Need to Do.* New York: Simon and Schuster.

Stephan, Cookie White, and Walter G. Stephan. 1985. *Two Social Psychologies: An Integrative Approach.* Homewood, Ill.: Dorsey.

Stephens, William N. 1963. *The Family in Cross-Cultural Perspective.* New York: Holt, Rinehart & Winston.

Stevens, Rosemary. 1971. *American Medicine and the Public Interest.* New Haven, Conn.: Yale University Press.

Stevens, Rosemary. 1989. *In Sickness and in Wealth: American Hospitals in the Twentieth Century.* New York: Basic.

Stinchcombe, Arthur L. 1968. *Constructing Social Theory.* New York: Harcourt Brace Jovanovich.

Stinchcombe, Arthur L. 1983. *Economic Sociology.* New York: Academic.

Stockard, Jean. 1985. "Education and Gender Equality: A Critical View." Pp. 293–321 in Alan Kerckhoff (ed.), *Research in Sociology of Education and Socialization* Vol. 5, Greenwich, Conn.: JAI.

Stockard, Jean. 1998 "Gender Socialization." In Janet S. Chafetz (ed.), *Handbook of Gender Sociology.* New York: Plenum.

Stockard, Jean, and Miriam M. Johnson. 1979. "The Social Origins of Male Dominance." *Sex Roles* 5: 199–218.

Stockard, Jean, and Miriam M. Johnson. 1992. *Sex and Gender in Society,* 2nd ed. Englewood Cliffs, N.J.: Prentice-Hall.

Stockard, Jean, and Maralee Mayberry. 1992. *Effective Educational Environments.* Newbury Park, Calif.: Corwin/Sage.

Stockard, Jean, and Robert M. O'Brien, 1999. "Changes in the Age Distribution of Suicide in the United States: A Simple Demographic Explanation." Paper presented at the 1999 Annual Meetings of the Pacific Sociological Association, April.

Stokes, Gale. 1993. *The Walls Came Tumbling Down: The Collapse of Communism in Eastern Europe.* New York: Oxford.

Stoll, David. 1990. *Is Latin America Turning Protestant? The Politics of Evangelical Growth.* Berkeley: University of California Press.

Straus, Murray A. 1990. "Measuring Intrafamily Conflict and Violence: The Conflict Tactics (CT) Scales." Pp. 29–47 in Murray A. Straus and Richard J. Gelles, *Physical Violence in American Families: Risk Factors and Adaptations to Violence in 8,145 Families.* New Brunswick, N.J.: Transaction.

Straus, Murray A., and Richard J. Gelles. 1990. "How Violent Are American Families? Estimates from the National Family Violence Resurvey and Other Studies." Pp. 95–112 in Murray A. Straus and Richard J. Gelles (eds.), *Physical Violence in American Families: Risk Factors and Adaptations to Violence in 8,145 Families.* New Brunswick, N.J.: Transaction.

Stryker, Sheldon. 1980. *Symbolic Interactionism: A Social Structural View.* Menlo Park, Calif.: Benjamin/Cummings.

Sugrue, Thomas J. 1996. *The Origins of the Urban Crisis: Race and Inequality in Postwar Detroit.* Princeton, N.J.: Princeton University Press.

Sutherland, Edwin H. 1939. *Principles of Criminology,* 3rd ed. Philadelphia: Lippincott.

Sutherland, Edwin H. 1983. *White-Collar Crime: The Uncut Version.* New Haven, Conn.: Yale University Press.

Sutton, John R., Frank Dobbin, John W. Meyer, and W. Richard Scott. 1994. "The Legalization of the Workplace." *American Journal of Sociology* 99: 944–971.

Swedberg, Richard. 1990. *Economics and Sociology: Redefining Their Boundaries: Conversations with Economists and Sociologists.* Princeton, N.J.: Princeton University Press.

Swedberg, Richard. 1991. "The Battle of Methods: Toward a Paradigm Shift." Pp. 13–34 in Amitai Etzioni and Paul R. Lawrence (eds.), *Socio-Economics: Toward a New Synthesis.* Armonk, N.Y.: Sharpe.

Swedberg, Richard (ed.). 1993. *Explorations in Economic Sociology.* New York: Russell Sage.

Swedberg, Richard. 1998. *Max Weber and the Idea of Economic Sociology.* Princeton, N.J.: Princeton University Press.

Sweet, James A., and Larry L. Bumpass. 1992. "Young Adults' Views of Marriage, Cohabitation, and Family." Pp. 143–170 in Scott J. South and Stewart E. Tolnay (eds.), *The Changing American Family: Sociological and Demographic Perspectives.* Boulder, Colo.: Westview.

Swoboda, Frank. 1992. "Incomes of Workers Lose Ground in 80s." *The Portland Oregonian,* 8 September, p. A14.

Szasz, Andrew. 1998. "Corporations, Organized Crime, and Hazardous Waste Disposal: Making a Criminogenic Regulatory Structure." Pp. 363–380 in Stuart Henry and Werner Einstadter (eds.), *The Criminology Theory Reader.* New York: New York University Press.

Tallman, Irving, and Louis N. Gray. 1990. "Choices, Decisions, and Problem-Solving." *Annual Review of Sociology* 16: 405–433.

Tang, Joyce, and Earl Smith (eds.). 1996. *Women and Minorities in American Professions.* Albany: State University of New York Press.

Tarrow, Sidney. 1996. "States and Opportunities: The Political Structuring of Social Movements." Pp. 41–61 in Doug McAdam, John D. McCarthy, and Mayer N. Zald. (eds.), 1996. *Comparative Perspectives on Social Movements: Political Opportunities, Mobilizing Structures, and Cultural Framings.* Cambridge: Cambridge University Press.

Tasker, Fiona L., and Susan Golombok. 1997. *Growing Up in a Lesbian Family: Effects on Child Development.* New York: Guilford.

Teachman, Jay D. 1992. "Intergenerational Resource Transfers Across Disrupted Households: Absent Fathers' Contributions to the Well-Being of Their Children." Pp. 224–246 in Scott J. South and Stewart E. Tolnay (eds.), *The Changing American Family: Sociological and Demographic Perspectives.* Boulder, Colo.: Westview.

Teitelbaum, Richard S. 1995. "The 500 Ranked by Performance." *Fortune,* 15 May, pp. F1–F35.

Testa, Mark, and Marilyn Krogh. 1995. "The Effect of Employment on Marriage Among Black Males in Inner-City Chicago." Pp. 59–95 in M. Belinda Tucker and Claudia Mitchell-Kernan (eds.), *The Decline in Marriage Among African Americans: Causes, Consequences, and Policy Implications.* New York: Russell Sage.

Thernstrom, Stephen (ed.). 1980. *Harvard Encyclopedia of American Ethnic Groups.* Cambridge, Mass.: Belknap.

Thernstrom, Stephan, and Abigail Thernstrom. 1997. *America in Black and White: One Nation, Indivisible.* New York. Simon and Schuster.

Theroux, Paul. 1989. "Malawi: Faces of a Quiet Land." *National Geographic* 176(3): 370–389.

Thomas, Sue, and Clyde Wilcox (eds). 1998. *Women and Elective Office: Past, Present and Future.* New York: Oxford University Press.

Thornberry, Terence P. 1996. "Empirical Support for Interactional Theory: A Review of the Literature." Pp. 198–235 in J. David Hawkins (ed.), *Delinquency and Crime: Current Theories.* New York: Cambridge University Press.

Thornberry, Terence P. (ed.). 1997. *Developmental Theories of Crime and Delinquency, Advances in Criminological Theory,* Vol. 7. New Brunswick: Transaction.

Thorne, Barrie. 1993. *Gender Play: Girls and Boys in School.* New Brunswick, N. J.: Rutgers University Press.

Thornton, Arland. 1991. "Influence of the Marital History of Parents on the Marital and Cohabitational Experiences of Children." *American Journal of Sociology* 96: 868–894.

Thornton, Arland. 1992. "The Influence of the Parental Family on the Attitudes and Behavior of Children." Pp. 247–266 in Scott J. South and Stewart E. Tolnay (eds.), *The Changing American Family: Sociological and Demographic Perspectives.* Boulder, Colo.: Westview.

Thrasher, Fredic M. 1927. *The Gang: A Study of 1,313 Gangs in Chicago.* Chicago: University of Chicago Press.

Tiger, Harriet (ed.). 1995. *Who's Who in America, 1996.* New Providence, N.J.: Marquis Who's Who.

Tilly, Chris. 1996. *Half a Job: Bad and Good Part-Time Jobs in a Changing Labor Market.* Philadelphia: Temple University Press.

Tiryakian, Edward A. 1993. "American Religious Exceptionalism: A Reconsideration." *Annals, AAPSS* 527: 40–53.

Toffler, Alvin. 1990. *Powershift: Knowledge, Wealth, and Violence at the Edge of the 21st Century.* New York: Bantam.

Toman, Walter. 1993. *Family Constellation: Its Effects on Personality and Social Behavior,* 4th ed. New York: Springer.

Tönnies, Ferdinand. 1963. *Gemeinschaft and Gesellschaft.* Trans. C. P. Loomis. New York: American. (Orig. pub. 1887.)

Tracy, Paul E., and Kimberly Kempf-Leonard. 1996. *Continuity and Discontinuity in Criminal Careers.* New York: Plenum.

Treas, Judith. 1995. "Older Americans in the 1990s and Beyond." *Population Bulletin* 50(2).

Trice, Harrison M., and Janice M. Beyer. 1993. *The Cultures of Work Organizations.* Englewood Cliffs, N.J.: Prentice-Hall.

Trow, Martin. 1961. "The Second Transformation of American Secondary Education." *International Journal of Comparative Sociology* 2: 144–161.

Tuan, Mia. 1998. *Forever Foreigners or Honorary Whites? The Asian Ethnic Experience Today.* New Brunswick, N.J.: Rutgers University Press.

Tuch, Steven A. 1987. "Urbanism, Region, and Tolerance Revisited: The Case of Racial Prejudice." *American Sociological Review* 52: 504–510.

Tucker, M. Belinda, and Claudia Mitchell-Kernan. 1995. "Trends in African American Family Formation: A Theoretical and Statistical Overview." Pp. 3–26 in M. Belinda Tucker and Claudia Mitchell-Kernan (eds.), *The Decline in Marriage Among African Americans: Causes, Consequences, and Policy Implications.* New York: Russell Sage.

Tumin, Melvin M. 1953. "Some Principles of Stratification: A Critical Analysis." *American Sociological Review* 18: 387–393.

Turner, Jonathan. 1982. *The Structure of Sociological Theory,* 3rd ed. Homewood, Ill.: Dorsey.

Turner, Jonathan H. 1984. *Societal Stratification: A Theoretical Analysis.* New York: Columbia University Press.

Turner, Jonathan H. 1989. "The Disintegration of American Sociology: Pacific Sociological Association 1988 Presidential Address." *Sociological Perspectives* 32: 419–433.

Turner, Jonathan H. 1998. "Must Sociological Theory and Sociological Practice Be So Far Apart?: A Polemical Answer." *Sociological Perspectives* 41: 243–258.

Turner, Jonathan H., Leonard Beeghley, and Charles H. Powers. 1997. *The Emergence of Sociological Theory,* 4th ed. Belmont, Calif.: Wadsworth.

Turner, Ralph H. 1967. "Introduction." Pp. ix–xlvi in Robert E. Park, *On Social Control and Collective Behavior.* Chicago: University of Chicago Press.

Turner, Ralph, and Lewis M. Killian. 1972. *Collective Behavior,* 2nd ed. Englewood Cliffs, N.J. Prentice Hall.

Tyack, David, and Elisabeth Hansot. 1990. *Learning Together: A History of Coeducation in American Schools.* New Haven, Conn.: Yale University Press.

Tyler, Tom R. 1990. *Why People Obey the Law.* New Haven, Conn.: Yale University Press.

UNESCO. 1998. *UNESCO Statistical Yearbook, 1998.* Paris: UNESCO. (http://unescostat.unesco.org/Yearbook/)

U.S. Bureau of the Census. 1975. *Historical Statistics of the United States, Colonial Times to 1970, Bicentennial Edition,* Parts 1 and 2. Washington, D.C.: U.S. Government Printing Office.

U.S. Bureau of the Census. 1985. *Estimates of the Population of the United States by Age, Sex, and Race, 1980–1984.* Current Population Reports, Series P-25, No. 965. Washington, D.C.: U.S. Government Printing Office.

U.S. Bureau of the Census. 1992. *Poverty in the United States: 1991.* Current Population Reports, Series P-60, No. 181. Washington, D.C.: U.S. Government Printing Office.

U.S. Bureau of the Census. 1993a. *1990 Census of the Population: Social and Economic Characteristics of the United States.* Washington, D.C.: U.S. Government Printing Office.

U.S. Bureau of the Census. 1993b. *Statistical Abstract of the United States, 1993,* 113th ed. Washington, D.C.: U.S. Government Printing Office.

U.S. Bureau of the Census. 1994. *Statistical Abstract of the United States, 1994,* 114th ed. Washington, D.C.: U.S. Government Printing Office.

U.S. Bureau of the Census. 1995. *Statistical Abstract of the United States, 1995,* 115th ed. Washington, D.C.: U.S. Government Printing Office.

U. S. Bureau of the Census. 1997. *Statistical Abstract of the United States: 1997,* 117th ed. Washington, D.C.: U.S. Government Printing Office.

U.S. Bureau of the Census. 1998a. *Measuring 50 Years of Economic Change Using the March Current Population Survey.* Current Population Reports, P60-203. Washington, D.C.: U.S. Government Printing Office.

U.S. Bureau of the Census. 1998b. *Money Income in the United States: 1997* (With Separate Data on Valuation of Noncash Benefits). Current Population Reports, P60-200, Washington, D.C.: U.S. Government Printing Office.

U.S. Bureau of the Census. 1998c. *Statistical Abstract of the United States 1998,* 118th ed. Washington, D.C.: U.S. Government Printing Office.

U.S. Department of Education, National Center for Education Statistics [E.D. Tabs]. 1998. *Degrees and Other Awards Conferred by Degree-Granting Institutions: 1995–96,* NCES 98-256, by Frank B. Morgan. Washington, D. C.: U.S. Government Printing Office.

Uhlenberg, Peter. 1992. "Death and the Family." Pp. 72–81 in Arlene S. Skolnick and Jerome H. Skolnick (eds.), *Family in Transition: Rethinking Marriage, Sexuality, Child Rearing, and Family Organization.* New York: Harper Collins.

Ullian, Dorothy Z. 1976. "The Development of Conceptions of Masculinity and Femininity." Pp. 25–48 in Barbara Lloyd and John Archer (eds.), *Exploring Sex Differences.* New York: Academic.

United Nations. 1993. *Demographic Yearbook, 1993.* New York: United Nations.

United Nations. 1997. *Demographic Yearbook, 1995.* New York: United Nations.

United Nations. 1998. *Demographic Yearbook, 1996.* New York: United Nations.

United Nations, Department for Economic and Social Information and Policy Analysis. 1997. *Report on the World Social Situation, 1997.* New York: United Nations.

United Nations High Commissioner for Refugees, Statistical Unit. 1998. "Refugees and Others of Concern to UNHCR: 1997 Statistical Overview." Geneva, Switzerland: UNHCR. (http://www.unhcr.ch/refworld/refbib/refstat/1998/98intro.htm)

United Nations Population Division. 1999. *World Population Prospects: The 1998 Revision*. New York: United Nations.

Vander Zanden, James W. 1987. *Social Psychology,* 4th ed. New York: Random House.

Vanneman, Reeve, and Lynn Weber Cannon. 1987. *The American Perception of Class*. Philadelphia: Temple University Press.

Van Zandt, David E. 1991. *Living in the Children of God*. Princeton, N.J.: Princeton University Press.

Ventura, Stephanie J., Robert N. Anderson, Joyce A. Martin, and Betty L. Smith. 1998. "Births and Deaths: Preliminary Date for 1997." *National Vital Statistics Reports* 47(4). Hyattsville, Md.: National Center for Health Statistics.

Ventura, Stephanie J., Joyce A. Martin, T. J. Mathews, and S. C. Clarke. 1996. "Advance Report of Final Natality Statistics, 1994." *Monthly Vital Statistics Report* 44(11), Supplement. Hyattsville, Md.: National Center for Health Statistics.

Ventura, Stephanie J., Kimberley D. Peters, Joyce A. Martin, and Jeffrey D. Maurer. 1997. "Births and Deaths: United States, 1996," *Monthly Vital Statistics Report* 46(1), Supp. 2. Hyattsville, Md.: National Center for Health Statistics.

Verba, Sidney, and Norman Nie. 1972. *Participation in America: Political Democracy and Social Equality*. New York: Harper & Row.

Verba, Sidney, Kay Lehman Schlozman, and Henry E. Brady. 1995. *Voice and Equality: Civic Voluntarism in American Politics*. Cambridge, Mass.: Harvard University Press.

Victora, Cesar G., Sharon R. A. Huttly, Fernando C. Barros, Cintia Lombardi, and J. Patrick Vaughan. 1992. "Maternal Education in Relation to Early and Late Child Health Outcomes: Findings from a Brazilian Cohort Study." *Social Science Medicine* 34: 899–905.

Visaria, Leela, and Pravin Visaria. 1995. "India's Population in Transition." *Population Bulletin* 50(3). Washington, D.C.: Population Reference Bureau, October, 1995.

Vogel, David. 1996. *Kindred Strangers: The Uneasy Relationship between Politics and Business in America*. Princeton, N.J.: Princeton University Press.

Vogel, Ezra F. 1991. *The Four Little Dragons: The Spread of Industrialization in East Asia*. Cambridge, Mass.: Harvard University Press.

Waite, Linda J. 1995. "Does Marriage Matter?" *Demography* 32: 483–507.

Waite, Linda J., Frances Kobrin Goldscheider, and Christian Witsberger. 1986. "Nonfamily Living and the Erosion of Traditional Family Orientations Among Young Adults." *American Sociological Review* 51: 541–554.

Waite, Linda J., and Lee A. Lillard. 1991. "Children and Marital Disruption." *American Journal of Sociology* 96: 930–953.

Wald, Roberta. 1992. *Factors Associated with Treatment Attendance and Treatment Completion for Substance Abusing Women*. Unpublished doctoral dissertation, Eugene, University of Oregon.

Waldinger, Roger. 1997. "Black/Immigrant Competition Re-assessed: New Evidence from Los Angeles." *Sociological Perspectives* 40: 365–386.

Waldinger, Roger, and Mehdi Bozorgmehr (eds.). 1996. *Ethnic Los Angeles*. New York: Russell Sage.

Walker, Jack L. 1990. "Political Mobilization in America." Pp. 163–187 in John E. Jackson (ed.), *Institutions in American Society: Essays in Market, Political, and Social Organizations*. Ann Arbor: University of Michigan Press.

Wallerstein, Immanuel. 1974. *The Modern World System: Capitalist Agriculture and the Origins of the European World-Economy in the Sixteenth Century*. New York: Academic.

Wallerstein, Immanuel. 1979. *The Capitalist World-Economy*. New York: Cambridge University Press.

Wallerstein, Immanuel. 1984. *The Politics of the World-Economy: The States, the Movements and the Civilizations*. Cambridge: Cambridge University Press.

Wallerstein, Immanuel. 1997. "Social Science and the Quest for a Just Society." *American Journal of Sociology* 102: 1241–1257.

Wallerstein, Judith S., and Sandra Blakeslee. 1989. *Second Chances: Women, Men and Children a Decade After Divorce*. New York: Ticknor & Fields.

Wallerstein, Judith S., and Joan B. Kelly. 1980. *Surviving the Breakup: How Children and Parents Cope with Divorce*. New York: Basic.

Wardhaugh, Ronald. 1985. *How Conversation Works*. Oxford: Basil Blackwell.

Warner, R. Stephen. 1988. *New Wine in Old Wineskins: Evangelicals and Liberals in a Small-Town Church*. Berkeley: University of California Press.

Warner, R. Stephen. 1993. "Work in Progress Toward a New Paradigm for the Sociological Study of Religion in the United States." *American Journal of Sociology* 98: 1044–1093.

Warner, R. Stephen. 1997. "A Paradigm is Not a Theory: Reply to Lechner." *American Journal of Sociology* 103: 192–198.

Warner, W. Lloyd. 1949a. *Democracy in Jonesville: A Study in Inequality*. New York: Harper & Row.

Warner, W. Lloyd. 1949b. *Social Class in America: A Manual of Procedure for the Measurement of Social Status*. New York: Harper & Row.

Warner, W. Lloyd, and Paul S. Lunt. 1942. *The Status System of a Modern Community*. New Haven, Conn.: Yale University Press.

Wartenberg, Thomas. 1990. *The Forms of Power: From Domination to Transformation*. Philadelphia: Temple University Press.

Wasburn, Philo C. 1982. *Political Sociology: Approaches, Concepts, Hypotheses*. Englewood Cliffs, N.J.: Prentice-Hall.

Waters, Mary C. 1990. *Ethnic Options: Choosing Identities in America*. Berkeley: University of California Press.

Weber, Max. 1946a. "Bureaucracy." Pp. 196–244 in H. H. Gerth and C. Wright Mills (eds. and trans.), *From Max Weber: Essays in Sociology*. New York: Oxford University Press. (Orig. pub. 1922.)

Weber, Max. 1946b. "The Protestant Sects and the Spirit of Capitalism." Pp. 302–322 in Hans H. Gerth and C. Wright Mills (eds. and trans.), *From Max Weber: Essays in Sociology*. New York: Oxford University Press.

Weber, Max. 1946c. "Class, Status, Party." Pp. 180–195 in H. H. Gerth and C. Wright Mills (eds. and trans.), *From Max Weber: Essays in Sociology.* New York: Oxford University Press.

Weber, Max. 1947. *The Theory of Social and Economic Organization.* A. M. Henderson and Talcott Parsons (eds. and trans.). New York: Oxford University Press.

Weber, Max. 1951. *The Religion of China.* Glencoe, Ill.: Free Press. (Orig. pub. 1916.)

Weber, Max. 1958. *The Protestant Ethic and the Spirit of Capitalism.* Trans. Talcott Parsons. New York: Scribner. (Orig. pub. 1905–6.)

Weber, Max. 1958. *The Religion of India.* Glencoe, Ill.: Free Press. (Orig. pub. 1916–17.)

Weber, Max. 1967a. "Religious Rejections of the World and Their Directions." Pp. 323–359 in H. H. Gerth and C. Wright Mills (eds. and trans.), *From Max Weber: Essays in Sociology.* New York: Oxford University Press. (Orig. pub. 1915.)

Weber, Max. 1967b. "The Social Psychology of the World Religions." Pp. 267–301 in H. H. Gerth and C. Wright Mills (eds. and trans.), *From Max Weber: Essays in Sociology.* New York: Oxford University Press. (Orig. pub. 1922.)

Weber, Max. 1968. *Economy and Society,* Vol. 1. New York: Bedminster. (Orig. pub. 1922.)

Weeks, John R. 1994. *Population: An Introduction to Concepts and Issues,* 5th ed., updated. Belmont, Calif.: Wadsworth.

Weeks, John R. 1996. *Population: An Introduction to Concepts and Issues,* 6th ed. Belmont, Calif.: Wadsworth.

Weinberg, Daniel H. 1996. "A Brief Look at Postwar U.S. Income Inequality." *Current Population Reports,* P60–191. U.S. Census Bureau. Washington, D.C.: U.S. Government Printing Office.

Weiner, Annette B. 1976. *Women of Value, Men of Renown: New Perspectives in Trobriand Exchange.* Austin: University of Texas Press.

Weiss, Lawrence D. 1997. *Private Medicine and Public Health: Profit, Politics, and Prejudice in the American Health Care Enterprise.* Boulder, Colo.: Westview.

Weitz, Rose. 1996. *The Sociology of Health, Illness, and Health Care: A Critical Approach.* Belmont, Calif.: Wadsworth.

Weitzman, Lenore J., and Mavis Maclean. 1992. *Economic Consequences of Divorce: The International Perspective.* Oxford: Clarendon.

Wellman, Barry. 1983. "Network Analysis: Some Basic Principles." *Sociological Theory* 1: 155–200.

Wellman, Barry, and Scot Wortley. 1990. "Different Strokes from Different Folks: Community Ties and Social Support." *American Journal of Sociology* 96: 558–588.

Wells, L. Edward, and Joseph H. Rankin. 1998. "Social Control Theories of Delinquency: Direct Parental Controls." Pp. 276–287 in Stuart Henry and Werner Einstadter (eds.), *The Criminology Theory Reader.* New York: New York University Press.

Wells, Robert V. 1982. *Revolutions in Americans' Lives: A Demographic Perspective on the History of Americans, Their Families, Their Society.* Westport, Conn.: Greenwood.

Wesolowski, Wlodzimierz. 1966. "Some Notes on the Functional Theory of Stratification." Pp. 28–38 in Reinhard Bendix and Seymour Martin Lipset (eds.), *Class, Status, and Power: Social Stratification in Comparative Perspective,* 2nd ed. New York: Free Press.

West, Candace. 1982. "Why Can't a Woman Be More Like a Man? An Interactional Note on Organizational Game-Playing for Managerial Women." *Work and Occupations* 9: 5–29.

West, Candace, and Don H. Zimmerman. 1983. "Small Insults: A Study of Interruptions in Cross-Sex Conversations Between Unacquainted Persons." Pp. 102–117 in Barrie Thorne, Cheris Kramarae, and Nancy Henley (eds.), *Language, Gender and Society.* Rowley, Mass.: Newbury.

Westerman, Marty. 1989. "Death of the Frito Bandito." *American Demographics* 11(March): 28–32.

Whalen, Jack. 1992. "Conversation Analysis." Pp. 303–310 in Edgar F. Borgatta and Marie L. Borgatta (eds.), *Encyclopedia of Sociology.* New York: Macmillan.

Whalen, Jack, Marilyn Whalen, and Kathryn Henderson. 1995. "On Line, on Paper, and on the Phone: The Choreography of Work in a Customer Call Center." Paper presented at the annual meeting of the American Sociological Association, August 1995, Washington, D.C.

Whalen, Marilyn R., and Don Zimmerman. 1987. "Sequential and Institutional Contexts in Calls for Help." *Social Psychology Quarterly* 50: 172–185.

White, James M. 1991. *Dynamics of Family Development: A Theoretical Approach.* New York: Guilford.

White, Leslie A. 1940. "Symbols, the Basis of Language and Culture." *Philosophy of Science* 7: 451–463.

White, Leslie A. 1949. *The Science of Culture: A Study of Man and Civilization.* New York: Farrar, Straus.

White, Michael J. 1987. *American Neighborhoods and Residential Differentiation.* New York: Russell Sage.

White House Fact Sheet. 1998. "Statement by the President: The Children's Health Insurance Program One Year Anniversay, October 1, 1998.")http://www.hcfa.gov/init/wh-chip9.htm)

Whiting, Beatrice Blyth, and Carolyn Pope Edwards. 1988. *Children of Different Worlds: The Formation of Social Behavior.* Cambridge, Mass.: Harvard University Press.

Whittier, Nancy. 1995. *Feminist Generations: The Persistence of the Radical Women's Movement.* Philadelphia: Temple University Press.

Whittier, Nancy. 1997. "Political Generations, Micro-Cohorts, and the Transformation of Social Movements." *American Sociological Review* 62: 760–778.

Whyte, Martin King. 1978. *The Status of Women in Preindustrial Societies.* Princeton, N.J.: Princeton University Press.

Whyte, Martin King. 1990. *Dating, Mating, and Marriage.* New York: Aldine de Gruyter.

Wickrama, K.A.S., Frederick O. Lorenz, and Rand D. Conger. 1997. "Parental Support and Adolescent Physical Health Status: A Latent Growth-Curve Analysis." *Journal of Health and Social Behavior* 38: 149–163.

Wickrama, K.A.S., Frederick O. Lorenz, Rand D. Conger, Lisa Matthews, and Glen H. Elder, Jr. 1997. "Linking Occupational Conditions to Physical Health through Marital, Social, and Intrapersonal Processes." *Journal of Health and Social Behavior* 38: 363–375.

Wilkie, Jane Riblett. 1993. "Changes in U.S. Men's Attitudes Toward the Family Provider Role, 1972–1989." *Gender and Society* 7: 261–279.

Wilkie, Raymond. 1956. "Navajo Folktales." Pp. 232–236 in Irwin T. Sanders, Richard B. Woodbury, Frank J. Essene, Thomas P. Field, Joseph R. Schwendeman, and Charles E. Snow (eds.), *Societies Around the World*. New York: Holt, Rinehart & Winston.

Wilkinson, Kenneth P. 1991. *The Community in Rural America*. New York: Greenwood.

Wilkinson, Richard G. (ed.). 1986. *Class and Health: Research and Longitudinal Data*. London and New York: Tavistock.

Wilkinson, Richard G. 1989. "Class Mortality Differentials, Income Distribution and Trends in Poverty 1921–1981." *Journal of Social Policy* 18: 307–335.

Wilkinson, Richard G. 1992. "Income Distribution and Life Expectancy." *British Medical Journal* 304: 165–168.

Wilkinson, Richard G.1996. *Unhealthy Societies: The Afflictions of Inequality*. London: Routledge.

Wilkinson, Richard G. 1997. "Comment: Income, Inequality, and Social Cohesion." *American Journal of Public Health* 87: 1504–1506.

Williams, Christine L. 1989. *Gender Differences at Work: Women and Men in Nontraditional Occupations*. Berkeley: University of California Press.

Willie, Charles Vert. 1979. *Caste and Class Controversy*. Bayside, N.Y.: General Hall.

Willie, Charles Vert. 1983. *Race, Ethnicity, and Socioeconomic Status: A Theoretical Analysis of Their Interrelationship*. Bayside, N.Y.: General Hall.

Wilmore, Gayraud S. 1972. *Black Religion and Black Radicalism*. Garden City, N.Y.: Doubleday.

Wilson, Bryan. 1976. *The Contemporary Transformation of Religion*. London and New York: Oxford University Press.

Wilson, Bryan. 1985. "Secularization: The Inherited Model." Pp. 9–20 in Phillip E. Hammond (ed.), *The Sacred in a Secular Age: Toward Revision in the Scientific Study of Religion*. Berkeley: University of California Press.

Wilson, Thomas C. 1985. "Urbanism and Tolerance: A Test of Some Hypotheses Drawn from Wirth and Stouffer." *American Sociological Review* 50: 117–123.

Wilson, Thomas C. 1991. "Urbanism, Migration, and Tolerance: A Reassessment." *American Sociological Review* 56: 117–123.

Wilson, William Julius. 1980. *The Declining Significance of Race: Blacks and Changing American Institutions*, 2nd ed. Chicago: University of Chicago Press.

Wilson, William Julius. 1987. *The Truly Disadvantaged: The Inner City, the Underclass, and Public Policy*. Chicago: University of Chicago Press.

Wilson, William Julius. 1991. "Studying Inner-City Social Dislocations: The Challenge of Public Agenda Research." *American Sociological Review* 56: 1–14.

Wilson, William Julius (ed.). 1989. "The Ghetto Underclass: Social Science Perspectives." *The Annals of the American Academy of Political and Social Science* 501 (January).

Wilson, William Julius. 1997. *When Work Disappears: The World of the New Urban Poor*. New York: Alfred A. Knopf.

Wirth, Louis. 1938. "Urbanism as a Way of Life." *American Journal of Sociology* 44: 1–24.

Wishman, S. 1981. "A Lawyer's Guilty Secrets." *Newsweek*, 9 November, p. 25.

Wolfgang, Marvin E., Terence P. Thornberry, and Robert M. Figlio. 1987. *From Boy to Man, from Delinquency to Crime*. Chicago: University of Chicago Press.

Wolfinger, Raymond S., and Steven J. Rosenstone. 1980. *Who Votes?* New Haven, Conn.: Yale University Press.

Wood, Julia T. 1994. *Gendered Lives: Communication, Gender, and Culture*. Belmont, Calif.: Wadsworth.

Wooffitt, Robin. 1990. "On the Analysis of Interaction: An Introduction to Conversation Analysis." Pp. 7–38 in Paul Luff, Nigel Gilbert, and David Frohlich (eds.), *Computers and Conversation*. San Diego, Calif.: Academic.

Woolf, Henry Bosley (ed.). 1975. *Webster's New Collegiate Dictionary*. Springfield, Mass.: G. and C. Merriam.

Worden, Robert E., and Alissa A. Pollitz. 1984. "Police Arrests in Domestic Disturbances: A Further Look." *Law and Society Review* 18: 105–119.

World Bank. 1995. *World Development Report, 1995: Workers in an Integrating World*. New York: Oxford University Press.

World Bank. 1998a. "Distribution of Income or Consumption." *World Development Indicators, 1998*, CD-ROM, Table 2.8.

World Bank. 1998b. "GNP per capita 1997, Atlas Method." (http://www.worldbank.org/data/databydopic/gnppc97.pdf) (12/18/98).

Wresch, William. 1996. *Disconnected: Haves and Have-Nots in the Information Age*. New Brunswick, N.J.: Rutgers University Press.

Wright, Erik Olin. 1985. *Classes*. London: Verso.

Wright, Erik Olin. 1989a. "A General Framework for the Analysis of Class Structure." Pp. 3–46 in Erik Olin Wright et al., *The Debate on Classes*. London: Verso.

Wright, Erik Olin. 1989b. "Rethinking, Once Again, the Concept of Class Structure." Pp. 269–348 in Erik Olin Wright et al., *The Debate on Classes*. London: Verso.

Wright, Erik Olin. 1997. *Class Counts: Comparative Studies in Class Analysis*. New York: Cambridge University Press.

Wright, Erik Olin, and Janeen Baxter, with Gunn Elizabeth Birkelund. 1995. "The Gender Gap in Workplace Authority: A Cross-National Study." *American Sociological Review* 60: 407–435.

Wright, John W. (ed.). 1998. *1999 New York Times Almanac*. New York: Penguin.

Wright, Richard T., and Scott H. Decker. 1997. *Armed Robbers in Action: Stickups and Street Culture*. Boston: Northeastern University Press.

Wrong, Dennis H. 1945. "The Functional Theory of Stratification: Some Neglected Considerations." *American Sociological Review* 10: 242–249.

Wu, Lawrence L. 1996. "Effects of Family Instability, Income, and Income Instability on the Risk of a Premarital Birth." *American Sociological Review* 61: 386–406.

Wuthnow, Robert J. 1988. "Sociology of Religion." Pp. 473–509 in Neil J. Smelser (ed.), *Handbook of Sociology.* Newbury Park, Calif.: Sage.

Wysocki, Diane Kholos. 1996. *Somewhere Over the Modem: Interpersonal Relationships Over Computer Bulletin Boards.* Unpublished Ph.D. Dissertation, University of California, Santa Barbara.

Wysocki, Diane Kholos. 1998. "Let Your Fingers Do the Talking: Sex on an Adult Chat Line." *Sexualities* 4: 425–452.

Yarrow, Leon J., et al. 1964. "Separation from Parents During Early Childhood." Pp. 89–136 in Martin L. Hoffman and Lois W. Hoffman (eds.), *Review of Child Development Research,* Vol. 1. New York: Russell Sage.

Ybarra, Lea. 1982. "When Wives Work: The Impact on the Chicano Family." *Journal of Marriage and the Family* 15: 169–178.

Yetman, Norman R. (ed.). 1985. *Majority and Minority: The Dynamics of Race and Ethnicity in American Life.* Boston: Allyn & Bacon.

Yinger, J. Milton. 1994. *Ethnicity: Source of Strength? Source of Conflict?* Albany: State University of New York Press.

Yinger, John. 1995. *Closed Doors, Opportunities Lost: The Continuing Costs of Housing Discrimination.* New York: Russell Sage.

Young, Lawrence A. (ed.). 1997. *Rational Choice Theory and Religion: Summary and Assessment.* New York: Routledge.

Zack, Naomi (ed.). 1995. *American Mixed Race: The Culture of Microdiversity.* Lanham, Md.: Rowman and Littlefield.

Zald, Mayer N. 1996. "Culture, Ideology, and Strategic Framing." Pp. 261–274 in Doug McAdam, John D. McCarthy, and Mayer N. Zald. (eds.), *Comparative Perspectives on Social Movements: Political Opportunities, Mobilizing Structures, and Cultural Framings.* Cambridge: Cambridge University Press.

Zellner, William W. 1995. *Countercultures: A Sociological Analysis.* New York: St. Martin's.

Zey, Mary. 1993. *Banking on Fraud: Drexel, Junk Bonds, and Buyouts.* New York: Aldine.

Zhao, Dingxin. 1998. "Ecologies of Social Movements: Student Mobilization during the 1989 Prodemocracy Movement in Beijing." *American Journal of Sociology* 103: 1493–1529.

Zipp, John F., and Joel Smith. 1979. "The Structure of Electoral Political Participation." *American Journal of Sociology* 85: 167–177.

Zippay, Allison. 1991. *From Middle Income to Poor: Downward Mobility Among Displaced Steelworkers.* New York: Praeger.

Ziskind, Samuel J. 1962. "Army." Pp. 602–614 in *The World Book Encyclopedia,* Vol. 1. Chicago: Field.

Zuckerman, Barry. 1991. "Drug-Exposed Infants: Understanding the Medical Risk." *The Future of Children* 1(1): 26–35.

Zwerling, Craig, and Hilary Silver. 1992. "Race and Job Dismissals in a Federal Bureaucracy." *American Sociological Review* 57: 651–660.

Photo and Text Credits

McVay/ Allstock/PNI; **p. 228** Frank Siteman/Stock, Boston/PNI; **p. 236** © Frank C. Dougherty/NYT Pictures ; **p. 246** © Peter Stephenson/Envision; **p. 249** © Paolo Koch/Photo Researchers; **p. 250** © (top left) Gale Zucker/Stock, Boston; (top right) Chuck Fishman/Woodfin Camp & Associates; (bottom right) Michael A. Schwarz/The Image Works; (bottom left) © Tom McKitterick/Impact Visuals; **p. 252** Courtesy State Historical Society of Iowa, Des Moines; **p. 261** © M. Philippot/Sgyma; **p. 262** © Gale Zucker/Stock, Boston; **p. 272** © Paolo Koch/Photo Researchers; **p. 276** Bob Kramer/Stock, Boston; **p. 279** © 1996 by The Metropolitan Museum of Art/Courtesy of DC Moore Gallery, NY; **p. 283** (left) Paul Harris/ AllStock/PNI; (right) Carlos Humberto T.D.C/ Contact Press Images/PNI; **p. 286** University of Berkeley; **p. 292** © Art Wolfe/Tony Stone Images; **p. 295** Courtesy Hewlett-Packard Company; **p. 305** Bob Daemmrich/Stock, Boston/PNI; **p. 311** Audrey Gottlieb/Monkmeyer Press; **p. 316** Courtesy of the Billy Graham Center Museum, Wheaton, Illinois; **p. 321** © Jana Birchum/Impact Visuals; **p. 330** Audrey Gottlieb/Monkmeyer Press; **p. 334** AP/ Wide World Photos; **p. 336** (left) Norman Lono/ NYT Pictures; (right) ©Liz Gilbert/Sygma; **p. 339** Lawrence Migdale/Stock, Boston/PNI; **p. 341** Peter Turnley/Black Star/PNI; **p. 348** AP/Wide World Photos; **p. 357** ©Liz Gilbert/Sygma; **p. 361** © George Olson/Photo 20-20; **p. 362** Christiana Dittmann/Rainbow/PNI; **p. 372** ©Ann Dowie; **p. 379** ©Archive Photos; **p. 390** Christiana Dittmann/Rainbow/PNI; **p. 395** Peter Menzel/ Stock, Boston/PNI; **p. 399** Courtesy State Historical Society of Iowa, Des Moines; **p. 401** Lawrence Migdale/Stock, Boston; **p. 410** CORBIS/Philip Gould; **p. 417** (top) Courtesy State Historical Society of Iowa, Des Moines; (bottom) CORBIS/ Philip Gould; **p. 422** Yoav Levy/Phototake/PNI; **p. 426** Charles Gutpon/AllStock, PNI; **p. 427** Michael Newmal/PhotoEdit/PNI; **p. 432** Culver Pictures/PNI; **p. 437** Thierry Boccon-Gibod/Black Star/PNI; **p. 446** (top) Thierry Boccon-Gibod/Black Star/PNI; (bottom) Michael Newmal/PhotoEdit/ PNI; **p. 452** (left) The Museum of the City of New York, 29.100.1692. The J. Clarence Davies Collection; (right) © Steve Goldberg/Monkmeyer Press; **p. 457** Dan Habib/Impact Visuals/PNI; **p. 461** ©Robert Caputo/Aurora; **p. 464** © Betty Press/Woodfin Camp & Associates; **p. 471** Fujihira/ Monkmeyer; **p. 482** ©Don L. Boroughs/The Image Works; **p. 487** © Robert Nickelsberg/Gamma-Liasion; **p. 489** ©Lawrence Migdale/Tony Stone Images; **p. 490** Copy print © The Museum of Modern Art, New York; **p. 492** © San Jose Mercury

News/Sipa Press; **p. 507** ©Lawrence Migdale/Tony Stone Images; **p. 511** © Gary John Norman/Tony Stone Images; **p. 514** Andrew Popper/Phototake/ PNI; **p. 524** © Werner Bischof/Magnum; **p. 534** Al Stephenson/Woodfin Camp & Associates; **p. 546** © Jeff Greenberg/Photo Researchers; **p. 554** Al Stephenson/Woodfin Camp & Associates.

Text Credits

p. 201 Eric Bryant Rhodes, "Chicago Hope: An Interview with William Julius Wilson," *Transition* 68, Vol. 5, No. 4 (Winter 1996). Copyright Duke University Press, 1996. Reprinted with permission. **pp. 205, 206** Excerpts from Joe R. Feagin, "The Cintinuing Significance of Race: Anti-Black Discrimination in Public Places," *American Sociological Review,* Vol. 56, 101–116, 1991. Reprinted by permission of the American Sociological Association and the author. **pp. 290** Figs. 11.4 a–g from *Dynamics of Organizational Populations: Density, Legitimation, and Competition* by Michael T. Hannan and Glenn R. Carroll. Copyright © 1992 by OxfordUniversity Press, Inc. Reprinted by permission. **p. 312** Table 12.1 from Rodney Stark and Laurence R. Iannaccone, "A Supply-Side Reinterpretation of the 'Secularization' of Europe," *Journal for the Scientific Study of Religion,* 33, 230–252, 1994. Reprinted by permission of the Society for the Scientific Study of Religion. **pp. 314, 315** Tables 12.4 and 12.5, from Laurence R. Iannaccone, "Why Stict Churches Are Strong," *American Journal of Sociology,* 99, 1080–1221, 1994. Used by permission of the University of Chicago Press. **p. 381** Figure 14.7 from *Musica on Demand: Composers and Careers in the Hollywood Film Industry* by Robert R. Faulkner, Transaction Books, 1983, p. 28. Reprinted by permission of the publisher. **p. 403** Table 15.1 from *Family Life and School Achievement: Why Poor Black Children Succeed or Fail* by Reginald Clark, University of Chicago Press, 1983, p. 200. Reprinted by permission of the publisher. **pp. 407, 411** Figs. 15.3, 15.4, and excerpts from *Catholic Schools and the Common Good* by Anthony Bryk, Valerie E. Lee, and Peter B. Holland, Cambridge, Mass.: Harvard University Press. Copyright © 1993 by the President and Fellows of Harvard College. Reprinted by permission of the publisher. **pp. 497, 499** Figs. 17.3 and 17.4 from *American Apartheid: Segregation and the Making of the Underclass* by Douglas S. Massey and Nancy A. Denton, Cambridge, Mass.: Harvard University Press, Copyright © 1993 by the President and Fellows of Harvard College. Reprinted by permission of the publisher.

Index